Course Tools

Here you'll find chapter-specific study outlines and a list of relevant websites, along with a link to Cladistics exercises. The Miller/Harley Zoology Online Learning Center also features career information and suggested readings.

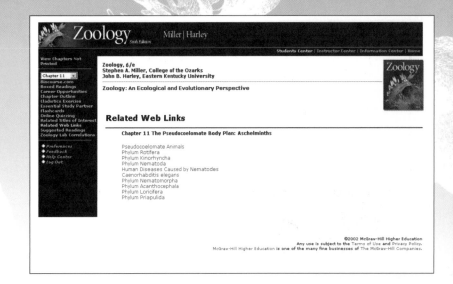

Access to Premium Learning Materials

Miller/Harley's Zoology Online Learning Center is your portal to exclusive interactive study tools like McGraw-Hill's Essential Study Partner.

Visit www.mhhe.com/zoology Today!

IMPORTANT:

HERE IS YOUR REGISTRATION CODE TO ACCESS
YOUR PREMIUM McGRAW-HILL ONLINE RESOURCES.

For key premium online resources you need THIS CODE to gain access. Once the code is entered, you will be able to use the Web resources for the length of your course.

If your course is using **WebCT** or **Blackboard**, you'll be able to use this code to access the McGraw-Hill content within your instructor's online course.

Access is provided if you have purchased a new book. If the registration code is missing from this book, the registration screen on our Website, and within your WebCT or Blackboard course, will tell you how to obtain your new code.

Registering for McGraw-Hill Online Resources

TO gain access to your McGraw-Hill web resources simply follow the steps below:

1. USE YOUR WEB BROWSER TO GO TO: **www.mhhe.com/zoology**

2. CLICK ON **FIRST TIME USER**.

3. ENTER THE REGISTRATION CODE* PRINTED ON THE TEAR-OFF BOOKMARK ON THE RIGHT.

4. AFTER YOU HAVE ENTERED YOUR REGISTRATION CODE, CLICK **REGISTER**.

5. FOLLOW THE INSTRUCTIONS TO SET-UP YOUR PERSONAL UserID AND PASSWORD.

6. WRITE YOUR UserID AND PASSWORD DOWN FOR FUTURE REFERENCE. KEEP IT IN A SAFE PLACE.

TO GAIN ACCESS to the McGraw-Hill content in your instructor's **WebCT** or **Blackboard** course simply log in to the course with the UserID and Password provided by your instructor. Enter the registration code exactly as it appears in the box to the right when prompted by the system. You will only need to use the code the first time you click on McGraw-Hill content.

Thank you, and welcome to your McGraw-Hill online Resources!

*YOUR REGISTRATION CODE CAN BE USED ONLY ONCE TO ESTABLISH ACCESS. IT IS NOT TRANSFERABLE.

0-07-293205-8 T/A MILLER/HARLEY: ZOOLOGY, 6E

MCGRAW-HILL
ONLINE RESOURCES

REGISTRATION CODE

ORTF-V3ZL-E7FM-GDMY-9IPC

Zoology

Zoology

Sixth Edition

Stephen A. Miller
College of the Ozarks

John P. Harley
Eastern Kentucky University

Boston   Burr Ridge, IL   Dubuque, IA   Madison, WI   New York   San Francisco   St. Louis
Bangkok   Bogotá   Caracas   Kuala Lumpur   Lisbon   London   Madrid   Mexico City
Milan   Montreal   New Delhi   Santiago   Seoul   Singapore   Sydney   Taipei   Toronto

Higher Education

ZOOLOGY, SIXTH EDITION

 This book is printed on recycled, acid-free paper containing 10% postconsumer waste.

1 2 3 4 5 6 7 8 9 0 VNH/VNH 0 9 8 7 6 5 4

ISBN 0–07–252836–2

Publishers: *Martin J. Lange/Colin H. Wheatley*
Developmental editor: *Fran Schreiber*
Marketing manager: *Heather K. Wagner*
Senior project manager: *Jayne Klein*
Production supervisor: *Sherry L. Kane*
Senior media project manager: *Tammy Juran*
Senior media technology producer: *John J. Theobald*
Senior designer: *David W. Hash*
Cover designer: *Rokusek Design*
Cover photo: *©2003 Nishina Masayoshi/Jun Imamoto* www.umiushi.info
Senior photo research coordinator: *Lori Hancock*
Photo research: *LouAnn K. Wilson*
Compositor: *The GTS Companies/Los Angeles, CA Campus*
Typeface: *10/12 Goudy*
Printer: *Von Hoffmann Corporation*

The credits section for this book begins on page 531 and is considered an extension of the copyright page.

Library of Congress Cataloging-in-Publication Data

Miller, Stephen A.
    Zoology / Stephen A. Miller, John P. Harley. — 6th ed.
        p. cm.
    Includes bibliographical references (p.   ) and index.
    ISBN 0–07–252836–2 (hard copy : alk. paper)
    1. Zoology. I. Harley, John P. II. Title.

QL47.2M55   2005
590—dc22   2003024621
                                            CIP

www.mhhe.com

BRIEF CONTENTS

*This chapter is available at www.mhhe.com/zoology (click on this book's cover).

CONTENTS

PART THREE
FORM AND FUNCTION: A COMPARATIVE PERSPECTIVE   359

*This chapter is available at www.mhhe.com/zoology (click on this book's cover).

PREFACE

As authors, we are honored to play a key role in the instruction of future generations of zoologists, ecologists, wildlife managers, and other life scientists. We undertook the revision for the sixth edition with this privilege, and the responsibility for content integrity, in mind.

The preparation of the sixth edition of *Zoology* involved careful evaluation of the previous editions and the features that contributed to the understanding of zoology as an exciting and dynamic scientific field. Our goal in preparing the sixth edition of *Zoology*, as in previous editions, was to prepare an introductory general zoology textbook that we believe is manageable in size and adaptable to a variety of course formats. We have retained the friendly, informative writing style that has attracted instructors and students to previous editions.

The shorter format of the fifth edition was well received by users as being less expensive and easily adapted to a one-semester course format. The sixth edition retains that format. The shorter format does mean that some general biological topics were eliminated from the book. These chapters are, however, still available, along with numerous other resources, in an electronic format on the book website and are free to adopters of the book. (Chapters found online only are indicated in the Table of Contents by an asterisk.)

CONTENT AND ORGANIZATION

We have maintained from the inception of this text that evolutionary and ecological perspectives captivate students and are fundamental to understanding the unifying principles of zoology. These perspectives are incorporated into *Zoology* in a number of ways. For example, animal structure and function are considered in the context of the environment, the animal phyla are described in the context of their roles in ecosystems, and most of the "Wildlife Alerts" that first appeared in the fourth edition, and were expanded in the fifth edition, have been retained. These boxed readings depict the plight of selected animal species or broader ecosystem issues relating to preserving various animal species.

We believe that the sixth edition of *Zoology* presents evolution as an exciting and dynamic field of study—a field of study that is vital for understanding all of biology. In addition, the continuing and expanding pseudoscientific attacks on biology make it a necessity that evolutionary concepts be presented clearly and convincingly throughout the biology curricula. We have attempted to do just that. A special font highlights important evolutionary concepts. Animal survey chapters begin with an "Evolutionary Perspective" and end with "Further Phylogenetic Considerations." These sections describe evolutionary relationships within each phylum and evolutionary connections to animals of previous and following chapters. Updated cladograms are used to depict taxonomic relationships. Evolutionary connections and animal adaptations are stressed in the structure and function sections.

To further explain and support evolutionary concepts, this sixth edition has a second set of themed boxed readings (in addition to "Wildlife Alerts") entitled "Evolutionary Insights." These boxes provide detailed examples of principles covered in a chapter and provide insight into how evolutionary biology works. For example, chapter 4 includes a reading on big-cat biogeography that illustrates how a variety of sources of evidence are used to paint a picture of the history of one group of animals. Chapter 5 has a reading on the speciation of Darwin's finches that illustrates how and why speciation occurs. Other readings describe ideas regarding animal origins and the debates that occur among taxonomists who try to sort out evolutionary relationships within animal groups.

Zoology is organized into three parts. Part One covers the common life processes, including cell and tissue structure and function, the genetic basis of evolution, and the evolutionary and ecological principles that unify all life.

Part Two is the survey of protists and animals, emphasizing evolutionary and ecological relationships, aspects of animal organization that unite major animal phyla, and animal adaptations. All of the chapters in Part Two have been updated. The presentation of taxonomic principles in chapter 7, and the taxonomic relationships in chapters 8 through 22, have been carefully revised and incorporate some of the flavor of the exciting changes occurring in the field of taxonomy. You will see some of these changes listed under "New to the Sixth Edition." Cladograms have been updated and, as in previous editions, full-color artwork, photographs, and lists of phylum characteristics are used to highlight each phylum.

Part Three covers animal form and function using a comparative approach. This approach includes descriptions and full-color artwork that depict evolutionary changes in the structure and function of selected organ systems. Part Three includes an appropriate balance between invertebrate and vertebrate descriptions.

NEW TO THE SIXTH EDITION

Major additions to the sixth edition focus on evolutionary principles and taxonomy. Evolutionary concepts must be presented clearly and convincingly in biology courses. We believe that changes we have made will help instructors accomplish that goal by providing more evidence of evolution, more examples to illustrate evolutionary principles, and more detail on evolutionary mechanisms. Recent, fast-paced changes in animal taxonomy require constant reevaluation of the presentation of evolutionary relationships between animal taxa. Because the taxonomy of many animal groups is unsettled, we have tried to take a conservative, yet up-to-date, position on taxonomic revisions. The following are major additions to this edition.

- "Evolutionary Insights" boxes appear in selected chapters. These readings present students with further information and examples of how evolutionary biology works.
- Chapter 4 is reorganized and presents new information on the distinction between microevolution and macroevolution. The coverage of the evidence for macroevolution includes an expanded discussion of paleontology, a reorganized and expanded discussion of homology and analogy from the perspectives of both comparative anatomy and molecular biology, and a new presentation of evidence from developmental biology. A new section on phylogeny and common descent caps this chapter.
- Chapter 5 begins with an expanded presentation of populations and gene pools. The sections on sources of variation and gene flow are enhanced with more information and new examples.
- Chapter 9 presents new information on the evolutionary relationships of the Porifera, Cnidaria, and Ctenophora.
- Chapter 13 provides an updated taxonomy of the Annelida, including the presentation of the oligochaetes and leeches as members of a single class Clitellata.
- Chapters 14 and 15 include an extensive update on arthropod taxonomy. Arthropods are presented as a monophyletic group, and recent thinking regarding crustacean ancestry for the phylum is discussed. There is expanded coverage of the hemocoel and insect nutrition and digestion.
- Chordate taxonomy in chapters 17 through 22 has been updated. Chapter 18 includes an expanded discussion of the evolution of jaws and paired appendages and the fish-to-amphibian transition. Chapter 19 introduces more coverage of the early evolution of the Stegocephalia and Tetrapoda. Chapter 21 has expanded coverage of bird evolution.

SUPPLEMENTARY MATERIALS

Supplementary materials are available to assist instructors with their presentations and course management and to augment student learning. The usefulness of these supplements is now greatly enhanced with the availability of online, digital, and printed resources. A Digital Content Manager is available as a CD-ROM. It contains PowerPoint slides of most line art and photographs from the textbook, which can be used in customizing classroom presentations.

ONLINE LEARNING CENTER

As with the previous edition, chapters on cell chemistry, energy and enzymes, embryology, and animal behavior—along with numerous boxed readings and pedagogical elements—have been moved to the Online Learning Center. This content-rich website is located at www.mhhe.com/zoology—just click on this book's

cover. Both instructors and students can take advantage of numerous teaching and learning aids within this book's Online Learning Center.

Instructor Resources:

- Instructor's Manual
- Instructor Resource Guide
- Link to Digital Zoology

Student and Instructor Resources:

- Interactive Cladistics Exercises
- Chapters on:
  - Chapter 30: The Chemical Basis of Animal Life
  - Chapter 31: Energy and Enzymes: Life's Driving and Controlling Forces
  - Chapter 32: How Animals Harvest Energy Stored in Nutrients
  - Chapter 33: Embryology
  - Chapter 34: Animal Behavior
- Quizzing
- Flashcards
- Suggested Readings
- Boxed Readings
- Animation Exercises
- Zoology Lab Correlations
- Zoology Essential Study Partner (ESP)

OTHER RESOURCES

The following items may accompany *Zoology*. Please consult your McGraw-Hill representative for policies, prices, and availability.

- An **Instructor's Manual,** prepared by Susan L. Keen, is available for instructors within the Online Learning Center. It provides items such as a lecture outline, lecture enrichments, research discussion topics, teaching suggestions, and/or suggested readings for each chapter.
- A **Zoology Test Item CD-ROM** is also available for instructors. This contains approximately 50 multiple-choice questions and the instructor's manual for each chapter.
- **General Zoology Study Guide,** prepared by Jane Aloi and Gina Erickson, contains subject-by-subject summaries, questions, and learning activities.
- A set of 100 full-color acetate transparencies is available to supplement classroom lectures.
- **General Zoology Laboratory Manual,** Fifth Edition, by Stephen A. Miller, is an excellent corollary to the text and incorporates many learning aids. It includes illustrations and photographs, plus activities on scientific method, cladistics, ecological and evolutionary principles, and animal structure and function. A Laboratory Resource Guide, available within the Online Learning Center, provides information about materials and procedures and answers to worksheet questions that accompany the lab exercises.

- **Digital Zoology CD-ROM** is an exciting interactive product designed to help you make the most of your zoology classes and laboratory sessions. This program contains interactive cladograms, laboratory modules, video, interactive quizzes, hundreds of photographs, a full glossary, and much detailed information about the diversity and evolution of the animals that we find on the planet. To find out the latest news on this ever-expanding product, log on to www.digitalzoology.com and find out how to incorporate this valuable resource into your course.

- **Study Aid/Poster: Chief Taxonomic Subdivisions & Organ Systems of the Animal Phyla**—This 30- × 36-inch poster is a great reference/study tool for students.

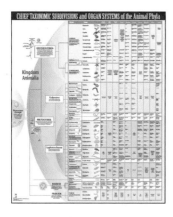

- Available through the Zoology Online Learning Center or on a free CD-ROM, the **Zoology Essential Study Partner** is a complete, interactive study tool offering animations and learning activities to help students understand complex zoology concepts. This valuable resource also includes self-quizzing to help students review each topic.
- PageOut® is the solution for professors who need to build a course website. The following features are now available to professors:

- The PageOut Library offers instant access to fully loaded course websites with no work required on the instructor's part.
- Courses can now be password protected.
- Professors can now upload, store, and manage up to 10 MB of data.
- Professors can copy their course and share it with colleagues or use it as a foundation for next semester. Short on time? Let us do the work. Our McGraw-Hill service team is ready to build your PageOut website and to provide content and any necessary training. Learn more about PageOut and other McGraw-Hill digital solutions at www.mhhe.com/solutions.

ACKNOWLEDGMENTS

We wish to thank the reviewers who provided detailed analysis of the text during development. In the midst of their busy teaching and research schedules, they took time to read our manuscript and offer constructive advice that greatly improved this sixth edition.

REVIEWERS

Barbara J. Abraham, Hampton University
Marc C. Albrecht, University of Nebraska at Kearney
Donald I. Anadu, South Carolina State University
Joe Arruda, Pittsburg State University
Amir M. Assadi-Rad, San Joaquin Delta College
Kris Burnell, Santa Barbara City College
Vincent A. Cobb, Northeastern State University
Marjorie M. Collier, Louisiana State University at Alexandria
D. Charles Dailey, Sierra College
Lewis Deaton, University of Louisiana at Lafayette
Elizabeth A. Desy, Southwest State University
Charles D. Dieter, South Dakota State University
Phillip Eichman, University of Rio Grande
David J. Eisenhour, Morehead State University
Bruce E. Felgenhauer, University of Louisiana at Lafayette
Jim Goetze, Laredo Community College
Glenn A. Gorelick, Citrus College
Edward J. Greding, Jr., Del Mar College
Ann Grens, Indiana University–South Bend
Leon E. Hallacher, University of Hawaii at Hilo
Jenna Jo Hellack, University of Central Oklahoma
J. L. Henriksen, Bellevue University
Gary W. Hunt, Tulsa Community College
Arthur B. Jantz, Western Oklahoma State College
Ritin Khan, Armstrong Atlantic State University
Scott L. Kight, Montclair State University
Shelley A. Kirkpatrick, Saint Francis University
Robert A. Krebs, Cleveland State University
Mary Jane Krotzer, Stillman College
Kevin Lyon, Jones County Junior College
Kelly M. Mack, University of Maryland Eastern Shore
Neal F. McCord, Stephen F. Austin State University

Bonnie McCormick, University of the Incarnate Word
Michael B. McDarby, Fulton-Montgomery Community College
Emily C. McDuffee, Freed-Hardeman University
William R. Miller, Chestnut Hill College
Devonna Sue Morra, Saint Francis University
Charles M. Page, El Camino College
Louis L. Pech, Carroll College
Polly K. Phillips, Miami-Dade Community College
Karen E. Plucinski, Defiance College
Michelle A. Priest, College of the Canyons
William M. Saidel, Rutgers University–Camden
Allen Sanborn, Barry University
Neil Schanker, College of the Siskiyous
Doug Schexnayder, Copiah-Lincoln County College
Fred Schindler, Indian Hills Community College
Shari Snitovsky, Skyline College
Ken Spitze, University of Miami
Annie M. Talley, Asheville-Buncombe Technical College
Michael E. Toliver, Eureka College
Richard E. Trout, Oklahoma City Community College
Elise M. Van Ginkel, Madison Area Technical College
Scott E. Walter, University of Wisconsin–Richland
Danny B. Wann, Carl Albert State College
MaryJo A. Witz, Monroe Community College
Jeffery S. Wooters, Pensacola Junior College
Robert W. Yost, Indiana University/Purdue University
    at Indianapolis
David D. Zeigler, University of North Carolina at Pembroke
Brenda Zink, Northeastern Junior College

The publication of a text requires the efforts of many people. We are grateful for the work of our colleagues at McGraw-Hill who have shown extraordinary patience, skill, and commitment to this textbook. Marge Kemp has helped shape *Zoology* from its earliest planning stages. Although she has moved up to other responsibilities within McGraw-Hill, her wisdom and skill is still evident in this sixth edition. Donna Nemmers, Senior Developmental Editor, has worked with this textbook through the last two revisions. We are grateful for her skill in coordinating many of the tasks involved with publishing previous editions of *Zoology*. Our Developmental Editor, Fran Schreiber, helped make the production of the sixth edition remarkably smooth. Fran kept us on schedule and the production moving in the plethora of directions that are nearly unimaginable to us. Jayne Klein served as Senior Project Manager for this edition. We appreciate her efficiency and organization. We also thank Rose Kramer for proofreading the entire textbook.

Finally, but most importantly, we wish to extend appreciation to out families for their patience and encouragement. Janice A. Miller lived with this text through many months of planning and writing. She died suddenly two months before the first edition was released. Our wives, Carol A. Miller and Donna Dailey Harley, have been supportive throughout the revision process. We appreciate the sacrifices that our families have made during the writing and revision of this text. We dedicate this book to the memory of Jan and to our families.

STEPHEN A. MILLER

JOHN P. HARLEY

ONLINE LEARNING CENTER

www.mhhe.com/zoology (click on this book's cover)

Students: You'll appreciate extensive self-quizzing opportunities; interactive activities; and related weblinks in addition to the Zoology Essential Study Partner—a web-based review of major zoology topics—hosted on this site.

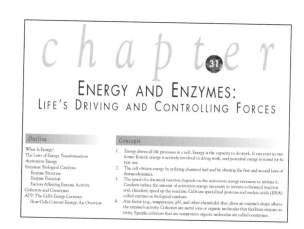

Instructor Resources:

- Instructor's Manual
- Instructor Resource Guide
- Link to Digital Zoology

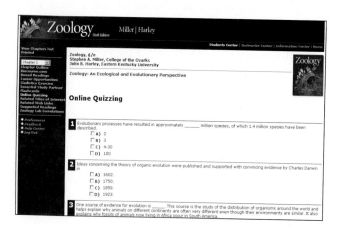

Student & Instructor Resources:

- Interactive Cladistics Exercises
- Quizzing
- Flashcards
- Suggested Readings
- Boxed Readings
- Animation Exercises
- Zoology Lab Correlations
- Zoology Essential Study Partner (ESP)

Additional Chapters:

- Chapter 30: The Chemical Basis of Animal Life
- Chapter 31: Energy and Enzymes: Life's Driving and Controlling Forces
- Chapter 32: How Animals Harvest Energy Stored in Nutrients
- Chapter 33: Descriptive Embryology
- Chapter 34: Animal Behavior

Digital Content Manager CD-ROM

For the first time, *Zoology* will have a **text-specific Digital Content Manager CD-ROM.** This powerful and easy-to-use tool is designed to help instructors easily incorporate text-specific illustrations and photos into lecture presentations and printed classroom materials. Organized by chapter, this cross-platform CD-ROM contains a collection of visual resources that can be imported and reproduced in multiple formats to create customized lectures, visually based tests and quizzes, dynamic course website content, or attractive printed materials.

Digital Zoology CD-ROM and Student Work Book

Digital Zoology is an interactive guide that will help reinforce what students are learning in their zoology lecture and laboratory sessions. Here is what you will find on this easy-to-use CD-ROM.

- Laboratory modules containing illustrations, photographs, annotations of the major structures of organisms, interactive quizzes, and over 70 video clips of animal diversity, habitats, and behaviors.
- Interactive cladograms and dendograms within lab modules, along with links to interactive synapomorphies of the various animal groups.
- Key terms with links to an interactive glossary of over 750 definitions.
- Taxon "Read Abouts" that include information on over 100 taxa, the number of living and extinct species, habitats, and how they function.
- Updated taxonomy of the animal phyla that allows for easy comparison of the differences between traditional morphological phylogenies and those created with molecular data.
- Image galleries available for many of the major phyla, providing photos that detail diversity within the phyla.
- An accompanying student workbook with self-tests, crossword puzzles, and a website to provide additional study tips, exercises, and phyla characteristics.

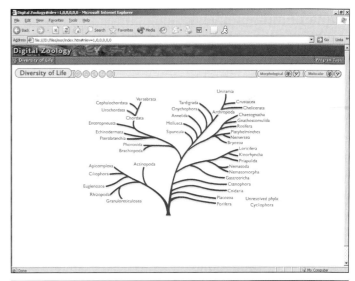

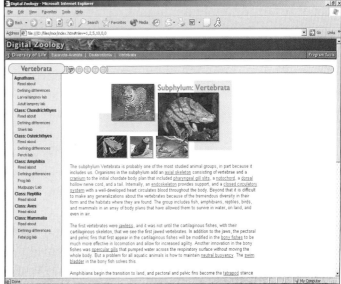

GUIDED TOUR

The organization and features of this book have been planned with students' learning and comprehension in mind.

CHAPTER CONCEPTS

The concepts most important to the understanding of each chapter are highlighted on the first page of each chapter.

CRITICAL THINKING QUESTIONS

Students can synthesize the chapter information by applying it to the Critical Thinking Questions in each chapter.

CHAPTER 24

COMMUNICATION I:

NERVOUS AND SENSORY SYSTEMS

Outline

Neurons: The Basic Functional Units of the
   Nervous System
   Neuron Structure: The Key to Function
Neuron Communication
   Resting Membrane Potential
   Mechanism of Neuron Action
   Transmission of the Action Potential
      Between Cells
Invertebrate Nervous Systems
Vertebrate Nervous Systems
   The Spinal Cord
   Spinal Nerves
   The Brain
   Cranial Nerves
   The Autonomic Nervous System
Sensory Reception
Invertebrate Sensory Receptors
   Baroreceptors
   Chemoreceptors
   Georeceptors
   Hygroreceptors
   Phonoreceptors
   Photoreceptors
   Proprioceptors
   Tactile Receptors
   Thermoreceptors
Vertebrate Sensory Receptors
   Lateral-Line System and Electrical Sensing
   Lateral-Line System and
      Mechanoreception
   Hearing and Equilibrium in Air
   Hearing and Equilibrium in Water
   Skin Sensors of Damaging Stimuli
   Skin Sensors of Heat and Cold
   Skin Sensors of Mechanical Stimuli
   Sonar
   Smell
   Taste
   Vision

Concepts

1. The nervous system helps to communicate, integrate, and coordinate the functions of the various organs and organ systems in the animal body.
2. Information flow through the nervous system has three main steps: (1) the collection of information from outside and inside the body (sensory activities), (2) the processing of this information in the nervous system, and (3) the initiation of appropriate responses.
3. Information is transmitted between neurons directly (electrically) or by means of chemicals called neurotransmitters.
4. The evolution of the nervous system in invertebrates has led to the elaboration of organized nerve cords and the centralization of responses in the anterior portion of the animal.
5. The vertebrate nervous system consists of the central nervous system, made up of the brain and spinal cord, and the peripheral nervous system, composed of the nerves in the rest of the body.
6. Nervous systems evolved through the gradual layering of additional nervous tissue over reflex pathways of more ancient origin.
7. Sensory receptors or organs permit an animal to detect changes in its body, as well as in objects and events in the world around it. Sensory receptors collect information that is then passed to the nervous system, which determines, evaluates, and initiates an appropriate response.
8. Sensory receptors initiate nerve impulses by opening channels in sensory neuron plasma membranes, depolarizing the membranes, and causing a generator potential. Receptors differ in the nature of the environmental stimulus that triggers an eventual nerve impulse.
9. Many kinds of receptors have evolved among invertebrates and vertebrates, and each receptor is sensitive to a specific type of stimulus.
10. The nature of its sensory receptors gives each animal species a unique perception of its body and environment.

The two forms of communication in an animal that integrate body functions to maintain homeostasis are: (1) neurons, which transmit electrical signals that report information or initiate a quick response in a specific tissue; and (2) hormones, which are slower, chemical signals that initiate a widespread, prolonged response, often in a variety of tissues. This chapter focuses on the function of the neuron, the anatomical organization of the nervous system in animals, and the ways in which the senses collect information and transmit it along nerves to the central nervous system. To conclude the study of communication, chapter 25 examines how hormones affect long-term changes in an animal's body.

This chapter contains evolutionary concepts, which are set off in this font.

379

---

126   PART TWO   Animal-like Protists and Animalia

SUMMARY

1. The kingdom Protista is a polyphyletic group that arose about 1.5 billion years ago from the Archaea. The evolutionary pathways leading to modern protozoa are uncertain.
2. Protozoa are both single cells and entire organisms. Organelles specialized for the unicellular lifestyle carry out many protozoan functions.
3. Many protozoa live in symbiotic relationships with other organisms, often in a host-parasite relationship.
4. Members of the phylum Sarcomastigophora possess pseudopodia and/or one or more flagella.
5. Members of the class Phytomastigophorea are photosynthetic and include the genera *Euglena* and *Volvox*. Members of the class Zoomastigophorea are heterotrophic and include *Trypanosoma*, which causes sleeping sickness.
6. Amoebae use pseudopodia for feeding and locomotion.
7. Members of the subphylum Sarcodina include the freshwater genera *Amoeba*, *Arcella*, and *Difflugia*, and the parasitic genus *Entamoeba*. Foraminiferans and radiolarians are common marine amoebae.
8. Members of the phylum Apicomplexa are all parasites. The phylum includes *Plasmodium* and *Toxoplasma*, which cause malaria and toxoplasmosis, respectively. Many apicomplexans have a three-part life cycle involving schizogony, gametogony, and sporogony.
9. The phylum Microspora consists of small protozoa that are intracellular parasites of every major animal group. They are transmitted from one host to the next as a spore, the form from which the group obtains its name.
10. The phylum Acetospora contains protozoa that produce spores lacking polar capsules. These protozoa are primarily parasitic in molluscs.
11. The phylum Myxozoa consists entirely of parasitic species, usually found in fishes. One to six polar filaments characterize the spore.
12. The phylum Ciliophora contains some of the most complex of all protozoa. Its members possess cilia, a macronucleus, and one or more micronuclei. Mechanical coupling of cilia coordinates their movements, and cilia can be specialized for different kinds of locomotion. Ciliates reproduce sexually by conjugation. Diploid ciliates undergo meiosis of the micronuclei to produce haploid pronuclei that two conjugants can exchange.
13. Precise evolutionary relationships are difficult to determine for the protozoa. The fossil record is sparse, and what does exist is not particularly helpful in deducing relationships. However, ribosomal RNA sequence comparisons indicate that each of the seven protozoan phyla probably had separate origins.

SELECTED KEY TERMS

ectoplasm (p. 112)   multiple fission (schizogony) (p. 113)
endoplasm (p. 112)   pellicle (p. 112)
host (p. 113)   protozoa (p. 112)
macronucleus (p. 123)   protozoologists (p. 114)
micronuclei (p. 123)   trichocysts (p. 123)

CRITICAL THINKING QUESTIONS

1. If knowing for certain the evolutionary pathways that gave rise to protozoa and animal phyla is impossible, is it worth constructing hypotheses about those relationships? Why or why not?
2. In what ways are protozoa similar to animal cells? In what ways are they different?
3. If sexual reproduction is unknown in *Euglena*, how do you think this lineage of organisms has survived through evolutionary time? (Recall that sexual reproduction provides the genetic variability that allows species to adapt to environmental changes.)
4. The use of DDT has been greatly curtailed for ecological reasons. In the past, it has proved to be an effective malaria deterrent. Many organizations would like to see this form of mosquito control resumed. Do you agree or disagree? Explain your reasoning.
5. If you were traveling out of the country and were concerned about contracting amoebic dysentery, what steps could you take to avoid contracting the disease? How would the precautions differ if you were going to a country where malaria is a problem?

ONLINE LEARNING CENTER

Visit our Online Learning Center (OLC) at www.mhhe.com/zoology (click on this book's title) to find the following chapter-related materials:

• CHAPTER QUIZZING
• RELATED WEB LINKS
   Phylum Sarcomastigophora
   Phylum Apicomplexa
   Phylum Ciliophora
   Other Protozoan Phyla
• BOXED READINGS ON
   Giardiasis: "Backpacker's Disease" in the Rocky Mountains
   Malaria Control—A Glimmer of Hope
• SUGGESTED READINGS
• LAB CORRELATIONS

   Check out the OLC to find specific information on these related lab exercises in the *General Zoology Laboratory Manual*, 5th edition, by Stephen A. Miller:

   Exercise 8   *Animal-like Protists*

ONLINE LEARNING CENTER

The Online Learning Center hosts specific study tools for each chapter, which are summarized at the end of each text chapter.

KEY TERMS

The most important terms from each chapter are linked to the text pages where they are defined, for further study.

WILDLIFE ALERT BOXES

These boxes feature issues related to endangered and threatened species of animals.

WILDLIFE ALERT
Coral Reefs

Coral reefs are among the most threatened marine habitats. Along with tropical rain forests, they are among the most diverse ecosystems on the earth. They are home to thousands of species of fish, and nearly 100,000 species of reef invertebrates have been described to date (box figure 9.1). This diversity gives coral reefs tremendous intrinsic and economic value. Their highly productive waters yield four to eight million tons of fish for commercial fisheries. This is one-tenth of the world's total fish harvest, from an area that represents only 0.17% of the ocean surface (box figure 9.2). Coral reefs attract billions of dollars' worth of tourist trade each year. The ecological, aesthetic, and economic reasons for preserving coral reefs are overwhelming.

Disturbances of coral reefs can be devastating, because reefs grow very slowly. Normally a coral reef is alive with color. A disturbed reef turns white as a result of the death of anthozoan polyps, zooxanthellae (dinoflagellate protists that live in a mutualistic relationship with the anthozoans), and coralline algae (box figure 9.3). This bleaching reaction of a coral reef, if it results from a local disturbance, can be reversed rather quickly. Large-scale disturbances, however, can result in the death of large expanses of coral reef, which requires thousands of years to recover. In recent years, massive bleaching has been reported in tropical waters of the Atlantic, Caribbean, Pacific, and Indian Oceans.

Reefs require clean, constantly warm, shallow water to support the growth of zooxanthellae, which sustain coral anthozoans. Changing water levels, water temperature, and turbidity can adversely affect reef growth. Sedimentation from mining, dredging, and logging, or clearing mangrove swamps that trap sediment from coastal runoff, can block sunlight and result in the death of zooxanthellae. Some island communities mine coral reefs to extract limestone for concrete. Coastal development results in sewage and industrial pollution, which have damaged coral

BOX FIGURE 9.2 Coral Reefs. Coral reefs are found throughout the tropical and subtropical regions of the world between 30°N and 30°S latitudes.

reefs. Oil spills are toxic to coral organisms. Ships that run aground damage large sections of coral reefs. Altered ecological relationships have resulted in the proliferation of the crown-of-thorns sea star (*Acanthaster planci*), which feeds on coral polyps and devastates reef communities of the South Pacific. Snorkelers and scuba divers who walk across reef surfaces, break off pieces of reef, or anchor their boats on reefs similarly threaten reef life. Recently, global warming (*see chapter 6*) has been implicated in reef bleaching. The results of global warming—changing water temperature, changing water levels, and increased frequency of tropical storms—have the potential to damage coral reefs by altering environmental conditions favorable for reef survival and growth.

The threats to coral reefs seem almost overwhelming. Fortunately, biologists are finding that coral reefs are resilient ecosystems. If water quality is good, coral reefs can recover from local disturbances. National and international policies are needed that will prevent disturbances,

BOX FIGURE 9.3 Coral Bleaching. The bleached portion of the coral is shown in the lower portion of this photograph. The polyps in the upper portion of the photograph are still alive.

X FIGURE 9.1 A Coral Reef Ecosystem.

EVOLUTIONARY INSIGHTS
Speciation of Darwin's Finches

When Charles Darwin visited the Galápagos Islands in 1835, he observed the dark-bodied finches whose adaptive radiation has become a classic example of speciation (box figure 5.1). Studies of these finches have provided insight into some of the ways in which speciation can occur. Peter R. and B. Rosemary Grant have been studying these finches for more than 30 years. They have directly observed microevolutionary change reflected in bill morphology in response to change in rainfall and food availability. Other molecular studies have also contributed to our knowledge of the adaptive radiation of this group of birds.

Molecular studies have identified the most likely South American relatives of Darwin's finches, members of the grassquit genus *Tiaris*. Comparisons of the mitochondrial DNA of this group with Darwin's finches suggest that the latter colonized some of the Galápagos Islands not more than 3 million years ago. A very rapid adaptive radiation occurred, with the number of finch species doubling approximately every 750,000 years. No other group of birds studied has undergone a more rapid evolutionary diversification (*see figure 4.4*). Darwin's finches have served as a model to answer questions of how and why species diverge.

The traditional explanation of speciation within Darwin's finches is based on the allopatric model discussed in this chapter. This explanation is based upon differences in food resources, and the observations of the Grants have provided support for this model. Geographic isolation

of populations of finches on different islands promoted speciation as these populations were influenced by natural selection and genetic drift. Each population adapted to the food resources available in their habitat (*see figure 4.6*). Most of these adaptations are reflected in bill morphology.

The Grants have discovered, however, that the allopatric model is not the entire explanation for finch adaptive radiation. Three million years ago, the Galápagos Islands were much simpler than they were today. In fact, there were fewer islands when they were first colonized by finches. Apparently the number of finch species increased as the number of islands increased as a result of volcanic activity. The increasing number of islands and oscillations in temperature and precipitation naturally affected vegetation. Habitats available for finches became more diverse and complex. The original warm, wet islands favored long, narrow bills that were used in gathering nectar and insects. The islands' moisture now fluctuates and the climate is more seasonal. The increasing diversity in habitats and food supply over 3 million years apparently promoted very rapid speciation among the finch populations.

The Grants have also discovered that sympatric forces probably have also promoted speciation. Different species of finches that live on the same island rarely hybridize. Lack of hybridization promotes isolation and speciation. Genetic incompatibility of gametes is apparently not the factor that discourages hybridization. Courtship behaviors of different species are similar, so courtship differences are not responsible.

BOX FIGURE 5.1 Speciation of Darwin's Finches. Speciation and adaptive radiation of Darwin's finches has been used as a classic example of allopatric speciation. Isolation of finches on different islands, and differences in food resources on those islands, selected for morphological differences in finch bills. For example, (a) the warbler finch (*Certhidea olivacea*) has a bill that is adapted for probing for insects, and (b) the large ground finch (*Geospiza magnirostris*) has a bill that is adapted for crushing seeds. Recent studies show that increasing numbers of islands over the last 3 million years and changes in temperature and precipitation resulted in very rapid speciation. In addition, sympatric influences regarding the role of the males' song and bill shape probably also promoted speciation.

EVOLUTIONARY INSIGHTS

These new boxes provide detailed examples of principles covered in a chapter and provide insight into how evolutionary biology works.

PART ONE

BIOLOGICAL PRINCIPLES

Animals are united with all other forms of life by the biological processes that they share with other organisms. Understanding these processes helps us to know how animals function and why animals are united with other forms of life from the evolutionary and ecological perspectives. Chapter 1 examines some of these unifying themes and sets the stage for the evolutionary and ecological perspectives that are developed throughout this book.

An understanding of the cell as the fundamental unit of life is key to understanding life on this planet. As you learn more about cell structure and function, you will find that many cellular components and processes are very similar in cells from a variety of organisms. One of the common functions of all cells is reproduction. Reproduction may involve individual cells within a multicellular organism, a single-celled organism, or the formation of single reproductive cells in multicelluar organisms. The processes involved in cellular reproduction, and the processes involved in determining the characteristics of the new cells and organisms that are produced, are based on common biological themes. Chapters 2 and 3

present cell structure and inheritance as an important, unifying framework within which biologists approach the diversity of organisms.

Principles of inheritance explain not only why offspring resemble their parents, but also why variation exists within populations. This variation is a key to understanding evolution. All organisms have an evolutionary history, and evolution helps us to understand the life-shaping experiences that all organisms share. Chapter 4 explores the work of pioneers of evolutionary theory, Charles Darwin and Alfred Russell Wallace, and how their work forms the nucleus for modern evolutionary theory. Chapter 5 examines the influence of modern genetics on evolutionary theory. This coverage of evolution will provide core knowledge for understanding the diversity of animal life presented in Part Two and how evolution has influenced the animal structure and function described in Part Three.

A fundamental unity of life also occurs at the environmental level. All animals are partners in the use of the earth's resources. Only by studying the interactions of organisms with one another and with their

environment can we appreciate the need for preserving resources for all organisms. Chapter 6 presents basic ecological principles that everyone must understand if we are to preserve the animal kingdom.

The tiger (Panthera tigris) is one of many animal species facing extinction. This solitary predator once roamed throughout Asia and numbered as many as 100,000. The results of poaching and habitat destruction have reduced its numbers to about 5,000 individuals. Part 1 emphasizes genetic and evolutionary principles that help explain the origin of diversity within the animal kingdom and the ecological principles that explain the interdependence of life. Understanding these principles helps us develop strategies for preserving species like the tiger and taking steps to reduce the likelihood that other species will follow in the tiger's paw-prints.

1

CHAPTER 1

ZOOLOGY:

AN EVOLUTIONARY AND ECOLOGICAL PERSPECTIVE

Outline

Concepts

1. The field of zoology is the study of animals. It is a very broad field with many subdisciplines.
2. An understanding of evolutionary processes is very important in zoology because evolution explains the family relationships among animals and how the great variety of animals arose.
3. An understanding of ecological principles is very important in zoology because it helps zoologists to understand the interrelationships among individual animals and groups of animals. Understanding ecological principles also helps zoologists to understand how human interference threatens animal populations and the human environment.

Zoology (Gr. *zoon*, animal + *logos*, to study) is the study of animals. It is one of the broadest fields in all of science because of the immense variety of animals and the complexity of the processes occurring within animals. There are, for example, over 20,000 described species of bony fishes and over 300,000 described (and many more undescribed) species of beetles! It is no wonder that zoologists usually specialize in one or more of the subdisciplines of zoology. They may study particular functional, structural, or ecological aspects of one or more animal groups (table 1.1), or they may choose to specialize in a particular group of animals (table 1.2).

Ichthyology, for example, is the study of fishes, and ichthyologists work to understand the structure, function, ecology, and evolution of fishes. These studies have uncovered an amazing diversity of fishes. One large group, the cichlids, is found in Africa (1,000 species), Central and South America (300 species), India (3 species) and North America (1 species). Members of this group have an enormous variety of color patterns (figure 1.1), habitats, and body forms. Ichthyologists have described a wide variety of feeding habits in cichlids. These fish include algae scrapers, like *Eretmodus*, that nip algae with chisel-like teeth; insect pickers, like *Tanganicodus*; and scale eaters, like *Perissodus*. All cichlids have two pairs of jaws. The mouth jaws are used for scraping or nipping food, and the throat jaws are used for crushing or macerating food before it is swallowed.

Many cichlids mouth brood their young. A female takes eggs into her mouth after the eggs are spawned. She then inhales sperm released by the male, and fertilization and development take place within the female's mouth! Even after the eggs hatch, young are taken back into the mouth of the female if danger threatens (figure 1.2). Hundreds of variations in color pattern, body form, and behavior in this family of fishes illustrate the remarkable diversity present in one relatively small branch of the animal kingdom. Zoologists are working around the world to understand and preserve this enormous diversity.

This chapter contains evolutionary concepts, which are set off in this font.

TABLE 1.1
EXAMPLES OF SPECIALIZATIONS IN ZOOLOGY

SUBDISCIPLINE	DESCRIPTION
Anatomy	Study of the structure of entire organisms and their parts
Cytology	Study of the structure and function of cells
Ecology	Study of the interaction of organisms with their environment
Embryology	Study of the development of an animal from the fertilized egg to birth or hatching
Genetics	Study of the mechanisms of transmission of traits from parents to offspring
Histology	Study of tissues
Molecular biology	Study of subcellular details of structure and function
Parasitology	Study of animals that live in or on other organisms at the expense of the host
Physiology	Study of the function of organisms and their parts
Systematics	Study of the classification of, and the evolutionary interrelationships among, animal groups

TABLE 1.2
EXAMPLES OF SPECIALIZATIONS IN ZOOLOGY BY TAXONOMIC CATEGORIES

Entomology	Study of insects
Herpetology	Study of amphibians and reptiles
Ichthyology	Study of fishes
Mammalogy	Study of mammals
Ornithology	Study of birds
Protozoology	Study of protozoa

ZOOLOGY: AN EVOLUTIONARY PERSPECTIVE

Animals share a common evolutionary past and evolutionary forces that influenced their history. Evolutionary processes are remarkable for their relative simplicity, yet they have had awesome effects on life-forms. These processes have resulted in an estimated 4 to 100 million species of animals living today. (Only 1 million animal species have been described.) Many more, about 90 percent, existed in the past and have become extinct. Zoologists must understand evolutionary processes if they are to understand what an animal is and how it originated.

(a)

(b)

FIGURE 1.1

Cichlids. Cichlids of Africa exist in an amazing variety of color patterns, habitats, and body forms. (*a*) This dogtooth cichlid (*Cynotilapia afra*) is native to Lake Malawi in Africa. The female of the species broods developing eggs in her mouth to protect them from predators. (*b*) The fontosa (*Cyphontilapia fontosa*) is native to Lake Tanganyika in Africa.

EVOLUTIONARY PROCESSES

Organic evolution (L. *evolutus*, unroll) is change in the genetic makeup of populations of organisms over time. It is the source of animal diversity, and it explains family relationships within animal groups. Charles Darwin published convincing evidence of evolution in 1859 and proposed a mechanism that could explain evolutionary change. Since that time, biologists have become convinced that evolution occurs. The mechanism proposed by Darwin has been confirmed and now serves as the nucleus of our broader understanding of evolutionary change (chapters 4 and 5).

Understanding how the diversity of animal structure and function arose is one of the many challenges faced by zoologists. For example, the cichlid scale eaters of Africa feed on the scales of other cichlids. They approach a prey cichlid from behind and bite

FIGURE 1.2

A Mouth-Brooding Cichlid.   *Nimbochromis livingstonii* is a mouth-brooding species. Eggs develop in the mouth of the female and, after hatching, young return to the female's mouth when danger threatens.

FIGURE 1.3

Lakes Victoria, Tanganyika, and Malawi.   These lakes have cichlid populations that have been traced by zoologists to an ancestry that is approximately 200,000 years old. Cichlid populations originated in Lake Tanganyika and then spread to the other two lakes.

a mouthful of scales from the body. The scales are then stacked and crushed by the second set of jaws and sent to the stomach and intestine for protein digestion. Michio Hori of Kyoto University found that there were two body forms within the species *Perissodus microlepis*. One form had a mouth that was asymmetrically curved to the right and the other form had a mouth that was asymmetrically curved to the left. The asymmetry allowed right-jawed fish to approach and bite scales from the left side of their prey and the left-jawed fish to approach and bite scales from the right side of their prey. Both right- and left-jawed fish have been maintained in the population; otherwise the prey would eventually become wary of being attacked from one side. The variety of color patterns within the species *Topheus duboisi* has also been explained in an evolutionary context. Different color patterns arose as a result of the isolation of populations among sheltering rock piles separated by expanses of sandy bottom. Breeding is more likely to occur within their isolated populations because fish that venture over the sand are exposed to predators.

ANIMAL CLASSIFICATION AND EVOLUTIONARY RELATIONSHIPS

Evolution not only explains why animals appear and function as they do, but it also explains family relationships within the animal kingdom. Zoologists have worked for many years to understand the evolutionary relationships among the hundreds of cichlid

species. Groups of individuals are more closely related if they share more of their genetic material (DNA) with each other than with individuals in other groups. (You are more closely related to your brother or sister than to your cousin for the same reason. Because DNA determines most of your physical traits, you will also more closely resemble your brother or sister.) Genetic studies have found that African cichlids originated in Lake Tanganyika, and from there the fish invaded African rivers and Lakes Malawi and Victoria. Lake Victoria's cichlids have been linked to an invasion by ancestral cichlids approximately 200,000 years ago (figure 1.3). That time period is long from the perspective of a human lifetime, but it is a blink of the eye from the perspective of evolutionary time. Even more remarkably, zoologists now believe that most of these species arose much more recently. Lake Victoria nearly dried out 12,000 years ago, and most of the original Victorian species would have been lost in the process. Over 300 cichlid species would have arisen in 12,000 years. Genetic studies have confirmed the close relationships of Lake Victorian species.

Like all organisms, animals are named and classified into a hierarchy of relatedness. Although Karl von Linne (1707–1778) is primarily remembered for collecting and classifying plants, his system of naming—**binomial nomenclature**—has also been adopted for animals. A two-part name describes each kind of organism. The first part indicates the genus, and the second part indicates the species to which the organism belongs. Each kind of organism—for example, the cichlid scale-eater *Perissodus microlepis*—is recognized throughout the world by its two-part

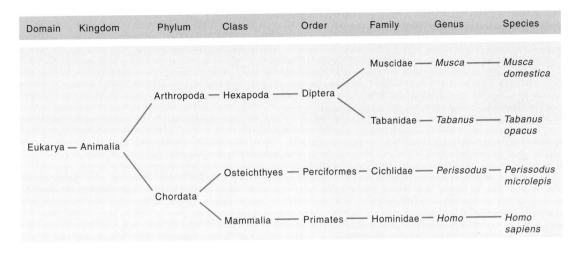

FIGURE 1.4

Hierarchy of Relatedness.   The classification of a housefly, horsefly, cichlid fish, and human illustrates how the classification system depicts degrees of relatedness.

name. Above the species and genus levels, organisms are grouped into families, orders, classes, phyla, kingdoms, and domains, based on a hierarchy of relatedness (figure 1.4). Organisms in the same species are more closely related than organisms in the same genus, and organisms in the same genus are more closely related than organisms in the same family, and so on. When zoologists classify animals into taxonomic groupings they are making hypotheses about the extent to which groups of animals share DNA, even when they study variations in traits like jaw structure, color patterns, and behavior, because these kinds of traits ultimately are based on the genetic material.

Evolutionary theory has affected zoology like no other single theory. It has impressed scientists with the fundamental unity of all of life. As the cichlids of Africa illustrate, evolutionary concepts hold the key to understanding why animals look and act in their unique ways, live in their particular geographical regions and habitats, and share characteristics with other related animals.

ZOOLOGY: AN ECOLOGICAL PERSPECTIVE

Just as important to zoology as an evolutionary perspective is an ecological perspective. **Ecology** (Gr. *okios*, house + *logos*, to study) is the study of the relationships between organisms and their environment (chapter 6). Throughout our history, humans have depended on animals, and that dependence too often has led to exploitation. We depend on animals for food, medicines, and clothing. We also depend on animals in other, more subtle ways. This dependence may not be noticed until human activities upset the delicate ecological balances that have evolved over hundreds of thousands of years. In the 1950s, the giant Nile perch (*Lates niloticus*) was introduced into Lake Victoria in an attempt to increase the lake's fishery. This voracious predator reduced the cichlid population from 99% to less than 1% of the total fish

population and has resulted in the extinction of many cichlid species. Because many of the cichlids fed on algae, the algae in the lake grew uncontrolled. When algae died and decayed, much of the lake became depleted of its oxygen. To make matters worse, when Nile perch are caught, their excessively oily flesh must be dried. Fishermen cut local forests for the wood needed to smoke the fish. This practice has resulted in severe deforestation around Lake Victoria. The resulting runoff of soil into the lake has caused further degradation.

Ecological problems also threaten Lake Tanganyika's cichlid populations. The area to the north of the lake has experienced nearly 100% deforestation. One-half of the forests on the Tanzania side of the lake are deforested to maintain a meager agricultural subsistence for human populations. Overfishing, agricultural runoff, and wastes from growing urban populations have led to some cichlid extinctions in the lake.

WORLD RESOURCES AND ENDANGERED ANIMALS

There is grave concern for the ecology of the entire world, not just Africa's greatest lakes. The problems, however, are most acute in developing countries, which are striving to attain the same wealth as industrialized nations. Two problems, global overpopulation and the exploitation of world resources, are the focus of our ecological concerns.

Population

Global overpopulation is at the root of virtually all other environmental problems. Human population growth is expected to continue in the twenty-first century. Most growth (92%) is in less developed countries, where 5 billion out of a total of 6.3 billion humans now live. Since a high proportion of the population is of childbearing age, the growth rate will increase in the twenty-first

(a)

(b)

FIGURE 1.5

Tropical Rain Forests: A Threatened World Resource.   A Brazilian tropical rain forest (*a*) before and (*b*) after clear-cutting to make way for agriculture. These soils quickly become depleted and are then abandoned for the richer soils of adjacent forests. Loss of tropical forests results in the extinction of many valuable forest species.

century. It is estimated that the world population will reach 10.4 billion by the year 2100. As the human population grows, the disparity between the wealthiest and poorest nations is likely to increase.

World Resources

Human overpopulation is stressing world resources. Although new technologies continue to increase food production, most food is produced in industrialized countries that already have a high per-capita food consumption. Maximum oil production is expected to continue in this millennium. Continued use of fossil fuels adds more carbon dioxide to the atmosphere, contributing to the greenhouse effect and global warming. Deforestation of large areas of the world results from continued demand for forest products and fuel. This trend contributes to the greenhouse effect, causes severe regional water shortages, and results in the extinction of many plant and animal species, especially in tropical forests. Forest preservation would result in the identification of new species of plants and animals that could be important human resources: new foods, drugs, building materials, and predators of pests (figure 1.5). Nature also has intrinsic value that is just as important as its provision of human resources. Recognition of this intrinsic worth provides important aesthetic and moral impetus for preservation.

Solutions

An understanding of basic ecological principles can help prevent ecological disasters like those we have described. Understanding how matter is cycled and recycled in nature, how populations grow, and how organisms in our lakes and forests use energy is fundamental to preserving the environment. There are no easy solutions to our ecological problems. Unless we deal with the problem of human overpopulation, however, solving the other problems will be impossible. We must work as a world community to prevent the spread of disease, famine, and other forms of suffering that accompany overpopulation. Bold and imaginative steps toward improved social and economic conditions and better resource management are needed.

"Wildlife Alerts" that appear at the end of selected chapters in the first two parts of this text remind us of the peril that an unprecedented number of species face around the world. Endangered or threatened species from a diverse group of animal phyla are highlighted.

WILDLIFE ALERT
An Overview of the Problems

Extinction has been the fate of most plant and animal species. It is a natural process that will continue. In recent years, however, the threat to the welfare of wild plants and animals has increased dramatically—mostly as a result of habitat destruction. Tropical rain forests, the most threatened areas on the earth, have been reduced to 44% of their original extent. In certain areas, such as Ecuador, forest coverage has been reduced by 95%. This decrease in habitat has resulted in tens of thousands of extinctions. Accurately estimating the number of extinctions is impossible in areas like rain forests, where taxonomists have not even described most species. We are losing species that we do not know exist, and we are losing resources that could lead to new medicines, foods, and textiles. Other causes of extinction include climate change, pollution, and invasions from foreign species. Habitats other than rain forests—grasslands, marshes, deserts, and coral reefs—are also being seriously threatened.

No one knows how many species living today are close to extinction. As of 2003, the U.S. Fish and Wildlife Service lists 1,263 species in the U.S. and 1,824 foreign species on its endangered or threatened species lists. An **endangered species** is in imminent danger of extinction throughout its range (where it lives). A **threatened species** is likely to become endangered in the near future. Box figure 1.1 shows the number of endangered and threatened species in different regions of the United States. Clearly, much work is needed to improve these alarming statistics.

In the chapters that follow, you will learn that saving species requires more than preserving a few remnant individuals. It requires a large diversity of genes within species groups to promote species survival in changing environments. This genetic diversity requires large populations of plants and animals.

Preservation of endangered species depends on a multifaceted conservation plan that includes the following components:

1. A global system of national parks to protect large tracts of land and wildlife corridors that allow movement between natural areas
2. Protected landscapes and multiple-use areas that allow controlled private activity but also retain value as a wildlife habitat
3. Zoos and botanical gardens to save species whose extinction is imminent

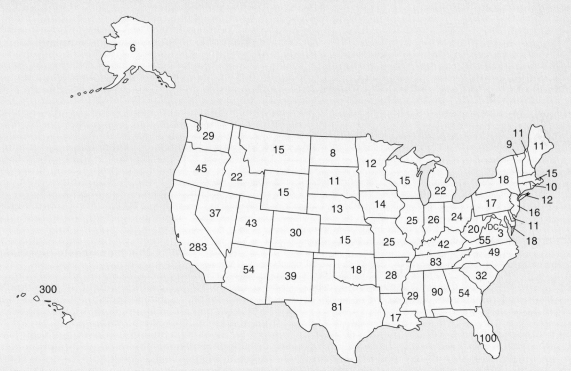

BOX FIGURE 1.1   **Map Showing Approximate Numbers of Endangered and Threatened Species in the United States.**   Because the ranges of some organisms overlap two or more states, the sum of all numbers is greater than the sum of all endangered and threatened species. The total number of endangered and threatened species in the United States is 1,263.

SUMMARY

1. Zoology is the study of animals. It is a broad field that requires zoologists to specialize in one or more subdisciplines.
2. Animals share a common evolutionary past and evolutionary forces that influenced their history.
3. Evolution explains how the diversity of animals arose.
4. Evolutionary relationships are the basis for the classification of animals into a hierarchical system. This classification system uses a two-part name for every kind of animal. Higher levels of classification denote more distant evolutionary relationships.
5. Animals share common environments, and ecological principles help us to understand how animals interact within those environments.
6. Human overpopulation is at the root of virtually all other environmental problems. It stresses world resources and results in pollution, global warming, deforestation, and the extinction of many plant and animal species.

SELECTED KEY TERMS

binomial nomenclature (p. 4)
ecology (p. 5)
endangered species (p. 7)
organic evolution (p. 3)
threatened species (p. 7)
zoology (p. 2)

CRITICAL THINKING QUESTIONS

1. How is zoology related to biology? What major biological concepts, in addition to evolution and ecology, are unifying principles shared between the two disciplines?
2. What are some current issues that involve both zoology and questions of ethics or public policy? What should be the role of zoologists in helping to resolve these issues?
3. Some people object to the teaching of evolution in public schools. What would be the effect on science education if evolution were banned from public school curricula?
4. Many of the ecological problems facing our world concern events and practices that occur in less developed countries. Many of these practices are the result of centuries of cultural evolution. What approach should people and institutions of developed countries take in helping to encourage ecologically minded resource use?
5. Why should people in all parts of the world be concerned with the extinction of cichlids in Lake Victoria?

ONLINE LEARNING CENTER

Visit our Online Learning Center (OLC) at www.mhhe.com/zoology (click on this book's title) to find the following chapter-related materials:

- CHAPTER QUIZZING
- RELATED WEB LINKS
  Introductory Materials
  Evolution
  Classification and Phylogeny of Animals
  Endangered Species
  Human Population Growth
- BOXED READINGS ON
  Science and Pseudoscience
  The Origin of Life on Earth—Life from Nonlife
  Continental Drift
  The Zebra Mussel
  Jaws from the Past
- SUGGESTED READINGS
- LAB CORRELATIONS

Check out the OLC to find specific information on these related lab exercises in the *General Zoology Laboratory Manual*, 5th edition, by Stephen A. Miller:

Exercise 6 *Adaptations of Stream Invertebrates—A Scavenger Hunt*
Exercise 7 *The Classification of Organisms*

CHAPTER 2

CELLS, TISSUES, ORGANS, AND ORGAN SYSTEMS OF ANIMALS

Outline

Concepts

1. Cells are the basic organizational units of life.
2. Eukaryotic cells exhibit a considerable degree of internal organization, with a dynamic system of membranes forming internal compartments called organelles.
3. The structure and function of a typical cell usually apply to all animals.
4. Different cell types organize into structural and functional units called tissues, organs, and organ systems.

Because all animals are made of cells, the cell is as fundamental to an understanding of zoology as the atom is to an understanding of chemistry. In the hierarchy of biological organization, the cell is the simplest organization of matter that exhibits all of the properties of life (figure 2.1). Some organisms are single celled; others are multicellular. An animal has a body composed of many kinds of specialized cells. A division of labor among cells allows specialization into higher levels of organization (tissues, organs, and organ systems). Yet, everything that an animal does is ultimately happening at the cellular level.

WHAT ARE CELLS?

Cells are the functional units of life, in which all of the chemical reactions necessary for the maintenance and reproduction of life take place. They are the smallest independent units of life. There are three basic types of cells, two of which lack nuclei and other membrane-bounded organelles. These simpler (**prokaryotic** or **prokaryotes**; "before nucleus") cells are in the domains Archaea and Eubacteria. The Archaea have unique characteristics but also share features with Eubacteria and the third domain, Eukarya. Eukaryotic cells are larger and more complex than prokaryotic cells. Since animals and protista are composed of eukaryotic cells, this cell type will be emphasized in this chapter. Table 2.1 compares prokaryotic and eukaryotic cells.

All **eukaryotes** ("true nucleus") have cells with a membrane-bound nucleus containing DNA. In addition, eukaryotic cells contain many other structures called **organelles** ("little organs") that perform specific functions. Eukaryotic cells also have a network of specialized structures called microfilaments and microtubules organized into the cytoskeleton, which gives shape to the cell and allows intracellular movement.

All eukaryotic cells have three basic parts (table 2.1):

1. The **plasma membrane** is the outer boundary of the cell. It separates the internal metabolic events from the environment and allows them to proceed in organized, controlled ways. The plasma membrane also has specific receptors for external molecules that alter the cell's function.

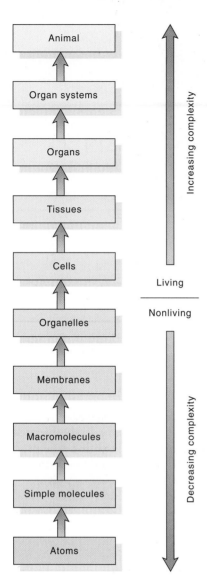

FIGURE 2.1

Structural Hierarchy in a Multicellular Animal. At each level, function depends on the structural organization of that level and those below it.

2. **Cytoplasm** (Gr. *kytos*, hollow vessel + *plasm*, fluid) is the portion of the cell outside the nucleus. The semifluid portion of the cytoplasm is called the cytosol. Suspended within the cytosol are the organelles.
3. The **nucleus** (pl., nuclei) is the cell control center. It contains the chromosomes and is separated from the cytoplasm by its own nuclear envelope. The nucleoplasm is the semifluid material in the nucleus.

Because cells vary so much in form and function, no "typical" cell exists. However, to help you learn as much as possible about cells, figure 2.2 shows an idealized version of a eukaryotic cell and most of its component parts.

WHY ARE MOST CELLS SMALL?

Most cells are small and can be seen only with the aid of a microscope. (Exceptions include the eggs of most vertebrates [fishes, amphibians, reptiles, and birds] and some long nerve cells.) One

TABLE 2.1
COMPARISON OF PROKARYOTIC AND EUKARYOTIC CELLS

COMPONENT	PROKARYOTE	EUKARYOTE
Organization of genetic material		
True membrane-bound nucleus	Absent	Present
DNA complexed with histones	No	Yes
Number of chromosomes	One	More than one
Nucleolus	Absent	Present
Mitosis occurs	No	Yes
Genetic recombination	Partial, unidirectional transfer of DNA	Meiosis and fusion of gametes
Mitochondria	Absent	Present
Chloroplasts	Absent	Present
Plasma membrane with sterols	Usually no	Yes
Flagella	Submicroscopic in size; composed of only one fiber	Microscopic in size; membrane bound; usually 20 microtubules in 9 + 2 pattern
Endoplasmic reticulum	Absent	Present
Golgi apparatus	Absent	Present
Cell walls	Usually chemically complex	Chemically simpler
Simpler organelles		
Ribosomes	70S	80S (except in mitochondria and chloroplasts)
Lysosomes and peroxisomes	Absent	Present
Microtubules	Absent or rare	Present
Cytoskeleton	May be absent	Present
Vacuoles	Present	Present
Vesicles	Present	Present
Differentiation	Rudimentary	Tissues and organs

reason for the small size of cells is that the ratio of the volume of the cell's nucleus to the volume of its cytoplasm must not be so small that the nucleus, the cell's major control center, cannot control the cytoplasm.

Another aspect of cell volume works to limit cell size. As the radius of a cell lengthens, cell volume increases more rapidly than cell surface area (figure 2.3). The need for nutrients and the rate of waste production are proportional to cell volume. The cell takes up nutrients and eliminates wastes through its surface

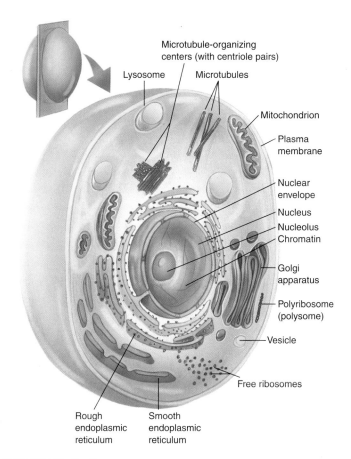

FIGURE 2.2

A Generalized Animal Cell. Understanding of the structures in this cell is based mainly on electron microscopy. The sizes of some organelles and structures are exaggerated to show detail.

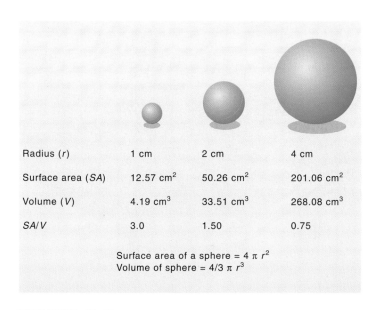

Radius (r)	1 cm	2 cm	4 cm
Surface area (SA)	12.57 cm²	50.26 cm²	201.06 cm²
Volume (V)	4.19 cm³	33.51 cm³	268.08 cm³
SA/V	3.0	1.50	0.75

Surface area of a sphere = $4 \pi r^2$
Volume of sphere = $4/3 \pi r^3$

FIGURE 2.3

The Relationship Between Surface Area and Volume. As the radius of a sphere increases, its volume increases more rapidly than its surface area. (SA/V = surface area-to-volume ratio.)

plasma membrane. If cell volume becomes too large, the surface area-to-volume ratio is too small for an adequate exchange of nutrients and wastes.

CELL MEMBRANES

The plasma membrane surrounds the cell. Other membranes inside the cell enclose some organelles and have properties similar to the plasma membrane.

STRUCTURE OF CELL MEMBRANES

In 1972, S. Jonathan Singer and Garth Nicolson developed the fluid-mosaic model of membrane structure. According to this model, a membrane is a double layer (bilayer) of proteins and phospholipids, and is fluid rather than solid. The phospholipid bilayer forms a fluid "sea" in which specific proteins float like icebergs (figure 2.4). Being fluid, the membrane is in a constant state of flux—shifting and changing, while retaining its uniform structure. The word *mosaic* refers to the many different kinds of proteins dispersed in the phospholipid bilayer.

The following are important points of the fluid-mosaic model:

1. The phospholipids have one polar end and one nonpolar end. The polar ends are oriented on one side toward the outside of the cell and into the fluid cytoplasm on the other side, and the nonpolar ends face each other in the middle of the bilayer. The "tails" of both layers of phospholipid molecules attract each other and are repelled by water (they are hydrophobic, "water dreading"). As a result, the polar spherical "heads" (the phosphate portion) are located over the cell surfaces (outer and inner) and are "water attracting" (they are hydrophilic).
2. Cholesterol is present in the plasma membrane and organelle membranes of eukaryotic cells. The cholesterol molecules are embedded in the interior of the membrane and help to make the membrane less permeable to water-soluble substances. In addition, the relatively rigid structure of the cholesterol molecules helps to stabilize the membrane (figure 2.5).
3. The membrane proteins are individual molecules attached to the inner or outer membrane surface (peripheral proteins) or embedded in it (intrinsic proteins) (*see figure 2.4*). Some intrinsic proteins are links to sugar-protein markers on the cell surface. Other intrinsic proteins help to move ions or molecules across the membrane, and still others attach the membrane to the cell's inner scaffolding (the cytoskeleton) or to various molecules outside the cell.
4. When carbohydrates unite with proteins, they form glycoproteins, and when they unite with lipids, they form glycolipids on the surface of a plasma membrane. Surface carbohydrates and portions of the proteins and lipids make up the **glycocalyx** ("cell coat") (figure 2.6). This arrangement of distinctively shaped groups of sugar molecules of the glycocalyx acts as a molecular "fingerprint" for each cell type. The glycocalyx is necessary for cell-to-cell recognition and the behavior of certain cells, and it is a key component in coordinating cell behavior in animals.

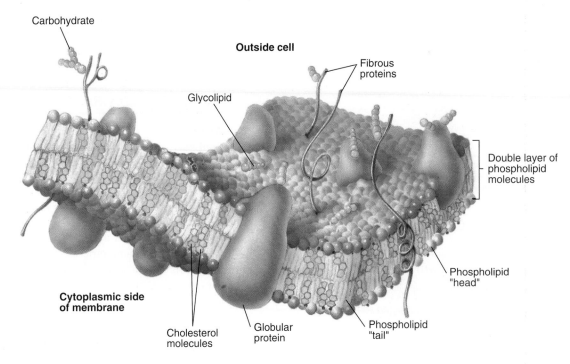

FIGURE 2.4

Fluid-Mosaic Model of Membrane Structure.　Intrinsic globular proteins may protrude above or below the lipid bilayer and may move about in the membrane. Peripheral proteins attach to either the inner or outer surfaces.

FUNCTIONS OF CELL MEMBRANES

Cell membranes (1) regulate material moving into and out of the cell, and from one part of the cell to another; (2) separate the inside of the cell from the outside; (3) separate various organelles within the cell; (4) provide a large surface area on which specific chemical reactions can occur; (5) separate cells from one another; and (6) are a site for receptors containing specific cell identification markers that differentiate one cell type from another.

The ability of the plasma membrane to let some substances in and keep others out is called **selective permeability** (L. *permeare* or *per*, through + *meare*, pass) and is essential for maintaining cellular homeostasis. **Homeostasis** (Gr. *homeo*, always the same + *stasis*, standing) is the maintenance of a relatively constant internal environment despite fluctuations in the external environment. However, before you can fully understand how substances pass into and out of cells and organelles, you must know how the molecules of those substances move from one place to another.

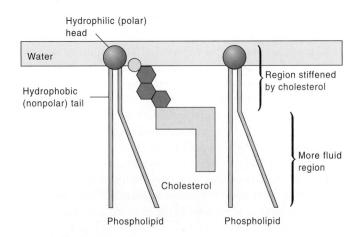

FIGURE 2.5

The Arrangement of Cholesterol Between Lipid Molecules of a Lipid Bilayer.　Cholesterol stiffens the outer lipid bilayer and causes the inner region of the bilayer to become slightly more fluid. Only half the lipid bilayer is shown; the other half is a mirror image.

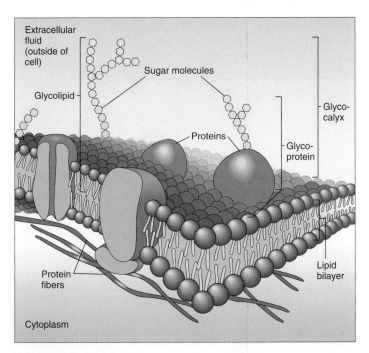

FIGURE 2.6

The Glycocalyx, Showing the Glycoproteins and Glycolipids.　Note that all of the attached carbohydrates are on the outside of the plasma membrane.

MOVEMENT ACROSS MEMBRANES

Molecules can cross membranes in a number of ways, both by using their own energy and by relying on an outside energy source. Table 2.2 summarizes the various kinds of transmembrane movement, and the sections that follow discuss them in more detail.

SIMPLE DIFFUSION

Molecules move randomly at all temperatures above absolute zero (−273° C) (due to spontaneous molecular motion) from areas where they are highly concentrated to areas of lower concentration, until they are evenly distributed in a state of dynamic equi-librium. This process is **simple diffusion** (L. *diffundere*, to spread). Simple diffusion accounts for most of the short-distance transport of substances moving into and out of cells. Figure 2.7 shows the diffusion of sugar particles away from a sugar cube placed in water.

FACILITATED DIFFUSION

Polar molecules (not soluble in lipids) may diffuse through protein channels (pores) in the lipid bilayer (figure 2.8). The protein channels offer a continuous pathway for specific molecules to move across the plasma membrane so that they never come into contact with the hydrophobic layer or the membrane's polar surface.

Large molecules and some of those not soluble in lipids require assistance in passing across the plasma membrane. These

TABLE 2.2
DIFFERENT TYPES OF MOVEMENT ACROSS PLASMA MEMBRANES

TYPE OF MOVEMENT	DESCRIPTION	EXAMPLE IN THE BODY OF A FROG
Simple diffusion	No cell energy is needed. Molecules move "down" a concentration gradient. Molecules spread out randomly from areas of higher concentration to areas of lower concentration until they are distributed evenly in a state of dynamic equilibrium.	A frog inhales air containing oxygen, which moves into the lungs and then diffuses into the bloodstream.
Facilitated diffusion	Carrier (transport) proteins in a plasma membrane temporarily bind with molecules and help them pass across the membrane. Other proteins form channels through which molecules move across the membrane.	Glucose in the gut of a frog combines with carrier proteins to pass through the gut cells into the bloodstream.
Osmosis	Water molecules diffuse across selectively permeable membranes from areas of higher concentration to areas of lower concentration.	Water molecules move into a frog's red blood cell when the concentration of water molecules outside the blood cell is greater than it is inside.
Filtration	Essentially protein-free plasma moves across capillary walls due to a pressure gradient across the wall.	A frog's blood pressure forces water and dissolved wastes into the kidney tubules during urine formation.
Active transport	Specific carrier proteins in the plasma membrane bind with molecules or ions to help them cross the membrane against a concentration gradient. Cellular energy is required.	Sodium ions move from inside the neurons of the sciatic nerve of a frog (the sodium-potassium pump) to the outside of the neurons.
Endocytosis	The bulk movement of material into a cell by the formation of a vesicle.	
Pinocytosis	The plasma membrane encloses small amounts of fluid droplets (in a vesicle) and takes them into the cell.	The kidney cells of a frog take in fluid to maintain fluid balance.
Phagocytosis	The plasma membrane forms a vesicle around a solid particle or other cell and draws it into the phagocytic cell.	The white blood cells of a frog engulf and digest harmful bacteria.
Receptor-mediated endocytosis	Extracellular molecules bind with specific receptor proteins on a plasma membrane, causing the membrane to invaginate and draw molecules into the cell.	The intestinal cells of a frog take up large molecules from the inside of the gut.
Exocytosis	The movement of material out of a cell. A vesicle (with particles) fuses with the plasma membrane and expels particles or fluids from the cell across the plasma membrane. The reverse of endocytosis.	The sciatic nerve of a frog releases a chemical (neurotransmitter).

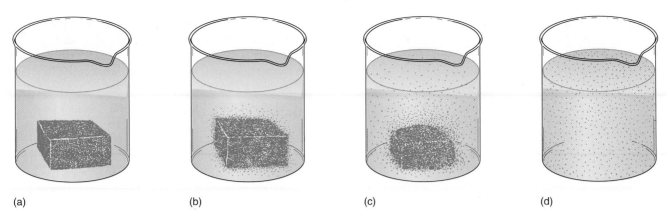

(a)   (b)   (c)   (d)

FIGURE 2.7

Simple Diffusion.   When a sugar cube is placed in water (*a*), it slowly dissolves (*b*) and disappears. As this happens, the sugar molecules diffuse from a region where they are more concentrated to a region (*c*) where they are less concentrated. Even distribution of the sugar molecules throughout the water is diffusion equilibrium (*d*).

molecules use **facilitated diffusion,** which, like simple diffusion, requires no energy input. To pass across the membrane, a molecule temporarily binds with a carrier (transport) protein in the plasma membrane and is transported from an area of higher concentration to one of lower concentration (figure 2.9).

OSMOSIS

The diffusion of water across a selectively permeable membrane from an area of higher concentration to an area of lower concentration is **osmosis** (Gr. *osmos*, pushing). Osmosis is just a special type of diffusion, not a different method (figure 2.10).

The term **tonicity** (Gr. *tonus*, tension) refers to the relative concentration of solutes in the water inside and outside the cell. For example, in an **isotonic** (Gr. *isos*, equal + *tonus*, tension) solution, the solute concentration is the same inside and outside a red blood cell (figure 2.11*a*). The concentration of water molecules is also the same inside and outside the cell. Thus, water molecules move across the plasma membrane at the same rate in both

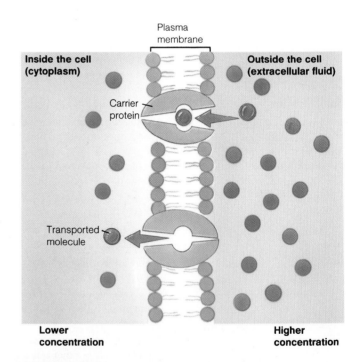

FIGURE 2.9

Facilitated Diffusion and Carrier (Transport) Proteins.   Some molecules move across the plasma membrane with the assistance of carrier proteins that transport the molecules down their concentration gradient, from a region of higher concentration to one of lower concentration. A carrier protein alternates between two configurations, moving a molecule across a membrane as the shape of the protein changes. The rate of facilitated diffusion depends on how many carrier proteins are available in the membrane and how fast they can move their specific molecule.

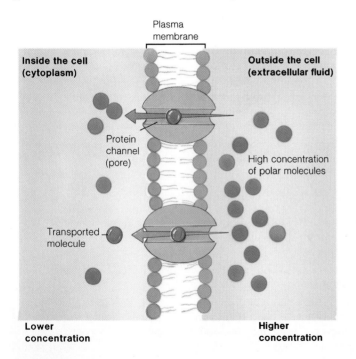

FIGURE 2.8

Transport Proteins.   Molecules can move into and out of cells through integrated protein channels (pores) in the plasma membrane without using energy.

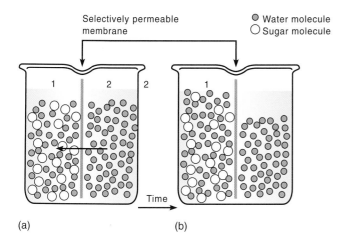

(a)   (b)

FIGURE 2.10

Osmosis. (*a*) A selectively permeable membrane separates the beaker into two compartments. Initially, compartment 1 contains sugar and water molecules, and compartment 2 contains only water molecules. Due to molecular motion, water moves down the concentration gradient (from compartment 2 to compartment 1) by osmosis. The sugar molecules remain in compartment 1 because they are too large to pass across the membrane. (*b*) At osmotic equilibrium, the number of water molecules in compartment 1 does not increase.

directions, and there is no net movement of water in either direction.

In a **hypertonic** (Gr. *hyper,* above) solution, the solute concentration is higher outside the red blood cell than inside. Because the concentration of water molecules inside the cell is higher than outside, water moves out of the cell, which shrinks (figure 2.11*b*). This condition is called crenation in red blood cells.

In a **hypotonic** (Gr. *hypo,* under) solution, the solute concentration is lower outside the red blood cell than inside. Conversely, the concentration of water molecules is higher outside the cell than inside. As a result, water moves into the cell, which swells and may burst (figure 2.11*c*).

FILTRATION

Filtration is a process that forces small molecules across selectively permeable membranes with the aid of hydrostatic (water) pressure (or some other externally applied force, such as blood pressure). For example, in the body of an animal such as a frog, filtration is evident when blood pressure forces water and dissolved molecules through the permeable walls of small blood vessels called capillaries (figure 2.12). In filtration, large molecules, such as proteins, do not pass through the smaller membrane pores. Filtration also takes place in the kidneys when blood pressure forces water and dissolved wastes out of the blood vessels and into the kidney tubules in the first step in urine formation.

ACTIVE TRANSPORT

Active-transport processes move molecules across a selectively permeable membrane against a concentration gradient—that is,

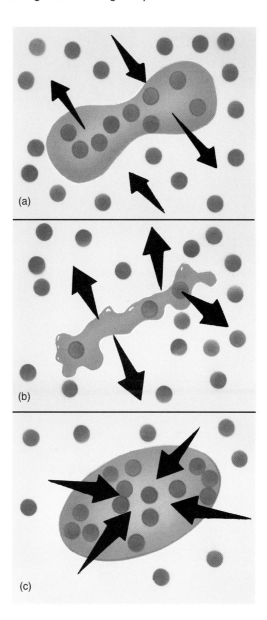

FIGURE 2.11

Effect of Salt Concentration on Cell Volumes. (*a*) An isotonic solution with the same salt concentration inside and outside the cell has no effect on the size of the red blood cell. (*b*) A hypertonic (high salt) solution causes water to leave the red blood cell, which shrinks. (*c*) A hypotonic (low salt) solution results in an inflow of water, causing the red blood cell to swell. Arrows indicate direction of water movement.

from an area of lower concentration to one of higher concentration. This movement against the concentration gradient requires ATP energy.

The active-transport process is similar to facilitated diffusion, except that the carrier protein in the plasma membrane must use energy to move the molecules against their concentration gradient (figure 2.13).

One active-transport mechanism, the sodium-potassium pump, helps maintain the high concentrations of potassium ions and low concentrations of sodium ions inside nerve cells that are necessary for the transmission of electrical impulses. Another

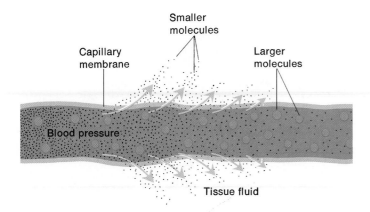

FIGURE 2.12

Filtration.   The high blood pressure in the capillary forces small molecules through the capillary membrane. Larger molecules cannot pass through the small openings in the capillary membrane and remain in the capillary. Arrows indicate the direction of small molecule movement.

active-transport mechanism, the calcium pump, keeps the calcium concentration hundreds of times lower inside the cell than outside.

ENDOCYTOSIS

Another process by which substances move across the plasma membrane is endocytosis. **Endocytosis** (Gr. *endon,* within) involves the bulk movement of materials across the plasma membrane, rather

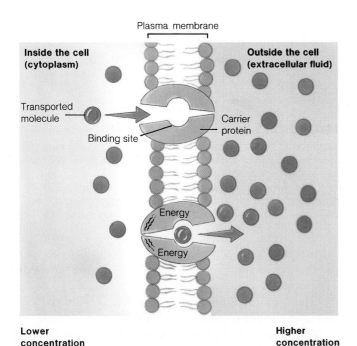

FIGURE 2.13

Active Transport.   During active transport, a molecule combines with a carrier protein whose shape is altered as a result of the combination. This change in configuration, along with ATP energy, helps move the molecule across the plasma membrane against a concentration gradient.

than movement of individual molecules. The three forms of endocytosis are pinocytosis, phagocytosis, and receptor-mediated endocytosis.

Pinocytosis ("cell drinking," from Gr. *pinein,* to drink + *cyto,* cell) is nonspecific uptake of small droplets of extracellular fluid. Any small solid dissolved in the fluid is also taken into the cell. Pinocytosis occurs when a small portion of the plasma membrane indents (invaginates). The open end of the invagination seals itself off, forming a small vesicle. This tiny vesicle detaches from the plasma membrane and moves into the cytoplasm (figure 2.14*a*).

Phagocytosis ("cell eating," from Gr. *phagein,* to eat + *cyto,* cell) is similar to pinocytosis, except that the cell takes in solid material rather than liquid. Commonly, an organelle called a lysosome combines with the vesicle to form a **phagolysosome** ("digestion vacuole"), and lysosomal digestive enzymes break down the vesicle's contents (figure 2.14*b*).

Receptor-mediated endocytosis involves a specific receptor protein on the plasma membrane that "recognizes" an extracellular molecule and binds with it (figure 2.14*c*). This reaction somehow stimulates the membrane to indent and create a vesicle containing the selected molecule. A variety of important molecules (such as cholesterol) are brought into cells in this manner.

EXOCYTOSIS

An organelle known as the Golgi apparatus (described in a later section) packages proteins and other molecules that the cell produces into vesicles for secretion. In the process of **exocytosis** (Gr. *exo,* outside), these secretory vesicles fuse with the plasma membrane and release their contents into the extracellular environment (figure 2.14*b*). This process adds new membrane material, which replaces the plasma membrane lost during endocytosis.

CYTOPLASM, ORGANELLES, AND CELLULAR COMPONENTS

Many cell functions that are performed in the cytoplasmic compartment result from the activity of specific structures called organelles. Organelles effectively compartmentalize a cell's activities, improving efficiency and protecting cell contents from harsh chemicals. Organelles also enable cells to secrete various substances, derive energy from nutrients, degrade debris and waste materials, and reproduce. Table 2.3 summarizes the structure and function of these organelles, and the sections that follow discuss them in more detail.

CYTOPLASM

The cytoplasm of a cell has two distinct parts: (1) The **cytomembrane system** consists of well-defined structures, such as the endoplasmic reticulum, Golgi apparatus, vacuoles, and vesicles. (2) The fluid **cytosol** suspends the structures of the cytomembrane system and contains various dissolved molecules.

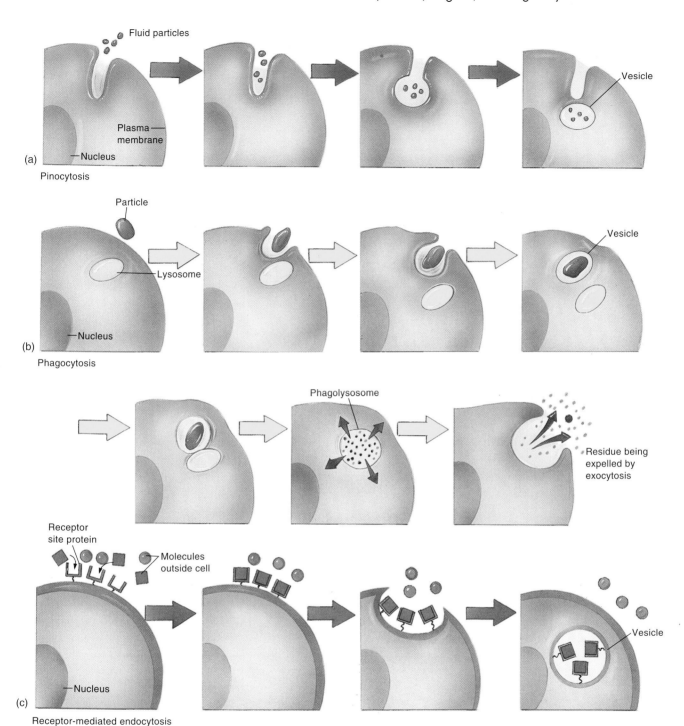

FIGURE 2.14

Endocytosis and Exocytosis. (*a*) Pinocytosis. A cell takes in small fluid particles and forms a vesicle. (*b*) Phagocytosis. A cell takes in a solid particle and forms a vesicle. A lysosome combines with a vesicle, forming a phagolysosome. Lysosomal enzymes digest the particle. The vesicle can also fuse with the plasma membrane and release its contents by exocytosis. (*c*) In receptor-mediated endocytosis, a specific molecule binds to a receptor protein, inducing the formation of a vesicle.

RIBOSOMES: PROTEIN WORKBENCHES

Ribosomes are non–membrane-bound structures that are the sites for protein synthesis. They contain almost equal amounts of protein and a special kind of ribonucleic acid called ribosomal RNA (rRNA). Some ribosomes attach to the endoplasmic reticulum (see next section), and some float freely in the cytoplasm. Whether ribosomes are free or attached, they usually cluster in groups connected by a strand of another kind of ribonucleic acid called messenger RNA (mRNA). These clusters are called polyribosomes or polysomes (*see figure 2.2*).

TABLE 2.3
STRUCTURE AND FUNCTION OF CELLULAR COMPONENTS

COMPONENT	STRUCTURE/DESCRIPTION	FUNCTION
Centriole	Located within microtuble-organizing center; contains nine triple microtubules	Forms basal body of cilia and flagella; functions in mitotic spindle formation
Chloroplast	Organelle that contains chlorophyll and is involved in photosynthesis	Traps, transforms, and uses light energy to convert carbon dioxide and water into glucose and oxygen
Chromosome	Made up of nucleic acid (DNA) and protein	Controls heredity and cellular activities
Cilia, flagella	Threadlike processes	Cilia and flagella move small particles past fixed cells and are a major form of locomotion in some cells
Cytomembrane system	The endoplasmic reticulum, Golgi apparatus, vacuoles, vesicles	Organelles, functioning as a system, to modify, package, and distribute newly formed proteins and lipids
Cytoplasm	Semifluid enclosed within plasma membrane; consists of fluid cytosol, organelles, and other structures	Dissolves substances; houses organelles and vesicles
Cytoskeleton	Interconnecting microfilaments and microtubules; flexible cellular framework	Assists in cell movement; provides support; site for binding of specific enzymes
Cytosol	Fluid part of cytoplasm; enclosed within plasma membrane; surrounds nucleus	Houses organelles; serves as fluid medium for metabolic reactions
Endoplasmic reticulum (ER)	Extensive membrane system extending throughout the cytoplasm from the plasma membrane to the nuclear envelope	Storage and internal transport; rough ER is a site for attachment of ribosomes; smooth ER makes lipids
Golgi apparatus	Stacks of disklike membranes	Sorts, packages, and routes cell's synthesized products
Lysosome	Membrane-bound sphere	Digests materials
Microfilament	Rodlike structure containing the protein actin	Gives structural support and assists in cell movement
Microtubule	Hollow, cylindrical structure	Assists in movement of cilia, flagella, and chromosomes; transport system
Microtubule-organizing center	Cloud of cytoplasmic material that contains centrioles	Dense site in the cytoplasm that gives rise to large numbers of microtubules with different functions in the cytoskeleton
Mitochondrion	Organelle with double, folded membranes	Converts energy into a form the cell can use
Nucleolus	Rounded mass within nucleus; contains RNA and protein	Preassembly point for ribosomes
Nucleus	Spherical structure surrounded by a nuclear envelope; contains nucleolus and DNA	Contains DNA that controls cell's genetic program and metabolic activities
Plasma membrane	The outer bilayered boundary of the cell; composed of protein, cholesterol, and phospholipid	Protection; regulation of material movement; cell-to-cell recognition
Ribosome	Contains RNA and protein; some are free, and some attach to ER	Site of protein synthesis
Vacuole	Membrane-surrounded, often large, sac in the cytoplasm	Storage site of food and other compounds; also pumps water out of a cell (e.g., contractile vacuole)
Vaults	Cytoplasmic ribonucleoproteins shaped like octagonal barrels.	Dock at nuclear pores, pick up molecules synthesized in the nucleus, and deliver their load to various places within the cell
Vesicle	Small, membrane-surrounded sac; contains enzymes or secretory products	Site of intracellular digestion, storage, or transport

ENDOPLASMIC RETICULUM: PRODUCTION AND TRANSPORT

The **endoplasmic reticulum (ER)** is a complex, membrane-bound labyrinth of flattened sheets, sacs, and tubules that branches and spreads throughout the cytoplasm. The ER is continuous from the nuclear envelope to the plasma membrane (*see figure 2.2*) and is a series of channels that helps various materials to circulate throughout the cytoplasm. It also is a storage unit for enzymes and other proteins and a point of attachment for ribosomes. ER with

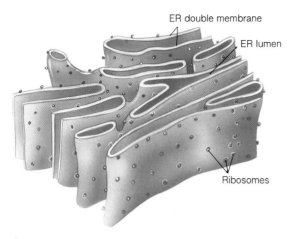

(a)   Rough endoplasmic reticulum

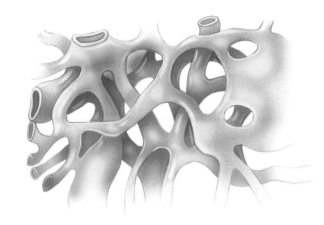

(b)   Smooth endoplasmic reticulum

FIGURE 2.15

Endoplasmic Reticulum (ER). (*a*) Ribosomes coat rough ER. Notice the double membrane and the lumen (space) between each membrane. (*b*) Smooth ER lacks ribosomes.

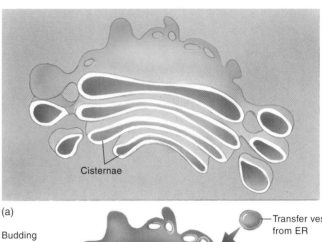

(a)

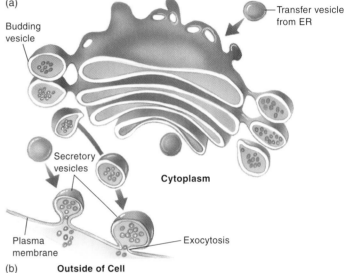

(b)

FIGURE 2.16

Golgi Apparatus. (*a*) The Golgi apparatus consists of a stack of cisternae. Notice the curved nature of the cisternae. (*b*) The Golgi apparatus stores, sorts, packages, and secretes cell products. Secretory vesicles move from the Golgi apparatus to the plasma membrane and fuse with it, releasing their contents to the outside of the cell via exocytosis.

attached ribosomes is rough ER (figure 2.15*a*), and ER without attached ribosomes is smooth ER (figure 2.15*b*). Smooth ER is the site for lipid production, detoxification of a wide variety of organic molecules, and storage of calcium ions in muscle cells. Most cells contain both types of ER, although the relative proportion varies among cells.

GOLGI APPARATUS: PACKAGING, SORTING, AND EXPORT

The **Golgi apparatus** or **complex** (named for Camillo Golgi, who discovered it in 1898) is a collection of membranes associated physically and functionally with the ER in the cytoplasm (figure 2.16*a*; *see also figure 2.2*). It is composed of flattened stacks of membrane-bound cisternae (sing., cisterna; closed spaces serving as fluid reservoirs). The Golgi apparatus sorts, packages, and secretes proteins and lipids.

Proteins that ribosomes synthesize are sealed off in little packets called transfer vesicles. Transfer vesicles pass from the ER to the Golgi apparatus and fuse with it (figure 2.16*b*). In the Golgi appara-

tus, the proteins are concentrated and chemically modified. One function of this chemical modification seems to be to mark and sort the proteins into different batches for different destinations. Eventually, the proteins are packaged into secretory vesicles, which are released into the cytoplasm close to the plasma membrane. When the vesicles reach the plasma membrane, they fuse with it and release their contents to the outside of the cell by exocytosis. Golgi apparatuses are most abundant in cells that secrete chemical substances (e.g., pancreatic cells secreting digestive enzymes and nerve cells secreting neurotransmitters). As noted in the next section, the Golgi apparatus also produces lysosomes.

LYSOSOMES: DIGESTION AND DEGRADATION

Lysosomes (Gr. *lyso*, dissolving + *soma*, body) are membrane-bound spherical organelles that contain enzymes called acid hydrolases, which are capable of digesting organic molecules

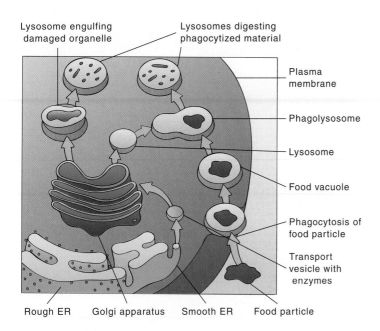

FIGURE 2.17

Lysosome Formation and Function. Lysosomes arise from the Golgi apparatus and fuse with vesicles that have engulfed foreign material to form digestive vesicles (phagolysosomes). These vesicles function in the normal recycling of cell constituents.

(lipids, proteins, nucleic acids, and polysaccharides) under acidic conditions. The enzymes are synthesized in the ER, transported to the Golgi apparatus for processing, and then secreted by the Golgi apparatus in the form of lysosomes or as vesicles that fuse with lysosomes (figure 2.17). Lysosomes fuse with phagocytic vesicles, thus exposing the vesicle's contents to lysosomal enzymes.

MITOCHONDRIA: POWER GENERATORS

Mitochondria (sing., mitochondrion) are double-membrane-bound organelles that are spherical to elongate in shape. A small space separates the outer membrane from the inner membrane. The inner membrane folds and doubles in on itself to form incomplete partitions called cristae (sing., crista; figure 2.18). The cristae increase the surface area available for the chemical reactions that trap usable energy for the cell. The space between the cristae is the matrix. The matrix contains ribosomes, circular DNA, and other material. Because they convert energy to a usable form, mitochondria are frequently called the "power generators" of the cell. Mitochondria usually multiply when a cell needs to produce more energy.

CYTOSKELETON: MICROTUBULES, INTERMEDIATE FILAMENTS, AND MICROFILAMENTS

In most cells, the microtubules, intermediate filaments, and microfilaments form the flexible cellular framework called the

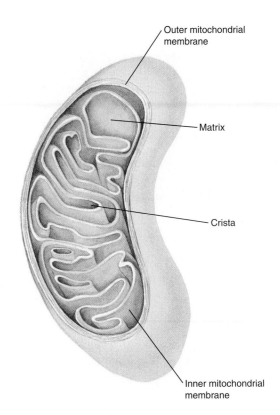

FIGURE 2.18

Mitochondrion. Mitochondrial membranes, cristae, and matrix. The matrix contains DNA, ribosomes, and enzymes.

cytoskeleton ("cell skeleton") (figure 2.19). This latticed framework extends throughout the cytoplasm, connecting the various organelles and cellular components.

Microtubules are hollow, slender, cylindrical structures in animal cells. Each microtubule is made of spiraling subunits of globular proteins called tubulin subunits (figure 2.20a). Microtubules function in the movement of organelles, such as secretory vesicles, and in chromosome movement during division of the cell nucleus. They are also part of a transport system within the cell. For example, in nerve cells, they help move materials through the long nerve processes. Microtubules are an important part of the cytoskeleton in the cytoplasm, and they are involved in the overall shape changes that cells undergo during periods of specialization.

Intermediate filaments are a chemically heterogeneous group of protein fibers, the specific proteins of which can vary with cell type (figure 2.20b). These filaments help to maintain cell shape and the spatial organization of organelles, as well as promote mechanical activities within the cytoplasm.

Microfilaments are solid strings of protein (actin) molecules (figure 2.20c). Actin microfilaments are most highly developed in muscle cells as myofibrils, which help muscle cells to shorten or contract. Actin microfilaments in nonmuscle cells provide mechanical support for various cellular structures and help form contractile systems responsible for some cellular movements (e.g., amoeboid movement in some protozoa).

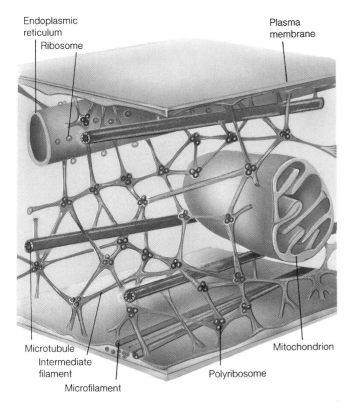

FIGURE 2.19

The Cytoskeleton.   Model of the cytoskeleton, showing the three-dimensional arrangement of the microtubules, intermediate filaments, and microfilaments.

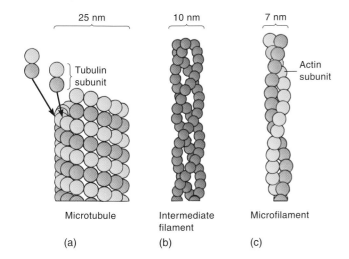

FIGURE 2.20

Three Major Classes of Protein Fibers Making Up the Cytoskeleton of Eukaryotic Cells.   (*a*) Microtubules consist of globular protein subunits (tubulins) linked in parallel rows. (*b*) Intermediate filaments in different cell types are composed of different protein subunits. (*c*) The protein actin is the key subunit in microfilaments.

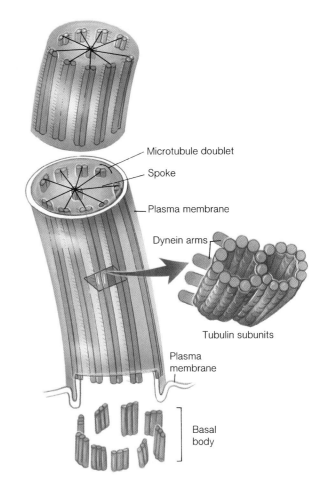

FIGURE 2.21

Internal Structure of Cilia and Flagella.   In cross section, the arms extend from each microtubule doublet toward a neighboring doublet, and spokes extend toward the central paired microtubules. The dynein arms push against the adjacent microtubule doublet to bring about movement.

CILIA AND FLAGELLA: MOVEMENT

Cilia (sing., cilium; L. "eyelashes") and **flagella** (sing., flagellum; L. "small whips") are elongated appendages on the surface of some cells by which the cells, including many unicellular organisms, propel themselves. In stationary cells, cilia or flagella move material over the cell's surface.

Although flagella are 5 to 20 times as long as cilia and move somewhat differently, cilia and flagella have a similar structure. Both are membrane-bound cylinders that enclose a matrix. In this matrix is an **axoneme** or **axial filament,** which consists of nine pairs of microtubules arranged in a circle around two central tubules (figure 2.21). This is called a 9 + 2 pattern of microtubules. Each microtubule pair (a doublet) also has pairs of dynein (protein) arms projecting toward a neighboring doublet and spokes extending toward the central pair of microtubules. Cilia and flagella move as a result of the microtubule doublets sliding along one another.

In the cytoplasm at the base of each cilium or flagellum lies a short, cylindrical **basal body,** also made up of microtubules and structurally identical to the centriole. The basal body controls the growth of microtubules in cilia or flagella. The microtubules in the basal body form a 9 + 0 pattern: nine sets of three with none in the middle.

CENTRIOLES AND MICROTUBULE-ORGANIZING CENTERS

The specialized nonmembranous regions of cytoplasm near the nucleus are the **microtubule-organizing centers.** These centers of dense material give rise to a large number of microtubules with different functions in the cytoskeleton. For example, one type of center gives rise to the **centrioles** (*see figure 2.2*) that lie at right angles to each other. Each centriole is composed of nine triplet microtubules that radiate from the center like the spokes of a wheel. The centrioles are duplicated before cell division, are involved with chromosome movement, and help to organize the cytoskeleton.

VACUOLES: CELL MAINTENANCE

Vacuoles are membranous sacs that are part of the cytomembrane system. Vacuoles occur in different shapes and sizes and have various functions. For example, some freshwater single-celled organisms (e.g., protozoa) and sponges have contractile vacuoles that collect water and pump it to the outside to maintain the organism's internal environment. Other vacuoles store food.

VAULTS: A NEWLY DISCOVERED ORGANELLE

In the 1990s, cell biologists discovered another organelle—the **vault.** Vaults are cytoplasmic ribonucleoproteins shaped like

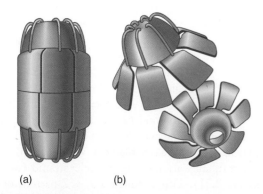

(a)   (b)

FIGURE 2.22

Vaults. (*a*) A three-dimensional drawing of the octagonal barrel-shaped organelle believed to transport messenger RNA from the nucleus to the ribosomes. (*b*) A vault opened to show its octagonal structure.

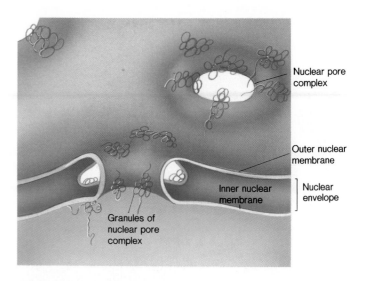

FIGURE 2.23

Nuclear Envelope. An artist's interpretation of pore structure, showing how the pore spans the two-layered nuclear envelope. The protein granules around the edge and in the center govern what passes through the pores.

octagonal barrels (figure 2.22). Their name is derived from their multiple arches that look like vaulted cathedral ceilings. One cell may contain thousands of vaults. The function of vaults may be related to their octagonal shape. Similarly, the nuclear pores (*see figure 2.23*) are also octagonally shaped and the same size as vaults, leading to speculation that vaults may be cellular "trucks." Vaults can dock at nuclear pores, pick up molecules synthesized in the nucleus, and deliver their load to various places within the cell. Since many vaults are always located near the nucleus, it is thought that they are picking up messenger RNA from the nucleus and transporting it to the ribosomes for protein synthesis.

THE NUCLEUS: INFORMATION CENTER

The nucleus contains the DNA and is the control and information center for the eukaryotic cell. It has two major functions. The nucleus directs chemical reactions in cells by transcribing genetic information from DNA into RNA, which then translates this specific information into proteins (e.g., enzymes) that determine the cell's specific activities. The nucleus also stores genetic information and transfers it during cell division from one cell to the next, and from one generation of organisms to the next.

NUCLEAR ENVELOPE: GATEWAY TO THE NUCLEUS

The **nuclear envelope** is a membrane that separates the nucleus from the cytoplasm and is continuous with the endoplasmic reticulum at a number of points. Over three thousand nuclear

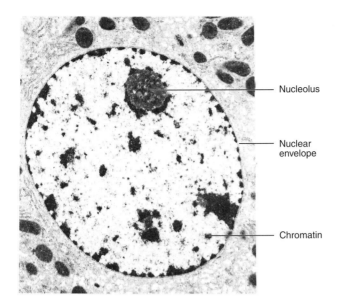

FIGURE 2.24

Nucleus.   The nucleolus, chromatin, and nuclear envelope are visible in this nucleus (TEM ×16,000).

pores penetrate the surface of the nuclear envelope (figure 2.23). These pores allow materials to enter and leave the nucleus, and they give the nucleus direct contact with the endoplasmic reticulum (*see figure 2.2*). Nuclear pores are not simply holes in the nuclear envelope; each is composed of an ordered array of globular and filamentous granules, probably proteins. These granules form the nuclear pore complex, which governs the transport of molecules into and out of the nucleus. The size of the pores prevents DNA from leaving the nucleus but permits RNA to be moved out.

CHROMOSOMES: GENETIC CONTAINERS

The nucleoplasm is the inner mass of the nucleus. In a nondividing cell, it contains genetic material called **chromatin.** Chromatin consists of a combination of DNA and protein and is the uncoiled, tangled mass of **chromosomes** ("colored bodies") containing the hereditary information in segments of DNA called genes. During cell division, each chromosome coils tightly, which makes the chromosome visible when viewed through a light microscope.

NUCLEOLUS: PREASSEMBLY POINT FOR RIBOSOMES

The **nucleolus** (pl., nucleoli) is a non–membrane-bound structure in the nucleoplasm that is present in nondividing cells (figure 2.24). Two or three nucleoli form in most cells, but some cells (e.g., amphibian eggs) have thousands. Nucleoli are preassembly points for ribosomes and usually contain proteins and RNA in many stages of synthesis and assembly. Assembly of ribosomes is completed after they leave the nucleus through the pores of the nuclear envelope.

LEVELS OF ORGANIZATION IN VARIOUS ANIMALS

Animals exhibit five major levels of organization. Each level is more complex than the one before and builds on it in a hierarchical fashion (*see figure 2.1*).

The first level exhibits **protoplasmic organization.** This level is found in unicellular organisms such as the protozoa where all life functions occur within the boundaries of a single cell. The second level exhibits **cellular organization.** Flagellates such as *Volvox* and some sponges can be placed at this level, where there is an aggregation of cells that are functionally differentiated and exhibit a division of labor. The third level is the **tissue level.** Jellyfishes have aggregations of cells organized into definite patterns or layers, which form a tissue. The fourth level is the **organ level.** Organs are composed of one or more tissues and have more specialized functions than tissues. This level first appears in the flatworms, where specific structures such as reproductive organs, eyespots, and feeding structures are present. The fifth and highest level of organization is the **system level.** At this level, organs work together to form systems such as the circulatory, digestive, reproductive, and respiratory system. This level first appears in the nemertean worms. Most animal phyla exhibit this level of organization.

TISSUES

In an animal, individual cells differentiate during development to perform special functions as aggregates called tissues. A **tissue** (Fr. *tissu*, woven) is a group of similar cells specialized for the performance of a common function. Animal tissues are classified as epithelial, connective, muscle, or nervous.

EPITHELIAL TISSUE: MANY FORMS AND FUNCTIONS

Epithelial tissue exists in many structural forms. In general, it either covers or lines something and typically consists of renewable sheets of cells that have surface specializations adapted for their specific roles. Usually, a basement membrane separates epithelial tissues from underlying, adjacent tissues. Epithelial tissues absorb (e.g., the lining of the small intestine), transport (e.g., kidney tubules), excrete (e.g., sweat glands), protect (e.g., the skin), and contain nerve cells for sensory reception (e.g., the taste buds in the tongue). The size, shape, and arrangement of epithelial cells are directly related to these specific functions.

Epithelial tissues are classified on the basis of shape and the number of layers present. Epithelium can be simple, consisting of only one layer of cells, or stratified, consisting of two or more stacked layers (figure 2.25e). Individual epithelial cells can be flat (squamous epithelium; figure 2.25a), cube shaped (cuboidal epithelium; figure 2.25b), or columnlike (columnar epithelium; figure 2.25c). The cells of pseudostratified ciliated columnar epithelium possess cilia and appear stratified or layered, but they are not; hence, the prefix pseudo. They look layered because their nuclei are at two or more levels within cells of the tissues (figure 2.25*d*) and they grow in height as old cells are replaced by new ones.

CONNECTIVE TISSUE: CONNECTION AND SUPPORT

Connective tissues support and bind. Unlike epithelial tissues, connective tissues are distributed throughout an extracellular matrix. This matrix frequently contains fibers that are embedded in a ground substance with a consistency anywhere from liquid to solid. To a large extent, the nature of this extracellular material determines the functional properties of the various connective tissues.

Connective tissues are of two general types, depending on whether the fibers are loosely or densely packed. In **loose connective tissue** strong, flexible fibers of the protein collagen are interwoven with fine, elastic, and reticular fibers, giving loose connective tissue its elastic consistency and making it an excellent binding tissue (e.g., binding the skin to underlying muscle tissue) (figure 2.25*g*). In **fibrous connective tissue,** the collagen fibers are densely packed and may lie parallel to one another, creating very strong cords, such as tendons (which connect muscles to bones or to other muscles) and ligaments (which connect bones to bones) (figure 2.25*h*).

Adipose tissue is a type of loose connective tissue that consists of large cells that store lipid (figure 2.25*f*). Most often, the cells accumulate in large numbers to form what is commonly called fat.

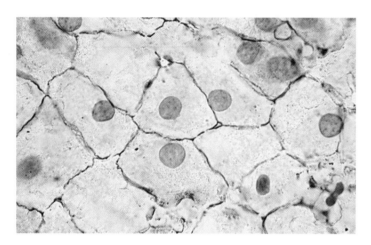

FIGURE 2.25

Tissue Types. (*a*) **Simple squamous epithelium** consists of a single layer of tightly packed, flattened cells with a disk-shaped central nucleus (LM × 250). **Location:** Air sacs of the lungs, kidney glomeruli, lining of heart, blood vessels, and lymphatic vessels. **Function:** Allows passage of materials by diffusion and filtration.

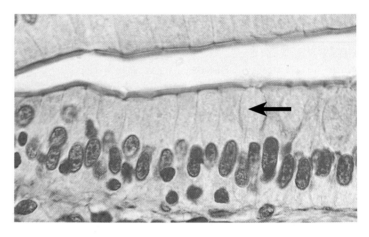

(*c*) **Simple columnar epithelium** consists of a single layer of elongated cells. The arrow points to a specialized goblet cell that secretes mucus (LM ×400). **Location:** Lines digestive tract, gallbladder, and excretory ducts of some glands. **Function:** Absorption, enzyme secretion.

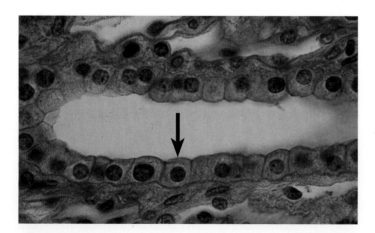

(*b*) **Simple cuboidal epithelium** consists of a single layer of tightly packed, cube-shaped cells. Notice the cell layer indicated by the arrow (LM ×250). **Location:** Kidney tubules, ducts and small glands, and surface of ovary. **Function:** Secretion and absorption.

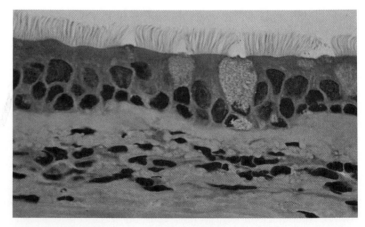

(*d*) **Pseudostratified ciliated columnar epithelium.** A tuft of cilia tops each columnar cell, except for goblet cells (LM ×500). **Location:** Lines the bronchi, uterine tubes, and some regions of the uterus. **Function:** Propels mucus or reproductive cells by ciliary action.

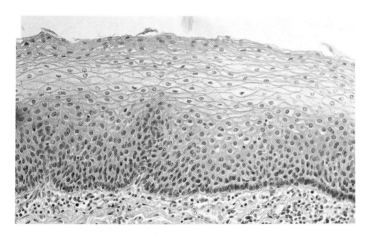

(*e*) **Stratified squamous epithelium** consists of many layers of cells (LM ×67). **Location:** Lines the esophagus, mouth, and vagina. Keratinized variety lines the surface of the skin. **Function:** Protects underlying tissues in areas subject to abrasion.

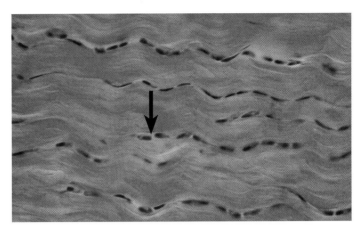

(*h*) **Fibrous connective tissue** consists largely of tightly packed collagenous fibers (LM ×250). The arrow points to a fibroblast. **Location:** Dermis of the skin, submucosa of the digestive tract, and fibrous capsules of organs and joints. **Function:** Provides structural strength.

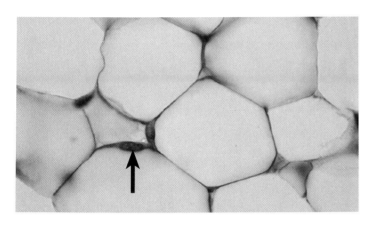

(*f*) **Adipose tissue** cells (adipocytes) contain large fat droplets that push the nuclei close to the plasma membrane. The arrow points to a nucleus (LM ×250). **Location:** Around kidneys, under skin, in bones, within abdomen, and in breasts. **Function:** Provides reserve fuel (lipids), insulates against heat loss, and supports and protects organs.

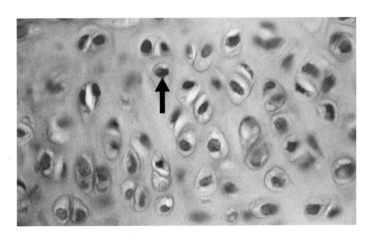

(*i*) **Hyaline cartilage** cells are located in lacunae (arrow) surrounded by intercellular material containing fine collagenous fibers (LM ×250). **Location:** Forms embryonic skeleton; covers ends of long bones; and forms cartilage of nose, trachea, and larynx. **Function:** Support and reinforcement.

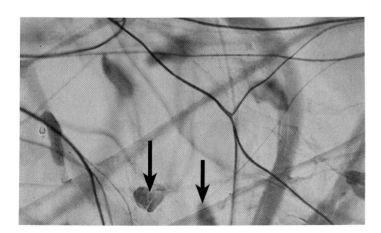

(*g*) **Loose connective tissue** contains numerous fibroblasts (arrows) that produce collagenous and elastic fibers (LM ×250). **Location:** Widely distributed under the epithelia of the human body. **Function:** Wraps and cushions organs.

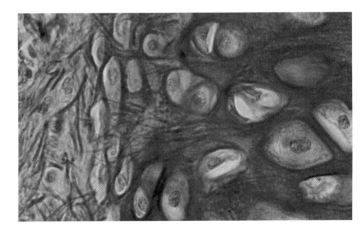

(*j*) **Elastic cartilage** contains fine collagenous fibers and many elastic fibers in its intercellular material (LM ×250). **Location:** External ear, epiglottis. **Function:** Maintains a structure's shape while allowing great flexibility.

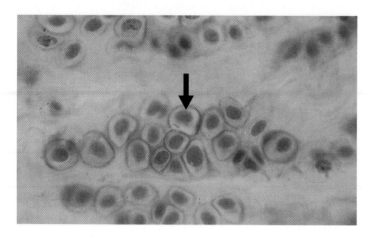

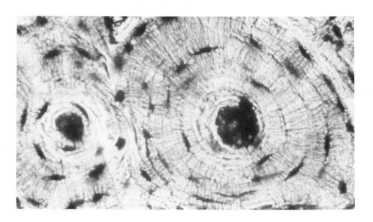

(*k*) **Fibrocartilage** contains many large, collagenous fibers in its intercellular material (LM ×195). The arrow points to a fibroblast. **Location:** Intervertebral disks, pubic symphysis, and disks of knee joint. **Function:** Absorbs compression shock.

(*l*) **Bone (osseous) tissue.** Bone matrix is deposited in concentric layers around osteonic canals (LM ×160). **Location:** Bones. **Function:** Supports, protects, provides lever system for muscles to act on, stores calcium and fat, and forms blood cells.

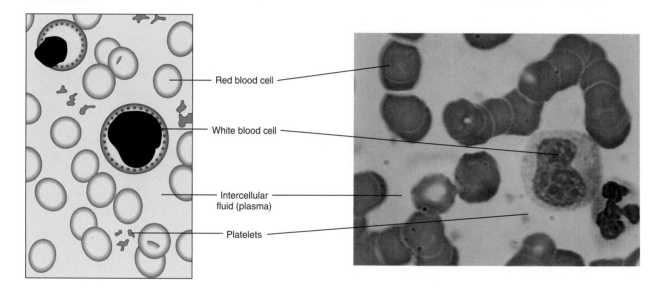

Red blood cell

White blood cell

Intercellular fluid (plasma)

Platelets

(*m*) **Blood** is a type of connective tissue. It consists of red blood cells, white blood cells, and platelets suspended in an intercellular fluid (plasma) (LM ×640). **Location:** Within blood vessels. **Function:** Transports oxygen, carbon dioxide, nutrients, wastes, hormones, minerals, vitamins, and other substances.

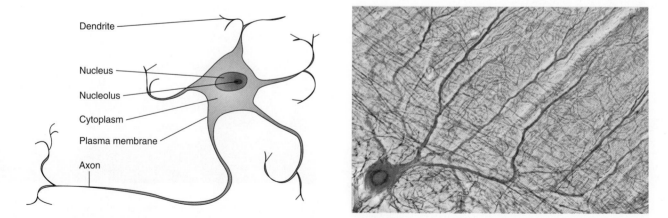

Dendrite

Nucleus

Nucleolus

Cytoplasm

Plasma membrane

Axon

(*n*) **Nervous tissue.** Neurons in nervous tissue transmit electrical signals to other neurons, muscles, or glands (LM ×450). **Location:** Brain, spinal cord, and nerves. **Function:** Transmits electrical signals from sensory receptors to the spinal cord or brain, and from the spinal cord or brain to effectors (muscles and glands).

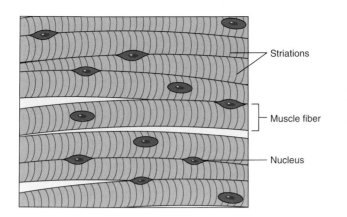
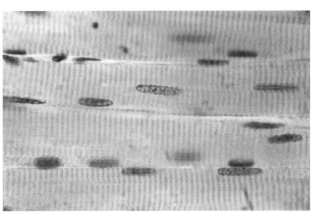

(*o*) **Skeletal muscle tissue** is composed of striated muscle fibers (cells) that are long and cylindrical and contain many peripheral nuclei (LM ×250). **Location:** In skeletal muscles attached to bones. **Function:** Voluntary movement, locomotion.

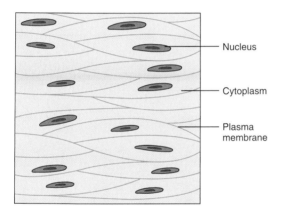
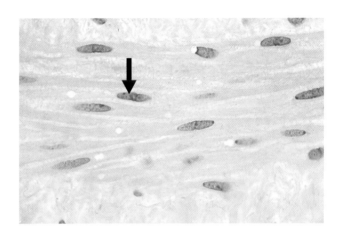

(*p*) **Smooth muscle tissue** is formed of spindle-shaped cells, each containing a single centrally located nucleus (arrow) (LM ×250). Cells are arranged closely to form sheets. Smooth muscle tissue is not striated. **Location:** Mostly in the walls of hollow organs. **Function:** Moves substances or objects (foodstuffs, urine, a baby) along internal passageways; involuntary control.

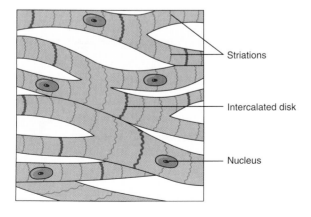
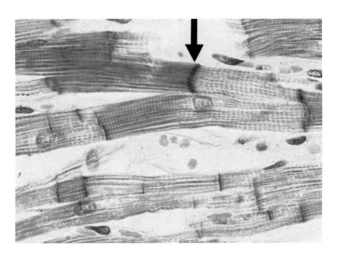

(*q*) **Cardiac muscle tissue** consists of branched striated cells, each containing a single nucleus and specialized cell junctions called intercalated disks (arrow) that allow ions (action potentials) to move quickly from cell to cell (LM ×400). **Location:** The walls of the heart. **Function:** As the walls of the heart contract, cardiac muscle tissue propels blood into the circulation; involuntary control.

EVOLUTIONARY INSIGHTS

The Origin of Eukaryotic Cells

The first cells were most likely very simple forms. The fossil record indicates that the earth is approximately 4 to 5 billion years old, and that the first cells may have arisen more than 3.5 billion years ago, whereas the eukaryotes are thought to have first appeared about 1.5 billion years ago.

The **endosymbiont theory** was first proposed by Lynn Margulis, a biologist at the University of Massachusetts at Amherst. She proposed that eukaryotes formed when large, nonnucleated cells engulfed smaller and simpler cells. (An **endosymbiont** is an organism that can live only inside another organism, forming a relationship that benefits both partners. **Symbiosis** is an intimate association between two organisms of different species.)

More recently, DNA evidence indicates that both Archaea and Eubacteria contributed to the origin of eukaryotic cells (box figure 2.1) as follows. About 2.5 billion years ago, bacteria and cyanobacteria occurred together in water environments. Over millions of years, the cyanobacteria pumped oxygen into the primitive atmosphere as a result of photosynthesis. As a result, those cellular organisms that could tolerate free oxygen began to flourish.

We now know that free oxygen reacts with other molecules, producing harmful byproducts (free radicals) that can disrupt normal biological functions. One way for a large cell, such as an archaeon, to survive in an oxygen-rich environment would be to engulf an aerobic (oxygen-utilizing) bacterium in an inward-budding vesicle of its plasma membrane. This captured bacterium would then contribute biological reactions to detoxify the free oxygen and radicals. Eventually the membrane of the enveloped vesicle became the outer membrane of the mitochondrion. The outer membrane of the engulfed aerobic bacterium became the inner membrane system of the mitochondrion. The small bacterium thus found a new home in the larger cell and as a result, the host cell could survive in the newly oxygenated atmosphere.

In a similar fashion, archaean cells that picked up cyanobacteria or a photosynthetic bacterium obtained the forerunners of chloroplasts and became the ancestors of the green plants. Once these ancient cells acquired their endosymbiont organelles, genetic changes impaired the ability of the captured cells to live on their own outside the host cells. Over many millions of years, the larger cells and the captured cells came to depend on one another for survival. The result of this interdependence is the compartmentalization in modern eukaryotic cells.

Although the exact mechanism for the evolution of the eukaryotic cell will never be known with certainty, the emergence of the eukaryotic cell led to a dramatic increase in the complexity and diversity of life forms on the earth. At first, these newly formed eukaryotic cells existed only by themselves. Later, however, some probably evolved into multicellular organisms in which various cells became specialized into tissues, which in turn led to the potential for many different functions. These multicellular forms would then be able to adapt to life in a greater variety of environments.

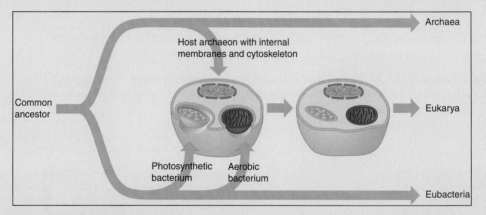

BOX FIGURE 2.1   **The Origin of Eukaryotic Cells.**   According to the endosymbiont theory, eukaryotic cells may have originated many millions of years ago from a joining of eubacterial cells with archaean cells. The captured bacteria eventually became the organelles called mitochondria and chloroplasts. The archaean host contributed the membranes and cytoskeleton elements, which enabled the new cell to move and engulf smaller cells in the watery environment in which they were living. When some of the genetic material of the captured cells moved to the nucleus-to-be (the archaeon's DNA), the smaller cells became dependent on their host.

Cartilage is a hard yet flexible tissue that supports such structures as the outer ear and forms the entire skeleton of such animals as sharks and rays (figure 2.25*i–k*). Cells called chondrocytes lie within spaces called lacunae that are surrounded by a rubbery matrix that chondroblasts secrete. This matrix, along with the collagen and/or elastin fibers, gives cartilage its strength and elasticity.

Bone cells (osteocytes) also lie within lacunae, but the matrix around them is heavily impregnated with calcium phosphate and calcium carbonate, making this kind of tissue hard and ideally suited for its functions of support and protection (figure 2.25*l*). Chapter 23 covers the structure and function of bone in more detail.

Blood is a connective tissue in which a fluid called plasma suspends specialized red and white blood cells plus platelets (figure 2.25*m*). Blood transports various substances throughout the bodies of animals. Chapter 26 covers blood in more detail.

MUSCLE TISSUE: MOVEMENT

Muscle tissue allows movement. The three kinds of muscle tissue are skeletal, smooth, and cardiac (figure 2.25*o*–*q*). Skeletal muscle is attached to bones and makes body movement possible in vertebrates. The rhythmic contractions of smooth muscle create a churning action (as in the stomach), help propel material through a tubular structure (as in the intestines), and control size changes in hollow organs such as the urinary bladder and uterus. The contractions of cardiac muscle result in the heart beating. Chapter 23 discusses the details of the contractile process in muscle tissue.

NERVOUS TISSUE: COMMUNICATION

Nervous tissue is composed of several different types of cells: Impulse-conducting cells are called neurons (figure 2.25*n*); cells involved with protection, support, and nourishment are called neuroglia; and cells that form sheaths and help protect, nourish, and maintain cells of the peripheral nervous system are called peripheral glial cells. Chapter 24 discusses nervous tissue in more detail.

ORGANS

Organs (Gr. *organnon*, an independent part of the body) are the functional units of an animal's body that are made up of more than one type of tissue. Examples include the heart, lungs, liver, spleen, and kidneys.

ORGAN SYSTEMS

The next higher level of structural organization in animals is the organ system. An **organ system** (Gr. *systema*, being together) is an association of organs that together performs an overall function. The organ systems in higher vertebrate animals are the integumentary, skeletal, muscular, nervous, endocrine, circulatory, lymphatic, respiratory, digestive, urinary, and reproductive systems. Chapters 23 through 29 discuss these systems in detail.

The highest level of organization in an animal body is the organismic level. All parts of the animal body function with one another to contribute to the total organism—a living entity or individual.

Animals need energy in order to survive. Many of the chemical reactions that produce energy are regulated by enzymes. Together, energy and enzymes are the driving and controlling forces in animals. All animals harvest energy from nutrients to fuel their metabolism with energy from ATP. To learn more about the fundamental chemistry that regulates energy production and storage, visit the web sites listed under **Related Web Links** at the end of this chapter. These are mini-chapters with illustrations that cover the basic biology of chemicals, enzymes, and energy in an animal's body.

SUMMARY

1. All animal cells have three basic parts: the nucleus, cytoplasm, and plasma membrane.
2. Cell membranes, composed mainly of phospholipids and proteins, allow certain materials to move across them. This quality is called selective permeability. The fluid-mosaic model is based on knowledge of the plasma membrane.
3. Some molecules use their own energy to pass across a cell membrane from areas of higher concentration to areas of lower concentration. Examples of these passive processes are simple diffusion, facilitated diffusion, osmosis, and filtration.
4. Active transport across cell membranes requires energy from the cell to move substances from areas of lower concentration to areas of higher concentration. Additional processes that move molecules across membranes are endocytosis and exocytosis. Three types of endocytosis are pinocytosis, phagocytosis, and receptor-mediated endocytosis.
5. The cytoplasm of a cell has two parts. The cytomembrane system consists of well-defined structures, such as the endoplasmic reticulum, Golgi apparatus, vacuoles, and vesicles. The aqueous part consists of the fluid cytosol.
6. Ribosomes are the sites of protein synthesis.
7. The endoplasmic reticulum (ER) is a series of channels that transports, stores enzymes and proteins, and provides a point of attachment for ribosomes. The two types of ER are smooth and rough.

8. The Golgi apparatus aids in the synthesis and secretion of glycoproteins, as well as processing and modifying other materials (e.g., enzymes).
9. Lysosomes containing digestive enzymes digest nutrients and clean away dead or damaged cell parts.
10. Mitochondria convert energy in food molecules to ATP, a form the cell can use.
11. Microtubules, intermediate filaments, and microfilaments make up the cytoskeleton of the cell. The cytoskeleton functions in transport, support, and the movement of structures in the cell, such as organelles and chromosomes.
12. Cilia and flagella are appendages on the surfaces of some cells and function in movement.
13. Centrioles assist in cell division and help move chromosomes during cell division.
14. Vacuoles are membranous sacs that are part of the cytomembrane system.
15. Vaults are octagonal, barrel-shaped organelles believed to transport messenger RNA from the nucleus to the ribosomes.
16. The nucleus of a cell contains DNA, which controls the cell's genetic program and other metabolic activities.
17. The nuclear envelope contains many pores that allow material to enter and leave the nucleus.
18. The chromosomes in the nucleus have DNA organized into genes, which are specific DNA sequences that control and regulate cell activities.

19. The nucleolus is a preassembly point for ribosomes.
20. Tissues are groups of cells with a common structure and function. The four types of tissues are epithelial, connective, muscle, and nervous.
21. An organ is composed of more than one type of tissue. An organ system is an association of organs.

SELECTED KEY TERMS

centrioles (p. 22)
cytomembrane system (p. 16)
cytoplasm (p. 10)
cytoskeleton (p. 20)
endoplasmic reticulum
    (ER) (p. 18)
epithelial tissue (p. 23)
Golgi apparatus (p. 19)

homeostasis (p. 12)
lysosomes (p. 19)
microtubule-organizing center
    (p. 22)
mitochondria (p. 20)
nucleolus (p. 23)
nucleus (p. 10)
vault (p. 22)

CRITICAL THINKING QUESTIONS

1. Why is the mitochondrion called the "power generator" of the cell?
2. One of the larger facets of modern zoology can be described as "membrane biology." What common principles unite the diverse functions of membranes?
3. Why is the current model of the plasma membrane called the "fluid-mosaic" model? What is the fluid, and in what sense is it fluid? What makes up the mosaic?
4. If you could visualize osmosis, seeing the solute and solvent particles as individual entities, what would an osmotic gradient look like?
5. Why can some animal cells transport materials against a concentration gradient? Could animals survive without this capability?

ONLINE LEARNING CENTER

Visit our Online Learning Center (OLC) at www.mhhe.com/zoology (click on this book's title) to find the following chapter-related materials:

- CHAPTER QUIZZING
- ANIMATIONS
    Diffusion
    Osmosis
    Active Transport
    Exocytosis, Endocytosis
- RELATED WEB LINKS
    Cell Theory
    Prokaryotes and Eukaryotes
    Cell Structure
    Cellular Organelles
    The Cell Membrane
    Cells and Tissues: Histology
- BOXED READINGS ON
    The Origin of Eukaryotic Cells
    Microscopes—Windows into the Cell
- SUGGESTED READINGS
- LAB CORRELATIONS

Check out the OLC to find specific information on these related lab exercises in the *General Zoology Laboratory Manual*, 5th edition, by Stephen A. Miller:

Exercise 1   *The Microscope and Scientific Method*
Exercise 2   *Cells and Tissues*
Exercise 3   *Aspects of Cell Function*

CELL DIVISION AND INHERITANCE

Outline

Concepts

1. The genetic material is organized into chromosomes. Chromosomes may be represented differently in males and females. However, the number of chromosomes is constant for a given species.
2. Mitosis is the form of nuclear division that results in growth and repair processes. It ensures an orderly and accurate distribution of chromosomes during the cell cycle. Cytokinesis results in the division of the cytoplasm.
3. Meiosis is the form of cell division that results in the formation of gametes. It reduces the chromosome number by half and allows for the random distribution of one member of each pair of parental chromosomes to the offspring.
4. Deoxyribonucleic acid (DNA) is the genetic material of the cell. Its double helix structure suggests how it can replicate itself and how it can code for the sequences of amino acids that make proteins.
5. Protein synthesis involves two processes. Transcription is the production of a messenger RNA (mRNA) molecule that is complementary to a gene in DNA. Translation is the assembly of proteins at the ribosomes based on the genetic information in the transcribed mRNA.
6. Changes in DNA and chromosomes increase the variation within a species and account for evolutionary change.
7. Principles of classical genetics explain the inheritance patterns of many animal traits, including dominance, segregation, and independent assortment.
8. Many alternative forms of a gene may exist in a population, and these alternative forms may interact in different ways.
9. Patterns of inheritance observed at an organismal level are explained at a molecular level by the presence or absence of functional enzymes.

Reproduction is essential for life. Each organism exists solely because its ancestors succeeded in producing progeny that could develop, survive, and reach reproductive age. At its most basic level, reproduction involves a single cell reproducing itself. For a unicellular organism, cellular reproduction also reproduces the organism. For multicellular organisms, cellular reproduction is involved in growth, repair, and the formation of sperm and egg cells that enable the organism to reproduce.

At the molecular level, reproduction involves the cell's unique capacity to manipulate large amounts of DNA, DNA's ability to replicate, and DNA's ability to carry information that will determine the characteristics of cells in the next generation. **Genetics** (Gr. *gennan*, to produce) is the study of how biological information is transmitted from one generation to the

This chapter contains evolutionary concepts, which are set off in this font.

next. Modern molecular genetics provides biochemical explanations of how this information is expressed in an organism. It holds the key to understanding the basis for inheritance. Information carried in DNA is manifested in the kinds of proteins that exist in each individual. Proteins contribute to observable traits, such as eye color and hair color, and they function as enzymes that regulate the rates of chemical reactions in organisms. Within certain environmental limits, animals are what they are by the proteins that they synthesize.

At the level of the organism, reproduction involves passing DNA from individuals of one generation to the next generation. The classical approach to genetics involves experimental manipulation of reproduction and observing patterns of inheritance between generations. This work began with Gregor Mendel (1822–1884) and continues today.

Gregor Mendel began a genetics revolution that has had a tremendous effect on biology and our society. *Genetic mechanisms explain how traits are passed between generations. They also help explain how species change over time. Genetic and evolutionary themes are interdependent in biology, and biology without either would be unrecognizable from its present form.* Genetic technologies have tremendous potential to improve crop production and health care, but society must deal with issues related to whole-organism cloning, the use of engineered organisms in biological warfare, and the application of genetic technologies to humans. This chapter introduces principles of cell division and genetics that are essential to understanding why animals function as they do, and it provides the background information to help you understand the genetic basis of evolutionary change that will be covered in chapters 4 and 5.

EUKARYOTIC CHROMOSOMES

DNA is the genetic material, and it exists with protein in the form of chromosomes in eukaryotic cells. During most of the life of a cell, chromosomes are in a highly dispersed state called chromatin. During these times, units of inheritance called **genes** (Gr. *genos*, race) may actively participate in the formation of protein. When a cell is dividing, however, chromosomes exist in a highly folded and condensed state that allows them to be distributed between new cells being produced. The structure of these chromosomes will be described in more detail in the discussion of cell division that follows.

Chromatin consists of DNA and histone proteins. This association of DNA and protein helps with the complex jobs of packing DNA into chromosomes (chromosome condensation) and regulating DNA activity.

There are five different histone proteins. Some of these proteins form a core particle. DNA wraps in a coil around the proteins, a combination called a **nucleosome** (figure 3.1). The fifth histone, sometimes called the linker protein, is not needed to form the nucleosome but may help anchor the DNA to the core and promote the winding of the chain of nucleosomes into a cylinder. Further folding and the addition of protective proteins result in the formation of chromosomes during mitosis and meiosis.

Not all chromatin is equally active. Some human genes, for example, are active only after adolescence. In other cases, entire

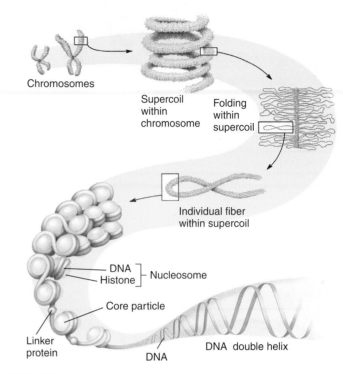

FIGURE 3.1

Organization of Eukaryotic Chromosomes.   Chromosomes consist of a supercoil of highly folded chromatin. Chromatin is a chain of nucleosomes. Each nucleosome is comprised of histone proteins wound by a strand of DNA. Linker histone proteins are associated with DNA between adjacent nucleosomes.

chromosomes may not function in particular cells. Inactive portions of chromosomes produce dark banding patterns with certain staining procedures and thus are called **heterochromatic regions,** whereas active portions of chromosomes are called **euchromatic regions.**

SEX CHROMOSOMES AND AUTOSOMES

In the early 1900s, attention turned to the cell to find a chromosomal explanation for the determination of maleness or femaleness. Some of the evidence for a chromosomal basis for sex determination came from work with the insect *Protenor.* One darkly staining chromosome of *Protenor,* called the X chromosome, is represented differently in males and females. All somatic (body) cells of males have one X chromosome (XO), and all somatic cells of females have two X chromosomes (XX). Similarly, half of all sperm contain a single X, and half contain no X, while all female gametes contain a single X. This pattern suggests that fertilization involving an X-bearing sperm will result in a female offspring and that fertilization involving a sperm with no X chromosome will result in a male offspring. As figure 3.2 illustrates, this sex determination system explains the approximately 50:50 ratio of females to males in this insect species. Chromosomes that are represented differently in females than in males and function in sex determination are **sex chromosomes.** Chromosomes that are alike and

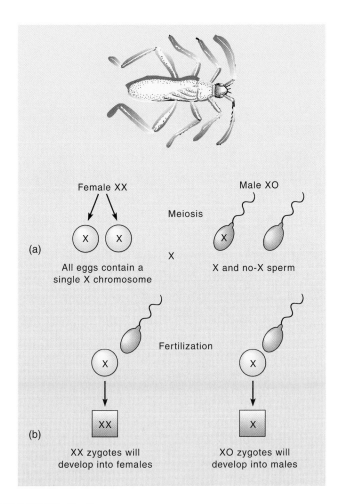

FIGURE 3.2

XO-System of Sex Determination for the Insect *Protenor*.   (*a*) In females, all cells except gametes possess two X chromosomes. During meiosis, homologous X chromosomes segregate, and all eggs contain one X chromosome. Males possess one X chromosome per cell. Meiosis results in half of all sperm cells having one X, and half of all sperm cells having no X. (*b*) Fertilization results in half of all offspring having one X chromosome (males), and half of all offspring having two X chromosomes (females).

not involved in determining sex are **autosomes** (Gr. *autos*, self + *soma*, body).

The system of sex determination described for *Protenor* is called the X-O system. It is the simplest system for determining sex because it involves only one kind of chromosome. Many other animals (e.g., humans and fruit flies) have an X-Y system of sex determination. In the X-Y system, males and females have an equal number of chromosomes, but the male is usually XY, and the female is XX. (In birds, the sex chromosomes are designated Z and W, and the female is ZW.) This mode of sex determination also results in approximately equal numbers of male and female offspring:

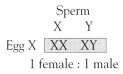

Sperm

	X	Y
Egg X	XX	XY

1 female : 1 male

NUMBER OF CHROMOSOMES

Even though the number of chromosomes is constant within a species, chromosome number varies greatly among species.

Chromosomes are present in sets, with the number in a set being characteristic of each kind of animal and expressed as "N." N identifies the number of different kinds of chromosomes. Most animals have two sets, or 2N chromosomes. This is the **diploid** (Gr. *di*, two + *eoides*, doubled) condition. Some animals have only one set, or N chromosomes (like gametes) and are **haploid** (Gr. *hapl*, single) (e.g., male honeybees and some rotifers).

Very few animals (e.g., brine shrimp, snout beetles, some flatworms, and some sow bugs) have more than the diploid number of chromosomes, a condition called **polyploidy** (Gr. *polys*, more). The upset in numbers of sex chromosomes apparently interferes with reproductive success. Asexual reproduction often accompanies polyploidy.

MITOTIC CELL DIVISION

Cell division occurs in all animals during growth and repair processes. Cells divide in two basic stages: **Mitosis** is division of the nucleus, and **cytokinesis** (Gr. *kytos*, hollow vessel + *kinesis*, motion) is division of the cytoplasm. Between divisions (interphase), the cell must grow and carry out its various metabolic processes. The **cell cycle** is that period from the time a cell is produced until it completes mitosis (figure 3.3).

The G₁ (first growth or gap) phase represents the early growth phase of the cell. During the S (DNA synthesis) phase, growth

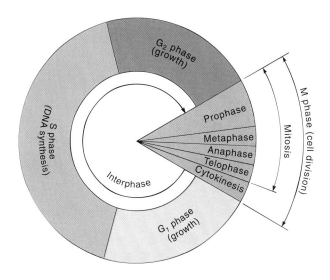

FIGURE 3.3

Life Cycle of a Eukaryotic Cell.   During the G₁ phase, cell components are synthesized and metabolism occurs, often resulting in cell growth. During the S (synthesis) phase, the chromosomes replicate, resulting in two identical copies called sister chromatids. During the G₂ phase, metabolism and growth continue until the mitotic phase is reached. This drawing is generalized, and the length of different stages varies greatly from one cell to the next.   *Source: Stuart Ira Fox, Human Physiology, 4th ed., copyright © 1993 Wm. C. Brown Communications, Inc., Dubuque, Iowa.*

continues, but this phase also involves DNA replication. The G_2 (second growth or gap) phase prepares the cell for division. It includes replication of the mitochondria and other organelles, synthesis of microtubules and protein that will make up the mitotic spindle fibers, and chromosome condensation. The M (mitotic) phase includes events associated with partitioning chromosomes between two daughter cells and the division of the cytoplasm (cytokinesis).

INTERPHASE: REPLICATING THE HEREDITARY MATERIAL

Interphase (L. *inter,* between) (includes the G_1, S, and G_2 phases) typically occupies about 90% of the total cell cycle. It is the period during which the normal activities of the cell take place. Interphase also sets the stage for cell division because DNA replication is completed during the S phase of interphase.

Before a cell divides, an exact copy of the DNA is made. This process is called replication, because the double-stranded DNA makes a replica, or duplicate, of itself. Replication is essential to ensure that each daughter cell receives identical genetic material as is present in the parent cell. The result is a pair of **sister chromatids** (figure 3.4). A **chromatid** is a copy of a chromosome produced by replication. Each chromatid attaches to its other copy, or sister, at a point of constriction called a centromere. The **centromere** is a specific DNA sequence of about 220 nucleotides and has a specific location on any given chromosome. Bound to each centromere is a disk of protein called a **kinetochore,** which eventually is an attachment site for the microtubules of the mitotic spindle.

As the cell cycle moves into the G_2 phase the chromosomes begin condensation. During the G_2 phase, the cell also begins to assemble the structures that it will later use to move the chromosomes to opposite poles (ends) of the cell. For example, centrioles replicate, and there is extensive synthesis of the proteins that make up the microtubules.

MITOSIS

Mitosis is divided into four phases: prophase, metaphase, anaphase, and telophase. In a dividing cell, however, the process is actually continuous, with each phase smoothly flowing into the next (figure 3.5).

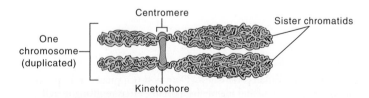

FIGURE 3.4

Duplicated Chromosomes.   Each parental chromosome replicates to make two genetically identical sister chromatids attached at a region of DNA called the centromere. The kinetochore is a disk of protein that is bound at the centromere and is an attachment site for microtubules.

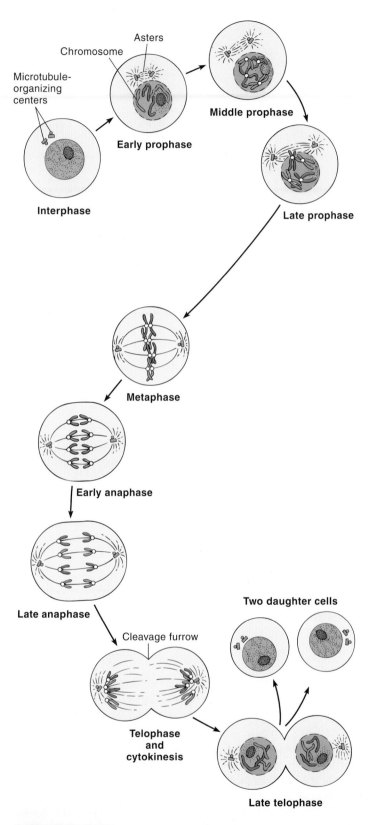

FIGURE 3.5

Continuum of Mitosis and Cytokinesis.   Mitosis is a continuous process during which the nuclear parts of a cell divide into two equal portions. Cytokinesis is the division of the cytoplasm of a cell.

The first phase of mitosis, **prophase** (Gr. *pro*, before + phase), begins when chromosomes become visible with the light microscope as threadlike structures. The nucleoli and nuclear envelope begin to break up, and the two centriole pairs move apart. By the end of prophase, the centriole pairs are at opposite poles of the cell. The centrioles radiate an array of microtubules called **asters** (L. *aster*, little star), which brace each centriole against the plasma membrane. Between the centrioles, the microtubules form a spindle of fibers that extends from pole to pole. The asters, spindle, centrioles, and microtubules are collectively called the **mitotic spindle** (or mitotic apparatus). As prophase continues, a second group of microtubules grows out from the kinetochore to the poles of the cell. These kinetochore microtubules connect each sister chromatid to the poles of the spindle.

As the dividing cell moves into **metaphase** (Gr. *meta*, after + phase), the chromatids (replicated chromosomes) begin to align in the center of the cell, along the spindle equator. Toward the end of metaphase, the centromeres divide and detach the two sister chromatids from each other, although the chromatids remain aligned next to each other. After the centromeres divide, the sister chromatids are considered full-fledged chromosomes (called daughter chromosomes).

During **anaphase** (Gr. *ana*, back again + phase), the shortening of the microtubules in the mitotic spindle pulls each daughter chromosome apart from its copy and toward its respective pole. Anaphase ends when all the daughter chromosomes have moved to the poles of the cell. Each pole now has a complete, identical set of chromosomes.

Telophase (Gr. *telos*, end + phase) begins once the daughter chromosomes arrive at the opposite poles of the cell. During telophase, the mitotic spindle disassembles. A nuclear envelope reforms around each set of chromosomes, which begin to uncoil for gene expression, and the nucleolus is resynthesized. The cell also begins to pinch in the middle. Mitosis is over, but cell division is not.

CYTOKINESIS: PARTITIONING THE CYTOPLASM

The final phase of cell division is cytokinesis, in which the cytoplasm divides. Cytokinesis usually starts sometime during late anaphase or early telophase. A contracting belt of microfilaments called the contractile ring pinches the plasma membrane to form the cleavage furrow. The furrow deepens, and two new, genetically identical daughter cells form.

MEIOSIS: THE BASIS OF SEXUAL REPRODUCTION

Sexual reproduction requires a genetic contribution from two different sex cells. Egg and sperm cells are specialized sex cells called **gametes** (Gr. *gamete*, wife; *gametes*, husband). In animals, a male gamete (sperm) unites with a female gamete (egg) during fertilization to form a single cell called a **zygote** (Gr. *zygotos*, yoked together). The zygote is the first cell of the new animal. The fusion of nuclei within the zygote brings together genetic information from the two parents, and each parent contributes half of the genetic information to the zygote.

To maintain a constant number of chromosomes in the next generation, animals that reproduce sexually must produce gametes with half the chromosome number of their ordinary body cells (called **somatic cells**). All of the cells in the bodies of most animals, except for the egg and sperm cells, have the diploid (2N) number of chromosomes. A type of cell division called **meiosis** (Gr. *meiosis*, diminution) occurs in specialized cells of the ovaries and testes and reduces the number of chromosomes to the haploid (1N) number. The nuclei of the two gametes combine during fertilization and restore the diploid number.

Meiosis begins after the G_2 phase in the cell cycle—after DNA replication. Two successive nuclear divisions, designated meiosis I and meiosis II, take place. The two nuclear divisions of meiosis result in four daughter cells, each with half the number of chromosomes of the parent cell. Moreover, these daughter cells are not genetically identical. Like mitosis, meiosis is a continuous process, and biologists divide it into the phases that follow only for convenience.

THE FIRST MEIOTIC DIVISION

In prophase I, chromatin folds and chromosomes become visible under a light microscope (figure 3.6a). Because a cell has a copy of each type of chromosome from each original parent cell, it contains the diploid number of chromosomes. **Homologous chromosomes** (homologues) carry genes for the same traits, are the same length, and have a similar staining pattern, making them identifiable as matching pairs. During prophase I, homologous chromosomes line up side-by-side in a process called **synapsis** (Gr. *synapsis*, conjunction), forming a **tetrad** of chromatids (also called a bivalent). The tetrad thus contains the two homologous chromosomes, each with its copy, or sister chromatid (figure 3.7). A network of protein and RNA is laid down between the sister chromatids of the two homologous chromosomes. This network holds the sister chromatids in a precise union so that each gene is directly across from its sister gene on the homologous chromosome.

Synapsis also initiates a series of events called **crossing-over,** whereby the nonsister chromatids of the two homologous chromosomes in a tetrad exchange DNA segments (figure 3.7). This process effectively redistributes genetic information among the paired homologous chromosomes and produces new combinations of genes on the various chromatids in homologous pairs. Thus, each chromatid ends up with new combinations of instructions for a variety of traits. *Crossing-over is a form of genetic recombination and is a major source of genetic variation in a population of a given species.*

In metaphase I, the microtubules form a spindle apparatus just as in mitosis (*see figures 3.4 and 3.5*). However, unlike mitosis, where homologous chromosomes do not pair, each pair of homologues lines up in the center of the cell, with centromeres on each side of the spindle equator.

Anaphase I begins when homologous chromosomes separate and begin to move toward each pole. Because the orientation of each pair of homologous chromosomes in the center of the cell

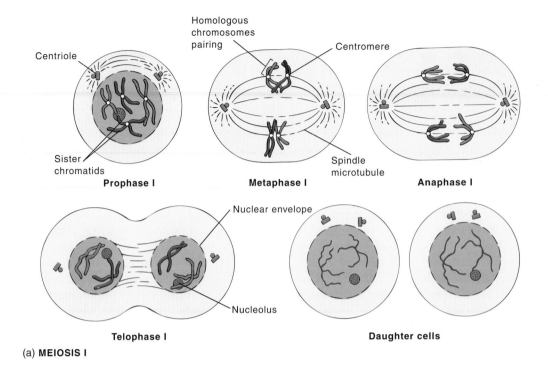

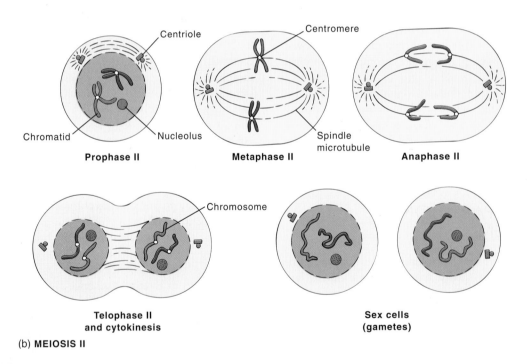

FIGURE 3.6

Meiosis and Cytokinesis. (*a*) Stages in the first meiotic division. (*b*) Stages in the second meiotic division.

is random, the specific chromosomes that each pole receives from each pair of homologues are also random.

Meiotic telophase I is similar to mitotic telophase. The transition to the second nuclear division is called interkinesis. Cells proceeding through interkinesis do not replicate their DNA. After a varying time period, meiosis II occurs.

THE SECOND MEIOTIC DIVISION

The second meiotic division (meiosis II) resembles an ordinary mitotic division (*see figure 3.6b*), except the number of chromosomes has been reduced by half. The phases are prophase II, metaphase II, anaphase II, and telophase II. At the end of

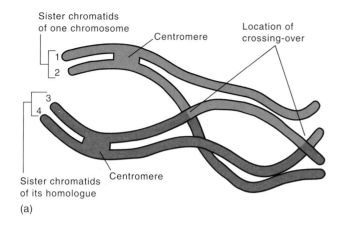

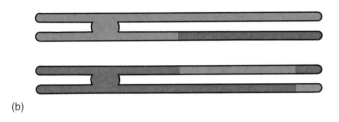

(b)

FIGURE 3.7

Synapsis and Crossing-Over. (*a*) Immediately after crossing-over, nonsister chromatids remain tightly associated in regions where crossing-over occurred. (*b*) The same chromosomes diagrammatically separated to show the result of crossing-over.

telophase II and cytokinesis, the final products of these two divisions of meiosis are four new "division products." In most animals, each of these "division products" is haploid and may function directly as a gamete (sex cell).

SPERMATOGENESIS AND OOGENESIS

The result of meiosis in most animals is the formation of sperm and egg cells. **Spermatogenesis** produces mature sperm cells and follows the sequence previously described. All four products of meiosis often acquire a flagellum for locomotion and a caplike structure that aids in the penetration of the egg. **Oogenesis** produces a mature ovum or egg. It differs from spermatogenesis in that only one of the four meiotic products develops into the functional gamete. The other products of meiosis are called polar bodies and eventually disintegrate. In some animals the mature egg is the product of the first meiotic division and only completes meiosis if it is fertilized by a sperm cell.

DNA: THE GENETIC MATERIAL

Twentieth-century biologists realized that a molecule that serves as the genetic material must have certain characteristics to explain the properties of life: First, the genetic material must be able

to code for the sequence of amino acids in proteins and control protein synthesis. Second, it must be able to replicate itself prior to cell division. Third, the genetic material must be in the nucleus of eukaryotic cells. Fourth, it must be able to change over time to account for evolutionary change. Only one molecule, DNA (deoxyribonucleic acid), fulfills all of these requirements.

THE DOUBLE HELIX MODEL

Two kinds of molecules participate in protein synthesis. Both are based on a similar building block, the nucleotide, giving them their name—nucleic acids. One of these molecules, **deoxyribonucleic acid** or **DNA,** is the genetic material, and the other, **ribonucleic acid** or **RNA,** is produced in the nucleus and moves to the cytoplasm, where it participates in protein synthesis. The study of how the information stored in DNA codes for RNA and protein is **molecular genetics.**

DNA and RNA are large molecules made up of subunits called nucleotides (figure 3.8). A nucleotide consists of a

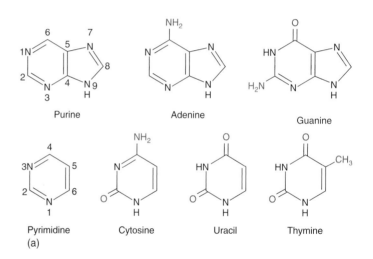

(a)

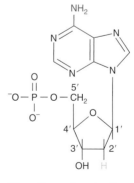

(b)   Deoxyadenosine-5′-monophosphate (dAMP)

FIGURE 3.8

Components of Nucleic Acids. (*a*) The nitrogenous bases in DNA and RNA. (*b*) Nucleotides form by attaching a nitrogenous base to the 1′ carbon of a pentose sugar and attaching a phosphoric acid to the 5′ carbon of the sugar. (Carbons of the sugar are numbered with primes to distinguish them from the carbons of the nitrogenous base.) The sugar in DNA is deoxyribose, and the sugar in RNA is ribose. In ribose, a hydroxyl group (−OH) would replace the hydrogen shaded yellow.

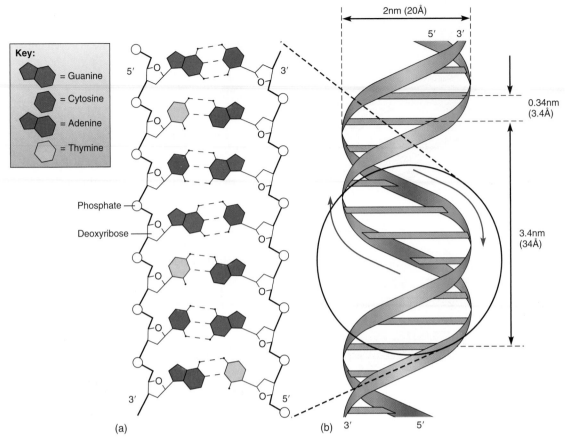

FIGURE 3.9

Structure of DNA. (*a*) Nucleotides of one strand of nucleic acid join by linking the phosphate of one nucleotide to the 3′ carbon of an adjacent nucleotide. Dashed lines between the nitrogenous bases indicate hydrogen bonds. Three hydrogen bonds are between cytosine and guanine, and two are between thymine and adenine. The antiparallel orientation of the two strands is indicated by using the 3′ and 5′ carbons at the ends of each strand. (*b*) Three-dimensional representation of DNA. The antiparallel nature of the strands is indicated by the curved arrows.

nitrogen-containing organic base, either in the form of a double ring **(purine)** or a single ring **(pyrimidine).** Nucleotides also contain a pentose (five-carbon) sugar and a phosphate ($-PO_4$) group. DNA and RNA molecules, however, differ in several ways. Both DNA and RNA contain the purine bases adenine and guanine, and the pyrimidine base cytosine. The second pyrimidine in DNA, however, is thymine, whereas in RNA it is uracil. A second difference between DNA and RNA involves the sugar present in the nucleotides. The pentose of DNA is deoxyribose, and in RNA it is ribose. A third important difference between DNA and RNA is that DNA is a double-stranded molecule and RNA is single stranded, although it may fold back on itself and coil.

The key to understanding the function of DNA is knowing how nucleotides link into a three-dimensional structure. The DNA molecule is ladderlike, with the rails of the ladder consisting of alternating sugar-phosphate groups (figure 3.9*a*). The phosphate of a nucleotide attaches at the fifth (5′) carbon of deoxyribose. Adjacent nucleotides attach to one another by a covalent bond between the phosphate of one nucleotide and the third (3′) carbon of deoxyribose. The pairing of nitrogenous bases between strands holds the two strands together. Adenine (a purine) is hy-

drogen bonded to its complement, thymine (a pyrimidine), and guanine (a purine) is hydrogen bonded to its complement, cytosine (a pyrimidine) (figure 3.9*a*). Each strand of DNA is oriented such that the 3′ carbons of deoxyribose in one strand are oriented in the opposite directions from the 3′ carbons in the other strand. The strands' terminal phosphates are, therefore, at opposite ends, and the DNA molecule is thus said to be **antiparallel** (Gr. *anti*, against + *para*, beside + *allelon*, of one another). The entire molecule is twisted into a right-handed helix, with one complete spiral every 10 base pairs (figure 3.9*b*).

DNA REPLICATION IN EUKARYOTES

During DNA replication, each DNA strand is a template for a new strand. The pairing requirements between purine and pyrimidine bases dictate the positioning of nucleotides in a new strand (figure 3.10). Thus, each new DNA molecule contains one strand from the old DNA molecule and one newly synthesized strand. Because half of the old molecule is conserved in the new molecule, DNA replication is said to be semiconservative.

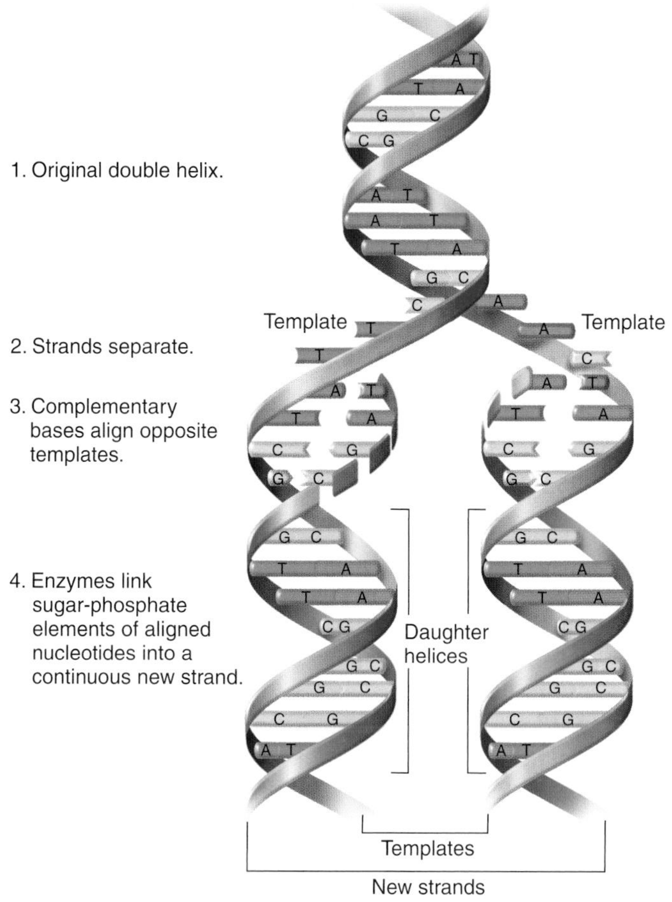

FIGURE 3.10

DNA Replication. (*1*) Replication begins simultaneously at many initiation sites along the length of a chromosome, and replication proceeds in both directions from the initiation site. Only one direction is shown here for simplicity. (*2*) An enzyme causes the double helix to unwind and the two strands to separate. Each original strand serves as a template for the synthesis of a new strand. (*3*) Other enzymes help align nucleotides with exposed, unpaired bases on the unwound portions of the original DNA and (*4*) link nucleotides into continuous new strands.

GENES IN ACTION

A gene can be defined as a sequence of bases in DNA that codes for the synthesis of one polypeptide, and genes must somehow transmit their information from the nucleus to the cytoplasm, where protein synthesis occurs. The synthesis of an RNA molecule from DNA is called **transcription** (L. *trans*, across + *scriba*, to write), and the formation of a protein from RNA at the ribosome is called **translation** (L. *trans*, across + *latere*, to remain hidden).

Three Major Kinds of RNA

Each of the three major kinds of RNA has a specific role in protein synthesis and is produced in the nucleus from DNA. **Messenger RNA (mRNA)** is a linear strand that carries a set of genetic instructions for synthesizing proteins to the cytoplasm. **Transfer RNA (tRNA)** picks up amino acids in the cytoplasm, carries them to ribosomes, and helps position them for incorporation into a polypeptide. **Ribosomal RNA (rRNA),** along with proteins, makes up ribosomes.

The Genetic Code

DNA must code for the 20 different amino acids found in all organisms. The information-carrying capabilities of DNA reside in the sequence of nitrogenous bases. The genetic code is a sequence of three bases—a triplet code. Figure 3.11 shows the genetic code as reflected in the mRNA that will be produced from DNA. Each three-base combination is a **codon.** More than one codon can specify the

Second position

First position	U	C	A	G	Third position
U	UUU Phe	UCU	UAU Tyr	UGU Cys	U
	UUC	UCC Ser	UAC	UGC	C
	UUA Leu	UCA	UAA STOP	UGA STOP	A
	UUG	UCG	UAG	UGG Trp	G
C	CUU	CCU	CAU His	CGU	U
	CUC Leu	CCC Pro	CAC	CGC Arg	C
	CUA	CCA	CAA Gln	CGA	A
	CUG	CCG	CAG	CGG	G
A	AUU	ACU	AAU Asn	AGU Ser	U
	AUC Ile	ACC Thr	AAC	AGC	C
	AUA	ACA	AAA Lys	AGA Arg	A
	AUG Met	ACG	AAG	AGG	G
G	GUU	GCU	GAU Asp	GGU	U
	GUC Val	GCC Ala	GAC	GGC Gly	C
	GUA	GCA	GAA Glu	GGA	A
	GUG	GCG	GAG	GGG	G

Ala =	Alanine	Leu =	Leucine
Arg =	Arginine	Lys =	Lysine
Asn =	Aparagine	Met =	Methionine
Asp =	Aspartic acid	Phe =	Phenylalanine
Cys =	Cysteine	Pro =	Proline
Gln =	Glutamine	Ser =	Serine
Glu =	Glutamic acid	Thr =	Threonine
Gly =	Glycine	Trp =	Tryptophan
His =	Histidine	Tyr =	Tyrosine
Ile =	Isoleucine	Val =	Valine

FIGURE 3.11

Genetic Code. Sixty-four messenger RNA codons are shown here. The first base of the triplet is on the left side of the figure, the second base is at the top, and the third base is on the right side. The abbreviations for amino acids are also shown. In addition to coding for the amino acid methionine, the AUG codon is the initiator codon. Three codons—UAA, UAG, and UGA—do not code for an amino acid but act as a signal to stop protein synthesis.

same amino acid because there are 64 possible codons, but only 20 amino acids. This characteristic of the code is referred to as **degeneracy.** Note that not all codons code for an amino acid. The base sequences UAA, UAG, and UGA are all stop signals that indicate where polypeptide synthesis should end. The base sequence AUG codes for the amino acid methionine, which is a start signal.

Transcription

The genetic information in DNA is not translated directly into proteins, but is first transcribed into mRNA. Transcription involves numerous enzymes that unwind a region of a DNA molecule, initiate and end mRNA synthesis, and modify the mRNA after transcription is complete. Unlike DNA replication, only one or a few genes are exposed, and only one of the two DNA strands is transcribed (figure 3.12).

One of the important enzymes of this process is RNA polymerase. After a section of DNA is unwound, RNA polymerase recognizes a specific sequence of DNA nucleotides. RNA polymerase attaches and begins joining ribose nucleotides, which are complementary to the 3′ end of the DNA strand. In RNA, the same complementary bases in DNA are paired, except that in RNA, the base uracil replaces the base thymine as a complement to adenine.

Newly transcribed mRNA, called the primary transcript, must be modified before leaving the nucleus to carry out protein synthesis. Some base sequences in newly transcribed mRNA do not code for proteins. RNA splicing involves cutting out noncoding regions so that the mRNA coding region can be read continuously at the ribosome.

Translation

Translation is protein synthesis at the ribosomes in the cytoplasm, based on the genetic information in the transcribed mRNA. Another type of RNA, called transfer RNA (tRNA), is important in the translation process (figure 3.13). It brings the different amino acids coded for by the mRNA into alignment so that a polypeptide can be made. Complementary pairing of bases across the molecule maintains tRNA's configuration. The presence of some unusual bases (i.e., other than adenine, cytosine, guanine, or uracil) disrupts the normal base pairing and forms loops in the molecule. The center loop (the "anticodon loop") has a sequence of three unpaired bases called the **anticodon.** During translation, pairing of the mRNA codon with its complementary anticodon of tRNA appropriately positions the amino acid that tRNA carries.

Ribosomes, the sites of protein synthesis, consist of large and small subunits that organize the pairing between the codon and the anticodon. Several sites on the ribosome are binding sites for mRNA and tRNA. At the initiation of translation, mRNA binds to a small, separate ribosomal subunit. Attachment of the mRNA requires that the initiation codon (AUG) of mRNA be aligned with the P (peptidyl) site of the ribosome. A tRNA with a complementary anticodon for methionine binds to the mRNA, and a large subunit joins, forming a complete ribosome.

Polypeptide formation can now begin. Another site, the A (aminoacyl) site, is next to the P site. A second tRNA, whose anticodon is complementary to the codon in the A site, is

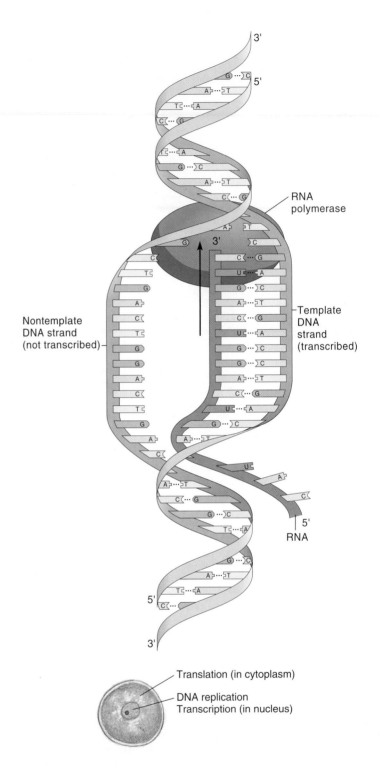

FIGURE 3.12

Transcription. Transcription involves the production of a messenger RNA molecule from the DNA segment. Note that transcription is similar to DNA replication in that the molecule is synthesized in the 5′ to 3′ direction.

positioned. Two tRNA molecules with their attached amino acids are now side-by-side in the P and A sites (figure 3.14). This step requires enzyme aid and energy, in the form of guanine triphosphate (GTP). An enzyme (peptidyl transferase), which is actually a part of

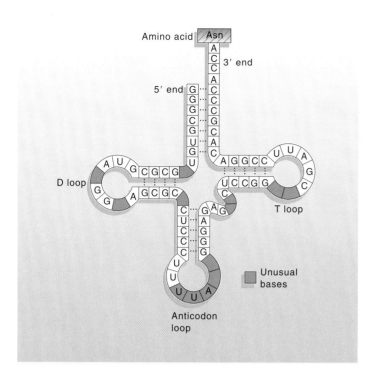

FIGURE 3.13

Structure of Transfer RNA. Diagrammatic representation of the secondary structure of transfer RNA (tRNA). An amino acid attaches to the 3′ end of the molecule. The anticodon is the sequence of three bases that pairs with the codon in mRNA, thus positioning the amino acid that tRNA carries. Other aspects of tRNA structure position the tRNA at the ribosome and in the enzyme that attaches the correct amino acid to the tRNA.

the larger ribosomal subunit, breaks the bond between the amino acid and tRNA in the P site, and catalyzes the formation of a peptide bond between that amino acid and the amino acid in the A site.

The mRNA strand then moves along the ribosome a distance of one codon. The tRNA with two amino acids attached to it that was in the A site is now in the P site. A third tRNA can now enter the exposed A site. This process continues until the entire mRNA has been translated, and a polypeptide chain has been synthesized. Translation ends when a termination codon (e.g., UAA) is encountered.

Protein synthesis often occurs on ribosomes on the surface of the rough endoplasmic reticulum. The positioning of ribosomes on the ER allows proteins to move into the ER as the protein is being synthesized. The protein can then be moved to the Golgi apparatus for packaging into a secretory vesicle or into a lysosome.

CHANGES IN DNA AND CHROMOSOMES

The genetic material of a cell can change, and these changes increase genetic variability and help increase the likelihood of survival in changing environments. These changes include alterations in the base sequence of DNA and changes that alter the structure or number of chromosomes.

Point Mutations

Genetic material must account for evolutionary change. **Point mutations** are changes in nucleotide sequences and may result from the replacement, addition, or deletion of nucleotides. Mutations are always random events. They may occur spontaneously as a result of base-pairing errors during replication, which result in a substitution of one base pair for another. Although certain environmental factors (e.g., electromagnetic radiation and many chemical mutagens) may change mutation rates, predicting what genes will be affected or what the nature of the change will be is impossible. *Some mutations may be unnoticed or even beneficial. However, the consequences of genetic changes are usually negative, because they disturb the structure of proteins that are the products of millions of years of evolution.*

Variation in Chromosome Number

Changes in chromosome number may involve entire sets of chromosomes, as in polyploidy, which was discussed earlier. **Aneuploidy** (Gr. *a*, without), on the other hand, involves the addition or deletion of one or more chromosomes, not entire sets. The addition of one chromosome to the normal 2N chromosome number (2N + 1) is a trisomy (Gr. *tri*, three + ME *some*, a group of), and the deletion of a chromosome from the normal 2N chromosome number (2N − 1) is a monosomy (Gr. *monos*, single).

Errors during meiosis usually cause aneuploidy. **Nondisjunction** occurs when a homologous pair fails to segregate during meiosis I or when chromatids fail to separate at meiosis II (figure 3.15). Gametes produced are either deficient in one chromosome or have an extra chromosome. If one of these gametes is involved in fertilization with a normal gamete, the monosomic or trisomic condition results. Aneuploid variations usually result in severe consequences involving mental retardation and sterility.

Variation in Chromosome Structure

Some changes may involve breaks in chromosomes. After breaking, pieces of chromosomes may be lost, or they may reattach, but not necessarily in their original position. The result is a chromosome that may have a different sequence of genes, multiple copies of genes, or missing genes. All of these changes can occur spontaneously. Various environmental agents, such as ionizing radiation and certain chemicals, can also induce these changes. The effects of changes in chromosome structure may be mild or severe, depending on the amount of genetic material duplicated or lost.

INHERITANCE PATTERNS IN ANIMALS

Classical genetics began with the work of Gregor Mendel and remains an important basis for understanding gene transfer between generations of animals. Understanding these genetics principles helps us to predict how traits will be expressed in offspring before these offspring are produced, something that has had profound implications in agriculture and medicine. One of the challenges of

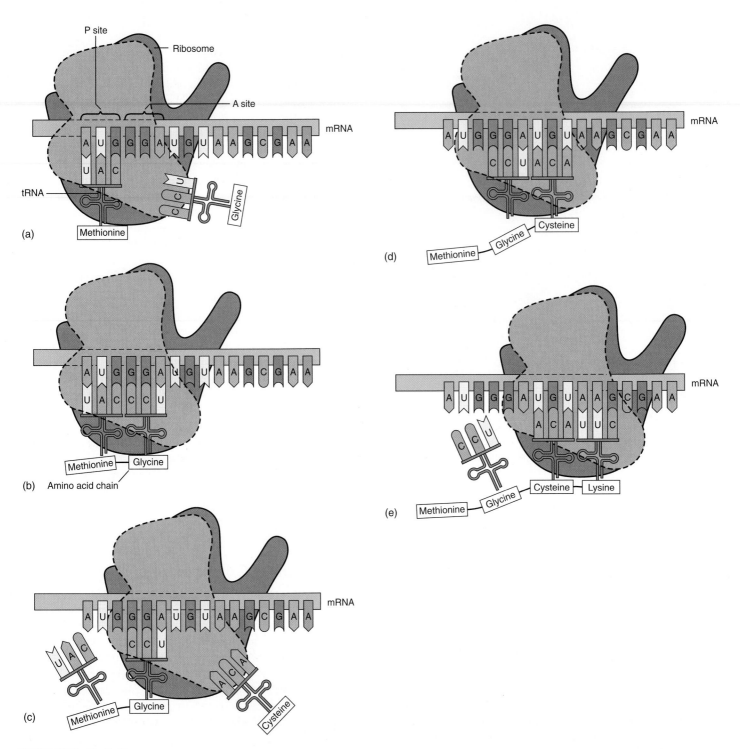

FIGURE 3.14

Events of Translation. (*a*) Translation begins when a methionine tRNA associates with the P site of the smaller ribosomal subunit and the initiation codon of mRNA associated with that subunit. The larger ribosomal subunit attaches to the small subunit/tRNA complex. (*b*) A second tRNA carrying the next amino acid enters the A site. A peptide bond forms between the two amino acids, freeing the first tRNA in the P site. (*c*) The mRNA, along with the second tRNA and its attached dipeptide, moves the distance of one codon. The first tRNA is discharged, leaving its amino acid behind. The second tRNA is now in the P site, and the A site is exposed and ready to receive another tRNA-amino acid. (*d*) A second peptide bond forms. (*e*) This process continues until an mRNA stop signal is encountered.

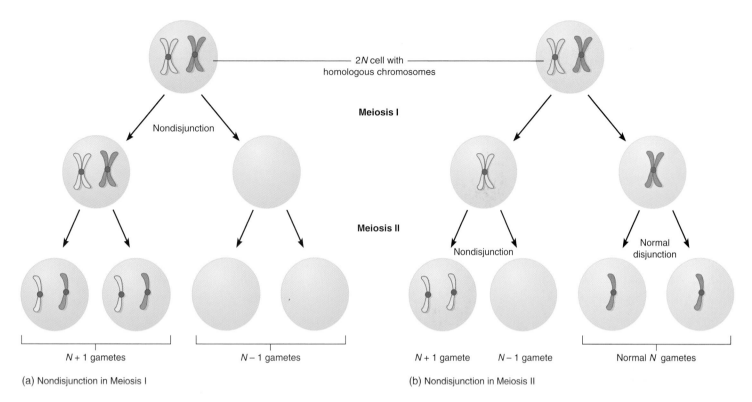

(a) Nondisjunction in Meiosis I

(b) Nondisjunction in Meiosis II

FIGURE 3.15

Results of Primary and Secondary Nondisjunction in Gamete Formation.   (*a*) Primary nondisjunction occurs in meiosis I and results from the failure of homologous pairs to separate normally. Both members of the homologous pair of chromosomes end up in one cell. A normal second meiotic division results in half of all gametes having both members of the homologous pair of chromosomes ($N + 1$). The other half of all gametes lacks members of this pair of homologous chromosomes ($N - 1$). (*b*) Secondary nondisjunction occurs after a normal first meiotic division. The failure of chromatids of one chromosome to separate in the second division means that a fourth of the gametes will be missing a member of one homologous pair ($N - 1$), and a fourth of the gametes will have an extra member of that homologous pair ($N + 1$). This illustration assumes that the second cell that resulted from meiosis I undergoes a normal second meiotic division.

modern genetics is to understand the molecular basis for these inheritance patterns.

The fruit fly, *Drosophila melanogaster,* is a classic tool for studying inheritance patterns. Its utility stems from its ease of handling, short life cycle, and easily recognized characteristics.

Studies of any fruit fly trait always make comparisons to a wild-type fly. If a fly has a characteristic similar to that found in wild flies, it is said to have the wild-type expression of that trait. (In the examples that follow, wild-type wings lay over the back at rest and extend past the posterior tip of the body, and wild-type eyes are red.) Numerous mutations from the wild-type body form, such as vestigial wings (reduced, shriveled wings) and sepia (dark brown) eyes have been described (figure 3.16).

SEGREGATION

During gamete formation, genes in each parent are incorporated into separate gametes. During anaphase I of meiosis, homologous chromosomes move toward opposite poles of the cell, and the resulting gametes have only one member of each chromosome pair. Genes carried on one member of a pair of homologous chromosomes end up in one gamete, and genes carried on the other member are segregated into a different gamete. The **principle of segregation** states that pairs of genes are distributed between gametes during gamete formation.

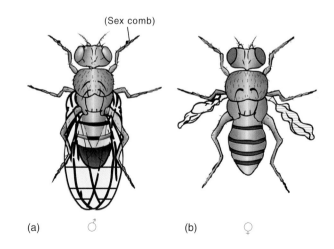

(Sex comb)

(a)   ♂   (b)   ♀

FIGURE 3.16

Distinguishing Sexes and Phenotypes of *Drosophila melanogaster.* (*a*) Male with wild-type wings and wild-type eyes. (*b*) Female with vestigial wings and sepia eyes. In contrast to the female, the posterior aspect of the male's abdomen has a wide, dark band and a rounded tip.

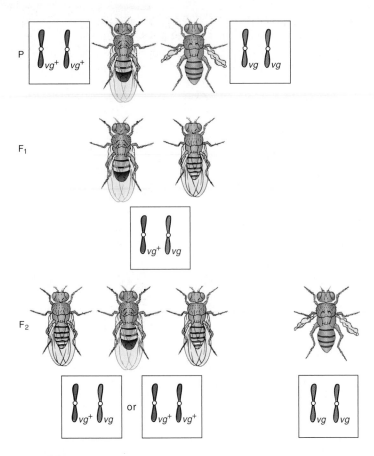

FIGURE 3.17

Cross Involving a Single Trait.   Cross between parental flies (P) with wild-type (vg^+) wings and vestigial (vg) wings, carried through two generations (F_1 and F_2).

Fertilization results in the random combination of gametes and brings homologous chromosomes together again.

A cross of wild-type fruit flies with flies having vestigial wings illustrates the principle of segregation. (The flies come from stocks that have been inbred for generations to ensure that they breed true for wild-type wings or vestigial wings.) The offspring (progeny) of this cross have wild-type wings and are the first generation of offspring, or the first filial (F_1) generation (figure 3.17). If these flies are allowed to mate with each other, their progeny are the second filial (F_2) generation. Approximately a fourth of these F_2 generation of flies have vestigial wings, and three-fourths have wild-type wings (figure 3.17). Note that the vestigial characteristic, although present in the parental generation, disappears in the F_1 generation and reappears in the F_2 generation. In addition, the ratio of wild-type flies to vestigial-winged flies in the F_2 generation is approximately 3:1. Reciprocal crosses, which involve the same characteristics but a reversal of the sexes of the individuals introducing a particular expression of the trait into the cross, yield similar results.

Genes that determine the expression of a particular trait can exist in alternative forms called **alleles** (Gr. *allelos,* each other). In the fruit-fly cross, the vestigial allele is present in the F_1 generation, and even though it is masked by the wild-type allele for wing shape, it retains its uniqueness because it is expressed again in some members of the F_2 generation. **Dominant** alleles hide the expression of another allele; **recessive** alleles are those whose

expression can be masked. In the fruit-fly example, the wild-type allele is dominant because it can mask the expression of the vestigial allele, which is therefore recessive.

The visual expression of alleles may not always indicate the underlying genetic makeup of an organism. This visual expression is the **phenotype,** and the genetic makeup is the **genotype.** In the example, the flies of the F_1 generation have the same phenotype as one of the parents, but they differ genotypically because they carry both a dominant and recessive allele. They are hybrids, and because this cross concerns only one pair of genes and a single trait, it is a **monohybrid cross** (Gr. *monos,* one + L. *hybrida,* offspring of two kinds of parents).

An organism is **homozygous** (L. *homo,* same + Gr. *zygon,* paired) if it carries two identical genes for a given trait and **heterozygous** (Gr. *heteros,* other) if the genes are different (alleles of each other). Thus, in the example, all members of the parental generation are homozygous because only true-breeding flies are crossed. All members of the F_1 generation are heterozygous.

Crosses are often diagrammed using a letter or letters descriptive of the trait in question. The first letter of the description of the dominant allele commonly is used. In fruit flies, and other organisms where all mutants are compared with a wild-type, the symbol is taken from the allele that was derived by a mutation from the wild condition. A superscript "$+$" next to the symbol represents the wild-type allele. A capital letter means that the mutant allele being represented is dominant, and a lowercase letter means that the mutant allele being represented is recessive.

Geneticists use the Punnett square to help predict the results of crosses. Figure 3.18 illustrates the use of a **Punnett square** to predict the results of the cross of two F_1 flies. The first step is to determine the kinds of gametes that each parent produces. One of the two axes of a square is designated for each parent, and the different kinds of gametes each parent produces are listed along the appropriate axis. Combining gametes in the interior of the square shows the results of random fertilization. As figure 3.18 indicates, the F_1 flies are heterozygous, with one wild-type allele and one vestigial allele. The two phenotypes of the F_2 generation are shown inside the Punnett square and are in a 3:1 ratio.

The **phenotypic ratio** expresses the results of a cross according to the relative numbers of progeny in each visually distinct class (e.g., 3 wild-type : 1 vestigial). The Punnett square has thus explained in another way the F_2 results in figure 3.17. It also shows that F_2 individuals may have one of three different genotypes. The **genotypic ratio** expresses the results of a cross according to the relative numbers of progeny in each genotypic category (e.g., 1 vg^+vg^+: 2 vg^+vg : 1 $vgvg$).

INDEPENDENT ASSORTMENT

It is also possible to make crosses using flies with two pairs of characteristics: flies with vestigial wings and sepia eyes, and flies that are wild for these characteristics. Sepia eyes are dark brown, and wild-type eyes are red. Figure 3.19 shows the results of crosses carried through two generations.

Note that flies in the parental generation are homozygous for the traits in question and that each parent produces only one kind of gamete. Gametes have one allele for each trait. Because each parent produces only one kind of gamete, fertilization results

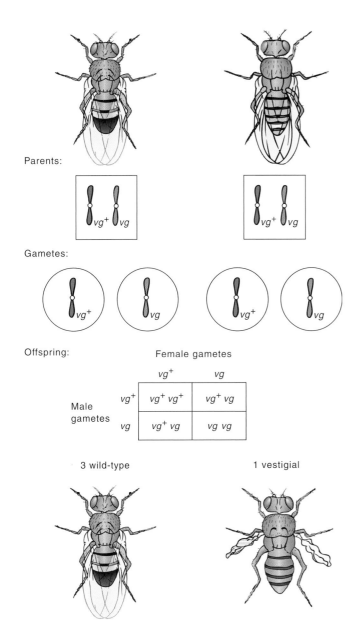

FIGURE 3.18
Use of a Punnett Square. A Punnett square helps predict the results of a cross. The kinds of gametes that each member of a cross produces are determined and placed along the axes of a square. Combining gametes in the interior of the square shows the results of mating: a phenotypic ratio of three flies with wild-type wings (vg^+) to one fly with vestigial wings (vg).

in offspring heterozygous for both traits. The F_1 flies have the wild-type phenotype; thus, wild-type eyes are dominant to sepia eyes. The F_1 flies are hybrids, and because the cross involves two pairs of genes and two traits, it is a **dihybrid cross** (Gr. *di*, two + L. *hybrida*, offspring of two kinds of parents).

The 9:3:3:1 ratio is typical of a dihybrid cross. During gamete formation, the distribution of genes determining one trait does not influence how genes determining the other trait are distributed. In the example, this means that an F_1 gamete with a vg^+ gene for wing condition may also have either the se or se^+ gene for eye color, as the F_1 gametes of figure 3.19 show. Note that all combinations of the eye color and wing condition genes are present, and that all combinations are equally likely. This illustrates the **principle of**

independent assortment, which states that, during gamete formation, pairs of factors segregate independently of one another.

The events of meiosis explain the principle of independent assortment (*see figure 3.6*). Cells produced during meiosis have one member of each homologous pair of chromosomes. Independent assortment simply means that when homologous chromosomes line up at metaphase I and then segregate, the behavior of one pair of chromosomes does not influence the behavior of any other pair (figure 3.20). After meiosis, maternal and paternal chromosomes are distributed randomly among cells.

OTHER INHERITANCE PATTERNS

The traits considered thus far have been determined by two genes, where one allele is dominant to a second. In this section, you learn that there are often many alleles in a population and that not all traits are determined by an interaction between a single pair of dominant or recessive genes.

Multiple Alleles

Two genes, one carried on each chromosome of a homologous pair, determine traits in one individual. A population, on the other hand, may have many different alleles with the potential to contribute to the phenotype of any member of the population. These are called **multiple alleles.**

Genes for a particular trait are at the same position on a chromosome. The gene's position on the chromosome is called its **locus** (L. *loca*, place). Numerous human loci have multiple alleles. Three alleles, symbolized I^A, I^B, and i, determine the familiar ABO blood types. Table 3.1 shows the combinations of alleles that determine a person's phenotype. Note that i is recessive to I^A and to I^B. I^A and I^B, however, are neither dominant nor recessive to each other. When I^A and I^B are present together, both are expressed.

Incomplete Dominance and Codominance

Incomplete dominance is an interaction between two alleles that are expressed more or less equally, and the heterozygote is different from either homozygote. For example, in cattle, the alleles for red coat color and for white coat color interact to produce an intermediate coat color called roan. Because neither the red nor the white allele is dominant, uppercase letters and a prime or a superscript are used to represent genes. Thus, red cattle are symbolized RR, white cattle are symbolized $R'R'$, and roan cattle are symbolized RR'.

TABLE 3.1
GENOTYPES AND PHENOTYPES IN THE ABO BLOOD GROUPS

GENOTYPE(S)	PHENOTYPE
$I^A I^A$, $I^A i$	A
$I^B I^B$, $I^B i$	B
$I^A I^B$	A and B
ii	O

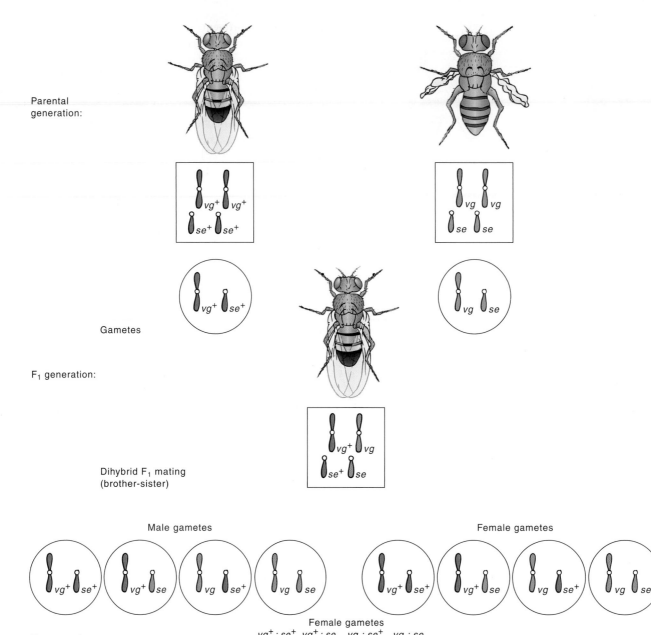

Parental generation:

Gametes

F₁ generation:

Dihybrid F₁ mating (brother-sister)

Male gametes

Female gametes

F₂ generation:

Male gametes

	Female gametes			
	vg^+ ; se^+	vg^+ ; se	vg ; se^+	vg ; se
vg^+ ; se^+	$vg^+ vg^+$; $se^+ se^+$	$vg^+ vg^+$; $se^+ se$	$vg^+ vg$; $se^+ se^+$	$vg^+ vg$; $se^+ se$
vg^+ ; se	$vg^+ vg^+$; $se se^+$	$vg^+ vg^+$; $se se$	$vg^+ vg$; $se se^+$	$vg^+ vg$; $se se$
vg ; se^+	$vg^+ vg$; $se^+ se^+$	$vg vg^+$; $se^+ se$	$vg vg$; $se^+ se^+$	$vg vg$; $se^+ se$
vg ; se	$vg^+ vg$; $se se^+$	$vg vg^+$; $se se$	$vg vg$; $se se^+$	$vg vg$; $se se$

Phenotypic ratio

9 wild-type wing (vg^+); wild-type eye (se^+)
3 wild-type wing; sepia eye (se)
3 vestigial wing (vg); wild-type eye
1 vestigial wing; sepia eye

FIGURE 3.19

Constructing a Punnett Square for a Cross Involving Two Characteristics. Note that every gamete has one allele for each trait and that all combinations of alleles for each trait are represented.

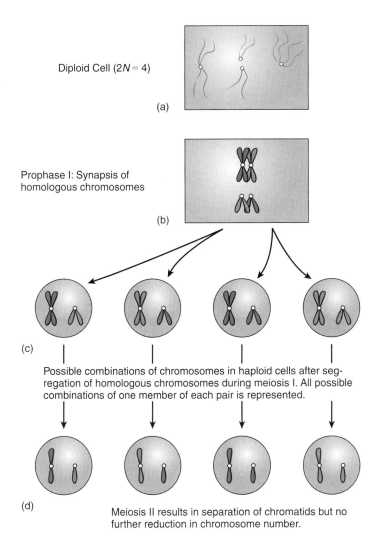

Diploid Cell (2*N* = 4)

(a)

Prophase I: Synapsis of
homologous chromosomes

(b)

(c)

Possible combinations of chromosomes in haploid cells after segregation of homologous chromosomes during meiosis I. All possible
combinations of one member of each pair is represented.

(d)

Meiosis II results in separation of chromatids but no
further reduction in chromosome number.

FIGURE 3.20

Independent Assortment of Chromosomes during Meiosis.   Color
distinguishes maternal and paternal chromosomes. Similar size and
shape indicate homologous chromosomes. (*a*) This cell has a diploid
(2*N*) chromosome number of four. (*b*) During the first meiotic division,
one homologous pair of chromosomes (and hence, the genes this pair
carries) is segregated without regard to the movements of any other homologous pair. (*c*) Thus, all combinations of large and small chromosomes in the cells are possible at the end of meiosis I. (*d*) Meiosis II
simply results in the separation of chromatids without further reduction
in chromosome number. Most organisms have more than two pairs of
homologous chromosomes in each cell. As the number of homologous
pairs increases, the number of different kinds of gametes also increases.

Codominance occurs when the heterozygote expresses the
phenotypes of both homozygotes. Thus, in the ABO blood types,
the I^AI^B heterozygote expresses both alleles.

THE MOLECULAR BASIS OF INHERITANCE PATTERNS

Just as the principles of segregation and independent assortment
can be explained based on our knowledge of the events of
meiosis, concepts related to dominance can be explained in

molecular terms. When we say that one allele is dominant to another, we do not mean that the recessive allele is somehow
"turned off" when the dominant allele is present. Instead, the
product of a gene's function is the result of a sequence of metabolic steps mediated by enzymes, which are encoded by the
gene(s) in question. A functional enzyme is usually encoded by a
dominant gene, and when that enzyme is present a particular
product is produced. A recessive allele usually arises by a mutation of the dominant gene, and the enzyme necessary for the production of the product is altered and does not function. In the
homozygous dominant state, both dominant genes code for the
enzyme that produces the product (figure 3.21*a*). In the heterozygous state, the activity of the single dominant allele is sufficient
to produce enough enzyme to form the product and the dominant
phenotype (figure 3.21*b*). In the homozygous recessive state,
no product can be formed and the recessive phenotype results
(figure 3.21*c*).

In the same way, one can explain incomplete dominance
and codominance. In these cases both alleles of a heterozygous individual produce approximately equal quantities of two enzymes
and products, and the phenotype that results would be either intermediate or show the products of both alleles.

Homozygous Dominant

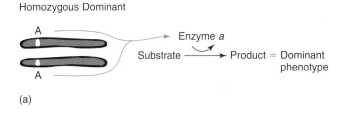

(a)

Heterozygous

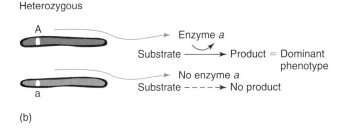

(b)

Homozygous Recessive

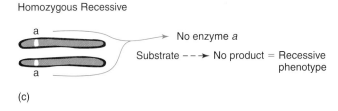

(c)

FIGURE 3.21

The Molecular Basis of Dominance.   (*a*) In a homozygous dominant
individual, both dominant genes code for enzymes that produce the
product and the dominant phenotype. (*b*) In the heterozygous state, the
single dominant allele is sufficient to produce enough enzyme to form
the product and the dominant phenotype. (*c*) In the homozygous recessive state, no product can be formed and the recessive phenotype results.

WILDLIFE ALERT
Preserving Genetic Diversity

One of the ways in which scientists evaluate the environmental health of a region is to assess the variety of organisms present in an area. Environments that have a great variety or diversity in species are usually considered healthier than environments with less diversity. Diversity can be reduced through habitat loss, the exploitation of animals or plants through hunting or harvesting, and the introduction of foreign species.

Another criterion used to evaluate environmental health is genetic diversity. Genetic diversity is the variety of alleles within a species. When a species on the brink of extinction is preserved, reduced genetic diversity within the species threatens the health of the species. Near-extinction events, in which many individuals die, eradicate many alleles from populations. Lowered numbers of individuals result in inbreeding, which also reduces genetic diversity. The result is that populations that survive near-extinction events tend to be genetically uniform. The effect of genetic uniformity on populations is nearly always detrimental because, when environmental conditions change, entire populations can be adversely affected. For example, if one individual in a genetically uniform population is susceptible to a particular disease, all individuals will be susceptible, and the disease will spread very quickly. High genetic diversity helps ensure that some individuals will survive the disease outbreak, and the species will be less likely to face extinction. Since mutation is the ultimate source of new variation within species, lost genetic diversity can only be replaced over evolutionary timescales. For all practical purposes, when genetic diversity is lost it is gone forever.

Conservation geneticists evaluate the genetic health of populations of organisms and try to preserve the genetic variation that exists within species. These efforts involve the use of virtually every genetic tool available to modern science, including the molecular techniques for studying DNA and the proteins of endangered organisms. Conservation geneticists search native populations for related individuals that could be used to enhance the genetic makeup of endangered organisms. They recommend breeding programs to preserve alleles that could easily be lost. Many zoos throughout the world cooperate in breeding programs that exchange threatened animals, or gametes from

threatened animals, to preserve alleles. Such a program is underway to help preserve the endangered snow leopard (*Panthera uncia*). There are between 3,000 and 7,500 snow leopards distributed throughout the mountains of central Asia, where they live at altitudes between 2,000 and 5,000 meters. Poaching the snow leopard to supply its coat for black-market trade is a serious threat to the cats that remain. Hunting wild prey and habitat destruction for farming and grazing livestock also threaten these cats (box figure 3.1). The American Zoo and Aquarium Association is coordinating an effort to maintain genetic diversity among the snow leopards in captivity in North America.

BOX FIGURE 3.1   Snow leopard (*Panthera uncia*).

SUMMARY

1. Eukaryotic chromosomes are complexly coiled associations of DNA and histone proteins.
2. The presence or absence of certain chromosomes that are represented differently in males and females determine the sex of an animal. The X-Y system of sex determination is most common.
3. The replication of DNA and its subsequent allocation to daughter cells during mitosis involves a number of phases collectively called the cell cycle. The cell cycle is that period from the time a cell is produced until it completes mitosis.
4. Mitosis maintains the parental number of chromosome sets in each daughter nucleus. It separates the sister chromatids of each (replicated) chromosome for distribution to daughter nuclei.
5. Interphase represents about 90% of the total cell cycle. It includes periods of cell growth and normal cell function. It also includes the time when DNA is replicated.
6. Mitosis is divided into four phases. During prophase, the mitotic spindle forms and the nuclear envelope disintegrates. During metaphase, the replicated chromosomes align along the spindle equator. During anaphase, the centromeres joining sister chromatids divide and microtubules pull sister chromatids to opposite

poles of the cell. During telophase, the mitotic spindle disassembles, the nuclear envelope reforms, and chromosomes unfold.

7. Cytokinesis, the division of the cytoplasm, begins in late anaphase and is completed in telophase.

8. Meiosis is a special form of nuclear division during gamete formation. It consists of a single replication of the chromosomes and two nuclear divisions that result in four daughter cells, each with half the original number of chromosomes.

9. In the life cycle of most animals, certain diploid cells undergo gametogenesis to form haploid gametes (sperm in males and eggs in females). Fusion of a sperm and an egg nucleus at fertilization produces a new diploid cell (zygote).

10. Deoxyribonucleic acid (DNA) is the hereditary material of the cell. Ribonucleic acid (RNA) participates in protein synthesis.

11. Nucleotides are nucleic acid building blocks. Nucleotides consist of a nitrogenous (purine or pyrimidine) base, a phosphate, and a pentose sugar.

12. DNA replication is semiconservative. During replication, the DNA strands separate, and each strand is a template for a new strand.

13. Protein synthesis is a result of two processes. Transcription involves the production of a messenger RNA molecule from a DNA molecule. Translation involves the movement of messenger RNA to the cytoplasm, where transfer RNA and ribosomes link amino acids in a proper sequence to produce a polypeptide.

14. Changes in DNA and chromosomes include point mutations, which alter the bases in DNA, and changes in chromosome number and structure. These changes are usually deleterious for the organism.

15. The principle of segregation states that pairs of genes are distributed between gametes during gamete formation when homologous chromosomes are distributed to different gametes during meiosis.

16. The principle of independent assortment states that, during gamete formation, pairs of genes segregate independently of one another. This is a result of meiotic processes in which members of one homologous pair of chromosomes are not influenced by the movements of any other pair of chromosomes.

17. Populations may have many alternative expressions of a gene at any locus. Human traits, like the ABO blood group, are traits determined by multiple alleles.

18. Incomplete dominance is an interaction between two alleles in which the alleles contribute more or less equally to the phenotype. Codominance is an interaction between two alleles in which both alleles are expressed in the heterozygote.

19. Patterns of inheritance observed at an organismal level are explained at a molecular level by the presence or absence of functional enzymes. A dominant allele usually encodes a functional enzyme, and a recessive allele usually encodes a nonfunctional enzyme.

SELECTED KEY TERMS

alleles (p. 44)
cell cycle (p. 33)
deoxyribonucleic acid (p. 37)
meiosis (p. 35)
mitosis (p. 33)
principle of independent
    assortment (p. 45)

principle of segregation (p. 43)
ribonucleic acid (p. 37)
transcription (p. 39)
translation (p. 39)

CRITICAL THINKING QUESTIONS

1. Which do you think evolved first—meiosis or mitosis? Why? What do you think may have been some of the stages in the evolution of one from the other?

2. Why is it important that all regions of chromosomes are not continually active?

3. In a laboratory experiment, you analyze the DNA from your own white blood cells. You determine that 15% of the nucleotide bases it contains is thymine. What percentage of the bases is adenine? cytosine? guanine?

4. A wild-type fruit fly, which had one vestigial-winged parent, is crossed with a vestigial-winged fruit fly. Use a Punnett square to predict the phenotypic and genotypic ratios in the offspring of this cross. What is this cross called?

5. A vestigial-winged, wild-eyed fruit fly is crossed with a wild-winged, sepia-eyed fruit fly. At least one parent of both of these flies had vestigial wings and sepia eyes. Use a Punnett square to predict the phenotypic ratio expected in the offspring of this cross.

6. Supply the genotypes, as completely as possible, for the parents and progeny of the following crosses:

   a. wild-type wing; sepia eye × wild-type wing; wild-type eye = 3 wild-type wing; sepia eye: 1 vestigial wing; sepia eye: 3 wild-type wing; wild-type eye: 1 vestigial wing; wild-type eye

   b. wild-type wing; sepia eye × wild-type wing; wild-type eye = 1 wild-type wing; sepia eye: 1 wild-type wing; wild-type eye

   c. vestigial wing; wild-type eye × wild-type wing; sepia eye = all wild-type wing; wild-type eye

   d. vestigial wing; wild-type eye × wild-type wing; sepia eye = 1 vestigial wing; wild-type eye: 1 vestigial wing; sepia eye: 1 wild-type wing; sepia eye: 1 wild-type wing; wild-type eye

7. The following progeny are the result of a cross between two fruit flies. Unfortunately, the phenotypes of the parental flies were not recorded. Formulate a hypothesis regarding the genotypes of the parental flies. (Hint: Consider the ratio between wing phenotypes separately from eye phenotypes in formulating your hypothesis.)
   Progeny:

      293 wild-type wing; wild-type eye

      310 wild-type wing; sepia eye

      97 vestigial wing; wild-type eye

      100 vestigial wing; sepia eye

8. Do you think that Mendel's conclusions regarding the assortment of genes for two traits would have been any different if he had used traits encoded by genes carried on the same chromosome? Explain.

9. To observe the laws of chance, toss a dime and a quarter simultaneously 10 times, and count the number of head-head, head-tail, and tail-tail combinations. How close are your results to a 1:2:1 ratio? Now toss the coins 90 times more for a total of 100 tosses. How close are your new results to a 1:2:1 ratio? Explain the difference.

ONLINE LEARNING CENTER

Visit our Online Learning Center (OLC) at www.mhhe.com/zoology (click on the book's title) to find the following chapter-related materials:

• CHAPTER QUIZZING

- ANIMATIONS
  DNA Structure
  DNA Replication
  Mitosis
  Meiosis
  Transcription
  Translation
  Mutations
- RELATED WEB LINKS
  Cell Division
  Mendelian Genetics
  Molecular Basis of Genetics
  Protein Synthesis

- BOXED READINGS ON
  Transgenic Animals
  Diabetes
  The Participation of the Fetus During Childbirth
- SUGGESTED READINGS
- LAB CORRELATIONS
  Check out the OLC to find specific information on these related lab exercises in the *General Zoology Laboratory Manual*, 5th edition, by Stephen A. Miller:

  Exercise 3   *Aspects of Cell Function*
  Exercise 4   *Genetics*

CHAPTER 4

EVOLUTION:

HISTORY AND EVIDENCE

Outline

Concepts

1. Organic evolution is the change of a population over time.
2. Although the concept of evolution is very old, Charles Darwin formulated the modern explanation of how change occurs. Darwin began gathering his evidence of evolution during a worldwide mapping expedition on the HMS *Beagle* and spent the rest of his life formulating and defending his ideas.
3. Darwin's theory of evolution by natural selection, although modified from its original form, is still a highly regarded account of how evolution occurs.
4. Modern evolutionary theorists apply principles of genetics, ecology, and geographic and morphological studies when investigating evolutionary mechanisms.
5. Adaptation may refer either to a process of evolutionary change or to the result of a change. In the latter sense, an adaptation is a structure or a process that increases an animal's potential to survive and reproduce in specific environmental conditions.
6. Microevolution refers to any change in the frequency of alleles in a population. Macroevolution refers to large-scale patterns and rates of change among groups of organisms. There is a wealth of evidence of macroevolution. This evidence comes from the study of biogeography, paleontology, comparative anatomy, molecular biology, and developmental biology.

Questions of the earth's origin and life's origin have been on the minds of humans since prehistoric times, when accounts of creation were passed orally from generation to generation. For many people, these questions centered around concepts of purpose. Religious and philosophical writings help provide answers to such questions as: Why are we here? What is human nature really like? How do we deal with our mortality?

Many of us are also concerned with other, very different, questions of origin: How old is the planet earth? How long has life been on the earth? How did life arise on the earth? How did a certain animal species come into existence? Answers for these questions come from a different authority—that of scientific inquiry.

This chapter presents the history of the study of organic evolution and introduces the theory of evolution by natural selection. **Organic evolution,** according to Charles Darwin, is "descent with modification." This simply means that populations change over time. Populations consist of individuals of the same species that occupy a given area at the same time. They share a unique set of genes. Population concepts are discussed in chapter 5. Evolution by itself does not imply any particular lineage or any particular mechanism, and virtually all scientists agree that the evidence for change in organisms over long time periods is overwhelming. Further, most scientists agree that natural selection, the mechanism for evolution that Charles Darwin outlined, is one explanation of how evolution occurs. In spite of the scientific certainty of evolution and an acceptance of a general mechanism, much is still to be learned about the details of evolutionary processes. Scientists will be debating these details for years to come.

PRE-DARWINIAN THEORIES OF CHANGE

The idea of evolution did not originate with Charles Darwin. Some of the earliest references to evolutionary change are from the ancient Greeks. The philosophers Empedocles (495–435 B.C.) and Aristotle (384–322 B.C.) described concepts of change in living organisms over time. Georges-Louis Buffon (1707–1788) spent many years studying comparative anatomy. His observations of structural variations in particular organs of related animals, and especially his observations of vestigial structures, convinced him that change must have occurred during the history of life on earth. Buffon attributed change in organisms to the action of the environment. He believed in a special creation of species and considered change as being degenerate—for example, he described apes as degenerate humans. Erasmus Darwin (1731–1802), a physician and the grandfather of Charles Darwin, was intensely interested in questions of origin and change. He believed in the common ancestry of all organisms.

Jean Baptiste Lamarck (1744–1829) was a distinguished French zoologist. His contributions to zoology include important studies of animal classification. Lamarck published a set of invertebrate zoology books. His theory was based on a widely accepted theory of inheritance that organisms develop new organs, or modify existing organs, as needs arise. (Charles Darwin also accepted this idea of inheritance.) Similarly, he believed that disuse resulted in the degeneration of organs. Lamarck believed that "need" was dictated by environmental change and that change involved movement toward perfection. The idea that change in a species is directed by need logically led Lamarck to the conclusion that species could not become extinct, they simply evolved into different species.

Lamarck illustrated his ideas of change with the often-quoted example of the giraffe. He contended that ancestral giraffes had short necks, much like those of any other mammal. Straining to reach higher branches during browsing resulted in their acquiring higher shoulders and longer necks. These modifications, produced in one generation, were passed on to the next generation. Lamarck published his theory in 1802 and included it in one of his invertebrate zoology books, *Philosophie Zoologique* (1809). He defended his ideas in spite of intense social criticism.

Lamarck's acceptance of a theory of inheritance that we now know is not correct led him to erroneous conclusions about how evolution occurs. There is no evidence that changes in the environment can initiate changes in organisms, which can be passed on to future generations. Instead, change originates in the process of gamete formation.

Random changes in the structure of DNA (mutation) and chance processes involved in the assortment of genes into gametes (e.g., independent assortment and crossing-over—chapter 3)

result in variation among offspring. The environment then plays a role in determining the survival of these variations in subsequent generations. Even though Larmarck's mechanism of change was incorrect, he should be remembered for his steadfastness in promoting the idea of evolutionary change and for his numerous accomplishments in zoology.

DARWIN'S EARLY YEARS AND HIS JOURNEY

Charles Robert Darwin (1809–1882) was born on February 12, 1809. His father, like his grandfather, was a physician. During Darwin's youth in Shrewsbury, England, his interests centered around dogs, collecting, and hunting birds—all popular pastimes in wealthy families of nineteenth-century England. These activities captivated him far more than the traditional education he received at boarding school. At the age of 16 (1825), he entered medical school in Edinburgh, Scotland. For two years, he enjoyed the company of the school's well-established scientists. Darwin, however, was not interested in a career in medicine because he could not bear the sight of pain. This prompted his father to suggest that he train for the clergy in the Church of England. With this in mind, Charles enrolled at Christ's College in Cambridge and graduated with honors in 1831. This training, like the medical training he received, was disappointing for Darwin. Again, his most memorable experiences were those with Cambridge scientists. During his stay at Cambridge, Darwin developed a keen interest in collecting beetles and made valuable contributions to beetle taxonomy.

VOYAGE OF THE HMS *BEAGLE*

One of his Cambridge mentors, a botanist by the name of John S. Henslow, nominated Darwin to serve as a naturalist on a mapping expedition that was to travel around the world. Darwin was commissioned as a naturalist on the HMS *Beagle*, which set sail on December 27, 1831 on a five-year voyage (figure 4.1). Darwin helped with routine seafaring tasks and made numerous collections, which he sent to Cambridge. The voyage gave him ample opportunity to explore tropical rain forests, fossil beds, the volcanic peaks of South America, and the coral atolls of the South Pacific. Most importantly, Darwin spent five weeks on the **Galápagos Islands,** a group of volcanic islands 900 km off the coast of Ecuador. Some of his most revolutionary ideas came from his observations of plant and animal life on these islands. At the end of the voyage, Darwin was just 27 years old.

By 1842, Darwin had developed the essence of his conclusions but delayed their publication because of uncertainty about how they would be received. His ideas were eventually presented before the Linnean Society in London in 1858, and *On the Origin of Species by Means of Natural Selection* was published in 1859 and revolutionized biology. In the years after his voyage, Darwin was an extremely prolific scientist. He published

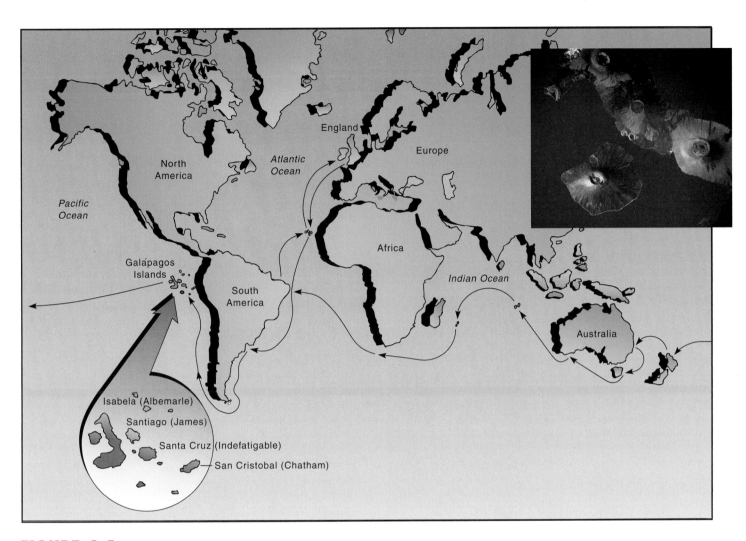

FIGURE 4.1

Voyage of the HMS Beagle.   Charles Darwin served as a naturalist on a five-year mapping expedition. Darwin's observations, especially those on the Galápagos Islands, were the basis for his theory of evolution by natural selection. The inset shows two of the islands as photographed from the space shuttle *Atlantis*. Fernandina is comprised of a single volcanic peak. Isabela is comprised of three volcanic peaks: Wolf (top), Darwin (center), and Alcedo (lower).

five volumes on *Zoology of the* Beagle *Voyage* (1843), *Fertilisation of Orchids* (1862), *The Variation of Plants and Animals under Domestication* (1873), *The Descent of Man* (1871), and numerous other works.

EARLY DEVELOPMENT OF DARWIN'S IDEAS OF EVOLUTION

The development of Darwin's theory of evolution by natural selection was a long, painstaking process. Darwin had to become convinced that change occurs over time. Before leaving on his voyage, Darwin accepted the prevailing opinion that the earth and its inhabitants had been created 6,000 years ago and had not changed since. Because his observations during his voyage suggested that change does occur, he realized that 6,000 years could not account for the diversity of modern species if they arose through gradual

change. Once ideas of change were established in Darwin's thinking, it took about 20 years of study to conceive, and thoroughly document, the mechanism by which change occurs. Darwin died without knowing the genetic principles that support his theory.

GEOLOGY

During his voyage, Darwin read Charles Lyell's (1779–1875) *Principles of Geology*. In this book, Lyell developed the ideas of another geologist, James Hutton, into the theory of **uniformitarianism.** His theory was based on the idea that the forces of wind, rain, rivers, volcanoes, and geological uplift shape the earth today, just as they have in the past. Lyell and Hutton contended that it was these forces, not catastrophic events, that shaped the face of the earth over hundreds of millions of years. This book planted two important ideas in Darwin's mind: (1) the earth could be much older than 6,000 years, and (2) if the face of the earth changed gradually over long periods, could not living forms also change during that time?

(a)

(b)

FIGURE 4.2

The Giant Sloth. (*a*) Charles Darwin found evidence of the existence of giant sloths in South America similar to this *Megantherium*. Giant sloths lived about 10,000 years ago and weighed in excess of 1,000 kg. They certainly did not move through tree branches like their only living relative, *Choloepus* (*b*). Instead, they probably fed on leaves of lower tree branches that they could reach from the ground. The similarity of giant sloths and modern-day sloths impressed Darwin with the fact that species change over time. Many species have become extinct. As in this case, they often leave descendants that provide evidence of evolutionary change.

FOSSIL EVIDENCE

Once the HMS *Beagle* reached South America, Darwin spent time digging in the dry riverbeds of the pampas (grassy plains) of Argentina. He found the fossil remains of an extinct hippopotamus-like animal, now called *Toxodon,* and fossils of a horselike animal, *Thoantherium*. Both of these fossils were from animals that were clearly different from any other animal living in the region. Modern horses were in South America, of course, but Spanish explorers had brought these horses to the Americas in the 1500s. The fossils suggested that horses had been present and had become extinct long before the 1500s. Darwin also found fossils of giant armadillos and giant sloths (figure 4.2). Except for their large size, these fossils were very similar to forms Darwin found living in the region.

Fossils were not new to Darwin. They were popularly believed to be the remains of animals that perished in catastrophic events, such as Noah's flood. In South America, however, Darwin understood them to be evidence that the species composition of the earth had changed. Some species became extinct without leaving any descendants. Others became extinct, but not before giving rise to new species.

GALÁPAGOS ISLANDS

On its trip up the western shore of South America, the HMS *Beagle* stopped at the Galápagos Islands, which are named after the large tortoises that inhabit them (Sp. *galápago,* tortoise). The tortoises weigh up to 250 kg, have shells up to 1.8 m in diameter, and live for 200 to 250 years. The islands' governor pointed out to Darwin that the shapes of the tortoise shells from different parts of Albemarle Island differed. Darwin noticed other differences as well. Tortoises from the drier regions had longer necks than tortoises from wetter habitats (figure 4.3). In spite of their differences, the tortoises were quite similar to each other and to the tortoises on the mainland of South America.

How could these overall similarities be explained? Darwin reasoned that the island forms were derived from a few ancestral animals that managed to travel from the mainland, across 900 km of ocean. Because the Galápagos Islands are volcanic (*see figure 4.1*) and arose out of the seabed, no land connection with the mainland ever existed. One modern hypothesis is that tortoises floated from the mainland on mats of vegetation that regularly break free from coastal riverbanks during storms. Without predators on the islands, tortoises gradually increased in number.

Darwin also explained some of the differences that he saw. In dryer regions, where vegetation was sparse, tortoises with longer necks would be favored because they could reach higher to get food. In moister regions, tortoises with longer necks would not necessarily be favored, and the shorter-necked tortoises could survive.

Darwin made similar observations of a group of dark, sparrowlike birds. Although he never studied them in detail, Darwin noticed that the Galápagos finches bore similarities suggestive of common ancestry. Scientists now think that Galápagos finches also descended from an ancestral species that originally inhabited the mainland of South America. The chance arrival of a few finches, in

(a)

(b)

FIGURE 4.3

Galápagos Tortoises.   (*a*) Shorter-necked subspecies of *Geochelone elephantopus* live in moister regions and feed on low-growing vegetation. (*b*) Longer-necked subspecies live in drier regions and feed on high-growing vegetation.

either single or multiple colonization events, probably set up the first bird populations on the islands. Early finches encountered many different habitats, all empty of other birds and predators. Ancestral finches, probably seed eaters, multiplied rapidly and filled the seed-bearing habitats most attractive to them. Fourteen species of finches arose from this ancestral group, including one species found on small Cocos Island northeast of the Galápagos Islands. Each species is adapted to a specific habitat on the islands. The most obvious difference between these finches relates to dietary adaptations and is reflected in the size and shape of their bills. The finches of the Galápagos Islands provide an example of **adaptive radiation**—the formation of new forms from an ancestral species, usually in response to the opening of new habitats (figure 4.4).

Darwin's experiences in South America and the Galápagos Islands convinced him that animals change over time. It took the remaining years of his life for Darwin to formulate and document his ideas, and to publish a description of the mechanism of evolutionary change.

THE THEORY OF EVOLUTION BY NATURAL SELECTION

On his return to England in 1836, Darwin worked diligently on the notes and specimens he had collected and made new observations. He was familiar with the obvious success of breeders in developing desired variations in plant and animal stocks (figure 4.5). He wondered if this artificial selection of traits could have a parallel in the natural world.

Ideas of how change occurred began to develop on his voyage. They took on their final form after 1838 when he read an essay by Thomas Malthus (1766–1834) entitled *Essay on the Principle of Population*. Malthus believed that the human population has the potential to increase geometrically. However, because resources cannot keep pace with the increased demands of a burgeoning population, population-restraining factors, such as poverty, wars, plagues, and famine, begin to have an influence. Darwin realized that a similar struggle to survive occurs in nature. This struggle, when viewed over generations, could be a means of **natural selection.** Traits that were detrimental for an animal would be eliminated by the failure of the animal containing them to reproduce.

NATURAL SELECTION

Charles Darwin had no knowledge of modern genetic concepts and, therefore, had no knowledge of the genetic principles that are the basis of evolutionary theory as it exists today. The modern version of his theory can be summarized as follows:

1. All organisms have a far greater reproductive potential than is ever realized. For example, a female oyster releases about 100,000 eggs with each spawning, a female sea star releases about 1 million eggs each season, and a female robin may lay four fertile eggs each season. What if all of these eggs were fertilized and developed to reproductive adults by the following year? A half million female sea stars (half of the million

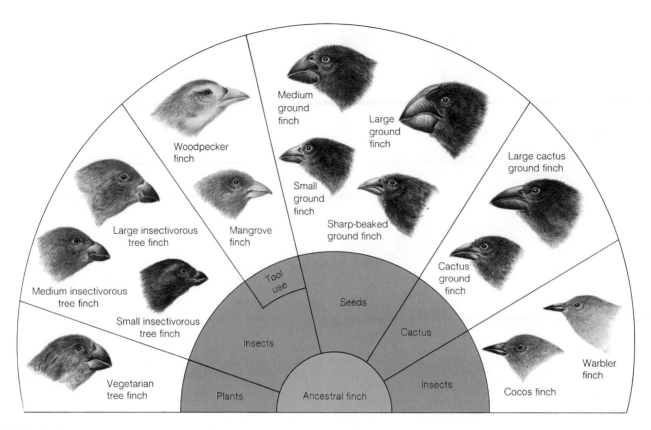

FIGURE 4.4

Adaptive Radiation of the Galápagos Finches. Ancestral finches from the South American mainland colonized the Galápagos Islands. Open habitats and few predators promoted the radiation of finches into 14 different species.

eggs would produce females and half would produce males), each producing another million eggs, repeated over just a few generations would soon fill up the oceans! Even the adult female robins, each producing four more robins, would result in unimaginable resource problems in just a few years.

2. Inherited variations exist. They arise from a variety of sources including mutation, genetic recombination (chapter 3), and random fertilization. Seldom are any two individuals exactly alike. Some of these genetic variations

may confer an advantage to the individual possessing them. In other instances, variations may be harmful to an individual. In still other instances, particular variations may be neither helpful nor harmful. (These are said to be neutral.) These variations can be passed on to offspring.

3. Because resources are limited, existence is a constant struggle. Many more offspring are produced than resources can support; therefore, many individuals die. Darwin reasoned that the individuals that die are those with the traits (variations) that

(a)   (b)   (c)

FIGURE 4.5

Artificial Selection. Artificial selection of domestic cattle has resulted in the diverse breeds seen today. (*a*) Angus cattle originated in Aberdeenshire and Angushire of Scotland and were imported into the United States in 1873. (*b*) Brahman cattle came to the United States from India and Brazil in the early 1990s. (*c*) The Brangus breed was developed in the United States around 1912 and resulted from crosses between Angus and Brahman cattle.

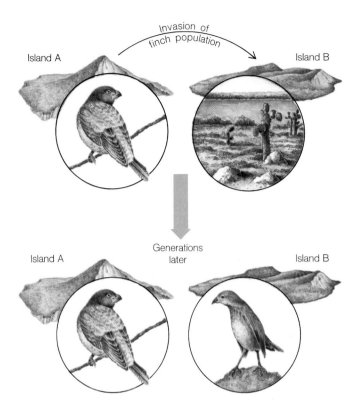

Invasion of
finch population

Island A   Island B

Generations
later
Island A   Island B

FIGURE 4.6

Natural Selection in Galápagos Finches.   Natural selection of finches on the Galápagos Islands resulted in changes in bill shape. In this illustration, tree-feeding birds on island A invade island B. The relatively treeless habitats of island B select against birds most adapted to living in trees, and individuals that can exploit seeds on the ground in low-growing vegetation are more likely to survive and reproduce. Subsequent generations of finches on island B show ground-feeding bill characteristics.

make survival and successful reproduction less likely. Traits that promote successful reproduction are said to be adaptive.

4. Adaptive traits are perpetuated in subsequent generations. Because organisms with maladaptive traits are less likely to reproduce, the maladaptive traits become less frequent in a population and eventually are eliminated.

With these ideas, Darwin formulated a theory that explained how the tortoises and finches of the Galápagos Islands changed over time (figure 4.6). In addition, Darwin's theory explained how some animals, such as the ancient South American horses, could become extinct. What if a group of animals is faced with a new environment to which it is ill-adapted? Climatic changes, food shortages, and other environmental stressors could lead to extinction.

ADAPTATION

Adaptation occurs when a change in a phenotype increases an animal's chance of successful reproduction. It is likely to be expressed when an organism encounters a new environment and may result in the evolution of multiple new groups if an environment can be exploited in different ways. No terms in evolution have been more

laden with confusion than adaptation and fitness or adaptedness. Adaptation is sometimes used to refer to a process of change in evolution. That use of the term is probably less confusing than when it is used to describe the result of the process of change. In this text, adaptations are defined as characteristics that increase the potential of an organism or species to successfully reproduce in a specified environment. In a similar fashion, adaptedness or fitness is a measure of the capacity for successful reproduction in a given environment relative to others of the same species. In other words, an adaptation is any characteristic that increases an individual's fitness.

Not every characteristic is an adaptation to some kind of environmental situation. Some have erroneously concluded that if a structure is now performing a specific function, it must have arisen for that purpose and is, therefore, an adaptation. An extreme extension of this incorrect view is that evolutionary adaptations lead to perfection.

ALFRED RUSSEL WALLACE

Alfred Russel Wallace (1823–1913) was an explorer of the Amazon Valley and led a zoological expedition to the Malay Archipelago, which is an area of great biogeographical importance. Wallace, like Darwin, was impressed with evolutionary change and had read the writings of Thomas Malthus on human populations. He synthesized a theory of evolution similar to Darwin's theory of evolution by natural selection. After writing the details of his theory, Wallace sent his paper to Darwin for criticism. Darwin recognized the similarity of Wallace's ideas and prepared a short summary of his own theory. Both Wallace's and Darwin's papers were published in the *Journal of the Proceedings of the Linnean Society* in 1858. Darwin's insistence on having Wallace's ideas presented along with his own shows Darwin's integrity. Darwin then shortened a manuscript he had been working on since 1856 and published it as *On the Origin of Species by Means of Natural Selection* in November 1859. The 1,250 copies prepared in the first printing sold out the day the book was released.

In spite of the similarities in the theories of Wallace and Darwin, there were also important differences. Wallace, for example, believed that every evolutionary modification was a product of selection and, therefore, had to be adaptive for the organism. Darwin, on the other hand, admitted that natural selection may not explain all evolutionary changes. He did not insist on finding adaptive significance for every modification. Further, unlike Darwin, Wallace stopped short of attributing human intellectual functions and the ability to make moral judgments to evolution. On both of these matters, Darwin's ideas are closer to the views of most modern scientists.

Wallace's work motivated Darwin to publish his own ideas. The theory of natural selection, however, is usually credited to Charles Darwin. Darwin's years of work and massive accumulations of evidence led even Wallace to attribute the theory to Darwin. Wallace wrote to Darwin in 1864:

I shall always maintain [the theory of evolution by natural selection] to be actually yours and yours only. You had worked it out in details I had never thought of years before I had a ray of light on the subject.

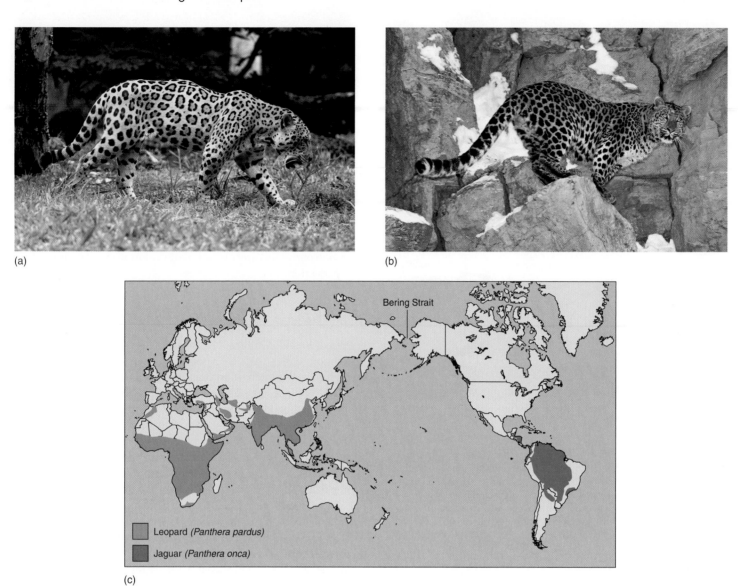

(a)

(b)

(c)

FIGURE 4.7

Biogeography as Evidence of Evolutionary Change.   (*a*) The jaguar (*Panthera onca*) of South America has a similar ecological role to that of the (*b*) leopard (*Panthera pardus*) of Africa and Asia. Their similar form suggests common ancestry, even though they are separated by apparently insurmountable oceanic barriers (*c*). Spotted varieties of these species are distinguished by the presence (jaguar) or absence (leopard) of small spots within dark rosette markings of their coats. Biogeographers have provided probable explanations for these observations. (*See Evolutionary Insights, page 65.*)

MICROEVOLUTION, MACROEVOLUTION, AND EVIDENCE OF MACROEVOLUTIONARY CHANGE

Organic evolution was defined earlier as a change in populations over time, or simply "descent with modification." The change must involve the genetic makeup of the population in order to be passed to future generations. These observations have led biologists to look for the mechanisms by which changes occur. There is no doubt that genetic changes in populations occur—they have been directly observed in the field and in the laboratory. These changes are the reason that bacteria gain resistance to antibiotics and agricultural pests

become resistant to pesticides. A change in the frequency of alleles in populations over time is called **microevolution.** The processes that result in microevolution are discussed in chapter 5.

Over longer timescales, microevolutionary processes result in large-scale changes. Large-scale changes that result in extinction and the formation of new species are called **macroevolution.** Macroevolutionary changes are difficult to observe in progress because of the geological timescales that are usually involved. Evidence that macroevolution occurs, however, is compelling. This evidence is in the form of patterns of plant and animal distribution, fossils, biochemical molecules, anatomical structures, and developmental processes. Just like a burglar cannot carry out a crime without leaving some kind of evidence behind, organisms leave evidence of what they looked like and how they lived. Evolutionary investigators piece this evidence together and provide

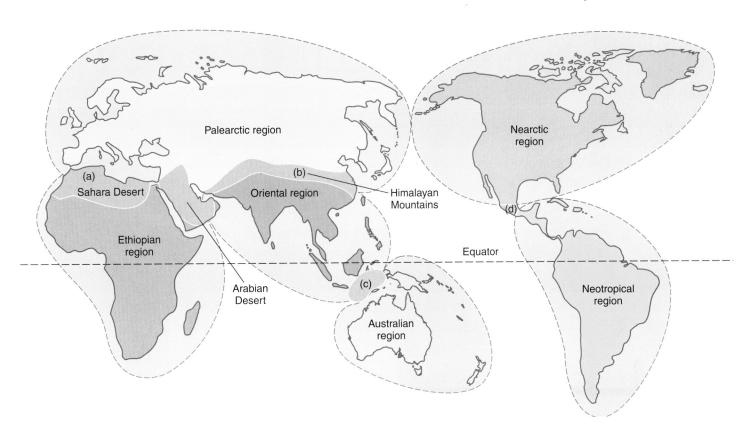

FIGURE 4.8

Biogeographic Regions of the World.   Barriers, such as oceans, mountain ranges, and deserts, separate biogeographic regions of the world. (*a*) The Sahara and Arabian Deserts separate the Ethiopian and Palearctic regions, (*b*) the Himalayan Mountains separate the Palearctic and Oriental regions, (*c*) deep ocean channels separate the Oriental and Australian regions, and (*d*) the mountains of southern Mexico and Mexico's tropical lowlands separate the Nearctic and Neotropical regions.

detailed accounts of the lives of extinct organisms and their relationships to modern forms. The sources of evidence for macroevolution are described in the next section.

BIOGEOGRAPHY

Biogeography is the study of the geographic distribution of plants and animals. Biogeographers try to explain why organisms are distributed as they are. Biogeographic studies show that life-forms in different parts of the world have distinctive evolutionary histories.

One of the distribution patterns that biogeographers try to explain is how similar groups of organisms have dispersed to places separated by seemingly impenetrable barriers. For example, native cats are inhabitants of most continents of the earth, yet they cannot cross expanses of open oceans. Obvious similarities suggest a common ancestry, but similarly obvious differences result from millions of years of independent evolution (figure 4.7 and Evolutionary Insights, page 65). Biogeographers also try to explain why plants and animals, separated by geographical barriers, are often very different in spite of similar environments. For example, why are so many of the animals that inhabit Australia and Tasmania so very different from animals in any other part of the world? The major native herbivores of Australia and Tasmania are the many species of kangaroos (*Macropus*). In other parts of the world, members of the deer and cattle groups fill these roles. Similarly, the Tasmanian wolf/tiger (*Thylacinus cyno-*

cephalus), now believed to be extinct, was a predatory marsupial that was unlike any other large predator. Finally, biogeographers try to explain why oceanic islands often have relatively few, but unique, resident species. They try to document island colonization and subsequent evolutionary events, which may be very different from the evolutionary events in ancestral, mainland groups. The discussion that follows will illustrate some of Charles Darwin's conclusions about the island biogeography of the Galápagos Islands.

Modern evolutionary biologists recognize the importance of geological events, such as volcanic activity, the movement of great landmasses, climatic changes, and geological uplift, in creating or removing barriers to the movements of plants and animals. Biogeographers divide the world into six major biogeographic regions (figure 4.8). As they observe the characteristic plants and animals in each of these regions and learn about the earth's geologic history, we understand more about animal distribution patterns and factors that played important roles in animal evolution. Only in understanding how the surface of the earth came to its present form can we understand its inhabitants.

PALEONTOLOGY

Paleontology (Gr. *palaios*, old + *on*, existing + *logos*, to study), which is the study of the fossil record, provides some of the most direct evidence for evolution. **Fossils** (L. *fossilis*, to dig) are evidence

FIGURE 4.9

Paleontological Evidence of Evolutionary Change. Fossils, such as this trilobite (*Griffithides*), are direct evidence of evolutionary change. Trilobites existed about 500 million years ago and became extinct about 250 million years ago. Fossils may form when an animal dies and is covered with sediments. Water dissolves calcium from hard body parts and replaces it with another mineral, forming a hard replica of the original animal. This process is called mineralization.

of plants and animals that existed in the past and have become incorporated into the earth's crust (e.g., as rock or mineral) (figure 4.9). Fossils are formed in sedimentary rock by a variety of methods. Most commonly, fossilization occurs when sediments (e.g., silt, sand, or volcanic ash) quickly cover an organism to prevent scavenging and in a way that seals out oxygen and slows decomposition. As sediments continue to be piled on top of the dead organism, pressures build. Water infiltrates the remains and inorganic compounds and ions replace organic components. Hard parts of the organism are most likely to be fossilized, but delicate structures are sometimes fossilized when silica is involved with replacement. These pressure and chemical changes transform the organism into a stony replica. Other fossilization processes form molds, casts of organisms, or carbon skeletons. Tracks and burrows, and even mummified remains, are sometimes found. Fossilization is most likely to occur in aquatic or semiaquatic environments. The fossil record is, therefore, more complete for those groups of organisms living in or around water and for organisms with hard parts.

The fossil record provides information regarding sequences in the appearance and disappearance of organisms. Different strata

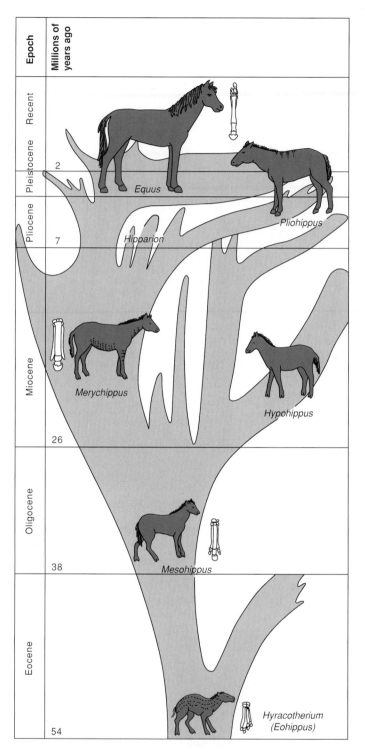

FIGURE 4.10

Reconstruction of an Evolutionary Lineage from Evidence in the Fossil Record. The fossil record allows horse evolution to be traced back about 60 million years. *Hyracotherium* (*Eohippus*) was a dog-sized animal with four prominent toes on each foot. A single middle digit of the toe and vestigial digits on either side of that remain in modern horses. About 17.7 million years ago, rapid evolutionary diversification resulted in the grazing lifestyle of modern horses. About 15 million years ago, 10–12 contemporaneous species of fossil horses lived in North America. Note that evolutionary lineages are seldom simple ladders of change. Instead, numerous evolutionary side branches often meet with extinction.

TABLE 4.1
THE HISTORY OF THE EARTH: GEOLOGICAL ERAS, PERIODS, AND MAJOR BIOLOGICAL EVENTS

ERA	PERIOD	AGE AT BEGINNING OF THE PERIOD (MILLIONS OF YEARS)	MAJOR BIOLOGICAL EVENTS
CENOZOIC	Quaternary	0.01	Subtropical forests gave way to cooler forests and grassland areas.
CENOZOIC	Tertiary	65	Modern orders of mammals evolved. Humans evolved in the last 5 million years.
MESOZOIC	Cretaceous	135	Continental seas and swamps spread. Extinction of ancient birds and reptiles.
MESOZOIC	Jurassic	195	Climate warm and stable. High reptilian diversity. Birds first appeared.
MESOZOIC	Triassic	240	Climate warm. Extensive deserts. Dinosaurs replace mammal-like reptiles. First true mammals.
PALEOZOIC	Permian	285	Climate cold early, but then warmed. Mammal-like reptiles common. Widespread extinction of amphibians.
PALEOZOIC	Carboniferous	375	Warm and humid with extensive coal-producing swamps. Arthropods and amphibians common. First reptiles appeared.
PALEOZOIC	Devonian	420	Land high and climate cool. Freshwater basins developed. Fish diversified. Early amphibians appeared.
PALEOZOIC	Silurian	450	Extensive shallow seas. Warm climate. First terrestrial arthropods. First jawed fish.
PALEOZOIC	Ordovician	520	Shallow extensive seas. Climate warmed. Many marine invertebrates. Jawless fish widespread.
PALEOZOIC	Cambrian	570	Extensive shallow seas and warm climate. Trilobites and brachiopods common. Earliest vertebrates found late in the Cambrian.
PROTEROZOIC		2,000	Multicellular organisms appeared and flourished. Many invertebrates. Eukaryotic organisms appeared (1,500 million years ago). Oxygen accumulated in the atmosphere.
ARCHEAN		4,600	Prokaryotic life appeared (3,500 million years ago). Origin of the earth (4,600 million years ago).

of rock result from differing rates of sedimentation. Climatic and geological events influence rates of sedimentation. When rates of sedimentation change, a break in deposition occurs, leaving a distinct layer or stratum. Successive strata are piled on top of each other with younger strata on top of older strata. Fossils in younger strata are of animals that lived more recently than fossils in older strata. Geologists can correlate strata around the world and deter-

mine when strata were formed using radioactive dating. Paleontologists use this information to provide an understanding of many evolutionary lineages. Many vertebrate lineages are very well documented in the fossil record (figure 4.10). Paleontologists have also used the fossil record to describe the history of life on earth (table 4.1). Evidence from paleontology is clearly some of the most convincing evidence of macroevolution.

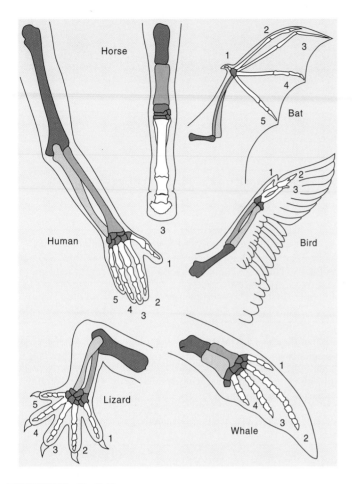

FIGURE 4.11

The Concept of Homology.   The forelimbs of vertebrates evolved from an ancestral pattern. Even vertebrates as dissimilar as whales and bats have the same basic arrangement of bones. The digits (fingers) are numbered 1 (thumb) to 5 (little finger). Color coding indicates homologous bones.

ANALOGY AND HOMOLOGY

Structures and processes of organisms may be alike. There are two reasons for similarities, and both cases provide evidence of evolution. Resemblance may occur when two unrelated organisms adapt to similar conditions. For example, adaptations for flight have produced flat, gliding surfaces in the wings of birds and insects. These similarities indicate that independent evolution in these two groups produced superficially similar structures and has allowed these two groups of animals to exploit a common aerial environment. The evolution of superficially similar structures in unrelated organisms is called **convergent evolution,** and the similar structures are said to be **analogous** (Gr. *analog* + *os*, proportionate).

Resemblance may also occur because two organisms share a common ancestry. Structures and processes in two kinds of organisms that are derived from common ancestry are said to be **homologous** (Gr. *homolog* + *os*, agreeing) (i.e., having the same or similar relation). Homology can involve aspects of an organism's structure, and these homologies are studied in the discipline called comparative anatomy. Homology can also involve aspects of func-

tion, and homologous processes are studied using techniques of molecular biology.

Comparative Anatomy

Comparative anatomy is the study of the structure of living and fossilized animals and the homologies that indicate evolutionarily close relationships. In many cases, homologies are readily apparent. For example, vertebrate appendages have a common arrangement of bones that can be traced back through primitive amphibians and certain groups of fish. Appendages have been modified for different functions like swimming, running, flying, and grasping, but the basic sets of bones and the relationships of bones to each other has been retained (figure 4.11). The similarities in structure of these bones reflect their common ancestry and the fact that vertebrate appendages, although modified in their details of structure, have retained their primary functions of locomotion.

Other structures may be homologous even though they differ in appearance and function. The origin of the middle ear bones of terrestrial vertebrates provides an example. Fish do not have middle and outer ears. Their inner ear provides for the senses of equilibrium, balance, and hearing, with sound waves being transmitted through bones of the skull. Terrestrial vertebrates evolved from primitive fish, and the evolution of life on land resulted in an ear that could detect airborne vibrations. These vibrations are transmitted to receptors of the inner ear through a middle ear and, in some cases, an outer ear. One (amphibians, reptiles, and birds) or more (mammals) small bones of the middle ear transmit vibrations from the eardrum (tympanic membrane) to the inner ear. Studies of the fossil record reveal the origin of these middle ear bones (figure 4.12). Small bones involved in jaw suspension in primitive fish are incorporated into the remnants of a pharyngeal (gill) slit to form the middle ear. In amphibians, reptiles, and birds, a single bone of ancient fish (the hyomandibular bone) forms the middle ear bone (the columella or stapes). In the evolution of mammals, two additional bones that contributed to jaw support of ancient fish (the quadrate and articular bones) are used in the middle ear (the incus and malleus, respectively).

There are many other examples of structures that have changed from an ancestral form. The human appendix functions as a small lymphatic structure. It originated as a large pouch involved with the fermentation and digestion of plant materials. Boa constrictors have minute remnants of hindlimb (pelvic) bones that are remnants of the appendages of their reptilian ancestors. These bones have no apparent function in modern boas. Changes in structure and function are clear indications of evolutionary change. These examples of the middle ear bones and appendix also illustrate an important evolutionary principle. Structures usually do not arise in evolution from nothing. Instead, previously existing structures are usually modified for new functions.

Molecular Biology

Recently, molecular biology has yielded a wealth of information on evolutionary relationships. Just as animals can have

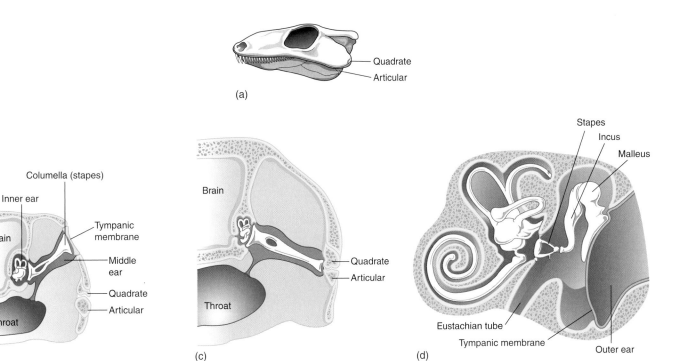

FIGURE 4.12

Evolution of the Vertebrate Ear Ossicles.   (*a*) Lateral view of the skull of a primitive amphibian showing the two bones (quadrate and articular) that function in jaw support and contribute to the middle ear bones of mammals. Diagrammatic sections of the heads of (*b*) a primitive amphibian, (*c*) a primitive reptile, and (*d*) a mammal showing the fate of three bones of primitive fish. The columella (stapes) is derived from a bone called the hyomandibular, the incus is derived from the quadrate bone, and the malleus is derived from the articular bone.

homologous structures, animals may also have homologous biochemical processes.

Ultimately, structure and function are based on the genetic blueprint found in all living animals: the DNA molecule. Related animals have similar DNA derived from their common ancestor. Because DNA carries the codes for proteins that make up each animal, related animals have similar proteins. With the modern laboratory technologies now available, zoologists can extract and analyze the structure of proteins from animal tissue, and compare the DNA of different animals. Molecular biologists have found that, within each kind of molecule, the rate of change is relatively constant. This observation has allowed scientists to estimate the approximate times when species diverged from a common ancestor. The molecular clock does not keep perfect time, but when data from these studies are compared with evidence from the fossil record, the conclusions are similar.

DEVELOPMENTAL PATTERNS

Evidence of evolution also comes from observing the developmental patterns of organisms. Changes in the genes that control the development of animals are usually harmful and are eliminated by natural selection. Therefore, the developmental stages of related animals often retain common features. Early embryonic stages of vertebrates are remarkably similar (figure 4.13*a*).

Many organ systems of vertebrates also show similar developmental patterns (figure 4.13*b*). These similarities are compelling evidence of evolutionary relationships within animal groups.

Adults in vertebrate groups are obviously different from one another. If developmental patterns are so similar, how did differences in adult stages arise? Evolution is again the answer. These differences are a result of evolutionary changes in the genes that control the onset of developmental stages and the rate at which development occurs. These changes result in differences in the size and proportions of organs. Differing growth rates of the bones of the skull, for example, can explain the proportional differences between bones of the human and ape skulls. Modern developmental biology is providing a growing appreciation of how evolution has conserved many genes that control the developmental similarities of animal groups. At the same time, it is helping to explain developmental changes that result in the great diversity of animal life.

INTERPRETING THE EVIDENCE: PHYLOGENY AND COMMON DESCENT

Scientists use the data gathered from the studies just described to understand how organisms are related to each other. **Phylogeny** refers to the evolutionary relationships among species. It includes the depiction of ancestral species and the relationships of modern descendants of a common ancestor. These depictions involve the

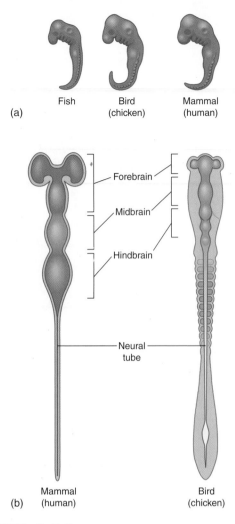

FIGURE 4.13

Developmental Patterns. (*a*) The early embryonic stages of various vertebrates are remarkably similar. These similarities result in the preservation of developmental sequences that evolved in early common ancestors of vertebrates. (*b*) Organ systems, like the nervous system, also show similar developmental patterns. Later developmental differences may result from evolutionary changes in the timing of developmental events.

use of tree diagrams showing lines of descent. Chapter 7 presents principles of animal classification and phylogeny, and chapters 8 through 22 show how the results of phylogenetic studies have provided a wealth of information on the evolutionary relationships among animal groups.

One example of how information from molecular biology is used to establish phylogenetic relationships concerns the evolu-

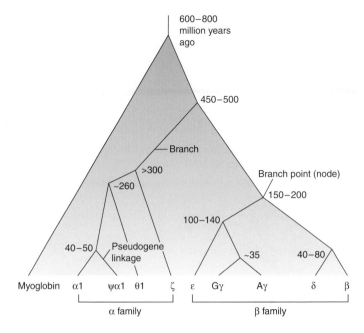

FIGURE 4.14

A Phylogenetic Tree of Hemoglobin. This phylogenetic tree is derived from molecular studies of the hemoglobin molecule. The numbers associated with each branch point indicate the approximate time in millions of years. The pseudogene in the alpha family is a gene that is apparently nonfunctional.

tionary history of the human hemoglobin genes. Hemoglobin genes are divided into alpha and beta families. The products of these genes are incorporated into the hemoglobin molecule that transports oxygen in blood cells. Another related gene codes for the oxygen storage pigment found in muscle, myoglobin. Molecular biologists have found that all of these modern genes are descendants of a single gene that was in existence between 600 and 800 million years ago. The phylogenetic tree that illustrates these relationships consists of branch points, which represent ancestral species, or in this case ancestral molecules, and branches that link species or molecules into related groups (figure 4.14). Evolutionary trees like this one, and the many that will follow, reinforce our ideas of common descent. All life is related, and the evidence of this relationship is overwhelming.

Evolution is the major unifying theme in biology because it helps explain both the similarities and the diversity of life. There is no doubt that it has occurred in the past and continues to occur today. Chapter 5 examines how the principles of population genetics have been combined with Darwinian evolutionary theory into what is often called the **modern synthesis** or **neo-Darwinism.**

EVOLUTIONARY INSIGHTS
An Example from Big-Cat Biogeography

All sources of evidence are used in piecing together the evolutionary relationships of animals. For example, paleontologists have established the common, but distant ancestry of all big-cat species. Earliest cats arose approximately 30 million years ago, and most recent lineages diverged between 2 and 10 million years ago (box figure 4.1). Molecular biologists are providing a wealth of information on genetic homologies that is helping to establish the phylogenetic relationships among modern wild-cat species, relationships that have been impossible to sort out using traditional methods.

Biogeographers have helped to explain the distribution of wild cats around the globe and the origin of variations within cat species. The leopard (*Panthera pardus*) and the jaguar (*Panthera onca*) are remarkably similar in appearance (*see figure 4.7*) and ecological roles. In the wild, their similar appearance presents no problem in the identification of these species because of their distribution. The leopard is found in Africa, the Middle East, and Asia. The jaguar is found only in Central America and northern and central South America. Paleontologists and biogeographers have found that these species evolved from a common ancestor in central Asia about two million years ago. Divergence of the two species began in Asia. Early jaguars were larger than their contemporaries and spread, along with leopards, into Europe. Jaguars also spread eastward, and about one million years ago they crossed the Bering land bridge that connected Asia and North America. Leopards continued to spread into Africa and other parts of their current range. Jaguars became extinct in Europe and Asia but survived in the Americas. Their current distribution in South America may be explained by the appearance of lions (*Felix concolor*) in North America. Competition may have driven jaguars southward.

Habitat differences may explain structural and behavioral differences between leopards and jaguars. Most jaguars are found in the densely forested areas of the Amazon Basin. Their smaller size is thought to be an adaptation to climatic and vegetational changes en-

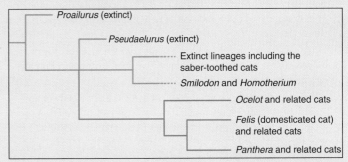

BOX FIGURE 4.1   **A Partial Cat Phylogeny.**   There are three lineages of modern cats: an *Ocelot* lineage, a *Felis* lineage, and a *Panthera* lineage. The latter includes the leopard and the jaguar. All cats trace their ancestry back about 30 million years. The earliest known cat fossil (*Proailurus*) is from a bobcat-sized animal that lived in Europe about 25 million years ago

countered as the cats moved south. Leopards, on the other hand, have evolved into a complex group of subspecies as they adapted to diverse environments across their range. One habit of African and Asian leopards is to use powerful neck and limb muscles to cache prey high in the boughs of trees. Antelope and other prey may be three times the body weight of the leopard. This behavior reduces competition from scavenging hyenas and opportunistic lions that may happen upon a leopard's kill.

Both of these species are threatened by habitat loss and hunting—both have been prized for their fur. Although the leopard has a very large range and diverse prey base, a number of subspecies are gone from many parts of their original range. Jaguars are severly threatened by deforestation. It is estimated that there are 15,000 individuals left in the wild.

SUMMARY

1. Organic evolution is the change of a species over time.

2. Ideas of evolutionary change can be traced back to the ancient Greeks.

3. Jean Baptiste Lamarck was an eighteenth-century proponent of evolution and proposed a mechanism—inheritance of acquired characteristics—to explain it.

4. Charles Darwin saw impressive evidence for evolutionary change while on a mapping expedition on the HMS *Beagle*. The theory of uniformitarianism, South American fossils, and observations of tortoises and finches on the Galápagos Islands convinced Darwin that evolution occurs.

5. After returning from his voyage, Darwin began formulating his theory of evolution by natural selection. In addition to his experiences on his voyage, later observations of artificial selection and Malthus's theory of human population growth helped shape his theory.

6. Darwin's theory of natural selection includes the following elements: (*a*) All organisms have a greater reproductive potential than is ever attained; (*b*) inherited variations arise by mutation; (*c*) in a constant struggle for existence, those organisms that are least suited to their environment die; (*d*) the adaptive traits present in the survivors tend to be passed on to subsequent generations, and the nonadaptive traits tend to be lost.

7. Adaptation may refer to a process of change or a result of change. An adaptation is a characteristic that increases an organism's potential to survive and reproduce in a given environment.

8. Not all evolutionary changes are adaptive, nor do all evolutionary changes lead to perfect solutions to environmental problems.

9. Alfred Russel Wallace outlined a theory similar to Darwin's but never accumulated significant evidence documenting his theory.

10. Microevolution is the change in the frequency of alleles in populations over time. Macroevolution is large-scale change that results in extinction and the formation of new species over geological timescales.

11. Evidence of macroevolutionary change comes from the study of biogeography, paleontology, comparative anatomy, molecular biology, and developmental biology.

12. All sources of evidence are used in studying the phylogeny of animals. These studies have resulted in the wealth of information on animal lineages that will be presented in chapters that follow.

SELECTED KEY TERMS

adaptation (p. 57)
biogeography (p. 59)
Galápagos Islands (p. 52)
homologous (p. 62)
macroevolution (p. 58)
microevolution (p. 58)

modern synthesis (p. 64)
natural selection (p. 55)
neo-Darwinism (p. 64)
organic evolution (p. 51)
paleontology (p. 59)

CRITICAL THINKING QUESTIONS

1. Outline a hypothesis and design a test of "inheritance of acquired characteristics," and define what is meant by the word *theory* in the theory of evolution by natural selection.

2. Describe the implications of inheritance of acquired characteristics for our modern concepts of how genes function.

3. Some opponents of evolutionary theory contend that evolutionary theory is not valid science because it concerns events of the past that cannot be observed or re-created in the laboratory. How would you respond to this criticism?

4. How would you explain the presence of gaps in the fossil record? Would you be more likely to see gaps in the fossil record of rodents, fish, molluscs, or segmented worms? Explain your answer.

5. Why is the stipulation of "a specified environment" included in the definition of adaptation?

6. Imagine that you could go back in time and meet simultaneously with Charles Darwin and Gregor Mendel. Construct a dialogue in which you explain to both the effect of their ideas on each other's theories and their theories on modern biology. Include their responses and questions throughout the dialogue.

ONLINE LEARNING CENTER

Visit our Online Learning Center (OLC) at www.mhhe.com/zoology (click on the book's title) to find the following chapter-related materials.

• CHAPTER QUIZZING

• CHAPTER 4. EVOLUTION—HISTORY AND EVIDENCE

• RELATED WEB LINKS
   Darwinian Evolutionary Theory
   Neo-Darwinism
   Evidences for Evolution

• BOXED READINGS ON
   Science and Pseudoscience
   The Origin of Eukaryotic Cells
   The Origin of Life on Earth—Life from Nonlife
   Continental Drift

• SUGGESTED READINGS

• LAB CORRELATIONS

   Check out the OLC to find specific information on these related lab exercises in the *General Zoology Laboratory Manual*, 5th edition, by Stephen A. Miller:

   Exercise 6   *Adaptations of Stream Invertebrates—A Scavenger Hunt*

EVOLUTION AND GENE FREQUENCIES

Outline

Concepts

1. In modern genetic theory, organic evolution is a change in the frequency of alleles in a population.
2. The principles of modern genetics help biologists understand how variation arises. This variation increases the chances of a population's survival in changing environments.
3. Population genetics is the study of events occurring in gene pools. The Hardy-Weinberg theorem helps scientists understand the circumstances under which evolution occurs. Evolution is a result of (a) genetic drift or neutral evolution, (b) gene flow, (c) mutations that introduce new genes into populations, and/or (d) natural selection.
4. Balanced polymorphism occurs when two or more body forms are maintained in a population without a range of phenotypes between them.
5. The fundamental unit of classification is the species, and the process by which new species form is speciation.
6. Speciation may occur when gene flow among populations is impeded.
7. Different organisms, as well as structures within organisms, evolve at different rates. Evolution may also proceed in jumps, rather than at a constant pace.
8. Molecular biologists study DNA and proteins to uncover evolutionary relationships.

Natural selection can be envisioned as operating in two ways, and both are important perspectives on evolution. One way (e.g., the focus of chapter 4) looks at characteristics of individual animals. When a population of birds acquires an adaptation through natural selection that permits its members to feed more efficiently on butterflies, the trait is described in terms of physical characteristics (e.g., bill shape) or inherited behaviors. This description of natural selection recognizes that natural selection must act in the context of living organisms.

The organism, however, must be viewed as a vehicle that permits the phenotypic expression of genes. This chapter examines the second way that natural selection operates—on genes. Birds and butterflies are not permanent—they die. The genes they carry, however, are potentially immortal. The result of natural selection (and evolution in general) is reflected in how common, or how rare, specific alleles are in a group of animals that are interbreeding—and, therefore, sharing genes.

POPULATIONS AND GENE POOLS

Individuals do not evolve. Evolution requires that genetic changes are passed from one generation to another within larger groups called populations. **Populations** are groups of individuals of the same species that occupy a given area at the same time and share a unique set of genes. With the possible exception of male and female differences, individuals of a population have the same number of genes and the same kinds of genes. A "kind" of gene would be a gene that codes for a given trait, such as hair length or the color of a mammal's coat. Differences within a population are based on variety within each trait, such as red or white hair in a mammal's coat. As described in chapter 3, this variety results from varying expressions of genes at each of the loci of an animal's chromosomes. Recall that these varying expressions of genes at each locus are called alleles. A population can be characterized by the frequency of alleles for a given trait, that is, the abundance of a particular allele in relation to the sum of all alleles at that locus. The sum of all the alleles for all traits in a sexually reproducing population is a pool of hereditary resources for the entire population and is called the **gene pool.**

Variety within individuals of a population results from having various combinations of alleles at each locus. Some of the sources of this variety have been discussed in chapter 3. These sources of variation include (1) the independent assortment of chromosomes that results in the random distribution of chromosomes into gametes, (2) the crossing-over that results in a shuffling of alleles between homologous chromosomes, and (3) the chance fertilization of an egg by a sperm cell. Variations also arise from (4) rearrangements in the number and structure of chromosomes and (5) mutations of existing alleles. Mutations are the only source of new alleles and will be discussed in greater detail later in this chapter. This chapter also describes how genetic variation may confer an advantage to individuals, leading to natural selection. It describes how other variations may become common in, or lost from, populations even though no particular advantage, or disadvantage, is derived from them.

The potential for genetic variation in individuals of a population is virtually unlimited. When generations of individuals of a population undergo sexual reproduction, there is a constant shuffling of alleles. The shuffling of alleles and the interaction of resulting phenotypes with the environment have one of two consequences for the population. Either the relative frequencies of alleles change across generations, or they do not. In the former case, evolution has occurred. In chapter 4, a change in the relative frequency of genes in a population across generations was defined as microevolution. In the next sections of this chapter, circumstances that favor microevolution are discussed in more detail.

MUST EVOLUTION HAPPEN?

Evolution is central to biology, but is evolution always occurring in a particular population? Sometimes the rate of evolution is slow, and sometimes it is rapid. But are there times when evolu-

tion does not occur at all? The answer to this question lies in the theories of **population genetics,** the study of the genetic events in gene pools.

THE HARDY-WEINBERG THEOREM

In 1908, English mathematician Godfrey H. Hardy and German physician Wilhelm Weinberg independently derived a mathematical model describing what happens to the relative frequency of alleles in a sexually reproducing population over time. Their combined ideas became known as the **Hardy-Weinberg theorem.** It states that the mixing of alleles at meiosis and their subsequent recombination do not alter the relative frequencies of the alleles in future generations, if certain assumptions are met. Stated another way, if certain assumptions are met, evolution will not occur because the relative allelic frequencies will not change from generation to generation, even though the specific mixes of alleles in individuals may vary.

The assumptions of the Hardy-Weinberg theorem are as follows:

1. The population size must be large. Large size ensures that gene frequency will not change by chance alone.
2. Sexual reproduction within the population must be random. Every individual must have an equal chance of mating with any other individual in the population. If this condition is not fulfilled, then some individuals are more likely to reproduce than others, and natural selection may occur.
3. Individuals cannot migrate into, or out of, the population. Migration may introduce new alleles into the gene pool or add or delete copies of existing alleles.
4. Mutations must not occur. If they do, mutational equilibrium must exist. Mutational equilibrium exists when mutation from the wild-type allele to a mutant form is balanced by mutation from the mutant form back to the wild type. In either case, no new genes are introduced into the population from this source.

These assumptions must be met if allelic frequencies are not changing—that is, if evolution is not occurring. Clearly, these assumptions are restrictive, and few, if any, real populations meet them. This means that most populations are evolving. The Hardy-Weinberg theorem, however, does provide a useful theoretical framework for examining changes in allelic frequencies in populations.

The next section explains how, when the assumptions are not met, microevolutionary change occurs.

EVOLUTIONARY MECHANISMS

Evolution is neither a creative force working for progress nor a dark force sacrificing individuals for the sake of the group. It is neither moral nor immoral. It has neither a goal nor a mind to conceive a goal. Such goal-oriented thinking is said to be teleological. Evolution is simply a result of some individuals in a population surviving and being more effective at reproducing than others in the population, leading to changes in relative allelic frequencies. This section examines some of the situations when the Hardy-Weinberg assumptions are not met—situations in which gene frequencies change from one generation to the next and evolution occurs.

POPULATION SIZE, GENETIC DRIFT, AND NEUTRAL EVOLUTION

Chance often plays an important role in the perpetuation of genes in a population, and the smaller the population, the more significant chance may be. Fortuitous circumstances, such as a chance encounter between reproductive individuals, may promote reproduction. Some traits of a population survive, not because they convey increased fitness, but because they happen to be in gametes involved in fertilization. Chance events influencing the frequencies of genes in populations result in **genetic drift.** Because gene frequencies are changing independently of natural selection, genetic drift is often called **neutral evolution.**

The process of genetic drift is analogous to flipping a coin. The likelihood of getting a head or a tail is equal. The 50:50 ratio of heads and tails is most likely in a large number of tosses. In only 10 tosses, for example, the ratio may be a disproportionate 7 heads and 3 tails. Similarly, the chance of one or the other of two equally adaptive alleles being incorporated into a gamete, and eventually into an individual in a second generation, is equal. Gamete sampling in a small population may show unusual proportions of alleles in any one generation of gametes because meiotic events, like tossing a coin, are random. Assuming that both alleles have equal fitness, these unusual proportions are reflected in the genotypes of the next generation. These chance events may result in a particular allele increasing or decreasing in frequency (figure 5.1a). In small populations, inbreeding is also common. Genetic drift and inbreeding are likely to reduce genetic variation within a population.

If a mutation introduces a new allele into a population and that allele is no more or less adaptive than existing alleles, genetic drift may permit the new allele to become established in the population (figure 5.1b), or the new allele may be lost because of genetic drift. The likelihood of genetic drift occurring in small populations suggests that a Hardy-Weinberg equilibrium will not occur.

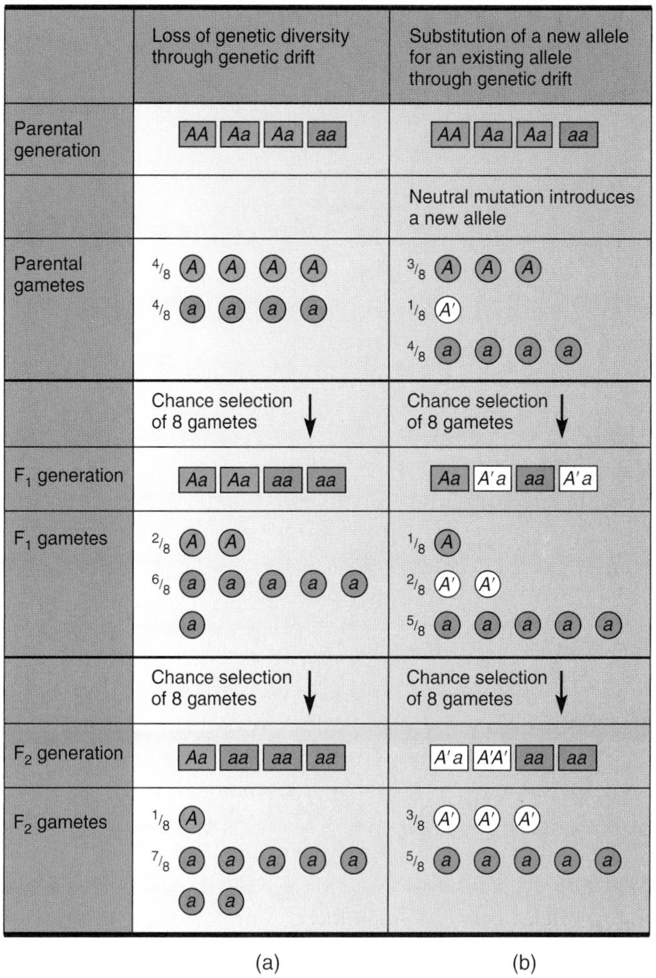

Two special cases of genetic drift have influenced the genetic makeup of some populations. When a few individuals from a parental population colonize new habitats, they seldom carry alleles in the same frequency as the alleles in the gene pool from which they came. The new colony that emerges from the founding individuals is likely to have a distinctive genetic makeup with far less variation than the larger population. This form of genetic drift is the **founder effect.**

An often-cited example of the founder effect concerns the genetic makeup of the Dunkers of eastern Pennsylvania. They emigrated from Germany to the United States early in the eighteenth century, and for religious reasons, have not married outside their sect. Examination of certain traits (e.g., ABO blood type) in their population reveals very different gene frequencies

	Loss of genetic diversity through genetic drift	Substitution of a new allele for an existing allele through genetic drift
Parental generation	AA Aa Aa aa	AA Aa Aa aa
		Neutral mutation introduces a new allele
Parental gametes	$^4/_8$ A A A A $^4/_8$ a a a a	$^3/_8$ A A A $^1/_8$ A' $^4/_8$ a a a a
	Chance selection of 8 gametes ↓	Chance selection of 8 gametes ↓
F₁ generation	Aa Aa aa aa	Aa A'a aa A'a
F₁ gametes	$^2/_8$ A A $^6/_8$ a a a a a a	$^1/_8$ A $^2/_8$ A' A' $^5/_8$ a a a a a
	Chance selection of 8 gametes ↓	Chance selection of 8 gametes ↓
F₂ generation	Aa aa aa aa	A'a A'A' aa aa
F₂ gametes	$^1/_8$ A $^7/_8$ a a a a a a a	$^3/_8$ A' A' A' $^5/_8$ a a a a a

(a)   (b)

FIGURE 5.1

Genetic Drift. (a) Genetic diversity may be lost as a result of genetic drift. Assume that alleles a and A are equally adaptive. Allele a might be incorporated into gametes more often than A, or it could be involved in more fertilizations. In either case, the frequency of a increases and the frequency of A decreases because of random events operating at the level of gametes. (b) A new, equally adaptive, allele may become established in a population as a result of genetic drift. In this example, the allele (A') substitutes for an existing allele (A). The same mechanisms could also account for the loss of the newly introduced allele or the establishment of A' as a third allele.

from the Dunker populations of Germany. These differences are attributed to the chance absence of certain genes in the individuals who founded the original Pennsylvania Dunker population.

A similar effect can occur when the number of individuals in a population is drastically reduced. For example, cheetah populations in South and East Africa are endangered. Their depleted populations have reduced genetic diversity to the point that even if population sizes are restored, they will have only a remnant of the original gene pool. This form of genetic drift is called the **bottleneck effect** (figure 5.2). A similar example concerns the northern elephant seal, which was hunted to near extinction in the late 1800s (figure 5.3). Legislation to protect the seal was enacted in 1922, and now the population is greater than 100,000 individuals.

(a)

(b)

Twenty-five different alleles of a particular gene in a population

a b c d e f g h j k
l m n o p q r s t
u v w x y z

d j l m

Bottleneck event (environmental upheaval)

d j l m j l
j m l j m d
m d j l d

Environmental upheaval lifts; population expands; only four different alleles remain

FIGURE 5.2

Bottleneck Effect. (*a*) Human activities have endangered cheetahs (*Acinonyx jubatus*) of South and East Africa. (*b*) Severe reduction in the original population has caused a bottleneck effect. Even if the population size recovers, genetic diversity has been significantly reduced.

FIGURE 5.3

Bottleneck Effect. The northern elephant seal (*Mirounga angustirostris*) was severely overhunted in the late 1800s. Even though its population numbers are now increasing, its genetic diversity is low.

In spite of this relatively large number, genetic variability in the population is low.

The effects of bottlenecks are somewhat controversial. The traditional interpretation is that decreases in genetic diversity make populations less likely to withstand environmental stress and more susceptible to extinction. That is, a population with high genetic diversity is more likely to have some individuals with a combination of genes that allows them to withstand environmental changes (*see Wildlife Alert*, chapter 3). In the case of the cheetah, recent evidence indicates that this cat's current problems may be more a result of lions and spotted hyenas preying on cheetah cubs than low genetic diversity. Most evolutionary biologists agree, however, that over evolutionary time frames, high genetic diversity makes extinction less likely.

GENE FLOW

The Hardy-Weinberg theorem assumes that no individuals enter a population from the outside (immigrate) and that no individuals leave a population (emigrate). Immigration or emigration upsets the Hardy-Weinberg equilibrium, resulting in changes in relative allelic frequency (evolution). Changes in relative allelic frequency from the migration of individuals are **gene flow.** Although some natural populations do not have significant gene flow, most populations do.

The effects of gene flow can be different, depending on the circumstance. The exchange of alleles between an island population and a neighboring continental population, for example, can change the genetic makeup of those populations. If gene flow continues and occurs in both directions, the two populations will become more similar. The absence of gene flow can make change in populations less likely. For example, recent evidence suggests that the African elephants should be divided into two species. Elephants of Africa's tropical forests are smaller, have straighter and thinner tusks, and distinctive skull morphology as compared to savannah elephants (figure 5.4). New molecular studies reveal marked genetic differences between these two groups. A lack of gene flow between these groups has helped to maintain the unique allelic frequencies within the groups.

MUTATION

Mutations are changes in the structure of genes and chromosomes (*see chapter 3*). The Hardy-Weinberg theorem assumes that no mutations occur or that mutational equilibrium exists. Mutations, however, are a fact of life. Most importantly, mutations are the origin of all new alleles and a source of variation that may prove adaptive for an animal. Mutation counters the loss of genetic material from natural selection and genetic drift, and it increases the likelihood that variations will be present that allow some individuals to survive future environmental shocks. Mutations make extinction less likely.

Mutations are random events, and the likelihood of a mutation is not affected by the mutation's usefulness. Organisms cannot filter harmful genetic changes from advantageous changes before they occur. The effects of mutations vary enormously. Most are deleterious. Some may be neutral or harmful in one environment and help an organism survive in another environment.

Mutational equilibrium exists when a mutation from the wild-type allele to a mutant form is balanced by a mutation from the mutant back to the wild type. This has the same effect on allelic frequency as if no mutation occurred. Mutational equilibrium rarely exists, however. **Mutation pressure** is a measure of the tendency for gene frequencies to change through mutation.

NATURAL SELECTION REEXAMINED

The theory of natural selection remains preeminent in modern biology. Natural selection occurs whenever some phenotypes are more successful at leaving offspring than other phenotypes. The tendency for natural selection to occur—and upset Hardy-Weinberg equilibrium—is **selection pressure.** Although natural selection is simple in principle, it is diverse in operation.

(a)

(b)

FIGURE 5.4

Gene Flow and African Elephants.   African elephants have been considered members of a single species, *Loxodonta africana*. Differences in morphology and habitats resulted in the description of two subspecies: (*a*) *L. africana africana* (savannah subspecies) and (*b*) *L. africana cyclotis* (tropical forest subspecies). Evidence from molecular studies indicates that little gene flow exists between these two groups of elephants and suggests that they should be considered separate species: *L. africana* and *L. cyclotis.*

Modes of Selection

For certain traits, many populations have a range of phenotypes, characterized by a bell-shaped curve that shows that phenotypic extremes are less common than the intermediate phenotypes. Natural selection may affect a range of phenotypes in three ways.

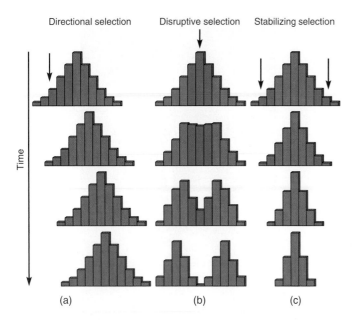

Directional selection | Disruptive selection | Stabilizing selection

Time

(a)   (b)   (c)

FIGURE 5.5

Modes of Selection. (*a*) Directional selection occurs when individuals at one phenotypic extreme are selected against. It shifts phenotypic distribution toward the advantageous phenotype. (*b*) Disruptive selection occurs when an intermediate phenotype is selected against. It produces distinct subpopulations. (*c*) Stabilizing selection occurs when individuals at both phenotypic extremes are selected against. It narrows at both ends of the range. Arrows indicate selection against one or more phenotypes. The X-axis of each graph indicates the range of phenotypes for the trait in question.

Directional selection occurs when individuals at one phenotypic extreme are at a disadvantage compared to all other individuals in the population (figure 5.5*a*). In response to this selection, the deleterious gene(s) decreases in frequency, and all other

genes increase in frequency. Directional selection may occur when a mutation gives rise to a new gene, or when the environment changes to select against an existing phenotype.

Industrial melanism, a classic example of directional selection, occurred in England during the Industrial Revolution. Museum records and experiments document how environmental changes affected selection against one phenotype of the peppered moth, *Biston betularia*.

In the early 1800s, a gray form made up about 99% of the peppered moth population. That form still predominates in nonindustrial northern England and Scotland. In industrial areas of England, a black form replaced the gray form over a period of about 50 years. In these areas, the gray form made up only about 5% of the population, and 95% of the population was black. The gray phenotype, previously advantageous, had become deleterious.

The nature of the selection pressure was understood when investigators discovered that birds prey more effectively on moths resting on a contrasting background. Prior to the Industrial Revolution, gray moths were favored because they blended with the bark of trees on which they rested. The black moth contrasted with the lighter, lichen-covered bark and was easily spotted by birds (figure 5.6*a*). Early in the Industrial Revolution, however, factories used soft coal, and spewed soot and other pollutants into the air. Soot covered the tree trunks and killed the lichens where the moths rested. Bird predators now could easily pick out gray moths against the black background of the tree trunk, while the black form was effectively camouflaged (figure 5.6*b*).

In the 1950s, the British Parliament enacted air pollution standards that have reduced soot in the atmosphere. As expected, the gray form of the moth has experienced a small but significant increase in frequency.

Another form of natural selection involves circumstances selecting against individuals of an intermediate phenotype (figure

(a)

(b)

FIGURE 5.6

Directional Selection of the Peppered Moth, *Biston betularia*. Each photo shows two forms of the moth: black and gray. (*a*) Prior to the Industrial Revolution, bird predators easily spotted the black form of moth, and the gray form was camouflaged. (*b*) In industrial regions after the Industrial Revolution, selection was reversed because pollution killed lichens that covered the bark of trees where moths rested. Note how clearly the gray form is seen, whereas the black form is almost invisible.

5.5b). **Disruptive selection** produces distinct subpopulations. Consider, for example, what could happen in a population of snails with a range of shell colors between white and dark brown and living in a marine tidepool habitat with two background colors. The sand, made up of pulverized mollusc shells, is white, and rock outcroppings are brown. In the face of shorebird predation, what phenotypes are going to be most common? Although white snails may not actively select a white background, those present on the sand are less likely to be preyed on than intermediate phenotypes on either sand or rocks. Similarly, brown snails on rocks are less likely to be preyed on than intermediate phenotypes on either substrate. Thus, disruptive selection could produce two distinct subpopulations, one white and one brown.

When both phenotypic extremes are deleterious, a third form of natural selection—**stabilizing selection**—narrows the phenotypic range (figure 5.5c). During long periods of environmental constancy, new variations that arise, or new combinations of genes that occur, are unlikely to result in more fit phenotypes than the genes that have allowed a population to survive for thousands of years, especially when the new variations are at the extremes of the phenotypic range.

A good example of stabilizing selection is the horseshoe crab (*Limulus*), which lives along the Atlantic coast of the United States (*see figure 14.8*). Comparison of the fossil record with living forms indicates that this body form has changed little over 200 million years. Apparently, the combination of characteristics present in this group of animals is adaptive for the horseshoe crab's environment.

Neutralist/Selectionist Controversy

Most biologists recognize that both natural selection and neutral evolution occur, but they may not be equally important in all circumstances. For example, during long periods when environments are relatively constant, and stabilizing selection is acting on phenotypes, neutral evolution may operate at the molecular level. Certain genes could be randomly established in a population. Occasionally, however, the environment shifts, and directional or disruptive selection begins to operate, resulting in gene frequency changes (often fairly rapid).

The relative importance of neutral evolution and natural selection in natural populations is debated and is an example of the kinds of debates that occur among evolutionists. These debates concern the mechanics of evolution and are the foundations of science. They lead to experiments that will ultimately present a clearer understanding of evolution.

BALANCED POLYMORPHISM AND HETEROZYGOTE SUPERIORITY

Polymorphism occurs in a population when two or more distinct forms exist without a range of phenotypes between them. **Balanced polymorphism** (Gr. *poly*, many + *morphe*, form) occurs when different phenotypes are maintained at relatively stable frequencies in the population and may resemble a population in which disruptive selection operates.

Sickle-cell anemia results from a change in the structure of the hemoglobin molecule. Some of the red blood cells of persons with the disease are misshapen, reducing their ability to carry oxygen. In the heterozygous state, the quantities of normal and sickled cells are roughly equal. Sickle-cell heterozygotes occur in some African populations with a frequency as high as 0.4. The maintenance of the sickle-cell heterozygotes and both homozygous genotypes at relatively unchanging frequencies makes this trait an example of a balanced polymorphism.

Why hasn't natural selection eliminated such a seemingly deleterious allele? The sickle-cell allele is most common in regions of Africa that are heavily infected with the malarial parasite, *Plasmodium falciparum*. Sickle-cell heterozygotes are less susceptible to malarial infections; if infected, they experience less severe symptoms than do homozygotes without sickled cells. Individuals homozygous for the normal allele are at a disadvantage because they experience more severe malarial infections, and individuals homozygous for the sickle-cell allele are at a disadvantage because they suffer from the severe anemia that the sickle cells cause. The heterozygotes, who usually experience no symptoms of anemia, are more likely to survive than either homozygote. This system is also an example of heterozygote superiority—when the heterozygote is more fit than either homozygote. Heterozygote superiority can lead to balanced polymorphism because perpetuation of the alleles in the heterozygous condition maintains both alleles at a higher frequency than would be expected if natural selection acted only on the homozygous phenotypes.

The processes that change relative allelic frequencies in populations can, over geological time periods (measured in thousands of years), result in the formation of new species. Species can become extinct very quickly when climatic or geological events cause abrupt enviromental changes. The formation of new species and the extinction of species were defined in chapter 4 as macroevolution. Even though the formation of new species is difficult to observe directly because of the long time frames involved, the evidence presented in chapter 4 has convinced the vast majority of scientists that macroevolution occurs. The rest of this chapter continues the theme of macroevolutionary change.

SPECIES AND SPECIATION

Taxonomists classify organisms according to their similarities and differences (*see chapters 1 and 7*). The fundamental unit of classification is the species. Unfortunately, formulating a universally applicable definition of species is difficult. According to a biological definition, a **species** is a group of populations in which genes are actually, or potentially, exchanged through interbreeding.

Although concise, this definition has problems associated with it. Taxonomists often work with morphological characteristics, and the reproductive criterion must be assumed based on morphological and ecological information. Also, some organisms do not reproduce sexually. Obviously, other criteria need to be applied in these cases. Another problem concerns fossil material. Paleontologists describe species of extinct organisms, but how can they test the reproductive criterion? Finally, populations of similar organisms may be so isolated from each other that gene exchange is geographically impossible. To test a reproductive criterion, biologists can transplant individuals to see if mating can occur, but mating of transplanted individuals does not necessarily prove what would happen if animals were together in a natural setting.

Rather than trying to establish a definition of a species that solves all these problems, it is probably better to simply understand the problems associated with the biological definition. In describing species, taxonomists use morphological, physiological, embryological, behavioral, molecular, and ecological criteria, realizing that all of these have a genetic basis.

Speciation is the formation of new species. A requirement of speciation is that subpopulations are prevented from interbreeding. For some reason, gene flow among populations or subpopulations does not occur. This is called **reproductive isolation.** When populations are reproductively isolated, natural selection and genetic drift can result in evolution taking a different course in each subpopulation. Reproductive isolation can occur in different ways.

Premating isolation prevents mating from taking place. For example, impenetrable barriers, such as rivers or mountain ranges, may separate subpopulations. Other forms of premating isolation are more subtle. If courtship behavior patterns of two animals are not mutually appropriate, mating does not occur. Similarly, individuals with different breeding periods or that occupy different habitats are unable to breed with each other.

Postmating isolation prevents successful fertilization and development, even though mating may have occurred. For example, conditions in the reproductive tract of a female may not support the sperm of another individual, which prevents successful fertilization. Postmating isolation also occurs because hybrids are usually sterile (e.g., the mule produced from a mating of a male donkey and a mare is a sterile hybrid). Mismatched chromosomes cannot synapse properly during meiosis, and any gametes produced are not viable. Other kinds of postmating isolation include developmental failures of the fertilized egg or embryo.

ALLOPATRIC SPECIATION

Allopatric (Gr. *allos*, other + *patria*, fatherland) **speciation** occurs when subpopulations become geographically isolated from one another. For example, a mountain range or river may permanently separate members of a population. Adaptations to different environments or genetic drift in these separate populations may result in members not being able to mate successfully with each other, even if experimentally reunited. Many biologists believe that allopatric speciation is the most common kind of speciation (figure 5.7).

The finches that Darwin saw on the Galápagos Islands are a classic example of allopatric speciation, as well as adaptive radiation (*see Evolutionary Insights, page 77, and chapter 4*). Adaptive radiation occurs when a number of new forms diverge from an ancestral form, usually in response to the opening of major new habitats.

Fourteen species of finches evolved from the original finches that colonized the Galápagos Islands. Ancestral finches, having emigrated from the mainland, probably were distributed among a few of the islands of the Galápagos. Populations became isolated on various islands over time, and though the original population probably

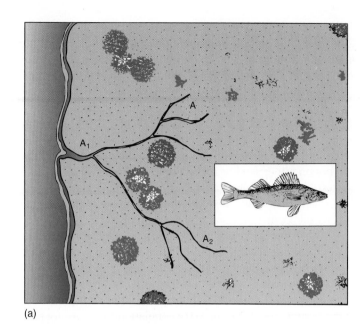

(a)

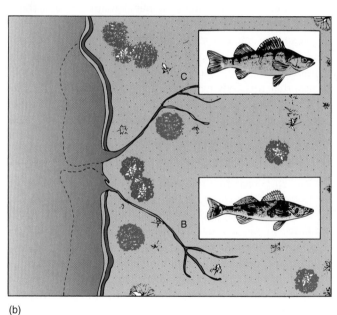

(b)

FIGURE 5.7

Allopatric Speciation.   Allopatric speciation can occur when a geographic barrier divides a population. (*a*) In this hypothetical example, a population of freshwater fish in a river drainage system is divided into three subpopulations: A, A_1, and A_2. Genetic exchanges may occur between A and A_1, and between A_1 and A_2. Exchanges are less common between A and A_2. (*b*) A rise in the level of the ocean forces the breakup of A_1 and makes A and A_2 separate populations. Genetic drift and different selection pressures in the two populations may eventually result in the formation of species B and C.

displayed some genetic variation, even greater variation arose. The original finches were seed eaters, and after their arrival, they probably filled their preferred habitats rapidly. Variations within the original finch population may have allowed some birds to exploit new islands and habitats where no finches had been. Mutations changed the genetic composition of the isolated finch populations, introducing further variations. Natural selection favored the retention of the variations that promoted successful reproduction.

The combined forces of isolation, mutation, and natural selection allowed the finches to diverge into a number of species with specialized feeding habits (*see figure 4.4*). Of the 14 species of finches, six have beaks specialized for crushing seeds of different sizes. Others feed on flowers of the prickly pear cactus or in the forests on insects and fruit.

PARAPATRIC SPECIATION

Another form of speciation, called **parapatric** (Gr. *para*, beside) **speciation,** occurs in small, local populations, called **demes.** For example, all of the frogs in a particular pond or all of the sea urchins in a particular tidepool make up a deme. Individuals of a deme are more likely to breed with one another than with other individuals in the larger population, and because they experience the same environment, they are subject to similar selection pressures. Demes are not completely isolated from each other because individuals, developmental stages, or gametes can move among demes of a population. On the other hand, the relative isolation of a deme may mean that its members experience different selection pressures than other members of the population. If so, speciation can occur. Although most evolutionists theoretically agree that parapatric speciation is possible, no certain cases are known. Parapatric speciation is therefore considered of less importance in the evolution of animal groups than allopatric speciation.

SYMPATRIC SPECIATION

A third kind of speciation, called **sympatric** (Gr. *sym*, together) **speciation,** occurs within a single population (*see Evolutionary Insights, page 77*). Even though organisms are sympatric, they still may be reproductively isolated from one another. Many plant species are capable of producing viable forms with multiple sets of chromosomes. Such events could lead to sympatric speciation among groups in the same habitat. While sympatric speciation in animals is uncommon, it has been documented in two species of bats and several species of insects and fish.

RATES OF EVOLUTION

Charles Darwin perceived evolutionary change as occurring gradually over millions of years. This concept, called **phyletic gradualism,** has been the traditional interpretation of the tempo, or rate, of evolution.

Some evolutionary changes, however, happen very rapidly. Studies of the fossil record show that many species do not change significantly over millions of years. These periods of stasis (Gr. *stasis*, standing still), or equilibrium, are interrupted when a group encounters an ecological crisis, such as a change in climate or a major geological event. Over the next 10,000 to 100,000 years, a variation that previously was selectively neutral or disadvanta-

geous might now be advantageous. Alternatively, geological events might result in new habitats becoming available. (Events that occur in 10,000 to 100,000 years are almost instantaneous in an evolutionary time frame.) This geologically brief period of change "punctuates" the previous million or so years of equilibrium and eventually settles into the next period of stasis. In this model, stabilizing selection characterizes the periods of stasis, and directional or disruptive selection characterizes the periods of change (figure 5.8). Long periods of stasis interrupted by brief

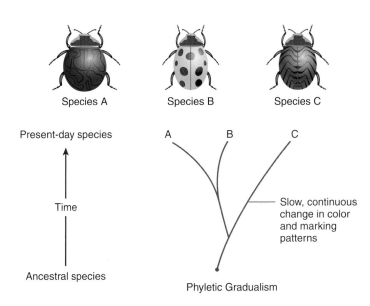

(a)
- Change is gradual over millions of years.
- Transitional fossils show a continuous range of phenotypes.

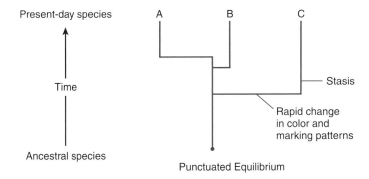

(b)
- Change is rapid over thousands of years.
- Transitional fossils are unlikely.

FIGURE 5.8

Rates of Evolution. A comparison of phyletic gradualism and punctuated equilibrium in three hypothetical beetle species. (*a*) In the phyletic gradualism model of evolution, changes are gradual over long time periods. Note that this tree implies a gradual change in color and marking patterns in the three beetle species. (*b*) In the punctuated equilibrium model of evolution, rapid periods of change interrupt long periods of stasis. This tree implies that the color and marking patterns in the beetles changed rapidly and did not change significantly during long periods of stabilizing selection (stasis).

periods of change is called the **punctuated equilibrium model** of evolution.

Biologists have observed such rapid evolutionary changes in small populations. Examples include the rapid pesticide and antibiotic resistance that insect pests and bacteria acquire. In addition, in a series of studies over a 20-year period, Peter R. Grant has shown that natural selection results in rapid morphological changes in the bills of Galápagos finches. A long, dry period from the middle of 1976 to early January 1978 resulted in birds with larger, deeper bills. Early in this dry period, birds quickly consumed smaller, easily cracked seeds. As they were forced to turn to larger seeds, birds with weaker bills were selected against, resulting in a measurable change in the makeup of the finch population of the island Daphne Major.

One advantage of the punctuated equilibrium model is its explanation for the fossil record not always showing transitional stages between related organisms. Gradualists attribute the absence of transitional forms to fossilization being an unlikely event; thus, many transitional forms disappeared without leaving a fossil record. Since punctuated equilibrium involves rapid changes in small, isolated populations, preservation of intermediate forms in the fossil record is even less likely. The rapid pace (geologically speaking) of evolution resulted in apparent "jumps" from one form to another.

MOLECULAR EVOLUTION

Many evolutionists study changes in animal structure and function that are observable on a large scale—for example, changes in the shape of a bird's bill or in the length of an animal's neck. All evolutionary change, however, results from changes in the base sequences in DNA and amino acids in proteins. Molecular evolutionists investigate evolutionary relationships among organisms by studying DNA and proteins. For example, cytochrome c is a protein present in the cellular respiration pathways in all eukaryotic organisms (table 5.1). Organisms that other research has shown to be closely related have similar cytochrome c molecules. That cytochrome c has changed so little during hundreds of millions of years suggests that mutations of the cytochrome c gene are nearly always detrimental, and are selected against. Because it has changed so little, cytochrome c is said to have been conserved evolutionarily.

Not all proteins are conserved as rigorously as cytochrome c. Although variations in highly conserved proteins can help establish evolutionary relationships among distantly related organisms, less-conserved proteins are useful for looking at relationships among more closely related animals. Because some proteins are conserved and others are not, the best information regarding evolutionary relationships requires comparing as many proteins as possible in any two species.

GENE DUPLICATION

Recall that most mutations are selected against. Sometimes, however, an extra copy of a gene is present. One copy may be modified, but as long as the second copy furnishes the essential protein,

TABLE 5.1 AMINO ACID DIFFERENCES IN CYTOCHROME C FROM DIFFERENT ORGANISMS	
ORGANISMS	**NUMBER OF VARIANT AMINO ACID RESIDUES**
Cow and sheep	0
Cow and whale	2
Horse and cow	3
Rabbit and pig	4
Horse and rabbit	5
Whale and kangaroo	6
Rabbit and pigeon	7
Shark and tuna	19
Tuna and fruit fly	21
Tuna and moth	28
Yeast and mold	38
Wheat and yeast	40
Moth and yeast	44

the organism is likely to survive. Gene duplication, the accidental duplication of a gene on a chromosome, is one way that extra genetic material can arise.

Vertebrate hemoglobin and myoglobin are believed to have arisen from a common ancestral molecule. Hemoglobin carries oxygen in red blood cells, and myoglobin is an oxygen storage molecule in muscle. The ancestral molecule probably carried out both functions. However, about 1 billion years ago, gene duplication followed by mutation of one gene resulted in the formation of two polypeptides: myoglobin and hemoglobin. Further gene duplications over the last 500 million years probably explain why most vertebrates, other than primitive fishes, have hemoglobin molecules consisting of four polypeptides.

MOSAIC EVOLUTION

As discussed earlier, rates of evolution can vary both in populations and in molecules and structures. A species is a mosaic of different molecules and structures that have evolved at different rates. Some molecules or structures are conserved in evolution; others change more rapidly. The basic design of a bird provides a simple example. All birds are easily recognizable as birds because of highly conserved structures, such as feathers, bills, and a certain body form. Particular parts of birds, however, are less conservative and have a higher rate of change. Wings have been modified for hovering, soaring, and swimming. Similarly, legs have been modified for wading, swimming, and perching. These are examples of **mosaic evolution.**

EVOLUTIONARY INSIGHTS
Speciation of Darwin's Finches

When Charles Darwin visited the Galápagos Islands in 1835, he observed the dark-bodied finches whose adaptive radiation has become a classic example of speciation (box figure 5.1). Studies of these finches have provided insight into some of the ways in which speciation can occur. Peter R. and B. Rosemary Grant have been studying these finches for more than 30 years. They have directly observed microevolutionary change reflected in bill morphology in response to change in rainfall and food availability. Other molecular studies have also contributed to our knowledge of the adaptive radiation of this group of birds.

Molecular studies have identified the most likely South American relatives of Darwin's finches, members of the grassquit genus *Tiaris*. Comparisons of the mitochondrial DNA of this group with Darwin's finches suggest that the latter colonized some of the Galápagos Islands not more than 3 million years ago. A very rapid adaptive radiation occurred, with the number of finch species doubling approximately every 750,000 years. No other group of birds studied has undergone a more rapid evolutionary diversification (*see figure 4.4*). Darwin's finches have served as a model to answer questions of how and why species diverge.

The traditional explanation of speciation within Darwin's finches is based on the allopatric model discussed in this chapter. This explanation is based upon differences in food resources, and the observations of the Grants have provided support for this model. Geographic isolation of populations of finches on different islands promoted speciation as these populations were influenced by natural selection and genetic drift. Each population adapted to the food resources available in their habitat (*see figure 4.6*). Most of these adaptations are reflected in bill morphology.

The Grants have discovered, however, that the allopatric model is not the entire explanation for finch adaptive radiation. Three million years ago, the Galápagos Islands were much simpler than they are today. In fact, there were fewer islands when they were first colonized by finches. Apparently the number of finch species increased as the number of islands increased as a result of volcanic activity. The increasing number of islands and oscillations in temperature and precipitation naturally affected vegetation. Habitats available for finches became more diverse and complex. The original warm, wet islands favored long, narrow bills that were used in gathering nectar and insects. The islands' moisture now fluctuates and the climate is more seasonal. The increasing diversity in habitats and food supply over 3 million years apparently promoted very rapid speciation among the finch populations.

The Grants have also discovered that sympatric forces probably have also promoted speciation. Different species of finches that live on the same island rarely hybridize. Lack of hybridization promotes isolation and speciation. Genetic incompatibility of gametes is apparently not the factor that discourages hybridization. Courtship behaviors of different species are similar, so courtship differences are not responsible.

(a)

(b)

BOX FIGURE 5.1   **Speciation of Darwin's Finches.**   Speciation and adaptive radiation of Darwin's finches has been used as a classic example of allopatric speciation. Isolation of finches on different islands, and differences in food resources on those islands, selected for morphological differences in finch bills. For example, (a) the warbler finch (*Certhidea olivacea*) has a bill that is adapted for probing for insects, and (b) the large ground finch (*Geospiza magnirostris*) has a bill that is adapted for crushing seeds. Recent studies show that increasing numbers of islands over the last 3 million years and changes in temperature and precipitation resulted in very rapid speciation. In addition, sympatric influences regarding the role of the males' song and bill shape probably also promoted speciation.

The Grants have discovered that visual and acoustic cues are used in mate choice. In these finches, only the males sing, and both male and female offspring respond to and learn the song of their father. The young associate the song and the bill shape of their father. Females tend to mate with males that have the bill shape and song of their father. The fact that learned behaviors are influencing speciation introduces new sets of variables that may influence speciation. Errors in learning, variations in the vocal apparatus of individuals, and characteristics of sound transmission through the environment could all result in changes in song characteristics and influence mate choice.

Observations of Darwin's finches have helped to revolutionize biology. They played an important role in the development of Darwin's theory of evolution by natural selection, and they continue to provide important evidence of how evolution occurs.

SUMMARY

1. Organic evolution is a change in the frequency of alleles in a population.
2. Virtually unlimited genetic variation, in the form of new alleles and new combinations of alleles, increases the chances that a population will survive future environmental changes.
3. Population genetics is the study of events occurring in gene pools. The Hardy-Weinberg theorem states that, if certain assumptions are met, gene frequencies of a population remain constant from generation to generation.
4. The assumptions of the Hardy-Weinberg theorem, when not met, define circumstances under which evolution will occur: (a) Fortuitous circumstances may allow only certain alleles to be carried into the next generation. Such chance variations in allelic frequencies are called genetic drift or neutral evolution. (b) Allelic frequencies may change as a result of individuals immigrating into, or emigrating from, a population. (c) Mutations are the source of new genetic material for populations. Mutational equilibrium rarely exists, and thus, mutations usually result in changing allelic frequencies. (d) The tendency for allelic frequencies to change, due to differing fitness, is called selection pressure.
5. Selection may be directional, disruptive, or stabilizing.
6. Balanced polymorphism occurs when two or more phenotypes are maintained in a population. Heterozygote superiority can lead to balanced polymorphism.
7. According to a biological definition, a species is a group of populations within which there is potential for the exchange of genes. Significant problems are associated with the application of this definition.
8. Speciation requires reproductive isolation. Speciation may occur sympatrically, parapatrically, or allopatrically, although most speciation events are believed to be allopatric.
9. Premating isolation prevents mating from occurring. Postmating isolation prevents the development of fertile offspring, if mating has occurred.
10. Phyletic gradualism is a model of evolution that depicts change as occurring gradually, over millions of years. Punctuated equilibrium is a model of evolution that depicts long periods of stasis interrupted by brief periods of relatively rapid change.
11. The study of rates of molecular evolution helps establish evolutionary interrelationships among organisms.
12. A mutation may modify a duplicated gene, which then may serve a function other than its original role.
13. A species is a mosaic of different molecules and structures that have evolved at differing rates.

SELECTED KEY TERMS

Hardy-Weinberg theorem (p. 68)
neutral evolution (p. 69)
phyletic gradualism (p. 75)
population genetics (p. 68)

punctuated equilibrium model (p. 76)
reproductive isolation (p. 74)
speciation (p. 74)
species (p. 73)

CRITICAL THINKING QUESTIONS

1. Can natural selection act on variations that are not inherited? (Consider, for example, deformities that arise from contracting a disease.) If so, what is the effect of that selection on subsequent generations?
2. In what way does overuse of antibiotics and pesticides increase the likelihood that these chemicals will eventually become ineffective? This is an example of which one of the three modes of natural selection?
3. What are the implications of the "bottleneck effect" for wildlife managers who try to help endangered species, such as the whooping crane, recover from near extinction?
4. What does it mean to think of evolutionary change as being goal-oriented? Explain why this way of thinking is wrong.
5. Imagine that two species of butterflies resemble one another closely. One of the species (the model) is distasteful to bird predators, and the other species (the mimic) is not. How could directional selection have resulted in the mimic species evolving a resemblance to the model species?

ONLINE LEARNING CENTER

Visit our Online Learning Center (OLC) at www.mhhe.com/zoology (click on the book's title) to find the following chapter-related materials:

- CHAPTER QUIZZING
- ANIMATIONS
    Hardy-Weinberg Equilibrium
    Molecular Clock
- RELATED WEB LINKS
    Speciation
    Punctuated versus Gradualist Evolution
    Macroevolution
    Microevolution

- BOXED READINGS ON
    Origin of Life on Earth—Life from Nonlife
    Genetic Equilibrium
    Malaria Control—A Glimmer of Hope
    Animal Origins—The Cambrian Explosion
    The Endostyle and the Vertebrate Thyroid Gland
- SUGGESTED READINGS

CHAPTER 6

ECOLOGY:

PRESERVING THE ANIMAL KINGDOM

Outline

Animals and Their Abiotic Environment
    Energy
    Temperature
    Other Abiotic Factors
Populations
    Population Growth
    Population Regulation
    Intraspecific Competition
Interspecific Interactions
    Herbivory and Predation
    Interspecific Competition
    Coevolution
    Symbiosis
    Other Interspecific Adaptations
Communities
    The Ecological Niche
    Community Stability
Trophic Structure of Ecosystems
Cycling Within Ecosystems
Ecological Problems
    Human Population Growth
    Pollution
    Resource Depletion and Biodiversity

Concepts

1. Ecology is the study of the relationships of organisms to their environment and to other organisms. In part, it involves the study of how abiotic factors, such as energy, temperature, light, and soils, influence individuals.
2. Ecology also involves the study of populations. Populations grow, and population density, the carrying capacity of the environment, and interactions among members of the same population regulate growth.
3. Ecology also involves the study of individuals interacting with members of their own and other species. Herbivory, predator–prey interactions, competition for resources, and other kinds of interactions influence the makeup of animal populations.
4. All populations living in an area make up a community. Communities have unique attributes that ecologists can characterize.
5. Communities and their physical surroundings are called ecosystems. Energy flowing through an ecosystem does not cycle. Energy that comes into an ecosystem must support all organisms living there before it is lost as heat. Nutrients and water, on the other hand, cycle through an ecosystem, and organisms reuse them.
6. Concepts of community and ecosystem ecology provide a basis for understanding many ecological problems, including human population growth, pollution, and resource depletion.

All animals have certain requirements for life. In searching out these requirements, animals come into contact with other organisms and their physical environment. These encounters result in a multitude of interactions among organisms and alter even the physical environment. **Ecology** is the study of the relationships of organisms to their environment and to other organisms. Understanding basic ecological principles helps us to understand why animals live in certain places, why animals eat certain foods, and why animals interact with other animals in specific ways. It is also the key to understanding how human activities can harm animal populations and what we must do to preserve animal resources. The following discussion focuses on ecological principles that are central to understanding how animals live in their environment.

*T*his chapter contains evolutionary concepts, which are set off in this font.

ANIMALS AND THEIR ABIOTIC ENVIRONMENT

An animal's **habitat** (environment) includes all living (biotic) and nonliving (abiotic) characteristics of the area in which the animal lives. Abiotic characteristics of a habitat include the availability of oxygen and inorganic ions, light, temperature, and current or wind velocity. Physiological ecologists who study abiotic influences have found that animals live within a certain range of values, called the **tolerance range,** for any environmental factor. At either limit of the tolerance range, one or more essential functions cease. A certain range of values within the tolerance range, called the **range of optimum,** defines the conditions under which an animal is most successful (figure 6.1).

Combinations of abiotic factors are necessary for an animal to survive and reproduce. When one of these is out of an animal's tolerance range, it becomes a **limiting factor.** For example, even though a stream insect may have the proper substrate for shelter, adequate current to bring in food and aid in dispersal, and the proper ions to ensure growth and development, inadequate supplies of oxygen make life impossible.

Often, an animal's response to an abiotic factor is to orient itself with respect to it; such orientation is called **taxis.** For example, a response to light is called phototaxis. If an animal favors well-lighted environments and moves toward a light source, it is displaying positive phototaxis. If it prefers low light intensities and moves away from a light source, it displays negative phototaxis.

ENERGY

Energy is the ability to do work. For animals, work includes everything from foraging for food to moving molecules around within cells. To supply their energy needs, animals ingest other organisms; that is, animals are **heterotrophic** (Gr. *hetero,* other + *tropho,* feeder). **Autotrophic** (Gr. *autos,* self + *tropho,* feeder) organisms (e.g., plants) carry on photosynthesis or other carbon-fixing activities that supply their food source. An accounting of an animal's total energy intake and a description of how that energy is used and lost is an **energy budget** (figure 6.2).

The total energy contained in the food an animal eats is the gross energy intake. Some of this energy is lost in feces and through excretion (excretory energy); some of this energy supports minimal maintenance activities, such as pumping blood, exchanging gases, and supporting repair processes (existence energy); and any energy left after existence and excretory functions can be devoted to growth, mating, nesting, and caring for young (productive energy). Survival requires that individuals acquire enough energy to supply these productive functions. Favorable energy budgets are sometimes difficult to attain, especially in temperate regions where winter often makes food supplies scarce.

TEMPERATURE

An animal expends part of its existence energy in regulating body temperature. Temperature influences the rates of chemical reactions in animal cells (metabolic rate) and affects the

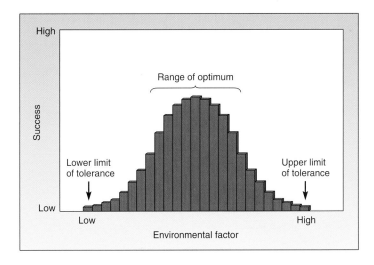

FIGURE 6.1

Tolerance Range of an Animal. Plotting changes in an environmental factor versus some index of success (perhaps egg production, longevity, or growth) shows an animal's tolerance range. The graphs that result are often, though not always, bell-shaped. The range of optimum is the range of values of the factor within which success is greatest. The range of tolerance and range of optimum may vary, depending on an animal's stage of life, health, and activity.

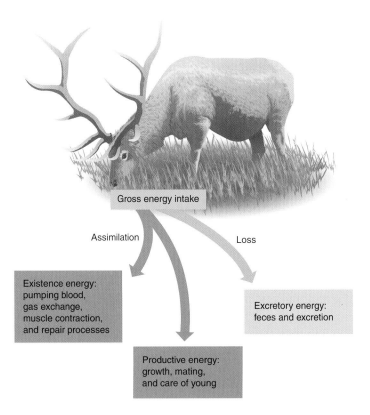

FIGURE 6.2

Energy Budgets of Animals. The gross energy intake of an animal is the sum of energy lost in excretory pathways plus energy assimilated for existence and productive functions. The relative sizes of the boxes in this diagram are not necessarily proportional to the amount of energy devoted to each function. An animal's gross energy intake, and thus, the amount of energy devoted to productive functions, depends on various internal and external factors (e.g., time of year and reproductive status).

animal's overall activity. The body temperature of an animal seldom remains constant because of an inequality between heat loss and heat gain. Heat energy can be lost to objects in an animal's surroundings as infrared and heat radiation, to the air around the animal through convection, and as evaporative heat. On the other hand, heat is gained from solar radiation, infrared and heat radiation from objects in the environment, and relatively inefficient metabolic activities that generate heat as a by-product of cellular functions. Thermoregulatory needs influence many habitat requirements, such as the availability of food, water, and shelter.

When food becomes scarce, or when animals are not feeding for other reasons, they are subject to starvation. Under these circumstances, metabolic activities may decrease dramatically.

Torpor is a time of decreased metabolism and lowered body temperature that occurs daily in bats, hummingbirds, and some other small birds and mammals who must feed almost constantly when they are active. Torpor allows these animals to survive brief periods when they do not feed.

Hibernation is a time of decreased metabolism and lowered body temperature that may last for weeks or months. True hibernation occurs in small mammals, such as rodents, shrews, and bats. The set point of a hibernator's thermoregulatory center drops to about 20° C, but thermoregulation is not suspended.

Winter sleep occurs in some larger animals. Large energy reserves sustain these mammals through periods of winter inactivity. Body temperatures do not drop substantially, and sleeping animals can wake and become active very quickly—as any rookie zoologist probing around the den of a sleeping bear quickly learns!

Aestivation is a period of inactivity in some animals that must withstand extended periods of drying. The animal usually enters a burrow as its environment begins to dry. It generally does not eat or drink and emerges again after moisture returns. Aestivation is common in many invertebrates, reptiles, and amphibians (*see figures 19.15b and 20.10*).

OTHER ABIOTIC FACTORS

Other important abiotic factors for animals include moisture, light, geology, and soils. All life's processes occur in the watery environment of the cell. Water that is lost must be replaced. The amount of light and the length of the light period in a 24-hour day is an accurate index of seasonal change. Animals use light for timing many activities such as reproduction and migration. Geology and soils often directly or indirectly affect organisms living in an area. Characteristics such as texture, amount of organic matter, fertility, and water-holding ability directly influence the number and kinds of animals living in or on the soil. These characteristics also influence the plants upon which animals feed.

POPULATIONS

Populations are groups of individuals of the same species that occupy a given area at the same time and have unique attributes. Two of the most important attributes involve the potential for

population growth and the limits that the environment places on population growth.

POPULATION GROWTH

Animal populations change over time as a result of birth, death, and dispersal. One way to characterize a population with regard to the death of individuals is with survivorship curves (figure 6.3). The Y-axis of a survivorship graph is a logarithmic plot of numbers of survivors, and the X-axis is a linear plot of age. There are three kinds of survivorship curves. Individuals in type 1 (convex) populations survive to an old age, then die rapidly. Environmental factors are relatively unimportant in influencing mortality, and most individuals live their potential life span. Some human populations approach type I survivorship. Individuals in type II (diagonal) populations have a constant probability of death throughout their lives. The environment has an important influence on death and is no harsher on the young than on the old. Populations of birds and rodents often have type II survivorship curves. Individuals in type III (concave) populations experience very high juvenile mortality. Those reaching adulthood, however, have a much lower mortality rate. Fishes and many invertebrates display type III survivorship curves.

A second attribute of populations concerns population growth. The potential for a population to increase in numbers of individuals is remarkable. Rather than increasing by adding a constant number of individuals to the population in every generation, the population increases by the same ratio per unit time. In other

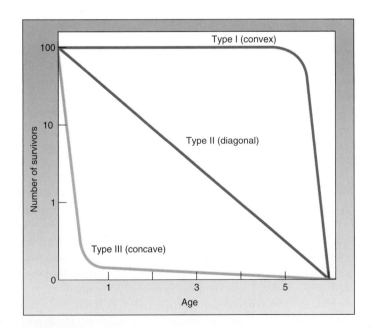

FIGURE 6.3

Survivorship. Survivorship curves are plots of the number of survivors (usually a logarithmic plot) versus age. Type I curves apply to populations in which individuals are likely to live out their potential life span. Type II curves apply to populations in which mortality rates are constant throughout age classes. Type III curves apply to populations in which mortality rates are highest for the youngest cohorts.

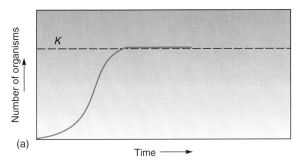

(a)

Time ⟶

(b)

FIGURE 6.4

Logistic Population Growth. (*a*) A logistic growth curve reflects limited resources placing an upper limit on population size. At carrying capacity (*K*), population growth levels off, creating an **S**-shaped curve. (*b*) The fur seal (*Callorhinus ursinus*) population on St. Paul Island, Alaska, was hunted nearly to extinction, and hunting was banned in 1911. Records of the recovery of this population show a logistic growth pattern that leveled off at around 10,000 individuals.

words, populations experience **exponential growth.** Not all populations display the same capacity for growth. Such factors as the number of offspring produced, the likelihood of survival to reproductive age, the duration of the reproductive period, and the length of time it takes to reach maturity all influence reproductive potential.

Exponential growth cannot occur indefinitely. The constraints that climate, food, space, and other environmental factors place on a population are called **environmental resistance.** The population size that a particular environment can support is the environment's **carrying capacity** and is symbolized by *K*. In these situations, growth curves assume a sigmoid, or flattened **S**, shape, and the population growth is referred to as **logistic population growth** (figure 6.4).

POPULATION REGULATION

The conditions that an animal must meet to survive are unique for every species. What many species have in common, however, is that population density and competition affect populations in predictable ways.

Population Density

Density-independent factors influence the number of animals in a population without regard to the number of individuals per unit space (density). For example, weather conditions often limit populations. An extremely cold winter with little snow cover may devastate a population of lizards sequestered beneath the litter of the forest floor. Regardless of the size of the population, a certain percentage of individuals will freeze to death. Human activities, such as construction and deforestation, often affect animal populations in a similar fashion.

Density-dependent factors are more severe when population density is high (or sometimes very low) than they are at other densities. Animals often use territorial behavior, song, and scent marking to tell others to look elsewhere for reproductive space. These actions become more pronounced as population density increases and are thus density dependent. Other density-dependent factors include competition for resources, disease, predation, and parasitism.

INTRASPECIFIC COMPETITION

Competition occurs when animals utilize similar resources and in some way interfere with each other's procurement of those resources. Competition among members of the same species, called **intraspecific competition,** is often intense because the resource requirements of individuals of a species are nearly identical. Intraspecific competition may occur without individuals coming into direct contact. (The "early bird that gets the worm" may not actually see later arrivals.) In other instances, the actions of one individual directly affect another. Territorial behavior and the actions of socially dominant individuals are examples of direct interference.

INTERSPECIFIC INTERACTIONS

Members of other species can affect all characteristics of a population. Interspecific interactions include herbivory, predation, competition, coevolution, and symbiosis. These artificial categories that zoologists create, however, rarely limit animals. Animals often do not interact with other animals in only one way. The nature of interspecific interactions may change as an animal matures, or as seasons or the environment changes.

HERBIVORY AND PREDATION

Animals that feed on plants by cropping portions of the plant, but usually not killing the plant, are herbivores. This conversion provides food for predators that feed by killing and eating other organisms. Interactions between plants and herbivores, and predators and prey, are complex, and many characteristics of the environment affect them. Many of these interactions are described elsewhere in this text.

INTERSPECIFIC COMPETITION

When members of different species compete for resources, one species may be forced to move or become extinct, or the two species may share the resource and coexist.

While the first two options (moving or extinction) have been documented in a few instances, most studies have shown that competing species can coexist. Coexistence can occur when species utilize resources in slightly different ways and when the effects of interspecific competition are less severe than the effects of intraspecific competition. Robert MacArthur studied five species of warblers that all used the same caterpillar prey. Warblers partitioned their spruce tree habitats by dividing a tree into preferred regions for foraging. Although foraging regions overlapped, competition was limited, and the five species coexisted (figure 6.5).

COEVOLUTION

The evolution of ecologically related species is sometimes coordinated such that each species exerts a strong selective influence on the other. This is **coevolution.**

Coevolution may occur when species are competing for the same resource or during predator–prey interactions. In the evolution of predator–prey relationships, for example, natural selection favors the development of protective characteristics in prey species. Similarly, selection favors characteristics in predators that allow them to become better at catching and immobilizing prey. Predator–prey relationships coevolve when a change toward greater predator efficiency is countered by increased elusiveness of prey.

Coevolution is obvious in the relationships between some flowering plants and their animal pollinators. Flowers attract pollinators with a variety of elaborate olfactory and visual

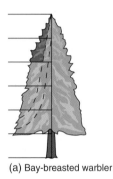

(a) Bay-breasted warbler

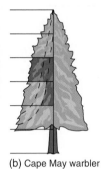
(b) Cape May warbler

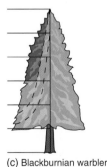

(c) Blackburnian warbler

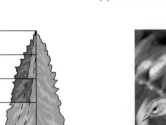

(d) Black-throated green warbler

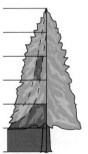
(e) Myrtle warbler

FIGURE 6.5

Coexistence of Competing Species.   Robert MacArthur found that five species of warblers (*a–e*) coexisted by partitioning spruce trees into preferred foraging regions (*shown in dark green*). (*a*) Bay-breasted warbler (*Dendroica castanea*). (*b*) Cape May warbler (*D. tigrina*). (*c*) Blackburnian warbler (*D. fusca*). (*d*) Black-throated green warbler (*D. virens*). (*e*) Myrtle warbler (*D. coronata*).

adaptations. Insect-pollinated flowers are usually yellow or blue because insects see these wavelengths of light best. In addition, petal arrangements often provide perches for pollinating insects. Flowers pollinated by hummingbirds, on the other hand, are often tubular and red. Hummingbirds have a poor sense of smell but see red very well. The long beak of hummingbirds is an adaptation that allows them to reach far into tubular flowers. Their hovering ability means that they have no need for a perch.

SYMBIOSIS

Some of the best examples of adaptations arising through coevolution come from two different species living in continuing, intimate associations, called **symbiosis** (Gr. *sym*, together + *bio*, life). Such interspecific interactions influence the species involved in dramatically different ways. In some instances, one member of the association benefits, and the other is harmed. In other cases, life without the partner would be impossible for both.

Parasitism is a common form of symbiosis in which one organism lives in or on a second organism, called a host. The host usually survives at least long enough for the parasite to complete one or more life cycles. The relationships between a parasite and its host(s) are often complex. Some parasites have life histories involving multiple hosts. The definitive or final host is the host that harbors the sexual stages of the parasite. A fertile female in a definitive host may produce and release hundreds of thousands of eggs in her lifetime. Each egg gives rise to an immature stage that may be a parasite of a second host. This second host is called an intermediate host, and asexual reproduction may occur in this host. Some life cycles may have more than one intermediate host and more than one immature stage. For the life cycle to be completed, the final immature stage must have access to a definitive host. Many examples of coevolutionary interactions between host and parasite are cited in Part Two of this text.

Commensalism is a symbiotic relationship in which one member of the relationship benefits, and the second is neither helped nor harmed. The distinction between parasitism and commensalism is somewhat difficult to apply in natural situations. Whether or not the host is harmed often depends on such factors as the host's nutritional state. Thus, symbiotic relationships may be commensalistic in some situations and parasitic in others.

Mutualism is a symbiotic relationship that benefits both members. Examples of mutualism abound in the animal kingdom, and many examples are described elsewhere in this text.

OTHER INTERSPECIFIC ADAPTATIONS

Interspecific interactions have shaped many other characteristics of animals. **Camouflage** occurs when an animal's color patterns help hide the animal, or a developmental stage, from another animal (figure 6.6). **Cryptic coloration** (L. *crypticus*, hidden) is a type of camouflage that occurs when an animal takes on color patterns in its environment to prevent the animal from being seen by other animals. **Countershading** is a kind of camouflage common in frog and toad eggs. These eggs are darkly pigmented on top and lightly pigmented on the bottom. When a bird or other predator

FIGURE 6.6

Camouflage.   The color pattern of this tiger (*Panthera tigris*) provides effective camouflage that helps when stalking prey.

views the eggs from above, the dark of the top side hides the eggs from detection against the darkness below. On the other hand, when fish view the eggs from below, the light undersurface blends with the bright air-water interface.

Some animals that protect themselves by being dangerous or distasteful to predators advertise their condition by conspicuous coloration. The sharply contrasting white stripe(s) of a skunk and bright colors of poisonous snakes give similar messages. These color patterns are examples of warning or **aposematic coloration** (Gr. *apo*, away from + *sematic*, sign).

Resembling conspicuous animals may also be advantageous. **Mimicry** (L. *mimus*, to imitate) occurs when a species resembles one, or sometimes more than one, other species and gains protection by the resemblance (figure 6.7).

FIGURE 6.7

Mimicry.   These six species of *Heliconius* are all distasteful to bird predators. A bird that consumes any member of the six species is likely to avoid all six species in the future.

COMMUNITIES

All populations living in an area make up a **community.** Communities are not just random mixtures of species; instead, they have a unique organization. Most communities have certain members that have overriding importance in determining community characteristics. For example, a stream community may have a large population of rainbow trout that helps determine the makeup of certain invertebrate populations on which the trout feed. Species that are responsible for establishing community characteristics are called **dominant species.**

Communities are also characterized by the variety of animals they contain. This variety is called **community (species) diversity** or richness. Factors that promote high diversity include a wide variety of resources, high productivity, climatic stability, moderate levels of predation, and moderate levels of disturbance from outside the community. Pollution often reduces the species diversity of ecosystems.

THE ECOLOGICAL NICHE

The **ecological niche** is an important concept of community structure. The niche of any species includes all the attributes of an animal's lifestyle: where it looks for food, what it eats, where it nests, and what conditions of temperature and moisture it requires (figure 6.8). Theoretically, competition results when the niches of two species overlap.

Although the niche concept is difficult to quantify, it is valuable for perceiving community structure. It illustrates that community members tend to complement each other in resource use. They tend to partition resources rather than compete for them. The niche concept is also helpful for visualizing the role of an animal in the environment.

COMMUNITY STABILITY

As with individuals, communities are born and they die. Between those events is a time of continual change. Some changes are the result of climatic or geological events. Members of the community may be responsible for others. In one model of community change, the dominant members of the community change a community in predictable ways in a process called **succession** (L. *successio*, to follow) (figure 6.9). Communities may begin in areas nearly devoid of life. The first community to become established in an area is called the **pioneer community.** Death, decay, and additional nutrients add to the community. Over thousands of years nutrients accumulate, and the characteristics of the ecosystem change. Each successional stage is called a **seral stage,** and the entire successional sequence is a **sere** (ME *seer*, to wither). Succession occurs because the dominant life-forms of a sere gradually make the area less favorable for themselves, but more favorable for organisms of the next successional stage. The final community is the **climax community.** It is different from the seral stages that preceded it because it can tolerate its own reactions. Accumulation of the products of life and death no longer make the area unfit for the individuals living there. Climax communities usually have complex structure and high species diversity.

Other models of community change take into consideration the observation that a nonclimax community may persist in a region because of the community's response to recurring disturbances, species interactions, and chance events. Similarly, changes in a community may result from patchy disturbances that occur repeatedly throughout the community and result in cyclic, nondirectional changes.

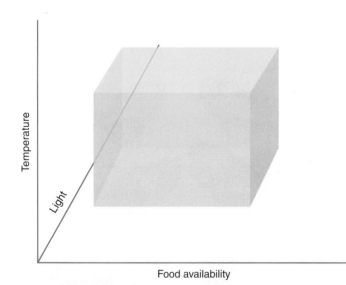

FIGURE 6.8

The Niche as a Multidimensional Space.   The niche of an animal includes all aspects of the animal's lifestyle. In this graph, three characteristics of a niche are plotted on a separate axis: food availability, the tolerance range for temperature, and the tolerance range for light. Adding other dimensions, such as food size, rainfall, and reproductive requirements, would better portray the complete niche.

FIGURE 6.9

Succession.   Primary succession on a sand dune. Beach grass is the first species to become established. It stabilizes the dune so that shrubs, and eventually trees, can grow.

TROPHIC STRUCTURE OF ECOSYSTEMS

Communities and their physical environment are called **ecosystems.** One important fact of ecosystems is that energy is constantly being used, and once it leaves the ecosystem this energy is never reused. Energy supports the activities of all organisms in the ecosystem. It usually enters the ecosystem in the form of sunlight, is incorporated into the chemical bonds of molecules within living and decaying tissues, and is eventually lost as heat.

The sequence of organisms through which energy moves in an ecosystem is a **food chain.** One relatively simple food chain might look like the following:

grass → grazing insects → shrews → owls

Complexly interconnected food chains, called **food webs,** that involve many kinds of organisms are more realistic (figure 6.10). Because food webs can be complex, it is convenient to group organisms according to the form of energy used. These groupings are called **trophic levels.**

Producers (autotrophs) obtain nutrition (complex organic compounds) from inorganic materials and an energy source. They form the first trophic level of an ecosystem. The most familiar producers are green plants that carry on photosynthesis. Other trophic levels are made up of consumers (heterotrophs). Consumers eat other organisms to obtain energy. Herbivores (primary consumers) eat producers. Some carnivores (secondary consumers) eat herbivores, and other carnivores (tertiary consumers) eat the carnivores that ate the herbivores. Consumers also include scavengers that feed on large chunks of dead and decaying organic matter. Decomposers break down dead organisms and feces by digesting organic matter extracellularly and absorbing the products of digestion.

The efficiency with which the animals of a trophic level convert food into new biomass depends on the nature of the food (figure 6.11). Biomass conversion efficiency averages 10%, although efficiencies range from less than 1% for herbivorous endotherms to 35% for carnivorous ectotherms.

CYCLING WITHIN ECOSYSTEMS

Did you ever wonder where the calcium atoms in your bones were 100 or even 100 million years ago? Perhaps they were in the bones of an ancient reptile or in the sediments of prehistoric seas. Unlike energy, all matter is cycled from nonliving reservoirs to living systems and then back to nonliving reservoirs. This is the second important lesson learned from the study of ecosystems—matter is constantly recycled within ecosystems. Matter moves through ecosystems in **biogeochemical cycles.**

A nutrient is any element essential for life. Approximately 97% of living matter is made of oxygen, carbon, nitrogen, and hydrogen. Gaseous cycles involving these elements use the atmosphere or oceans as a reservoir. Elements such as sulfur, phosphorus, and calcium are less abundant in living tissues than are those with gaseous cycles, but they are no less important in sustaining life. The nonliving reservoir for these nutrients is the earth, and the cycles involving these elements are called sedimentary cycles. Water also cycles through ecosystems. Its cycle is called the hydrological cycle.

To help you understand the concept of a biogeochemical cycle, study the carbon cycle in figure 6.12. Carbon is plentiful on the earth and is rarely a limiting factor. The reservoir for carbon is carbon dioxide (CO_2) in the atmosphere or water. Carbon enters the reservoir when organic matter is oxidized to CO_2. CO_2 is released to the atmosphere or water where autotrophs incorporate it into organic compounds. In aquatic systems, some of the CO_2 combines with water to form carbonic acid ($CO_2 + H_2O \rightleftharpoons H_2CO_3$). Because this reaction is reversible, carbonic acid can supply CO_2 to aquatic plants for photosynthesis when CO_2 levels in the water decrease. Carbonic acid can also release CO_2 to the atmosphere. Some of the carbon in aquatic systems is tied up as calcium carbonate ($CaCO_3$) in the shells of molluscs and the skeletons of echinoderms. Accumulations of mollusc shells and echinoderm skeletons have resulted in limestone formations that are the bedrock of much of the United States. Geological uplift, volcanic activities, and weathering return much of this carbon to the earth's surface and the atmosphere. Other carbon is tied up in fossil fuels. Burning fossil fuels returns large quantities of this carbon to the atmosphere as CO_2 (figure 6.12).

ECOLOGICAL PROBLEMS

In the last few hundred years, humans have tried to provide for the needs and wants of their growing population. In the search for longer and better lives, however, humans have lost a sense of being a part of the world's ecosystems. Now that you have studied some general ecological principles, it should be easier to understand many of the ecological problems.

HUMAN POPULATION GROWTH

An expanding human population is the root of virtually all environmental problems. Human populations, like those of other animals, tend to grow exponentially. The earth, like any ecosystem on it, has a carrying capacity and a limited supply of resources. When human populations achieve that carrying capacity, populations should stabilize. If they do not stabilize in a fashion that limits human misery, then war, famine, and/or disease are sure to take care of the problem.

What is the earth's carrying capacity? The answer is not simple. In part, it depends on the desired standard of living and on whether or not resources are distributed equally among all populations. Currently, the earth's population stands at 6.1 billion people. Virtually all environmentalists agree that number is too high if all people are to achieve the affluence of developed countries.

Efforts are being made to curb population growth in many countries, and these efforts have met with some success. Looking at the age characteristics of world populations helps to explain why control measures are needed. The **age structure** of a population

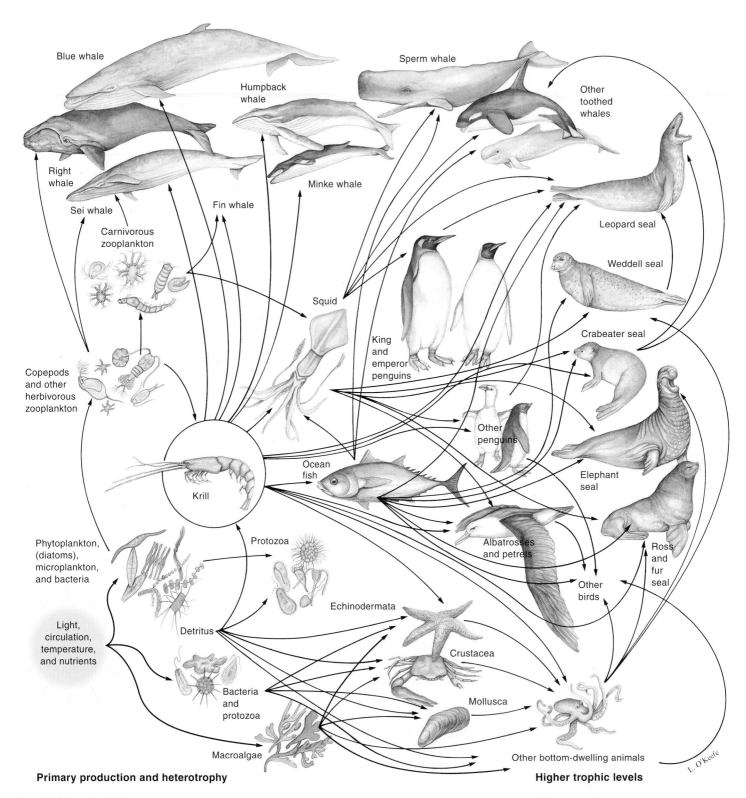

Blue whale

Sperm whale

Other toothed whales

Humpback whale

Right whale

Sei whale

Minke whale

Fin whale

Carnivorous zooplankton

Squid

Leopard seal

Weddell seal

Copepods and other herbivorous zooplankton

King and emperor penguins

Crabeater seal

Other penguins

Elephant seal

Phytoplankton, (diatoms), microplankton, and bacteria

Krill

Ocean fish

Albatrosses and petrels

Ross and fur seal

Light, circulation, temperature, and nutrients

Protozoa

Echinodermata

Other birds

Detritus

Crustacea

Bacteria and protozoa

Mollusca

Macroalgae

Other bottom-dwelling animals

L. O'Keefe

Primary production and heterotrophy

Higher trophic levels

FIGURE 6.10

Food Webs. An Antarctic food web. Small crustaceans called krill support nearly all life in Antarctica. Six species of baleen whales, 20 species of squid, over 100 species of fish, 35 species of birds, and 7 species of seals eat krill. Krill feed on algae, protozoa, other small crustaceans, and various larvae. To appreciate the interconnectedness of food webs, trace the multiple paths of energy from light (lower left), through krill, to the leopard seal.

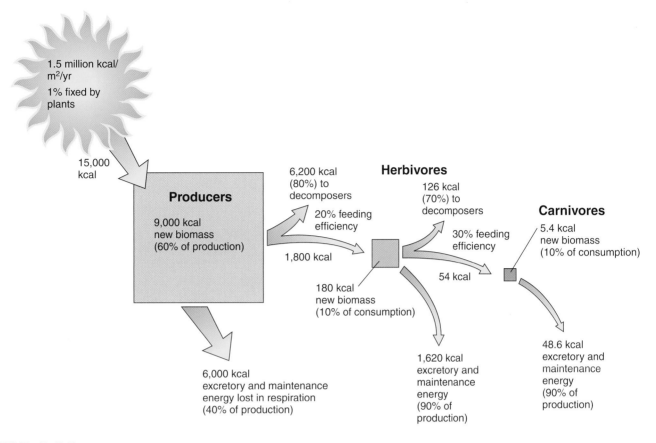

FIGURE 6.11

Energy Flow through Ecosystems.   Approximately 1.5 million kcal of radiant energy strike a square meter of the earth's surface each year. Plants convert less than 1% (15,000 kcal/m^2/yr) into chemical energy. Of this, approximately 60% is converted into new biomass, and 40% is lost in respiration. The herbivore trophic level harvests approximately 20% of net primary production, and decomposers get the rest. Of the 1,800 kcal moving into the herbivore trophic level, 10% (180 kcal) is converted to new biomass, and 90% (1,620 kcal) is lost in respiration. Carnivores harvest about 30% of the herbivore biomass, and 10% of that is converted to carnivore biomass. At subsequent trophic levels, harvesting efficiencies of about 30% and new biomass production of about 10% can be assumed. All of these percentages are approximations. Absolute values depend on the nature of the primary production (e.g., forest versus grassland) and characteristics of the herbivores and carnivores (e.g., ectothermic versus endothermic).

shows the proportion of a population in prereproductive, reproductive, and postreproductive classes. Age structure is often represented by an age pyramid. Figure 6.13 shows an age pyramid for a developed country and for a developing country. In developing countries, the age pyramid has a broad base, indicating high birthrates. As in many natural populations, high infant mortality offsets these high birthrates. However, what happens when developing countries begin accumulating technologies that reduce prereproductive mortality and prolong the life of the elderly? Unless reproductive practices change, a population explosion occurs.

The United Nations population division reports that birthrates are falling in some countries around the world. A birthrate of 2.1 children per woman would replace the current population. The U.N. population division reports an average birthrate of 1.85 for a key group of major developing nations. This average includes rates of 1.2 for Estonia and 1.8 for China. Italy has Europe's lowest birthrate of 1.19. The reasons for the declining birthrate are complex. In China, stringent family planning strategies have been implemented. In Brazil, government-subsidized sterilizations are common. Throughout the world, more

women are joining a workforce and enjoying higher standards of living. In the hardest-hit countries, an increased death rate from the AIDS epidemic is partly responsible for slower growth. In light of these observations, former predictions that the world population would double to 12 billion by 2050 have been revised downward to approximately 9 billion. Virtually all of the projected growth is expected to occur in the world's poorest countries.

Interestingly, the U.S. population is growing, largely because of permissive immigration policies. The current population of the United States is 280 million and is likely to grow to 400 million by 2050. Given the current standard of living, even 200 million people living in the United States is too many. Homelessness, hunger, resource depletion, and pollution all stem from trying to support too many people at the current standard of living.

POLLUTION

Pollution is any detrimental change to an ecosystem. Most kinds of pollution are the results of human activities. Large human populations and demands for increasing goods and services compound

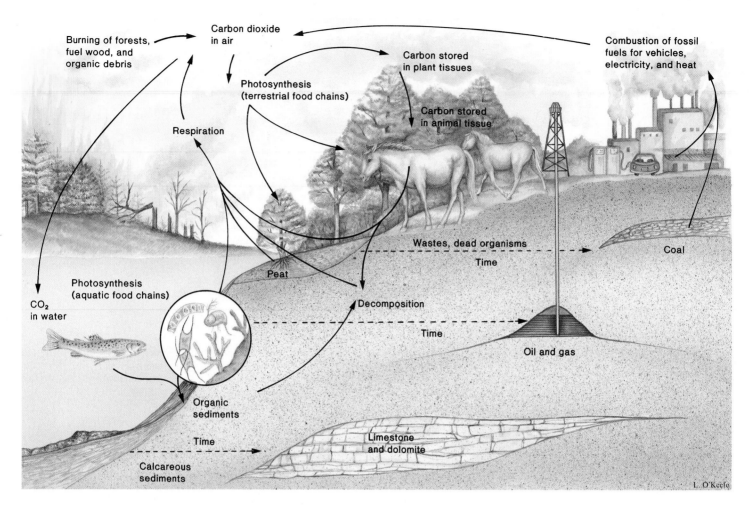

FIGURE 6.12

Carbon Cycle.   Carbon cycles between its reservoir in the atmosphere, living organisms, fossil fuels, and limestone bedrock.

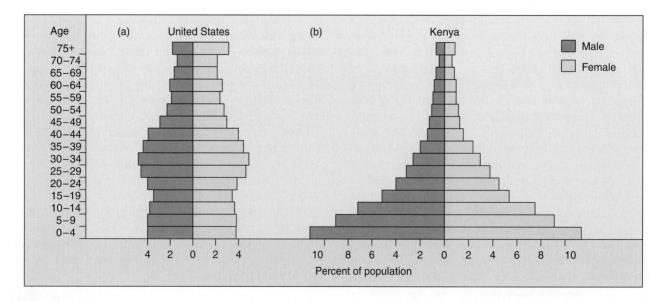

FIGURE 6.13

Human Age Pyramids from 1990.   (a) In developed countries, the age structure is rectangular because mortality in all age classes is relatively low. In this example the widening of the pyramid in the 25 to 50 age range is because of the "baby boom" that occurred between 1945 and 1965. (b) In developing countries, a greater proportion of the population is in the prereproductive age classes. High mortality compensates for high birth rates, and the pyramid is triangular. As technologies reduce infant mortality and prolong the life span of the elderly, populations increase rapidly.

pollution problems. Previous sections on ecosystem productivity and nutrient cycling in this chapter should clarify why these problems exist.

Water pollution takes many forms. Industries generate toxic wastes, heat, and plastics, some of which will persist for centuries. Every household in the world generates human wastes for disposal. All too often, industrial and human wastes find their ways into groundwater, streams, lakes, and oceans. When they do, water becomes unfit for human consumption and unfit for wildlife.

Air pollution also presents serious problems. Burning fossil fuels releases sulfur dioxide and nitrogen oxides into the atmosphere. Sulfur dioxide and water combine to produce sulfuric acid, which falls as **acid deposition.** Acid deposition lowers the pH of lakes, often many kilometers from the site of sulfur dioxide production. Carbon dioxide released in burning fuels is accumulating in the atmosphere, causing the **greenhouse effect.** Carbon dioxide reflects solar radiation back to the earth. This reflection of solar radiation is predicted to cause an increase in world temperature, polar ice caps to melt, and ocean levels to rise. The release of chlorinated fluorocarbons from aerosol cans, air conditioners, and refrigerators contributes to the depletion of the earth's ultraviolet filter—the atmospheric ozone layer. As a result, the incidence of skin cancer is likely to increase.

When wastes and poisons enter food webs, organisms at the highest trophic levels usually suffer the most. Tiny amounts of a toxin incorporated into primary production can quickly build up as carnivores feed on herbivores that have concentrated toxins in their tissues. This problem is especially severe when the material is not biodegradable (not broken down by biological processes). The accumulation of matter in food webs is called **biological magnification.**

RESOURCE DEPLETION AND BIODIVERSITY

Other environmental problems arise because humans have been too slow to realize that an ecosystem's energy is used only once. When a quantity of energy is lost to outer space as heat, it is gone from the earth forever.

As with energy, human populations are also squandering other resources. Our burgeoning population stresses water resources. Nearly two-thirds of the world's people face water shortages. About 70% of water from surface and underground sources is used for irrigation. Water tables in agricultural regions of China are dropping three meters per year. In the United States, water tables in parts of Texas have dropped 30 meters. In the poorest countries, providing clean drinking water and water for sanitation is a severe problem—nearly 2.5 billion people do not have adequate sanitation facilities. Overgrazing and deforestation have led to the spread of deserts. Exploitation of tropical rain forests has contributed to the extinction of many plant and animal species.

The variety of living organisms in an ecosystem is called **biodiversity.** No one knows the number of species in the world. About 1.4 million species have been described, but taxonomists estimate that there are 4 to 30 million more. Much of this unseen, or unnoticed, biodiversity is unappreciated for the free services it performs. Forests hold back flood waters and recycle CO_2 and nutrients. Insects pollinate crops and control insect pests, and subterranean organisms promote soil fertility through decomposition.

FIGURE 6.14

Tropical Deforestation.   Severe erosion quickly follows the removal of the tropical rain forest in Belize. Forests are often cut and burned to make land available for agriculture, leading to loss of biodiversity.

Many of these undescribed species would, when studied, provide new food crops, petroleum substitutes, new fibers, and pharmaceuticals. All of these functions require not just remnant groups but large, healthy populations. Large populations promote the genetic diversity required for surviving environmental changes. When genetic diversity is lost, it is lost forever. Heroic attempts to save endangered species come far too late. Even when they succeed, they salvage only a tiny portion of an original gene pool.

The biodiversity of all natural areas of the world is threatened. Acid rain, pollution, urban development, and agriculture know no geographic or national boundaries. The main threats to biological diversity arise from habitat destruction by expanding human populations. Humans are either directly or indirectly exploiting about 40% of the earth's net primary production. Often, this involves converting natural areas to agricultural uses, frequently substituting less efficient crop plants for native species. Habitat loss displaces thousands of native plants and animals.

Some of the most important threatened natural areas include tropical rain forests, coastal wetlands, and coral reefs. Of these, tropical rain forests have probably received the most attention. Tropical rain forests cover only 7% of the earth's land surface, but they contain more than 50% of the world's species. Tropical rain forests are being destroyed rapidly, mostly for agricultural production. About 76,000 km² (an area larger than Costa Rica) is being cleared each year (*see figure 1.5*). At current rates of destruction, most tropical rain forests will disappear in this century. According to some estimates, 17,500 rain forest species are lost each year. Clearing of tropical rain forests achieves little, because the thin, nutrient-poor soils of tropical rain forests are exhausted within two years (figure 6.14). Sadly, rain forests are a nonrenewable resource. Seeds of rain forest plants germinate rapidly, but

seedlings are unprotected on sterile, open soils. Even if a forest were able to become reestablished, it would take many centuries to return to a climax rain forest.

The problems of threatened biodiversity have solutions, but none of the solutions is quick and easy. First, more money needs to be appropriated for training taxonomists and ecologists, and for supporting their work. Second, all countries of the world need to realize that biodiversity, when preserved and managed properly, is a source of economic wealth. Third, the world needs a system of international ethics that values natural diversity for the beauty it brings to human lives. Anything short of these steps will surely lead to severe climatic changes and mass starvation.

WILDLIFE ALERT
Kirtland's Warbler (*Dendroica kirtlandii*)

VITAL STATISTICS

Classification: Phylum Chordata, class Aves, order Passeriformes

Range: Northern lower peninsula of Michigan

Habitat: Stands of jack-pine between 2 and 6 m high

Number remaining: 400

Status: Endangered

NATURAL HISTORY AND ECOLOGICAL STATUS

Kirtland's warbler is a small bird that breeds in Michigan (box figure 6.1) and winters in the Bahamas and other Caribbean islands. It is bluegray

BOX FIGURE 6.2 Male Kirtland's Warbler (*Dendroica kirtlandii*).

with a yellow breast and black streaks on its back. The male has a black mask (box figure 6.2). Kirtland's warbler feeds on insects, pine pitch, grass, and berries.

During its reproductive period in late spring and early summer, the loud, nearly constant singing of Kirtland's warbler can be heard over distances of 0.5 km. Kirtland's warbler nests on the ground in jack-pine stands that are larger than 80 acres and that have trees between 2 and 6 m tall. A jack-pine stand achieves this status 9 to 13 years after fire destroys an earlier stand. Kirtland's warblers abandon older tree stands because the jack-pines become too tall for lower branches to form protective cover below the trees. Each female produces four or five eggs, and both parents take responsibility for feeding the young chicks. By September, all birds are ready for the migration to the Caribbean.

Fire control and habitat destruction are the major threats to Kirtland's warbler. This bird's very specific habitat requirements mean that changes in the jack-pine habitat interrupt nesting activities. Jack-pine reproduction depends on fire because fire opens jack-pine cones for seed release. Human fire-control measures have prevented the burning of old jack-pine forests and the subsequent

BOX FIGURE 6.1 Breeding Range of Kirtland's Warbler (*Dendroica kirtlandii*).

reestablishment of the younger forests that Kirtland's warbler needs for its reproductive activities.

Cowbirds have also been a significant threat to Kirtland's warbler. Cowbirds are often called nest parasites because they lay their eggs in the nests of other birds and do not rear their young. Parents in the host nest incubate the cowbird eggs along with their own. Because cowbird eggs develop quickly and hatch early into large chicks, cowbird chicks usually are fed most often. Cowbird chicks often push host chicks from the nest. Unfortunately, cowbirds' parasitism of Kirtland's warbler nests has contributed to this warbler's endangered status.

Finally, zoologists do not know much about the overwintering conditions that Kirtland's warbler faces. Of the eight hundred to nine hundred birds that leave Michigan in the fall for the Caribbean, only about four hundred return the following spring.

Extensive efforts are underway to save Kirtland's warbler. Cutting, burning, and planting are restoring jack-pine habitats. Careful monitoring of human development, control of human intrusion into jack-pine forests during the breeding season, and cowbird control measures may provide for the recovery of this species.

SUMMARY

1. Many abiotic factors influence where an animal may live. Animals have a tolerance range and a range of the optimum for environmental factors.
2. Energy for animal life comes from consuming autotrophs or other heterotrophs. Energy is expended in excretory, existence, and productive functions.
3. Temperature, water, light, geology, and soils are important environmental factors that influence animal lifestyles.
4. Animal populations change in size over time. Changes can be characterized using survivorship curves.
5. Animal populations grow exponentially until the carrying capacity of the environment is achieved, at which point constraints such as food, chemicals, climate, and space restrict population growth.
6. Interspecific interactions influence animal populations. These interspecific interactions include herbivory, predator–prey interactions, interspecific competition, coevolution, mimicry, and symbiosis.
7. All populations living in an area make up a community.
8. Organisms have roles in their communities. The ecological niche concept helps ecologists to visualize those roles.
9. Communities often change in predictable ways. Successional changes often lead to a stable climax community.
10. Energy in an ecosystem is not recyclable. Energy that is fixed by producers is eventually lost as heat.
11. Nutrients are cycled through ecosystems. Cycles involve movements of material from nonliving reservoirs in the atmosphere or earth to biological systems and back to the reservoirs again.
12. Human population growth is the root of virtually all of our environmental problems. Trying to support too many people at the standard of living found in developed countries has resulted in air and water pollution and resource depletion.
13. Pollution and resource depletion are important environmental problems that threaten life as we know it. Biodiversity is an important resource that is threatened by human activities.

SELECTED KEY TERMS

biodiversity (p. 91)
biogeochemical cycles (p. 87)
camouflage (p. 85)
carrying capacity (p. 83)
coevolution (p. 84)
community (p. 86)
ecological niche (p. 86)
ecosystems (p. 87)
exponential growth (p. 83)
symbiosis (p. 85)

CRITICAL THINKING QUESTIONS

1. Assuming a starting population of 10 individuals, a doubling time of one month, and no mortality, how long would it take a hypothetical population to achieve 10,000 individuals?
2. Why do you think that winter inactivity of many small mammals takes the form of hibernation, whereas winter inactivity in large mammals takes the form of winter sleep?
3. Which of the following would be a more energy-efficient strategy for supplying animal protein for human diets? Explain your answers.
   a. Feeding people grain-fed beef raised in feed lots, or feeding people beef that has been raised in pastures.
   b. Feeding people sardines and herrings, or processing sardines and herrings into fishmeal that is subsequently used to raise poultry, which is used to feed people.
4. Explain why the biomass present at one trophic level of an ecosystem decreases at higher trophic levels.

ONLINE LEARNING CENTER

Visit our Online Learning Center (OLC) at www.mhhe.com/zoology (click on the book's title) to find the following chapter-related materials:
• CHAPTER QUIZZING
• ANIMATIONS
   Exponential Population Growth

- RELATED WEB LINKS
    Animal Population Ecology
    Community Ecology
    Nutrient Cycling
    Biodiversity
- BOXED READINGS ON
    Hydrothermal Vent Communities
    Coral Reefs
    Ecology of Soil Nematodes
    Soil Conditioning by Earthworms

- SUGGESTED READINGS
- LAB CORRELATIONS

Check out the OLC to find specific information on these related lab exercises in the *General Zoology Laboratory Manual,* 5th edition, by Stephen A. Miller:

Exercise 6 *Adaptations of Stream Invertebrates—A Scavenger Hunt*

PART TWO

ANIMAL-LIKE PROTISTS AND ANIMALIA

One person can never fully appreciate the amazing diversity in the animal kingdom. Zoologists, therefore, must specialize and study particular animal groups. Knowledge of all aspects of animal biology is invaluable, because it reveals the delicate balances in nature and gives clues to how to preserve those balances. Furthermore, as zoologists learn more about animal groups, information that directly affects human welfare emerges. For example, from research on the nervous systems of squid and cockroaches comes much of what is known about nerve cells and many human nervous disorders. In spite of the work of generations of zoologists, many questions about animals remain unanswered. Many species, especially in the tropics, have not been described. Some species may hold keys to unlocking the secrets of cancer, AIDS, and other diseases. Other species may provide insight into managing world resources.

Chapters 7 to 22 present an overview of the known animal-like protists and animal phyla. Specifically, chapter 7 is an introduction to animal taxonomy, which is the study of the naming of organisms and their evolutionary relationships, and to the basic organization of animal bodies. Chapter 8 then covers the animal-like protists, and chapters 9 to 22 survey the animal phyla. These chapters are the beginning of an exciting journey into the diversity of the animal kingdom. Perhaps one of these chapters will captivate your attention, and you will join the thousands of zoologists who spend their lives studying a portion of the animal kingdom.

The animal kingdom has astounding variety and beauty. These feather duster worms (Spirobranchus spinosus) are members of the phylum Annelida and the class Polychaeta. The tentacles are used for feeding and gas exchange. Part Two describes the diversity found within the Protista and the Animalia.

CHAPTER 7

ANIMAL CLASSIFICATION, PHYLOGENY, AND ORGANIZATION

Outline

Concepts

1. Order in nature allows systematists to name animals and discern evolutionary relationships among them.
2. Traditional grouping resulted in the classification of organisms into five kingdoms. Recent information from molecular studies has challenged this concept with three major evolutionary lineages or domains: Eubacteria, Archaea, and Eukarya.
3. Animal systematists use a variety of methods to discern evolutionary relationships. Evolutionary systematics and phylogenetic systematics (cladistics) are two widely used approaches to the study of evolutionary relationships.
4. Animal relationships are represented by branching evolutionary-tree diagrams.
5. Animal body plans can be categorized according to how cells are organized into tissues and how body parts are distributed within and around an animal.
6. Animals are traditionally considered monophyletic. Higher taxonomic levels between kingdom and phylum are used to represent hypotheses of relatedness between animal phyla. Recent cladistic analyses are resulting in reinterpretations of traditional higher groupings.

Biologists have identified approximately 1.4 million species, more than three-fourths of which are animals. Many zoologists spend their lives grouping animals according to shared characteristics. These groupings reflect the order found in living systems that is a natural consequence of shared evolutionary histories. Often, the work of these zoologists involves describing new species and placing them into their proper relationships with other species. Obviously, much work remains in discovering and classifying the world's 4 to 30 million undescribed species.

Rarely do zoologists describe new taxa above the species level (*see figure 1.4*). In 1995, however, R. M. Kirstensen and P. Funch of the University of Copenhagen described a new animal species—*Symbion pandora*—on the mouthparts of lobsters. This species is so different that it has been assigned to a new phylum—the broadest level of animal classification (figure 7.1). The description of this new phylum, Cycliophora, is a remarkable event that brings the total number of recognized extant animal phyla to 36. This chapter describes the principles that zoologists use as they group and name animals.

This chapter contains evolutionary concepts, which are set off in this font.

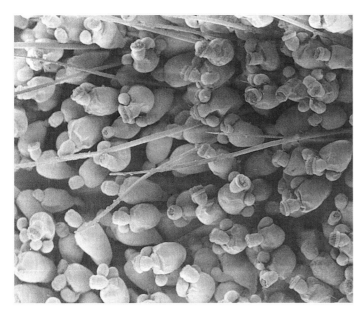

TAXON	HUMAN	DOMESTIC DOG
Domain	Eukarya	Eukarya
Kingdom	Animalia	Animalia
Phylum	Chordata	Chordata
Class	Mammalia	Mammalia
Order	Primates	Carnivora
Family	Hominidae	Canidae
Genus	*Homo*	*Canis*
Species	*sapiens*	*familiaris*

TABLE 7.1
TAXONOMIC CATEGORIES OF A HUMAN AND A DOG

FIGURE 7.1

The Most Recently Described Phylum, Cycliophora. Systematists group animals according to evolutionary relationships. Usually, the work of systematists results in newly described species (or species for which new information has been uncovered) being placed in higher taxonomic categories along with previously studied species. They rarely describe new higher taxonomic groups because finding an organism so different from any previously known organism is unlikely. *Symbion pandora* (shown here) was discovered in 1995 and was distinctive enough for the description of an entirely new phylum, Cycliophora. The individuals shown here are covering the mouthparts of a lobster and are about 0.3 mm long.

CLASSIFICATION OF ORGANISMS

One of the characteristics of modern humans is our ability to communicate with a spoken language. Language not only allows us to communicate but also helps us encode and classify concepts, objects, and organisms that we encounter. To make sense out of life's diversity, we need more than just names for organisms. A potpourri of over a million animal names is of little use to anyone. To be useful, a naming system must reflect the order and relationships that arise from evolutionary processes. The study of the kinds and diversity of organisms and of the evolutionary relationships among them is called **systematics** (Gr. *systema*, system + *ikos*, body of facts) or **taxonomy** (Gr. *taxis*, arrangement + L. *nominalis*, belonging to a name). These studies result in the description of new species and the organization of animals into groups (taxa) based on degree of evolutionary relatedness. (Some biologists distinguish between systematics and taxonomy, preferring to think of taxonomy as the work involved with the original description of species, and systematics as the assignment of species into evolutionary groups. This text does not make this distinction because of the extensive overlap between the two tasks.) **Nomenclature** (L. *nominalis*, belonging to a name + *calator*, to call) is the assignment of a distinctive name to each species.

A TAXONOMIC HIERARCHY

The modern classification system is rooted in the work of Karl von Linné (Carolus Linnaeus) (1707–1778). His binomial system (*see chapter 1*) is still used today. Von Linné also recognized that different species could be grouped into broader categories based on shared characteristics. Any grouping of animals that shares a particular set of characteristics forms an assemblage called a **taxon** (pl., taxa). For example, a housefly (*Musca domestica*), although obviously unique, shares certain characteristics with other flies (the most important of these being a single pair of wings). Based on these similarities, all true flies form a logical, more inclusive taxon. Further, all true flies share certain characteristics with bees, butterflies, and beetles. Thus, these animals form an even more inclusive taxon. They are all insects.

Von Linné recognized five taxonomic categories. Modern taxonomists use those five and have added three others. The categories are arranged hierarchically (from broad to specific): **domain, kingdom, phylum, class, order, family, genus,** and **species** (table 7.1). (Domain, the broadest taxonomic category, was added recently, is not yet universally accepted, and is discussed under "Kingdoms of Life.")

Even though von Linné did not accept evolution, many of his groupings reflect evolutionary relationships. Morphological similarities between two animals have a genetic basis and are the result of a common evolutionary history. Thus, in grouping animals according to shared characteristics, von Linné grouped them according to their evolutionary relationships. Ideally, members of the same taxonomic group are more closely related to each other than to members of different taxa (*see figure 1.4*)

Above the species level, the definitions of what constitutes a particular taxon are not precise. (The species concept is discussed in chapter 5.) Disagreements as to whether two species should be grouped in the same taxon or different taxa are common.

NOMENCLATURE

Do you call certain freshwater crustaceans crawdads, crayfish, or crawfish? Do you call a common sparrow an English sparrow, a barn sparrow, or a house sparrow? The binomial system of nomenclature brings order to a chaotic world of common names. Common names have two problems. First, they vary from country to country, and from region to region within a country. Some species have literally hundreds of different common names. Biology transcends regional and nationalistic boundaries, and so must the names of what biologists study. Second, many common names refer to taxonomic categories higher than the species level. Most different kinds of pillbugs (class Crustacea, order Isopoda) or most different kinds of crayfish (class Crustacea, order Decapoda) cannot be distinguished from a superficial examination. A common name, even if you recognize it, often does not specify a particular species.

The binomial system of nomenclature is universal and clearly indicates the level of classification involved in any description. No two kinds of animals have the same binomial name, and every animal has only one correct name, as required by the *International Code of Zoological Nomenclature*, thereby avoiding the confusion that common names cause. The genus of an animal begins with a capital letter, the species designation begins with a lowercase letter, and the entire scientific name is italicized or underlined because it is derived from Latin or is latinized. Thus, the scientific name of humans is written *Homo sapiens*. When the genus is understood, the binomial name can be abbreviated *H. sapiens*.

MOLECULAR APPROACHES TO ANIMAL SYSTEMATICS

In recent years, molecular biological techniques have provided important information for taxonomic studies. The relatedness of animals is reflected in the gene products (proteins) animals produce and in the genes themselves (the sequence of nitrogenous bases in DNA). Related animals have DNA derived from a common ancestor. Genes and proteins of related animals, therefore, are more similar than genes and proteins from distantly related animals. Molecular clocks (*see chapter 4*), although limited by their differing rates of change, depending on the molecule studied, can provide estimates of the time elapsed since divergence from a common ancestor. Sequencing the nuclear DNA and the mitochondrial DNA of animals has become commonplace. Mitochondrial DNA is useful in taxonomic studies because mitochondria have their own genetic systems and are inherited cytoplasmically. That is, mitochondria are transmitted from parent to offspring through the egg cytoplasm and can be used to trace maternal lineages. Using mitochondrial DNA involves relatively small quantities of DNA that change at a relatively constant rate. As you will see in the next section, the sequencing of ribosomal RNA has been used extensively in studying taxonomic relationships.

Although molecular techniques have proved to be extremely valuable to animal taxonomists, they will not replace traditional taxonomic methods. Molecular and traditional methods of investigation will probably always be used to complement each other in taxonomic studies.

KINGDOMS OF LIFE

In 1969, Robert H. Whittaker described a system of classification that distinguished between kingdoms according to cellular organization and mode of nutrition (figure 7.2a). According to this system, members of the kingdom **Monera** are the bacteria and the cyanobacteria. They are distinguished from all other organisms by being prokaryotic (*see table 2.1*). Members of the kingdom **Protista** are eukaryotic and consist of single cells or colonies of cells. This kingdom includes *Amoeba*, *Paramecium*, and many others. Members of the kingdom **Plantae** are eukaryotic, multicellular, and photosynthetic. Plants have cell walls and are usually nonmotile. Members of the kingdom **Fungi** are also eukaryotic and multicellular. They also have cell walls and are usually nonmotile. Mode of nutrition distinguishes fungi from plants. Fungi digest organic matter extracellularly and absorb the breakdown products. Members of the kingdom **Animalia** are eukaryotic and multicellular, and they usually feed by ingesting other organisms or parts of other organisms. Their cells lack walls and they are usually motile.

In recent years, new information has challenged the five-kingdom classification system. For the first 2 billion years of life on the earth, the only living forms were prokaryotic microbes. Fossil evidence from this early period is scanty; however, molecular studies of variations in base sequences (*see chapter 3*) of ribosomal RNA from more than two thousand organisms are providing evidence of relationships rooted within this 2-billion-year period. The emerging picture is that the five previously described kingdoms do not represent distinct evolutionary lineages.

Ribosomal RNA is excellent for studying the evolution of early life on earth. It is an ancient molecule, and it is present and retains its function in virtually all organisms. In addition, ribosomal RNA changes very slowly. This slowness of change, called **evolutionary conservation,** *indicates that the protein-producing machinery of a cell can tolerate little change and still retain its vital function. Evolutionary conservation of this molecule means that closely related organisms (recently diverged from a common ancestor) are likely to have similar ribosomal RNAs. Distantly related organisms are expected to have ribosomal RNAs that are less similar, but the differences are small enough that the relationships to some ancestral molecule are still apparent.*

Molecular systematists compare the base sequences in ribosomal RNA of different organisms to find the number of positions in the RNAs where bases are different. They enter these data into computer programs and examine all possible relationships among the different organisms. The systematists then decide which arrangement of the organisms best explains the data.

Studies of ribosomal RNA have led systematists to the conclusions that all life shares a common ancestor and that

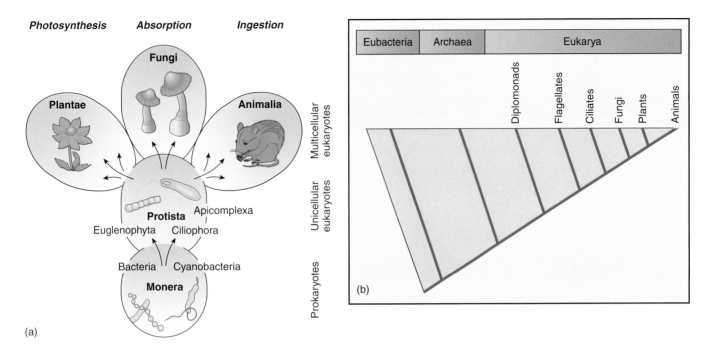

FIGURE 7.2

Classification of Organisms. (*a*) In 1969, Robert H. Whittaker described a five-kingdom classification system based on cellular organization and mode of nutrition. (*b*) Recent studies of ribosomal RNA indicate that a grouping into three domains more accurately portrays early evolutionary relationships.

*there are three major evolutionary lineages (figure 7.2 b). Each of these lineages is called a domain. The domain thus supersedes the kingdom as the broadest taxonomic grouping. The **Archaea** are prokaryotic microbes that live in extreme environments, such as high-temperature rift valleys on the ocean floor, or high-salt or acidic environments. All members of the Archaea inhabit anaerobic environments. These environments may reflect the conditions on the earth at the time of life's origin. The Archaea are the most primitive life-forms known. Ancient archaeans gave rise to two other domains of organisms: Eubacteria and Eukarya. The **Eubacteria** are true bacteria and are prokaryotic microorganisms. The **Eukarya** include all eukaryotic organisms. Interestingly, the Eukarya diverged more recently than the Eubacteria from the Archaea. Thus, the Eukarya are more closely related than the Eubacteria to the Archaea.*

The Eukarya arose about 1.5 billion years ago. The photosynthetic accumulation of oxygen in the atmosphere probably resulted in the production of ozone, which shielded the planet from deadly ultraviolet light. The Eukarya then underwent a late, but rapid, evolutionary diversification into the modern lineages of protists, fungi, plants, and animals.

Many biologists find the relatively close relationships of all eukaryotic organisms depicted in this taxonomic scheme strange and unsettling. The relationships depicted in figure 7.2b not only are new but also are based on cellular characteristics, rather than the whole-organism characteristics that taxonomists traditionally use. As a result, many questions remain to be answered before zoologists accept this system of higher taxonomy.

One of these questions involves how to deal with the eukaryotic kingdom lineages. Should the Animalia, Fungi, and Plantae still be considered kingdoms? Most zoologists would answer, "Yes." If so, molecular studies indicate that the separate protist lineages shown in figure 7.2b should also be elevated to kingdom status. Questions such as this make systematics a lively and challenging field of study, and future editions of this text will almost certainly contain taxonomic revisions that center around the relationships within the Eukarya.

This text is primarily devoted to the animals. Chapter 8, however, covers the animal-like protists (protozoa). The inclusion of protozoa is part of a tradition that originated with an old two-kingdom classification system. Animal-like protists (e.g., *Amoeba* and *Paramecium*) were once considered a phylum (Protozoa) in the animal kingdom, and general zoology courses usually include them.

ANIMAL SYSTEMATICS

*The goal of animal systematics is to arrange animals into groups that reflect evolutionary relationships. Ideally, these groups should include a single ancestral species and all of its descendants. Such a group is called a **monophyletic group** (figure 7.3). In searching out monophyletic groups, taxonomists look for animal attributes called characters that indicate relatedness. A **character** is virtually anything that has a genetic basis and can be measured—from an anatomical feature to a sequence of nitrogenous bases in DNA or RNA. **Polyphyletic***

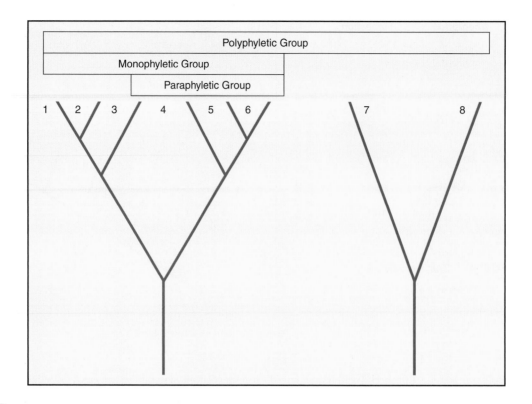

FIGURE 7.3

Evolutionary Groups.   An assemblage of species 1–8 is a polyphyletic group because species 1–6 have a different ancestor than species 7 and 8. An assemblage of species 3–6 is a paraphyletic group because species 1 and 2 share the same ancestor as 3–6, but they have been left out of the group. An assemblage of species 1–6 is a monophyletic group because it includes all of the descendants of a single ancestor.

groups have members that can be traced to separate ancestors. Since each group should have a single ancestor, a polyphyletic group reflects insufficient knowledge of the group. A **paraphyletic group** includes some, but not all, members of a lineage. Paraphyletic groups also result when knowledge of the group is insufficient (see figure 7.3).

As in any human endeavor, disagreements have arisen in animal systematics. These disagreements revolve around methods of investigation and whether or not data may be used in describing distant evolutionary relationships. Three contemporary schools of systematics exist: evolutionary systematics, numerical taxonomy, and phylogenetic systematics (cladistics).

Evolutionary systematics is the oldest of the three approaches. It is sometimes called the "traditional approach," although it has certainly changed since the beginnings of animal systematics. *A basic assumption of evolutionary systematists is that organisms closely related to an ancestor will resemble that ancestor more closely than they resemble distantly related organisms.*

Two kinds of similarities between organisms are recognized: homologies and analogies (*see also the discussion of homology and analogy in chapter 4*). Homologies are resemblances that result from common ancestry and are useful in classifying animals. An example is the similarity in the arrangement of bones in the wing of a bird and the arm of a human (*see figure 4.11*). Analogies are resemblances that result from organisms adapting under similar evolutionary pressures. The latter process is sometimes called convergent evolution. Analogies do not reflect common ancestry and are not used in animal taxonomy. The similarity between the wings of birds and insects is an analogy.

Evolutionary systematists often portray the results of their work on phylogenetic trees, where organisms are grouped according to their evolutionary relationships. Figure 7.4 is a phylogenetic tree showing vertebrate evolutionary relationships, as well as timescales and relative abundance of animal groups. These diagrams reflect judgments made about rates of evolution and the relative importance of certain key characters (e.g., feathers in birds).

Numerical taxonomy emerged during the 1950s and 1960s and represents the opposite end of the spectrum from evolutionary systematics. The founders of numerical taxonomy believed that the criteria for grouping taxa had become too arbitrary and vague. They tried to make taxonomy more objective. Numerical taxonomists use mathematical models and computer-aided techniques to group samples of organisms according to overall similarity. They do not attempt to distinguish between homologies and analogies. Numerical taxonomists admit that analogies exist. They contend, however, that telling one from the other is sometimes impossible and that the numerous homologies used in data analysis overshadow the analogies. A second major difference between evolutionary systematics and numerical taxonomy is that numerical taxonomists limit discussion

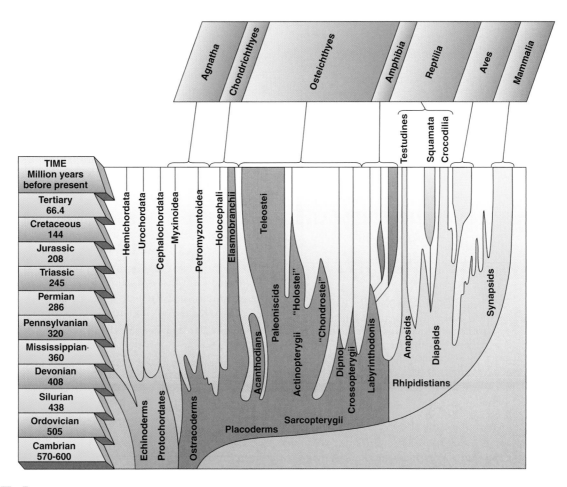

FIGURE 7.4

Phylogenetic Tree Showing Vertebrate Phylogeny. A phylogenetic tree derived from evolutionary systematics depicts the degree of divergence since branching from a common ancestor, which is indicated by the time periods on the vertical axis. The width of the branches indicates the number of recognized genera for a given time period. Note that this diagram shows the birds (Aves) as being closely related to the reptiles (Reptilia), and both groups as having class-level status.

of evolutionary relationships to closely related taxa. Numerical taxonomy is the least popular of the three taxonomic schools; however, all taxonomists use the computer programs that numerical taxonomists developed.

Phylogenetic systematics (cladistics) is a third approach to animal systematics. The goal of cladistics is similar to that described for evolutionary systematics—the generation of hypotheses of genealogical relationships among monophyletic groups of organisms. Cladists contend, however, that their methods are more open to analysis and testing, and thus are more scientific, than those of evolutionary systematists.

As do evolutionary systematists, cladists differentiate between homologies and analogies. They believe, however, that homologies of recent origin are most useful in phylogenetic studies. Characters that all members of a group share are referred to as **symplesiomorphies** (Gr. *sym*, together + *plesio*, near + *morphe*, form). These characters are homologies that may indicate a shared ancestry, but they are useless in describing relationships within the group.

To decide what character is ancestral for a group of organisms, cladists look for a related group of organisms, called an **out-**

group, that is not included in the study group. Figure 7.5 shows a hypothetical lineage for five taxa. Notice that taxon 5 is the outgroup for taxa 1–4. Character A is symplesiomorphic (ancestral) for the outgroup and the study group.

Characters that have arisen since common ancestry with the outgroup are called **derived characters.** Derived characters shared by members of a group are called **synapomorphies** (Gr. *syn*, together + *apo*, away + *morphe*, form). Taxa 1–4 in figure 7.5 share derived character B. This character separates taxa 1–4 from the outgroup. Similarly, derived characters C and D arose more recently than character B. Closely related taxa 1 and 2 share character C. Closely related taxa 3 and 4 share character D. The presence of a synapomorphy, a shared derived character, provides evidence that taxa form a related subset called a **clade** (Gr. *klados*, branch). Taxa 1 and 2 form a clade characterized by C.

The hypothetical lineage shown in figure 7.5 is called a **cladogram.** Cladograms depict a sequence in the origin of derived characters. A cladogram is interpreted as a family tree depicting a hypothesis regarding monophyletic lineages. New data in the form of newly investigated characters or reinterpretations of old data are used to test the hypothesis the cladogram describes.

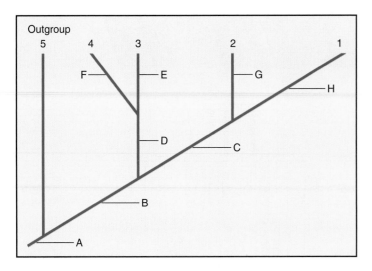

FIGURE 7.5

Interpreting Cladograms. This hypothetical cladogram shows five taxa (1–5) and the characters (A–H) used in deriving the taxonomic relationships. Character A is symplesiomorphic for the entire group. Taxon 5 is the outgroup because it shares only that ancestral character with taxa 1–4. All other characters are more recently derived. What single character is a synapomorphy for taxa 1 and 2, separating them from all other taxa?

Figure 7.6 is a cladogram depicting the evolutionary relationships among the vertebrates. The tunicates and cephalochordates are an outgroup for the entire vertebrate lineage. Derived characters are listed on the right side of the cladogram. Notice that extraembryonic membranes is a synapomorphy used to define the clade containing the reptiles, birds, and mammals. These extraembryonic membranes are a shared character for these groups and are not present in any of the fish taxa or the amphibians. Distinguishing between reptiles, birds, and mammals requires looking at characters that are even more recently derived than extraembryonic membranes. A derived character, the shell, distinguishes turtles from all other members of the clade; skull characters distinguish the lizard/crocodile/bird lineage from the mammal lineage; and hair, mammary glands, and endothermy is a unique mammalian character combination. Note that a synapomorphy at one level of taxonomy may be a symplesiomorphy at a different level of taxonomy. Extraembryonic membranes is a synapomorphic character within the vertebrates that distinguishes the reptile/bird/mammal clade. It is symplesiomorphic for reptiles, birds, and mammals because it is ancestral for the clade and cannot be used to distinguish among members of these three groups.

Zoologists widely accept cladistics. This acceptance has resulted in some nontraditional interpretations of animal phylogeny. A comparison of figures 7.4 and 7.6 shows one example of different interpretations derived through evolutionary systematics and cladistics. Generations of taxonomists have assigned class-level status (Aves) to birds. Reptiles also have had class-level status (Reptilia). Cladistic analysis has shown, however, that birds are more closely tied by common ancestry to the alligators and crocodiles than to any other group of living vertebrates. According to the cladistic interpretation, birds and crocodiles should be

assigned to a group that reflects this close common ancestry. Birds would become a subgroup within a larger group that included both birds and reptiles. Crocodiles would be depicted more closely related to the birds than they would be to snakes and lizards. Traditional evolutionary systematists maintain that the traditional interpretation is still correct because it takes into account the greater importance of key characters of birds (e.g., feathers and endothermy) that make the group unique. Cladists support their position by pointing out that the designation of "key characters" involves value judgments that cannot be tested.

As debates between cladists and evolutionary systematists continue, our knowledge of evolutionary relationships among animals will become more complete. Debates like these are the fuel that force scientists to examine and reexamine old assumptions. Animal systematics is certain to be a lively and exciting field in future years.

Chapters 8 through 22 are a survey of the animal kingdom. The organization of these chapters reflects the traditional taxonomy that makes most zoologists comfortable. Cladograms are usually included in "Further Phylogenetic Considerations" at the end of most chapters, and any different interpretations of animal phylogeny implicit in these cladograms are discussed.

EVOLUTIONARY RELATIONSHIPS AND TREE DIAGRAMS

Although evolutionary-tree diagrams can help clarify evolutionary relationships and timescales, they are often a source of misunderstanding because they illustrate relationships among levels of classification above the species (*see figure 7.6*). Depicting phyla or classes as ancestral is misleading because evolution occurs in species groups (populations), not at higher taxonomic levels. Also, even though phyla or classes are depicted as ancestral, modern representatives of these "ancestral phyla" have had just as long an evolutionary history as animals in other taxonomic groups that may have descended from the common ancestor. All modern representatives of any group of animals should be visualized at the tips of a "tree branch," and they may be very different from ancestral species. Zoologists use modern representatives to help visualize general characteristics of an ancestral species, but never to specify details of the ancestor's structure, function, or ecology.

In addition to these problems of interpretation, evolutionary trees often imply a ladderlike progression of increasing complexity. This is misleading, because evolution has often resulted in reduced complexity and body forms that are evolutionary failures. In many cases, evolution does not lead to phenotypes that permit survival under changing conditions, and extinction occurs. Further, the common representation of a phylogeny as an inverted cone, or a tree with a narrow trunk and many higher branches, is often misleading. This implies that evolution is a continuous process of increasing diversification. The fossil records show that this is often wrong. For example, 20 to 30 groups of echinoderms (sea stars and their relatives) are in the fossil record, but there are only five modern groups. This evolutionary lineage, like many

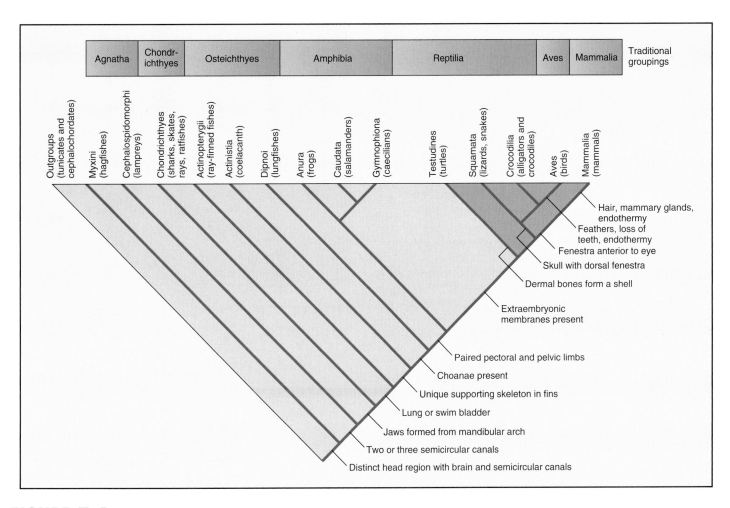

FIGURE 7.6

Cladogram Showing Vertebrate Phylogeny. A cladogram is constructed by identifying points at which two groups diverged. Animals that share a branching point are included in the same taxon. Notice that timescales are not given or implied. The relative abundance of taxa is also not shown. Notice that this diagram shows the birds and crocodilians sharing a common branch, and that these two groups are more closely related to each other than either is to any other group of animals.

others, underwent rapid initial evolutionary diversification. After the initial diversification, extinction—not further diversification—was the rule. Paleontologist Stephen J. Gould used the term *contingency* to refer to rapid evolutionary explosion followed by a high likelihood of extinction.

In spite of these problems, tree diagrams persist in scientific literature and are used in this text. As long as you keep their limitations in mind, they can help you visualize evolutionary relationships.

PATTERNS OF ORGANIZATION

One of the most strikingly ordered series of changes in evolution is reflected in body plans in the animal kingdom and the protists. Evolutionary changes in animal body plans might be likened to a road map through a mountain range. What is most easily depicted are the starting and ending points and a few of the "attractions" along the route. What cannot be seen from this perspective are the tortuous curves and grades that must be navigated and the extra miles that must be traveled to chart back roads. *Evolutionary*

changes do not always mean "progress" and increased complexity. Evolution frequently results in backtracking, in failed experiments, and in inefficient or useless structures. Evolution results in frequent dead ends, even though the route to that dead end may be filled with grandeur. The account that follows is a look at patterns of animal organization. As far as evolutionary pathways are concerned, this account is an inexplicit road map through the animal kingdom. On a grand scale, it portrays evolutionary trends, but it does not depict an evolutionary sequence.

SYMMETRY

The bodies of animals and protists are organized into almost infinitely diverse forms. Within this diversity, however, are certain patterns of organization. The concept of symmetry is fundamental to understanding animal organization. **Symmetry** describes how the parts of an animal are arranged around a point or an axis (table 7.2).

TABLE 7.2
ANIMAL SYMMETRY

TERM	DEFINITION
Asymmetry	The arrangement of body parts without a central axis or point (e.g., the sponges).
Bilateral symmetry	The arrangement of body parts such that a single plane passing between the upper and lower surfaces and through the longitudinal axis divides the animal into right and left mirror images (e.g., the vertebrates).
Radial symmetry	The arrangement of body parts such that any plane passing through the oral-aboral axis divides the animal into mirror images (e.g., the cnidarians). Radial symmetry can be modified by the arrangement of some structures in pairs, or other combinations, around the central axis (e.g., biradial symmetry in the ctenophorans and some anthozoans, and pentaradial symmetry in the echinoderms).

Asymmetry, which is the absence of a central point or axis around which body parts are equally distributed, characterizes most protists and many sponges (figure 7.7). Asymmetry cannot be said to be an adaptation to anything or advantageous to an organism. Asymmetrical organisms do not develop complex communication, sensory, or locomotor functions. Clearly, however, protists and animals whose bodies consist of aggregates of cells have flourished.

A sea anemone can move along a substrate, but only very slowly. How does it gather food? How does it detect and protect it-

FIGURE 7.7

Asymmetry. Sponges display a cell-aggregate organization, and as this red encrusting sponge (*Monochora barbadensis*) shows, many are asymmetrical.

FIGURE 7.8

Radial Symmetry. Planes that pass through the oral-aboral axis divide radially symmetrical animals, such as this tube coral polyp (*Tubastraea* sp.), into equal halves. Certain arrangements of internal structures modify the radial symmetry of sea anemones.

self from predators? For this animal, a blind side would leave it vulnerable to attack and cause it to miss many meals. The sea anemone, as is the case for most sedentary animals, has sensory and feeding structures uniformly distributed around its body. Sea anemones do not have distinct head and tail ends. Instead, one point of reference is the end of the animal that possesses the mouth (the oral end), and a second point of reference is the end opposite the mouth (the aboral end). Animals such as the sea anemone are radially symmetrical. **Radial symmetry** is the arrangement of body parts such that any plane passing through the central oral-aboral axis divides the animal into mirror images (figure 7.8). Radial symmetry is often modified by the arrangement of some structures in pairs, or in other combinations, around the central oral-aboral axis. The paired arrangement of some structures in radially symmetrical animals is called biradial symmetry. The arrangement of structures in fives around a radial animal is called pentaradial symmetry.

Although the sensory, feeding, and locomotor structures in radially symmetrical animals could never be called "simple," they are not comparable to the complex sensory, locomotor, and feeding structures in many other animals. The evolution of such complex structures in radially symmetrical animals would require repeated distribution of specialized structures around the animal.

Bilateral symmetry is the arrangement of body parts such that a single plane, passing between the upper and lower surfaces and through the longitudinal axis of an animal, divides the animal into right and left mirror images (figure 7.9). Bilateral symmetry is characteristic of active, crawling, or swimming animals. Because bilateral animals move primarily in one direction, one end of the animal is continually encountering the environment. The end that meets the environment is usually where complex sensory, nervous, and feeding structures evolve and develop.

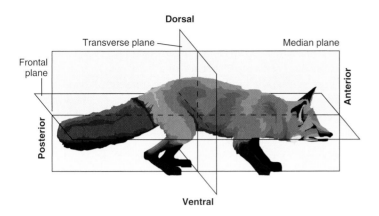

FIGURE 7.9

Bilateral Symmetry. Planes and terms of direction useful in locating parts of a bilateral animal. A bilaterally symmetrical animal, such as this fox, has only one plane of symmetry. An imaginary median plane is the only plane through which the animal could be cut to yield mirror-image halves.

These developments result in the formation of a distinct head and are called **cephalization** (Gr. *kephale*, head). Cephalization occurs at an animal's anterior end. Posterior is opposite anterior; it is the animal's tail end. Other important terms of direction and terms describing body planes and sections apply to bilateral animals. These terms are for locating body parts relative to a point of reference or an imaginary plane passing through the body (tables 7.2 and 7.3; figure 7.9).

OTHER PATTERNS OF ORGANIZATION

In addition to body symmetry, other patterns of animal organization are recognizable. In a broad context, these patterns may reflect evolutionary trends. As explained earlier, however, these trends are not exact sequences in animal evolution.

The Unicellular (Cytoplasmic) Level of Organization

Organisms whose bodies consist of single cells or cellular aggregates display the unicellular level of organization. Unicellular body plans are characteristic of the Protista. Some zoologists prefer to use the designation *cytoplasmic* to emphasize that all living functions are carried out within the confines of a single plasma membrane. Unicellular organization is not "simple." All unicellular organisms must provide for the functions of locomotion, food acquisition, digestion, water and ion regulation, sensory perception, and reproduction in a single cell.

Cellular aggregates (colonies) consist of loose associations of cells that exhibit little interdependence, cooperation, or coordination of function—therefore, cellular aggregates cannot be considered tissues (*see chapter 2*). In spite of the absence of interdependence, these organisms show some division of labor. Some cells may be specialized for reproductive, nutritive, or structural functions.

TABLE 7.3
TERMS OF DIRECTION

TERM	DESCRIPTION
Aboral	The end opposite the mouth
Oral	The end containing the mouth
Anterior	The head end; usually the end of a bilateral animal that meets its environment
Posterior	The tail end
Caudal	Toward the tail
Cephalic	Toward the head
Distal	Away from the point of attachment of a structure on the body (e.g., the toes are distal to the knee)
Proximal	Toward the point of attachment of a structure on the body (e.g., the hip is proximal to the knee)
Dorsal	The back of an animal; usually the upper surface; synonymous with posterior for animals that walk upright
Ventral	The belly of an animal; usually the lower surface; synonymous with anterior for animals that walk upright
Inferior	Below a point of reference (e.g., the mouth is inferior to the nose in humans)
Superior	Above a point of reference (e.g., the neck is superior to the chest)
Lateral	Away from the plane that divides a bilateral animal into mirror images
Medial (median)	On or near the plane that divides a bilateral animal into mirror images

Diploblastic Organization

Cells are organized into tissues in most animal phyla. **Diploblastic** (Gr. *diplóos*, twofold + *blaste*, to sprout) organization is the simplest tissue-level organization (figure 7.10). Body parts are organized into layers derived from two embryonic tissue layers. **Ectoderm** (Gr. *ektos*, outside + *derm*, skin) gives rise to the epidermis, the outer layer of the body wall. **Endoderm** (Gr. *endo*, within) gives rise to the gastrodermis, the tissue that lines the gut cavity. Between the epidermis and the gastrodermis is a middle layer called mesoglea. This mesoglea may or may not contain cells, but when cells occur they are always derived from ectoderm or endoderm. When cells are present, this middle layer is sometimes referred to as mesenchyme. The term "mesoglea" is then reserved for the acellular condition. This text uses the term "mesoglea" for the middle layer and specifies the acellular or cellular condition.

The cells in each tissue layer are functionally interdependent. The gastrodermis consists of nutritive (digestive) and muscular cells, and the epidermis contains epithelial and muscular cells. The feeding movements of *Hydra* or the swimming

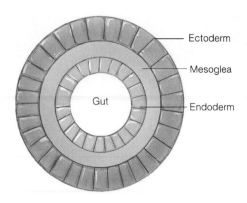

FIGURE 7.10

Diploblastic Body Plan.   Diploblastic animals have tissues derived from ectoderm and endoderm. Between these two layers is a noncellular mesoglea.

movements of a jellyfish are only possible when groups of cells co-operate, showing tissue-level organization.

Triploblastic Organization

Animals described in chapters 10 to 22 are **triploblastic** (Gr. *treis*, three + *blaste*, to sprout); that is, their tissues are derived from three embryological layers. As with diploblastic animals, ectoderm forms the outer layer of the body wall, and endoderm lines the gut. A third embryological layer is sandwiched between the ectoderm and endoderm. This layer is **mesoderm** (Gr. *meso*, in the middle), which gives rise to supportive, contractile, and blood cells. Most triploblastic animals have an organ-system level of organization. Tissues are organized to form excretory, nervous, digestive, reproductive, circulatory, and other systems. Triploblastic

animals are usually bilaterally symmetrical (or have evolved from bilateral ancestors) and are relatively active.

Triploblastic animals are organized into several subgroups based on the presence or absence of a body cavity and, for those that possess one, the kind of body cavity present. A body cavity is a fluid-filled space in which the internal organs can be suspended and separated from the body wall. Body cavities are advantageous because they

1. Provide more room for organ development.
2. Provide more surface area for diffusion of gases, nutrients, and wastes into and out of organs.
3. Provide an area for storage.
4. Often act as hydrostatic skeletons.
5. Provide a vehicle for eliminating wastes and reproductive products from the body.
6. Facilitate increased body size.

Of these, the hydrostatic skeleton deserves further comment. Body-cavity fluids give support, while allowing the body to remain flexible. Hydrostatic skeletons can be illustrated with a water-filled balloon, which is rigid yet flexible. Because the water in the balloon is incompressible, squeezing one end causes the balloon to lengthen. Compressing both ends causes the middle of the balloon to become fatter. In a similar fashion, body-wall muscles, acting on coelomic fluid, are responsible for movement and shape changes in many animals.

The Triploblastic Acoelomate Pattern   Triploblastic
animals whose mesodermally derived tissues form a relatively solid mass of cells between ectodermally and endodermally derived tissues are called **acoelomate** (Gr. *a*, without + *kilos*, hollow) (figure 7.11*a*). Some cells between the ectoderm and endoderm of acoelomate animals are densely packed cells called parenchyma. Parenchymal cells are not specialized for a particular function.

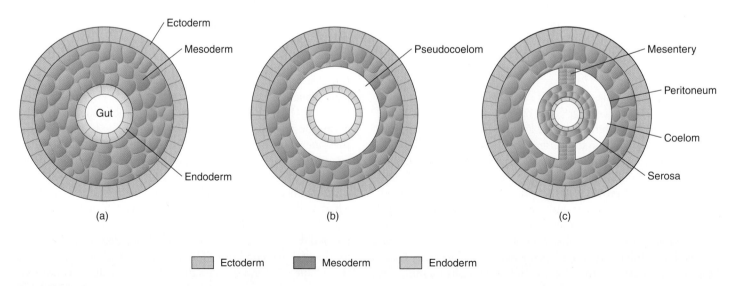

FIGURE 7.11

Triploblastic Body Plans.   Triploblastic animals have tissues derived from ectoderm, mesoderm, and endoderm. (*a*) Triploblastic acoelomate pattern. (*b*) Triploblastic pseudocoelomate pattern. Note the absence of mesodermal lining on the gut tract. (*c*) Triploblastic coelomate pattern. Mesodermally derived tissues completely surround the coelom.

The Triploblastic Pseudocoelomate Pattern A pseudo-coelom (Gr. *pseudes*, false) is a body cavity not entirely lined by mesoderm (figure 7.11*b*). No muscular or connective tissues are associated with the gut tract, no mesodermal sheet covers the inner surface of the body wall, and no membranes suspend organs in the body cavity.

The Triploblastic Coelomate Pattern A **coelom** is a body cavity completely surrounded by mesoderm (figure 7.11*c*). A thin mesodermal sheet, the peritoneum, lines the inner body wall and is continuous with the serosa, which lines the outside of visceral organs. The peritoneum and the serosa are continuous and suspend visceral structures in the body cavity. These suspending sheets are called mesenteries. Having mesodermally derived tissues, such as muscle and connective tissue, associated with internal organs enhances the function of virtually all internal body systems. The chapters that follow show many variations on the triploblastic coelomate pattern.

HIGHER ANIMAL TAXONOMY

Traditionally, the Animalia have been considered monophyletic (having a single ancestry) because of the impressive similarities in animal cellular organization. About 0.6 billion years ago, at the beginning of the Cambrian period, an evolutionary explosion occurred that resulted in the origin of all modern phyla (along with other animals that are now extinct) (*see Evolutionary Insights, page 109*). This rapid origin and diversification of animals is called "the Cambrian explosion" because it occurred over a relatively brief one hundred million–year period. This rapid appearance of all animal phyla leads some scientists to believe that animals may be polyphyletic because such a rapid divergence of many kinds of animals from a single ancestor seems unlikely. This idea remains a minority view. This textbook takes the traditional approach and assumes a monophyletic origin of the animals. The following brief description of higher taxonomy helps one to visualize possible relationships among animal phyla.

Taxonomic levels between kingdom and phylum are used to represent hypotheses of relatedness between animal phyla. Many zoologists recognize three major groups within the animal kingdom. These groups are often called branches and include the Mesozoa (phylum Mesozoa, the mesozoans), Parazoa (phylum Porifera, the sponges), and Eumetazoa (all other phyla). The Eumetazoa are further divided into two groups, Radiata and Bilateria, based on body symmetry and embryological characteristics (table 7.4). In this system, the bilaterally symmetrical animals are divided into two large groups, Protostomia and Deuterostomia, based on similarities and differences in the early development of the animals. The study of **comparative embryology** is based on the observation that embryological events may be similar because of shared ancestry. As with other comparative studies, however, embryologists must sort homologous developmental sequences from analogous developmental sequences (*see chapter 4*).

Protostomes traditionally include animals in the phyla Platyhelminthes, Nematoda, Mollusca, Annelida, Arthropoda, and others. Figure 7.12*a–d* shows the developmental characteristics

TABLE 7.4
TWO INTERPRETATIONS OF THE HIGHER TAXONOMY OF SELECTED ANIMAL PHYLA

PHYLUM	HIGHER TAXONOMIC DESIGNATION	
	Traditional Categories	*Alternate Categories*
Mesozoa	**Mesozoa**	
Porifera	**Parazoa**	**Diploblasts**
Cnidaria and Ctenophora	**Eumetazoa** Radiata	
Platyhelminthes and Nemertea	Bilateria Protostomia	**Lophotrochozoa**
Mollusca		
Annelida		
Arthropoda		**Ecdysozoa**
Nematoda		
Echinodermata	Deuterostomia	**Deuterostomia**
Hemichordata		
Chordata		

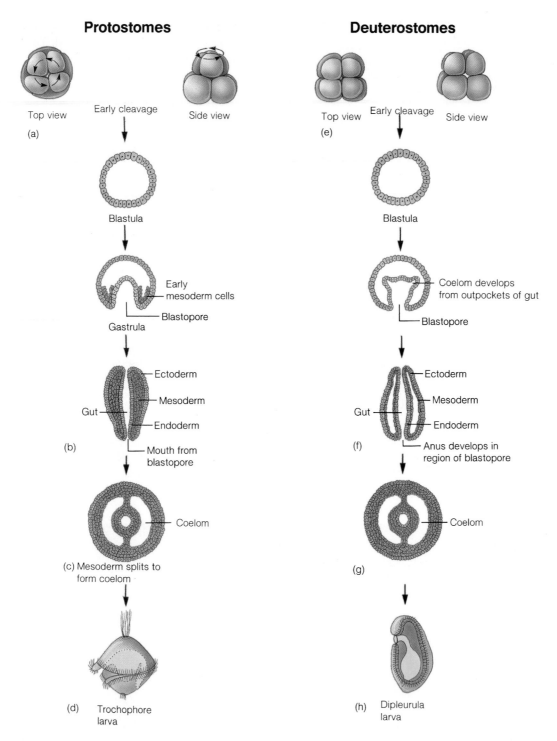

FIGURE 7.12

Developmental Characteristics of Protostomes and Deuterostomes.   Protostomes are characterized by (*a*) spiral and determinate cleavage, (*b*) a mouth that forms from an embryonic blastopore, (*c*) schizocoelous coelom formation, and (*d*) a trochophore larva. Deuterostomes are characterized by (*e*) radial and indeterminate cleavage, (*f*) an anus that forms in the region of the embryonic blastopore, and (*g*) enterocoelous coelom formation. A dipleurula larva is present in some echinoderms; however, this larval stage (*h*) is often absent in other deuterostomes.

that unite these phyla. One characteristic is the pattern of early cleavages of the zygote. In spiral cleavage, the mitotic spindle is oriented obliquely to the axis of the zygote. This orientation produces an eight-celled embryo in which the upper tier of cells is

twisted out of line with the lower cells. A second characteristic common to many protostomes is that early cleavage is determinate, meaning that the fate of the cells is established very early in development. If cells of the two- or four-celled embryo are

EVOLUTIONARY INSIGHTS
Animal Origins

The geological timescale is marked by significant geological and biological events, including the origin of the earth about 4.6 billion years ago, the origin of life about 3.5 billion years ago, the origin of eukaryotic life-forms about 1.5 billion years ago, and the origin of animals about 0.6 billion years ago. The latter event marks the beginning of the Cambrian period (*see table 4.1*). Animals originated relatively late in the history of life on earth. During a geologically brief 100-million-year period, all modern animal phyla (along with other animals that are now extinct) evolved. This rapid origin and diversification of animals is often referred to as the "Cambrian explosion."

Scientists have asked important questions about this explosion since Charles Darwin. Why did it occur so late in the history of the earth? The origin of multicellularity seems a relatively simple step compared to the origin of life itself. Why do few fossil records document the series of evolutionary changes during the evolution of the animal phyla? Why did animal life evolve so quickly? Paleontologists continue to search the fossil records for answers to these questions.

The Ediacaran fossil formation contains the oldest known animal fossils. Although it is named after a site in Australia, the Ediacaran formation is worldwide in distribution and dates to 600 million years before the present. This formation includes some modern phyla, including Porifera (sponges), Cnidaria (jellyfishes and their relatives), Mollusca (bivalves and their relatives), and Echinodermata (sea stars and their relatives). Many other animals are not clearly members of any extant phyla. The earliest Ediacaran fossils reveal animals that were radially symmetrical. Later fossils show the advent of bilateral symmetry. The animals of the Ediacara fossil formation are all soft-bodied forms. Even the molluscs and arthropod-like organisms had soft, nonmineralized skeletons. This fact may provide a clue to understanding the apparent sudden appearance of animal life. The earliest animals were probably so small and soft-bodied that fossilization was highly unlikely, and they left no record of their existence.

Another fossil formation provides evidence of the results of the Cambrian explosion. This fossil formation, called the Burgess Shale, is in Yoho National Park in the Canadian Rocky Mountains of British Columbia (box figure 7.1). Shortly after the Cambrian explosion, mudslides rapidly buried thousands of marine animals under conditions that favored fossilization. These fossil beds provide evidence of virtually all the 34 extant animal phyla, plus about 20 other animal body forms that are so different from any modern animals that they cannot be assigned to any one of the modern phyla. These unassignable animals include a large swimming predator called *Anomalocaris* and a

BOX FIGURE 7.1   The Burgess Shale.   An artist's reconstruction of the Burgess Shale is shown here. The Burgess Shale contained numerous unique forms of animal life as well as representatives of the animal phyla described in this textbook. A trilobite (Arthropoda) is shown on the lower left. Tall sponges (Porifera) are shown on the right and left in the foreground. *Sidneyia* (Arthropoda) is shown on the seafloor in the middle foreground and in the middle left.

soft-bodied detritus- or algae-eating animal called *Wiwaxia*. The Burgess Shale formation also has fossils of many extinct representatives of modern phyla. For example, a well-known Burgess Shale animal called *Sidneyia* is a representative of a previously unknown group of arthropods (insects, spiders, mites, crabs).

Fossil formations like the Ediacara and Burgess Shale show that evolution cannot always be thought of as a slow progression. The Cambrian explosion involved rapid evolutionary diversification, followed by the extinction of many unique animals. Why was this evolution so rapid? No one really knows. Many zoologists believe that it was because so many ecological niches were available with virtually no competition from existing species. Others suggest that the presence of mineralized skeletons and predatory lifestyles that mark the beginning of the Cambrian period promoted rapid animal evolution. Will zoologists ever know the evolutionary sequences in the Cambrian explosion? Perhaps another ancient fossil bed of soft-bodied animals from 600-million-year-old seas is awaiting discovery.

---

separated, none develops into a complete organism. Other characteristics of protostome development include the manner in which the embryonic gut tract and the coelom form. Many protostomes have a top-shaped larva, called a **trochophore larva.**

The other group, the **deuterostomes,** includes animals in the phyla Echinodermata, Hemichordata, Chordata, and others. Figure 7.12*e–h* shows the developmental characteristics that unite these phyla. Radial cleavage occurs when the mitotic spindle is oriented perpendicular to the axis of the zygote and results in embryonic cells directly over one another. Cleavage is indeterminate, meaning that the fate of blastomeres is determined late in development, and if embryonic cells are separated, they can develop into entire individuals. The manner of gut tract and coelom formation differs from that of protostomes. Some deuterostomes possess a kidney-bean-shaped larva called a dipleurula. There is, however, no single kind of deuterostome larval stage.

Even though these higher taxonomic groupings are traditional, a number of phyla (e.g., Mollusca, Nematoda, and others) do not fit well into either deuterostome or protostome groups. Cladistic analyses and molecular approaches to taxonomy are resulting in alternative groupings and are gaining widespread acceptance (*see table 7.4*). This textbook will rely on the traditional interpretations of higher taxonomy until a consensus opinion is established.

SUMMARY

1. Systematics is the study of the evolutionary history and classification of organisms. The binomial system of classification originated with von Linné and is used throughout the world in classifying organisms.
2. Organisms are classified into broad categories called kingdoms. The five-kingdom classification system used in recent years is being challenged as new information regarding evolutionary relationships among the monerans and protists is discovered.
3. The three modern approaches to systematics are evolutionary systematics, numerical taxonomy, and phylogenetic systematics (cladistics). The ultimate goal of systematics is to establish evolutionary relationships in monophyletic groups. Evolutionary systematists use homologies and rank the importance of different characteristics in establishing evolutionary relationships. These taxonomists take into consideration differing rates of evolution in taxonomic groups. Phylogenetic systematists (cladists) look for shared, derived characteristics that can be used to investigate evolutionary relationships. Cladists do not attempt to weigh the importance of different characteristics. Wide acceptance of cladistic methods has resulted in some nontraditional taxonomic groupings of animals.
4. Evolutionary-tree diagrams are useful for depicting evolutionary relationships, but their limitations must be understood.
5. The bodies of animals are organized into almost infinitely diverse forms. Within this diversity, however, are certain patterns of organization. Symmetry describes how the parts of an animal are arranged around a point or an axis.
6. Other patterns of organization reflect how cells associate into tissues, and how tissues organize into organs and organ systems.
7. Many zoologists recognize three higher taxonomic groups within the animal kingdom. These groups include the Mesozoa, Parazoa, and Eumetazoa. The Eumetazoa are further divided based on body symmetry and embryological characteristics. Recent cladistic analyses are resulting in reinterpretations of traditional higher groupings.

SELECTED KEY TERMS

bilateral symmetry (p. 104)
class (p. 97)
coelom (p. 107)
family (p. 97)
genus (p. 97)

kingdom (p. 97)
order (p. 97)
phylum (p. 97)
radial symmetry (p. 104)
species (p. 97)

CRITICAL THINKING QUESTIONS

1. In one sense, the animal classification system above the species level is artificial. In another sense, however, it is real. Explain this paradox.
2. Give proper scientific names to six hypothetical animal species. Assume that you have three different genera represented in your group of six. Be sure your format for writing scientific names is correct.
3. Describe hypothetical synapomorphies that would result in an assemblage of one order and two families (in addition to the three genera and six species from question 2).
4. Construct a cladogram, similar to that shown in figure 7.6, using your hypothetical animals from questions 2 and 3. Make drawings of your animals.
5. Describe the usefulness of evolutionary-tree diagrams in zoological studies. Describe two problems associated with their use.

ONLINE LEARNING CENTER

Visit our Online Learning Center (OLC) at www.mhhe.com/zoology (click on the book's title) to find the following chapter-related materials:

- CHAPTER QUIZZING
- ANIMATIONS
    Symmetry in Nature
- RELATED WEB LINKS
    Architectural Pattern and Diversity of Animals
    Classification and Phylogeny of Animals
    Theories of Evolutionary Patterns
- BOXED READINGS ON
    The Origin of Life on Earth—Life from Nonlife
    Hydrothermal Vent Communities
    Animal Origins—The Cambrian Explosion
    Jaws from the Past
    Collision or Coincidence?
- READINGS ON LESSER-KNOWN INVERTEBRATES
    The Lophophrates, Entoprocts, Cycliophores, and Chaetognaths
- SUGGESTED READINGS
- LAB CORRELATIONS

Check out the OLC to find specific information on these related lab exercises in the *General Zoology Laboratory Manual*, 5th edition, by Stephen A. Miller:

Exercise 7   *The Classification of Organisms*

CHAPTER 8

ANIMAL-LIKE PROTISTS:

THE PROTOZOA

Outline

Concepts

1. The kingdom Protista is a polyphyletic group with origins in ancestral Archaea.
2. Protozoa display unicellular or colonial organization. All functions must be carried out within the confines of a single plasma membrane.
3. According to the most widely accepted classification scheme, the seven protozoan phyla are Sarcomastigophora, Labyrinthomorpha, Apicomplexa, Microspora, Acetospora, Myxozoa, and Ciliophora.
4. Certain members of these phyla have had, and continue to have, important influences on human health and welfare.

EVOLUTIONARY PERSPECTIVE

Where are your "roots"? Although most people are content to go back into their family tree several hundred years, scientists look back millions of years to the origin of all life-forms. The fossil record indicates that virtually all protist and animal phyla living today were present during the Cambrian period, about 550 million years ago. Unfortunately, fossil evidence of the evolutionary pathways that gave rise to these phyla is scant. Instead, scientists gather evidence by examining the structure and function of living species. The "Evolutionary Perspective" sections in chapters 8 to 22 present hypotheses regarding the origin of protist and animal phyla. These hypotheses seem reasonable to most zoologists; however, they are untestable, and alternative interpretations are in the scientific literature.

Ancient members of the Archaea were the first living organisms on this planet. The Archaea and Eubacteria probably contributed to the origin of the kingdom Protista (also called Protoctista) about 1.5 billion years ago. The endosymbiont hypothesis is one of a number of explanations of how this could have occurred (see Evolutionary Insights, page 125). Most scientists agree that the protists probably arose from more than one ancestral Archaean group. Depending on the classification system, zoologists recognize between 7 and 45 phyla of protists. These phyla represent numerous evolutionary lineages. Groups of organisms believed to have had separate origins are said to be **polyphyletic** *(Gr. polys, many + phylon, race).* Some protists are plantlike because they are primarily autotrophic (they produce their own food). Others are animal-like because they are primarily heterotrophic (they feed on other organisms). This chapter covers seven phyla of protists commonly called the protozoa (figures 8.1, 8.2).

This chapter contains evolutionary concepts, which are set off in this font.

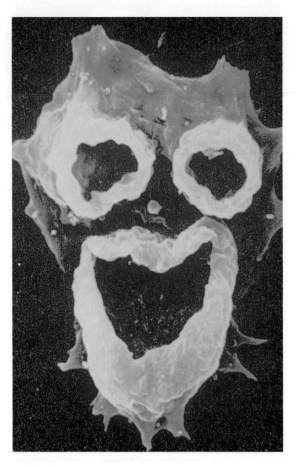

FIGURE 8.1

A Protozoan. *Naegleria fowleri,* the cause of primary amebic meningo-encephalitis in humans. The facelike structures attack and engulf food sources (SEM ×600).

LIFE WITHIN A SINGLE PLASMA MEMBRANE

Protozoa (Gr. *proto,* first + *zoa,* animal) display unicellular (cytoplasmic) organization, which does not necessarily imply that they are simple organisms. Often, they are more complex than any particular cell in higher organisms. In some protozoan phyla, individuals group to form colonies, associations of individuals that are not dependent on one another for most functions. Protozoan colonies, however, can become complex, with some individuals becoming so specialized that differentiating between a colony and a multicellular organism becomes difficult.

MAINTAINING HOMEOSTASIS

Organelles that are similar to the organelles of other eukaryotic cells carry out specific functions in protozoa (figure 8.3; *see also figure 2.2*). Some protozoan organelles, however, reflect specializations for unicellular lifestyles.

A regular arrangement of microtubules, called the **pellicle,** underlies the plasma membrane of many protozoa. The pellicle is rigid enough to maintain the shape of the protozoan, but it is also flexible.

The cytoplasm of a protozoan is differentiated into two regions. The portion of the cytoplasm just beneath the pellicle is called **ectoplasm** (Gr. *ectos,* outside + *plasma,* to form). It is relatively clear and firm. The inner cytoplasm, called **endoplasm** (Gr. *endon,* within), is usually granular and more fluid. The conversion of cytoplasm between these two states is important in one kind of protozoan locomotion and is discussed later in the chapter.

Most marine protozoa have solute concentrations similar to that of their environments. Freshwater protozoa, however, must regulate the water and solute concentrations of their cytoplasm. Water enters freshwater protozoa by osmosis because of higher solute concentrations in the protozoan than in the environment. **Contractile vacuoles** or **water expulsion vacuoles** remove this excess water (figure 8.3). In some protozoa, contractile vacuoles form by the coalescence of smaller vacuoles. In others, the vacuoles are permanent organelles that collecting tubules radiating into the cytoplasm fill. Contracting microfilaments (*see figure 2.20*) have been implicated in the emptying of contractile vacuoles.

Most protozoa absorb dissolved nutrients by active transport or ingest whole or particulate food through endocytosis (*see figure 2.14*). Some protozoa ingest food in a specialized region analogous to a mouth, called the **cytopharynx.** Digestion and transport of food occurs in **food vacuoles** that form during endocytosis. Enzymes and acidity changes mediate digestion. Food vacuoles fuse with enzyme-containing lysosomes and circulate through the cytoplasm, distributing the products of digestion. After digestion is complete, the vacuoles are called **egestion vacuoles.** They release their waste contents by exocytosis, sometimes at a specialized region of the plasma membrane or pellicle called the **cytopyge.**

Because protozoa are small, they have a large surface area in proportion to their volume (*see figure 2.3*). This high surface-area-to-volume ratio facilitates two other maintenance functions: gas exchange and excretion. Gas exchange involves acquiring oxygen for cellular respiration and eliminating the carbon dioxide produced as a by-product. Excretion is the elimination of the nitrogenous by-products of protein metabolism. The primary by-product in protozoa is ammonia. Both gas exchange and excretion occur by diffusion across the plasma membrane.

REPRODUCTION

Both asexual and sexual reproduction occur among the protozoa. One of the simplest and most common forms of asexual reproduction is **binary fission.** In binary fission, mitosis produces two nuclei that are distributed into two similar-sized individuals when the cytoplasm divides. During cytokinesis, some organelles duplicate to ensure that each new protozoan has the needed organelles to continue life. Depending on the group of protozoa, cytokinesis may be longitudinal or transverse (figures 8.4 and 8.5).

Other forms of asexual reproduction are common. During **budding,** mitosis is followed by the incorporation of one nucleus

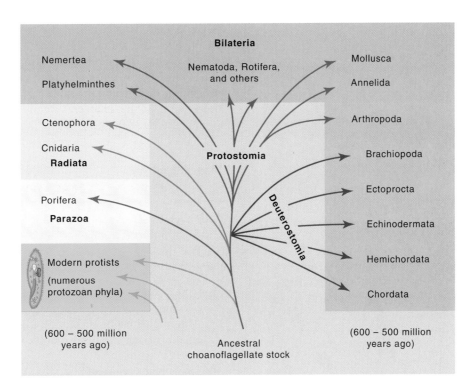

FIGURE 8.2

Animal-Like Protists: The Protozoa.   A generalized evolutionary tree depicting the major events and possible lines of descent for the protozoa (shaded in orange).

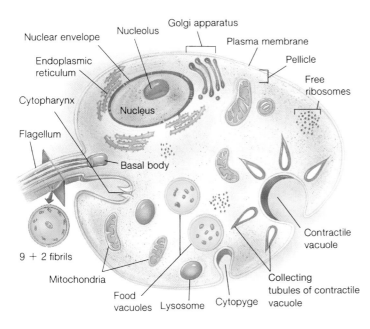

FIGURE 8.3

A Protozoan Protist.   This drawing of a stylized protozoan with a flagellum illustrates the basic protozoan morphology.   *From: "A LIFE OF INVERTEBRATES" © 1979 W. D. Russell-Hunter.*

into a cytoplasmic mass that is much smaller than the parent cell. **Multiple fission** or **schizogony** (Gr. *schizein*, to split) occurs when a large number of daughter cells form from the division of a single protozoan. Schizogony begins with multiple mitotic nuclear

divisions in a mature individual. When a certain number of nuclei have been produced, cytoplasmic division results in the separation of each nucleus into a new cell.

Sexual reproduction requires gamete formation and the subsequent fusion of gametes to form a zygote. In most protozoa, the sexually mature individual is haploid. Gametes are produced by mitosis, and meiosis follows the union of the gametes. Ciliated protozoa are an exception to this pattern. Specialized forms of sexual reproduction are covered as individual protozoan groups are discussed.

SYMBIOTIC LIFESTYLES

Many protozoa have symbiotic lifestyles. **Symbiosis** (Gr. *syn*, with + *bios*, life) is an intimate association between two organisms. For many protozoa, these interactions involve a form of symbiosis called **parasitism,** in which one organism lives in or on a second organism, called a **host.** The host is harmed but usually survives, at least long enough for the parasite to complete one or more life cycles.

The relationships between a parasite and its host(s) are often complex. Some parasites have life cycles involving multiple hosts. The **definitive host** harbors the sexual stages of the parasite. The sexual stages may produce offspring that enter another host, called an **intermediate host,** where they reproduce asexually. Some life cycles require more than one intermediate host and more than one immature stage. For the life cycle to be complete, the final, asexual stage must have access to a definitive host.

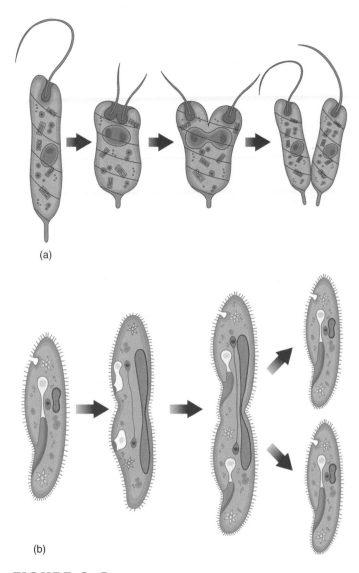

(a)

(b)

FIGURE 8.4

Asexual Reproduction in Protozoa. Binary fission begins with mitosis. Cytoplasmic division (cytokinesis) divides the organelles between the two cells and results in two similarly sized protozoa. Binary fission is (a) longitudinal in some protozoa (e.g., mastigophorans) and (b) transverse in other protozoa (e.g., ciliates).

Other kinds of symbiosis involve relationships that do not harm the host. **Commensalism** is a symbiotic relationship in which one member of the relationship benefits, and the second member is neither benefited nor harmed. **Mutualism** is a symbiotic relationship in which both species benefit.

PROTOZOAN TAXONOMY

Zoologists who specialize in the study of protozoa are called **protozoologists.** Most protozoologists now regard the protozoa as a subkingdom, consisting of seven separate phyla within the kingdom Protista. These are the phylum Sarcomastigophora, consisting of flagellates and amoebae with a single type of nucleus; the phyla Labyrinthomorpha, Apicomplexa, Microspora, Acetospora, and Myxozoa, consisting of either saprozoic or parasitic species; and the phylum Ciliophora, containing ciliated protozoa with two types of nuclei. The classification of this subkingdom into phyla is based primarily on types of nuclei, mode of reproduction, and mechanism of locomotion (table 8.1). These taxa are discussed further later in this chapter. Overall, the number of protozoan species exceeds 38,000.

PHYLUM SARCOMASTIGOPHORA

With over 18,000 described species, Sarcomastigophora (sar"ko-mas-ti-gof'o-rah) (Gr. *sarko*, fleshy + *mastigo*, whip + *phoros*, to bear) is the largest protozoan phylum and has the following characteristics:

1. Unicellular or colonial
2. Locomotion by flagella, pseudopodia, or both
3. Autotrophic (self-nourishing), saprozoic (living in decaying organic matter), or heterotrophic (obtains energy from organic compounds)
4. Single type of nucleus
5. Sexual or asexual reproduction

SUBPHYLUM MASTIGOPHORA: FLAGELLAR LOCOMOTION

Members of the subphylum Mastigophora (mas"ti-gof'o-rah) use flagella in locomotion. Flagella may produce two-dimensional, whiplike movements or helical movements that push or pull the protozoan through its aquatic medium.

CLASS PHYTOMASTIGOPHOREA

The subphylum Mastigophora has two classes. Members of the class Phytomastigophorea (fi"to-mas-ti-go-for-ee'-ah) (Gr. *phytos*, plant) possess chlorophyll and one or two flagella. Phytomastigophoreans produce a large portion of the food in marine food webs. Much of the oxygen used in aquatic habitats comes from photosynthesis by these marine organisms.

Marine phytomastigophoreans include the dinoflagellates (figure 8.6). Dinoflagellates have one flagellum that wraps around the organism in a transverse groove. The primary action of this flagellum causes the organism to spin on its axis. A second flagellum is a trailing flagellum that pushes the organism forward. In addition to chlorophyll, many dinoflagellates contain xanthophyll pigments, which give them a golden brown color. At times, dinoflagellates become so numerous that they color the water. Several genera, such as *Gymnodinium*, have representatives that produce toxins. Periodic "blooms" of these organisms are called "red tides" and result in fish kills along the continental shelves. Humans who consume tainted molluscs or fish may die. The Bible reports that the first plague Moses visited upon the Egyptians was a blood-red tide that killed fish and fouled water. Indeed, the Red Sea is probably named after these toxic dinoflagellate blooms.

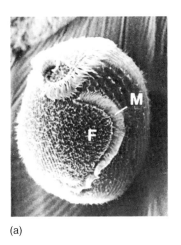

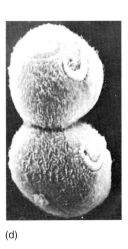

(a)　　　　　(b)　　　　　(c)　　　　　(d)

FIGURE 8.5

Binary Fission of the Ciliate *Stentor coeruleus*. (*a,b*) Fission includes the division of some surface features—in this case, cilia modified into a band-like structure called a membranelle (M). F designates the frontal field. The arrow in (*b*) shows the beginning of a fission furrow. (*c,d*) Fission is completed by division of the cytoplasm. (SEMs ×250.)

TABLE 8.1
SUMMARY OF PROTOZOAN CLASSIFICATION*

Kingdom Protista (pro-tees′ta) Single-celled eukaryotes.

Subkingdom Protozoa (pro″to-zo′ah) Animal-like protistans.

Phylum Sarcomastigophora (sar″ko-mas-ti-gof′o-rah)

Protozoa that possess flagella, pseudopodia, or both for locomotion and feeding; single type of nucleus. About 18,000 species.

Subphylum Mastigophora (mas″ti-gof′o-rah)

One or more flagella for locomotion; autotrophic, heterotrophic, or saprozoic.

Class Phytomastigophorea (fi″to-mas-ti-go-for-ee′ah)

Chloroplasts usually present; mainly autotrophic, some heterotrophic. *Euglena, Volvox, Chlamydomonas.*

Class Zoomastigophorea (zo″o-mas-ti-go-for-ee′ah)

Lack chloroplasts; heterotrophic or saprozoic. *Trypanosoma, Trichonympha, Trichomonas, Giardia, Leishmania.*

Subphylum Sarcodina (sar″ko-din′ah)

Pseudopodia for movement and food gathering; naked or with shell or test; mostly free living.

Superclass Rhizopoda (ri-zop′o-dah)

Lobopodia, filopodia, reticulopodia, or no distinct pseudopodia. *Amoeba, Entamoeba, Naegleria, Arcella, Difflugia;* foraminiferans (*Gumbelina*). About 4,000 species.

Superclass Actinopoda (ak″ti-nop′o-dah)

Spherical, planktonic; axopodia supported by microtubules; includes marine radiolarians with siliceous tests and freshwater heliozoans (*Actinophrys*). About 3,000 species.

Subphylum Opalinata (op″ah-li-not′ah)

Cylindrical; covered with cilia. *Opalina, Zelleriella.*

Phylum Labyrinthomorpha (la″brinth-o-mor′fa)

Trophic stage as ectoplasmic network with spindle-shaped or spherical, nonamoeboid cells; saprozoic and parasitic on algae and seagrass; mostly marine and estuarine. *Labyrinthula.*

Phylum Apicomplexa (a″pi-kom-plex′ah)

Parasitic with an apical complex used for penetrating host cells; cilia and flagella lacking, except in certain reproductive stages. The gregarines (*Monocystis*), coccidians (*Eimeria, Isospora, Sarcocystis, Toxoplasma*), *Plasmodium.* About 5,500 species.

Phylum Microspora (mi″cro-spor′ah)

Unicellular spores; intracellular parasites in nearly all major animal groups. The microsporeans (*Nosema*). About 850 species.

Phylum Acetospora (ah-sēt-o-spor′ah)

Multicellular spore; all parasitic in invertebrates. The acetosporans (*Paramyxa, Haplosporidium*).

Phylum Myxozoa (mix″o-zoo′ah)

Spores of multicellular origin; all parasitic. The myxozoans (*Myxosoma*). About 1,250 species.

Phylum Ciliophora (sil″i-of′or-ah)

Protozoa with simple or compound cilia at some stage in the life history; heterotrophs with a well-developed cytostome and feeding organelles; at least one macronucleus and one micronucleus present. *Paramecium, Stentor, Euplotes, Vorticella, Balantidium.* About 9,000 species.

*For many years the four main groups of protozoa were amoebae, ciliates, flagellates, and spore formers. The following taxonomy continues to follow the principles of evolutionary taxonomy rather than cladistic taxonomy. It should be noted that the true evolutionary relationships of many protozoan groups have yet to be elucidated, and some of the following taxa may be either para- or polyphyletic. The reason for this approach is that it is a compromise between reasonably current evolutionary thinking and the practical need for a system of nomenclatures that allows faculty and students to communicate with one another and retrieve information from the older literature.

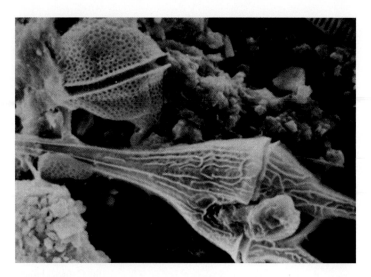

FIGURE 8.6

Class Phytomastigophorea: Dinoflagellates.   Two species of dinoflagellates—*Peridinium* in the upper left and *Ceratium* in the lower half. The transverse groove in the center of each dinoflagellate is the location for one of the two flagella (SEM ×800).

Euglena is a freshwater phytomastigophorean (figure 8.7). Each chloroplast has a **pyrenoid,** which synthesizes and stores polysaccharides. If cultured in the dark, euglenoids feed by absorption and lose their green color. Some euglenoids (e.g., *Peranema*) lack chloroplasts and are always heterotrophic.

Euglena orients toward light of certain intensities. A pigment shield **(stigma)** covers a photoreceptor at the base of the flagellum. The stigma permits light to strike the photoreceptor from only one direction, allowing *Euglena* to orient and move in relation to a light source.

Euglenoid flagellates are haploid and reproduce by longitudinal binary fission (*see figure 8.4a*). Sexual reproduction in these species is unknown.

Volvox is a colonial flagellate consisting of up to 50,000 cells embedded in a spherical, gelatinous matrix (figure 8.8a). Individual cells possess two flagella, which cause the colony to roll and turn gracefully through the water (figure 8.8b).

Although most *Volvox* cells are relatively unspecialized, reproduction depends on certain specialized cells. Asexual reproduction occurs in the spring and summer when certain cells withdraw to the watery interior of the parental colony and form daughter colonies. When the parental colony dies, it ruptures and releases daughter colonies.

Sexual reproduction in *Volvox* occurs during autumn. Some species are **dioecious** (having separate sexes); other species are **monoecious** (having both sexes in the same colony). In autumn, specialized cells differentiate into macrogametes or microgametes. Macrogametes are large, filled with nutrient reserves, and nonmotile. Microgametes form as a packet of flagellated cells that leaves the parental colony and swims to a colony containing macrogametes. The packet then breaks apart, and fertilization (syngamy) occurs between macro- and microgametes. The zygote, an overwintering stage, secretes a resistant wall around itself and

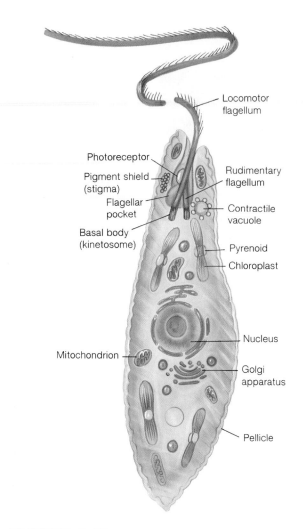

FIGURE 8.7

Phytomastigophorean Anatomy: The Structure of *Euglena*.   Note the large, well-organized chloroplasts. The photoreceptor allows the organism to swim toward light. The organism is about 50 μm long.

is released when the parental colony dies. Because the parental colony consists of haploid cells, the zygote must undergo meiosis to reduce the chromosome number from the diploid zygotic condition. One of the products of meiosis then undergoes repeated mitotic divisions to form a colony consisting of just a few cells. The other products of meiosis degenerate. This colony is released from the protective zygotic capsule in the spring.

CLASS ZOOMASTIGOPHOREA

Members of the class Zoomastigophorea (zo″o-mas-ti-go-for-ee′-ah) (Gr. *zoion,* animal) lack chloroplasts and are heterotrophic. Some members of this class are important parasites of humans.

One of the most important species of zoomastigophoreans is *Trypanosoma brucei.* This species is divided into three subspecies: *T. b. brucei, T. b. gambiense,* and *T. b. rhodesiense,* often referred to as the *Trypanosoma brucei* complex. The first of these three subspecies is a parasite of nonhuman mammals of Africa. The latter

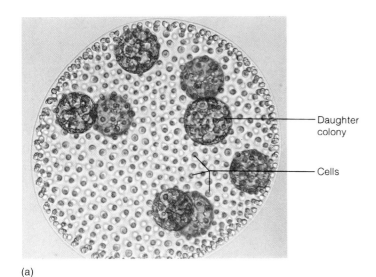

(a)

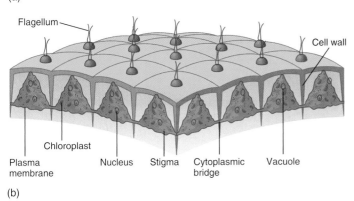

(b)

FIGURE 8.8

Class Phytomastigophorea: *Volvox,* a Colonial Flagellate. (*a*) A *Volvox* colony, showing asexually produced daughter colonies (LM ×400). (*b*) An enlargement of a portion of the colony wall.

two cause African sleeping sickness in humans. Tsetse flies (*Glossina* spp.) are intermediate hosts and vectors of all three subspecies. When a tsetse fly bites an infected human or mammal, it picks up parasites in addition to its meal of blood. Trypanosomes multiply asexually in the gut of the fly for about 10 days, then migrate to the salivary glands. While in the fly, trypanosomes transform, in 15 to 35 days, through a number of body forms. When the infected tsetse fly bites another vertebrate host, the parasites travel with salivary secretions into the blood of a new definitive host. The parasites multiply asexually in the new host and again transform through a number of body forms. Parasites may live in the blood, lymph, spleen, central nervous system, and cerebrospinal fluid (figure 8.9).

When trypanosomes enter the central nervous system, they cause general apathy, mental dullness, and lack of coordination. "Sleepiness" develops, and the infected individual may fall asleep during normal daytime activities. Death results from the pathology occurring in the nervous system, as well as from heart failure, malnutrition, and other weakened conditions. If detected early, sleeping sickness is curable. However, if an infection has advanced to the central nervous system, recovery is unlikely.

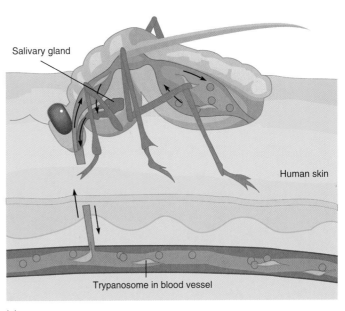

(a)

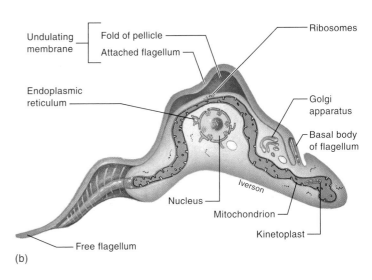

(b)

FIGURE 8.9

Class Zoomastigophorea: The Life Cycle of *Trypanosoma brucei.* (*a*) When a tsetse fly feeds on a vertebrate host, trypanosomes enter the vertebrate's circulatory system (first arrow on right) with the fly's saliva. Trypanosomes multiply in the vertebrate's circulatory and lymphatic systems by binary fission. When another tsetse fly bites this vertebrate host again, trypanosomes move into the gut of the fly and undergo binary fission. Trypanosomes then migrate to the fly's salivary glands, where they are available to infect a new host. (*b*) Structure of the flagellate, *Trypanosoma brucei rhodesiense.* This flagellate is about 25 μm long.

SUBPHYLUM SARCODINA: PSEUDOPODIA AND AMOEBOID LOCOMOTION

Members of the subphylum Sarcodina (sar″ko-din′ah) are the amoebae (sing., amoeba). When feeding and moving, they form temporary cell extensions called **pseudopodia** (sing., pseudopodium)

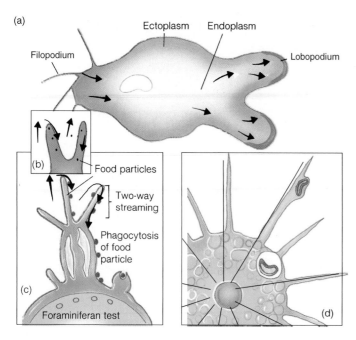

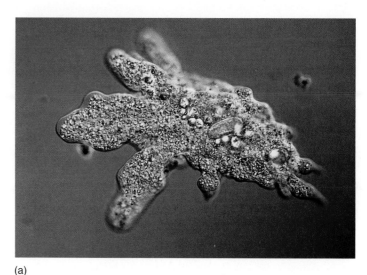

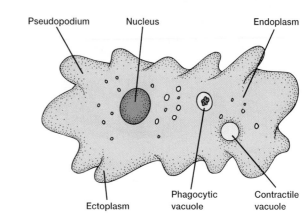

FIGURE 8.10

Variations in Pseudopodia. (*a*) Lobopodia of *Amoeba* contain both ectoplasm and endoplasm and are used for locomotion and engulfing food. (*b*) Filopodia of a shelled amoeba contain ectoplasm only and provide constant two-way streaming that delivers food particles to this protozoan in a conveyor-belt fashion. (*c*) Reticulopodia are similar to filopodia except that they branch and rejoin to form a netlike series of cell extensions. They occur in foraminiferans such as *Globigerina*. (*d*) Axopodia on the surface of a heliozoan such as *Actinosphaerium* deliver food to the central cytoplasm.

FIGURE 8.11

Subphylum Sarcodina: Superclass Rhizopoda, Class Lobosea. (*a*) *Amoeba proteus*, showing blunt lobopodia (LM ×160). (*b*) Anatomy of *Amoeba proteus*.

(Gr. *pseudes*, false + *podion*, little foot). Pseudopodia exist in a variety of forms. **Lobopodia** (sing., lobopodium) (Gr. *lobos*, lobe) are broad cell processes containing ectoplasm and endoplasm and are used for locomotion and engulfing food (figure 8.10*a*). **Filopodia** (sing., filopodium) (L. *filum*, thread) contain ectoplasm only and provide a constant two-way streaming that delivers food in a conveyor-belt fashion (figure 8.10*b*). **Reticulopodia** (sing., reticulopodium) (L. *reticulatus*, netlike) are similar to filopodia, except that they branch and rejoin to form a netlike series of cell extensions (figure 8.10*c*). **Axopodia** (sing., axopodium) (L. *axis*, axle) are thin, filamentous, and supported by a central axis of microtubules. The cytoplasm covering the central axis is adhesive and movable. Food caught on axopodia can be delivered to the central cytoplasm of the amoeba (figure 8.10*d*).

SUPERCLASS RHIZOPODA, CLASS LOBOSEA

The most familiar amoebae belong to the superclass Rhizopoda (ri-zop′o-dah), class Lobosea (lo-bo′sah) (Gr. *lobos*, lobe), and the genus *Amoeba* (figure 8.11). These amoebae are naked (they have no test or shell) and are normally found on shallow-water

substrates of freshwater ponds, lakes, and slow-moving streams, where they feed on other protists and bacteria. They engulf food by phagocytosis, a process that involves the cytoplasmic changes described earlier for amoeboid locomotion (*see figure 2.14*). In the process, food is incorporated into food vacuoles. Binary fission occurs when an amoeba reaches a certain size limit. As with other amoebae, no sexual reproduction is known to occur.

Other members of the superclass Rhizopoda possess a test (shell). **Tests** are protective structures that the cytoplasm secretes. They may be calcareous (made of calcium carbonate), proteinaceous (made of protein), siliceous (made of silica [SiO_2]), or chitinous (made of chitin—a polysaccharide). Other tests may be composed of sand or other debris cemented into a secreted matrix. Usually, one or more openings in the test allow pseudopodia to be extruded. *Arcella* is a common freshwater, shelled amoeba. It has a brown, proteinaceous test that is flattened on one side and domed on the other. Pseudopodia project from an opening on the flattened side. *Difflugia* is another

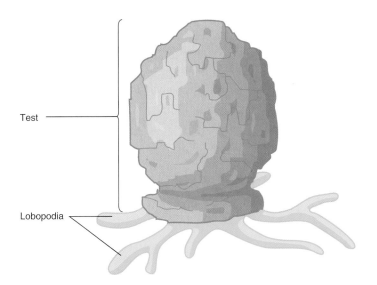

FIGURE 8.12

Subphylum Sarcodina. *Difflugia oblongata*, a common freshwater, shelled amoeba. The test consists of cemented mineral particles.

FIGURE 8.13

Subphylum Sarcodina: Foraminiferan Test (*Polystomella*). As this foraminiferan grows, it secretes new, larger chambers that remain attached to older chambers, making this protozoan look like a tiny snail (LM ×63).

common freshwater, shelled amoeba (figure 8.12). Its test is vase shaped and is composed of mineral particles embedded in a secreted matrix.

All free-living amoebae are particle feeders, using their pseudopodia to capture food; a few are pathogenic. For example, *Entamoeba histolytica* causes one form of dysentery in humans. Inflammation and ulceration of the lower intestinal tract and a debilitating diarrhea that includes blood and mucus characterize dysentery. Amoebic dysentery is a worldwide problem that plagues humans in crowded, unsanitary conditions.

A significant problem in the control of *Entamoeba histolytica* is that an individual can be infected and contagious without experiencing symptoms of the disease. Amoebae live in the folds of the intestinal wall, feeding on starch and mucoid secretions. They pass from one host to another in the form of cysts transmitted by fecal contamination of food or water. After ingestion by a new host, amoebae leave their cysts and take up residence in the host's intestinal wall, causing a multitude of problems.

SUPERCLASS ACTINOPODA: FORAMINIFERANS, HELIOZOANS, AND RADIOLARIANS

Foraminiferans (commonly called forams) are primarily a marine group of amoebae. Foraminiferans possess reticulopodia and secrete a test that is primarily calcium carbonate. As foraminiferans grow, they secrete new, larger chambers that remain attached to the older chambers (figure 8.13). Test enlargement follows a symmetrical pattern that may result in a straight chain of chambers or a spiral arrangement that resembles a snail shell. Many of these tests become relatively large; for example,

"Mermaid's pennies," found in Australia, may be several centimeters in diameter.

Foram tests are abundant in the fossil record since the Cambrian period. They make up a large component of marine sediments, and their accumulation on the floor of primeval oceans resulted in limestone and chalk deposits. The white cliffs of Dover in England are one example of a foraminiferan-chalk deposit. Oil geologists use fossilized forams to identify geologic strata when taking exploratory cores.

Heliozoans are aquatic amoebae that are either planktonic or live attached by a stalk to some substrate. (The plankton of a body of water consists of those organisms that float freely in the water.) Heliozoans are either naked or enclosed within a test that contains openings for axopodia (figure 8.14*a*).

Radiolarians are planktonic marine and freshwater amoebae. They are relatively large; some colonial forms may reach several centimeters in diameter. They possess a test (usually siliceous) of long, movable spines and needles or of a highly sculptured and ornamented lattice (figure 8.14*b*). When radiolarians die, their tests drift to the ocean floor. Some of the oldest known fossils of eukaryotic organisms are radiolarians.

PHYLUM LABYRINTHOMORPHA

The very small phylum Labyrinthomorpha (la"brinth-o-mor'fa) consists of protozoa with spindle-shaped, nonamoeboid, vegetative cells. In some genera, amoeboid cells use a typical gliding motion to move within a network of mucous tracks. Most members are marine, and either saprozoic or parasitic on algae or seagrass. Several years ago, *Labyrinthula* killed most of the "eel grass" (a grasslike marine flowering plant) on the Atlantic coast, starving many ducks that feed on the grass.

(a)

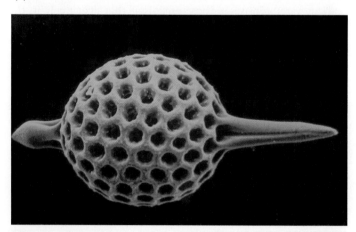

(b)

FIGURE 8.14

Subphylum Sarcodina: Heliozoan and Radiolarian Tests. (a) *Actinosphaerium sol* has a spherical body covered with fine, long axopodia made of numerous microtubules and surrounded by streaming cytoplasm. Following phagocytosis by an axopodium, waves of cytoplasmic movement carry trapped food particles into the main body of this protozoan (LM ×450). (b) The radiolarian *Spaerostylus* is typically spherical with a highly sculptured test (LM ×480).

PHYLUM APICOMPLEXA

Members of the phylum Apicomplexa (a″pi-kom-plex′ah) (L. *apex*, point + *com*, together, + *plexus*, interweaving) are all parasites. Characteristics of the phylum include:

1. Apical complex for penetrating host cells
2. Single type of nucleus
3. No cilia and flagella, except in certain reproductive stages
4. Life cycles that typically include asexual (schizogony, sporogony) and sexual (gametogony) phases

CLASS SPOROZOEA

The most important species in the phylum Apicomplexa are members of the class Sporozoea (spor″o-zo′e). The class name derives from most sporozoeans producing a resistant spore or oocyst following sexual reproduction. Some members of this class, including *Plasmodium* and coccidians, cause a variety of diseases in domestic animals and humans.

Although the life cycles of sporozoeans vary considerably, certain generalizations are possible. Many are intracellular parasites, and their life cycles have three phases. **Schizogony** is multiple fission of an asexual stage in host cells to form many more (usually asexual) individuals, called merozoites, that leave the host cell and infect many other cells. (Schizogony to produce merozoites is also called **merogony.**)

Some of the merozoites undergo **gametogony,** which begins the sexual phase of the life cycle. The parasite forms either microgametocytes or macrogametocytes. Microgametocytes undergo multiple fission to produce biflagellate microgametes that emerge from the infected host cell. The macrogametocyte develops directly into a single macrogamete. A microgamete fertilizes a macrogamete to produce a zygote that becomes enclosed in a membranous cyst called an oocyst.

The zygote undergoes meiosis, and the resulting cells divide repeatedly by mitosis. This process, called **sporogony,** produces many rodlike sporozoites in the oocyst. Sporozoites infect the cells of a new host after the new host ingests and digests the oocyst, or sporozoites are otherwise introduced (e.g., by a mosquito bite).

One sporozoean genus, *Plasmodium*, causes malaria and has a long recorded history of devastating effects on humans. Accounts of the disease go back as far as 1550 B.C. Malaria contributed significantly to the failure of the Crusades during the medieval era and, along with typhus, has devastated more armies than has actual combat. Recently (since the early 1970s), malaria has resurged throughout the world. Over 100 million humans are estimated to annually contract the disease.

The *Plasmodium* life cycle involves vertebrate and mosquito hosts (figure 8.15). Schizogony occurs first in liver cells and later in red blood cells, and gametogony also occurs in red blood cells. A mosquito takes in gametocytes during a meal of blood, and the gametocytes subsequently fuse. The zygote penetrates the gut of the mosquito and transforms into an oocyst. Sporogony forms haploid sporozoites that may enter a new host when the mosquito bites the host.

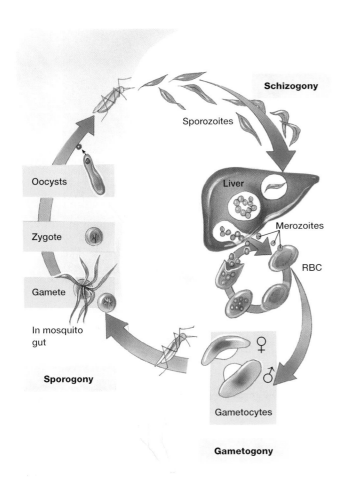

FIGURE 8.15

Phylum Apicomplexa: The Life Cycle of *Plasmodium*. Schizogony (merogony) occurs in liver cells and, later, in the red blood cells (RBCs) of humans. Gametogony occurs in RBCs. During a blood meal, the mosquito takes in micro- and macrogametes, which fuse to form zygotes. Zygotes penetrate the gut of the mosquito and form oocysts. Meiosis and sporogony form many haploid sporozoites that may enter a new host when the mosquito bites the host.

The symptoms of malaria recur periodically and are called paroxysms. Chills and fever correlate with the maturation of parasites, the rupture of red blood cells, and the release of toxic metabolites.

Four species of *Plasmodium* are the most important human malarial species. *P. vivax* causes malaria in which the paroxysms recur every 48 hours. This species occurs in temperate regions and has been nearly eradicated in many parts of the world. *P. falciparum* causes the most virulent form of malaria in humans. Paroxysms are more irregular than in the other species. *P. falciparum* was once worldwide, but is now mainly tropical and subtropical in distribution. It remains one of the greatest killers of humanity, especially in Africa. *P. malariae* is worldwide in distribution and causes malaria with paroxysms that recur every 72 hours. *P. ovale* is the rarest of the four human malarial species and is primarily tropical in distribution.

Other members of the class Sporozoea also cause important diseases. Coccidiosis is primarily a disease of poultry, sheep, cattle, and rabbits. Two genera, *Isospora* and *Eimeria*, are particularly im-

portant parasites of poultry. Yearly losses to the U.S. poultry industry from coccidiosis have approached $35 million. Another coccidian, *Cryptosporidium*, has become more well known with the advent of AIDS since it causes chronic diarrhea in AIDS patients, is the only known protozoan to resist chlorination, and is most virulent in immunosuppressed individuals. Toxoplasmosis is a disease of mammals, including humans, and birds. Sexual reproduction of *Toxoplasma* occurs primarily in cats. Infections occur when oocysts are ingested with food contaminated by cat feces, or when meat containing encysted merozoites is eaten raw or poorly cooked. Most infections in humans are asymptomatic, and once infection occurs, an effective immunity develops. However, if a woman is infected near the time of pregnancy, or during pregnancy, congenital toxoplasmosis may develop in a fetus. Congenital toxoplasmosis is a major cause of stillbirths and spontaneous abortions. Fetuses that survive frequently show signs of mental retardation and epileptic seizures. Congenital toxoplasmosis has no cure. Toxoplasmosis also ranks high among the opportunistic diseases afflicting AIDS patients. Steps to avoid infections by *Toxoplasma* include keeping stray and pet cats away from children's sandboxes; using sandbox covers; and awareness, on the part of couples considering having children, of the potential dangers of eating raw or very rare pork, lamb, and beef.

PHYLUM MICROSPORA

Members of the phylum Microspora (mi"cro-spor'ah), commonly called microsporidia, are small, obligatory intracellular parasites. Included in this phylum are several species that parasitize beneficial insects. *Nosema bombicus* parasitizes silkworms (figure 8.16), causing the disease **pebrine,** and *N. apis* causes serious dysentery (foul brood) in honeybees. These parasites have a possible role as biological control agents for insect pests. For example, the U.S. Environmental Protection Agency has approved and registered *N. locustae* for use in residual control of rangeland grasshoppers. Recently, four microsporidian genera have been implicated in secondary infections of immunosuppressed and AIDS patients.

PHYLUM ACETOSPORA

Acetospora (ah-sēt-o-spor'ah) is a relatively small phylum that consists exclusively of obligatory extracellular parasites characterized by spores lacking polar caps or polar filaments. The acetosporeans (e.g., *Haplosporidium*) primarily are parasitic in the cells, tissues, and body cavities of molluscs.

PHYLUM MYXOZOA

The phylum Myxozoa (myx"o-zoo'ah), commonly called myxosporeans, are all obligatory extracellular parasites in freshwater and marine fish. They have a resistant spore with one to six coiled polar filaments. The most economically important myxosporean is *Myxosoma cerebralis*, which infects the nervous system

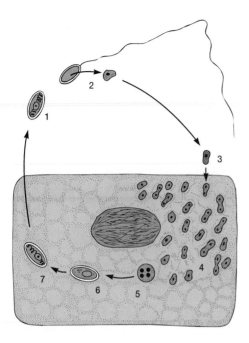

FIGURE 8.16

Phylum Microspora: The Microsporean *Nosema bombicus,* Which Is Fatal to Silkworms.   The cell of a silkworm infected with *N. bombicus* is shown here. (*1*) A typical spore with one coiled filament. (*2*) When ingested, the spore extrudes the filament, which is used in locomotion. (*3*) The parasite enters an epithelial cell in the silkworm intestine and (*4*) divides many times to form small amoebae that eventually fill the cell and kill it. (*5–7*) During this phase, some of the amoebae with four nuclei become spores. Silkworms are infected by eating leaves contaminated with the feces of infected worms.

and auditory organs of trout and salmon, causing whirling or tumbling disease.

PHYLUM CILIOPHORA

The phylum Ciliophora (sil″i-of′or-ah) includes some of the most complex protozoa (*see table 8.1*). Ciliates are widely distributed in freshwater and marine environments. A few ciliates are symbiotic. Characteristics of the phylum Ciliophora include:

1. Cilia for locomotion and for the generation of feeding currents in water
2. Relatively rigid pellicle and more or less fixed shape
3. Distinct cytostome (mouth) structure
4. Dimorphic nuclei, typically a larger macronucleus and one or more smaller micronuclei

CILIA AND OTHER PELLICULAR STRUCTURES

Cilia are generally similar to flagella, except that they are much shorter, more numerous, and widely distributed over the surface of the protozoan (figure 8.17). Ciliary movements are coordinated so that ciliary waves pass over the surface of the ciliate. Many ciliates

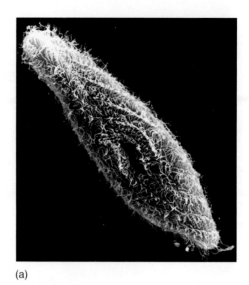

(a)

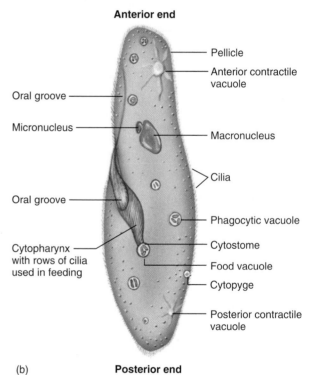

FIGURE 8.17

Phylum Ciliophora.   (*a*) The ciliate, *Paramecium sonneborn*. This paramecium is 40 μm in length. Note the oral groove near the middle of the body that leads into the cytopharynx (SEM ×1,600). (*b*) The structure of a typical ciliate such as *Paramecium.*

can reverse the direction of ciliary beating and the direction of cell movement.

Basal bodies (kinetosomes) of adjacent cilia are interconnected with an elaborate network of fibers that are believed to anchor the cilia and give shape to the organism.

Some ciliates have evolved specialized cilia. Cilia may cover the outer surface of the protozoan. They may join to form **cirri,** which are used in movement. Alternatively, cilia may be lost from large regions of a ciliate.

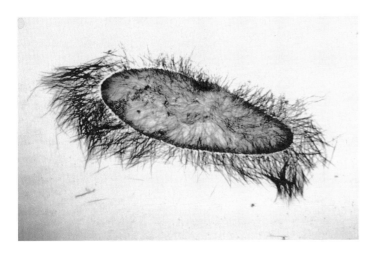

FIGURE 8.18

Discharged Trichocysts of *Paramecium*. Each trichocyst transforms itself into a long, sticky, proteinaceous thread when discharged (LM ×150).

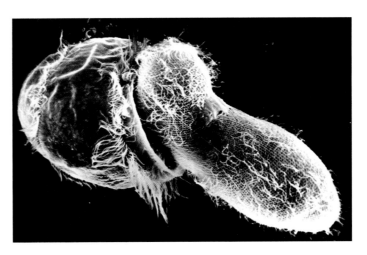

FIGURE 8.19

A Single-Celled Hunter and Its Prey. The juglike *Didinium* (left) swallowing a slipper-shaped *Paramecium* (right) (SEM ×550).

Trichocysts are pellicular structures primarily used for protection. They are rodlike or oval organelles oriented perpendicular to the plasma membrane. In *Paramecium*, they have a "golf tee" appearance. The pellicle can discharge trichocysts, which then remain connected to the body by a sticky thread (figure 8.18).

NUTRITION

Some ciliates, such as *Paramecium*, have a ciliated oral groove along one side of the body (*see* figure 8.17). Cilia of the oral groove sweep small food particles toward the end of the cytopharynx, where a food vacuole forms. When the food vacuole reaches an upper size limit, it breaks free and circulates through the endoplasm. Indigestible material is voided either through a temporary opening or through a permanent cytopyge which is found in many ciliates.

Some free-living ciliates prey upon other protists or small animals. Prey capture is usually a case of fortuitous contact. The ciliate *Didinium* feeds principally on *Paramecium*, a prey that is bigger than itself. *Didinium* forms a temporary opening that can greatly enlarge to consume its prey (figure 8.19).

Suctorians are ciliates that live attached to their substrate. They possess tentacles whose secretions paralyze prey, often ciliates or amoebae. The tentacles digest an opening in the pellicle of the prey, and prey cytoplasm is drawn into the suctorian through tiny channels in the tentacle. The mechanism for this probably involves tentacular microtubules (figure 8.20).

GENETIC CONTROL AND REPRODUCTION

Ciliates have two kinds of nuclei. A large, polyploid **macronucleus** regulates daily metabolic activities. One or more smaller **micronuclei** are the genetic reserve of the cell.

Ciliates reproduce asexually by transverse binary fission and, occasionally, by budding. Budding occurs in suctorians and results

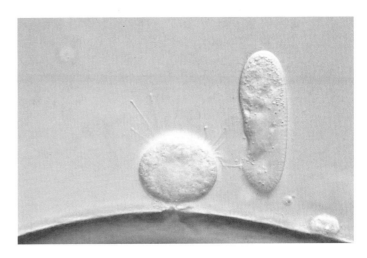

FIGURE 8.20

Suctorian (*Tokophrya* spp.) Feeding. The knobbed tip of a tentacle holds a ciliate (right). Tentacles discharge enzymes that immobilize the prey and dissolve the pellicle. Pellicles of tentacle and prey fuse, and the tentacle enlarges and invaginates to form a feeding channel. Prey cytoplasm moves down the feeding channel and incorporates into endocytic vacuoles at the bottom of the tentacle (LM ×181).

in the formation of ciliated, free-swimming organisms that attach to the substrate and take the form of the adult.

Ciliates reproduce sexually by **conjugation** (figure 8.21). The partners involved are called conjugants. Many species of ciliates have numerous mating types, not all of which are mutually compatible. Initial contact between individuals is apparently random, and sticky secretions of the pellicle facilitate adhesion. Ciliate plasma membranes then fuse and remain that way for several hours.

The macronucleus does not participate in the genetic exchange that follows. Instead, the macronucleus breaks up during or after micronuclear events, and reforms from micronuclei of the daughter ciliates.

After separation, the exconjugants undergo a series of nuclear divisions to restore the nuclear characteristics of the particular

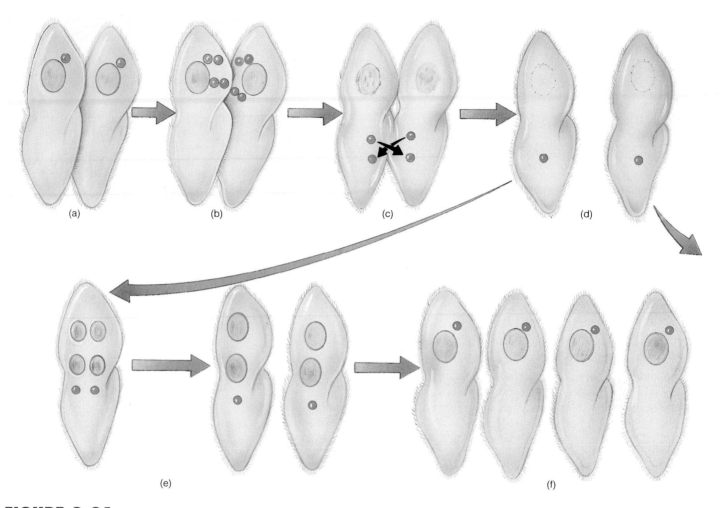

FIGURE 8.21

Conjugation in *Paramecium*. (*a*) Random contact brings individuals of opposite mating types together. (*b*) Meiosis results in four haploid pronuclei. (*c*) Three pronuclei and the macronucleus degenerate. Mitosis and mutual exchange of pronuclei is followed by fusion of pronuclei. (*d–f*) Conjugants separate. Nuclear divisions that restore nuclear characteristics of the species follow. Cytoplasmic divisions may accompany these events.

species, including the formation of a macronucleus from one or more micronuclei. Cytoplasmic divisions that form daughter cells accompany these events.

SYMBIOTIC CILIATES

Most ciliates are free living; however, some are commensalistic or mutualistic, and a few are parasitic. *Balantidium coli* is an important parasitic ciliate that lives in the large intestines of humans, pigs, and other mammals. At times, it is a ciliary feeder; at other times, it produces proteolytic enzymes that digest host epithelium, causing a flask-shaped ulcer. (Its pathogenicity resembles that of *Entamoeba histolytica*.) *B. coli* is passed from one host to another in cysts that form as feces begin to dehydrate in the large intestine. Fecal contamination of food or water is the most common form of transmission. Its distribution is potentially worldwide, but it is most common in the Philippines.

Large numbers of different species of ciliates also inhabit the rumen of many ungulates (hoofed animals). These ciliates contribute to the digestive processes of their hosts.

FURTHER PHYLOGENETIC CONSIDERATIONS

Protozoa probably originated about 1.5 billion years ago. Although known fossil species exceed 30,000, they are of little use in investigations of the origin and evolution of the various protozoan groups. Only protozoa with hard parts (tests) have left much of a fossil record, and only the foraminiferans and radiolarians have well-established fossil records in Precambrian rocks. Recent evidence from the study of base sequences in ribosomal RNA indicates that each of the seven protozoan phyla probably had separate origins and that each is sufficiently different from the others to warrant elevating all of these groups to phylum status, as in this chapter (figure 8.22). Additional modifications to the present scheme of protozoan classification are continually being proposed as the results of new ultrastructural and molecular studies are published. For example, in 1993, T. Cavalier-Smith proposed that the protozoa be elevated to kingdom status with 18 phyla. Whether protozoologists will accept this new classification, however, remains to be determined. (*See the footnote at the end of table 8.1.*)

EVOLUTIONARY INSIGHTS
The Animal-Like Protists May Lie at the Crossroads
Between the Simpler and the Complex

Between the unicellular microorganisms (Eubacteria and Archaea) and the multicellular eukaryotes lies the kingdom Protista. The protists may represent a bridge from simple to complex life-forms. As noted in this chapter, most Protista are single eukaryotic cells that lack mitochondria and provide some insight as to what the earliest eukaryotes might have been like.

Along these lines, protists are of interest to evolutionary biologists because current organisms may retain clues to important milestones in eukaryotic evolution. For example, a group of protists called jakobids have mitochondria that resemble bacteria more than those of any other type of eukaryote. Therefore, jackobids may resemble those microorganisms that lived shortly after cells acquired aerobic bacteria as endosymbionts (*see box in chapter 2*). At the other end of the evolutionary spectrum are the choanoflagellates (a group of free-living zooflagellates

found primarily in freshwater). Many choanoflagellate species are sessile, being permanently attached to a substrate (box figure 8.1). Each individual bears a single flagellum that bears an uncanny resemblance to the "collar cells" of sponges (*see figure 9.5*)—the simplest animals. Commonly, individuals are stalked and/or embedded in a gelatinous secretion. Most species are colonial and immobile. Members of the genus *Proterospongia* form (planktonic) colonies of up to several hundred cells and bear a striking resemblance to primitive sponges. Whether this simple relationship reflects a true phylogenetic relationship and crossroads between the unicellular flagellates and the more complex multicellular sponges or whether the similarities are a product of independent, convergent evolution is not certain. Definitive answers will have to await nucleic acid sequencing, which provides a more objective measure of relatedness than comparing possible superficial appearances.

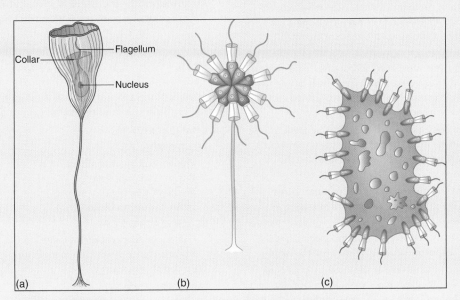

BOX FIGURE 8.1 **Zooflagellate Diversity.** Choanoflagellates: (*a*) *Stephanoeca*. (*b*) *Codosiga*, a colonial species. (*c*) *Proterospongia*, another colonial species, with individuals embedded in a thick, gelatinous matrix.

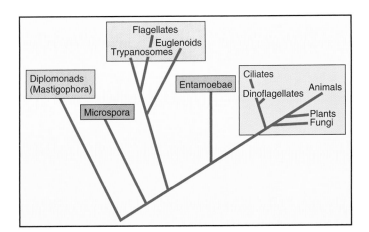

FIGURE 8.22

Cladogram Showing the Evolutionary Relationships of Protozoa and Other Eukaryotes Based on 18S Ribosomal RNA Sequence Comparisons. This cladogram suggests that evolution along the nuclear line of descent was not continuous but instead occurred in major epochs (an epoch is a particular time period marked by distinctive features and events). Five major evolutionary radiations (colored boxes) are apparent for the protozoa. The Mastigophora (e.g., *Giardia*) and Microspora (e.g., *Nosema*) are modern relatives of the earliest major eukaryotic cell lines. Following the development of these protozoa, the other groups of protozoa radiated off the nuclear line of descent.

SUMMARY

1. The kingdom Protista is a polyphyletic group that arose about 1.5 billion years ago from the Archaea. The evolutionary pathways leading to modern protozoa are uncertain.

2. Protozoa are both single cells and entire organisms. Organelles specialized for the unicellular lifestyle carry out many protozoan functions.

3. Many protozoa live in symbiotic relationships with other organisms, often in a host-parasite relationship.

4. Members of the phylum Sarcomastigophora possess pseudopodia and/or one or more flagella.

5. Members of the class Phytomastigophorea are photosynthetic and include the genera *Euglena* and *Volvox*. Members of the class Zoomastigophorea are heterotrophic and include *Trypanosoma*, which causes sleeping sickness.

6. Amoebae use pseudopodia for feeding and locomotion.

7. Members of the subphylum Sarcodina include the freshwater genera *Amoeba*, *Arcella*, and *Difflugia*, and the parasitic genus *Entamoeba*. Foraminiferans and radiolarians are common marine amoebae.

8. Members of the phylum Apicomplexa are all parasites. The phylum includes *Plasmodium* and *Toxoplasma*, which cause malaria and toxoplasmosis, respectively. Many apicomplexans have a three-part life cycle involving schizogony, gametogony, and sporogony.

9. The phylum Microspora consists of small protozoa that are intracellular parasites of every major animal group. They are transmitted from one host to the next as a spore, the form from which the group obtains its name.

10. The phylum Acetospora contains protozoa that produce spores lacking polar capsules. These protozoa are primarily parasitic in molluscs.

11. The phylum Myxozoa consists entirely of parasitic species, usually found in fishes. One to six polar filaments characterize the spore.

12. The phylum Ciliophora contains some of the most complex of all protozoa. Its members possess cilia, a macronucleus, and one or more micronuclei. Mechanical coupling of cilia coordinates their movements, and cilia can be specialized for different kinds of locomotion. Ciliates reproduce sexually by conjugation. Diploid ciliates undergo meiosis of the micronuclei to produce haploid pronuclei that two conjugants can exchange.

13. Precise evolutionary relationships are difficult to determine for the protozoa. The fossil record is sparse, and what does exist is not particularly helpful in deducing relationships. However, ribosomal RNA sequence comparisons indicate that each of the seven protozoan phyla probably had separate origins.

SELECTED KEY TERMS

ectoplasm (p. 112)	multiple fission (schizogony) (p. 113)
endoplasm (p. 112)	pellicle (p. 112)
host (p. 113)	protozoa (p. 112)
macronucleus (p. 123)	protozoologists (p. 114)
micronuclei (p. 123)	trichocysts (p. 123)

CRITICAL THINKING QUESTIONS

1. If knowing for certain the evolutionary pathways that gave rise to protozoa and animal phyla is impossible, is it worth constructing hypotheses about those relationships? Why or why not?

2. In what ways are protozoa similar to animal cells? In what ways are they different?

3. If sexual reproduction is unknown in *Euglena*, how do you think this lineage of organisms has survived through evolutionary time? (Recall that sexual reproduction provides the genetic variability that allows species to adapt to environmental changes.)

4. The use of DDT has been greatly curtailed for ecological reasons. In the past, it has proved to be an effective malaria deterrent. Many organizations would like to see this form of mosquito control resumed. Do you agree or disagree? Explain your reasoning.

5. If you were traveling out of the country and were concerned about contracting amoebic dysentery, what steps could you take to avoid contracting the disease? How would the precautions differ if you were going to a country where malaria is a problem?

ONLINE LEARNING CENTER

Visit our Online Learning Center (OLC) at www.mhhe.com/zoology (click on this book's title) to find the following chapter-related materials:

- CHAPTER QUIZZING
- RELATED WEB LINKS
  Phylum Sarcomastigophora
  Phylum Apicomplexa
  Phylum Ciliophora
  Other Protozoan Phyla
- BOXED READINGS ON
  Giardiasis: "Backpacker's Disease" in the Rocky Mountains
  Malaria Control—A Glimmer of Hope
- SUGGESTED READINGS
- LAB CORRELATIONS

Check out the OLC to find specific information on these related lab exercises in the *General Zoology Laboratory Manual*, 5th edition, by Stephen A. Miller:

Exercise 8   *Animal-like Protists*

CHAPTER 9

MULTICELLULAR AND TISSUE LEVELS OF ORGANIZATION

Outline

Concepts

1. How multicellularity originated in animals is largely unknown. Most zoologists consider the animal kingdom to be monophyletic, having arisen from ancestral choanoflagellate protists.
2. Animals whose bodies consist of aggregations of cells, but whose cells do not form tissues, are in the phylum Porifera, as well as some lesser known phyla.
3. Animals that show diploblastic, tissue-level organization are in the phylum Cnidaria.
4. Members of the phylum Cnidaria are important in zoological research because of their relatively simple organization and their contribution to coral reefs.
5. Members of the phylum Ctenophora are called sea walnuts or comb jellies. They are diploblastic or triploblastic and biradially symmetrical.

EVOLUTIONARY PERSPECTIVE

Animals with multicellular and tissue levels of organization have captured the interest of scientists and laypersons alike. A description of some members of the phylum Cnidaria, for example, could fuel a science fiction writer's imagination:

> From a distance I was never threatened, in fact I was infatuated with its beauty. A large, inviting, bright blue float lured me closer. As I swam nearer I could see that hidden from my previous view was an infrastructure of tentacles, some of which dangled nearly nine meters below the water's surface! The creature seemed to consist of many individuals and I wondered whether or not each individual was the same kind of being because, when I looked closely, I counted eight different body forms!
>
> I was drawn closer and the true nature of this creature was painfully revealed. The beauty of the gasfilled float hid some of the most hideous weaponry imaginable. When I brushed against those silky tentacles I experienced the most excruciating pain. Had it not been for my life vest, I would have drowned. Indeed, for some time, I wished that had been my fate.

Swimmers of tropical waters who have come into contact with *Physalia physalis*, the Portuguese man-of-war, know that this fictitious account rings true (figure 9.1). In organisms such as *Physalia physalis*, cells are grouped, specialized for various functions, and interdependent. This chapter covers three animal phyla whose multicellular organization varies from an association of cells lacking tissue organization to cells organized into two

This chapter contains evolutionary concepts, which are set off in this font.

FIGURE 9.1

Physalia physalis, **the Portuguese Man-of-War.**   The tentacles shown here can be up to 9 m long and are laden with nematocysts that are lethal to small vertebrates and dangerous to humans. A bluish float at the surface of the water is about 12 cm long. It is not shown in this photograph. The entire organism is actually a colony of polypoid and medusoid individuals.

distinct tissue layers. These phyla are the Porifera, Cnidaria, and Ctenophora. The Porifera are members of the branch Parazoa (*see table 7.4*) and, although not simple, are evolutionarily the most primitive animals. The Cnidaria and Ctenophora, by virtue of the presence of embryological tissue layers, are the first animals covered in this textbook that are members of the branch Eumetazoa. The radial symmetry of these latter two phyla is often used to group them into a division of the Eumetazoa called Radiata (figure 9.2).

ORIGINS OF MULTICELLULARITY

Multicellular life has been a part of the earth's history for approximately 550 million years. Although this seems a very long time, it represents only 10% of the earth's geological history.

Multicellular life arose quickly in the 100 million years prior to the Precambrian/Cambrian boundary, in what scientists view as an evolutionary explosion. These evolutionary events resulted not only in the appearance of all of the animal phyla recognized today, but also 15 to 20 animal groups that are now extinct. Since this initial evolutionary explosion, most of the history of multicellular life has been one of extinction.

The evolutionary events leading to multicellularity are shrouded in mystery. Many zoologists believe that multicellularity could have arisen as dividing cells remained together, in the fashion of many colonial protists. Although variations of this hypothesis exist, they are all treated here as the **colonial hypothesis** (figure 9.3*a*).

A second proposed mechanism is called the **syncytial hypothesis** (figure 9.3*b*). A syncytium is a large, multinucleate cell. The formation of plasma membranes in the cytoplasm of a syncytial protist could have produced a small, multicellular organism. Both the colonial and syncytial hypotheses are supported by the colonial and syncytial organization that occurs in some protist phyla.

ANIMAL ORIGINS

Figure 9.2 depicts the animal kingdom as being monophyletic—derived from a common ancestor. This hypothesis is supported by impressive similarities within the Animalia as regards certain cellular structures that are common to animals. The presence of flagellated cells, especially monoflagellated cells, is characteristic of animals. Asters (*see figure 3.5*) form during mitosis in most animals, certain cell junctions are similar in all animal cells, and the proteins that accomplish movement are similar in most animal cells.

Not only is the animal kingdom most likely monophyletic, but the most likely candidate group of ancestral protists has also been identified. The choanoflagellates are a group of protists that possess a basket-like collar surrounding the base of a flagellum that is used in feeding. These cells are virtually identical to one kind of sponge cell used in feeding—the choanocyte. Evidence from molecular biology also supports the hypothesis that animals arose from ancestral choanoflagellate stock.

PHYLUM PORIFERA

The Porifera (po-rif'er-ah) (L. *porus*, pore + *fera*, to bear), or sponges, are primarily marine animals consisting of loosely organized cells (figure 9.4; table 9.1). The approximately nine thousand species of sponges vary in size from less than a centimeter to a mass that would more than fill your arms.

Characteristics of the phylum Porifera include:

1. Asymmetrical or superficially radially symmetrical
2. Three cell types: pinacocytes, mesenchyme cells, and choanocytes
3. Central cavity, or a series of branching chambers, through which water circulates during filter feeding
4. No tissues or organs

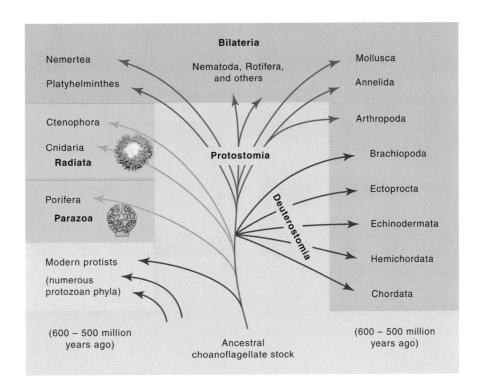

FIGURE 9.2

Evolutionary Relationships of the Poriferans and the Radiate Phyla.   Members of the phylum Porifera are derived from ancestral choanoflagellate stocks. The Radiata include the Cnidaria and Ctenophora. Evidence for a common protistan ancestry for all animals comes from the common elements of cellular organization and from molecular biology.

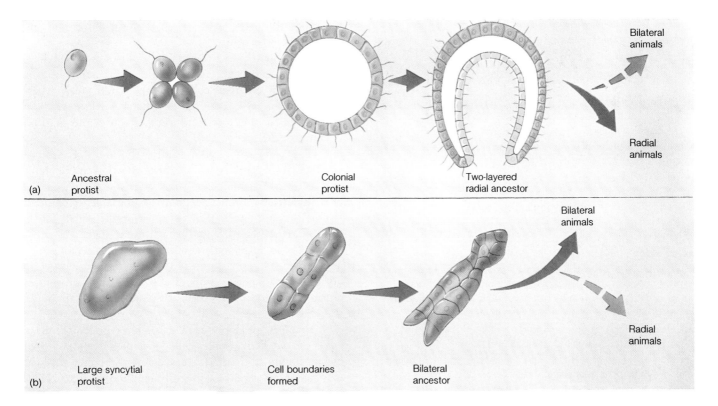

FIGURE 9.3

Two Hypotheses Regarding the Origin of Multicellularity.   (*a*) The colonial hypothesis. Multicellularity may have arisen when cells that a dividing protist produced remained together. Cell invagination could have formed a second cell layer. This hypothesis is supported by the colonial organization of some Sarcomastigophora. (The colonial protist and the two-layered radial ancestor are shown in sectional views.) (*b*) The syncytial hypothesis. Multicellularity could have arisen when plasma membranes formed within the cytoplasm of a large, multinucleate protist. Multinucleate, bilateral ciliates support this hypothesis.

(a)

(b)

FIGURE 9.4

Phylum Porifera. Many sponges are brightly colored with hues of red, orange, green, or yellow. (*a*) *Verongia* sp. (*b*) *Axiomella* sp.

CELL TYPES, BODY WALL, AND SKELETONS

In spite of their relative simplicity, sponges are more than colonies of independent cells. As in all animals, sponge cells are specialized

---

TABLE 9.1
CLASSIFICATION OF THE PORIFERA

Phylum Porifera (po-rif′er-ah)*
The animal phylum whose members are sessile and either asymmetrical or radially symmetrical; body organized around a system of water canals and chambers; cells not organized into tissues or organs. Approximately 9,000 species.

Class Calcarea (kal-kar′e-ah)
Spicules composed of calcium carbonate; spicules are needle shaped or have three or four rays; ascon, leucon, or sycon body forms; all marine. Calcareous sponges. *Grantia* (*Scypha*), *Leucosolenia*.

Class Hexactinellida (hex-act″ in-el′id-ah)
Spicules composed of silica and six rayed; spicules often fused into an intricate lattice; cup or vase shaped; sycon or leucon body form; found at 450 to 900 m depths in tropical West Indies and eastern Pacific. Glass sponges. *Euplectella* (Venus flower-basket).

Class Demospongiae (de-mo-spun′je-e)
Brilliantly colored sponges with needle-shaped or four-rayed siliceous spicules or spongin or both; leucon body form; up to 1 m in height and diameter. Includes one family of freshwater sponges, Spongillidae, and the bath sponges. *Cliona, Spongilla*.

---

*The class Sclerospongiae has been recently abandoned and its members assigned to Calcarea and Demospongiae.

---

for particular functions. This organization is often referred to as division of labor.

Thin, flat cells, called **pinacocytes,** line the outer surface of a sponge. Pinacocytes may be mildly contractile, and their contraction may change the shape of some sponges. In a number of sponges, some pinacocytes are specialized into tubelike, contractile **porocytes,** which can regulate water circulation (figure 9.5*a*). Openings through porocytes are pathways for water moving through the body wall.

Just below the pinacocyte layer of a sponge is a jellylike layer called the **mesohyl** (Gr. *meso*, middle + *hyl*, matter). Amoeboid cells called **mesenchyme cells** move about in the mesohyl and are specialized for reproduction, secreting skeletal elements, transporting and storing food, and forming contractile rings around openings in the sponge wall.

Below the mesohyl and lining the inner chamber(s) are choanocytes, or collar cells. **Choanocytes** (Gr. *choane*, funnel + *cyte*, cell) are flagellated cells that have a collarlike ring of microvilli surrounding a flagellum. Microfilaments connect the microvilli, forming a netlike mesh within the collar. The flagellum creates water currents through the sponge, and the collar filters microscopic food particles from the water (figure 9.5*b*). The presence of choanocytes in sponges suggests an evolutionary link between the sponges and a group of protists called choanoflagellates. This link is discussed further at the end of this chapter.

Sponges are supported by a skeleton that may consist of microscopic needlelike spikes called **spicules.** Spicules are formed by amoeboid cells, are made of calcium carbonate or silica, and may take on a variety of shapes (figure 9.6). Alternatively, the skeleton may be made of **spongin** (a fibrous protein made of collagen). A commercial sponge is prepared by drying, beating, and washing a spongin-supported sponge until all cells are removed. The nature of the skeleton is an important characteristic in sponge taxonomy.

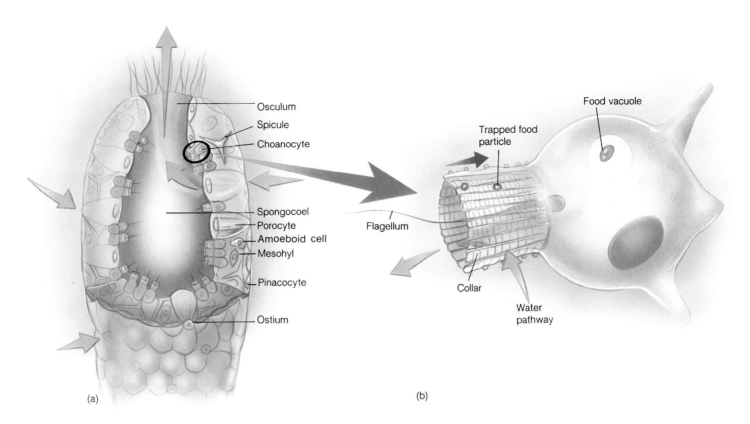

FIGURE 9.5

Morphology of a Simple Sponge. (*a*) In this example, pinacocytes form the outer body wall, and mesenchyme cells and spicules are in the mesohyl. Porocytes that extend through the body wall form ostia. (*b*) Choanocytes are cells with a flagellum surrounded by a collar of microvilli that traps food particles. Food moves toward the base of the cell, where it is incorporated into a food vacuole and passed to amoeboid mesenchyme cells, where digestion takes place. Blue arrows show water flow patterns. The brown arrow shows the direction of movement of trapped food particles.

FIGURE 9.6

Sponge Spicules. Photomicrograph of a variety of sponge spicules (×150).

WATER CURRENTS AND BODY FORMS

The life of a sponge depends on the water currents that choanocytes create. Water currents bring food and oxygen to a sponge and carry away metabolic and digestive wastes. Methods of food filtration and circulation reflect the body forms in the phylum. Zoologists have described three sponge body forms.

The simplest and least common sponge body form is the **ascon** (figure 9.7*a*). Ascon sponges are vaselike. Ostia are the outer openings of porocytes and lead directly to a chamber called the spongocoel. Choanocytes line the spongocoel, and their flagellar movements draw water into the spongocoel through the ostia. Water exits the sponge through the osculum, which is a single, large opening at the top of the sponge.

In the **sycon** body form, the sponge wall appears folded (figure 9.7*b*). Water enters a sycon sponge through openings called dermal pores. Dermal pores are the openings of invaginations of the body wall, called incurrent canals. Pores in the body wall connect incurrent canals to radial canals, and the radial canals lead to the spongocoel. Choanocytes line radial canals (rather than the spongocoel), and the beating of choanocyte flagella moves water from the ostia, through incurrent and radial canals, to the spongocoel, and out the osculum.

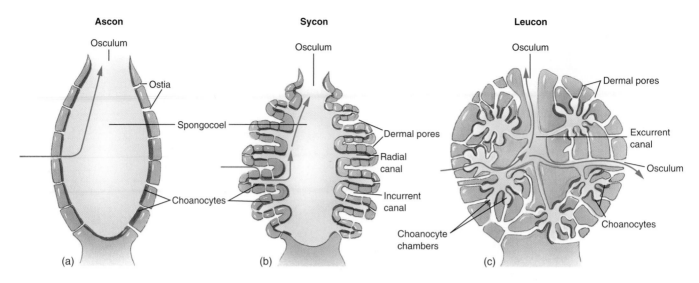

FIGURE 9.7

Sponge Body Forms.   (*a*) An ascon sponge. Choanocytes line the spongocoel in ascon sponges. (*b*) A sycon sponge. The body wall of sycon sponges appears folded. Choanocytes line radial canals that open into the spongocoel. (*c*) A leucon sponge. The proliferation of canals and chambers results in the loss of the spongocoel as a distinct chamber. Multiple oscula are frequently present. Blue arrows show the direction of water flow.

Leucon sponges have an extensively branched canal system (figure 9.7*c*). Water enters the sponge through ostia and moves through branched incurrent canals, which lead to choanocyte-lined chambers. Canals leading away from the chambers are called excurrent canals. Proliferation of chambers and canals has resulted in the absence of a spongocoel, and often, multiple exit points (oscula) for water leaving the sponge.

In complex sponges, an increased surface area for choanocytes results in large volumes of water being moved through the sponge and greater filtering capabilities. Although the evolutionary pathways in the phylum are complex and incompletely described, most pathways have resulted in the leucon body form.

MAINTENANCE FUNCTIONS

Sponges feed on particles that range in size from 0.1 to 50 μm. Their food consists of bacteria, microscopic algae, protists, and other suspended organic matter. The prey are slowly drawn into the sponge and consumed. Large populations of sponges play an important role in reducing the turbidity of coastal waters. A single leucon sponge, 1 cm in diameter and 10 cm high, can filter in excess of 20 liters of water every day! Recent investigations have discovered that a few sponges are carnivorous. These deep-water sponges (*Asbestopluma*) can capture small crustaceans using spicule-covered filaments.

Choanocytes filter small, suspended food particles. Water passes through their collar near the base of the cell and then moves into a sponge chamber at the open end of the collar. Suspended food is trapped on the collar and moved along microvilli to the base of the collar, where it is incorporated into a food vacuole (*see figure 9.5b*). Digestion begins in the food vacuole by lysosomal enzymes and pH changes. Partially digested food is passed to amoeboid cells, which distribute it to other cells.

Filtration is not the only way that sponges feed. Pinacocytes lining incurrent canals may phagocytize larger food particles (up to 50 μm). Sponges also may absorb by active transport nutrients dissolved in seawater.

Because of extensive canal systems and the circulation of large volumes of water through sponges, all sponge cells are in close contact with water. Thus, nitrogenous waste (principally ammonia) removal and gas exchange occur by diffusion.

Sponges do not have nerve cells to coordinate body functions. Most reactions result from individual cells responding to a stimulus. For example, water circulation through some sponges is at a minimum at sunrise and at a maximum just before sunset because light inhibits the constriction of porocytes and other cells surrounding ostia, keeping incurrent canals open. Other reactions, however, suggest some communication among cells. For example, the rate of water circulation through a sponge can drop suddenly without any apparent external cause. This reaction can be due only to choanocytes ceasing activities more or less simultaneously, and this implies some form of internal communication. The nature of this communication is unknown. Amoeboid cells transmitting chemical messages and ion movement over cell surfaces are possible control mechanisms.

REPRODUCTION

Most sponges are monoecious (both sexes occur in the same individual) but do not usually self-fertilize because individual sponges produce eggs and sperm at different times. Certain choanocytes lose their collars and flagella and undergo meiosis to form flagellated sperm. Other choanocytes (and amoeboid cells in some sponges) probably undergo meiosis to form eggs. Sperm and eggs are released from sponge oscula. Fertilization occurs in the ocean water, and planktonic larvae develop. In a few sponges, eggs are retained in the mesohyl of the parent. Sperm cells exit one sponge

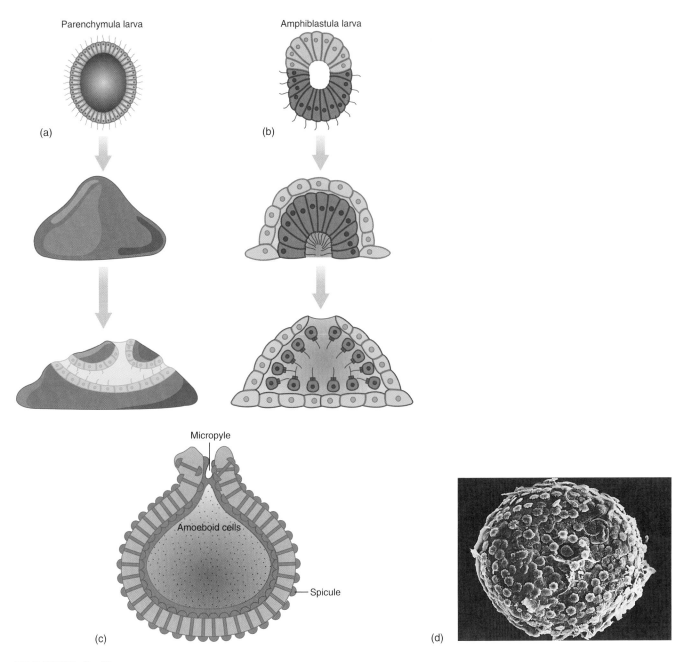

Parenchymula larva

Amphiblastula larva

(a)

(b)

Micropyle

Amoeboid cells

Spicule

(c)

(d)

FIGURE 9.8

Development of Sponge Larval Stages.   (*a*) Most sponges have a parenchymula larva (0.2 mm). Flagellated cells cover most of the larva's outer surface. After the larva settles and attaches, the outer cells lose their flagella, move to the interior, and form choanocytes. Interior cells move to the periphery and form pinacocytes. (*b*) Some sponges have an amphiblastula larva (0.2 mm), which is hollow and has half of the larva composed of flagellated cells. On settling, the flagellated cells invaginate into the interior of the embryo and form choanocytes. Nonflagellated cells overgrow the choanocytes and form the pinacocytes. (*c*) Gemmules (0.9 mm) are resistant capsules containing masses of amoeboid cells. Gemmules are released when a parent sponge dies (e.g., in the winter), and amoeboid cells form a new sponge when favorable conditions return. (*d*) Scanning electron micrograph of a gemmule of the freshwater sponge (*Dosilia brouni*). It is 0.5 mm in diameter and covered with a shell of spicules and spongin.

through the osculum and enter another sponge with the incurrent water. Sperm are trapped by choanocytes and incorporated into a vacuole. The choanocytes lose their collar and flagellum, become amoeboid, and transport sperm to the eggs.

In some sponges, early development occurs in the mesohyl. Cleavage of a zygote results in the formation of a flagellated larval stage. (A **larva** is an immature stage that may undergo a dramatic change in structure before attaining the adult body form.) The larva breaks free, and water currents carry the larva out of the parent sponge. After no more than two days of a free-swimming existence, the larva settles to the substrate and begins to develop into the adult body form (figure 9.8*a,b*).

Asexual reproduction of freshwater and some marine sponges involves the formation of resistant capsules, called

gemmules, containing masses of amoeboid cells. When the parent sponge dies in the winter, it releases gemmules, which can survive both freezing and drying (figure 9.8c,d). When favorable conditions return in the spring, amoeboid cells stream out of a tiny opening, called the micropyle, and organize into a sponge.

Some sponges possess remarkable powers of regeneration. Portions of a sponge that are cut or broken from one individual regenerate new individuals.

PHYLUM CNIDARIA

Members of the phylum Cnidaria (ni-dar′e-ah) (*Gr. knide*, nettle) possess radial or biradial symmetry. Biradial symmetry is a modification of radial symmetry in which a single plane, passing through a central axis, divides the animal into mirror images. It results from the presence of a single or paired structure in a basically radial animal and differs from bilateral symmetry in that dorsal and ventral surfaces are not differentiated. Radially symmetrical animals have no anterior or posterior ends. Thus, terms of direction are based on the position of the mouth opening. The end of the animal that contains the mouth is the oral end, and the opposite end is the aboral end. Radial symmetry is advantageous for sedentary animals because sensory receptors are evenly distributed around the body. These organisms can respond to stimuli from all directions.

The Cnidaria include over nine thousand species, are mostly marine, and are important in coral reef ecosystems (table 9.2).

Characteristics of the phylum Cnidaria include:

1. Radial symmetry or modified as biradial symmetry
2. Diploblastic, tissue-level organization
3. Gelatinous mesoglea between the epidermal and gastrodermal tissue layers
4. Gastrovascular cavity
5. Nervous system in the form of a nerve net
6. Specialized cells, called cnidocytes, used in defense, feeding, and attachment

THE BODY WALL AND NEMATOCYSTS

Cnidarians possess diploblastic, tissue-level organization (*see figure 7.10*). Cells organize into tissues that carry out specific functions, and all cells are derived from two embryological layers. The ectoderm of the embryo gives rise to an outer layer of the body wall, called the **epidermis,** and the inner layer of the body wall, called the **gastrodermis,** is derived from endoderm (figure 9.9). Cells of the epidermis and gastrodermis differentiate into a number of cell types for protection, food gathering, coordination, movement, digestion, and absorption. Between the epidermis and gastrodermis is a jellylike layer called **mesoglea.** Cells are present in the middle layer of some cnidarians, but they have their origin in either the epidermis or the gastrodermis.

TABLE 9.2
CLASSIFICATION OF THE CNIDARIA

Phylum Cnidaria (ni-dar′e-ah)
Radial or biradial symmetry, diploblastic organization, a gastrovascular cavity, and cnidocytes. Over 9,000 species.

Class Hydrozoa (hi ″dro-zo′ah)
Cnidocytes present in the epidermis; gametes produced epidermally and always released to the outside of the body; mesoglea is largely acellular; medusae usually with a velum; many polyps colonial; mostly marine with some freshwater species. *Hydra, Obelia, Gonionemus, Physalia.*

Class Scyphozoa (si ″fo-zo′ah)
Medusa prominent in the life history; polyp small; gametes gastrodermal in origin and released into the gastrovascular cavity; cnidocytes present in the gastrodermis as well as epidermis; medusa lacks a velum; mesoglea with wandering mesenchyme cells of epidermal origin, marine. *Aurelia.*

Class Cubozoa (ku ″bo-zo′ah)
Medusa prominent in life history; polyp small; gametes gastrodermal in origin; medusa cuboidal in shape with tentacles that hang from each corner of the bell; marine. *Chironex.*

Class Anthozoa (an ″tho-zo′ah)
Colonial or solitary polyps; medusae absent; cnidocytes present in the gastrodermis; cnidocils absent; gametes gastrodermal in origin; gastrovascular cavity divided by mesenteries that bear nematocysts; internal biradial or bilateral symmetry present; mesoglea with wandering mesenchyme cells; marine. Anemones and corals. *Metridium.*

One kind of cell is characteristic of the phylum. Epidermal and/or gastrodermal cells called **cnidocytes** produce structures called cnidae, which are used for attachment, defense, and feeding. A **cnida** is a fluid-filled, intracellular capsule enclosing a coiled, hollow tube (figure 9.10). A lidlike operculum caps the capsule at one end. The cnidocyte usually has a modified cilium, called a cnidocil. Stimulation of the cnidocil forces open the operculum, discharging the coiled tube—as you would evert a sweater sleeve that had been turned inside out.

Zoologists have described nearly 30 kinds of cnidae. **Nematocysts** are a type of cnida used in food gathering and defense that may discharge a long tube armed with spines that penetrates the prey. The spines have hollow tips that deliver paralyzing toxins. Other cnidae contain unarmed tubes that wrap around prey or a substrate. Still other cnidae have sticky secretions that help the animal anchor itself. Six or more kinds of cnidae may be present in one individual.

ALTERNATION OF GENERATIONS

Most cnidarians possess two body forms in their life histories (figure 9.11). The **polyp** is usually asexual and sessile. It attaches to a substrate at the aboral end, and has a cylindrical body, called the column, and a mouth surrounded by food-gathering tentacles.

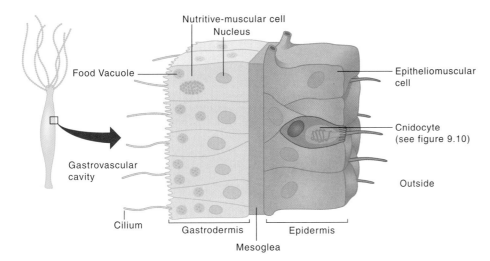

FIGURE 9.9

Body Wall of a Cnidarian (*Hydra*). Cnidarians are diploblastic (two tissue layers). The epidermis is derived embryologically from ectoderm, and the gastrodermis is derived embryologically from endoderm. Between these layers is mesoglea. Mesoglea is normally acellular in the Hydrozoa, but it contains wandering mesenchyme cells in members of the other classes. In the Hydrozoa, cnidocytes are present only in the epidermis. In members of other classes, they are present in both the epidermis and endodermis.

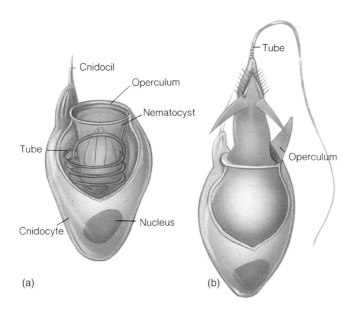

FIGURE 9.10

Cnidocyte Structure and Nematocyst Discharge. (*a*) A nematocyst develops in a capsule in the cnidocyte. The capsule is capped at its outer margin by an operculum (lid) that is displaced upon discharge of the nematocyst. The triggerlike cnidocil is responsible for nematocyst discharge. (*b*) A discharged nematocyst. When the cnidocil is stimulated, a rapid (osmotic) influx of water causes the nematocyst to evert, first near its base, and then progressively along the tube from base to tip. The tube revolves at enormous speeds as the nematocyst is discharged. In nematocysts armed with barbs, the advancing tip of the tube is aided in its penetration of the prey as barbs spring forward from the interior of the tube and then flick backward along the outside of the tube.

The **medusa** (pl., medusae) is dioecious and free swimming. It is shaped like an inverted bowl, and tentacles dangle from its margins. The mouth opening is centrally located facing downward, and the medusa swims by gentle pulsations of the body wall. The mesoglea is more abundant in a medusa than in a polyp, giving the former a jellylike consistency. When a cnidarian life cycle involves both polyp and medusa stages, the phrase "alternation of generations" is often applied.

MAINTENANCE FUNCTIONS

The gastrodermis of all cnidarians lines a blind-ending **gastrovascular cavity.** This cavity functions in digestion, the exchange of respiratory gases and metabolic wastes, and the discharge of gametes. Food, digestive wastes, and reproductive stages enter and leave the gastrovascular cavity through the mouth.

The food of most cnidarians consists of very small crustaceans, although some cnidarians feed on small fish. Nematocysts entangle and paralyze prey, and contractile cells in the tentacles cause the tentacles to shorten, which draws food toward the mouth. As food enters the gastrovascular cavity, gastrodermal gland cells secrete lubricating mucus and enzymes, which reduce food to a soupy broth. Certain gastrodermal cells, called nutritive-muscular cells, phagocytize partially digested food and incorporate it into food vacuoles, where digestion is completed. Nutritive-muscular cells also have circularly oriented contractile fibers that help move materials into or out of the gastrovascular cavity by peristaltic contractions. During peristalsis, ringlike contractions move along the body wall, pushing contents of the gastrovascular cavity ahead of them, expelling undigested material through the mouth.

Cnidarians derive most of their support from the buoyancy of water around them. In addition, a hydrostatic skeleton aids in support and movement. A **hydrostatic skeleton** is water or body fluids confined in a cavity of the body and against which contractile elements of the body wall act. In the Cnidaria, the water-filled gastrovascular cavity acts as a hydrostatic skeleton. Certain cells of the body wall, called epitheliomuscular cells, are contractile and aid in movement. When a polyp closes its mouth (to prevent water from escaping) and contracts longitudinal epitheliomuscular cells on one side of the body, the polyp bends toward that side. If these cells contract while the mouth is open, water escapes from the gastrovascular cavity, and the polyp collapses. Contraction of circular epitheliomuscular cells causes constriction of a part of the body and, if the mouth is closed, water in the gastrovascular cavity is compressed, and the polyp elongates.

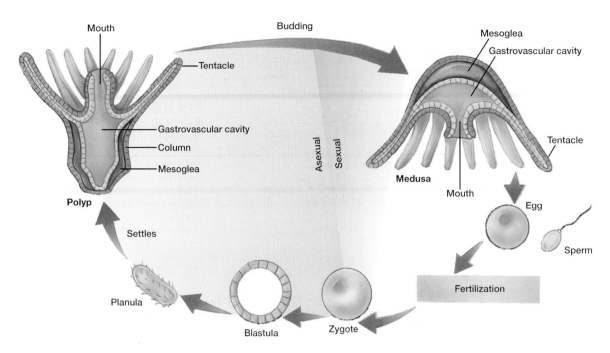

FIGURE 9.11

Generalized Cnidarian Life Cycle. This figure shows alternation between medusa and polyp body forms. Dioecious medusae produce gametes that may be shed into the water for fertilization. Early in development, a ciliated planula larva forms. After a brief free-swimming existence, the planula settles to the substrate and forms a polyp. Budding of the polyp produces additional polyps and medusa buds. Medusae break free of the polyp and swim away. The polyp or medusa stage of many species is either lost or reduced, and the sexual and asexual stages have been incorporated into one body form.

Polyps use a variety of forms of locomotion. They may move by somersaulting from base to tentacles and from tentacles to base again, or move in an inchworm fashion, using their base and tentacles as points of attachment. Polyps may also glide very slowly along a substrate while attached at their base or walk on their tentacles.

Medusae move by swimming and floating. Water currents and wind are responsible for most horizontal movements. Vertical movements are the result of swimming. Contractions of circular and radial epitheliomuscular cells cause rhythmic pulsations of the bell and drive water from beneath the bell, propelling the medusa through the water.

Cnidarian nerve cells have been of interest to zoologists for many years because they may be the most primitive nervous elements in the animal kingdom. By studying these cells, zoologists may gain insight into the evolution of animal nervous systems. Nerve cells are located below the epidermis, near the mesoglea, and interconnect to form a two-dimensional nerve net. This net conducts nerve impulses around the body in response to a localized stimulus. The extent to which a nerve impulse spreads over the body depends on stimulus strength. For example, a weak stimulus applied to a polyp's tentacle may cause the tentacle to be retracted. A strong stimulus at the same point may cause the entire polyp to withdraw.

Sensory structures of cnidarians are distributed throughout the body and include receptors for perceiving touch and certain chemicals. More specialized receptors are located at specific sites on a polyp or medusa.

Because cnidarians have large surface-area-to-volume ratios, all cells are a short distance from the body surface, and oxygen, carbon dioxide, and nitrogenous wastes exchange by diffusion.

REPRODUCTION

Most cnidarians are dioecious. Sperm and eggs may be released into the gastrovascular cavity or to the outside of the body. In some instances, eggs are retained in the parent until after fertilization.

A blastula forms early in development, and migration of surface cells to the interior fills the embryo with cells that will eventually form the gastrodermis. The embryo elongates to form a ciliated, free-swimming larva, called a **planula.** The planula attaches to a substrate, interior cells split to form the gastrovascular cavity, and a young polyp develops (*see figure 9.11*).

Medusae nearly always form by budding from the body wall of a polyp, and polyps may form other polyps by budding. Buds may detach from the polyp, or they may remain attached to the parent to contribute to a colony of individuals. Variations on this general pattern are discussed in the survey of cnidarian classes that follows.

CLASS HYDROZOA

Hydrozoans (hi"dro-zo'anz) are small, relatively common cnidarians. The vast majority are marine, but this is the one cnidarian class with freshwater representatives. Most hydrozoans have life cycles that display alternation of generations; however, in some, the medusa stage is lost, while in others, the polyp stage is very small.

Three features distinguish hydrozoans from other cnidarians: (1) nematocysts are only in the epidermis; (2) gametes are epidermal and released to the outside of the body rather than into the gastrovascular cavity; and (3) the mesoglea is largely acellular (*see table 9.2*).

Most hydrozoans have colonial polyps in which individuals may be specialized for feeding, producing medusae by budding, or defending the colony. In *Obelia,* a common marine cnidarian, the planula develops into a feeding polyp, called a **gastrozooid** (gas′tra-zo′oid) or hydranth (hi″dranth)(figure 9.12). The gastrozooid has tentacles, feeds on microscopic organisms in the water, and secretes a skeleton of protein and chitin, called the perisarc, around itself.

Growth of an *Obelia* colony results from budding of the original gastrozooid. Rootlike processes grow into and horizontally along the substrate. They anchor the colony and give rise to branch colonies. The entire colony has a continuous gastrovascular cavity, body wall, and perisarc, and is a few centimeters high. Gastrozooids are the most common type of polyp in the colony; however, as an *Obelia* colony grows, gonozooids are produced. A **gonozooid** (gon′o-zo′oid) or gonangium (go′nanj″ium) is a reproductive polyp that produces medusae by

budding. *Obelia*'s small medusae form on a stalklike structure of the gonozooid. When medusae mature, they break free of the stalk and swim out an opening at the end of the gonozooid. Medusae reproduce sexually to give rise to more colonies of polyps.

Gonionemus is a hydrozoan in which the medusa stage predominates (figure 9.13*a*). It lives in shallow marine waters, where it often clings to seaweeds by adhesive pads on its tentacles. The biology of *Gonionemus* is typical of most hydrozoan medusae. The margin of the *Gonionemus* medusa projects inward to form a shelflike lip, called the velum. A velum is present on most hydrozoan medusae but is absent in all other cnidarian classes. The velum concentrates water expelled from beneath the medusa to a smaller outlet, creating a jet-propulsion system. The mouth is at the end of a tubelike **manubrium** that hangs from the medusa's oral surface. The gastrovascular cavity leads from the inside of the manubrium into four radial canals that extend to the margin of the medusa. An encircling ring canal connects the radial canals at the margin of the medusa (figure 9.13*b*).

In addition to a nerve net, *Gonionemus* has a concentration of nerve cells, called a nerve ring, that encircles the margin of the medusa. The nerve ring coordinates swimming movements. Embedded in the mesoglea around the margin of the medusa are

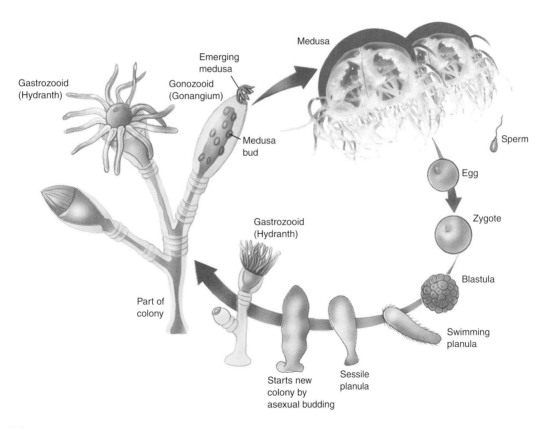

FIGURE 9.12

Obelia **Structure and Life Cycle.**   *Obelia* alternates between polyp and medusa stages. An entire polyp colony stands about 1 cm tall. A mature medusa is about 1 mm in diameter, and the planula is about 0.2 mm long. Unlike *Obelia,* the majority of colonial hydrozoans have medusae that remain attached to the parental colony, and they release gametes or larval stages through the gonozooid. The medusae often degenerate and may be little more than gonadal specializations in the gonozooid.

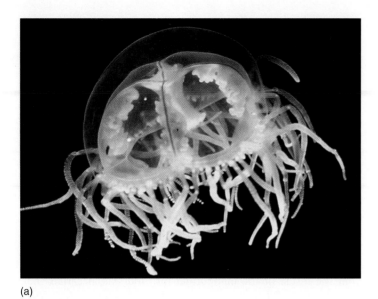

(a)

FIGURE 9.13

A Hydrozoan Medusa.   (*a*) A *Gonionemus* medusa. (*b*) Structure of *Gonionemus*.

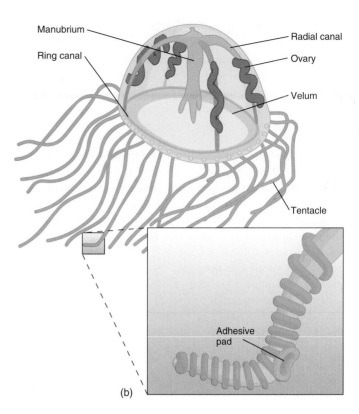

(b)

sensory structures called statocysts. A **statocyst** consists of a small sac surrounding a calcium carbonate concretion called a statolith. When *Gonionemus* tilts, the statolith moves in response to the pull of gravity. This initiates nerve impulses that may change the animal's swimming behavior.

Gonads of *Gonionemus* medusae hang from the oral surface, below the radial canals. *Gonionemus* is dioecious and sheds gametes directly into seawater. A planula larva develops and attaches to the substrate, eventually forming a polyp (about 5 mm tall). The polyp reproduces by budding to make more polyps and medusae.

Hydra is a common freshwater hydrozoan that hangs from the underside of floating plants in clean streams and ponds. *Hydra* lacks a medusa stage and reproduces both asexually by budding from the side of the polyp and sexually. Hydras are somewhat unusual hydrozoans because sexual reproduction occurs in the polyp stage. Testes are conical elevations of the body surface that form from the mitosis of certain epidermal cells, called interstitial cells. Sperm form by meiosis in the testes. Mature sperm exit the testes through temporary openings. Ovaries also form from interstitial cells. One large egg forms per ovary. During egg formation, yolk is incorporated into the egg cell from gastrodermal cells. As ovarian cells disintegrate, a thin stalk of tissue attaches the egg to the body wall. After fertilization and early development, epithelial cells lay down a resistant chitinous shell. The embryo drops from the parent, overwinters, hatches in the spring, and develops into an adult.

Large oceanic hydrozoans belong to the order Siphonophora. These colonies are associations of numerous polypoid and medusoid individuals. Some polyps, called dactylozooids, possess a single, long (up to 9 m) tentacle armed with cnidocytes for capturing prey. Other polyps are specialized for digesting prey. Various medusoid individuals form swimming bells, sac floats, oil floats, leaflike defensive structures, and gonads.

Physalia physalis, commonly called the Portuguese man-of-war, is a large, colonial siphonophore. It lacks swimming capabilities and moves at the mercy of wind and waves. Its cnidocyte-laden dactylozooids are lethal to small vertebrates and dangerous to humans.

CLASS SCYPHOZOA

Members of the class Scyphozoa (si″fo-zo′ah) are all marine and are "true jellyfish" because the dominant stage in their life history is the medusa (figure 9.14). Unlike hydrozoan medusae, scyphozoan medusae lack a velum, the mesoglea contains amoeboid mesenchyme cells, cnidocytes occur in the gastrodermis as well as the epidermis, and gametes are gastrodermal in origin (*see table 9.2*).

Many scyphozoans are harmless to humans; others can deliver unpleasant and even dangerous stings. For example, *Mastigias quinquecirrha,* the so-called stinging nettle, is a common Atlantic scyphozoan whose populations increase in late summer and become hazardous to swimmers (figure 9.14*a*). A rule of thumb for swimmers is to avoid helmet-shaped jellyfish with long tentacles and fleshy lobes hanging from the oral surface.

Aurelia is a common scyphozoan in both Pacific and Atlantic coastal waters of North America (figure 9.14*b*). The margin of its medusa has a fringe of short tentacles and is divided by notches. The mouth of *Aurelia* leads to a stomach with four gastric pouches, which contain cnidocyte-laden gastric filaments. Radial canals lead from gastric pouches to the margin of the bell. In *Aurelia,* but not all scyphozoans, the canal system is extensively branched and leads to a ring canal around the margin of the

(a)

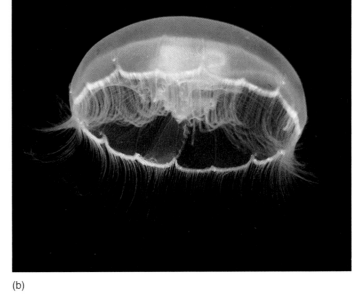

(b)

FIGURE 9.14

Representative Scyphozoans.   (a) *Mastigias quinquecirrha*. (b) *Aurelia* sp.

medusa. Gastrodermal cells of all scyphozoans possess cilia to continuously circulate seawater and partially digested food.

Aurelia is a plankton feeder. At rest, it sinks slowly in the water and traps microscopic animals in mucus on its epidermal surfaces. Cilia carry this food to the margin of the medusa. Four fleshy lobes, called oral lobes, hang from the manubrium and scrape food from the margin of the medusa (figure 9.15a). Cilia on the oral lobes carry food to the mouth.

In addition to sensory receptors on the epidermis, *Aurelia* has eight specialized structures, called rhopalia, in the notches at the margin of the medusa. Each **rhopalium** consists of sensory structures surrounded by rhopalial lappets. Two sensory pits (presumed to be olfactory) are associated with sensory lappets. A statocyst and photoreceptors, called ocelli, are associated with rhopalia (figure 9.15b). *Aurelia* displays a distinct negative phototaxis, coming to the surface at twilight and descending to greater depths during bright daylight.

Scyphozoans are dioecious. *Aurelia's* eight gonads are in gastric pouches, two per pouch. Gametes are released into the gastric pouches. Sperm swim through the mouth to the outside of the medusa. In some scyphozoans, eggs are fertilized in the female's gastric pouches, and early development occurs there. In *Aurelia*, eggs lodge in the oral lobes, where fertilization and development to the planula stage occur.

The planula develops into a polyp called a **scyphistoma** (figure 9.16). The scyphistoma lives a year or more, during which time budding produces miniature medusae, called **ephyrae.** The budding scyphistoma is often called a strobila. Repeated budding of the scyphistoma results in ephyrae being stacked on the polyp—as you might pile saucers on top of one another. After ephyrae are released, they gradually attain the adult form.

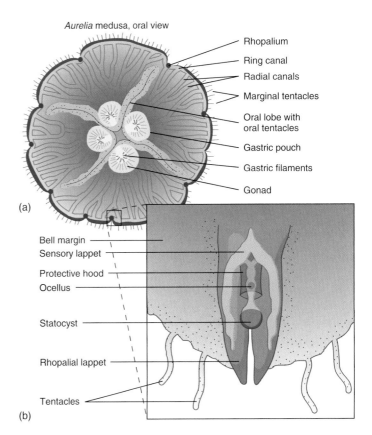

Aurelia medusa, oral view

- Rhopalium
- Ring canal
- Radial canals
- Marginal tentacles
- Oral lobe with oral tentacles
- Gastric pouch
- Gastric filaments
- Gonad

(a)

- Bell margin
- Sensory lappet
- Protective hood
- Ocellus
- Statocyst
- Rhopalial lappet
- Tentacles

(b)

FIGURE 9.15

Structure of a Scyphozoan Medusa.   (a) Internal structure of *Aurelia*. (b) A section through a rhopalium of *Aurelia*. Each rhopalium consists of two sensory (olfactory) lappets, a statocyst, and a photoreceptor called an ocellus.   (b) *Source: After L. H. Hyman, Biology of the Invertebrates, copyright 1940 McGraw-Hill Publishing Co.*

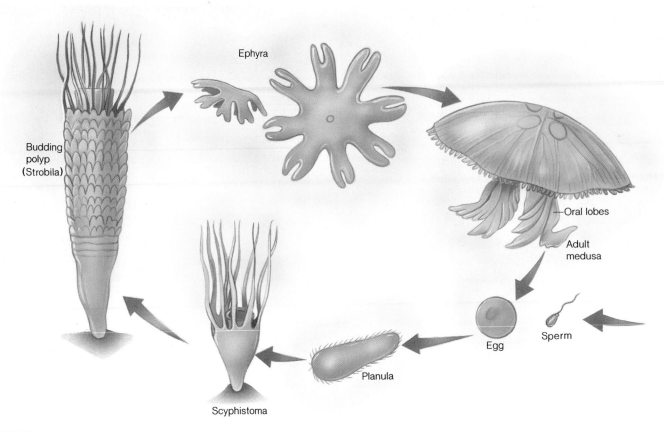

FIGURE 9.16

Aurelia **Life History.**   *Aurelia* is dioecious, and like all scyphozoans, the medusa (10 cm) predominates in the organism's life history. The planula (0.3 mm) develops into a polyp called a scyphistoma (4 mm), which produces young medusae, or ephyrae, by budding.

CLASS CUBOZOA

The class Cubozoa (ku″bo-zo′ah) was formerly classified as an order in the Scyphozoa. The medusa is cuboidal, and tentacles hang from each of its corners. Polyps are very small and, in some species, are unknown. Cubozoans are active swimmers and feeders in warm tropical waters. Some possess dangerous nematocysts (figure 9.17).

CLASS ANTHOZOA

Members of the class Anthozoa (an″tho-zo′ah) are colonial or solitary, and lack medusae. Their cnidocytes lack cnidocils. They include anemones and stony and soft corals. Anthozoans are all marine and are found at all depths.

Anthozoan polyps differ from hydrozoan polyps in three respects: (1) the mouth of an anthozoan leads to a pharynx, which is an invagination of the body wall that leads into the gastrovascular cavity; (2) mesenteries (membranes) that bear cnidocytes and gonads on their free edges divide the gastrovascular cavity into sections; and (3) the mesoglea contains amoeboid mesenchyme cells (*see table 9.2*).

Externally, anthozoans appear to show perfect radial symmetry. Internally, the mesenteries and other structures convey biradial symmetry to members of this class.

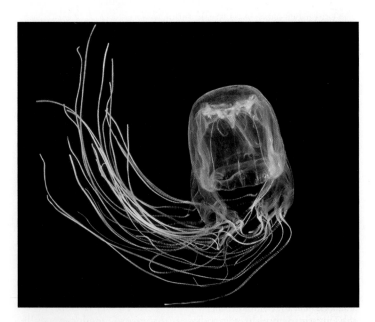

FIGURE 9.17

Class Cubozoa.   The sea wasp, *Chironex fleckeri*. The medusa is cuboidal, and tentacles hang from the corners of the bell. *Chironex fleckeri* has caused more human suffering and death off Australian coasts than the Portuguese man-of-war has in any of its home waters. Death from heart failure and shock is not likely unless the victim is stung repeatedly.

Sea anemones are solitary, frequently large, and colorful (figure 9.18a). Some attach to solid substrates, some burrow in soft substrates, and some live in symbiotic relationships (figure 9.18b). The polyp attaches to its substrate by a pedal disk (figure 9.19). An oral disk contains the mouth and hollow, oral tentacles. At one or both ends of the slitlike mouth is a siphonoglyph, which is a ciliated tract that moves water into the gastrovascular cavity to maintain the hydrostatic skeleton.

Mesenteries are arranged in pairs. Some attach at the body wall at their outer margin and to the pharynx along their inner margin. Other mesenteries attach to the body wall but are free along their entire inner margin. Openings in mesenteries near the

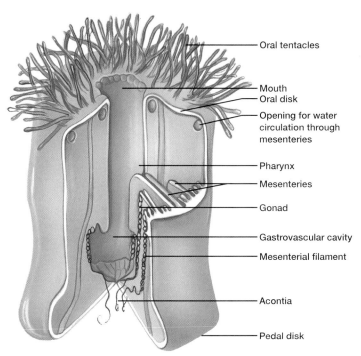

FIGURE 9.19
Class Anthozoa. Structure of the anemone, *Metridium* sp.

(a)

(b)

FIGURE 9.18
Representative Sea Anemones. (a) Giant sea anemone (*Anthopleura xanthogrammica*). Symbiotic algae give this anemone its green color. (b) This sea anemone (*Calliactis parasitical*) lives in a mutualistic relationship with a hermit crab (*Eupargurus*). Hermit crabs lack a heavily armored exoskeleton over much of their bodies and seek refuge in empty snail shells. When this crab outgrows its present home, it will take its anemone with it to a new snail (whelk) shell. This anemone, riding on the shell of the hermit crab, has an unusual degree of mobility. In turn, the anemone's nematocysts protect the crab from predators.

oral disk permit water to circulate between compartments the mesenteries set off. The free lower edges of the mesenteries form a trilobed mesenterial filament. Mesenterial filaments bear cnidocytes, cilia that aid in water circulation, gland cells that secrete digestive enzymes, and cells that absorb products of digestion. Threadlike acontia at the ends of mesenterial filaments bear cnidocytes. Acontia subdue live prey in the gastrovascular cavity and can be extruded through small openings in the body wall or through the mouth when an anemone is threatened.

Muscle fibers are largely gastrodermal. Longitudinal muscle bands are restricted to the mesenteries. Circular muscles are in the gastrodermis of the column. When threatened, anemones contract their longitudinal fibers, allowing water to escape from the gastrovascular cavity. This action causes the oral end of the column to fold over the oral disk, and the anemone appears to collapse. Reestablishment of the hydrostatic skeleton depends on gradual uptake of water into the gastrovascular cavity via the siphonoglyphs.

Anemones have limited locomotion. They glide on their pedal disks, crawl on their sides, and walk on their tentacles. When disturbed, some "swim" by thrashing their bodies or tentacles. Some anemones float using a gas bubble held within folds of the pedal disk.

Anemones feed on invertebrates and fishes. Tentacles capture prey and draw it toward the mouth. Radial muscle fibers in the mesenteries open the mouth to receive the food.

Anemones show both sexual and asexual reproduction. In asexual reproduction, a piece of pedal disk may break away from the polyp and grow into a new individual in a process called pedal

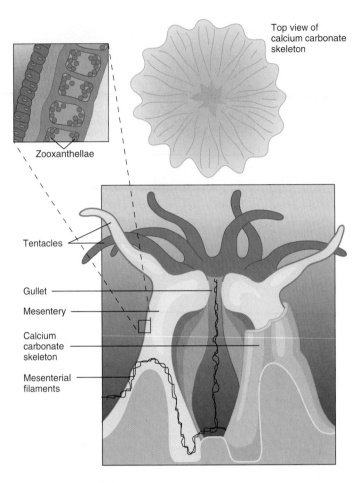

FIGURE 9.20

Class Anthozoa.   A stony coral polyp in its calcium carbonate skeleton (longitudinal section).

(a)

(b)

FIGURE 9.21

Representative Octacorallian Corals.   (a) Fleshy sea pen (*Ptilosaurus gurneyi*). (b) Purple sea fan (*Gorgonia ventalina*).

laceration. Alternatively, longitudinal or transverse fission may divide one individual into two, with missing parts being regenerated. Unlike other cnidarians, anemones may be either monoecious or dioecious. In monoecious species, male gametes mature earlier than female gametes so that self-fertilization does not occur. This is called **protandry** (Gr. *protos*, first + *andros*, male). Gonads occur in longitudinal bands behind mesenterial filaments. Fertilization may be external or within the gastrovascular cavity. Cleavage results in the formation of a planula, which develops into a ciliated larva that settles to the substrate, attaches, and eventually forms the adult.

Other anthozoans are corals. Stony corals form coral reefs and, except for lacking siphonoglyphs, are similar to the anemones. Their common name derives from a cuplike calcium carbonate exoskeleton that epithelial cells secrete around the base and the lower portion of the column (figure 9.20). When threatened, polyps retract into their protective exoskeletons. Sexual reproduction is similar to that of anemones, and asexual budding produces other members of the colony.

Many cnidarians have developed close symbiotic relationships with unicellular algae. In marine cnidarians these algae usually reside in the epidermis or gastrodermis and are called zooxanthellae (*see figure 9.20*). Stony corals have large populations of these algae. Photosynthesis by dinoflagellate (*see figure 8.6*) zooxanthellae often provides a significant amount of organic carbon for the coral polyps, and metabolism by the polyps provides algae with nitrogen and phosphorus by-products. Studies suggest that zooxanthellae aid in building coral reefs by promoting exceptionally high rates of calcium carbonate deposition. As zooxanthellae remove CO_2 from the environment of the polyp, associated pH changes induce the precipitation of dissolved calcium carbonate as aragonite (coral limestone). It is thought that the 90 m depth limit for reef building corresponds to the limits to which sufficient light penetrates to support zooxanthellae photosynthesis. Environmental disturbances, such as increased water temperature, can stress and kill zooxanthellae and result in coral bleaching (Wildlife Alert, page 144).

The colorful octacorallian corals are common in warm waters. They have eight pinnate (featherlike) tentacles, eight mesenteries, and one siphonoglyph. The body walls of members of a colony are connected, and mesenchyme cells secrete an internal skeleton of protein or calcium carbonate. Sea fans, sea pens, sea whips, red corals, and organ-pipe corals are members of this group (figure 9.21).

PHYLUM CTENOPHORA

Animals in the phylum Ctenophora (ti-nof'er-ah) (Gr. *kteno*, comb + *phoros*, to bear) are called sea walnuts or comb jellies (table 9.3). The approximately 90 described species are all marine (figure 9.22). Most ctenophorans have a spherical form, although several groups are flattened and/or elongate.

Characteristics of the phylum Ctenophora include:

1. Diploblastic or possibly triploblastic, tissue-level organization
2. Biradial symmetry
3. Gelatinous, cellular mesoglea between the epidermal and gastrodermal tissue layers
4. True muscle cells develop within the mesoglea

TABLE 9.3
CLASSIFICATION OF THE CTENOPHORA

Phylum Ctenophora (ti-nof'er-ah)
The animal phylum whose members are biradially symmetrical, diploblastic or possibly triploblastic, usually ellipsoid or spherical in shape, possess colloblasts, and have meridionally arranged comb rows. The ctenophors have been divided into the following classes. Recent information suggests that these two class designations do not reflect monophyletic groups. Higher-level taxanomy is likely to change in the near future.

Class Tentaculata (ten-tak'u-lah-tah)
With tentacles that may or may not be associated with sheaths, into which the tentacles can be retracted. *Pleurobranchia.*

Class Nuda (nu'dah)
Without tentacles; flattened; a highly branched gastrovascular cavity. *Beroë.*

(a)

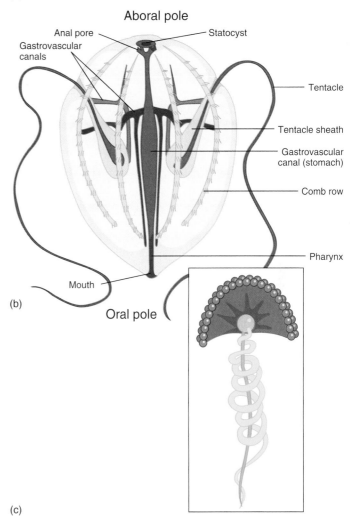

(b)

(c)

FIGURE 9.22

Phylum Ctenophora. (*a*) The ctenophore *Mnemiopsis* sp. Ctenophorans are well known for their bioluminescence. Light-producing cells are in the walls of their digestive canals, which are beneath comb rows. (*b*) The structure of *Pleurobranchia* sp. The animal usually swims with the oral end forward or upward. (*c*) Colloblasts consist of a hemispherical sticky head that connects to the core of the tentacle by a straight filament. A contractile spiral filament coils around the straight filament. Straight and spiral filaments prevent struggling prey from escaping.

WILDLIFE ALERT
Coral Reefs

Coral reefs are among the most threatened marine habitats. Along with tropical rain forests, they are among the most diverse ecosystems on the earth. They are home to thousands of species of fish, and nearly 100,000 species of reef invertebrates have been described to date (box figure 9.1). This diversity gives coral reefs tremendous intrinsic and economic value. Their highly productive waters yield four to eight million tons of fish for commercial fisheries. This is one-tenth of the world's total fish harvest, from an area that represents only 0.17% of the ocean surface (box figure 9.2). Coral reefs attract billions of dollars' worth of tourist trade each year. The ecological, aesthetic, and economic reasons for preserving coral reefs are overwhelming.

Disturbances of coral reefs can be devastating, because reefs grow very slowly. Normally a coral reef is alive with color. A disturbed reef turns white as a result of the death of anthozoan polyps, zooxanthellae (dinoflagellate protists that live in a mutualistic relationship with the anthozoans), and coralline algae (box figure 9.3). This bleaching reaction of a coral reef, if it results from a local disturbance, can be reversed rather quickly. Large-scale disturbances, however, can result in the death of large expanses of coral reef, which requires thousands of years to recover. In recent years, massive bleaching has been reported in tropical waters of the Atlantic, Caribbean, Pacific, and Indian Oceans.

Reefs require clean, constantly warm, shallow water to support the growth of zooxanthellae, which sustain coral anthozoans. Changing water levels, water temperature, and turbidity can adversely affect reef growth. Sedimentation from mining, dredging, and logging, or clearing mangrove swamps that trap sediment from coastal runoff, can block sunlight and result in the death of zooxanthellae. Some island communities mine coral reefs to extract limestone for concrete. Coastal development results in sewage and industrial pollution, which have damaged coral

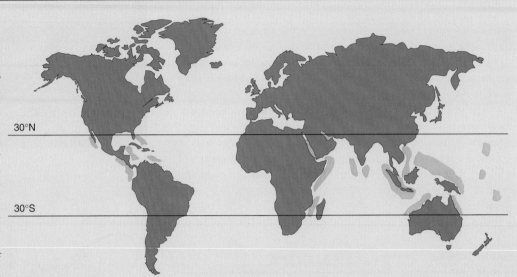

BOX FIGURE 9.2 **Coral Reefs.** Coral reefs are found throughout the tropical and subtropical regions of the world between 30°N and 30°S latitudes.

reefs. Oil spills are toxic to coral organisms. Ships that run aground damage large sections of coral reefs. Altered ecological relationships have resulted in the proliferation of the crown-of-thorns sea star (*Acanthaster planci*), which feeds on coral polyps and devastates reef communities of the South Pacific. Snorklers and scuba divers who walk across reef surfaces, break off pieces of reef, or anchor their boats on reefs similarly threaten reef life. Recently, global warming (*see chapter 6*) has been implicated in reef bleaching. The results of global warming—changing water temperature, changing water levels, and increased frequency of tropical storms—have the potential to damage coral reefs by altering environmental conditions favorable for reef survival and growth.

The threats to coral reefs seem almost overwhelming. Fortunately, biologists are finding that coral reefs are resilient ecosystems. If water quality is good, coral reefs can recover from local disturbances. National and international policies are needed that will prevent disturbances,

BOX FIGURE 9.1 **A Coral Reef Ecosystem.**

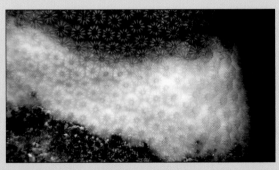

BOX FIGURE 9.3 **Coral Bleaching.** The bleached portion of the coral is shown in the lower portion of this photograph. The polyps in the upper portion of the photograph are still alive.

including those from coastal sources, and manage coral reefs as resources. The Great Barrier Reef Marine Park off the Australian coast is the largest reef preserve in the world. It includes 2,900 reef formations and is managed by the Great Barrier Reef Marine Park Authority (GBRMPA). The goal of this management is to regulate the use of the reef. GBRMPA

monitors water quality and coastal development with the purpose of conserving the reef biodiversity while sustaining commercial, educational, and recreational uses. While other reef systems would require other approaches to management, the preservation steps taken by the GBRMPA should serve as a model for managing other reefs.

5. Gastrovascular cavity
6. Nervous system in the form of a nerve net
7. Adhesive structures called colloblasts
8. Eight rows of ciliary bands, called comb rows, for locomotion

Ctenophorans have traditionally been classified as having diploblastic organization. Their "mesoglea," however, is always highly cellular and contains true muscle cells. This observation has led some zoologists to conclude that the ctenophoran middle layer should be considered a true tissue layer, which would make these animals triploblastic acoelomates.

Pleurobranchia has a spherical or ovoid, transparent body about 2 cm in diameter. It occurs in the colder waters of the Atlantic and Pacific Oceans (figure 9.22b). *Pleurobranchia*, like most ctenophorans, has eight meridional bands of cilia, called **comb rows,** between the oral and aboral poles. Comb rows are locomotor structures that are coordinated through a statocyst at the aboral pole. *Pleurobranchia* normally swims with its aboral pole oriented downward. The statocyst detects tilting, and the comb rows adjust the animal's orientation. Two long, branched tentacles arise from pouches near the aboral pole. Tentacles possess contractile fibers that retract the tentacles, and adhesive cells, called **colloblasts,** that capture prey (figure 9.22c).

Ingestion occurs as the tentacles wipe the prey across the mouth. The mouth leads to a branched gastrovascular canal system. Some canals are blind; however, two small anal canals open to the outside near the apical sense organ. Thus, unlike the cnidarians, ctenophores have an anal opening. Some undigested wastes are eliminated through these canals, and some are probably also eliminated through the mouth (*see figure 9.22b*).

Pleurobranchia is monoecious, as are all ctenophores. Two bandlike gonads are associated with the gastrodermis. One of these is an ovary, and the other is a testis. Gametes are shed through the mouth, fertilization is external, and a slightly flattened larva develops.

FURTHER PHYLOGENETIC CONSIDERATIONS

The evolutionary relationships of the phyla covered in this chapter are subject to debate. As described earlier, the animal kingdom is usually considered monophyletic, with the sponges

being closest to the choanoflagellate ancestors. As would be expected, the sponges are represented in the oldest fossil deposits—the Ediacaran formation (*see Evolutionary Insights, page 109*). As a colonial choanoflagellate increased in size, a bottom-dwelling existence would have been favored. Increased size would have selected for increased complexity of cell types and the unique system of water canals and chambers, which is the most important synapomorphy of sponges. The increased surface-to-volume ratio found in the syconoid and leuconoid body forms probably evolved in response to these selection pressures. Evolutionary relationships between poriferan classes have not been clearly defined.

Members of the phylum Cnidaria arose very early and are also represented in the Ediacaran formation. They may have arisen from a radially symmetrical ancestor, which, in turn, may have been similar in form to *Volvox* (*see figures 8.8 and 9.3a*). This interpretation places the Cnidaria very close to the base of the evolutionary tree. Another, probably minority, interpretation is that bilateral symmetry is the ancestral body form (*see figure 9.3b*) and that a bilateral ancestor gave rise to both the radiate phyla and bilateral phyla. In this interpretation, the Cnidaria are further removed from the base of the evolutionary tree.

The classical interpretation of cnidarian phylogeny is that primitive Hydrozoa were the ancestral radial animals and that the medusoid body form is the primitive body form. The polyp may have evolved secondarily as a larval stage. Recent molecular data suggests that this interpretation is probably not correct and that primitive anthozoans were the ancestral cnidarian stock. This interpretation is shown in figure 9.23. The absence of synapomorphies for the Anthozoa probably indicates that members of this class originated after the origin of the first cnidarian. The presence of cnidocils distinguishes members of the other three classes from the Anthozoa. The Scyphozoa and Cubozoa are distinguished from the Anthozoa by the evolutionary reduction of the polyp stage. The Hydrozoa are distinguished by the loss of mesenteries, the loss of nematocysts in the gastrodermis, and the presence of gamete-forming cells in the epidermis.

The relationships of the Ctenophora to any other group of animals are uncertain. The Ctenophora and Cnidaria share important characteristics, such as radial (biradial) symmetry, diploblastic organization, nerve nets, and gastrovascular cavities. In spite of these similarities, differences in adult body forms and embryological development make it difficult to derive the Ctenophora from any group of cnidarians. Relationships between the Cnidaria and Ctenophora are probably distant.

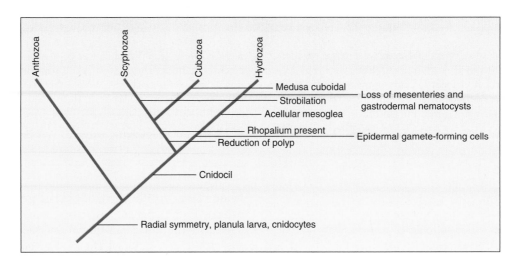

FIGURE 9.23

Cladogram Showing Cnidarian Taxonomy.   Selected synapomorphic characters are shown. Most zoologists believe anthozoans to be ancestral to other cnidarians.

SUMMARY

1. Although the origin of multicellularity in animals is unknown, the colonial hypothesis and the syncytial hypothesis are explanations of how animals could have arisen. The animal kingdom is most likely monophyletic.
2. Animals in the phylum Porifera are the sponges. Cells of sponges are specialized to create water currents, filter food, produce gametes, form skeletal elements, and line the sponge body wall.
3. Sponges circulate water through their bodies to bring in food and oxygen and to carry away wastes and reproductive products. Evolution has resulted in most sponges having complex canal systems and large water-circulating capabilities.
4. Members of the phylum Cnidaria are radially symmetrical and possess diploblastic, tissue-level organization. Cells are specialized for food gathering, defense, contraction, coordination, digestion, and absorption.
5. Hydrozoans differ from other cnidarians in having ectodermal gametes, mesoglea without mesenchyme cells, and nematocysts only in their epidermis. Most hydrozoans have well-developed polyp and medusa stages.
6. The class Scyphozoa contains the jellyfish. The polyp stage of scyphozoans is usually very small.
7. Members of the class Cubozoa live in warm, tropical waters. Some possess dangerous nematocysts.
8. The Anthozoa lack the medusa stage. They include sea anemones and corals.
9. Members of the phylum Ctenophora are biradially symmetrical and diploblastic or possibly triploblastic. Bands of cilia, called comb rows, characterize the ctenophores.
10. The Porifera probably evolved from ancestral choanoflagellate protists. The Cnidaria and Ctenophora are distantly related phyla. Within the Cnidaria, the ancient anthozoans are believed to be the stock from which modern anthozoans and other cnidarians evolved.

SELECTED KEY TERMS

choanocytes (p. 130)
cnidocytes (p. 134)
epidermis (p. 134)
gastrodermis (p. 134)
gastrovascular cavity (p. 135)
hydrostatic skeleton (p. 135)
medusa (p. 135)
mesoglea (p. 134)
polyp (p. 134)
statocyst (p. 138)

CRITICAL THINKING QUESTIONS

1. If most animals are derived from a single ancestral stock, and if that ancestral stock was radially symmetrical, would the colonial hypothesis or the syncytial hypothesis of animal origins be more attractive to you? Explain.
2. Colonies are defined in chapter 7 as "loose associations of independent cells." Why are sponges considered to have surpassed that level of organization? In your answer, compare a sponge with a colonial protist like *Volvox*.
3. Evolution is often viewed as a continuous process of increasing diversification of body forms and species. What evidence in early animal evolution contradicts this viewpoint?
4. Most sponges and sea anemones are monoecious, yet separate individuals usually reproduce sexually. Why is this advantageous for these animals? What ensures that sea anemones do not self-fertilize?
5. Do you think that the polyp stage or the medusa stage predominated in the ancestral cnidarian? Support your answer. What implications does your answer have when interpreting the evolutionary relationships among the cnidarian classes?

ONLINE LEARNING CENTER

Visit our Online Learning Center (OLC) at www.mhhe.com/zoology (click on this book's title) to find the following chapter-related materials:

- CHAPTER QUIZZING
- RELATED WEB LINKS
  - Phylum Porifera
  - Phylum Cnidaria
  - Class Hydrozoa
  - Class Scyphozoa
  - Class Anthozoa
  - Coral Reefs
  - Class Cubozoa
  - Phylum Ctenophora
- BOXED READINGS ON
  - Hydrothermal Vent Communities
  - Animal Origins—The Cambrian Explosion

  - Coral Reefs
  - A Thorny Problem for Australia's Barrier Reef
  - Planktonic Tunicates
- READINGS ON LESSER-KNOWN INVERTEBRATES
  - Mesozoa and Placozoa
- SUGGESTED READINGS
- LAB CORRELATIONS

Check out the OLC to find specific information on these related lab exercises in the *General Zoology Laboratory Manual*, 5th edition, by Stephen A. Miller:

Exercise 9 *Porifera*
Exercise 10 *Cnidaria (Coelenterata)*

THE TRIPLOBLASTIC, ACOELOMATE BODY PLAN

Outline

Concepts

1. The acoelomates are represented by the phyla Platyhelminthes, Nemertea, and Gastrotricha. These phyla are phylogenetically important because they are transitional between radial animals and the more complex bilateral animals. Three important characteristics appeared initially in this group: bilateral symmetry, a true mesoderm that gives rise to muscles and other organs, and a nervous system with a primitive brain and nerve cords.
2. Acoelomates lack a body cavity because the mesodermal mass completely fills the area between the outer epidermis and digestive tract.
3. The phylum Platyhelminthes is a large group of dorsoventrally flattened animals commonly called the flatworms.
4. Members of the class Turbellaria are mostly free-living. Representatives of the classes Monogenea, Trematoda, and Cestoidea are parasitic.
5. The phylum Nemertea (proboscis or ribbon worms) contains predominantly marine, elongate, burrowing worms with digestive and vascular systems.
6. The gastrotrichs are members of a small phylum (Gastrotricha) of free-living and freshwater species that inhabit the space between bottom sediments.
7. Acoelomates may have evolved from a primitive organism resembling a modern turbellarian.

EVOLUTIONARY PERSPECTIVE

Members of the phyla Platyhelminthes, Nemertea, and Gastrotricha are the first animals to exhibit bilateral symmetry and a body organization more complex than that of the cnidarians. All of the animals covered in this chapter are triploblastic (have three primary germ layers), acoelomate (without a coelom), and classified into three phyla: The phylum Platyhelminthes includes the flatworms (figure 10.1) that are either free living (e.g., turbellarians) or parasitic (e.g., flukes and tapeworms); the phylum Nemertea includes a small group of elongate, unsegmented, softbodied worms that are mostly marine and free-living; and the phylum Gastrotricha includes members that inhabit the space between bottom sediments.

The evolutionary relationship of the major phylum in this chapter (Platyhelminthes) to other phyla is controversial. One interpretation is that the triploblastic acoelomate body plan is an important intermediate between the radial, diploblastic plan and the triploblastic coelomate plan. The flatworms would thus represent an evolutionary side branch from a hypothetical triploblastic acoelomate ancestor. Evolution from radial ancestors could have

This chapter contains evolutionary concepts, which are set off in this font.

FIGURE 10.1

Flatworms: Animals with Primitive Organ Systems. The marine tiger flatworm (*Prostheceraeus bellustriatus*) shows brilliant markings and bilateral symmetry. This flatworm inhabits the warm, shallow waters around the Hawaiian Islands and is about 25 mm long.

involved a larval stage that became sexually mature in its larval body form. Sexual maturity in a larval body form is called **paedomorphosis** (Gr. *pais*, child + *morphe*, form) and has occurred many times in animal evolution.

Other zoologists envision the evolution of the triploblastic, acoelomate plan from a bilateral ancestor. Primitive acoelomates,

similar to flatworms, would have preceded the radiate phyla, and the radial, diploblastic plan would be secondarily derived.

The recent discovery of a small group of worms (Lobatocercebridae, Annelida) that shows both flatworm and annelid characteristics (annelids are a group of coelomate animals, such as the earthworm) suggests that the acoelomate body plan is a secondary characteristic. In this case, the flatworms would represent a side branch that resulted from the loss of a body cavity (figure 10.2)

PHYLUM PLATYHELMINTHES

The phylum Platyhelminthes (plat"e-hel-min'thez) (Gr. *platys*, flat + *helmins*, worm) contains over 20,000 animal species. Flatworms range in adult size from 1 mm or less to 25 m (*Taeniarhynchus saginatus*; see figure 10.18), in length. Their mesodermally derived tissues include a loose tissue called **parenchyma** (Gr. *parenck*, anything poured in beside) that fills spaces between other more specialized tissues, organs, and the body wall. *Depending on the species, parenchyma may provide skeletal support, nutrient storage, motility, reserves of regenerative cells, transport of materials, structural interactions with other tissues, modifiable tissue for morphogenesis, oxygen storage, and perhaps other functions yet to be determined. This is the first phylum covered that has an organ-system level of organization—a significant evolutionary advancement over the tissue level of organization.* The phylum is divided into four classes (table 10.1): (1) the Turbellaria consist of mostly free-living flatworms, whereas the (2) Monogenea, (3) Trematoda, and (4) Cestoidea contain solely parasitic species.

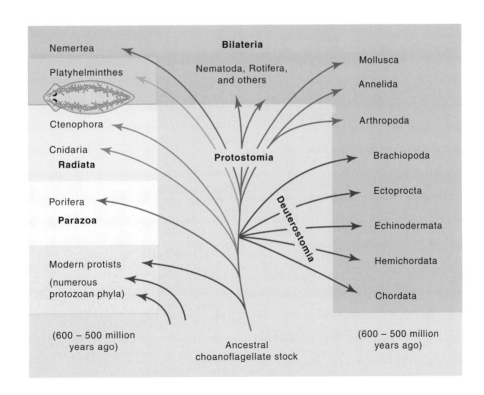

FIGURE 10.2

Acoelomate Phyla. A generalized evolutionary tree depicting the major events and possible lines of descent for the acoelomates (shaded in orange).

TABLE 10.1
CLASSIFICATION OF THE PLATYHELMINTHES*

Phylum Platyhelminthes (plat″e-hel-min′thez)

Flatworms; bilateral acoelomates. Over 20,000 species.

   Class Turbellaria*(tur″bel-lar′e-ah)

   Mostly free-living and aquatic; external surface usually ciliated; predaceous; possess rhabdites, protrusible proboscis, frontal glands, and many mucous glands: mostly hermaphroditic, *Convoluta, Notoplana, Dugesia.* Over 3,000 species.

   Class Monogenea (mon″oh-gen′e-ah)

   Monogenetic flukes; mostly ectoparasites on vertebrates (usually on fishes; occasionally on turtles, frogs, copepods, squids); one life-cycle form in only one host; bear opisthaptor. *Disocotyle, Gyrodactylus, Polystoma.* About 1,100 species.

   Class Trematoda (trem″ah-to′dah)

   Trematodes; all are parasitic; several holdfast devices present; have complicated life cycles involving both sexual and asexual reproduction. Over 10,000 species.

      Subclass Aspidogastrea (= Aspidobothrea)

      Mostly endoparasites of molluscs; possess large opisthaptor; most lack an oral sucker. *Aspidogaster, Cotylaspis, Multicotyl.* About 32 species.

      Subclass Digenea

      Adults endoparasites in vertebrates; at least two different life-cycle forms in two or more hosts; have oral sucker and acetabulum. *Schistosoma, Fasciola, Clonorchis.* About 1,350 species.

   Class Cestoidea (ses-toid′e-ah)

   All parasitic with no digestive tract; have great reproductive potential; tapeworms. About 3,500 species.

      Subclass Cestodaria

      Body not subdivided into proglottids; larva in crustaceans, adult in fishes. *Amphilina, Gyrocotyle.* About 15 species.

      Subclass Eucestoda

      True tapeworms; body divided into scolex, neck, and strobila; strobila composed of many proglottids; both male and female reproductive systems in each proglottid; adults in digestive tract of vertebrates. *Protocephalus, Taenia, Echinococcus, Taeniarhynchus, Diphyllobothrium.* About 1,000 species.

*In some of the current literature, the class Turbellaria has been abandoned as a formal taxonomic category. This is based, in part, on ultrastructural studies and cladistic analyses. There is also much uncertainty about the interrelationships among other platyhelminth groups. Until there is greater stability, we retain here the older, simpler classification scheme.

Characteristics of the phylum Platyhelminthes include:

1. Usually flattened dorsoventrally, triploblastic, acoelomate, bilaterally symmetrical
2. Unsegmented worms (members of the class Cestoidea are strobilated)
3. Incomplete gut usually present; gut absent in Cestoidea
4. Somewhat cephalized, with an anterior cerebral ganglion and usually longitudinal nerve cords
5. Protonephridia as excretory/osmoregulatory structures
6. Most forms monoecious; complex reproductive systems
7. Nervous system consists of a pair of anterior ganglia with longitudinal nerve cords connected by transverse nerves and located in the mesenchyme

CLASS TURBELLARIA: THE FREE-LIVING FLATWORMS

Members of the class Turbellaria (tur″bel-lar′e-ah) (L. *turbellae*, a commotion + *aria*, like) are mostly free-living bottom dwellers in freshwater and marine environments, where they crawl on stones, sand, or vegetation. Turbellarians are named for the turbulence that their beating cilia create in the water. Over three thousand species have been described. Turbellarians are predators and scavengers. The few terrestrial turbellarians known live in the humid tropics and subtropics. Although most turbellarians are less than 1 cm long, the terrestrial, tropical ones may reach 60 cm in length. Coloration is mostly in shades of black, brown, and gray, although some groups display brightly colored patterns.

Body Wall

As in the Cnidaria, the ectodermal derivatives include an epidermis that is in direct contact with the environment (figure 10.3). Some epidermal cells are ciliated, and others contain microvilli. A basement membrane of connective tissue separates the epidermis from mesodermally derived tissues. An outer layer of circular muscle and an inner layer of longitudinal muscle lie beneath the basement membrane. Other muscles are located dorsoventrally and obliquely between the dorsal and ventral surfaces. Between the longitudinal muscles and the gastrodermis are the loosely organized parenchymal cells.

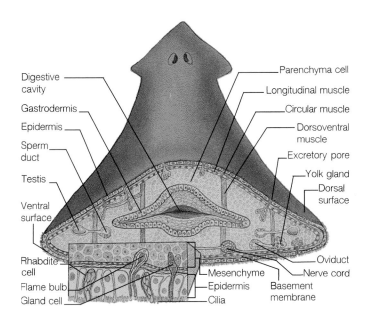

FIGURE 10.3

Phylum Platyhelminthes: Class Turbellaria.   Cross section through the body wall of a sexually mature turbellarian (the planarian, *Dugesia*), showing the relationships of the various body structures.

The innermost tissue layer is the endodermally derived gastrodermis. It consists of a single layer of cells that comprise the digestive cavity. The gastrodermis secretes enzymes that aid in digestion, and it absorbs the end products of digestion.

On the ventral surface of the body wall are several types of glandular cells of epidermal origin. **Rhabdites** are rodlike cells that swell and form a protective mucous sheath around the body, possibly in response to attempted predation or desiccation. **Adhesive glands** open to the epithelial surface and produce a chemical that attaches part of the turbellarian to a substrate. **Releaser glands** secrete a chemical that dissolves the attachment as needed.

Locomotion

Turbellarians are the first group of bilaterally symmetrical animals. Bilateral symmetry is usually characteristic of animals with an active lifestyle. Turbellarians are primarily bottom dwellers that glide over the substrate. They move using cilia and muscular undulations. As they move, turbellarians lay down a sheet of mucus that aids in adhesion and helps the cilia gain traction. The densely ciliated ventral surface and the flattened body of turbellarians enhance the effectiveness of this locomotion.

Digestion and Nutrition

Some marine turbellarians lack the digestive cavity characteristic of other turbellarians (figure 10.4*a*). This blind cavity varies from a simple, unbranched chamber (figure 10.4*b*) to a highly branched system of digestive tubes (figure 10.4*d,e*). Other turbellarians have digestive tracts that are lobed (figure 10.4*c*). *From an evolutionary perspective, highly branched digestive*

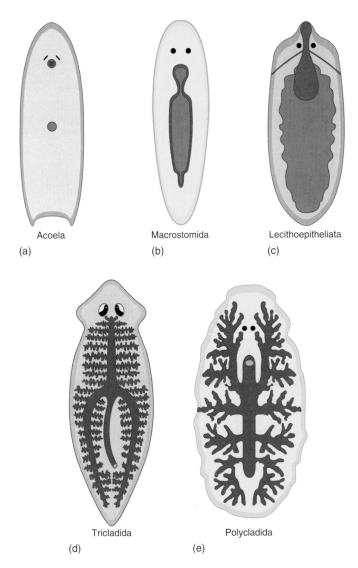

FIGURE 10.4

Digestive Systems in Some Orders of Turbellarians.   (*a*) No pharynx and digestive cavity. (*b*) A simple pharynx and straight digestive cavity. (*c*) A simple pharynx and unbranched digestive cavity. (*d*) A branched digestive cavity. (*e*) An extensively branched digestive cavity in which the branches reach almost all parts of the body.

systems are an advancement that results in more gastrodermis closer to the sites of digestion and absorption, reducing the distance nutrients must diffuse. This aspect of digestive tract structure is especially important in some of the larger turbellarians and partially compensates for the absence of a circulatory system.

The turbellarian pharynx functions as an ingestive organ. It varies in structure from a simple, ciliated tube to a complex organ developed from the folding of muscle layers. In the latter, the free end of the tube lies in a pharyngeal sheath and can project out of the mouth when feeding (figure 10.5).

Most turbellarians, such as the common planarian, are carnivores and feed on small, live invertebrates or scavenge on larger, dead animals; some are herbivores and feed on algae that they

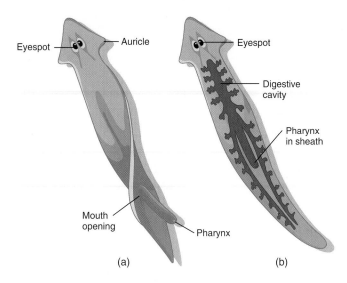

FIGURE 10.5

The Turbellarian Pharynx. A planarian turbellarian with its pharynx (*a*) extended in the feeding position and (*b*) retracted within the pharyngeal sheath.

scrape from rocks. Sensory cells (chemoreceptors) on their heads help them detect food from a considerable distance.

Food digestion is partially extracellular. Pharyngeal glands secrete enzymes that help break down food into smaller units that can be taken into the pharynx. In the digestive cavity, phagocytic cells engulf small units of food, and digestion is completed in intracellular vesicles.

Exchanges with the Environment

The turbellarians do not have respiratory organs; thus, respiratory gases (CO_2 and O_2) are exchanged by diffusion through the body wall. Most metabolic wastes (e.g., ammonia) are also removed by diffusion through the body wall.

In marine environments, invertebrates are often in osmotic equilibrium with their environment. In freshwater, invertebrates are hypertonic to their aquatic environment and thus must regulate the osmotic concentration (water and ions) of their body tissues. *The evolution of osmoregulatory structures in the form of protonephridia enabled turbellarians to invade freshwater.*

Protonephridia (Gr. *protos*, first + *nephros*, kidney) (sing., protonephridium) are networks of fine tubules that run the length of the turbellarian, along each of its sides (figure 10.6a). Numerous, fine side branches of the tubules originate in the parenchyma as tiny enlargements called **flame cells** (figure 10.6b). Flame cells (so named because, in the living organism, they resemble a candle flame) have numerous cilia that project into the lumen of the tubule. Slitlike fenestrations (openings) perforate the tubule wall surrounding the flame cell. The beating of the cilia drives fluid down the tubule, creating a negative pressure in the tubule. As a result, fluid from the surrounding tissue is sucked through the fenestrations into the tubule. The tubules eventually merge and open to the outside of the body wall through a minute opening called a **nephridiopore.**

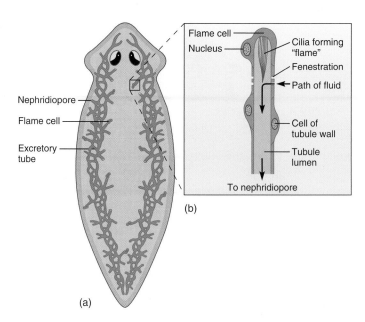

FIGURE 10.6

Protonephridial System in a Turbellarian. (*a*) The protonephridial system lies in the parenchyma and consists of a network of fine tubules that run the length of the animal on each side and open to the surface by minute nephridiopores. (*b*) Numerous, fine side branches from the tubules originate in the parenchyma in enlargements called flame cells. The black arrows indicate the direction of fluid movement.

Nervous System and Sense Organs

The most primitive type of flatworm nervous system, found in worms in the order Acoela (e.g., *Convoluta* spp.), is a subepidermal nerve plexus (figure 10.7a). This plexus resembles the nerve net of cnidarians. A statocyst in the anterior end functions as a mechanoreceptor (a receptor excited by pressure) that detects the turbellarian's body position in reference to the direction of gravity.

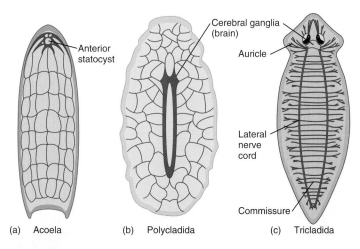

FIGURE 10.7

Nervous Systems in Three Orders of Turbellaria. (*a*) *Convoluta* has a nerve net with a statocyst. (*b*) The nerve net in a turbellarian in the order Polycladida has cerebral ganglia and two lateral nerve cords. (*c*) The cerebral ganglia and nerve cords in the planarian, *Dugesia*.

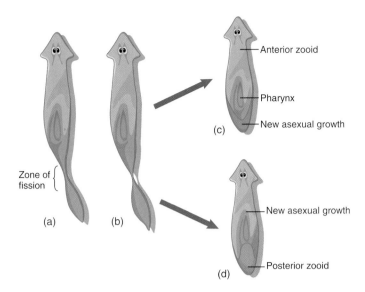

FIGURE 10.8

Asexual Reproduction in a Turbellarian. (*a*) Just before division and (*b*) just after. The posterior zooid soon develops a head, pharynx, and other structures. (*c,d*) Later development.

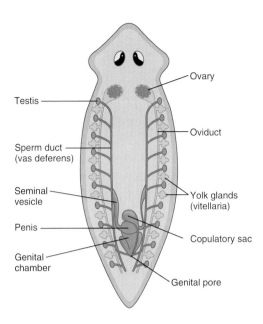

FIGURE 10.9

Triclad Turbellarian Reproductive System. Note that this single individual has both male and female reproductive organs.

Some turbellarians have a more centralized nerve net with cerebral ganglia (figure 10.7*b*). The nervous system of most other turbellarians, such as the planarian *Dugesia*, consists of a subepidermal nerve net and several pairs of long nerve cords (figure 10.7*c*). Lateral branches called commissures (points of union) connect the nerve cords. Nerve cords and their commissures give a ladderlike appearance to the turbellarian nervous system. *Neurons are organized into sensory (going to the primitive brain), motor (going away from the primitive brain), and association (connecting) types—an important evolutionary advance with respect to the nervous system. Anteriorly, the nervous tissue concentrates into a pair of cerebral ganglia (sing., ganglion) called a primitive brain.*

Turbellarians respond to a variety of stimuli in their external environment. Many tactile and sensory cells distributed over the body detect touch, water currents, and chemicals. **Auricles** (sensory lobes) may project from the side of the head (figure 10.7*c*). Chemoreceptors that aid in food location are especially dense in these auricles.

Most turbellarians have two simple eyespots called **ocelli** (sing., ocellus). These ocelli orient the animal to the direction of light. (Most turbellarians are negatively phototactic and move away from light.) Each ocellus consists of a cuplike depression lined with black pigment. Photoreceptor nerve endings in the cup are part of the neurons that leave the eye and connect with a cerebral ganglion.

Reproduction and Development

Many turbellarians reproduce asexually by transverse fission. Fission usually begins as a constriction behind the pharynx (figure 10.8). The two (or more) animals that result from fission are called **zooids** (Gr., *zoon*, living being or animal), and they regenerate missing parts after separating from each other. Sometimes, the zooids remain attached until they have attained a fairly complete degree of development, at which time they detach as independent individuals.

Turbellarians are monoecious, and reproductive systems arise from the mesodermal tissues in the parenchyma. Numerous paired testes lie along each side of the worm and are the sites of sperm production. Sperm ducts (vas deferens) lead to a seminal vesicle (a sperm storage organ) and a protrusible penis (figure 10.9). The penis projects into a genital chamber.

The female system has one to many pairs of ovaries. Oviducts lead from the ovaries to the genital chamber, which opens to the outside through the genital pore.

Even though turbellarians are monoecious, reciprocal sperm exchange between two animals is usually the rule. This cross-fertilization ensures greater genetic diversity than does self-fertilization. During cross-fertilization, the penis of each individual is inserted into the copulatory sac of the partner. After copulation, sperm move from the copulatory sac to the genital chamber and then through the oviducts to the ovaries, where fertilization occurs. Yolk may either be directly incorporated into the egg during egg formation, or yolk cells may be laid around the zygote as it passes down the female reproductive tract past the vitellaria (yolk glands).

Eggs are laid with or without a gel-like mass. A hard capsule called a **cocoon** (L., *coccum*, eggshell) encloses many turbellarian eggs. These cocoons attach to the substrate by a stalk and contain several embryos per capsule. Two kinds of capsules are laid. Summer capsules hatch in two to three weeks, and immature animals emerge. Autumn capsules have thick walls that can resist freezing and drying, and they hatch after overwintering.

Development of most turbellarians is direct—a gradual series of changes transforms embryos into adults. A few turbellarians have a free-swimming stage called a **Müller's larva.** It has ciliated extensions for feeding and locomotion. The larva eventually settles to the substrate and develops into a young turbellarian.

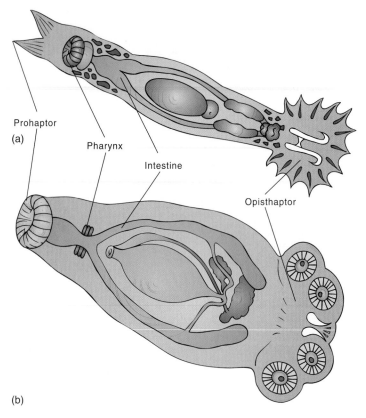

(a) Prohaptor
Pharynx
Intestine
Opisthaptor
(b)

FIGURE 10.10

Class Monogenea.   Two monogeneid trematodes. (*a*) *Gyrodactylus.*
(*b*) *Sphyranura*. Note the opisthaptors by which these ectoparasites cling to
the gills of their fish hosts. Both of these monogenea are about 1 cm long.

CLASS MONOGENEA

Monogenetic flukes are so named because they have only one generation in their life cycle; that is, one adult develops from one egg.
Monogeneans are mostly external parasites (ectoparasites) of
freshwater and marine fishes, where they attach to the gill filaments and feed on epithelial cells, mucus, or blood. A large, posterior attachment organ called an opisthaptor facilitates attachment (figure 10.10). Adult monogeneans produce and release eggs
that have one or more sticky threads that attach the eggs to the
fish gill. Eventually, a ciliated larva called an **oncomiracidium**
hatches from the egg and swims to another host fish, where it attaches by its opisthaptor and develops into an adult.

CLASS TREMATODA

The approximately eight thousand species of parasitic flatworms
in the class Trematoda (trem″ah-to′dah) (Gr. *trematodes*, perforated form) are collectively called **flukes,** which describes their
wide, flat shape. Almost all adult flukes are parasites of vertebrates,
whereas immature stages may be found in vertebrates or invertebrates, or encysted on plants. Many species are of great economic
and medical importance.

Most flukes are flat and oval to elongate, and range from less
than 1 mm to 6 cm in length (figure 10.11). They feed on host

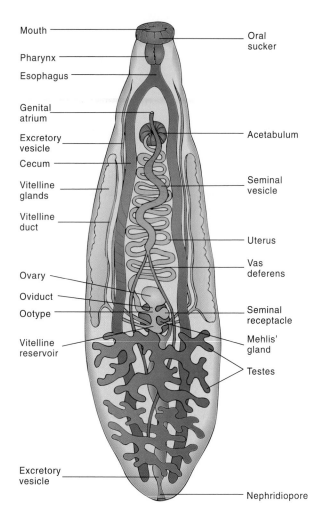

Mouth
Pharynx
Esophagus
Genital atrium
Excretory vesicle
Cecum
Vitelline glands
Vitelline duct
Ovary
Oviduct
Ootype
Vitelline reservoir
Excretory vesicle
Oral sucker
Acetabulum
Seminal vesicle
Uterus
Vas deferens
Seminal receptacle
Mehlis' gland
Testes
Nephridiopore

FIGURE 10.11

Generalized Fluke (Digenetic Trematode).   Note the large
percentage of the body devoted to reproduction. The Mehlis' gland is a
conspicuous feature of the female reproductive tract; its function in
trematodes is uncertain.

cells and cell fragments. The digestive tract includes a mouth and
a muscular, pumping pharynx. Posterior to the pharynx, the digestive tract divides into two blind-ending, variously branched
pouches called cecae (sing., cecum). Some flukes supplement their
feeding by absorbing nutrients across their body walls.

*Body-wall structure is similar for all flukes and represents
an evolutionary adaptation to the parasitic way of life.* The epidermis consists of an outer layer called the **tegument** (figure
10.12), which forms a syncytium (a continuous layer of fused
cells). The outer zone of the tegument consists of an organic layer
of proteins and carbohydrates called the glycocalyx. The glycocalyx aids in the transport of nutrients, wastes, and gases across the
body wall, and protects the fluke against enzymes and the host's
immune system. Also found in this zone are microvilli that facilitate nutrient exchange. Cytoplasmic bodies that contain the nuclei and most of the organelles lie below the basement membrane.
Slender cell processes called cytoplasmic bridges connect the cytoplasmic bodies with the outer zone of the tegument.

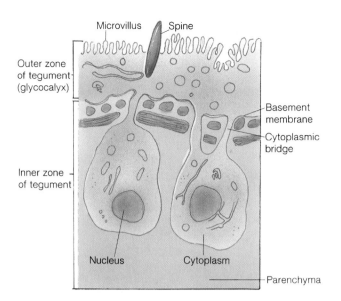

FIGURE 10.12

Trematode Tegument.   The fine structure of the tegument of a fluke. The tegument is an evolutionary adaptation that is highly efficient at absorbing nutrients and effective for protection.

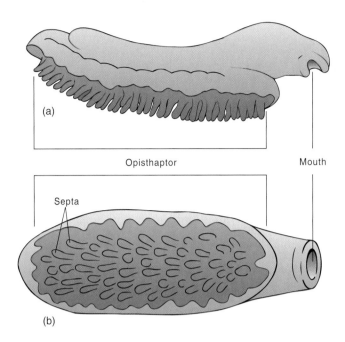

FIGURE 10.13

Class Trematoda: Subclass Aspidogastrea.   A representative aspidogastrean fluke. (*a*) Lateral and (*b*) ventral views show the large opisthaptor and numerous septa. These flukes are about 3 mm long.

Subclass Aspidogastrea

Subclass Aspidogastrea consists of a small group of flukes that are primarily internal parasites (endoparasites) of molluscs. A large, oval holdfast organ called the **opisthaptor** covers the entire ventral surface of the animal and characterizes all aspidogastreans (figure 10.13). The opisthaptor is an extremely strong attachment organ, and ridges or septa usually subdivide it. The oral sucker, characteristic of most other trematode mouths, is absent. The life cycle of aspidogastreans may involve only one host (a mollusc) or two hosts. In the latter case, the final host is usually a vertebrate (fishes or turtles) that becomes infected by ingesting a mollusc that contains immature aspidogastreans.

Subclass Digenea

The vast majority of all flukes belong to the subclass Digenea (Gr. *di*, two + *genea*, birth). In this subclass, at least two different forms, an adult and one or more larval stages, develop—a characteristic from which the name of the subclass was derived. Because digenetic flukes require at least two different hosts to complete their life cycles, these animals possess the most complex life cycles in the entire animal kingdom. As adults, they are all endoparasites in the bloodstreams, digestive tracts, ducts of the digestive organs, or other visceral organs in a wide variety of vertebrates that serve as definitive, or final, hosts. The one or more intermediate hosts (the hosts that harbor immature stages) may harbor several different larval stages. The adhesive organs are two large suckers. The anterior sucker is the **oral sucker** and surrounds the mouth. The other sucker, the **acetabulum,** is located below the oral sucker on the middle portion of the body (*see figure 10.11*).

The eggs of digenetic trematodes are oval and usually have a lidlike hatch called an **operculum** (figure 10.14*a*). When an egg

reaches freshwater, the operculum opens, and a ciliated larva called a **miracidium** (pl., miracidia) swims out (figure 10.14*b*). The miracidium swims until it finds a suitable first intermediate host (a snail) to which it is chemically attracted. The miracidium penetrates the snail, loses its cilia, and develops into a **sporocyst** (figure 10.14*c*). (Alternately, the miracidium may remain in the egg and hatch after a snail eats it.) Sporocysts are baglike and contain embryonic cells that develop into either **daughter sporocysts** or **rediae** (sing., redia) (figure 10.14*d*). At this point in the life cycle, asexual reproduction first occurs. From a single miracidium, hundreds of daughter sporocysts, and in turn, hundreds of rediae can form by asexual reproduction. Embryonic cells in each daughter sporocyst or redia produce hundreds of the next larval stage, called **cercariae** (sing., cercaria) (figure 10.14*e*). (This phenomenon of producing many cercariae is called polyembryony. It greatly enhances the chances that one or two of these cercaria will further the life cycle.) A cercaria has a digestive tract, suckers, and a tail. Cercariae leave the snail and swim freely until they encounter a second intermediate or final host, which may be a vertebrate, invertebrate, or plant. The cercaria penetrates this host and encysts as a **metacercaria** (pl., metacercariae) (figure 10.14*f*). When the definitive host eats the second intermediate host, the metacercaria excysts and develops into an adult (figure 10.14*g*).

Some Important Trematode Parasites of Humans

The Chinese liver fluke, *Clonorchis sinensis*, is a common parasite of humans in Asia, where over 30 million people are infected. The adult lives in the bile ducts of the liver, where it feeds on epithelial tissue and blood (figure 10.15*a*). The adults release

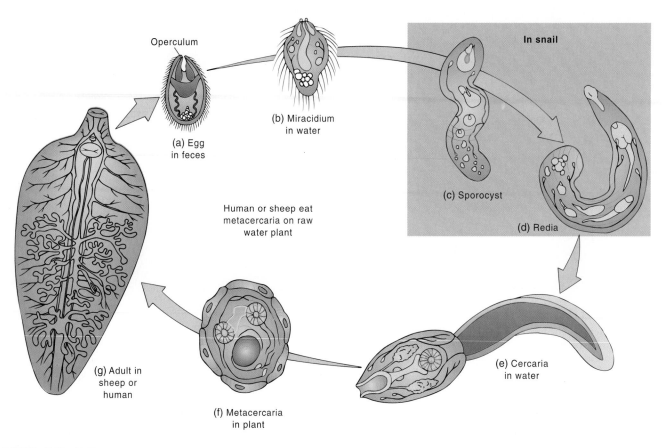

FIGURE 10.14

Class Trematoda: Subclass Digenea. The life cycle of the digenetic trematode, *Fasciola hepatica* (the common liver fluke). The adult is about 30 mm long and 13 mm wide. The cercaria is about 0.5 mm long.

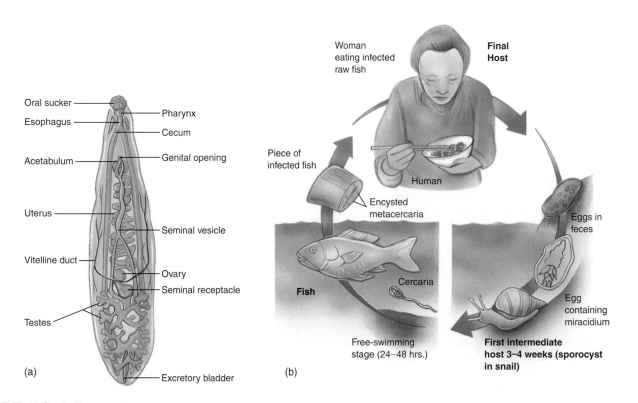

FIGURE 10.15

Chinese Liver Fluke, *Clonorchis sinensis*. (*a*) Dorsal view. (*b*) Life cycle. The adult worm is 10 to 25 mm long and 1 to 5 mm wide.

embryonated eggs into the common bile duct. The eggs make their way to the intestine and are eliminated with feces (figure 10.15*b*). The miracidia are released when a snail ingests the eggs. Following the sporocyst and redial stages, cercariae emerge into the water. If a cercaria contacts a fish (the second intermediate host), it penetrates the epidermis of the fish, loses its tail, and encysts. The metacercaria develops into an adult in a human who eats raw or poorly cooked fish, a delicacy in Asian countries and gaining in popularity in the Western world.

Fasciola hepatica is called the sheep liver fluke (*see figure 10.14*a) because it is common in sheep-raising areas and uses sheep or humans as its definitive host. The adults live in the bile duct of the liver. Eggs pass via the common bile duct to the intestine, from which they are eliminated. Eggs deposited in freshwater hatch, and the miracidia must locate the proper species of snail. If a snail is found, miracidia penetrate the snail's soft tissue and develop into sporocysts that develop into rediae and give rise to cercariae. After the cercariae emerge from the snail, they encyst on

aquatic vegetation. Sheep or other animals become infected when they graze on the aquatic vegetation. Humans may become infected with *Fasciola hepatica* by eating a freshwater plant called watercress that contains the encysted metacercaria.

Schistosomes are blood flukes with vast medical significance. The impact these flukes have had on history is second only to that of *Plasmodium*. They infect over 200 million people throughout the world. Infections are most common in Africa (*Schistosoma haematobium* and *S. mansoni*), South and Central America (*S. mansoni*), and Southeast Asia (*S. japonicum*). The adult dioecious worms live in the human bloodstream (figure 10.16*a*). The male fluke is shorter and thicker than the female, and the sides of the male body curve under to form a canal along the ventral surface (schistosoma means "split body"). The female fluke is long and slender, and is carried in the canal of the male (figure 10.16*b*). Copulation is continuous, and the female produces thousands of eggs over her lifetime. Each egg contains a spine that mechanically aids it in moving through host tissue until

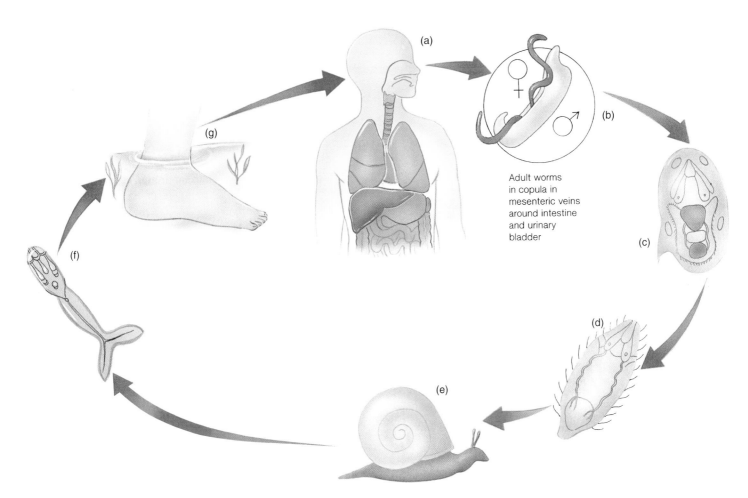

Adult worms in copula in mesenteric veins around intestine and urinary bladder

FIGURE 10.16

Representative Life Cycle of a Schistosome Fluke.   The cycle begins in a human (*a*) when the female fluke lays eggs (*b, c*) in the thin-walled, small vessels of the large or small intestine (*S. mansoni* and *S. japonicum*) or urinary bladder (*S. haematobium*). Secretions from the eggs weaken the walls, and the blood vessels rupture, releasing eggs into the intestinal lumen or urinary bladder. From there, the eggs leave the body. If they reach freshwater, the eggs hatch into ciliated, free-swimming larvae called miracidia (*d*). A miracidium burrows into the tissues of an aquatic snail (*e*), losing its cilia in the process, and develops into a sporocyst, then daughter sporocysts. Eventually, fork-tailed larvae (cercariae) are produced (*f*). After the cercariae leave the snail, they actively swim about. If they encounter human skin (*g*), they attach to it and release tissue-degrading enzymes. The larvae enter the body and migrate to the circulatory system, where they mature. They end up at the vessels of the intestines or urinary bladder, where sexual reproduction takes place, and the cycle begins anew. The adult worms are 10 to 20 mm long.

it is eliminated in either the feces or urine (figure 10.16c). Unlike other flukes, schistosome eggs lack an operculum. The miracidium escapes through a slit that develops in the egg when the egg reaches freshwater (figure 10.16d). The miracidium seeks, via chemotaxis, a snail (figure 10.16e). The miracidium penetrates it, and develops into a sporocyst, then daughter sporocysts, and finally fork-tailed cercariae (figure 10.16f). There is no redial generation. The cercariae leave the snail and penetrate the skin of a human (figure 10.16g). Anterior glands that secrete digestive enzymes aid penetration. Once in a human, the cercariae lose their tails and develop into adults in the intestinal veins, skipping the metacercaria stage.

CLASS CESTOIDEA: THE TAPEWORMS

The most highly specialized class of flatworms are members of the class Cestoidea (ses-toid'e-ah) (Gr. *kestos*, girdle + *eidos*, form), commonly called either tapeworms or cestodes. All of the approximately 3,500 species are endoparasites that usually reside in the vertebrate digestive system. Since they lack pigment as adults, their color is often white with shades of yellow or gray. Adult tapeworms range from 1 mm to 25 m in length.

Two unique adaptations to a parasitic lifestyle characterize tapeworms: (1) Tapeworms lack a mouth and digestive tract in all of their life-cycle stages; they absorb nutrients directly across their body wall. (2) Most adult tapeworms consist of a long series of repeating units called **proglottids.** Each proglottid contains one or two complete sets of reproductive structures.

As with most endoparasites, adult tapeworms live in a very stable environment. The vertebrate intestinal tract has very few environmental variations that would require the development of great anatomical or physiological complexity in any single tapeworm body system. The physiology of the tapeworm's host maintains the tapeworm's homeostasis (internal constancy). *In adapting to such a specialized environment, tapeworms have lost some of the structures believed to have been present in ancestral turbellarians. Tapeworms are, therefore, a good example of evolution not always resulting in greater complexity.*

Subclass Cestodaria

Representatives of the subclass Cestodaria are all endoparasites in the intestine and coelom of primitive fishes. Zoologists have identified about 15 species. Cestodarians possess some digenetic trematode features in that only one set of both reproductive systems is present in each animal, some bear suckers, and their bodies are not divided into proglottids like other cestodes. Yet, the absence of a digestive system, the presence of larval stages similar to those of cestodes, and the presence of parenchymal muscle cells, which are not present in any other platyhelminth, all suggest strong phylogenetic affinities with other cestodes.

Subclass Eucestoda

Almost all of the cestodes belong to the subclass Eucestoda and are called true tapeworms. They represent the ultimate degree of specialization of any parasitic animal. The body is divided into three regions (figure 10.17a). At one end is a hold-fast structure called the **scolex** that contains circular or leaflike suckers and sometimes a rostellum of hooks (figure 10.17b). With the scolex, the tapeworm firmly anchors itself to the intestinal wall of its definitive vertebrate host. No mouth is present.

Posteriorly, the scolex narrows to form the neck. Transverse constrictions in the neck give rise to the third body region, the **strobila** (Gr. *strobilus*, a linear series) (pl., strobilae). The strobila consists of a series of linearly arranged proglottids, which function primarily as reproductive units. As a tapeworm grows, new proglottids are added in the neck region, and older proglottids are gradually pushed posteriorly. As they move posteriorly, proglottids mature and begin producing eggs. Thus, anterior proglottids are said to be immature, those in the midregion of the strobila are mature, and those at the posterior end that have accumulated eggs are gravid (L., *gravida*, heavy, loaded, pregnant).

The outer body wall of tapeworms consists of a tegument similar in structure to that of trematodes (*see figure 10.12*). It plays a vital role in nutrient absorption because tapeworms have no digestive system. The tegument even absorbs some of the host's own enzymes to facilitate digestion.

With the exception of the reproductive systems, the body systems of tapeworms are reduced in structural complexity. The nervous system consists of only a pair of lateral nerve cords that arise from a nerve mass in the scolex and extend the length of the strobila. A protonephridial system also runs the length of the tapeworm (*see figure 10.6*).

Tapeworms are monoecious, and most of their physiology is devoted to producing large numbers of eggs. Each proglottid contains one or two complete sets of male and female reproductive organs (figure 10.17a). Numerous testes are scattered throughout the proglottid and deliver sperm via a duct system to a copulatory organ called a cirrus. The cirrus opens through a genital pore, which is an opening shared with the female system. The male system of a proglottid matures before the female system, so that copulation usually occurs with another mature proglottid of the same tapeworm or with another tapeworm in the same host. As previously mentioned, the avoidance of self-fertilization leads to hybrid vigor.

A single pair of ovaries in each proglottid produce eggs. Sperm stored in a seminal receptacle fertilize eggs as the eggs move through the oviduct. Vitelline cells from the vitelline gland are then dumped onto the eggs in the ootype. The ootype is an expanded region of the oviduct that shapes capsules around the eggs. The ootype is also surrounded by the Mehlis' gland, which aids in the formation of the egg capsule. Most tapeworms have a blind-ending uterus, where eggs accumulate (figure 10.17a). As eggs accumulate, the reproductive organs degenerate; thus, gravid proglottids can be thought of as "bags of eggs." Eggs are released when gravid proglottids break free from the end of the tapeworm and pass from the host with the host's feces. In a few tapeworms, the uterus opens to the outside of the worm, and eggs are released into the host's intestine. Sometimes proglottids are not continuously lost, and some adult tapeworms become very long, such as the beef tapeworm (*Taeniarhynchus saginatus*).

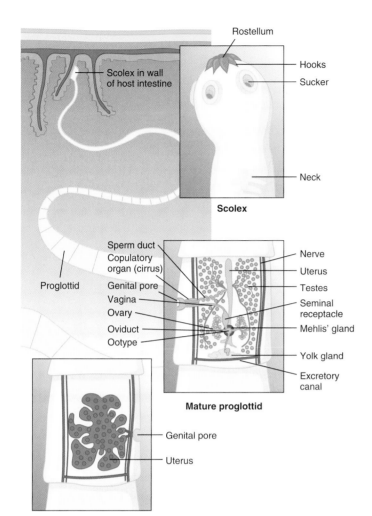

(a)   **Gravid proglottid**

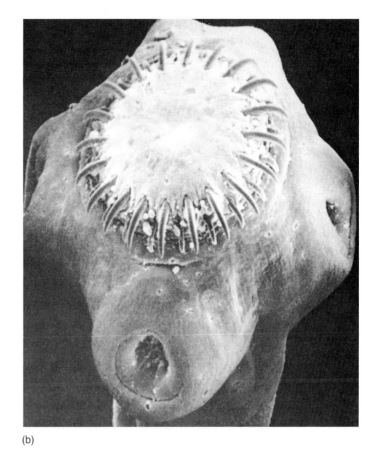

(b)

FIGURE 10.17

Class Cestoidea: A Tapeworm.   (*a*) The scolex, neck, and proglottids of the pork tapeworm, *Taenia solium*. The adult worm attains a length of 2 to 7 m. Included is a detailed view of a mature proglottid with a complete set of male and female reproductive structures. (*b*) The scolex of the cestode *Taenia solium* (SEM ×100). Notice the rostellum with two circles of hooks.

Some Important Tapeworm Parasites of Humans

One medically important tapeworm of humans is the beef tapeworm *Taeniarhynchus saginatus* (figure 10.18). Adults live in the small intestine and may reach lengths of 25 m. About 80,000 eggs per proglottid are released as proglottids break free of the adult worm. As an egg develops, it forms a six-hooked (hexacanth) larva called the **onchosphere.** As cattle (the intermediate host) graze in pastures contaminated with human feces, they ingest oncospheres (or proglottids). Digestive enzymes of the cattle free the oncospheres, and the larvae use their hooks to bore through the intestinal wall into the bloodstream. The bloodstream carries the larvae to skeletal muscles, where they encyst and form a fluid-filled bladder called a **cysticercus** (pl., cysticerci) or **bladder worm.** When a human eats infected meat (termed "measly beef") that is raw or improperly cooked, the cysticercus is released from the meat, the scolex attaches to the human intestinal wall, and the tapeworm matures.

A closely related tapeworm, *Taenia solium* (the pork tapeworm), has a life cycle similar to that of *Taeniarhynchus saginatus*, except that the intermediate host is the pig. The strobila has been reported as being 10 m long, but 2 to 3 m is more common. The pathology is more serious in the human than in the pig. Gravid proglottids frequently release oncospheres before the proglottids have had a chance to leave the small intestine of the human host. When these larvae hatch, they move through the intestinal wall, enter the bloodstream, and are distributed throughout the body, where they eventually encyst in human tissue as cysticerci. The disease that results is called **cysticercosis** and can be fatal if the cysticerci encyst in the brain.

The broad fish tapeworm, *Diphyllobothrium latum,* is relatively common in the northern parts of North America, in the Great Lakes area of the United States and throughout northern Europe. This tapeworm has a scolex with two longitudinal grooves (bothria; sing., bothrium) that act as hold-fast structures (figure 10.19). The adult worm may attain a length of 10 m and shed up to a million

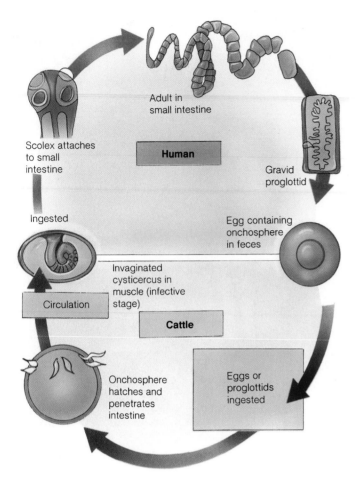

FIGURE 10.18

Life Cycle of the Beef Tapeworm, *Taeniarhynchus saginatus.*   Adult worms may attain a length of 25 m.   *Source: Redrawn from Centers for Disease Control, Atlanta, GA.*

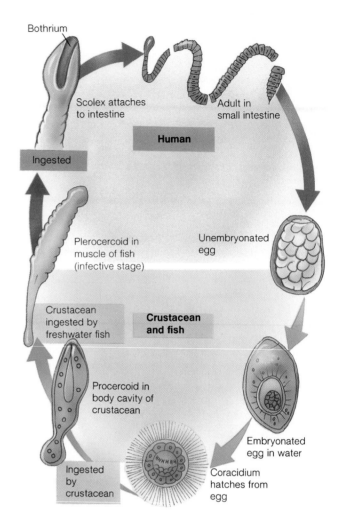

FIGURE 10.19

Life Cycle of the Broad Fish Tapeworm, *Diphyllobothrium latum.* Adult worms may be 3 to 10 m long.   *Source: Redrawn from Centers for Disease Control, Atlanta, GA.*

eggs a day. Many proglottids release eggs through uterine pores. When eggs are deposited in freshwater, they hatch, and ciliated larvae called **coracidia** (sing., coracidium) emerge. These coracidia swim about until small crustaceans called copepods ingest them. The larvae shed their ciliated coats in the copepods and develop into **procercoid larvae.** When fish eat the copepods, the procercoids burrow into the muscle of the fish and become **plerocercoid larvae.** Larger fishes that eat smaller fishes become similarly infected with plerocercoids. When humans (or other carnivores) eat infected, raw, or poorly cooked fishes, the plerocercoids attach to the small intestine and grow into adult worms.

PHYLUM NEMERTEA

Most of the approximately nine hundred species of nemerteans (nem-er′te-ans) (Gr. *Nemertes,* a Mediterranean sea nymph; the daughter of Nereus and Doris) are elongate, flattened worms found in marine mud and sand. Due to a long proboscis, nemerteans are commonly called proboscis worms. Adult worms

range in size from a few millimeters to several centimeters in length. Most nemerteans are pale yellow, orange, green, or red. Characteristics of the phylum Nemertea include:

1. Triploblastic, acoelomate, bilaterally symmetrical, unsegmented worms possessing a ciliated epidermis containing mucous glands
2. Complete digestive tract with an anus
3. Protonephridia
4. Cerebral ganglion, longitudinal nerve cords, and transverse commissures
5. Closed circulatory system
6. Body musculature organized into two or three layers

The most distinctive feature of nemerteans is a long proboscis held in a sheath called a **rhynchocoel** (figure 10.20). The proboscis may be tipped with a barb called a **stylet.** Carnivorous species use the proboscis to capture annelid (segmented worms) and crustacean prey.

Unlike the platyhelminths, nemerteans have a complete one-way digestive tract. They have a mouth for ingesting food and

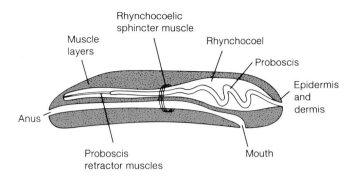

FIGURE 10.20

Phylum Nemertea.   Longitudinal section of a nemertean, showing the tubular gut and proboscis.   *Source: Modified from Turbeville and Ruppert, 1983, Zoomorphology, 103:103, Copyright 1983, Springer-Verlag, Heidelberg, Germany.*

an anus for eliminating digestive wastes. This enables mechanical breakdown of food, digestion, absorption, and feces formation to proceed sequentially in an anterior to posterior direction—a major evolutionary innovation found in all higher bilateral animals.

Another major innovation found in all higher animals evolved first in the nemerteans—a circulatory system consisting of two lateral blood vessels and, often, tributary vessels that branch from lateral vessels. However, no heart is present, and contractions of the walls of the large vessels propel blood. Blood does not circulate but simply moves forward and backward through the longitudinal vessels. Blood cells are present in some species. *This combination of blood vessels with their capacity to serve local tissues and a one-way digestive system with its greater efficiency at processing nutrients allows nemerteans to grow much larger than most flatworms.*

Nemerteans are dioecious. Male and female reproductive structures develop from parenchymal cells along each side of the body. External fertilization results in the formation of a helmet-shaped, ciliated **pilidium larva.** After a brief free-swimming existence, the larva develops into a young worm that settles to the substrate and begins feeding.

When they move, adult nemerteans glide on a trail of mucus. Cilia and peristaltic contractions of body muscles provide the propulsive forces.

PHYLUM GASTROTRICHA

The gastrotrichs (gas-tro-tri′ks) (Gr. *gastros,* stomach + *trichos,* hair) are members of a small phylum of about five hundred free-living marine and freshwater species that inhabit the space between bottom sediments. They range from 0.01 to 4 mm in length. Gastrotrichs use cilia on their ventral surface to move over the substrate. The phylum contains a single class divided into two orders.

The dorsal cuticle often contains scales, bristles, or spines, and a forked tail is often present (figure 10.21). A syncytial epidermis is beneath the cuticle. Sensory structures include tufts of long cilia and bristles on the rounded head. The nervous system includes a brain and a pair of lateral nerve trunks. The digestive system is a straight tube with a mouth, a muscular pharynx, a

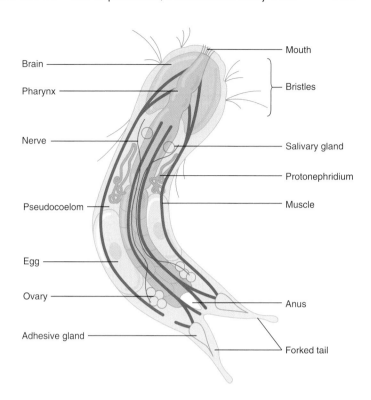

FIGURE 10.21

Phylum Gastrotricha.   The internal anatomy of a freshwater gastrotrich. This animal is about 3 mm long.

stomach-intestine, and an anus. The action of the pumping pharynx allows the ingestion of microorganisms and organic detritus from the bottom sediment and water. Digestion is mostly extracellular. Adhesive glands in the forked tail secrete materials that anchor the animal to solid objects. Paired protonephridia occur in freshwater species, rarely in marine ones. Gastrotrich protonephridia, however, are morphologically different from those found in other acoelomates. Each protonephridium possesses a single flagellum instead of the cilia found in flame cells.

Most of the marine species reproduce sexually and are hermaphroditic. Most of the freshwater species reproduce asexually by parthenogenesis; the females can lay two kinds of unfertilized eggs. Thin-shelled eggs hatch into females during favorable environmental conditions, whereas thick-shelled resting eggs can withstand unfavorable conditions for long periods before hatching into females. There is no larval stage; development is direct, and the juveniles have the same form as the adults.

FURTHER PHYLOGENETIC CONSIDERATIONS

Zoologists who believe that the platyhelminth body form is central to animal evolution envision an ancestral flatworm similar to a turbellarian. Figure 10.22 is a cladogram emphasizing the uniqueness of the tegument as a synapomorphy (a shared, evolutionarily derived character used to describe common descent

EVOLUTIONARY INSIGHTS
The Flatworms Were the First Carnivorous Hunters

Since the first animals inhabited the surface of the Precambrian seabed, what would they have had available to them as potential food? The answer can only be unicellular or colonial protists and bacteria, which in all but the shallowest of waters, would mainly have also been heterotrophic.

Hunters are mobile animals that attack, kill, and consume individual prey items one at a time. Since the current flatworms are exclusively feeders on animal tissue, whether as predators, scavengers, or parasites, they can be considered the first hunters. They prey on organisms ranging in size from bacteria to small animals, which they can locate with their sense organ–bearing head.

It is therefore not surprising to find that the simplest surviving small flatworms, and presumably the ancestral forms, are and were essentially consumers of protists and bacteria associated in some way with the bottom sediments and rocks. What is surprising is that in direct and indirect ways this ancestral diet has dominated the nutritional lives of all of the flatworm descendants, even including the terrestrial ones.

Protists and bacteria are abundant in marine sediments, but they are individually small and widely dispersed in space. They rarely occur in dense clumps. This has had two fundamental repercussions on animal lifestyles. Firstly, animals must be mobile in order to find new food supplies when local food has been exhausted. Secondly, a consumer of small, widely scattered microorganisms must itself be relatively small. The larger an animal is, the greater will be its metabolic requirements and the more food it must obtain per unit time. Thus, a large animal could not subsist on a diet of protists and bacteria because it could not find enough of them to eat.

However, to increase in size is itself likely to offer selective advantages for three reasons: (1) larger animals tend to be able to produce more offspring than smaller animals; (2) a larger size can confer greater protection from consumption by other organisms; and (3) larger animals tend to be able to displace smaller animals from limited shared resources.

Considering these advantages, it is not surprising to find that most surviving flatworms are larger than the protists and bacteria-consuming species and that they have turned to the hunting and consumption of larger food items—other animals—rather than protists and bacteria. Today, the two more significant lines of structurally simplest animals, the radial cnidarians and the bilaterally symmetrical flatworms, are both essentially carnivorous.

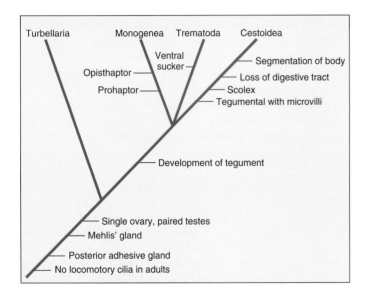

FIGURE 10.22

Cladogram Showing Evolutionary Relationships among the Classes of Platyhelminthes. The absence of synapomorphies for the Turbellaria suggests that the ancestral platyhelminth was itself a turbellarian and that some member of that class was ancestral to the three living parasitic groups. *Source: Data from D. R. Brooks,* Journal of Parasitology, *1989, pp. 606–616.*

among two or more species) uniting the Monogenea, Trematoda, and Cestoidea. Recent molecular data suggest that the acoelomate flatworms should not be considered members of the Phylum Platyhelminthes and that they are very close to the first ancestral bilateral animals.

More conclusive evidence links the parasitic flatworms to ancient, free-living ancestors. The free-living and parasitic ways of life probably diverged in the Cambrain period, 600 million years ago. The first flatworm parasites were probably associated with primitive molluscs, arthropods, and echinoderms. They must have acquired the vertebrate hosts and complex life cycles described in this chapter much later.

The gastrotrichs show some distant relationships to the acoelomates. For example, many gastrotrichs lack a body cavity and are monoecious and small, and their ventral cilia may have been derived from the same ancestral sources as those of the turbellarian flatworms.

SUMMARY

1. The free-living platyhelminths, members of the class Turbellaria, are small, bilaterally symmetrical, acoelomate animals with some cephalization.
2. Most turbellarians move entirely by cilia and are predators and scavengers. Digestion is initially extracellular and then intracellular.
3. Protonephridia are present in many flatworms and are involved in osmoregulation. A primitive brain and nerve cords are present.
4. Turbellarians are monoecious with the reproductive systems adapted for internal fertilization.
5. The monogenetic flukes (class Monogenea) are mostly ectoparasites of fishes.
6. The class Trematoda is divided into two subclasses (Aspidogastrea and Digenea), and most are external or internal parasites of vertebrates. A gut is present, and most of these flukes are monoecious.
7. Cestodes, or tapeworms, are gut parasites of vertebrates. They are structurally more specialized than flukes, having a scolex with attachment organs, a neck region, and a strobila, which consists of a chain of segments (proglottids) budded off from the neck region. A gut is absent, and the reproductive system is repeated in each proglottid.
8. Nemerteans are similar to platyhelminths but can be much larger. They prey on other invertebrates, which they capture with a unique proboscis. They have a one-way digestive tract and a blood-vascular system, and the sexes are separate.
9. Gastrotrichs are microscopic, aquatic animals with a head, neck, and trunk. Numerous adhesive glands are present. The group is generally hermaphroditic, although males are rare and female parthenogenesis is common in freshwater species.

SELECTED KEY TERMS

acetabulum (p. 155)
cercariae (p. 155)
coracidia (p. 160)
procercoid larvae (p. 160)
rediae (p. 155)
sporocyst (p. 155)

CRITICAL THINKING QUESTIONS

1. Describe the morphological and developmental similarities and differences between nemerteans and turbellarians.
2. How do parasitic flatworms evade their host's immune system?
3. How could a zoologist document the complex life cycle of a digenetic trematode?
4. Describe some of the key features of acoelomate animals.

ONLINE LEARNING CENTER

Visit our Online Learning Center (OLC) at www.mhhe.com/zoology (click on this book's title) to find the following chapter-related materials:

- CHAPTER QUIZZING
- RELATED WEB LINKS
    Phylum Platyhelminthes
    Class Turbellaria
    Class Trematoda
    Class Cestoda
    Class Monogenea
    Phylum Nemertea
    Phylum Gastrotricha
- BOXED READING ON SWIMMER'S ITCH
- SUGGESTED READINGS
- LAB CORRELATIONS

Check out the OLC to find specific information on these related lab exercises in the *General Zoology Laboratory Manual*, 5th edition, by Stephen A. Miller:

Exercise 11   *Platyhelminthes*

CHAPTER 11

THE PSEUDOCOELOMATE BODY PLAN:

ASCHELMINTHS

Outline

Concepts

1. The aschelminths comprise seven phyla: Rotifera, Kinorhyncha, Nematoda, Nematomorpha, Acanthocephala, Loricifera, and Priapulida. Because most of these phyla have had a separate evolutionary history, this grouping is mostly one of convenience.

2. The major unifying aschelminth feature is a pseudocoelom. The pseudocoelom is a type of body cavity that develops from the blastocoel (the primitive cavity in the embryo) and is not fully lined by mesoderm, as in the true coelomates. In the pseudocoelomates, the muscles and other structures of the body wall and internal organs are in direct contact with fluid in the pseudocoelom.

3. Other common aschelminth features include a complete digestive tract, a muscular pharynx, constant cell numbers (eutely), protonephridia, a cuticle, and adhesive glands.

EVOLUTIONARY PERSPECTIVE

The seven different phyla grouped for convenience as the **aschelminths** (Gr. *askos*, bladder + *helmins,* worm) are very diverse animals. They have obscure phylogenetic affinities, and their fossil record is meager. Two hypotheses have been proposed for their phylogeny. The first hypothesis contends that the phyla are related based on the presence of the following structures: a pseudocoelom, a cuticle, a muscular pharynx, and adhesive glands. The second hypothesis contends that the various aschelminth phyla are not related to each other; thus, they are probably polyphyletic. The absence of any single unique feature found in all groups strongly suggests independent evolution of each phylum. The similarities among the living aschelminths may simply be the result of convergent evolution as these various animals adapted to similar environments.

    The correct phylogeny may actually be something between the two hypotheses. All phyla are probably distantly related to each other based on the few anatomical and physiological features they share (figures 11.1, 11.2). True convergent evolution may have also produced some visible analogous similarities, but each phylum probably arose from a common acoelomate ancestor and diverged very early in evolutionary history (figure 11.3). Such an ancestor might have been a primitive, ciliated, acoelomate turbellarian (*see figure 10.4*a), from which it would follow that the first ancestor was ciliated, acoelomate, marine, and probably monoecious, and lacked a cuticle.

*T*his chapter contains evolutionary concepts, which are set off in this font.

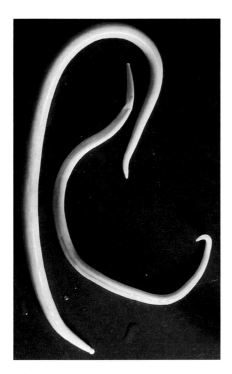

FIGURE 11.1

Roundworm Characteristics: A Fluid-Filled Body Cavity and a Complete Digestive System.   *Ascaris* species inhabit the intestines of both pigs and humans. Male ascarid worms measure 15 to 31 cm in length and are smaller than females (20 to 35 cm in length). Males also have a curved posterior end.

GENERAL CHARACTERISTICS

The aschelminths are the first assemblage of animals to possess a distinct body cavity, but they lack the peritoneal linings and membranes, called mesenteries, found in more advanced animals. As a result, the various internal (visceral) organs lie free in the cavity. Such a cavity is called a pseudocoelom or **pseudocoel** (*see figure 7.11*b), and the animals are called **pseudocoelomates**. The pseudocoelom is often fluid filled or may contain a gelatinous substance with mesenchyme cells, serves as a cavity for circulation, aids in digestion, and acts as an internal hydrostatic skeleton that functions in locomotion.

Most aschelminths (the acanthocephalans and nematomorphs are exceptions) have a complete tubular digestive tract that extends from an anterior mouth to a posterior anus. This complete digestive tract was first encountered in the nemerteans (*see figure 10.20*) and is characteristic of almost all other higher animals. *It permits, for the first time, the mechanical breakdown of food, digestion, absorption, and feces formation to proceed sequentially and continually from an anterior to posterior direction—an evolutionary advancement over the blind-ending digestive system.* Most aschelminths also have a specialized muscular pharynx that is adapted for feeding.

Many aschelminths show eutely (Gr. *euteia*, thrift), a condition in which the number of cells (or nuclei in syncytia) is constant both for the entire animal and for each given organ in all the animals of that species. For example, the number of body (somatic) cells in all adult *Caenorhabditis elegans* nematodes is 959,

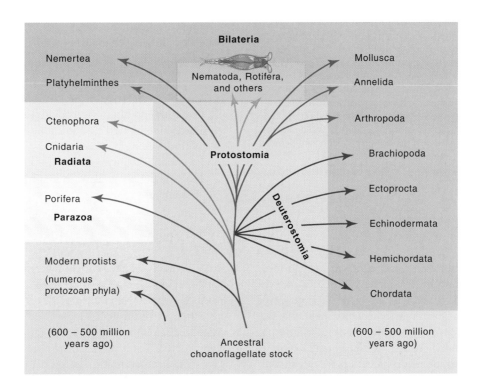

FIGURE 11.2

Pseudocoelomate Phyla.   A generalized evolutionary tree depicting the major events and possible lines of descent for the pseudocoelomates (shaded in orange).

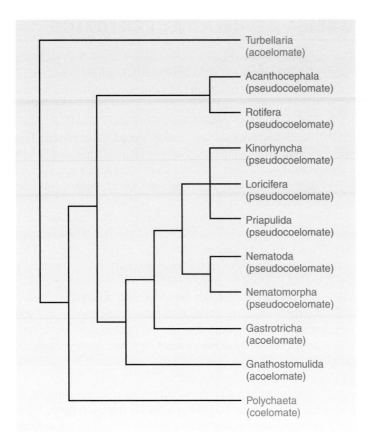

FIGURE 11.3

Cladogram of the Seven Pseudocoelomate (Aschelminth) Phyla.   For comparative purposes, one coelomate (Polychaeta) and three acoelomate (Turbellaria, Gastrotricha, and Gnathostomulida) phyla are included. Notice that in this cladogram the Kinorhyncha, Loricifera, and Priapulida are closely related, the Acanthocephala and Rotifera are closely related, and the Nematoda and Nematomorpha are closely related. Thus, the pseudocoelomates comprise two main clades and a minor clade: comprising the Acanthocephala + Rotifera defined principally by a syncytial epidermis, an intracytoplasmic lamina, and sperm with anterior flagella; and a major clade composed of two subclades—Loricifera + Priapulida + Kinorhyncha and Nematoda + Nematomorpha—defined principally by an extracellular cuticle that is molted. This consensus cladogram permits, but does not require, a polyphyletic origin of the pseudocoelom.   *Source: Data from Robert L. Wallace, Claudia Ricci, and Giulio Melone, Invertebrate Biology, 1996, pp. 104–12.*

and the number of cells in the pharynx of every worm in the species is precisely 80.

Most aschelminths are microscopic, although some grow to over a meter in length. They are bilaterally symmetrical, unsegmented, triploblastic, and cylindrical in cross section. Most aschelminths have an osmoregulatory system of protonephridia (*see figure 10.6*). This system is best developed in freshwater forms, in which osmotic problems are the greatest. No separate blood or gas exchange systems are present. Some cephalization is evident, with the anterior end containing a primitive brain, sensory organs, and a mouth. The vast majority of aschelminths are dioecious. Both reproductive systems are relatively uncomplicated; life cycles are usually simple, except in parasites. Cilia are generally absent from the

external surface, but a thin, tough, external cuticle is present. The **cuticle** (L. *cutis*, skin) may bear spines, scales, or other forms of ornamentation that protect the animal and are useful to taxonomists. Some aschelminths shed this cuticle in a process called **molting** or **ecdysis** (Gr. *ekdysis*, to strip off) in order to grow. Beneath the cuticle is a syncytial epidermis that actively secretes the cuticle. Several longitudinal muscle layers lie beneath the epidermis.

Most aschelminths are freshwater animals; only a few live in marine environments. The nematomorphs and acanthocephalans, and many of the nematodes, are parasitic. The rest of the aschelminths are mostly free-living; some rotifers are colonial.

PHYLUM ROTIFERA

The rotifers (ro′tif-ers) (L. *rota*, wheel + *fera*, to bear) derive their name from the characteristic ciliated organ, the **corona** (Gr. *krowe*, crown), around lobes on the heads of these animals (figure 11.4a). The corona functions in locomotion and food gathering. The cilia of the corona do not beat in synchrony; instead, each cilium is at a slightly earlier stage in the beat cycle than the next cilium in the sequence. A wave of beating cilia thus appears to pass around the periphery of the ciliated lobes and gives the impression of a pair of spinning wheels. (Interestingly, the rotifers were first called "wheel animalicules.")

Rotifers are small animals (0.1 to 3 mm in length) that are abundant in most freshwater habitats; a few (less than 10%) are marine. The approximately two thousand species are divided into three classes (table 11.1). The body has approximately a thousand cells, and the organs are eutelic. Rotifers are usually solitary, free-swimming animals, although some colonial forms are known. Others occur between grains of sand.

Characteristics of the phylum Rotifera include:

1. Triploblastic, bilateral, unsegmented, pseudocoelomate
2. Complete digestive system, regionally specialized
3. Anterior end often has a ciliated organ called a corona
4. Posterior end with toes and adhesive glands
5. Well-developed cuticle
6. Protonephridia with flame cells
7. Males generally reduced in number or absent; parthenogenesis common

EXTERNAL FEATURES

An epidermally secreted cuticle covers a rotifer's external surface. In many species, the cuticle thickens to form an encasement called a **lorica** (L. *corselet*, a loose-fitting case). The cuticle or lorica provides protection and is the main supportive element, although fluid in the pseudocoelom also provides hydrostatic support. The epidermis is syncytial; that is, no plasma membranes are between nuclei.

The head contains the corona, mouth, sensory organs, and brain (figure 11.4b). The corona surrounds a large, ciliated area called the buccal field. The trunk is the largest part of a rotifer, and is elongate and saclike. The anus occurs dorsally on the posterior trunk. The posterior narrow portion is called the foot. The

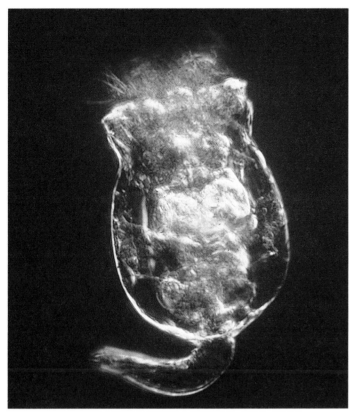

(a)

FIGURE 11.4

Phylum Rotifera.   (*a*) A rotifer, *Brachionus* (LM ×150). (*b*) Internal anatomy of a typical rotifer, *Philodina*. This rotifer is about 2 mm long.

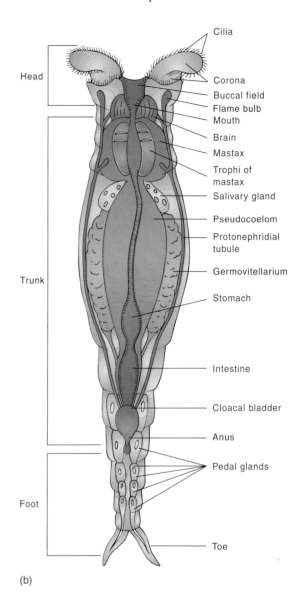

(b)

<table>
<tr><td colspan="2">**TABLE 11.1**
CLASSIFICATION OF THE ROTIFERA</td></tr>
</table>

Phylum Rotifera (ro′tif-er-ah)

A cilated corona surrounding a mouth; muscular pharynx (mastax) present with jawlike features; nonchitinous cuticle; parthenogenesis is common; both freshwater and marine species. About 2,000 species.

Class Seisonidea (sy″son-id′e-ah)

A single genus of marine rotifers that are commensals of crustaceans; large and elongate body with reduced corona. *Seison*. Only 2 species.

Class Bdelloidea (del-oid′e-ah)

Anterior end retractile and bearing two trochal disks; mastax adapted for grinding; paired ovaries; cylindrical body; males absent. *Adineta, Philodina, Rotaria*. About 590 species.

Class Monogononta (mon″o-go-non′tah)

Rotifers with one ovary; mastax not designed for grinding; produce mictic and amictic eggs. Males appear only sporadically. *Conochilus, Collotheca, Notommata*. About 1,400 species.

terminal portion of the foot usually bears one or two toes. At the base of the foot are many pedal glands whose ducts open on the toes. Secretions from these glands aid in the temporary attachment of the foot to a substratum.

FEEDING AND THE DIGESTIVE SYSTEM

Most rotifers feed on small microorganisms and suspended organic material. The coronal cilia create a current of water that brings food particles to the mouth. The pharynx contains a unique structure called the **mastax** (jaws). The mastax is a muscular organ that grinds food. The inner walls of the mastax contain several sets of jaws called trophi (figure 11.4*b*). The trophi vary in morphological detail, and taxonomists use them to distinguish species.

From the mastax, food passes through a short, ciliated esophagus to the ciliated stomach. Salivary and digestive glands secrete digestive enzymes into the pharynx and stomach.

Complete extracellular digestion and absorption of food occur in the stomach. In some species, a short, ciliated intestine extends posteriorly and becomes a cloacal bladder, which receives water from the protonephridia and eggs from the ovaries, as well as digestive waste. The cloacal bladder opens to the outside via an anus at the junction of the foot with the trunk.

OTHER ORGAN SYSTEMS

All visceral organs lie in a pseudocoelom filled with fluid and interconnecting amoeboid cells. Protonephridia that empty into the cloacal bladder function in osmoregulation. Rotifers, like other pseudocoelomates, exchange gases and dispose of nitrogenous wastes across body surfaces. The nervous system is composed of two lateral nerves and a bilobed, ganglionic brain on the dorsal surface of the mastax. Sensory structures include numerous ciliary clusters and sensory bristles concentrated on either one or more short antennae or the corona. One to five photosensitive eyespots may be on the head.

REPRODUCTION AND DEVELOPMENT

Some rotifers reproduce sexually, although several types of parthenogenesis occur in most species. Smaller males appear only sporadically in one class (Monogononta), and no males are known in another class (Bdelloidea). In the class Seisonidea, fully developed males and females are equally common in the population. Most rotifers have a single ovary and an attached syncytial vitellarium, which produces yolk that is incorporated into the eggs. The ovary and vitellarium often fuse to form a single germovitellarium (figure 11.4b). After fertilization, each egg travels through a short oviduct to the cloacal bladder and out its opening.

In males, the mouth, cloacal bladder, and other digestive organs are either degenerate or absent. A single testis produces sperm that travel through a ciliated vas deferens to the gonopore. Male rotifers typically have an eversible penis that injects sperm, like a hypodermic needle, into the pseudocoelom of the female (hypodermic impregnation).

In one class (Seisonidea), the females produce haploid eggs that must be fertilized to develop into either males or females. In another class (Bdelloidea), all females are parthenogenetic and produce diploid eggs that hatch into diploid females. The third class (Monogononta) produces two different types of eggs (figure 11.5). **Amictic** (Gr. *a*, without + *miktos*, mixed or blended; thin-shelled summer eggs) **eggs** are produced by mitosis, are diploid, cannot be fertilized, and develop directly into amictic females. Thin-shelled, **mictic** (Gr. *miktos*, mixed or blended) **eggs** are haploid. If the mictic egg is not fertilized, it develops parthenogenetically into a male; if fertilized, mictic eggs secrete a thick, heavy shell and become dormant or resting winter eggs. Dormant eggs always hatch with melting snows and spring rains into amictic females, which begin a first amictic cycle, building up large populations quickly. By early

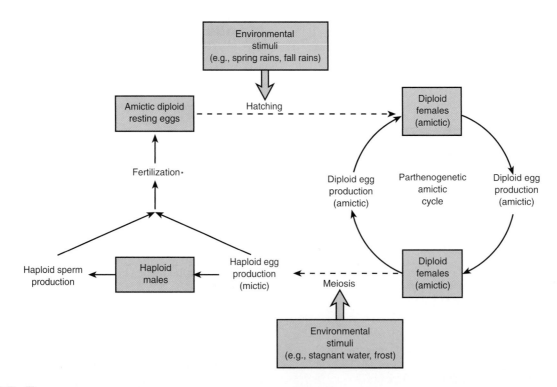

FIGURE 11.5

Life Cycle of a Monogonont Rotifer.   Dormant, diploid, resting eggs hatch in response to environmental stimuli (e.g., melting snows and spring rains) to begin a first amictic cycle. Other environmental stimuli (e.g., population density, stagnating water) later stimulate the production of haploid mictic eggs that lead to the production of dormant eggs that carry the species through the summer (e.g., when the pond dries up). The autumn rains initiate a second amictic cycle. Frost stimulates the production of mictic eggs again and the eventual dormant resting eggs that allow the population of rotifers to overwinter.

summer, some females have begun to produce mictic eggs, males appear, and dormant eggs are produced. Another amictic cycle, as well as the production of more dormant eggs, occurs before the yearly cycle is over. Winds or birds often disperse dormant eggs, accounting for the unique distribution patterns of many rotifers. Most females lay either amictic or mictic eggs, but not both. Apparently, during oocyte development, the physiological condition of the female determines whether her eggs will be amictic or mictic.

PHYLUM KINORHYNCHA

Kinorhynchs (kin'o-rinks) are small (less than 1 mm long), elongate, bilaterally symmetrical worms found exclusively in marine environments, where they live in mud and sand. Because they have no external cilia or locomotor appendages, they simply burrow through the mud and sand with their snouts. In fact, the phylum takes its name (Kinorhyncha, Gr. *kinein*, motion + *rhynchos*, snout) from this method of locomotion. The phylum Kinorhyncha contains about 150 known species.

The body surface of a kinorhynch is devoid of cilia and is composed of 13 or 14 definite units called **zonites** (figure 11.6). The head, represented by zonite 1, bears the mouth, an oral cone, and spines. The neck, represented by zonite 2, contains spines called **scalids** and plates called **placids.** The head can be retracted into the neck. The trunk consists of the remaining 11 or 12 zonites and terminates with the anus. Each trunk zonite bears a pair of lateral spines and one dorsal spine.

The body wall consists of a cuticle, epidermis, and two pairs of muscles: dorsolateral and ventrolateral. The pseudocoelom is large and contains amoeboid cells.

A complete digestive system is present, consisting of a mouth, buccal cavity, muscular pharynx, esophagus, stomach, intestine (where digestion and absorption take place), and anus. Most kinorhynchs feed on diatoms, algae, and organic matter.

A pair of protonephridia is in zonite 11. The nervous system consists of a brain and single ventral nerve cord with a ganglion (a mass of nerve cells) in each zonite. Some species have eyespots and sensory bristles. Kinorhynchs are dioecious with paired gonads. Several spines that may be used in copulation surround the male gonopore. The young hatch into larvae that do not have all of the zonites. As the larvae grow and molt, the adult morphology appears. Once adulthood is attained, molting no longer occurs.

PHYLUM NEMATODA

Nematodes (nem'a-todes) (Gr. *nematos*, thread) or roundworms are some of the most abundant animals on earth—some five billion may be in every acre (4,046 square meters) of fertile garden soil. Zoologists estimate that the number of roundworm species ranges from 16,000 to 500,000. Roundworms feed on every conceivable source of organic matter—from rotting substances to the living tissues of other invertebrates, vertebrates, and plants. They range in size from microscopic to several meters long. Many nematodes are parasites of plants or animals; most others are free-living in marine, freshwater, or soil habitats. Some nematodes play an important role in recycling nutrients in soils and bottom sediments.

Except in their sensory structures, nematodes lack cilia, a characteristic they share with arthropods. Also in common with some arthropods, the sperm of nematodes is amoeboid. Zoologists recognize two classes of nematodes (table 11.2).

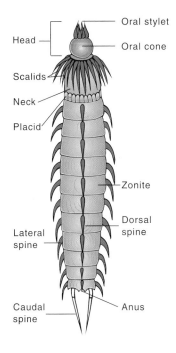

FIGURE 11.6
Phylum Kinorhyncha. External anatomy of an adult kinorhynch (dorsal view).

TABLE 11.2
CLASSIFICATION OF THE NEMATODES*

Phylum Nematoda (nem-a-to'dah)
Nematodes, or roundworms. About 16,000 species have been described to date.

Class Secernentea (Phasmidea) (ses-er-nen'te-ah)
Paired glandular or sensory structures called phasmids in the tail region; similar pair of structures (amphids) poorly developed in anterior end; excretory system present; both free-living and parasitic species. *Ascaris, Enterobius, Rhabditis, Turbutrix, Necator, Wuchereria*. About 5,000 described species.

Class Adenophorea (Aphasmidia) (a-den"o-for'e-ah) Phasmids absent; most free-living, but some parasitic species occur. *Dioctophyme, Trichinella, Trichuris*. About 3,000 species.

*A recent cladistic analysis using molecular data suggests that many of the morphological similarities among the different nematode groups evolved through convergence, and that one of the two nematode classes (the Secernentea) may have evolved from within the other class (the Adenophorea). If these data are confirmed, the long-accepted classification scheme presented in this table and in the text will be undergoing revision over the next few years.

Characteristics of the phylum Nematoda include:

1. Triploblastic, bilateral, vermiform (resembling a worm in shape; long and slender), unsegmented, pseudocoelomate
2. Body round in cross section and covered by a layered elastic cuticle; molting usually accompanies growth in juveniles
3. Complete digestive tract; mouth usually surrounded by lips bearing sense organs
4. Most with unique excretory system comprised of one or two renette cells or a set of collecting tubules
5. Body wall has only longitudinal muscles

EXTERNAL FEATURES

A typical nematode body is slender, elongate, cylindrical, and tapered at both ends (figure 11.7a,b). Much of the success of nematodes is due to their outer, noncellular, collagenous cuticle (figure 11.7c) that is continuous with the foregut, hindgut, sense organs, and parts of the female reproductive system. The cuticle may be smooth, or it may contain spines, bristles, papillae (small, nipplelike projections), warts, or ridges, all of which are of taxonomic significance. Three primary layers make up the cuticle: cortex, matrix layer, and basal layer. The cuticle maintains internal hydrostatic pressure, provides mechanical protection, and in parasitic species of nematodes, resists digestion by the host. The cuticle is usually molted four times during maturation.

Beneath the cuticle is the epidermis, or hypodermis, which surrounds the pseudocoelom (figure 11.7d). The epidermis may be syncytial, and its nuclei are usually in the four epidermal cords (one dorsal, one ventral, and two lateral) that project inward. The longitudinal muscles are the principal means of locomotion in nematodes. Contraction of these muscles results in undulatory waves

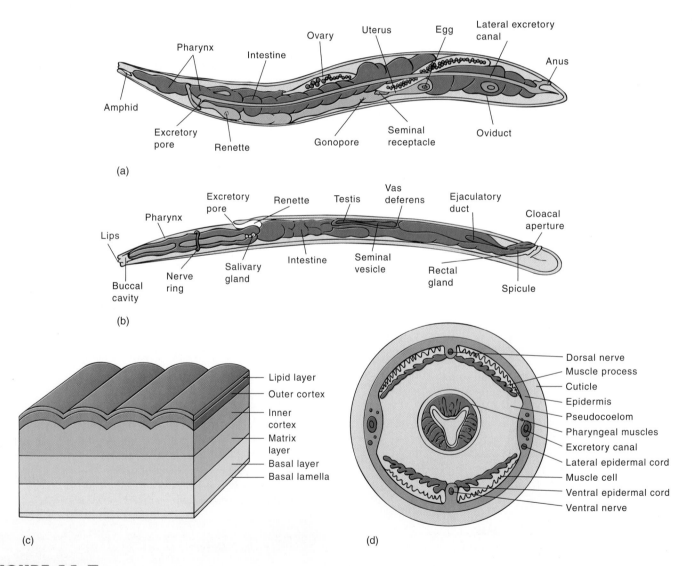

FIGURE 11.7

Phylum Nematoda.   Internal anatomical features of an (a) female and (b) male *Rhabditis*. (c) Section through a nematode cuticle, showing the various layers. (d) Cross section through the region of the muscular pharynx of a nematode. The hydrostatic pressure in the pseudocoelom maintains the rounded body shape of a nematode and also collapses the intestine, which helps move food and waste material from the mouth to the anus.

that pass from the anterior to the posterior end of the animal, creating characteristic thrashing movements. Nematodes lack circular muscles and therefore cannot crawl as do worms with more complex musculature.

Some nematodes have lips surrounding the mouth, and some species bear spines or teeth on or near the lips. In others, the lips have disappeared. Some roundworms have head shields that afford protection. Sensory organs include amphids, phasmids, or ocelli. **Amphids** are anterior depressions in the cuticle that contain modified cilia and function in chemoreception. **Phasmids** are near the anus and also function in chemoreception. The presence or absence of these organs determines the taxonomic class to which nematodes belong (*see table 11.2*). Paired ocelli (eyes) are present in aquatic nematodes.

INTERNAL FEATURES

The nematode pseudocoelom is a spacious, fluid-filled cavity that contains the visceral organs and forms a hydrostatic skeleton. All nematodes are round because the body muscles contracting against the pseudocoelomic fluid generate an equal outward force in all directions (figure 11.7d).

FEEDING AND THE DIGESTIVE SYSTEM

Depending on the environment, nematodes are capable of feeding on a wide variety of foods; they may be carnivores, herbivores, omnivores, or saprobes (saprotrophs) that consume decomposing organisms, or parasitic species that feed on blood and tissue fluids of their hosts.

Nematodes have a complete digestive system consisting of a mouth, which may have teeth, jaws, or stylets (sharp, pointed structures); buccal cavity; muscular pharynx; long, tubular intestine where digestion and absorption occur; short rectum; and anus. Hydrostatic pressure in the pseudocoelom and the pumping action of the pharynx push food through the alimentary canal.

OTHER ORGAN SYSTEMS

Nematodes accomplish osmoregulation and excretion of nitrogenous waste products (ammonia, urea) with two unique systems. The glandular system is in aquatic species and consists of ventral gland cells, called **renettes,** posterior to the pharynx (figure 11.8a). Each gland absorbs wastes from the pseudocoelom and empties them to the outside through an excretory pore. Parasitic nematodes have a more advanced system, called the tubular system, that develops from the renette system (figure 11.8b). In this system, the renettes unite to form a large canal, which opens to the outside via an excretory pore.

The nervous system consists of an anterior nerve ring (*see figure 11.7b*). Nerves extend anteriorly and posteriorly; many

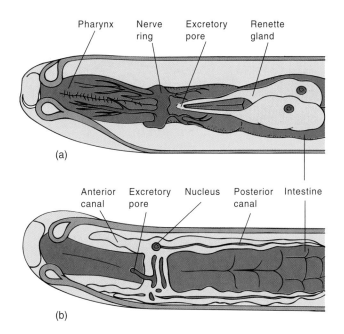

FIGURE 11.8
Nematode Excretory Systems.   (*a*) Glandular, as in *Rhabditis*. (*b*) Tubular, as in *Ascaris*.

connect to each other via commissures. Certain neuroendocrine secretions are involved in growth, molting, cuticle formation, and metamorphosis.

REPRODUCTION AND DEVELOPMENT

Most nematodes are dioecious and dimorphic, with the males being smaller than the females. The long, coiled gonads lie free in the pseudocoelom.

The female system consists of a pair of convoluted ovaries (figure 11.9a). Each ovary is continuous with an oviduct whose proximal end is swollen to form a seminal receptacle. Each oviduct becomes a tubular uterus; the two uteri unite to form a vagina that opens to the outside through a genital pore.

The male system consists of a single testis, which is continuous with a vas deferens that eventually expands into a seminal vesicle (figure 11.9b). The seminal vesicle connects to the cloaca. Males are commonly armed with a posterior flap of tissue called a bursa. The bursa aids the male in the transfer of sperm to the female genital pore during copulation.

After copulation, hydrostatic forces in the pseudocoelom (*see figure 11.7d*) move each fertilized egg to the gonopore (genital pore). The number of eggs produced varies with the species; some nematodes produce only several hundred, whereas others may produce hundreds of thousands daily. Some nematodes give birth to larvae (ovoviviparity). External factors, such as temperature and moisture, influence the development and hatching of the eggs. Hatching produces a larva (some parasitologists refer to it as a juvenile) that has most adult structures. The larva (juvenile) undergoes four molts, although in some species, the first one or two molts may occur before the eggs hatch.

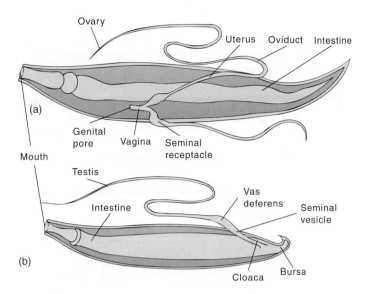

FIGURE 11.9

Nematode Reproductive Systems. The reproductive systems of (*a*) female and (*b*) male nematodes, such as *Ascaris*. The sizes of the reproductive systems are exaggerated to show details.

SOME IMPORTANT NEMATODE PARASITES OF HUMANS

Parasitic nematodes show a number of evolutionary adaptations to their way of life. These include a high reproductive potential, life cycles that increase the likelihood of transmission from one host to another, an enzyme-resistant cuticle, resistant eggs, and encysted larvae. Nematode life cycles are not as complicated as those of cestodes or trematodes because only one host is usually involved. Discussions of the life cycles of five important human parasites follow.

Ascaris lumbricoides: *The Giant Intestinal Roundworm of Humans*

As many as 800 million people throughout the world may be infected with *Ascaris lumbricoides*. Adult *Ascaris* (Gr. *askaris*, intestinal worm) live in the small intestine of humans. They produce large numbers of eggs that exit with the feces (figure 11.10). A first-stage larva develops rapidly in the egg, molts, and matures into a second-stage larva, the infective stage. When a human ingests embryonated eggs, they hatch in the intestine. The larvae penetrate the intestinal wall and are carried via the circulation to the lungs. They molt twice in the lungs, migrate up the trachea, and are swallowed. The worms attain sexual maturity in the intestine, mate, and begin egg production.

Enterobius vermicularis: *The Human Pinworm*

Pinworms (*Enterobius*; Gr. *enteron*, intestine + *bios*, life) are the most common roundworm parasites in the United States. Adult *Enterobius vermicularis* become established in the lower region of

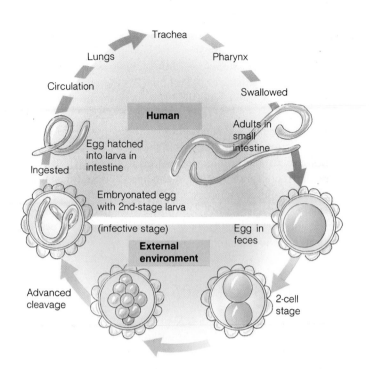

FIGURE 11.10

Life Cycle of *Ascaris lumbricoides*. (See text for details.)
Source: Redrawn from Centers for Disease Control, Atlanta, GA.

the large intestine. At night, gravid females migrate out of the rectum to the perianal area, where they deposit eggs containing a first-stage larva (figure 11.11) and then die. When humans ingest these eggs, the eggs hatch. The larvae molt four times in the small intestine and migrate to the large intestine. Adults mate, and females soon begin egg production.

Necator americanus: *The New World Hookworm*

The New World or American hookworm, *Necator americanus* (L. *necator*, killer), is found in the southern United States. The adults live in the small intestine, where they hold onto the intestinal wall with teeth and feed on blood and tissue fluids (figure 11.12). Individual females may produce as many as 10,000 eggs daily, which pass out of the body in the feces. An egg hatches on warm, moist soil and releases a small rhabditiform (the first- and second-stage juveniles of some nematodes) larva. It molts and becomes the infective filariform (the infective third-stage larva of some nematodes) larva. Humans become infected when the filariform larva penetrates the skin, usually between the toes. (Outside defecation and subsequent walking barefoot through the immediate area maintains the life cycle in humans.) The larva burrows through the skin to reach the circulatory system. The rest of its life cycle is similar to that of *Ascaris* (*see figure 11.10*).

Trichinella spiralis: *The Porkworm*

Adult *Trichinella* (Gr. *trichinos*, hair) *spiralis* live in the mucosa of the small intestine of humans and other carnivores and

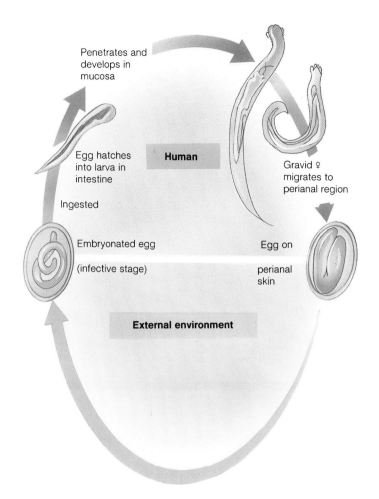

FIGURE 11.11

Life Cycle of *Enterobius vermicularis.*   (See text for details.)
Source: Redrawn from Centers for Disease Control, Atlanta, GA.

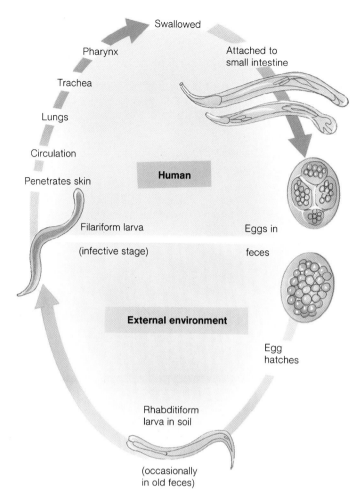

FIGURE 11.12

Life Cycle of *Necator americanus.*   (See text for details.)
Source: Redrawn from Centers for Disease Control, Atlanta, GA.

omnivores (e.g., the pig). In the intestine, adult females give birth to young larvae that then enter the circulatory system and are carried to skeletal (striated) muscles of the same host (figure 11.13). The young larvae encyst in the skeletal muscles and remain infective for many years. The disease this nematode causes is called **trichinosis.** Another host must ingest infective meat (muscle) to continue the life cycle. Humans most often become infected by eating improperly cooked pork products. Once ingested, the larvae excyst in the stomach and make their way to the small intestine, where they molt four times and develop into adults.

Wuchereria *spp.: The Filarial Worms*

In tropical countries, over 250 million humans are infected with filarial (L. *filium*, thread) worms. Two examples of human filarial worms are *Wuchereria bancrofti* and *W. malayi.* These elongate, threadlike nematodes live in the lymphatic system, where they block the vessels. Because lymphatic vessels return tissue fluids to the circulatory system, when the filiarial nematodes block these

vessels, fluids and connective tissue tend to accumulate in peripheral tissues. This fluid and connective tissue accumulation causes the enlargement of various appendages, a condition called **elephantiasis** (figure 11.14).

In the lymphatic vessels, filarial nematode adults copulate and produce larvae called **microfilariae** (figure 11.15). The microfilariae are released into the bloodstream of the human host and migrate to the peripheral circulation at night. When a mosquito feeds on a human, it ingests the microfilariae. The microfilariae migrate to the mosquito's thoracic muscles, where they molt twice and become infective. When the mosquito takes another meal of blood, the mosquito's proboscis injects the infective third-stage larvae into the blood of the human host. The final two molts take place as the larvae enter the lymphatic vessels.

A filarial worm prevalent in the United States is *Dirofilaria immitis,* a parasite of dogs. Since the adult worms live in the heart and large arteries of the lungs, the infection is called **heartworm disease.** Once established, these filarial worms are difficult to

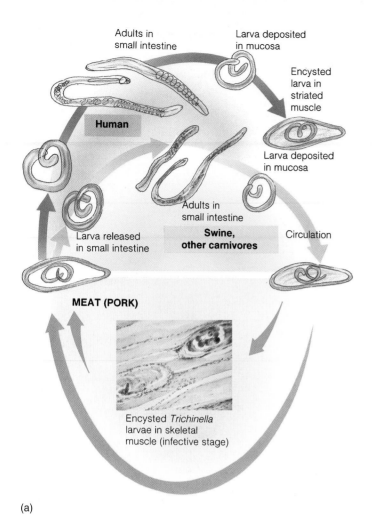

(a)

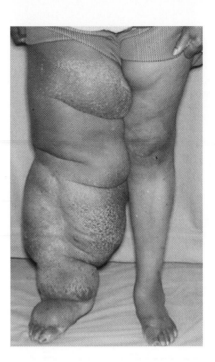

FIGURE 11.14

Elephantiasis Caused by the Filarial Worm *Wuchereria bancrofti*. It takes years for elephantiasis to advance to the degree shown in this photograph.

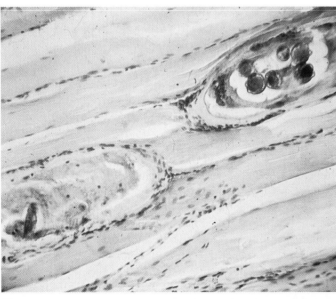

(b)

FIGURE 11.13

Life Cycle of *Trichinella spiralis*. (*a*) (See text for details.) (*b*) An enlargement of the insert in (*a*), showing two encysted larvae in skeletal muscle (LM ×450). (*a*) *Source: Redrawn from Centers for Disease Control, Atlanta, GA.*

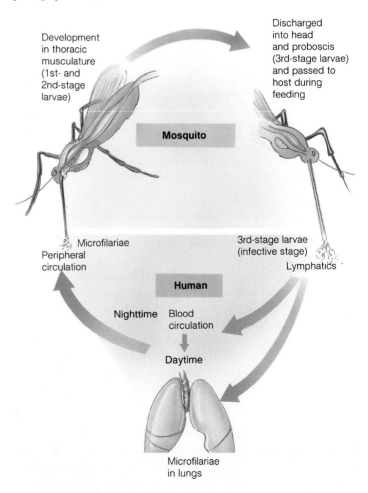

FIGURE 11.15

Life Cycle of *Wuchereria* spp. (See text for details.) *Source: Redrawn from Centers for Disease Control, Atlanta, GA.*

eliminate, and the condition can be fatal. Prevention with heart-worm medicine is thus advocated for all dogs.

PHYLUM NEMATOMORPHA

Nematomorphs (nem″a-to-mor′fs) (Gr. *nema*, thread + *morphe*, form) are a small group (about 250 species) of elongate worms commonly called either **horsehair worms** or **Gordian worms.** The hairlike nature of these worms is so striking that they were formerly thought to arise spontaneously from the hairs of a horse's tail in drinking troughs or other stock-watering places. The adults are free-living, but the juveniles are all parasitic in arthropods. They have a worldwide distribution and can be found in both running and standing water.

The nematomorph body is extremely long and threadlike and has no distinct head (figure 11.16). The body wall has a thick cuticle, a cellular epidermis, longitudinal cords, and a muscle layer of longitudinal fibers. The nervous system contains an anterior nerve ring and a ventral cord.

Nematomorphs have separate sexes; two long gonads extend the length of the body. After copulation, the eggs are deposited in water. A small larva with a protrusible proboscis armed with spines hatches from the egg. Terminal stylets are also present on the pro-

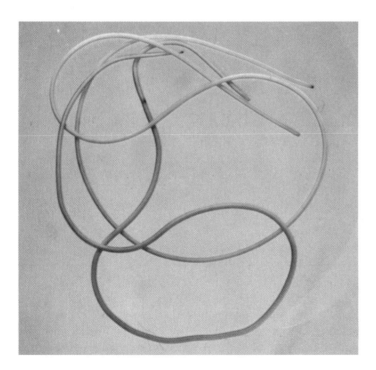

FIGURE 11.16

Phylum Nematomorpha. Photomicrograph of two adult worms, each about 25 cm long. These worms tend to twist and turn upon themselves, giving the appearance of complicated knots—thus, the name "Gordian worms." (Legend has it that King Gordius of Phrygia tied a formidable knot—the Gordian knot—and declared that whoever might undo it would be the ruler of all Asia. No one could accomplish this until Alexander the Great cut through it with his broadsword.)

boscis. The larva must quickly enter an arthropod (e.g., a beetle, cockroach) host, either by penetrating the host or by being eaten. Lacking a digestive system, the larva feeds by absorbing material directly across its body wall. Once mature, the worm leaves its host only when the arthropod is near water. Sexual maturity is attained during the free-living adult phase of the life cycle.

PHYLUM ACANTHOCEPHALA

Adult acanthocephalans (a-kan″tho-sef′a-lans) (Gr. *akantha*, spine or thorn + *kephale*, head) are endoparasites in the intestinal tract of vertebrates (especially fishes). Two hosts are required to complete the life cycle. The juveniles are parasites of crustaceans and insects. Acanthocephalans are generally small (less than 40 mm long), although one important species, *Macracanthorhynchus hirudinaceus*, which occurs in pigs can be up to 80 cm long. The body of the adult is elongate and composed of a short anterior proboscis, a neck region, and a trunk (figure 11.17*a*). The proboscis is covered with recurved spines (figure 11.17*b*); hence, the name "spiny-headed worms." The retractible proboscis provides the means of attachment in the host's intestine. Females are always larger than males, and zoologists have identified about a thousand species.

A living syncytial tegument that has been adapted to the parasitic way of life covers the body wall of acanthocephalans. A glycocalyx (*see figure 2.6*) consisting of mucopolysaccharides and glycoproteins covers the tegument and protects against host enzymes and immune defenses. No digestive system is present; acanthocephalans absorb food directly through the tegument from the host by specific membrane transport mechanisms and pinocytosis. Protonephridia may be present. The nervous system is composed of a ventral, anterior ganglionic mass from which anterior and posterior nerves arise. Sensory organs are poorly developed.

The sexes are separate, and the male has a protrusible penis. Fertilization is internal, and eggs develop in the pseudocoelom. The biotic potential of certain acanthocephalans is great; for example, a gravid female *Macroacanthorhynchus hirudinaceus* may contain up to 10 million embryonated eggs. The eggs pass out of the host with the feces and must be eaten by certain insects (e.g., cockroaches or grubs [beetle larvae]) or by aquatic crustaceans (e.g., amphipods, isopods, ostracods). Once in the invertebrate, the larva emerges from the egg and is now called an **acanthor.** It burrows through the gut wall and lodges in the hemocoel, where it develops into an **acanthella** and, eventually, into a **cystacanth.** When a mammal, fish, or bird eats the intermediate host the cystacanth excysts and attaches to the intestinal wall with its spiny proboscis.

PHYLUM LORICIFERA

The phylum Loricifera (lor′a-sif-er-ah) (L. *lorica*, clothed in armor + *fero*, to bear) is a recently described animal phylum. Its first members were identified and named in 1983. Loriciferans live in

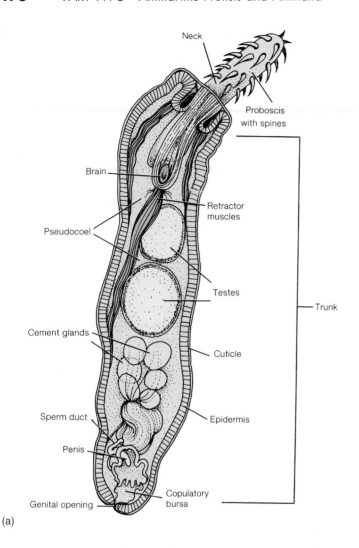

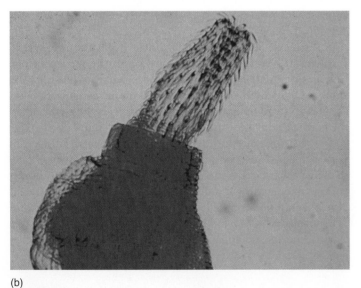

FIGURE 11.17

Phylum Acanthocephala. (*a*) Dorsal view of an adult male. (*b*) The proboscis of a spiny-headed worm (LM ×50).

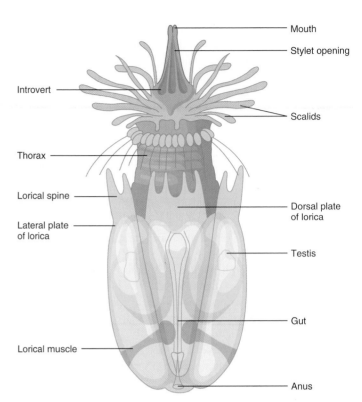

FIGURE 11.18

Phylum Loricifera. Dorsal view of the anatomy of an adult male *Nanaloricus*.

spaces between marine gravel. A characteristic species is *Nanaloricus mysticus*. It is a small, bilaterally symmetrical worm with a spiny head called an **introvert,** a thorax, and an abdomen surrounded by a lorica (figure 11.18). Loriciferans can retract both the introvert and thorax into the anterior end of the lorica. The introvert bears eight oral stylets that surround the mouth. The lorical cuticle is periodically molted. A pseudocoelom is present and contains a short digestive system, brain, and several ganglia. Loriciferans are dioecious with paired gonads. Zoologists have described about 14 species.

PHYLUM PRIAPULIDA

The priapulids (pri′a-pyu-lids) (Gr. *priapos,* phallus + *ida,* pleural suffix; from *Priapos,* the Greek god of reproduction, symbolized by the penis) are a small group (only 16 species) of marine worms found in cold waters. They live buried in the mud and sand of the seafloor, where they feed on small annelids and other invertebrates.

The priapulid body is cylindrical in cross section and ranges in length from 2 mm to about 8 cm (figure 11.19). The anterior part of the body is an introvert (proboscis), which priapulids can draw into the longer, posterior trunk. The introvert functions in burrowing, and spines surround it. A thin cuticle that bears spines covers the muscular body, and the trunk bears superficial annuli. A straight digestive tract is suspended in a large pseudocoelom

EVOLUTIONARY INSIGHTS
What Are Worms?

We used the term "worm" in the previous chapter in a very generic sense; for example, free-living flatworms, parasitic flatworms, soft-bodied worms, tapeworms, and bladder worms. This generic use of the word "worm" also occurs in this chapter: roundworms, porkworm, filarial worms, horsehair or Gordian worms. In other chapters you will encounter the terms acorn-worms, arrow-worms, tongue-worms, and others. Therefore, what are worms? Worms covered in this zoology book can be defined as animals that do not possess legs, are not covered by a protective shell, are not deuterostomes, do not bear a lophophore, are soft-bodied, have bodies from two or three to over 15,000 times longer than wide, and are either flattened or rounded in section. This means that *worm* is an indefinite though suggestive term popularly applied to any animal that is not obviously something else. The worms therefore do not form a natural group, as might be expected from a group largely defined on several negative characteristics. They are more of a group for convenience, distinguished more by the common absence of the various features that characterize other groups of phyla than by anything else.

If the ancestral animal was wormlike, as many phylogenetic schemes postulate, it would follow that a wide variety of worms might evolve from a vermiform (L. *vermis*, a worm, wormlike) ancestor, and that ultimately some of these worms might give rise to non-worms. There are other schemes that read the sequence in the opposite direction and derive at least some worms from non-worms.

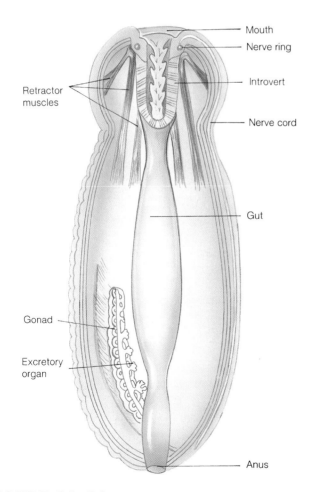

FIGURE 11.19

Phylum Priapulida.   Internal anatomy of the priapulid, *Priapulus caudatus*, with the introvert withdrawn into the body.

that acts as a hydrostatic skeleton. In some species, the pseudocoelom contains amoeboid cells that probably function in gas transport. The nervous system consists of a nerve ring around the pharynx and a single midventral nerve cord. The sexes are separate but not superficially distinguishable. A pair of gonads is suspended in the pseudocoelom and shares a common duct with the protonephridia. The duct opens near the anus, and gametes are shed into the sea. Fertilization is external, and the eggs eventually sink to the bottom, where the larvae develop into adults. The cuticle is repeatedly molted throughout life. The most commonly encountered species is *Priapulus caudatus*.

FURTHER PHYLOGENETIC CONSIDERATIONS

The aschelminths are clearly a diverse assemblage of animals. Despite the common occurrence of a cuticle, pseudocoelom, muscular pharynx, and adhesive glands, no distinctive features occur in every phylum.

The rotifers have certain features in common with the acoelomates. The protonephridia of rotifers closely resemble those of some freshwater turbellarians, and zoologists generally believe that rotifers originated in freshwater habitats. Both flatworms and rotifers have separate ovaries and vitellaria. Rotifers probably had their origins from the earliest acoelomates and may have had a common bilateral, metazoan ancestor.

The kinorhynchs, acanthocephalans, loriciferans, and priapulids all have a spiny anterior end that can be retracted; thus, they are probably related. Loriciferans and kinorhynchs appear to be most closely related.

The affinities of the nematodes to other phyla are vague. No other living group is believed to be closely related to these worms. Nematodes probably evolved in freshwater habitats and then colonized the oceans and soils. The ancestral nematodes may have been sessile, attached at the posterior end, with the anterior end protruding upward into the water. The nematode cuticle, feeding structures, and food habits probably preadapted these worms for parasitism. In fact, free-living species could become parasitic without substantial anatomical or physiological changes.

Nematomorphs may be more closely related to nematodes than to any other group by virtue of both groups being cylindrical in shape, having a cuticle, and being dioecious and sexually dimorphic. However, because the larval form of some nematomorphs resembles priapulids, the exact affinity to the nematodes is questionable.

SUMMARY

1. The aschelminths are seven phyla grouped for convenience. Most have a well-defined pseudocoelom, a constant number of body cells or nuclei (eutely), protonephridia, and a complete digestive system with a well-developed pharynx. No organs are developed for gas exchange or circulation. A cuticle that may be molted covers the body. Only longitudinal muscles are often present in the body wall.

2. The phylogenetic affinities among the seven phyla and with other phyla are uncertain.

3. The majority of rotifers inhabit freshwater. The head of these animals bears a unique ciliated corona used for locomotion and food capture. Males are smaller than females and unknown in some species. Females may develop parthenogenetically.

4. Kinorhynchs are minute worms living in marine habitats. Their bodies are comprised of 13 zonites, which have cuticular scales, plates, and spines.

5. Nematodes live in aquatic and terrestrial environments. Many are parasitic and of medical and agricultural importance. They are all elongate, slender, and circular in cross section. Two sexes are present.

6. Nematomorpha are threadlike and free-living in freshwater. They lack a digestive system.

7. Acanthocephalans are also known as spiny-headed worms because of their spiny proboscis. All are endoparasites in vertebrates.

8. The phylum Loricifera was described in 1983. These microscopic animals have a spiny head and thorax, and they live in gravel in marine environments.

9. The phylum Priapulida contains only 16 known species of cucumber-shaped, wormlike animals that live buried in the bottom sand and mud in marine habitats.

SELECTED KEY TERMS

amictic eggs (p. 168)
aschelminths (p. 164)
corona (p. 166)
cuticle (p. 166)

mastax (p. 167)
mictic eggs (p. 168)
trichinosis (p. 173)
zonites (p. 169)

CRITICAL THINKING QUESTIONS

1. Discuss how the structure of the body wall places limitations on shape changes in nematodes.

2. What characteristics set the Nematomorpha apart from the Nematoda? What characteristics do the Nematomorpha share with the Nematoda?

3. In what respects are the kinorhynchs like nematodes? How are they like rotifers?

4. How are nematodes related to the rotifers?

5. What environmental factors appear to trigger the production of mictic females in monogonont rotifers?

ONLINE LEARNING CENTER

Visit our Online Learning Center (OLC) at www.mhhe.com/zoology (click on this book's title) to find the following chapter-related materials:

- CHAPTER QUIZZING
- RELATED WEB LINKS
  Phylum Rotifera
  Phylum Kinorhyncha
  Phylum Loricifera
  Phylum Priapulida
  Phylum Nematoda
  Human Diseases Caused by Nematodes
  Caenorhabditis elegans
  Phylum Nematomorpha
  Phylum Acanthocephala
- BOXED READINGS ON
  An Application of Eutely
  The Ecology of Soil Nematodes
- SUGGESTED READINGS
- LAB CORRELATIONS

Check out the OLC to find specific information on these related lab exercises in the *General Zoology Laboratory Manual,* 5th edition, by Stephen A. Miller:

Exercise 12   *The Pseudocoelomate Body Plan: Aschelminths*

CHAPTER 12

MOLLUSCAN SUCCESS

Outline

Concepts

1. Molluscs are protostomes. Relationships to other protostomes are distant and evolutionary pathways are speculative.
2. Molluscs have a coelom, as well as a head-foot, visceral mass, mantle, and mantle cavity. Most also have a radula.
3. Members of the class Gastropoda are the snails and slugs. They include the only terrestrial molluscs. Torsion and shell coiling modify their bodies.
4. Clams, oysters, mussels, and scallops are members of the class Bivalvia. They are all aquatic filter feeders and often burrow in soft substrates or attach to hard substrates.
5. The class Cephalopoda includes the octopuses, squids, cuttlefish, and nautiluses. They are the most complex of all invertebrates and are adapted for predatory lifestyles.
6. Other molluscs include members of the classes Scaphopoda (tooth shells), Monoplacophora, Aplacophora (solenogasters), and Polyplacophora (chitons). Members of these classes are all marine.
7. Specializations of molluscs have obscured evolutionary relationships among molluscan classes.

EVOLUTIONARY PERSPECTIVE

Octopuses, squids, and cuttlefish (the cephalopods) are some of the invertebrate world's most adept predators (figure 12.1). Predatory lifestyles have resulted in the evolution of large brains (by invertebrate standards), complex sensory structures (by any standards), rapid locomotion, grasping tentacles, and tearing mouthparts. In spite of these adaptations, cephalopods rarely make up a major component of any community. Once numbering about nine thousand species, the class Cephalopoda now includes only about 550 species (figure 12.1).

Zoologists do not know why the cephalopods have declined so dramatically. Vertebrates may have outcompeted cephalopods because the vertebrates were also making their appearance in prehistoric seas, and some vertebrates (e.g., bony fish) acquired active, predatory lifestyles.

This evolutionary decline has not been the case for all molluscs. Overall, this group has been very successful. If success is measured by numbers of species, the molluscs are twice as successful as vertebrates! The vast majority of the nearly 100,000 living species of molluscs belongs to two classes: Gastropoda, the snails and slugs; and Bivalvia, the clams and their close relatives.

This chapter contains evolutionary concepts, which are set off in this font.

FIGURE 12.1

Phylum Mollusca. The phylum Mollusca includes nearly 100,000 living species, including members of the class Cephalopoda—some of the invertebrate world's most adept predators. A member of the genus *Octopus* is shown here.

Molluscs are triploblastic, as are all the remaining animals covered in this text. In addition, they are the first animals described in this text that possess a coelom, although the coelom of molluscs is only a small cavity (the pericardial cavity) surrounding the heart and gonads. A coelom is a body cavity that arises in mesoderm and is lined by a sheet of mesoderm called the peritoneum (*see figure 7.11c*).

RELATIONSHIPS TO OTHER ANIMALS

Molluscs are protostomes (figure 12.2). The similarities in the embryological development of the molluscs and other protostomes, especially the annelids (segmented worms), are striking. Some embryological stages, for example the trochophore larvae (*see figure 7.12*), are virtually indistinguishable in molluscs and annelids. Certain adult structures of molluscs and annelids, for example the excretory organs and their duct systems, are very similar in structure. Even though most zoologists accept the protosome affiliation of the molluscs, the relationships between members of this phylum and other protostomes are distant, and the ancestral evolutionary pathways are speculative.

ORIGIN OF THE COELOM

A number of hypotheses focus on the origin of the coelom. These hypotheses influence how zoologists picture the evolutionary relationships among triploblastic phyla.

The schizocoel hypothesis (Gr. schizen, to split + koilos, hollow) is patterned after the method of mesoderm development and coelom formation in many protostomes (see figure 7.12a). Mesoderm fills the area between ectoderm and endoderm. The coelom arises from a splitting of this mesoderm. If the coelom formed in this way during evolution, mesodermally derived tissues would have preceded the coelom, implying that a triploblastic, acoelomate (flatworm) body form could be the forerunner of the coelomate body form (*see figure 7.11a*).

The enterocoel hypothesis (Gr. enteron, gut + koilos, hollow) suggests that the coelom may have arisen as outpocketings of a primitive gut tract. This hypothesis is patterned after the method of coelom formation in deuterostomes (other than vertebrates; *see figure 7.12b*). The implication of this hypothesis is that mesoderm and the coelom formed from the gut of a diploblastic animal. If this is true, the triploblastic, acoelomate body form would have been secondarily derived by mesoderm filling the body cavity of a coelomate animal.

The coelom usually performs multiple functions, including one or more of the following: providing room for organ development; surfaces for diffusion of gases, nutrients, and wastes; storage; elimination of reproductive products; and hydrostatic support (*see chapter 7*). These functions of the coelom are important in animals of both the protostomate and deuterostomate lineages. The differences in coelom development in these lineages have led some zoologists to suggest that the coelom may have arisen independently in each lineage; thus more than one explanation of coelom origin could be correct.

Molluscs are protostomate animals, and their coelom forms by splitting blocks of mesoderm in the usual protostomate fashion. The result of molluscan development, however, is that the coelom is evidenced only as relatively small cavities around the heart, the pericardial cavity, the nephridia, and the gonads. The evolution of the unique and very successful body form of the molluscs resulted in reduced size and importance of the coelom.

MOLLUSCAN CHARACTERISTICS

Molluscs range in size and body form from the giant squid, measuring 18 m in length, to the smallest garden slug, less than 1 cm long. In spite of this diversity, the phylum Mollusca (mol-lus'kah) (L. *molluscus*, soft) is not difficult to characterize (table 12.1).

Characteristics of the phylum Mollusca include:

1. Body of two parts: head-foot and visceral mass
2. Mantle that secretes a calcareous shell and covers the visceral mass
3. Mantle cavity functions in excretion, gas exchange, elimination of digestive wastes, and release of reproductive products
4. Bilateral symmetry

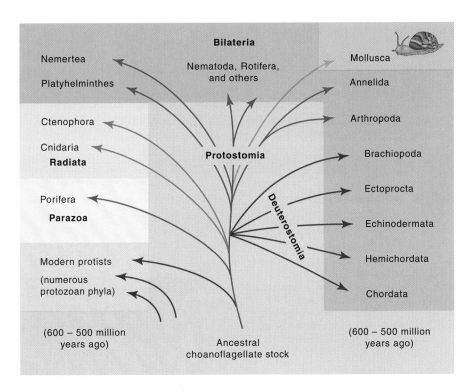

FIGURE 12.2

Evolutionary Relationships of the Molluscs. Molluscs are protostomates. They are distantly related to other protostomes. The earliest molluscan fossils are found in Ediacaran fossil beds that date to about 600 million years ago. Evolutionary pathways linking the Mollusca with other protostomate phyla are largely unknown.

TABLE 12.1
CLASSIFICATION OF THE MOLLUSCA

Phylum Mollusca (mol-lus′kah)

The coelomate animal phylum whose members possess a head-foot, visceral mass, mantle, and mantle cavity. Most molluscs also possess a radula and a calcareous shell. Nearly 100,000 species.

Class Caudofoveata (kaw′do-fo″ve-a″ta)

Wormlike molluscs with a cylindrical, shell-less body and scale-like, calcareous spicules; lack eyes, tentacles, statocysts, crystalline style, foot, and nephridia. Deep-water, marine burrowers. *Chaetoderma.* Approximately 70 species.

Class Aplacophora (a″pla-kof′o-rah)

Shell, mantle, and foot lacking; wormlike; head poorly developed; burrowing molluscs. Marine. *Neomenia.* Approximately 250 species.

Class Polyplacophora (pol″e-pla-kof′o-rah)

Elongate, dorsoventrally flattened; head reduced in size; shell consisting of eight dorsal plates. Marine, on rocky intertidal substrates. *Chiton.*

Class Monoplacophora (mon″o-pla-kof′o-rah)

Molluscs with a single arched shell; foot broad and flat; certain structures serially repeated. Marine. *Neopilina.*

Class Scaphopoda (ska-fop′o-dah)

Body enclosed in a tubular shell that is open at both ends; tentacles used for deposit feeding; no head. Marine. *Dentalium.* Over 300 species.

Class Bivalvia (bi″val′ve-ah)

Body enclosed in a shell consisting of two valves, hinged dorsally; no head or radula; wedge-shaped foot. Marine and freshwater. *Anodonta, Mytilus, Venus.* Approximately 30,000 species.

Class Gastropoda (gas-trop′o-dah)

Shell, when present, usually coiled; body symmetry distorted by torsion; some monoecious species. Marine, freshwater, terrestrial. *Nerita, Orthaliculus, Helix.* Over 35,000 species.

Class Cephalopoda (sef″ah-lah″po′dah)

Foot modified into a circle of tentacles and a siphon; shell reduced or absent; head in line with the elongate visceral mass. Marine. *Octopus, Loligo, Sepia, Nautilus.*

This taxonomic listing reflects a phylogenetic sequence. The discussions that follow, however, begin with molluscs that are familiar to most students.

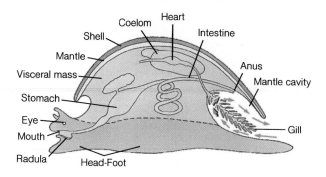

FIGURE 12.3

Molluscan Body Organization. This illustration of a hypothetical mollusc shows three features unique to the phylum. The head-foot is a muscular structure usually used for locomotion and sensory perception. The visceral mass contains organs of digestion, circulation, reproduction, and excretion. The mantle is a sheet of tissue that enfolds the rest of the body and secretes the shell. Blue arrows indicate the flow of water through the mantle cavity.

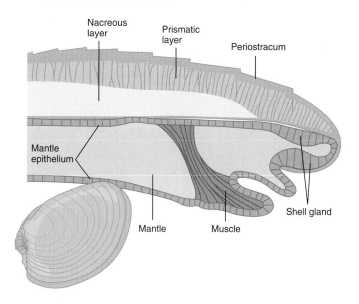

FIGURE 12.4

Molluscan Shell and Mantle. A transverse section of a bivalve shell and mantle shows the three layers of the shell and the portions of the mantle responsible for shell secretion.

5. Protostome characteristics, including trochophore larvae, spiral cleavage, and schizocoelous coelom formation
6. Coelom reduced to cavities surrounding the heart, nephridia, and gonads
7. Open circulatory system in all but one class (Cephalopoda)
8. Radula usually present and used in scraping food

The body of a mollusc has three main regions—the head-foot, the visceral mass, and the mantle (figure 12.3). The **head-foot** is elongate with an anterior head, containing the mouth and certain nervous and sensory structures, and an elongate foot, used for attachment and locomotion. The **visceral mass** contains the organs of digestion, circulation, reproduction, and excretion and is positioned dorsal to the head-foot.

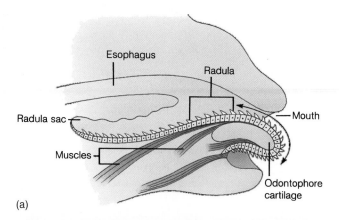

(a)

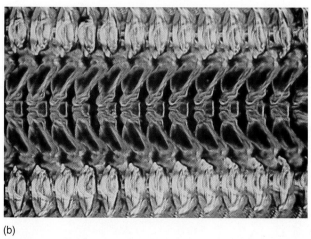

(b)

FIGURE 12.5

Radular Structure. (a) The radular apparatus lies over the cartilaginous odontophore. Muscles attached to the radula move the radula back and forth over the odontophore (see arrows). (b) Micrograph of radular teeth arrangement of the marine snail, *Nerita.* Tooth structure is an important taxonomic characteristic for zoologists who study molluscs. (a) From: "A LIFE OF INVERTEBRATES" © 1979 W. D. Russel-Hunter.

The **mantle** of a mollusc usually attaches to the visceral mass, enfolds most of the body, and may secrete a shell that overlies the mantle. The shell of a mollusc is secreted in three layers (figure 12.4). The outer layer of the shell is called the periostracum. Mantle cells at the mantle's outer margin secrete this protein layer. The middle layer of the shell, called the prismatic layer, is the thickest of the three layers and consists of calcium carbonate mixed with organic materials. Cells at the mantle's outer margin also secrete this layer. The inner layer of the shell, the nacreous layer, forms from thin sheets of calcium carbonate alternating with organic matter. Cells along the entire epithelial border of the mantle secrete the nacreous layer. Nacre secretion thickens the shell.

Between the mantle and the foot is a space called the **mantle cavity.** The mantle cavity opens to the outside and functions in gas exchange, excretion, elimination of digestive wastes, and release of reproductive products.

The mouth of most molluscs possesses a rasping structure called a **radula,** which consists of a chitinous belt and rows of posteriorly curved teeth (figure 12.5). The radula overlies a fleshy, tonguelike structure supported by a cartilaginous **odontophore.**

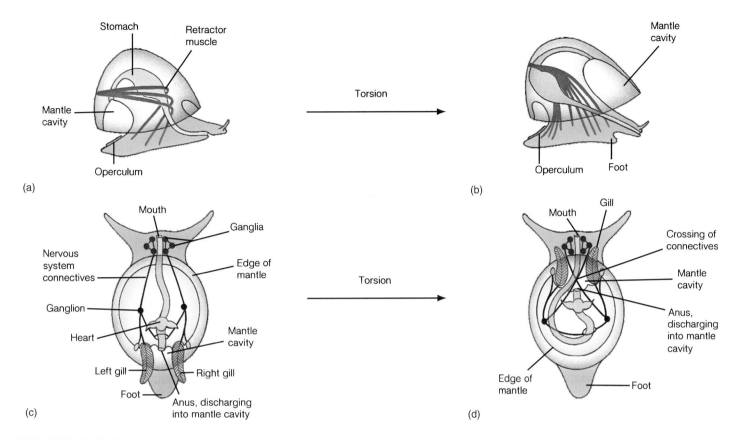

FIGURE 12.6

Torsion in Gastropods. (*a*) A pretorsion gastropod larva. Note the posterior opening of the mantle cavity and the untwisted digestive tract. (*b*) After torsion, the digestive tract is looped, and the mantle cavity opens near the head. The foot is drawn into the shell last, and the operculum closes the shell opening. (*c*) A hypothetical adult ancestor, showing the arrangement of internal organs prior to torsion. (*d*) Modern adult gastropods have an anterior opening of the mantle cavity and the looped digestive tract. *Redrawn from L. Hyman,* The Invertebrates, Volume VI. *Copyright © 1967 McGraw-Hill, Inc. Used by permission.*

Muscles associated with the odontophore permit the radula to be protruded from the mouth. Muscles associated with the radula move the radula back and forth over the odontophore. Food is scraped from a substrate and passed posteriorly to the digestive tract.

CLASS GASTROPODA

The class Gastropoda (gas-trop'o-dah) (Gr. *gaster*, gut + *podos*, foot) includes the snails, limpets, and slugs. With over 35,000 living species (*see table 12.1*), Gastropoda is the largest and most varied molluscan class. Its members occupy a wide variety of marine, freshwater, and terrestrial habitats. Most people give gastropods little thought unless they encounter *Helix pomatia* (escargot) in a French restaurant or are pestered by garden slugs and snails. One important impact of gastropods on humans is that gastropods are intermediate hosts for some medically important trematode parasites of humans (*see chapter 10*).

TORSION

One of the most significant modifications of the molluscan body form in the gastropods occurs early in development. **Torsion** is a 180°, counterclockwise twisting of the visceral mass, mantle, and mantle cavity. Torsion positions the gills, anus, and openings from the excretory and reproductive systems just behind the head and nerve cords, and twists the digestive tract into a **U** shape (figure 12.6).

The adaptive significance of torsion is speculative; however, three advantages are plausible. First, without torsion, withdrawal into the shell would proceed with the foot entering first and the more vulnerable head entering last. With torsion, the head enters the shell first, exposing the head less to potential predators. In some snails, a proteinaceous, and in some calcareous, covering, called an **operculum,** on the dorsal, posterior margin of the foot enhances protection. When the gastropod draws the foot into the mantle cavity, the operculum closes the opening of the shell, thus preventing desiccation when the snail is in drying habitats. A second advantage of torsion concerns an anterior opening of the mantle cavity that allows clean water from in front of the snail to enter the mantle cavity, rather than water contaminated with silt stirred up by the snail's crawling. The twist in the mantle's sensory organs around to the head region is a third advantage of torsion because it makes the snail more sensitive to stimuli coming from the direction in which it moves.

(a)

(b)

FIGURE 12.7

Gastropod Structure.   (*a*) A land (pulmonate) gastropod (*Orthaliculus*).
(*b*) Internal structure of a generalized gastropod.

Note in figure 12.6*d* that, after torsion, the anus and nephridia empty dorsal to the head and create potential fouling problems. However, a number of evolutionary adaptations seem to circumvent this problem. Various modifications allow water and the wastes it carries to exit the mantle cavity through notches or openings in the mantle and shell posterior to the head. Some gastropods undergo detorsion, in which the embryo undergoes a full 180° torsion and then untwists approximately 90°. The mantle cavity thus opens on the right side of the body, behind the head.

SHELL COILING

The earliest fossil gastropods had a shell that was coiled in one plane. This arrangement is not common in later fossils, probably because growth resulted in an increasingly cumbersome shell. (Some modern snails, however, have secondarily returned to this shell form.)

Most modern snail shells are asymmetrically coiled into a more compact form, with successive coils or whorls slightly larger than, and ventral to, the preceding whorl (figure 12.7*a*).

This pattern leaves less room on one side of the visceral mass for certain organs, which means that organs that are now single were probably paired ancestrally. This asymmetrical arrangement of internal organs is described further in the descriptions of particular body systems.

LOCOMOTION

Nearly all gastropods have a flattened foot that is often ciliated, covered with gland cells, and used to creep across the substrate (figure 12.7*b*). The smallest gastropods use cilia to propel themselves over a mucous trail. Larger gastropods use waves of muscular contraction that move over the foot. The foot of some gastropods is modified for clinging, as in abalones and limpets, or for swimming, as in sea butterflies and sea hares.

FEEDING AND DIGESTION

Most gastropods feed by scraping algae or other small, attached organisms from their substrate using their radula. Others are herbivores that feed on larger plants, scavengers, parasites, or predators.

The anterior portion of the digestive tract may be modified into an extensible proboscis, which contains the radula. This structure is important for some predatory snails that must extract animal flesh from hard-to-reach areas. The digestive tract of gastropods, like that of most molluscs, is ciliated. Food is trapped in mucous strings and incorporated into a mucoid mass called the **protostyle,** which extends to the stomach and is rotated by cilia. A digestive gland in the visceral mass releases enzymes and acid into the stomach, and food trapped on the protostyle is freed and digested. Wastes form fecal pellets in the intestine.

OTHER MAINTENANCE FUNCTIONS

Gas exchange always involves the mantle cavity. Primitive gastropods had two gills; modern gastropods have lost one gill because of coiling. Some gastropods have a rolled extension of the mantle, called a **siphon,** that serves as an inhalant tube. Burrowing species extend the siphon to the surface of the substrate to bring in water. Gills are lost or reduced in land snails (pulmonates), but these snails have a richly vascular mantle for gas exchange between blood and air. Mantle contractions help circulate air and water through the mantle cavity.

Gastropods, like most molluscs, have an **open circulatory system.** During part of its circuit around the body, blood leaves the vessels and directly bathes cells in tissue spaces called sinuses. Molluscs typically have a heart consisting of a single, muscular ventricle and two auricles. Most gastropods have lost one member of the pair of auricles because of coiling and thus have a single auricle and a single ventricle (*see figure 12.7b*).

In addition to transporting nutrients, wastes, and gases, the blood of molluscs acts as a hydraulic skeleton. A **hydraulic skeleton** consists of fluid under pressure that may be confined to tissue spaces to extend body structures and to support the body. Molluscs contract muscles to force fluid, in this case blood, into a distant structure to push it forward. For example, snails have sensory tentacles on their heads, and if a tentacle is touched, retractor muscles can rapidly withdraw it. However, no antagonistic muscles exist to extend the tentacle. The snail slowly extends the tentacle by contracting distant muscles to squeeze blood into the tentacle from adjacent blood sinuses.

The nervous system of primitive gastropods is characterized by six ganglia located in the head-foot and visceral mass. In primitive gastropods, torsion twists the nerves that link these ganglia. *The evolution of the gastropod nervous system has resulted in the untwisting of nerves and the concentration of nervous tissues into fewer, larger ganglia, especially in the head (see figure 12.7b).*

Gastropods have well-developed sensory structures. Eyes may be at the base or at the end of tentacles. They may be simple pits of photoreceptor cells or they may consist of a lens and cornea. Statocysts are in the foot. Osphradia are chemoreceptors in the anterior wall of the mantle cavity that detect sediment and chemicals in inhalant water or air. The osphradia of predatory gastropods help detect prey.

Primitive gastropods possessed two nephridia. In modern species, the right nephridium has disappeared, probably because of shell coiling. The nephridium consists of a sac with highly folded walls and connects to the reduced coelom, the pericardial cavity. Excretory wastes are derived largely from fluids filtered and secreted into the coelom from the blood. The nephridium modifies this waste by selectively reabsorbing certain ions and organic molecules. The nephridium opens to the mantle cavity or, in land snails, on the right side of the body adjacent to the mantle cavity and anal opening. Aquatic gastropod species excrete ammonia because they have access to water in which the toxic ammonia is diluted. Terrestrial snails must convert ammonia to a less-toxic form—uric acid. Because uric acid is relatively insoluble in water and less toxic, it can be excreted in a semisolid form, which helps conserve water.

REPRODUCTION AND DEVELOPMENT

Many marine snails are dioecious. Gonads lie in spirals of the visceral mass (*see figure 12.7b*). Ducts discharge gametes into the sea for external fertilization.

Many other snails are monoecious, and internal, cross-fertilization is the rule. Copulation may result in mutual sperm transfer, or one snail may act as the male and the other as the female. A penis has evolved from a fold of the body wall, and portions of the female reproductive tract have become glandular and secrete mucus, a protective jelly, or a capsule around the fertilized egg. Some monoecious snails are protandric in that testes develop first, and after they degenerate, ovaries mature.

Eggs are shed singly or in masses for external fertilization. Internally fertilized eggs are deposited in gelatinous strings or masses. The large, yolky eggs of terrestrial snails are deposited in moist environments, such as forest-floor leaf litter, and a calcareous shell may encapsulate them. In most marine gastropods, spiral cleavage results in a free-swimming **trochophore larva** that develops into another free-swimming larva with foot, eyes, tentacles, and shell, called a **veliger larva.** Sometimes, the trochophore is suppressed, and the veliger is the primary larva. Torsion occurs during the veliger stage, followed by settling and metamorphosis to the adult.

GASTROPOD DIVERSITY

The largest group of gastropods is the subclass Prosobranchia. Its 20,000 species are mostly marine, but a few are freshwater or terrestrial. Most members of this subclass are herbivores or deposit feeders; however, some are carnivorous. Some carnivorous species inject venom into their fish, mollusc, or annelid prey with a radula modified into a hollow, harpoonlike structure. Prosobranch gastropods include most of the familiar marine snails and the abalone. This subclass also includes the heteropods. Heteropods are voracious predators, with very small shells or no shells. Their foot is modified into an undulating "fin" that propels the animal through the water (figure 12.8a).

Members of the subclass Opisthobranchia include sea hares, sea slugs, and their relatives (figure 12.8b). They are mostly marine and include fewer than two thousand species. The shell, mantle cavity, and gills are reduced or lost in these animals, but they are not defenseless. Many acquire undischarged nematocysts from their cnidarian prey, which they use to ward off predators. The pteropods have a foot modified into thin lobes for swimming.

The subclass Pulmonata contains about 17,000 predominantly freshwater or terrestrial species (*see figure 12.7*). These snails are mostly herbivores and have a long radula for scraping plant material. The mantle cavity of pulmonate gastropods is highly vascular and serves as a lung. Air or water moves in or out of the opening of the mantle cavity, the **pneumostome.** In addition to typical freshwater or terrestrial snails, the pulmonates include terrestrial slugs (figure 12.8c).

CLASS BIVALVIA

With close to 30,000 species, the class Bivalvia (bi′val″ve-ah) (L. *bis,* twice + *valva,* leaf) is the second largest molluscan class. This class includes the clams, oysters, mussels, and scallops (*see table 12.1*). A sheetlike mantle and a shell consisting of two valves (hence, the class name) cover these laterally compressed animals. Many bivalves are edible, and some form pearls. Because most bivalves are filter feeders, they are valuable in removing bacteria from polluted water.

SHELL AND ASSOCIATED STRUCTURES

The two convex halves of the shell are called **valves.** Along the dorsal margin of the shell is a proteinaceous hinge and a series of tongue-and-groove modifications of the shell, called teeth, that prevent the valves from twisting (figure 12.9). The oldest part of the shell is the **umbo,** a swollen area near the shell's anterior

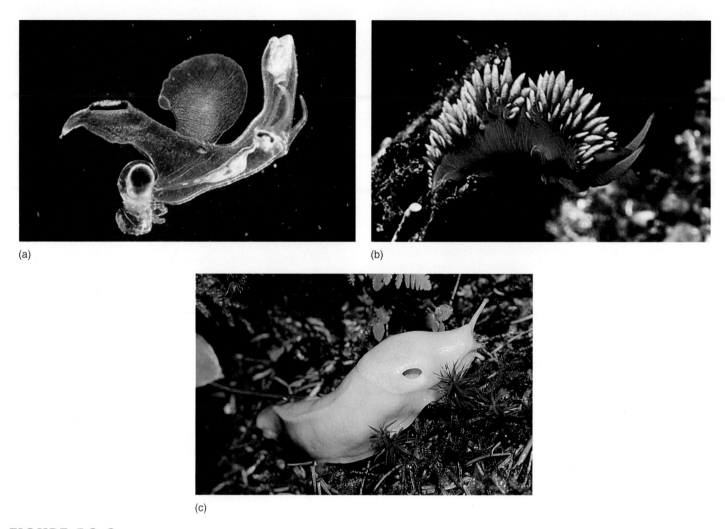

(a)

(b)

(c)

FIGURE 12.8

Variations in the Gastropod Body Form.   (*a*) Subclass Prosobranchia. This heteropod (*Carinaria*) is a predator that swims upside down in the open ocean. Its body is nearly transparent. The head is at the left, and the shell is below and to the right. Heteropoda is a superfamily of prosobranchs comprised of open-ocean, swimming snails with a finlike foot and reduced shell. (*b*) Subclass Opisthobranchia. Colorful nudibranchs have no shell or mantle cavity. The projections on the dorsal surface are used in gas exchange. In some nudibranchs, the dorsal projections are armed with nematocysts for protection. Nudibranchs prey on sessile animals, such as soft corals and sponges. (*c*) Subclass Pulmonata. Terrestrial slugs like this one (*Ariolimax columbianus*) lack a shell. Note the opening to the lung (pneumostome).

margin. Although bivalves appear to have two shells, embryologically, the shell forms as a single structure. The shell is continuous along its dorsal margin, but the mantle, in the region of the hinge, secretes relatively greater quantities of protein and relatively little calcium carbonate. The result is an elastic hinge ligament. The elasticity of the hinge ligament opens the valves when certain muscles relax.

Adductor muscles at either end of the dorsal half of the shell close the shell. Anyone who has tried to force apart the valves of a bivalve mollusc knows the effectiveness of these muscles. This is important for bivalves because their primary defense against predatory sea stars is to tenaciously refuse to open their shells. Chapter 16 explains how sea stars have adapted to meet this defense strategy.

The bivalve mantle attaches to the shell around the adductor muscles and near the shell margin. If a sand grain or a parasite lodges between the shell and the mantle, the mantle secretes

nacre around the irritant, gradually forming a pearl. The Pacific oysters, *Pinctada margaritifera* and *Pinctada mertensi*, form the highest-quality pearls.

GAS EXCHANGE, FILTER FEEDING, AND DIGESTION

Bivalve adaptations to sedentary, filter-feeding lifestyles include the loss of the head and radula and, except for a few bivalves, the expansion of cilia-covered gills. Gills form folded sheets (lamellae), with one end attached to the foot and the other end attached to the mantle. The mantle cavity ventral to the gills is the inhalant region, and the cavity dorsal to the gills is the exhalant region (figure 12.10*a*). Cilia move water into the mantle cavity through an incurrent opening of the mantle. Sometimes, this

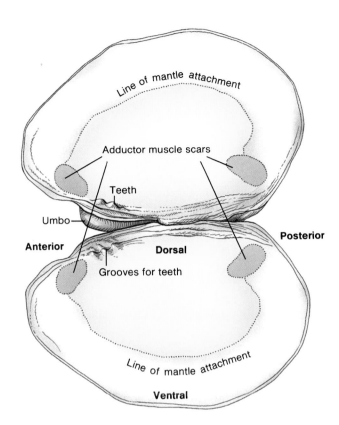

FIGURE 12.9
Inside View of a Bivalve Shell. The umbo is the oldest part of the bivalve shell. As the bivalve grows, the mantle lays down more shell in concentric lines of growth.

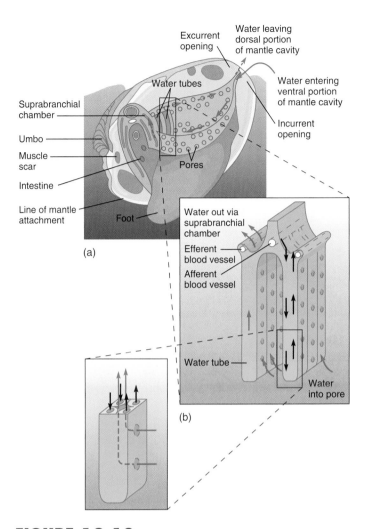

FIGURE 12.10
Lamellibranch Gill of a Bivalve. (*a*) Blue arrows indicate incurrent and excurrent water currents. Food is filtered as water enters water tubes through pores in the gills. (*b*) Cross section through a portion of a gill. Water passing through a water tube is in close proximity to blood. Water and blood exchange gases in the water tubes. Blue arrows show the path of water. Black arrows show the path of blood.

opening is at the end of a siphon, which is an extension of the mantle. A bivalve buried in the substrate can extend its siphon to the surface and still feed and exchange gases. Water moves from the mantle cavity into small pores in the surface of the gills, and from there, into vertical channels in the gills, called water tubes. In moving through water tubes, blood and water are in close proximity, and gases exchange by diffusion (figure 12.10*b*). Water exits the bivalve through a part of the mantle cavity at the dorsal aspect of the gills, called the suprabranchial chamber, and through an excurrent opening in the mantle (figure 12.10*a*).

The gills trap food particles brought into the mantle cavity. Zoologists originally thought that ciliary action was responsible for the trapping. However, the results of a recent study indicate that cilia and food particles have little contact. The food-trapping mechanism is unclear, but once food particles are trapped, cilia move them to the gills' ventral margin. Cilia along the ventral margin of the gills then move food toward the mouth (figure 12.11). Cilia covering leaflike **labial palps** on either side of the mouth also sort filtered food particles. Cilia carry small particles into the mouth and move larger particles to the edges of the palps and gills. This rejected material, called pseudofeces, falls, or is thrown, onto the mantle, and a ciliary tract on the mantle transports the pseudofeces posteriorly. Water rushing out when valves are forcefully closed washes pseudofeces from the mantle cavity.

The digestive tract of bivalves is similar to that of other molluscs (figure 12.12*a*). Food entering the esophagus entangles in a

mucoid food string, which extends to the stomach and is rotated by cilia lining the digestive tract. A consolidated mucoid mass, the **crystalline style,** projects into the stomach from a diverticulum, called the style sac (figure 12.12*b*). Enzymes for carbohydrate and fat digestion are incorporated into the crystalline style. Cilia of the style sac rotate the style against a chitinized **gastric shield.** This abrasion and acidic conditions in the stomach dislodge enzymes. The mucoid food string winds around the crystalline style as it rotates, which pulls the food string farther into the stomach from the esophagus. This action and the acidic pH in the stomach dislodge food particles in the food string. Further sorting separates fine particles from undigestible coarse materials. The latter are sent on to the intestine. Partially digested food from the stomach enters a digestive gland for intracellular digestion. Cilia carry undigested wastes in the digestive gland back to the stomach and then to the intestine. The intestine empties through the anus near the excurrent opening, and excurrent water carries feces away.

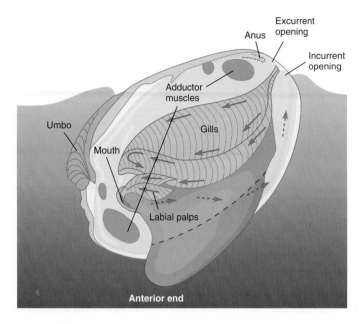

FIGURE 12.11

Bivalve Feeding. Solid blue arrows show the path of food particles after the gills filter them. Dashed blue arrows show the path of particles that the gills and the labial palps reject.

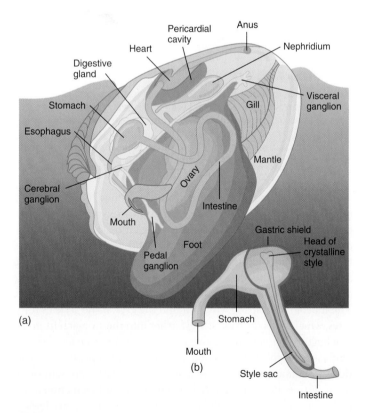

FIGURE 12.12

Bivalve Structure. (*a*) Internal structure of a bivalve. (*b*) Bivalve stomach, showing the crystalline style and associated structures.

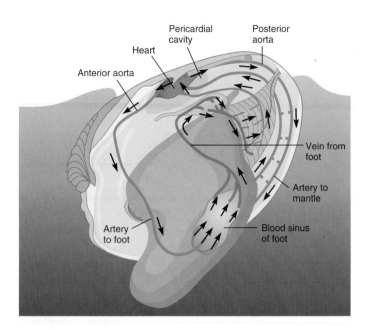

FIGURE 12.13

Bivalve Circulation. Blood flows (black arrows) from the single ventricle of the heart to tissue sinuses through anterior and posterior aortae. Blood from tissue sinuses flows to the nephridia, to the gills, and then to the auricles of the heart. In all bivalves, the mantle is an additional site for oxygenation. In some bivalves, a separate aorta delivers blood to the mantle. This blood returns directly to the heart. The ventricle of bivalves is always folded around the intestine. Thus, the pericardial cavity (the coelom) encloses the heart and a portion of the digestive tract.

OTHER MAINTENANCE FUNCTIONS

Bivalves have an open circulatory system. Blood flows from the heart to tissue sinuses, nephridia, gills, and back to the heart (figure 12.13). The mantle is an additional site for oxygenation. In some bivalves, a separate aorta delivers blood directly to the mantle. Two nephridia are below the pericardial cavity (the coelom). Their duct system connects to the coelom at one end and opens at nephridiopores in the anterior region of the suprabranchial chamber (*see figure 12.12*).

The nervous system of bivalves consists of three pairs of interconnected ganglia associated with the esophagus, the foot, and the posterior adductor muscle. The margin of the mantle is the principal sense organ. It always has sensory cells, and it may have sensory tentacles and photoreceptors. In some species (e.g., scallops), photoreceptors are in the form of complex eyes with a lens and a cornea. Other receptors include statocysts near the pedal ganglion and an osphradium in the mantle, beneath the posterior adductor muscle.

REPRODUCTION AND DEVELOPMENT

Most bivalves are dioecious. A few are monoecious, and some of these species are protandric. Gonads are in the visceral mass, where they surround the looped intestine. Ducts of these gonads open directly to the mantle cavity or by the nephridiopore to the mantle cavity.

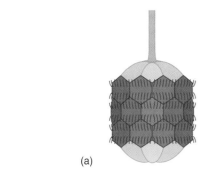

(a)

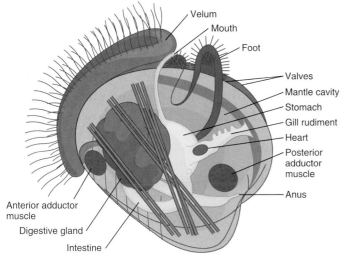

(b)

(c)

FIGURE 12.14

Larval Stages of Bivalves.   (*a*) Trochophore larva (0.4 mm) of *Yoldia limatula*. (*b*) Veliger (0.5 mm) of an oyster. (*c*) Glochidium (1.0 mm) of a freshwater clam. Note the tooth used to attach to fish gills.

Most bivalves exhibit external fertilization. Gametes exit through the suprabranchial chamber of the mantle cavity and the exhalant opening. Development proceeds through trochophore and veliger stages (figure 12.14*a,b*). When the veliger settles to the substrate, it assumes the adult form.

FIGURE 12.15

Class Bivalvia.   This photograph shows a modification of the mantle of a freshwater bivalve (*Lampsilis reeviana*) into a lure. The edge of the bivalve shell is shown in the lower right corner of the photograph. When a fish approaches and bites at the lure, glochidia are released onto the fish.

Most freshwater bivalves brood their young. Fertilization occurs in the mantle cavity by sperm brought in with inhalant water. Some brood their young in maternal gills through reduced trochophore and veliger stages. Young clams are shed from the gills. Others brood their young to a modified veliger stage called a **glochidium,** which is parasitic on fishes (figure 12.14*c*). These larvae possess two tiny valves, and some species have tooth-like hooks. Larvae exit through the exhalant aperture and sink to the substrate. Most die. If a fish contacts a glochidium, however, the larva attaches to the gills, fins, or another body part and may begin to feed on host tissue. The fish may form a cyst around the larva. The mantles of some freshwater bivalves have elaborate modifications that present a fishlike lure to entice predatory fish. When a fish attempts to feed on the lure, the bivalve ejects glochidia onto the fish (figure 12.15). After several weeks of larval development, during which a glochidium begins acquiring its adult structures, the miniature clam falls from its host and takes up its filter-feeding lifestyle. The glochidium is a dispersal stage for an otherwise sedentary animal and probably has little effect on the fish.

BIVALVE DIVERSITY

Bivalves live in nearly all aquatic habitats (figure 12.16). They may completely or partially bury themselves in sand or mud, attach to solid substrates, or bore into submerged wood, coral, or limestone.

The mantle margins of burrowing bivalves are frequently fused to form distinct openings in the mantle cavity (siphons). This fusion helps to direct the water washed from the mantle cavity during burrowing and helps keep sediment from accumulating in the mantle cavity.

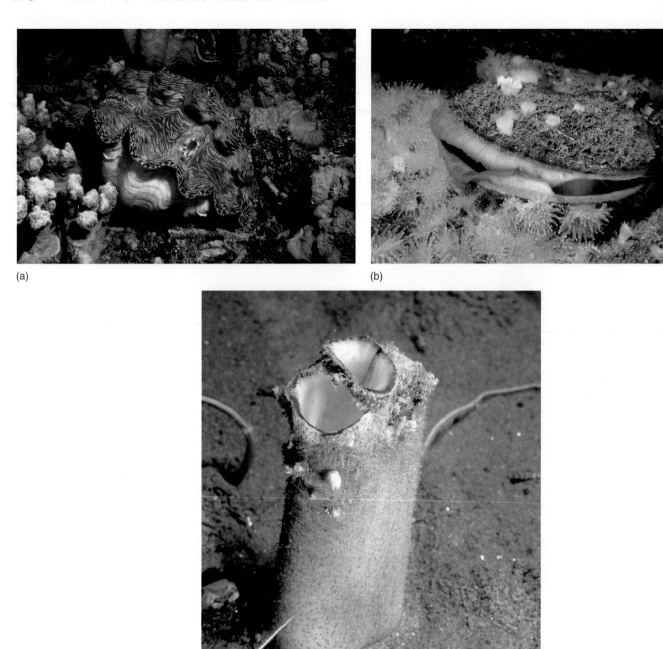

(a)

(b)

(c)

FIGURE 12.16

Bivalve Diversity. (*a*) This giant clam (*Tridacna derasa*) is one of two genera and nine species of giant clams. Giant clams occur in association with coral reefs throughout the tropical Indo-Pacific region. Giant clams are unusual in that they derive a substantial portion of their nutrition from a relationship with photosynthetic, symbiotic algae (zooxanthelle) that live in their large fleshy mantle. Their mantles are typically brightly colored as a result of this association. (*b*) The giant rock scallop (*Hinnites giganteus*) occurs on the Pacific coast from British Columbia to central Baja California. As a mature adult, it is large (up to 25 cm) and is attached to a hard substrate by mantle secretions. Before attachment, however, a young rock scallop can swim in a jet-propulsion fashion by opening its valves and clapping them closed. The brightly colored mantle is highly sensory. (*c*) The geoduck (pronounced "gooey duck") (*Panopea generosa*) is the largest burrowing bivalve. It can attain a mass of about 4 kg and has a siphon that is commonly 10 cm long. It burrows in soft mud and extends its siphon to the surface for filter feeding.

Some surface-dwelling bivalves attach to the substrate either by proteinaceous strands called byssal threads, which a gland in the foot secretes, or by cementation to the substrate. The common marine mussel, *Mytilus,* uses the former method, while oysters employ the latter.

Boring bivalves live beneath the surface of limestone, clay, coral, wood, and other substrates. Boring begins when the larvae settle to the substrate, and the anterior margin of their valves mechanically abrades the substrate. Acidic secretions from the mantle margin that dissolve limestone sometimes

(a)

(b)

FIGURE 12.17

Class Cephalopoda. (*a*) Chambered nautilus (*Nautilus*). (*b*) A cuttle-fish (*Sepia*).

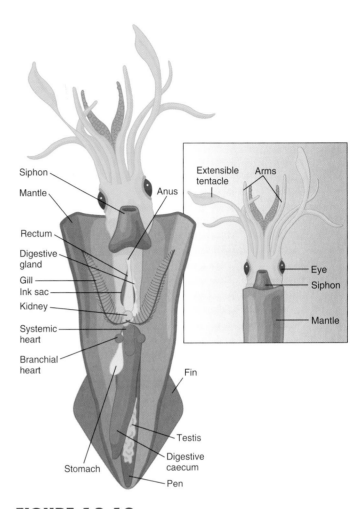

FIGURE 12.18

Internal Structure of the Squid, *Loligo*. The shell of most cephalopods is reduced or absent, and the foot is modified into a funnel and a circle of tentacles and/or arms that encircle the head. The inset shows the undissected anatomy of the squid.

accompany physical abrasion. As the bivalve grows, it is often imprisoned in its rocky burrow because the most recently bored portions of the burrow are larger in diameter than portions bored earlier.

CLASS CEPHALOPODA

The class Cephalopoda (sef'ah-lop"o-dah) (Gr. *kephale*, head + *podos*, foot) includes the octopuses, squid, cuttlefish, and nautiluses (figure 12.17; *see table 12.1; see figure 12.1*). They are the most complex molluscs and, in many ways, the most complex invertebrates. The anterior portion of their foot has been modified into a circle of tentacles or arms used for prey capture, attachment, locomotion, and copulation (figure 12.18). The foot is also incorporated into a funnel associated with the mantle cavity and is used for jetlike locomotion. The molluscan body plan is further modified in that the cephalopod head is in line with the visceral mass. Cephalopods have a highly muscular mantle that encloses all of the body except the head and tentacles. The

mantle acts as a pump to bring large quantities of water into the mantle cavity.

SHELL

Ancestral cephalopods probably had a conical shell. The only living cephalopod that possesses an external shell is the nautilus (*see figure 12.17a*). Septa subdivide its coiled shell. As the nautilus grows, it moves forward, secreting new shell around itself and leaving an empty septum behind. Only the last chamber is occupied. When formed, these chambers are fluid filled. A cord of tissue called a siphuncle perforates the septa, absorbing fluids by osmosis and replacing them with metabolic gases. The amount of gas in the chambers is regulated to alter the buoyancy of the animal.

In all other cephalopods, the shell is reduced or absent. In cuttlefish, the shell is internal and laid down in thin layers, leaving small, gas-filled spaces that increase buoyancy. Cuttlefish shell, called cuttlebone, is used to make powder for polishing and is fed to pet birds to supplement their diet with calcium. The shell of a squid is reduced to an internal, chitinous structure called the pen.

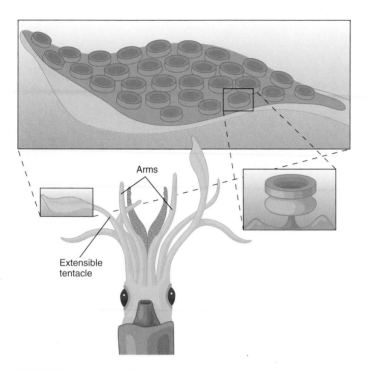

Arms

Extensible
tentacle

FIGURE 12.19

Cephalopod Tentacle. Cephalopods use suction cups for prey capture and as holdfast structures.

In addition, squid also have cartilaginous plates in the mantle wall, neck, and head that support the mantle and protect the brain. The shell is absent in octopuses.

LOCOMOTION

As predators, cephalopods depend on their ability to move quickly using a jet-propulsion system. The mantle of cephalopods contains radial and circular muscles. When circular muscles contract, they decrease the volume of the mantle cavity and close collarlike valves to prevent water from moving out of the mantle cavity between the head and the mantle wall. Water is thus forced out of a narrow funnel. Muscles attached to the funnel control the direction of the animal's movement. Radial mantle muscles bring water into the mantle cavity by increasing the cavity's volume. Posterior fins act as stabilizers in squid and also aid in propulsion and steering in cuttlefish. "Flying squid" (family Onycoteuthidae) have been clocked at speeds of 30 km/hr. Octopuses are more sedentary animals. They may use jet propulsion in an escape response, but normally, they crawl over the substrate using their tentacles. In most cephalopods, the use of the mantle in locomotion coincides with the loss of an external shell, since a rigid external shell would preclude the jet-propulsion method of locomotion described.

FEEDING AND DIGESTION

Most cephalopods locate their prey by sight and capture prey with tentacles that have adhesive cups. In squid, the margins of these

cups are reinforced with tough protein and sometimes possess small hooks (figure 12.19).

All cephalopods have jaws and a radula. The jaws are powerful, beaklike structures for tearing food, and the radula rasps food, forcing it into the mouth cavity.

Cuttlefish and nautiluses feed on small invertebrates on the ocean floor. Octopuses are nocturnal hunters and feed on snails, fishes, and crustaceans. Octopuses have salivary glands that inject venom into prey. Squid feed on fishes and shrimp, which they kill by biting across the back of the head.

The digestive tract of cephalopods is muscular, and peristalsis (coordinated muscular waves) replaces ciliary action in moving food. Most digestion occurs in a stomach and a large cecum. Digestion is primarily extracellular, with large digestive glands supplying enzymes. An intestine ends at the anus, near the funnel, and exhalant water carries wastes out of the mantle cavity.

OTHER MAINTENANCE FUNCTIONS

Cephalopods, unlike other molluscs, have a **closed circulatory system.** Blood is confined to vessels throughout its circuit around the body. Capillary beds connect arteries and veins, and exchanges of gases, nutrients, and metabolic wastes occur across capillary walls. In addition to having a heart consisting of two auricles and one ventricle, cephalopods have contractile arteries and structures called branchial hearts. The latter are at the base of each gill and help move blood through the gill. These modifications increase blood pressure and the rate of blood flow—necessary for active animals with relatively high metabolic rates. Large quantities of water circulate over the gills at all times. *Cephalopods exhibit greater excretory efficiency because of the closed circulatory system. A close association of blood vessels with nephridia allows wastes to filter and secrete directly from the blood into the excretory system.*

The cephalopod nervous system is unparalleled in any other invertebrate. Cephalopod brains are large, and their evolution is directly related to cephalopod predatory habits and dexterity. The brain forms by a fusion of ganglia. Large areas are devoted to controlling muscle contraction (e.g., swimming movements and sucker closing), sensory perception, and functions such as memory and decision making. Research on cephalopod brains has provided insight into human brain functions.

The eyes of octopuses, cuttlefish, and squid are similar in structure to vertebrate eyes (figure 12.20). *(This similarity is an excellent example of convergent evolution.)* In contrast to the vertebrate eye, nerve cells leave the eye from the outside of the eyeball, so that no blind spot exists. (The blind spot of the vertebrate eye is a region of the retina where no photoreceptors exist because of the convergence of nerve cells into the optic nerve. When light falls on the blind spot, no image is perceived.) Like many aquatic vertebrates, cephalopods focus by moving the lens back and forth. Cephalopods can form images, distinguish shapes, and discriminate some colors. The nautiloid eye is less complex. It lacks a lens, and the interior is open to seawater: thus, it acts as a pinhole camera.

Cephalopod statocysts respond to gravity and acceleration, and are embedded in cartilages next to the brain. Osphradia

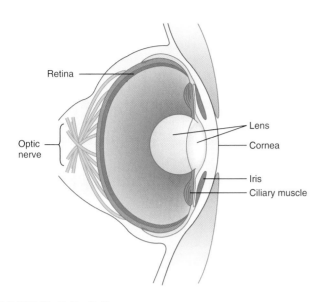

FIGURE 12.20

Cephalopod Eye.   The eye is immovable in a supportive and protective socket of cartilages. It contains a rigid, spherical lens. An iris in front of the lens forms a slitlike pupil that can open and close in response to varying light conditions. Note that the optic nerve comes off the back of the retina.

are present only in *Nautilus*. Tactile receptors and additional chemoreceptors are widely distributed over the body.

Cephalopods have pigment cells called **chromatophores.** When tiny muscles attached to these pigment cells contract, the chromatophores quickly expand and change the color of the animal. Color changes, in combination with ink discharge, function in alarm responses. In defensive displays, color changes may spread in waves over the body to form large, flickering patterns. Color changes may also help cephalopods to blend with their background. The cuttlefish, *Sepia*, can even make a remarkably good impression of a checkerboard background. Color changes are also involved with courtship displays. Some species combine chromatophore displays with bioluminescence.

All cephalopods possess an ink gland that opens just behind the anus. Ink is a brown or black fluid containing melanin and other chemicals. Discharged ink confuses a predator, allowing the cephalopod to escape. For example, *Sepiola* reacts to danger by darkening itself with chromatophore expansion prior to releasing ink. After ink discharge, *Sepiola* changes to a lighter color again to assist its escape.

REPRODUCTION AND DEVELOPMENT

Cephalopods are dioecious with gonads in the dorsal portion of the visceral mass. The male reproductive tract consists of testes and structures for encasing sperm in packets called **spermatophores.** The female reproductive tract produces large, yolky eggs and is modified with glands that secrete gel-like cases around eggs. These cases frequently harden on exposure to seawater.

One tentacle of male cephalopods, called the **hectocotylus,** is modified for spermatophore transfer. In *Loligo* and *Sepia*, the

hectocotylus has several rows of smaller suckers capable of picking up spermatophores. During copulation, male and female tentacles intertwine, and the male removes spermatophores from his mantle cavity. The male inserts his hectocotylus into the mantle cavity of the female and deposits a spermatophore near the opening to the oviduct. Spermatophores have an ejaculatory mechanism that frees sperm from the baseball-bat-shaped capsule. Eggs are fertilized as they leave the oviduct, and are deposited singly or in stringlike masses. They usually attach to some substrate, such as the ceiling of an octopus's den. Octopuses clean developing eggs of debris with their arms and squirts of water.

Cephalopods develop in the confines of the egg membranes, and the hatchlings are miniature adults. Young are never cared for after hatching.

CLASS POLYPLACOPHORA

The class Polyplacophora (pol″e-pla-kof′o-rah) (Gr. *polys*, many + *plak*, plate + *phoros*, to bear) contains the chitons. Chitons are common inhabitants of hard substrates in shallow marine water. Early Native Americans ate chitons. Chitons have a fishy flavor but are tough to chew and difficult to collect.

Chitons have a reduced head, a flattened foot, and a shell that divides into eight articulating dorsal valves (figure 12.21*a*). A muscular mantle that extends beyond the margins of the shell and foot covers the broad foot (figure 12.21*b*). The mantle cavity is restricted to the space between the margin of the mantle and the foot. Chitons crawl over their substrate in a manner similar to gastropods. The muscular foot allows chitons to securely attach to a substrate and withstand strong waves and tidal currents. When chitons are disturbed, the edges of the mantle tightly grip the substrate, and foot muscles contract to raise the middle of the foot, creating a powerful vacuum that holds the chiton in place. Articulations in the shell allow chitons to roll into a ball when dislodged from the substrate.

A linear series of gills is in the mantle cavity on each side of the foot. Cilia on the gills create water currents that enter below the anterior mantle margins and exit posteriorly. The digestive, excretory, and reproductive tracts open near the exhalant area of the mantle cavity, and exhalant water carries products of these systems away.

Most chitons feed on attached algae. A chemoreceptor, the subradular organ, extends from the mouth to detect food, which the radula rasps from the substrate. Mucus traps food, which then enters the esophagus by ciliary action. Extracellular digestion and absorption occur in the stomach, and wastes move on to the intestine (figure 12.21*c*).

The nervous system is ladderlike, with four anteroposterior nerve cords and numerous transverse nerves. A nerve ring encircles the esophagus. Sensory structures include osphradia, tactile receptors on the mantle margin, chemoreceptors near the mouth, and statocysts in the foot. In some chitons, photoreceptors dot the surface of the shell.

Sexes are separate in chitons. External fertilization and development result in a swimming trochophore that settles and metamorphoses into an adult without passing through a veliger stage.

(a)

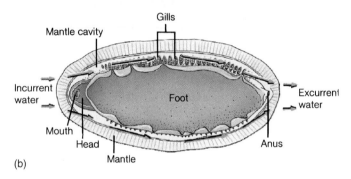

(b)

Gills
Mantle cavity
Incurrent water
Mouth
Head
Mantle
Foot
Excurrent water
Anus

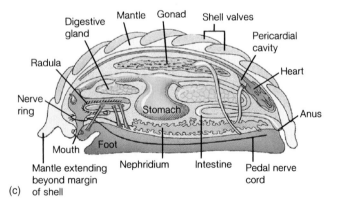

(c)

Digestive gland
Mantle
Gonad
Shell valves
Pericardial cavity
Radula
Heart
Nerve ring
Stomach
Anus
Mouth
Foot
Mantle extending beyond margin of shell
Nephridium
Intestine
Pedal nerve cord

FIGURE 12.21

Class Polyplacophora. (*a*) Dorsal view of a chiton (*Tonicella lineata*). Note the shell consisting of eight valves and the mantle extending beyond the margins of the shell. (*b*) Ventral view of a chiton. The mantle cavity is the region between the mantle and the foot. Arrows show the path of water moving across gills in the mantle cavity. (*c*) Internal structure of a chiton.

CLASS SCAPHOPODA

Members of the class Scaphopoda (ska-fop'o-dah) (Gr. *skaphe*, boat + *podos*, foot) are called tooth shells or tusk shells. The over three hundred species are all burrowing marine animals that inhabit moderate depths. Their most distinctive characteristic is a conical shell that is open at both ends. The head and foot project from the wider end of the shell, and the rest of the body, including

FIGURE 12.22

Class Scaphopoda. This conical shell is open at both ends. In its living state, the animal is mostly buried, with the apex of the shell projecting into the water.

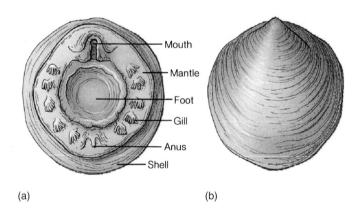

(a)

Mouth
Mantle
Foot
Gill
Anus
Shell

(b)

FIGURE 12.23

Class Monoplacophora. (*a*) Ventral and (*b*) dorsal views of *Neopilina*.

the mantle, is greatly elongate and extends the length of the shell (figure 12.22). Scaphopods live mostly buried in the substrate with head and foot oriented down and with the apex of the shell projecting into the water above. Incurrent and excurrent water enters and leaves the mantle cavity through the opening at the apex of the shell. Functional gills are absent, and gas exchange occurs across mantle folds. Scaphopods have a radula and tentacles, which they use in feeding on foraminiferans. Sexes are separate, and trochophore and veliger larvae are produced.

CLASS MONOPLACOPHORA

Members of the class Monoplacophora (mon"o-pla-kof'o-rah) (Gr. *monos*, one + *plak*, plate + *phoros*, to bear) have an undivided, arched shell; a broad, flat foot; and serially repeated pairs of gills and foot-retractor muscles. They are dioecious; however, nothing is known of their embryology. This group of molluscs was known only from fossils until 1952, when a limpetlike monoplacophoran, named *Neopilina*, was dredged up from a depth of 3,520 m off the Pacific coast of Costa Rica (figure 12.23).

WILDLIFE ALERT
Fat Pocketbook Mussel (*Potamilus capax*)

VITAL STATISTICS

Classification: Phylum Mollusca, class Bivalvia, order Lamellibranchia

Range: Upper Mississippi River

Habitat: Shallow, stable, sandy, or gravelly substrates of freshwater rivers

Number Remaining: Unknown

Status: Endangered

NATURAL HISTORY AND ECOLOGICAL STATUS

The fat pocketbook mussel lives in the sand, mud, and fine-gravel bottoms of large rivers in the upper Mississippi River drainage (box figure 12.1). It buries itself, leaving the posterior margin of its shell and mantle exposed so that its siphons can accommodate the incurrent and excurrent water that circulates through its mantle cavity. The fat pocketbook mussel is a filter feeder, and like other members of its family (Unionidae), it produces glochidia larvae (*see figure 12.14*c). Sperm released by a male are taken into the female's mantle cavity. Eggs are fertilized and develop into glochidia larvae in the gills of the female. Females release glochidia, which live as parasitic hitchhikers on fish gills. The freshwater drum (*Aplodinotus grunniens*) is probably the primary host fish. After a short parasitic existence, the young clams drop to the substrate, where they may live up to 50 years.

The fat pocketbook mussel (box figure 12.2) is one of the more than 60 freshwater mussels that the U.S. Fish and Wildlife Service lists as threatened or endangered. The good news is that researchers from Southwest Missouri State University unexpectedly found an apparently healthy, reproducing population of the fat pocketbook mussel in Southeast Missouri.

The problems that freshwater mussels face stem from economic exploitation, habitat destruction, pollution, and invasion of a foreign mussel. The pearl-button industry began harvesting freshwater mussels in the late 1800s. In the early 1900s, 196 pearl-button factories were

BOX FIGURE 12.2 Fat Pocketbook Mussel (*Potamilus capax*).

operating along the Mississippi River. After harvesting and cleaning, bivalve shells were drilled to make buttons (box figure 12.3). In 1912, this industry produced more than $6 million in buttons. Since that year, however, the harvest has declined.

In the 1940s, plastic buttons replaced pearl buttons, but mussel harvesters soon discovered a new market for freshwater bivalve shells. Small pieces of mussel shell placed into pearl oysters are a nucleus for the formation of a cultured pearl. A renewed impetus for harvesting freshwater mussels in the 1950s resulted in overharvesting. In 1966, harvesters took 3,500 tons of freshwater mussels.

Habitat destruction, pollution, and invasion by the zebra mussel also threaten freshwater mussels. Freshwater mussels require shallow, stable, sandy or gravelly substrates, although a few species prefer mud. Channelization for barge traffic and flood control purposes has destroyed many mussel habitats. Siltation from erosion has replaced stable substrates with soft, mucky river bottoms. A variety of pollutants, especially agricultural runoff containing pesticides, also threaten mussel conservation. In addition, the full impact of the recently introduced zebra mussel (*Dreissena polymorpha*) on native mussels has yet to be determined. However, the zebra mussel has the unfortunate habit of using native mussel shells as a substrate for attachment. They can cover a native mussel so densely that the native mussel cannot feed.

BOX FIGURE 12.1 Distribution of the Fat Pocketbook Mussel (*Potamilus capax*).

BOX FIGURE 12.3 Shell of a Freshwater Mussel After Having Been Drilled to Make Pearl Buttons.

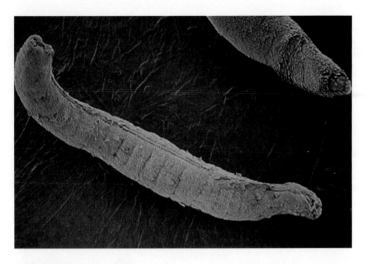

FIGURE 12.24

Class Aplacophora.   Scanning electron micrograph of the solenogaster, *Meiomenia*. Flattened, spinelike, calcareous spicules cover the body. The ventral groove shown here may form from a rolling of the mantle margins. *Meiomenia* is approximately 2 mm long.

CLASS APLACOPHORA

Members of the class Aplacophora (a″pla-kof′o-rah) (Gr. *a*, without + *plak*, plate + *phoros*, to bear) are divided into two subclasses. Members of the subclass Neomeniomorpha are called solenogasters (figure 12.24). The approximately 250 species of these cylindrical molluscs lack a shell and crawl on their ventral surface. Their nervous system is ladderlike and reminiscent of the flatworm body form, causing some to suggest that this group may be closely related to the ancestral molluscan stock. One small group of aplacophorans contains burrowing species that feed on microorganisms and detritus, and possess a radula and nephridia. Most aplacophorans, however, lack nephridia and a radula, are surface dwellers on corals and other substrates, and are carnivores, frequently feeding on cnidarian polyps.

Members of the subclass Chaetodermomorpha (Caudo-foveata) are wormlike molluscs that range in size from 2 mm to 14 cm and live in vertical burrows on the deepsea floor. They have scalelike spicules on the body wall and lack the following typical

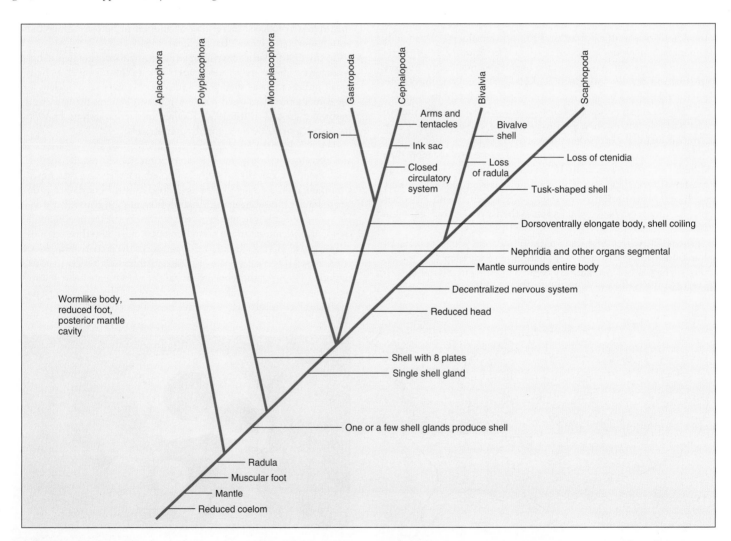

FIGURE 12.25

Molluscan Phylogeny.   Cladogram showing possible evolutionary relationships among the molluscs.

molluscan characteristics: shell, crystalline style, statocysts, foot, and nephridia. Zoologists have described approximately 120 species, but little is known of their ecology.

FURTHER PHYLOGENETIC CONSIDERATIONS

Fossil records of molluscan classes indicate that the phylum is over 500 million years old. Although molluscs have protostome affinities, zoologists do not know the exact relationship of this phylum to other animal phyla. The discovery of *Neopilina* (class Monoplacophora) in 1952 seemed to revolutionize ideas regarding the position of the molluscs in the animal phyla. The most striking feature of *Neopilina* is a segmental arrangement of gills, excretory structures, and nervous system. Because annelids and arthropods (*see chapters 13 to 15*) also have a segmental arrangement of body parts, monoplacophorans were considered a "missing link" between other molluscs and the annelid-arthropod evolutionary line. This idea was further supported by molluscs, annelids, and arthropods sharing certain protostome characteristics (*see figure 7.12a*), and chitons also showing a repetition of some body parts.

Most zoologists now agree that the segmentation in some molluscs is very different from that of annelids and arthropods. Although information on the development of the serially repeat-ing structures in *Neopilina* is not available, no serially repeating structure in any other mollusc develops in an annelid-arthropod fashion. *Segmentation is probably not an ancestral molluscan characteristic. In spite of the absence of annelid-like segmentation in molluscs, protostomate affinities still tie the molluscs to the annelid-arthropod line. Whichever is the case, the relationship of the molluscs to other animal phyla is distant.*

The diversity of body forms and lifestyles in the phylum Mollusca is an excellent example of adaptive radiation. Molluscs probably began in Precambrian times as slow-moving, marine bottom dwellers. The development of unique molluscan features allowed them to diversify relatively quickly. By the end of the Cambrian period, some were filter feeders, some were burrowers, and others were swimming predators. Later, some molluscs became terrestrial and invaded many habitats, from tropical rain forests to arid deserts.

Figure 12.25 shows one interpretation of molluscan phylogeny. Zoologists believe that the lack of a shell in the class Aplacophora is a primitive character. All other molluscs have a shell or are derived from shelled ancestors. The multipart shell distinguishes the Polyplacophora from other classes. Other selected synapomorphies, discussed earlier in this chapter, are noted in the cladogram. There are, of course, other interpretations of molluscan phylogeny. The extensive adaptive radiation of this phylum has made higher taxonomic relationships difficult to discern.

SUMMARY

1. Molluscs are protostomes. They share embryological stages with other protostomes, especially the annelids. In spite of these similarities, relationships between molluscs and other protostomes are distant.

2. Theories regarding the origin of the coelom influence how zoologists interpret evolutionary relationships among triploblastic animals.

3. Molluscs are characterized by a head-foot, a visceral mass, a mantle, and a mantle cavity. Most molluscs also have a radula.

4. Members of the class Gastropoda are the snails and slugs. They are characterized by torsion and often have a coiled shell. Like most molluscs, they use cilia for feeding and have an open circulatory system, well-developed sensory structures, and nephridia. Gastropods may be either monoecious or dioecious.

5. The class Bivalvia includes the clams, oysters, mussels, and scallops. Bivalves lack a head and are covered by a sheetlike mantle and a shell consisting of two valves. Most bivalves use expanded gills for filter feeding, and most are dioecious.

6. Members of the class Cephalopoda are the octopuses, squids, cuttlefish, and nautiluses. Except for the nautiluses, they have a reduced shell. The anterior portion of their foot has been modified into a circle of tentacles. Cephalopods have a closed circulatory system and highly developed nervous and sensory systems. They are efficient predators.

7. Other molluscs include tooth shells (class Scaphopoda), *Neopilina* (class Monoplacophora), caudofoveates and solenogasters (class Aplacophora), and chitons (class Polyplacophora).

8. Some zoologists believe that the molluscs are derived from the annelid-arthropod lineage. Others believe that they arose from triploblastic stocks independent of any other phylum. Adaptive radiation in the molluscs has resulted in their presence in most ecosystems of the earth.

SELECTED KEY TERMS

closed cirulatory system (p. 192)
glochidium (p. 189)
head-foot (p. 182)
mantle (p. 182)
open circulatory system (p. 184)

radula (p. 182)
torsion (p. 183)
trochophore larva (p. 185)
veliger larva (p. 185)
visceral mass (p. 182)

CRITICAL THINKING QUESTIONS

1. Compare and contrast hydraulic skeletons of molluscs with the hydrostatic skeletons of cnidarians and pseudocoelomates.

2. Review the functions of body cavities presented in chapter 7. Which of those functions apply to the coelom of molluscs? What additional function(s), if any, does the coelom of molluscs serve?

3. Students often confuse torsion and shell coiling. Describe each and its effect on gastropod structure and function.

4. Why do you think nautiloids have retained their external shell, while other cephalopods have a reduced internal shell or no shell at all?

5. Bivalves are often used as indicators of environmental quality. Based on your knowledge of bivalves, describe why bivalves are useful environmental indicator organisms.

ONLINE LEARNING CENTER

Visit our Online Learning Center (OLC) at www.mhhe.com/zoology (click on this book's title) to find the following chapter-related materials:

- CHAPTER QUIZZING
- RELATED WEB LINKS
    Phylum Mollusca
    Primitive Classes of Molluscs

    Class Polyplacophora
    Class Gastropoda
    Class Bivalvia
    Class Cephalopoda

- BOXED READINGS ON
    Hydrothermal Vent Communities
    Animal Origins—The Cambrian Explosion
    The Coral Reefs
    The Zebra Mussel—Another Biological Invasion

- SUGGESTED READINGS
- LAB CORRELATIONS

    Check out the OLC to find specific information on these related lab exercises in the *General Zoology Laboratory Manual*, 5th edition, by Stephen A. Miller:

    Exercise 13 *Mollusca*

CHAPTER 13

ANNELIDA:

THE METAMERIC BODY FORM

Concepts

1. Members of the phylum Annelida are the segmented worms. Zoologists debate the relationships of annelids to lower triploblastic animals.
2. Metamerism has important influences on virtually every aspect of annelid structure and function.
3. Members of the class Polychaeta are annelids that have adapted to a variety of marine habitats. Some live in or on marine substrates; others live in burrows or are free-swimming. Parapodia and numerous, long setae characterize the polychaetes.
4. Members of the class Clitellata include earthworms and their relatives and the leeches. A clitellum and few or no setae characterize members of this class.
5. Members of the subclass Oligochaeta live in freshwater and terrestrial habitats. They lack parapodia and have fewer, short setae.
6. The subclass Hirudinea contains the leeches. They are predators in freshwater, marine, and terrestrial environments. Body-wall musculature and the coelom are modified from the pattern found in the other annelid classes. These differences influence locomotor and other functions of leeches.
7. The ancient polychaetes are traditionally considered the ancestral stock from which modern polychaetes, oligochaetes, and leeches evolved.

EVOLUTIONARY PERSPECTIVE

At the time of the November full moon on islands near Samoa in the South Pacific, people rush about preparing for one of their biggest yearly feasts. In just one week, the sea will yield a harvest that can be scooped up in nets and buckets (figure 13.1). Worms by the millions transform the ocean into what one writer called "vermicelli soup!" Celebrants gorge themselves on worms that have been cooked or wrapped in breadfruit leaves. The Samoan palolo worm (*Palola viridis*) spends its entire adult life in coral burrows at the sea bottom. Each November, one week after the full moon, this worm emerges from its burrow, and specialized body segments devoted to sexual reproduction break free and float to the surface, while the rest of the worm is safe on the ocean floor. The surface water is discolored as gonads release their countless eggs and sperm. The natives' feast is short-lived, however; these reproductive swarms last only two days and do not recur for another year.

The Samoan palolo worm is a member of the phylum Annelida (ah-nel'i-dah) (L. *annellus*, ring). Other members of this phylum include countless marine worms (figure 13.2), the soil-building earthworms, and predatory leeches (table 13.1).

This chapter contains evolutionary concepts, which are set off in this font.

(a)

(b)

FIGURE 13.1

Palolo Feasts. (*a*) Indigenous Samoans use nets to collect the reproductive segments of *Palola viridis* that surface each November, one week after the full moon. (*b*) The results of a night of collecting are shown here.

Characteristics of the phylum Annelida include:

1. Body metameric, bilaterally symmetrical, and wormlike
2. Protostome characteristics include spiral cleavage, trochophore larvae (when larvae are present), and schizocoelous coelom formation
3. Paired, epidermal setae (chaetae)
4. Closed circulatory system
5. Dorsal suprapharyngeal ganglia and ventral nerve cord(s) with ganglia
6. Metanephridia (usually) or protonephridia

RELATIONSHIPS TO OTHER ANIMALS

Annelids are protostomes (*see figure 7.12*). Protostome characteristics, such as spiral cleavage, a mouth derived from an embryonic blastopore, schizocoelous coelom formation, and trochophore larvae, are present in most members of the phylum (figure 13.3). (Certain exceptions are discussed later.) This diverse phylum, like most other phyla, originated at least as early as Precambrian times, more than six hundred million years ago. Unfortunately, little

FIGURE 13.2

Phylum Annelida. The phylum Annelida includes over ten thousand species of segmented worms. Most of these are marine members of the class Polychaeta. The fanworm (*Sriro branchus*) is shown here. The fan of this tube-dwelling polychaete is specialized for feeding and gas exchange.

evidence documents the evolutionary pathways that resulted in the first annelids.

A number of hypotheses account for annelid origins. These hypotheses are tied into hypotheses regarding the origin of the coelom (*see chapter 12*). *If a schizocoelous origin of the coelom is correct, as many zoologists believe, then the annelids evolved from ancient flatworm stock. On the other hand, if an enterocoelous coelom origin is correct, then annelids evolved from ancient diploblastic animals, and the triploblastic, acoelomate body may have been derived from a triploblastic, coelomate ancestor.* The recent discovery of a worm, *Lobatocerebrum*, that shares annelid and flatworm characteristics has lent support to the enterocoelous origin hypothesis. *Lobatocerebrum* is classified as an annelid based on the presence of certain segmentally arranged excretory organs, an annelid-like body covering, a complete digestive tract, and an annelid-like nervous system. However, it has a ciliated epidermis and is acoelomate like flatworms. Some

TABLE 13.1
CLASSIFICATION OF THE PHYLUM ANNELIDA

Phylum Annelida (ah-nel'i-dah)

The phylum of triploblastic, coelomate animals whose members are metameric (segmented), elongate, and cylindrical or oval in cross section. Annelids have a complete digestive tract; paired, epidermal setae; and a ventral nerve cord. The phylum is traditionally divided into three classes. Cladistic analysis has resulted in other interpretations, which will be discussed later.

Class Polychaeta (pol"e-kēt'ah)

The largest annelid class; mostly marine; head with eyes and tentacles; parapodia bear numerous setae; monoecious or dioecious; development frequently involves a trochophore larval stage. *Nereis, Arenicola, Sabella.* More than 5,300 species.

Class Clitellata (kli-te'lah-tah)

No parapodia; few or no setae; clitellum functions in cocoon formation; monoecious with direct development. Earthworms and leeches.

Subclass Oligochaeta (ol" i-go-kēt'ah)

Few setae; no distinct head; primarily freshwater or terrestrial. *Lumbricus, Tubifex.* Over 3,000 species.

Subclass Hirudinea (hi'ru-din" e-ah)

Leeches; bodies with 34 segments; each segment subdivided into annuli; posterior and usually anterior suckers present; setae reduced or absent. Freshwater, marine, and terrestrial. *Hirudo.* Approximately 500 species.

zoologists believe that *Lobatocerebrum* illustrates how the triploblastic, acoelomate design could have been derived from the annelid lineage.

METAMERISM AND TAGMATIZATION

Earthworm bodies are organized into a series of ringlike segments. What is not externally obvious, however, is that the body is divided internally as well. Segmental arrangement of body parts in an animal is called **metamerism** (Gr. *meta*, after + *mere*, part).

Metamerism profoundly influences virtually every aspect of annelid structure and function, such as the anatomical arrangement of organs that are coincidentally associated with metamerism. For example, the compartmentalization of the body has resulted in each segment having its own excretory, nervous, and circulatory structures. As will be discussed later, there is evidence that metamerism may have developed in conjunction with blood circulation to segmental appendages of marine worms (class Polychaeta). *In most modern annelids, however, two related functions are probably the primary adaptive features of metamerism: flexible support and efficient locomotion.* These functions depend on the metameric arrangement of the coelom and can be understood by examining the development of the coelom and the arrangement of body-wall muscles.

During embryonic development, the body cavity of annelids arises by a segmental splitting of a solid mass of mesoderm that occupies the region between ectoderm and endoderm on either side of the embryonic gut tract. Enlargement of each cavity forms a

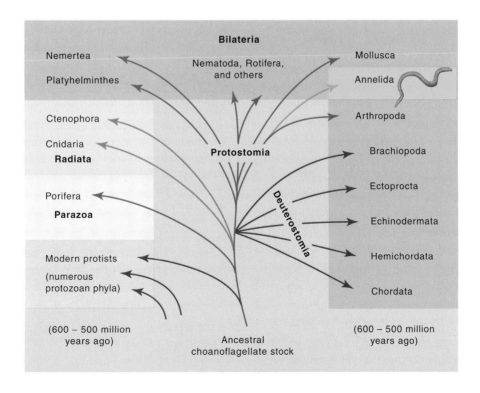

FIGURE 13.3
Evolutionary Relationships of the Annelida.   Annelids (shaded in orange) are protostomes with close evolutionary ties to the Arthropods.

double-membraned septum on the anterior and posterior margin of each coelomic space and dorsal and ventral mesenteries associated with the digestive tract (figure 13.4).

Muscles also develop from the mesodermal layers associated with each segment. A layer of circular muscles lies below the epidermis, and a layer of longitudinal muscles, just below the circular muscles, runs between the septa that separate each segment. In addition, some polychaetes have oblique muscles, and the leeches have dorsoventral muscles.

One advantage of the segmental arrangement of coelomic spaces and muscles is the creation of hydrostatic compartments, which allow a variety of advantageous locomotor and supportive functions not possible in nonmetameric animals that use a hydro-

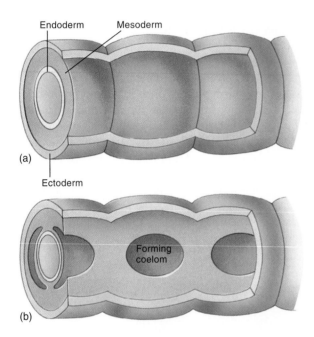

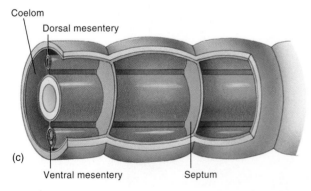

FIGURE 13.4

Development of Metameric, Coelomic Spaces in Annelids. (*a*) A solid mesodermal mass separates ectoderm and endoderm in early embryological stages. (*b*) Two cavities in each segment form from the mesoderm splitting on each side of the endoderm (schizocoelous coelom formation). (*c*) These cavities spread in all directions. Enlargement of the coelomic sacs leaves a thin layer of mesoderm applied against the outer body wall (the parietal peritoneum) and the gut tract (the visceral peritoneum), and dorsal and ventral mesenteries form. Anterior and posterior expansion of the coelom in adjacent segments forms the double-membraned septum that separates annelid metameres.

static skeleton. Each segment can be controlled independently of distant segments, and muscles can act as antagonistic pairs within a segment. The constant volume of coelomic fluid provides a hydrostatic skeleton against which muscles operate. Resultant localized changes in the shape of groups of segments provide the basis for swimming, crawling, and burrowing.

A second advantage of metamerism is that it lessens the impact of injury. If one or a few segments are injured, adjacent segments, set off from injured segments by septa, may be able to maintain nearly normal functions, which increases the likelihood that the worm, or at least a part of it, will survive the trauma.

A third advantage of metamerism is that it permits the modification of certain regions of the body for specialized functions, such as feeding, locomotion, and reproduction. The specialization of body regions in a metameric animal is called **tagmatization** (Gr. *tagma*, arrangement). Although it is best developed in the arthropods, some annelids also display tagmatization. (The arthropods include animals such as insects, spiders, mites, ticks, and crayfish.)

Because of similarities in the development of metamerism in the two groups, annelids and arthropods are thought to be closely related. Other common features include triploblastic coelomate organization, bilateral symmetry, a complete digestive tract, and a ventral nerve cord. As usual, fossil evidence documenting ancestral pathways that led from a common ancestor to the earliest representatives of these two phyla is scant. Annelids and arthropods may have evolved from a marine, wormlike, bilateral ancestor that possessed metameric design. On the other hand, recent data from molecular studies has led some zoologists to suggest that the relationship between the annelids and arthropods is more distant. This hypothesis has fueled considerable debate and research and will be briefly discussed in chapter 15.

CLASS POLYCHAETA

Members of the class Polychaeta (pol″-e-kēt′ah) (Gr. *polys*, many + *chaite*, hair) are mostly marine, and are usually between 5 and 10 cm long (*see table 22.1*). With more than 5,300 species, Polychaeta is the largest of the annelid classes. Polychaetes have adapted to a variety of habitats. Many live on the ocean floor, under rocks and shells, and within the crevices of coral reefs. Other polychaetes are burrowers and move through their substrate by peristaltic contractions of the body wall. A bucket of intertidal sand normally yields vast numbers and an amazing variety of these burrowing annelids. Other polychaetes construct tubes of cemented sand grains or secreted organic materials. Mucus-lined tubes serve as protective retreats and feeding stations.

EXTERNAL STRUCTURE AND LOCOMOTION

In addition to metamerism, the most distinctive feature of polychaetes is the presence of lateral extensions called **parapodia** (Gr. *para*, beside + *podion*, little foot) (figure 13.5). Chitinous rods

FIGURE 13.5

Class Polychaeta.   External structure of *Nereis virens.* Note the numerous parapodia.

support the parapodia, and numerous setae project from the parapodia, giving them their class name. **Setae** (L. *saeta,* bristle) **(chaetae)** are bristles secreted from invaginations of the distal ends of parapodia. They aid locomotion by digging into the substrate and also hold a worm in its burrow or tube.

The **prostomium** (Gr. *pro,* before + *stoma,* mouth) of a polychaete is a lobe that projects dorsally and anteriorly to the mouth and contains numerous sensory structures, including eyes, antennae, palps, and ciliated pits or grooves, called nuchal organs. The first body segment, the **peristomium** (Gr. *peri,* around), surrounds the mouth and bears sensory tentacles or cirri.

The epidermis of polychaetes consists of a single layer of columnar cells that secrete a protective, nonliving **cuticle.** Some polychaetes have epidermal glands that secrete luminescent compounds.

Various species of polychaetes are capable of walking, fast crawling, or swimming. To enable them to do so, the longitudinal muscles on one side of the body act antagonistically to the longitudinal muscles on the other side of the body so that undulatory waves move along the length of the body from the posterior end toward the head. The propulsive force is the result of parapodia and setae acting against the substrate or water. Parapodia on opposite sides of the body are out of phase with one another. When longitudinal muscles on one side of a segment contract, the parapodial muscles on that side also contract, stiffening the parapodium and protruding the setae for the power stroke (figure 13.6a). As a polychaete changes from a slow crawl to swimming, the period and amplitude of undulatory waves increase (figure 13.6b).

Burrowing polychaetes push their way through sand and mud by contractions of the body wall or by eating their way through the substrate. In the latter, polychaetes digest organic matter in the substrate and eliminate absorbed and undigestible materials via the anus.

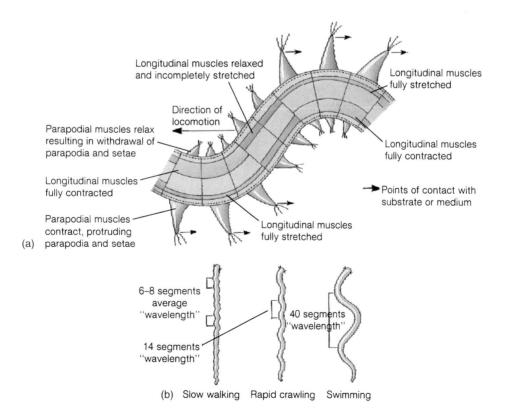

FIGURE 13.6

Polychaete Locomotion.   (*a*) Dorsal view of a primitive polychaete, showing the antagonism of longitudinal muscles on opposite sides of the body and the resultant protrusion and movement of parapodia. (*b*) Both the period and amplitude of locomotor waves increase as a polychaete changes from a "slow walk" to a swimming mode.   *From: "A LIFE OF INVERTEBRATES" © 1979 W. D. Russell-Hunter.*

FEEDING AND THE DIGESTIVE SYSTEM

The digestive tract of polychaetes is a straight tube that mesenteries and septa suspend in the body cavity. The anterior region of the digestive tract is modified into a proboscis that special protractor muscles and coelomic pressure can evert through the mouth. Retractor muscles bring the proboscis back into the peristomium. In some, when the proboscis is everted, paired jaws are opened and may be used for seizing prey. Predatory polychaetes may not leave their burrow or coral crevice. When prey approaches a burrow entrance, the worm quickly extends its anterior portion, everts the proboscis, and pulls the prey back into the burrow. Some polychaetes have poison glands at the base of the jaw. Other polychaetes are herbivores and scavengers and use jaws for tearing food. Deposit-feeding polychaetes (e.g., *Arenicola*, the lugworm) extract organic matter from the marine sediments they ingest. The digestive tract consists of a pharynx that, when everted, forms the proboscis; a storage sac, called a crop; a grinding gizzard; and a long, straight intestine. These are similar to the digestive organs of earthworms (*see figure 13.13*). Organic matter is digested extracellularly, and the inorganic particles are passed through the intestine and released as "castings."

Many sedentary and tube-dwelling polychaetes are filter feeders. They usually lack a proboscis but possess other specialized feeding structures. Some tube dwellers, called fanworms, possess radioles that form a funnel-shaped fan. Cilia on the radioles circulate water through the fan, trapping food particles. Trapped particles are carried along a food groove at the axis of the radiole. During transport, a sorting mechanism rejects the largest particles and transports the finest particles to the mouth. Another filter feeder, *Chaetopterus*, lives in a U-shaped tube and secretes a mucous bag that collects food particles, which may be as small as 1 μm. The parapodia of segments 14 through 16 are modified into fans that create filtration currents. When full, the entire mucous bag is ingested.

Elimination of digestive waste products can be a problem for tube-dwelling polychaetes. Those that live in tubes that open at both ends simply have wastes carried away by water circulating through the tube. Those that live in tubes that open at one end must turn around in the tube to defecate, or they may use ciliary tracts along the body wall to carry feces to the tube opening.

Polychaetes that inhabit substrates rich in dissolved organic matter can absorb as much as 20 to 40% of their energy requirements across their body wall as sugars and other organic compounds. This method of feeding occurs in other animal phyla, too, but rarely accounts for more than 1% of their energy needs.

GAS EXCHANGE AND CIRCULATION

The respiratory gases of most annelids simply diffuse across the body wall, and parapodia increase the surface area for these exchanges. In many polychaetes, parapodial gills further increase the surface area for gas exchange.

Polychaetes have a closed circulatory system. Oxygen is usually carried in combination with molecules called respiratory pigments, which are usually dissolved in the plasma rather than contained in blood cells. Blood may be colorless, green, or red, depending on the presence and/or type of respiratory pigment.

Contractile elements of polychaete circulatory systems consist of a dorsal aorta that lies just above the digestive tract and propels blood from rear to front, and a ventral aorta that lies ventral to the digestive tract and propels blood from front to rear. Running between these two vessels are two or three sets of segmental vessels that receive blood from the ventral aorta and break into capillary beds in the gut and body wall. Capillaries coalesce again into segmental vessels that deliver blood to the dorsal aorta (figure 13.7).

NERVOUS AND SENSORY FUNCTIONS

Nervous systems are similar in both classes of annelids. The annelid nervous system includes a pair of suprapharyngeal ganglia, which connect to a pair of subpharyngeal ganglia by circumpharyngeal connectives that run dorsoventrally along either side of the pharynx. A double ventral nerve cord runs the length of the worm along the ventral margin of each coelomic space, and a paired segmental ganglion is in each segment. The double ventral nerve cord and paired segmental ganglia may fuse to varying extents in different taxonomic groups. Lateral nerves emerge from each segmental ganglion, supplying the body-wall musculature and other structures of that segment (figure 13.8a).

Segmental ganglia coordinate swimming and crawling movements in isolated segments. (Anyone who has used portions of worms as live fish bait can confirm that the head end—with the pharyngeal ganglia—is not necessary for coordinated movements.) Each segment acts separately from, but is closely coordinated with, neighboring segments. The subpharyngeal ganglia help mediate locomotor functions requiring coordination of

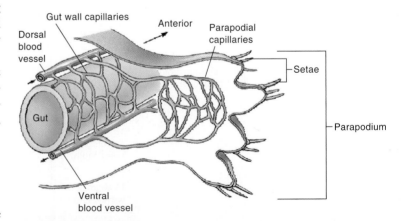

FIGURE 13.7

Circulatory System of a Polychaete. Cross section through the body and a parapodium. In the closed circulatory system shown here, blood passes posterior to anterior in the dorsal vessel and anterior to posterior in the ventral vessel. Capillary beds interconnect dorsal and ventral vessels.

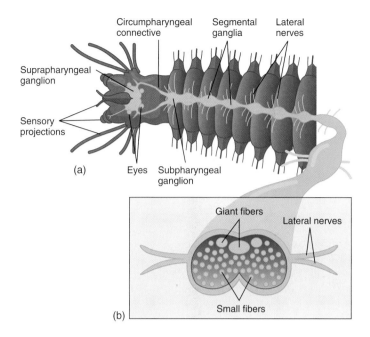

FIGURE 13.8

Nervous System of a Polychaete.   (*a*) Connectives link suprapharyngeal and subpharyngeal ganglia. Segmental ganglia and lateral nerves occur along the length of the worm. (*b*) Cross section of the ventral nerve cord, showing giant fibers.

distant segments. The suprapharyngeal ganglia probably control motor and sensory functions involved with feeding, and sensory functions associated with forward locomotion.

In addition to small-diameter fibers that help coordinate locomotion, the ventral nerve cord also contains giant fibers involved with escape reactions (figure 13.8*b*). For example, a harsh stimulus, such as a fishhook, at one end of a worm causes rapid withdrawal from the stimulus. Giant fibers are approximately 50 μm in diameter and conduct nerve impulses at 30 m/second (as opposed to 0.5 m/second in the smaller, 4 μm diameter annelid fibers).

Polychaetes have various sensory structures. Two to four pairs of eyes are on the surface of the prostomium. They vary in complexity from a simple cup of receptor cells to structures made up of a cornea, lens, and vitreous body. Most polychaetes react negatively to increased light intensities. Fanworms, however, react negatively to decreasing light intensities. If shadows cross them, fanworms retreat into their tubes. This response is believed to help protect fanworms from passing predators. Nuchal organs are pairs of ciliated sensory pits or slits in the head region. Nerves from the suprapharyngeal ganglia innervate nuchal organs, which are thought to be chemoreceptors for food detection. Statocysts are in the head region of polychaetes, and ciliated tubercles, ridges, and bands, all of which contain receptors for tactile senses, cover the body wall.

EXCRETION

Annelids excrete ammonia, and because ammonia diffuses readily into the water, most nitrogen excretion probably occurs across the body wall. Excretory organs of annelids are more active in regulating water and ion balances, although these abilities are limited. *Most marine polychaetes, if presented with extremely diluted seawater, cannot survive the osmotic influx of water and the resulting loss of ions. The evolution of efficient osmoregulatory abilities has allowed only a few polychaetes to invade freshwater.*

The excretory organs of annelids, like those of many invertebrates, are called nephridia. Annelids have two types of nephridia. A protonephridium consists of a tubule with a closed bulb at one end and a connection to the outside of the body at the other end. Protonephridia have a tuft of flagella in their bulbular end that drives fluids through the tubule (figure 13.9*a*; *see also figure 10.6*). Some primitive polychaetes possess paired, segmentally arranged protonephridia that have their bulbular end projecting through the anterior septum into an adjacent segment and the opposite end opening through the body wall at a nephridiopore.

Most polychaetes possess a second kind of nephridium, called a metanephridium. A **metanephridium** consists of an open, ciliated funnel, called a nephrostome, that projects through an anterior septum into the coelom of an adjacent segment. At the opposite end, a tubule opens through the body wall at a nephridiopore or, occasionally, through the intestine (figure 13.9*b,c*). There is usually one pair of metanephridia per segment, and tubules may be extensively coiled, with one portion dilated into a bladder. A capillary bed is usually associated with the tubule of a metanephridium for active transport of ions between the blood and the nephridium (figure 13.9*d*).

Some polychaetes also have chloragogen tissue associated with the digestive tract. This tissue functions in amino acid metabolism in all annelids and is described further in the discussion of oligochaetes.

REGENERATION, REPRODUCTION, AND DEVELOPMENT

All polychaetes have remarkable powers of regeneration. They can replace lost parts, and some species have break points that allow worms to sever themselves when a predator grabs them. Lost segments are later regenerated.

Some polychaetes reproduce asexually by budding or by transverse fission; however, sexual reproduction is much more common. Most polychaetes are dioecious. Gonads develop as masses of gametes and project from the coelomic peritoneum. Primitively, gonads occur in every body segment, but most polychaetes have gonads limited to specific segments. Gametes are shed into the coelom, where they mature. Mature female worms are often packed with eggs. Gametes may exit worms by entering nephrostomes of metanephridia and exiting through the nephridiopore, or they may be released, in some polychaetes, after the worm ruptures. In these cases, the adult soon dies. Only a few polychaetes have separate gonoducts, a condition believed to be primitive (*see figure 13.9a–c*).

Fertilization is external in most polychaetes, although a few species copulate. A unique copulatory habit has been reported in *Platynereis megalops* from Woods Hole, Massachusetts. Toward the

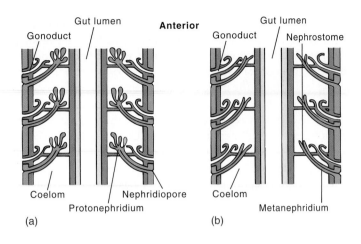

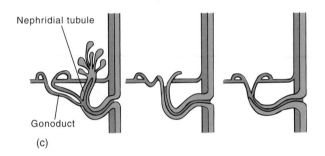

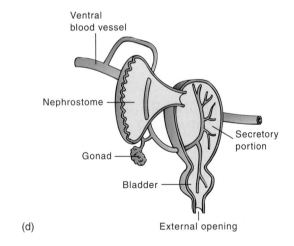

FIGURE 13.9

Annelid Nephridia. (*a*) Protonephridium. The bulbular ends of this nephridium contain a tuft of flagella that drives wastes to the outside of the body. In primitive polychaetes, a gonoduct (coelomoduct) carries reproductive products to the outside of the body. (*b*) Metanephridium. An open ciliated funnel (the nephrostome) drives wastes to the outside of the body. (*c*) In modern annelids, the gonoduct and the nephridial tubules undergo varying degrees of fusion. (*d*) Nephridia of modern annelids are closely associated with capillary beds for secretion, and nephridial tubules may have an enlarged bladder. *From: "A LIFE OF INVERTEBRATES" © 1979 W. D. Russell-Hunter.*

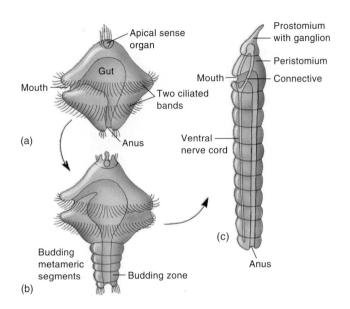

FIGURE 13.10

Polychaete Development. (*a*) Trochophore. (*b*) A later planktonic larva, showing the development of body segments. As more segments develop, the larva settles to the substrate. (*c*) Juvenile worm. *From: "A LIFE OF INVERTEBRATES" © 1979 W. D. Russell-Hunter.*

end of their lives, male and female worms cease feeding, and their intestinal tracts begin to degenerate. At this time, gametes have accumulated in the body cavity. During sperm transfer, male and female worms coil together, and the male inserts his anus into the mouth of the female. Because the digestive tracts of the worms have degenerated, sperm transfer directly from the male's coelom to the egg-filled coelom of the female. This method ensures fertilization of most eggs, after which the female sheds eggs from her anus. Both worms die soon after copulation.

Epitoky is the formation of a reproductive individual (an epitoke) that differs from the nonreproductive form of the species (an atoke). Frequently, an epitoke has a body that is modified into two body regions. Anterior segments carry on normal maintenance functions, and posterior segments are enlarged and filled with gametes. The epitoke may have modified parapodia for more efficient swimming.

This chapter begins with an account of the reproductive swarming habits of *Palola viridis* (the Samoan palolo worm) and one culture's response to those swarms. Similar swarming occurs in other species, usually in response to changing light intensities and lunar periods. The Atlantic palolo worm, for example, swarms at dawn during the first and third quarters of the July lunar cycle.

Zoologists believe that swarming of epitokes accomplishes at least three things. First, because nonreproductive individuals remain safe below the surface waters, predators cannot devastate an entire population. Second, external fertilization requires that individuals become reproductively active at the same time and in close proximity to one another. Swarming ensures that large numbers of individuals are in the right place at the proper time. Third, swarming of vast numbers of individuals for brief periods provides a banquet for predators. However, because vast numbers of prey

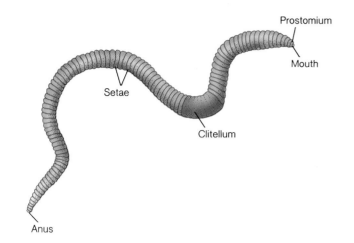

FIGURE 13.11

Subclass Oligochaeta.   External structures of the earthworm, *Lumbricus terrestris*.

are available for only short periods during the year, predator populations cannot increase beyond the limits of their normal diets. Therefore, predators can dine gluttonously and still leave epitokes that will yield the next generation of animals.

Spiral cleavage of fertilized eggs may result in planktonic trochophore larvae that bud segments anterior to the anus. Larvae eventually settle to the substrate (figure 13.10). As growth proceeds, newer segments continue to be added posteriorly. Thus, the anterior end of a polychaete is the oldest end. Many other polychaetes lack a trochophore and display direct development or metamorphosis from another larval stage.

CLASS CLITELLATA

Members of the class Clitellata (kli-te'la-tah) (L. *clitellae*, saddle) include the earthworms and their relatives and the leeches. The subclasses described in the sections that follow formerly had class-level status. Cladistic studies have established that the presence of a clitellum used in cocoon formation, monoecious direct development, and few or no setae are symplesiomorphic characters for a clade containing these two groups. Most zoologists now accept the class-level status for Clitellata.

SUBCLASS OLIGOCHAETA

The subclass Oligochaeta (ol"i-go-kēt'ah) has over three thousand species found throughout the world in freshwater and terrestrial habitats (*see table 13.1*). A few oligochaetes are estuarine, and some are marine. Aquatic species live in shallow water, where they burrow in mud and debris. Terrestrial species live in soils with high organic content and rarely leave their burrows. In hot, dry weather, they may retreat to depths of 3 m below the surface. The soil-conditioning habits of earthworms are well known. *Lumbricus terrestris* is commonly used in zoology laboratories because of its large size. It was introduced to the United States from northern Europe and has flourished. Common native species like *Eisenia foetida* and various species of *Allolobophora* are smaller.

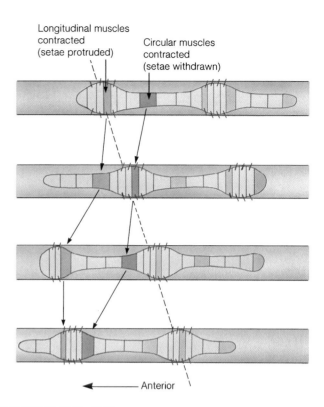

FIGURE 13.12

Earthworm Locomotion.   Arrows designate activity in specific segments of the body, and broken lines indicate regions of contact with the substrate.   *From: "A LIFE OF INVERTEBRATES" © 1979 W. D. Russell-Hunter.*

External Structure and Locomotion

Oligochaetes (Gr. *oligos*, few + *chaite*, hair) have setae, but fewer than are found in polychaetes (thus, the derivation of the class name). Oligochaetes lack parapodia because parapodia and long setae would interfere with their burrowing lifestyles, although they do have short setae on their integument. The prostomium consists of a small lobe or cone in front of the mouth and lacks sensory appendages. A series of segments in the anterior half of an oligochaete is usually swollen into a girdlelike structure called the **clitellum** that secretes mucus during copulation and forms a cocoon (figure 13.11). As in the polychaetes, a nonliving, secreted cuticle covers the body.

Oligochaete locomotion involves the antagonism of circular and longitudinal muscles in groups of segments. Neurally controlled waves of contraction move from rear to front.

Segments bulge and setae protrude when longitudinal muscles contract, providing points of contact with the burrow wall. In front of each region of longitudinal muscle contraction, circular muscles contract, causing the setae to retract, and the segments to elongate and push forward. Contraction of longitudinal muscles in segments behind a bulging region pulls those segments forward. Thus, segments move forward relative to the burrow as waves of muscle contraction move anteriorly on the worm (figure 13.12).

Burrowing is the result of coelomic hydrostatic pressure being transmitted toward the prostomium. As an earthworm pushes its way through the soil, it uses expanded posterior segments and

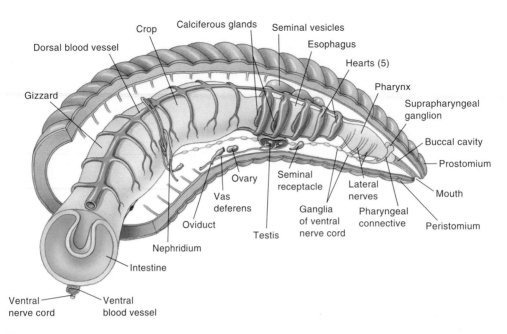

FIGURE 13.13

Earthworm Structure.   Lateral view of the internal structures in the anterior third of an earthworm.

protracted setae to anchor itself to its burrow wall. Any person pursuing fishing worms experiences the effectiveness of this anchor when trying to extract a worm from its burrow. Contraction of circular muscles transforms the prostomium into a conical wedge, 1 mm in diameter at its tip. Contraction of body-wall muscles generates coelomic pressure that forces the prostomium through the soil. During burrowing, earthworms swallow considerable quantities of soil.

Feeding and the Digestive System

Oligochaetes are scavengers and feed primarily on fallen and decaying vegetation, which they drag into their burrows at night. The digestive tract of oligochaetes is tubular and straight (figure 13.13). The mouth leads to a muscular pharynx. In the earthworm, pharyngeal muscles attach to the body wall. The pharynx acts as a pump for ingesting food. The mouth pushes against food, and the pharynx pumps the food into the esophagus. The esophagus is narrow and tubular, and frequently expands to form a stomach, crop, or gizzard; the latter two are common in terrestrial species. A crop is a thin-walled storage structure, and a gizzard is a muscular, cuticle-lined grinding structure. Calciferous glands are evaginations of the esophageal wall that rid the body of excess calcium absorbed from food. They also help regulate the pH of body fluids. The intestine is a straight tube and is the principal site of digestion and absorption. A dorsal fold of the lumenal epithelium called the typhlosole substantially increases the surface area of the intestine (figure 13.14). The intestine ends at the anus.

Gas Exchange and Circulation

Oligochaete respiratory and circulatory functions are as described for polychaetes. Some segmental vessels expand and may be contractile. In the earthworm, for example, expanded segmental

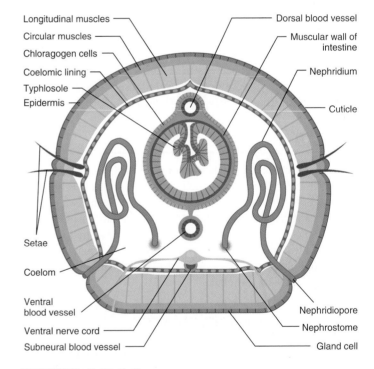

FIGURE 13.14

Earthworm Cross Section.   The nephrostomes shown here would actually be associated with the next anterior segment.

vessels surrounding the esophagus propel blood between dorsal and ventral blood vessels and anteriorly in the ventral vessel toward the mouth. Even though these are sometimes called "hearts," the main propulsive structures are the dorsal and ventral vessels (*see figure 13.13*). Branches from the ventral vessel supply the intestine and body wall.

Nervous and Sensory Functions

The ventral nerve cords and all ganglia of oligochaetes have undergone a high degree of fusion. Other aspects of nervous structure and function are essentially the same as those of polychaetes. As with polychaetes, giant fibers mediate escape responses. An escape response results from the stimulation of either the anterior or posterior end of a worm. An impulse conducted to the opposite end of the worm initiates the formation of an anchor, and longitudinal muscles contract to quickly pull the worm away from the stimulus.

Oligochaetes lack well-developed eyes, which is not surprising, given their subterranean lifestyle. Animals living in perpetual darkness often do not have well-developed eyes. Other oligochaetes have simple pigment-cup ocelli, and all have a "dermal light sense" that arises from photoreceptor cells scattered over the dorsal and lateral surfaces of the body. Scattered photoreceptor cells mediate a negative phototaxis in strong light (evidenced by movement away from the light source) and a positive phototaxis in weak light (evidenced by movement toward the light source).

Oligochaetes are sensitive to a wide variety of chemical and mechanical stimuli. Receptors for these stimuli are scattered over the body surface, especially around the prostomium.

Excretion

Oligochaetes use metanephridia for excretion and for ion and water regulation. As with polychaetes, funnels of metanephridia are associated with the segment just anterior to the segment containing the tubule and the nephridiopore. Nitrogenous wastes include ammonia and urea. Oligochaetes excrete copious amounts of very dilute urine, although they retain vital ions, which is important for organisms living in environments where water is plentiful but essential ions are limited.

Oligochaetes (as well as other annelids) possess chloragogen tissue that surrounds the dorsal blood vessel and lies over the dorsal surface of the intestine (*see figure 13.14*). **Chloragogen tissue** acts similarly to the vertebrate liver. It is a site of amino acid metabolism. Chloragogen tissue deaminates amino acids and converts ammonia to urea. It also converts excess carbohydrates into energy-storage molecules of glycogen and fat.

Reproduction and Development

All oligochaetes are monoecious and exchange sperm during copulation. One or two pairs of testes and one pair of ovaries are located on the anterior septum of certain anterior segments. Both the sperm ducts and the oviducts have ciliated funnels at their proximal ends to draw gametes into their respective tubes.

Testes are closely associated with three pairs of **seminal vesicles,** which are sites for maturation and storage of sperm prior to their release. **Seminal receptacles** receive sperm during copulation. A pair of very small ovisacs, associated with oviducts, are sites for the maturation and storage of eggs prior to egg release (figure 13.15).

During copulation of *Lumbricus,* two worms line up facing in opposite directions, with the ventral surfaces of their anterior ends in contact with each other. This orientation lines up the

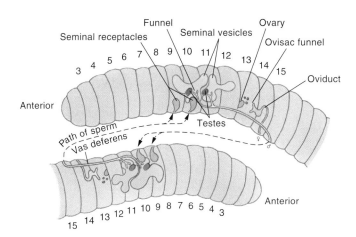

FIGURE 13.15

Earthworm Reproduction.   Mating earthworms, showing arrangements of reproductive structures and the path sperm take during sperm exchange (shown by arrows).

clitellum of one worm with the genital segments of the other worm. A mucous sheath that the clitellum secretes envelops the anterior halves of both worms and holds the worms in place. Some species also have penile structures and genital setae that help maintain contact between worms. In *Lumbricus,* the sperm duct releases sperm, which travel along the external, ventral body wall in sperm grooves formed when special muscles contract. Muscular contractions along this groove help propel sperm toward the openings of the seminal receptacles. In other oligochaetes, copulation results in the alignment of sperm duct and seminal receptacle openings, and sperm transfer is direct. Copulation lasts two to three hours, during which both worms give and receive sperm.

Following copulation, the clitellum forms a cocoon for the deposition of eggs and sperm. The cocoon consists of mucoid and chitinous materials that encircle the clitellum. The clitellum secretes a food reserve, albumen, into the cocoon, and the worm begins to back out of the cocoon. Eggs are deposited in the cocoon as the cocoon passes the openings to the oviducts, and sperm are released as the cocoon passes the openings to the seminal receptacles. Fertilization occurs in the cocoon, and as the worm continues backing out, the ends of the cocoon are sealed, and the cocoon is deposited in moist soil.

Spiral cleavage is modified, and no larva forms. Hatching occurs in one to a few weeks, depending on the species, when young worms emerge from one end of the cocoon.

Freshwater oligochaetes also reproduce asexually. Asexual reproduction involves transverse division of a worm, followed by the regeneration of missing segments.

SUBCLASS HIRUDINEA

The subclass Hirudinea (hi"ru-din'e-ah) (L. *hirudin,* leech) contains approximately five hundred species of leeches (*see table 13.1*). Most leeches are freshwater; others are marine or completely terrestrial. Leeches prey on small invertebrates or feed on the body fluids of vertebrates.

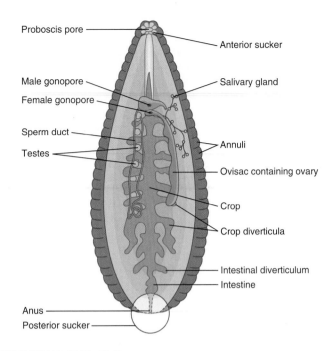

FIGURE 13.16

Internal Structure of a Leech. Annuli subdivide each true segment. Septa do not subdivide the coelom.

External Structure and Locomotion

Leeches lack parapodia and head appendages. Setae are absent in most leeches. In a few species, setae occur only on anterior segments. Leeches are dorsoventrally flattened and taper anteriorly. They have 34 segments, but the segments are difficult to distinguish externally because they have become secondarily divided. Several secondary divisions, called **annuli,** are in each true segment. Anterior and posterior segments are usually modified into suckers (figure 13.16).

Modifications of body-wall musculature and the coelom influence patterns of leech locomotion. The musculature of leeches is more complex than that of other annelids. A layer of oblique muscles is between the circular and longitudinal muscle layers. In addition, dorsoventral muscles are responsible for the typical leech flattening. The leech coelom has lost its metameric partitioning. Septa are lost, and connective tissue has invaded the coelom, resulting in a series of interconnecting sinuses.

These modifications have resulted in altered patterns of locomotion. Rather than being able to utilize independent coelomic compartments, the leech has a single hydrostatic cavity and uses it in a looping type of locomotion. Figure 13.17 describes the mechanics of this locomotion. Leeches also swim using undulations of the body.

Feeding and the Digestive System

Many leeches feed on body fluids or the entire bodies of other invertebrates. Some feed on the blood of vertebrates, including human blood. Leeches are sometimes called parasites; however, the association between a leech and its host is relatively brief. Therefore, describing leeches as predatory is probably more accurate. Leeches are also not species specific, as are most parasites.

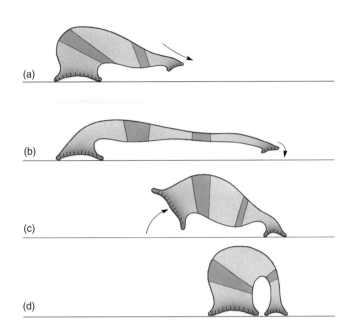

FIGURE 13.17

Leech Locomotion. (*a,b*) Attachment of the posterior sucker causes reflexive release of the anterior sucker, contraction of circular muscles, and relaxation of longitudinal muscles. This muscular activity compresses fluids in the single hydrostatic compartment, and the leech extends. (*c,d*) Attachment of the anterior sucker causes reflexive release of the posterior sucker, the relaxation of circular muscles, and the contraction of longitudinal muscles, causing body fluids to expand the diameter of the leech. The leech shortens, and the posterior sucker again attaches. *From: "A LIFE OF INVERTEBRATES" © 1979 W. D. Russell-Hunter.*

(Leeches are, however, class specific. That is, a leech that preys on a turtle may also prey on an alligator, but probably would not prey on a fish or a frog.)

The mouth of a leech opens in the middle of the anterior sucker. In some leeches, the anterior digestive tract is modified into a protrusible proboscis, lined inside and outside by a cuticle. In others, the mouth is armed with three chitinous jaws. While feeding, a leech attaches to its prey by the anterior sucker and either extends its proboscis into the prey or uses its jaws to slice through host tissues. Salivary glands secrete an anticoagulant called hirudin that prevents blood from clotting.

Behind the mouth is a muscular pharynx that pumps body fluids of the prey into the leech. The esophagus follows the pharynx and leads to a large stomach with lateral cecae. Most leeches ingest large quantities of blood or other body fluids and gorge their stomachs and lateral cecae, increasing their body mass 2 to 10 times. After engorgement, a leech can tolerate periods of fasting that may last for months. The digestive tract ends in a short intestine and anus (*see figure 13.16*).

Gas Exchange and Circulation

Leeches exchange gases across the body wall. Some leeches retain the basic annelid circulatory pattern, but in most leeches, it is highly modified, and coelomic sinuses replace vessels. Coelomic fluid has taken over the function of blood and, except in two orders, respiratory pigments are lacking.

Nervous and Sensory Functions

The leech nervous system is similar to that of other annelids. Ventral nerve cords are unfused, except at the ganglia. The suprapharyngeal and subpharyngeal ganglia and the pharyngeal connectives all fuse into a nerve ring that surrounds the pharynx. Ganglia at the posterior end of the animal fuse in a similar way.

A variety of epidermal sense organs are widely scattered over the body. Most leeches have photoreceptor cells in pigment cups (2 to 10) along the dorsal surface of the anterior segments. Normally, leeches are negatively phototactic, but when they are searching for food, the behavior of some leeches changes, and they become positively phototactic, which increases the likelihood of contacting prey that happen to pass by.

Hirudo medicinalis, the medicinal leech, has a well-developed temperature sense, which helps it to detect the higher body temperature of its mammalian prey. Other leeches are attracted to extracts of prey tissues.

All leeches have sensory cells with terminal bristles in a row along the middle annulus of each segment. These sensory cells, called sensory papillae, are of uncertain function but are taxonomically important.

Excretion

Leeches have 10 to 17 pairs of metanephridia, one per segment in the middle segments of the body. Their metanephridia are highly modified and possess, in addition to the nephrostome and tubule, a capsule believed to be involved with the production of coelomic fluid. Chloragogen tissue proliferates through the body cavity of most leeches.

Reproduction and Development

All leeches reproduce sexually and are monoecious. None are capable of asexual reproduction or regeneration. They have a single pair of ovaries and from four to many testes. Leeches have a clitellum that includes three body segments. The clitellum is present only in the spring, when most leeches breed.

Sperm transfer and egg deposition usually occur in the same manner as described for oligochaetes. A penis assists sperm transfer between individuals. A few leeches transfer sperm by expelling a spermatophore from one leech into the integument of another, a form of hypodermic impregnation. Special tissues within the integument connect to the ovaries by short ducts. Cocoons are deposited in the soil or are attached to underwater objects. There are no larval stages, and the offspring are mature by the following spring.

FURTHER PHYLOGENETIC CONSIDERATIONS

Taxonomic relationships within the Annelida are the subject of intense research and debate (Evolutionary Insights, page 212). Figure 13.18 shows a traditional interpretation of the evolutionary relationships with the phylum divided into the two classes discussed in this chapter. According to this interpretation, the ancestral

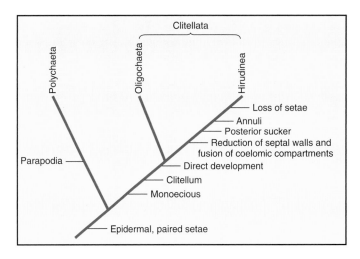

FIGURE 13.18

Annelid Phylogeny. Cladogram illustrating a traditional interpretation of the evolutionary relationships of the two annelid classes. The ancestors of the annelids and arthropods were metameric coelomate animals in the protostome lineage. Paired epidermal setae are diagnostic of the phylum. The polychaetes were the first modern annelids to be derived from the ancestral annelids. The oligochaetes and leeches were derived from a second major lineage of annelids. The absence of derived characters that distinguishes the oligochaetes from the leeches indicates that the oligochaetes were ancestral and that the Hirudinea were derived from them.

polychaetes gave rise to modern polychaetes through adaptive radiation and a group of annelids that invaded freshwater. This freshwater invasion required the ability to regulate the salt and water content of body fluids. In addition, direct development inside of a cocoon rather than as free-swimming larval stages promoted the invasion of terrestrial environments. The absence of synapomorphic characters in the oligochaete lineage indicates that the oligochaetes are the ancestral members of this clade and that the Hirudinea were derived from them.

The evolutionary history of the polychaetes is a story of impressive adaptive radiation into the variety of body forms and habits described earlier in this chapter. A few polychaetes adapted to freshwater environments.

During the Cretaceous period, approximately 100 million years ago, oligochaetes invaded moist, terrestrial environments. This period saw the climax of the giant land reptiles, but more importantly, it was a time of proliferation of flowering plants. The reliance of modern oligochaetes on deciduous vegetation can be traced back to their ancestors' exploitation of this food resource. A few oligochaetes secondarily invaded marine environments. Some of the early freshwater oligochaetes gave rise to leeches. As with oligochaetes, the leeches colonized marine and terrestrial habitats from freshwater.

The systematics of members of the phylum Annelida is the subject of intensive research. Molecular and morphological studies are slowly sorting out the relationships among the 25 orders and 87 families of polychaetes and 6 orders of Clitellata. It is likely to be some time before a widely accepted picture of these evolutionary relationships emerges.

EVOLUTIONARY INSIGHTS
Annelida: A Taxonomist's Vermicelli Soup

The original classification of the Annelida can be traced back to Lamarck (1809), who established the taxon. He recognized the similarities between the oligochaetes and the polychaetes, but the leeches were left out of the group until the mid-1800s. Since that time there have been many taxonomic studies that attempt to sort out relationships within the phylum and to other phyla. These studies have resulted in radically different, and sometimes conflicting, interpretations of annelid phylogenetics. The following summarizes a few of the points of contention regarding the taxonomy and evolutionary relationships within the phylum.

1. Recent cladistic studies suggest that the phylum Annelida is not a monophyletic grouping and that the polychaetes arose from a metameric ancestor independently of the oligochaetes and leeches. If this conclusion is true, then the Polychaeta, Clitellata, and Arthropoda probably all had their origins in an ancestral metameric species, and the phylum name "Annelida" should be abandoned.

2. Some authorities unite the Annelida and the Arthropoda under a taxon originally established by Cuvier (1816) called Articulata.

3. Members of three other smaller groups—echiurans, pogonophorans, and vestimeniferans—were formerly described as members of a separate phylum. Recent work suggests that they are polychaetes.

4. Within the annelid classes there remain many questions regarding taxonomic relationships. There are 10,000 described species of polychaetes. They were formerly divided into two subclasses: Errantia and Sedentaria. The former included motile forms and the latter included nonmotile forms. These subclass designations are now abandoned because they reflect convergent evolution of distantly related groups. The names are now used only as nontaxonomic references. The polychaetes are now divided into two clades: Palpata and Scolecida. The former includes most polychaetes, and the latter includes a small group of burrowing worms.

5. There continue to be differences of opinion whether or not Polychaeta is a monophyletic taxon. Some believe that Clitellata is a clade within Polychaeta.

6. Most systematists view the polychaetes as ancestral annelids, but details of annelid origins are debated. There are two opposing hypotheses regarding the body form and habits of the ancestral annelid. One hypothesis suggests that the ancestral annelid was a crawling worm with well-developed parapodia and numerous setae. Modern ultrastructural studies have been interpreted to suggest that metamerism and parapodia evolved together. Septa, the basic structure of annelid metamerism, are required for the development of transverse blood vessels associated with parapodia. Assuming metamerism is an ancestral character, then parapodia—a synapomorphic character of the polychaetes—would have been present in ancestral annelids and would have been lost during the evolution of the Clitellata. A second hypothesis suggests that the ancestral annelid was a burrowing form that may have lacked parapodia, which are most useful in crawling across the substrate. Support for this hypothesis is in the form of paleontological evidence that burrowing coelomate organisms existed 1 billion years ago. It also rests on the notion that metamerism may have originated in conjunction with a burrowing lifestyle. The usefulness of metamerism in burrowing and locomotion was discussed earlier (*see figure 13.12*). In this scenario, parapodia would have evolved very early, but would not be an ancestral character for the entire phylum.

SUMMARY

1. The origin of the Annelida is largely unknown. A diagnostic characteristic of the annelids is metamerism.

2. Metamerism allows efficient utilization of separate coelomic compartments as a hydrostatic skeleton for support and movement. Metamerism also lessens the impact of injury and makes tagmatization possible.

3. Members of the class Polychaeta are mostly marine and possess parapodia with numerous setae. Locomotion of polychaetes involves the antagonism of longitudinal muscles on opposite sides of the body, which creates undulatory waves along the body wall and causes parapodia to act against the substrate.

4. Polychaetes may be predators, herbivores, scavengers, or filter feeders.

5. The nervous system of polychaetes usually consists of a pair of suprapharyngeal ganglia, subpharyngeal ganglia, and double ventral nerve cords that run the length of the worm.

6. Polychaetes have a closed circulatory system. Respiratory pigments dissolved in blood plasma carry oxygen.

7. Polychaetes use either protonephridia or metanephridia in excretion.

8. Most polychaetes are dioecious, and gonads develop from coelomic epithelium. Fertilization is usually external. Epitoky occurs in some polychaetes.

9. Development of polychaetes usually results in a planktonic trochophore larva that buds off segments near the anus.

10. The class Clitellata includes earthworms and their relatives (subclass Oligochaeta) and leeches (subclass Hirudinea). They possess a

clitellum used in cocoon formation, monoecious direct development, and few or no setae.

11. The subclass Oligochaeta includes primarily freshwater and terrestrial annelids. Oligochaetes possess few setae, and they lack a head and parapodia.

12. Oligochaetes are scavengers that feed on dead and decaying vegetation. Their digestive tract is tubular and straight, and frequently has modifications for storing and grinding food and for increasing the surface area for secretion and absorption.

13. Oligochaetes use metanephridia for excretion and for ion and water regulation. Chloragogen tissue is a site for urea formation from protein metabolism, and for the synthesis and storage of glycogen and fat.

14. Oligochaetes are monoecious and exchange sperm during copulation.

15. Members of the subclass Hirudinea are the leeches. Complex arrangements of body-wall muscles and the loss of septa influence patterns of locomotion.

16. Leeches are predatory and feed on body fluids, the entire bodies of other invertebrates, and the blood of vertebrates.

17. Leeches are monoecious, and reproduction and development occur as in oligochaetes.

18. The traditional interpretation of annelid phylogeny is that ancestral polychaetes gave rise to modern polychaetes and a group of annelids that invaded freshwater. The oligochaetes and then the leeches were derived from this freshwater lineage.

SELECTED KEY TERMS

chloragogen tissue (p. 209)
clitellum (p. 207)
epitoky (p. 206)
metamerism (p. 201)
metanephridium (p. 205)

parapodia (p. 202)
peristomium (p. 203)
prostomium (p. 203)
tagmatization (p. 202)

CRITICAL THINKING QUESTIONS

1. What evidence links the annelids and arthropods in the same evolutionary line?
2. Distinguish between a protonephridium and a metanephridium. Name a class of annelids whose members may have protonephridia. What other phylum have we studied whose members also had protonephridia? Do you think that metanephridia would be more useful for a coelomate or an acoelomate animal? Explain.
3. In what annelid groups are septa between coelomic compartments lost? What advantages does this loss give each group?
4. What differences in nephridial structure might you expect in freshwater and marine annelids?
5. Few polychaetes have invaded freshwater. Can you think of a reasonable explanation for this?

ONLINE LEARNING CENTER

Visit our Online Learning Center (OLC) at www.mhhe.com/zoology (click on this book's title) to find the following chapter-related materials:

- CHAPTER QUIZZING
- RELATED WEB LINKS
   Phylum Annelida
   Class Polychaeta
   Class Clitellata
- BOXED READINGS ON
   Soil Conditioning by Earthworms
   Leeches and Science
- SUGGESTED READINGS
- LAB CORRELATIONS

Check out the OLC to find specific information on these related lab exercises in the *General Zoology Laboratory Manual*, 5th edition, by Stephen A. Miller:

   Exercise 14   *Annelida*

CHAPTER 14

THE ARTHROPODS:

BLUEPRINT FOR SUCCESS

Concepts

1. Arthropods have been successful in almost all habitats on the earth. Some ancient arthropods were the first animals to live most of their lives in terrestrial environments.
2. Metamerism with tagmatization, a chitinous exoskeleton, and metamorphosis have contributed to arthropod success.
3. Members of the subphylum Trilobitomorpha are extinct arthropods that were a dominant life-form in the oceans between 345 and 600 million years ago.
4. The bodies of members of the subphylum Chelicerata are divided into two regions and have chelicerae. The class Merostomata contains the horseshoe crabs. The class Arachnida contains the spiders, mites, ticks, and scorpions. Some ancient arachnids were among the earliest terrestrial arthropods, and modern arachnids have numerous adaptations for terrestrial life. The class Pycnogonida contains the sea spiders.
5. Animals in the subphylum Crustacea have biramous appendages and two pairs of antennae. The class Malacostraca includes the crabs, lobsters, crayfish, and shrimp. The class Branchiopoda includes the fairy shrimp, brine shrimp, and water fleas. The class Maxillopoda includes the copepods and barnacles.

EVOLUTIONARY PERSPECTIVE

What animal species has the greatest number of individuals? The answer can only be an educated guess; however, many zoologists would argue that one of the many species of small (1 to 2 mm) crustaceans, called copepods, that drift in the open oceans must have this honor. Copepods have been very successful, feeding on the vast photosynthetic production of the open oceans (figure 14.1). After only 20 minutes of towing a plankton net behind a slowly moving boat (at the right location and during the right time of year), you can collect over three million copepods—enough to solidly pack a two-gallon pail! Copepods are food for fish, such as herring, sardines, and mackerel, as well as for whale sharks and the largest mammals, the blue whale and its relatives. Humans benefit from copepod production by eating fish that feed on copepods. (Unfortunately, we use only a small fraction of the total food energy in these animals. In spite of half of the earth's human inhabitants lacking protein in their diets, humans process most of the herring and sardines caught into fish meal, which is then fed to poultry and hogs. In eating the poultry and hogs, we lose over 99% of the original energy present in the copepods!)

This chapter contains evolutionary concepts, which are set off in this font.

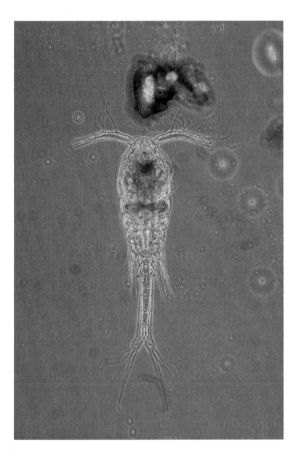

FIGURE 14.1

The Most Abundant Animal?   Copepods, such as *Calanus* sp. (shown here), are abundant in the oceans of the world and form important links in oceanic food webs.

Copepods are one of many groups of animals belonging to the phylum Arthropoda (ar"thrah-po'dah) (Gr. *arthro,* joint + *podos,* foot). Crayfish, lobsters, spiders, mites, scorpions, and insects are also arthropods. Zoologists have described about one million species of arthropods, and recent studies estimate that 30 to 50 million species may yet be undescribed. In this chapter and chapter 15, you will discover the many ways in which some arthropods are considered among the most successful of all animals.

Characteristics of the phylum Arthropoda include:

1. Metamerism modified by the specialization of body regions for specific functions (tagmatization)
2. Chitinous exoskeleton that provides support and protection and is modified to form sensory structures
3. Paired, jointed appendages
4. Growth accompanied by ecdysis or molting
5. Ventral nervous system
6. Coelom reduced to cavities surrounding gonads and sometimes excretory organs
7. Open circulatory system in which blood is released into tissue spaces (hemocoel) derived from the blastocoel
8. Complete digestive tract
9. Metamorphosis often present; reduces competition between immature and adult stages

CLASSIFICATION AND RELATIONSHIPS TO OTHER ANIMALS

As discussed in chapter 13, the traditional interpretation is that arthropods and annelids are closely related. Shared protostome characteristics, such as schizocoelous coelom formation and the development of the mouth from the blastopore, as well as other common characteristics, such as the presence of a paired ventral nerve cord and metamerism, are evidence of a common ancestry (figure 14.2).

The last decade has brought a wealth of compelling evidence from molecular biology that is stimulating new research and debate regarding evolutionary relationships within the Protostomia. Some of this evidence has led some zoologists to suggest that the protostomes should be divided into two major clades: Lophotrochozoa and Ecdysozoa (*see table 7.4*). The latter includes the Nematoda, Nematomorpha, Kinorhyncha (*see chapter 11*), and Arthropoda. Proposed synapomorphies for this clade include a cuticle, loss of epidermal cilia, and shedding the cuticle in a process called ecdysis. The separation of Arthropoda and Annelida into two separate clades requires that their common form of metamerism would have resulted from convergent evolution. The convergence of this character seems unlikely to many zoologists, and this debate is likely to continue for some time.

There has also been an explosion of new information regarding the evolutionary relationships within the phylum Arthropoda that is causing zoologists to reexamine current and older hypotheses regarding arthropod phylogeny. There is now general agreement that the Arthropoda is a monophyletic taxon, and living arthropods are divided into four subphyla: Chelicerata, Crustacea, Hexapoda, and Myriapoda. All members of a fifth subphylum, Trilobitomorpha, are extinct (table 14.1). Ideas regarding the evolutionary relationships among these subphyla are discussed at the end of chapter 15. This chapter examines Trilobitomorpha, Chelicerata, and Crustacea, and chapter 15 covers Myriapoda and Hexapoda.

Finally, the subphylum Chelicerata contains a group of organisms known as sea spiders (class Pycnogonida). As we will see, this group is unusual among the Chelicerata, and its placement within the arthropods has been debated. Some authorities believe that pycnogonids are distinct enough to warrant separating them from other chelicerates. In this case, the subphylum name Cheliceriformes is used, and it contains two classes, Pycnogonida and Chelicerata. Merostomata and Arachnida then become subclasses within the Chelicerata.

METAMERISM AND TAGMATIZATION

Four aspects of arthropod biology have contributed to their success. One of these is metamerism. Metamerism of arthropods is most evident externally because the arthropod body is often composed of a series of similar segments, each bearing a pair of appendages (*see figure 13.4c*). Internally, however, septa do not

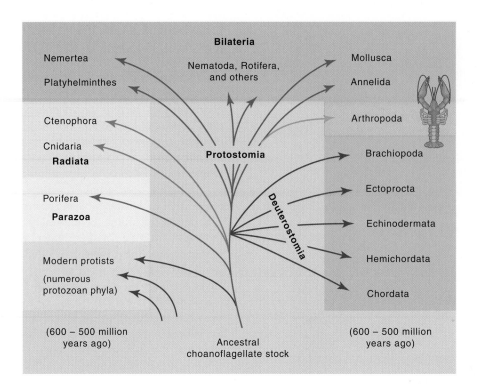

FIGURE 14.2

Evolutionary Relationships of the Arthropods. The traditional view is that the arthropods (shaded in orange) are protostomes with close evolutionary ties to the annelids. A paired ventral nerve cord and metamerism in both groups are evidence of common ancestry. Recent information from molecular studies is causing zoologists to reexamine this traditional view.

divide the body cavity of an arthropod, and most organ systems are not metamerically arranged. *The reason for the loss of internal metamerism is speculative; however, the presence of metamerically arranged hydrostatic compartments would be of little value in the support or locomotion of animals enclosed by an external skeleton (discussed under "The Exoskeleton").*

As discussed in chapter 13, metamerism permits the specialization of regions of the body for specific functions. This regional specialization is called tagmatization. In arthropods, body regions, called tagmata (sing., tagma), are specialized for feeding and sensory perception, locomotion, and visceral functions.

THE EXOSKELETON

An external, jointed skeleton, called an **exoskeleton** *or* **cuticle,** *encloses arthropods. The exoskeleton is often cited as the major reason for arthropod success.* It provides structural support, protection, impermeable surfaces for the prevention of water loss, and a system of levers for muscle attachment and movement.

The exoskeleton covers all body surfaces and invaginations of the body wall, such as the anterior and posterior portions of the gut tract. It is nonliving and is secreted by a single layer of epidermal cells (figure 14.3). The epidermal layer is sometimes called the hypodermis because, unlike other epidermal tissues, it is covered on the outside by an exoskeleton, rather than being directly exposed to air or water.

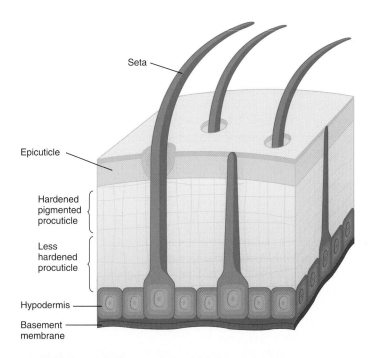

FIGURE 14.3

Arthropod Exoskeleton. The epicuticle is made of a waxy lipoprotein and is impermeable to water. Calcium carbonate deposition and/or sclerotization harden the outer layer of the procuticle. Chitin, a tough, leathery polysaccharide, and several kinds of proteins make up the bulk of the procuticle. The hypodermis secretes the entire exoskeleton.

TABLE 14.1
CLASSIFICATION OF THE PHYLUM ARTHROPODA

Phylum Arthropoda (ar″thrah-po′dah)
Animals that show metamerism with tagmatization, a jointed exoskeleton, and a ventral nervous system.

Subphylum Trilobitomorpha (tri″lo-bit″o-mor′fah)
Marine, all extinct; lived from Cambrian to Carboniferous periods; bodies divided into three longitudinal lobes; head, thorax, and abdomen present; one pair of antennae and biramous appendages.

Subphylum Chelicerata (ke-lis″er-ah′tah)
Body usually divided into prosoma and opisthosoma; first pair of appendages piercing or pincerlike (chelicerae) and used for feeding.

Class Merostomata (mer″o-sto′mah-tah)
Marine, with book gills on opisthosoma. Two subclasses: Eurypterida, a group of extinct arthropods called giant water scorpions, and Xiphosura, the horseshoe crabs. *Limulus.*

Class Arachnida (ah-rak′nĭ′-dah)
Mostly terrestrial, with book lungs, tracheae, or both; usually four pairs of walking legs in adults. Spiders, scorpions, ticks, mites, harvestmen, and others.

Class Pycnogonida (pik″no-gon′ĭ′-dah)
Reduced abdomen; no special respiratory or excretory structures; four to six pairs of walking legs; common in all oceans. Sea spiders.

Subphylum Crustacea (krus-tās′e-ah)
Most aquatic, head with two pairs of antennae, one pair of mandibles, and two pairs of maxillae; biramous appendages.

Class Remipedia (re-mi-pe′de-ah)
Cave-dwelling crustaceans from the Caribbean basin, Indian Ocean, Canary Islands, and Australia; body with approximately 30 segments that bear uniform, biramous appendages.

Class Cephalocarida (sef″ah-lo-kar′ĭ-dah)
Small (3 mm) marine crustaceans with uniform, leaflike, triramous appendages.

Class Branchiopoda (brang″ke-o-pod′ah)
Flattened, leaflike appendages used in respiration, filter feeding, and locomotion, found mostly in freshwater. Fairy shrimp, brine shrimp, clam shrimp, water fleas.

Class Malacostraca (mal-ah-kos′trah-kah)
Appendages possibly modified for crawling, feeding, swimming. Lobsters, crayfish, crabs, shrimp, isopods (terrestrial).

Class Maxillopoda (maks″il-ah-pod′ah)
Five head, six thoracic, and four abdominal somites plus a telson; thoracic segments variously fused with the head; abdominal segments lack typical appendages; abdomen often reduced. Barnacles and copepods.

Subphylum Hexapoda (hex″sah-pod′ah)
Body divided into head, thorax, and abdomen; five pairs of head appendages; three pairs of uniramous appendages on the thorax. Insects and their relatives.

Subphylum Myriapoda (mir″e-a-pod′ah)
Body divided into head and trunk; four pairs of head appendages; uniramous appendages. Millipedes and centipedes.

The exoskeleton has two layers. The epicuticle is the outermost layer. Made of a waxy lipoprotein, it is impermeable to water and a barrier to microorganisms and pesticides. The bulk of the exoskeleton is below the epicuticle and is called the procuticle. (In crustaceans, the procuticle is sometimes called the endocuticle.) The procuticle is composed of **chitin,** a tough, leathery polysaccharide, and several kinds of proteins. The procuticle hardens through a process called sclerotization and sometimes by impregnation with calcium carbonate. Sclerotization is a tanning process in which layers of protein are chemically cross-linked with one another—hardening and darkening the exoskeleton. In insects and most other arthropods, this bonding occurs in the outer portion of the procuticle. The exoskeleton of crustaceans hardens by sclerotization and by the deposition of calcium carbonate in the middle regions of the procuticle. Some proteins give the exoskeleton resiliency. Distortion of the exoskeleton stores energy for such activities as flapping wings and jumping. The inner portion of the procuticle does not harden.

Hardening in the procuticle provides armorlike protection for arthropods, but it also necessitates a variety of adaptations to allow arthropods to live and grow within their confines. Invaginations of the exoskeleton form firm ridges and bars for muscle attachment. Another modification of the exoskeleton is the formation of joints. A flexible membrane, called an articular membrane, is present in regions where the procuticle is thinner and less hardened (figure 14.4). Other modifications of the exoskeleton include sensory receptors, called sensilla, in the form of pegs, bristles, and lenses, and modifications of the exoskeleton that permit gas exchange.

The growth of an arthropod would be virtually impossible unless the exoskeleton were periodically shed, such as in the molting process called **ecdysis** (Gr. *ekdysis*, getting out). Ecdysis is divided into four stages: (1) Enzymes, secreted from hypodermal glands, begin digesting the old procuticle to separate the hypodermis and the exoskeleton (figure 14.5a,b); (2) new procuticle and epicuticle are secreted (figure 14.5c,d); (3) the old exoskeleton splits open along predetermined ecdysal lines when the animal stretches by air or water intake; pores in the procuticle secrete additional epicuticle (figure 14.5e); and (4) finally, calcium carbonate deposits and/or sclerotization harden the new exoskeleton (figure 14.5f). During the few hours or days of the hardening process, the arthropod is vulnerable to predators and remains hidden. The nervous and endocrine systems control all of these changes; the controls are discussed in more detail later in the chapter.

THE HEMOCOEL

A third arthropod characteristic is the presence of a **hemocoel.** The hemocoel is derived from an embryonic cavity called the blastocoel that forms in the blastula (*see figure 7.12*). The hemocoel provides an internal cavity for the open circulatory system of arthropods. Internal organs are bathed by body fluids in the hemocoel to provide for the exchange of nutrients, wastes, and sometimes gases. The coelom, which forms by splitting blocks of

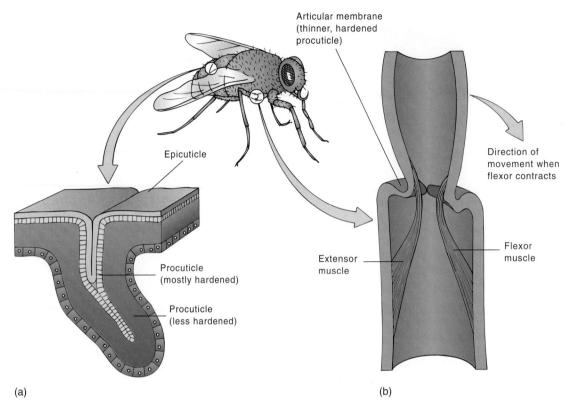

FIGURE 14.4

Modifications of the Exoskeleton. (*a*) Invaginations of the exoskeleton result in firm ridges and bars when the procuticle in the region of the invagination remains thick and hard. These are muscle attachment sites. (*b*) Regions where the procuticle is thinned are flexible and form membranes and joints. *From: "A LIFE OF INVERTEBRATES" © 1979 W. D. Russell-Hunter.*

mesoderm later in the development of most protostomes, was reduced in ancestral arthropods. The presence of the rigid exoskeleton and body wall means that the coelom is no longer used as a hydrostatic compartment. In modern arthropods, the coelom forms small cavities around the gonads and sometimes the excretory structures.

METAMORPHOSIS

A fourth characteristic that has contributed to arthropod success is a reduction of competition between adults and immature stages because of metamorphosis. Metamorphosis is a radical change in body form and physiology as an immature stage, usually called a larva, becomes an adult. The evolution of arthropods has resulted in an increasing divergence of body forms, behaviors, and habitats between immature and adult stages. Adult crabs, for example, usually prowl the sandy bottoms of their marine habitats for live prey or decaying organic matter, whereas larval crabs live and feed in the plankton. Similarly, the caterpillar that feeds on leafy vegetation eventually develops into a nectar-feeding adult butterfly or moth. Having different adult and immature stages means that the stages do not compete with each other for food or living space. In some arthropod and other animal groups, larvae also serve as the dispersal stage.

SUBPHYLUM TRILOBITOMORPHA

Members of the subphylum Trilobitomorpha (tri″lo-bit″o-mor′fah) (Gr. *tri,* three + *lobos,* lobes) were a dominant form of life in the oceans from the Cambrian period (600 million years ago) to the Carboniferous period (345 million years ago). They crawled along the substrate feeding on annelids, molluscs, and decaying organic matter. The trilobite body was oval, flattened, and divided into three tagmata: head (cephalon), thorax, and pygidium (figure 14.6). All body segments articulated so that the trilobite could roll into a ball to protect its soft ventral surface. Most fossilized trilobites are found in this position. Trilobite appendages consisted of two lobes or rami, and are called **biramous** (L. *bi,* twice + *ramus,* branch) **appendages.** The inner lobe was a walking leg, and the outer lobe bore spikes or teeth that may have been used in digging or swimming or as gills in gas exchange.

SUBPHYLUM CHELICERATA

One arthropod lineage, the subphylum Chelicerata (ke-lis″er-ah′tah) (Gr. *chele,* claw + *ata,* plural suffix), includes familiar animals, such as spiders, mites, and ticks, and less familiar animals, such as horseshoe crabs and sea spiders. These animals have two

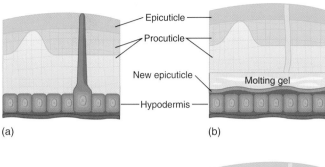

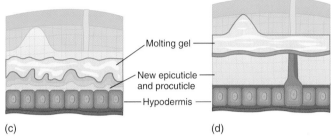

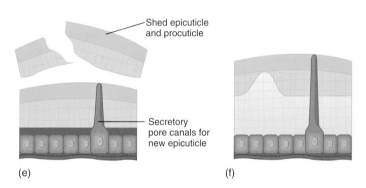

FIGURE 14.5

Events of Ecdysis. (*a,b*) During preecdysis, the hypodermis detaches from the exoskeleton, and the space between the old exoskeleton and the hypodermis fills with a fluid called molting gel. (*c,d*) The hypodermis begins secreting a new epicuticle, and a new procuticle forms as the old procuticle is digested. The products of digestion are incorporated into the new procuticle. Note that the new epicuticle and procuticle are wrinkled beneath the old exoskeleton to allow for increased body size after ecdysis. (*e*) Ecdysis occurs when the animal swallows air or water, and the exoskeleton splits along predetermined ecdysal lines. The animal pulls out of the old exoskeleton. (*f*) After ecdysis, the new exoskeleton hardens by calcium carbonate deposition and/or sclerotization, and pigments are deposited in the outer layers of the procuticle. Additional material is added to the epicuticle.

tagmata. The **prosoma** or **cephalothorax** is a sensory, feeding, and locomotor tagma. It usually bears eyes, but unlike other arthropods, never has antennae. Paired appendages attach to the prosoma. The first pair, called **chelicerae,** are often pincerlike or chelate, and are most often used in feeding. They may also be specialized as hollow fangs or for a variety of other functions. The second pair, called **pedipalps,** are usually sensory but may also be used in feeding, locomotion, or reproduction. Paired walking legs follow pedipalps. Posterior to the prosoma is the **opisthosoma,** which contains digestive, reproductive, excretory, and respiratory organs.

FIGURE 14.6

Trilobite Structure. The trilobite body had three longitudinal sections (thus, the subphylum name). It was also divided into three tagmata. A head, or cephalon, bore a pair of antennae and eyes. The trunk, or thorax, bore appendages for swimming or walking. A series of posterior segments formed the pygidium, or tail.

CLASS MEROSTOMATA

Members of the class Merostomata (mer″o-sto′mah-tah) are divided into two subclasses. The Xiphosura are the horseshoe crabs, and the Eurypterida are the giant water scorpions (figure 14.7). The latter are extinct, having lived from the Cambrian period (600 million years ago) to the Permian period (280 million years ago).

Only four species of horseshoe crabs are living today. One species, *Limulus polyphemus*, is widely distributed in the Atlantic Ocean and the Gulf of Mexico (figure 14.8*a*). Horseshoe crabs scavenge sandy and muddy substrates for annelids, small molluscs, and other invertebrates. Their body form has remained virtually unchanged for over 200 million years, and they were cited in chapter 5 as an example of stabilizing selection.

A hard, horseshoe-shaped carapace covers the prosoma of horseshoe crabs. The chelicerae, pedipalps, and first three pairs of walking legs are chelate and are used for walking and food handling. The last pair of appendages has leaflike plates at their tips and are used for locomotion and digging (figure 14.8*b*).

The opisthosoma of a horseshoe crab includes a long, unsegmented telson. If wave action flips a horseshoe crab over, the crab arches its opisthosoma dorsally, which helps it to roll to its side and flip right side up again. The first pair of opisthosomal appendages cover genital pores and are called genital opercula. The remaining five pairs of appendages are **book gills.** The name is derived from the resemblance of these platelike gills to the pages of a closed book. Gases are exchanged between the blood and water as blood circulates through the book gills. Horseshoe crabs have an open circulatory system, as do all arthropods. Blood circulation in horseshoe crabs is similar to that described later in this chapter for arachnids and crustaceans.

Horseshoe crabs are dioecious. During reproductive periods, males and females congregate in intertidal areas. The male

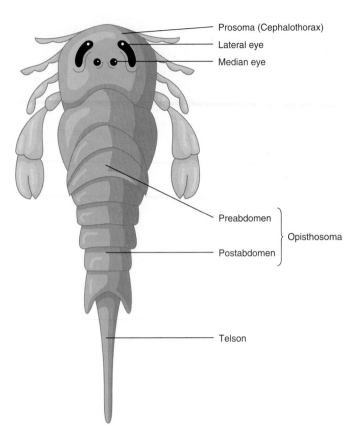

FIGURE 14.7

Class Merostomata. A eurypterid, *Euripterus remipes*.

mounts the female and grasps her with his pedipalps. The female excavates shallow depressions in the sand, and as she sheds eggs into the depressions, the male fertilizes them. Fertilized eggs are covered with sand and develop unattended.

CLASS ARACHNIDA

Members of the class Arachnida (ah-rak'nĭ-dah) (Gr. *arachne*, spider) are some of the most misrepresented members of the animal kingdom. Their reputation as fearsome and grotesque creatures is vastly exaggerated. The majority of spiders, mites, ticks, scorpions, and related forms are either harmless or very beneficial to humans.

Most zoologists believe that arachnids arose from the eurypterids and were early terrestrial inhabitants. The earliest fossils of aquatic scorpions date from the Silurian period (405 to 425 million years ago), fossils of terrestrial scorpions date from the Devonian period (350 to 400 million years ago), and fossils of all other arachnid groups are present by the Carboniferous period (280 to 345 million years ago).

Water conservation is a major concern for any terrestrial organism, and their relatively impermeable exoskeleton preadapted ancestral arachnids for terrestrialism. **Preadaptation** occurs when a structure present in members of a species proves useful in promoting reproductive success when an individual encounters new environmental situations. Later adaptations included the evolution of efficient excretory structures,

(a)

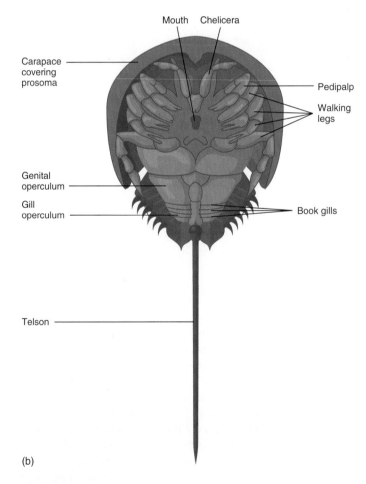

(b)

FIGURE 14.8

Class Merostomata. (*a*) Dorsal view of the horseshoe crab, *Limulus polyphemus*. (*b*) Ventral view.

internal surfaces for gas exchange, appendages modified for locomotion on land, and greater deposition of wax in the epicuticle.

Form and Function

Most arachnids are carnivores. They hold small arthropods with their chelicerae while enzymes from the gut tract pour over the prey. Partially digested food is then taken into the mouth. Others inject enzymes into prey through hollow chelicerae (e.g., spiders) and suck partially digested animal tissue. The gut tract of arachnids is divided into three regions. The anterior portion is the foregut, and the posterior portion is the hindgut. Both develop as infoldings of the body wall and are lined with cuticle. A portion of the foregut is frequently modified into a pumping stomach, and the hindgut is frequently a site of water reabsorption. The midgut between the foregut and hindgut is noncuticular and lined with secretory and absorptive cells. Lateral diverticula increase the area available for absorption and storage.

Arachnids use coxal glands and/or Malpighian tubules for excreting nitrogenous wastes. **Coxal glands** are paired, thin-walled, spherical sacs bathed in the blood of body sinuses. Nitrogenous wastes are absorbed across the wall of the sacs, transported in a long, convoluted tubule, and excreted through excretory pores at the base of the posterior appendages. Arachnids that are adapted to dry environments possess blind-ending diverticula of the gut tract that arise at the juncture of the midgut and hindgut. These tubules, called **Malpighian tubules,** absorb waste materials from the blood and empty them into the gut tract. Excretory wastes are then eliminated with digestive wastes. The major excretory product of arachnids is uric acid. As discussed in chapter 28, uric acid excretion is advantageous for terrestrial animals because uric acid is excreted as a semisolid with little water loss.

Gas exchange also occurs with minimal water loss because arachnids have few exposed respiratory surfaces. Some arachnids possess structures, called **book lungs,** that are assumed to be modifications of the book gills in the Merostomata. Book lungs are paired invaginations of the ventral body wall that fold into a series of leaflike lamellae (figure 14.9). Air enters the book lung through a slitlike opening and circulates between lamellae. Respiratory gases diffuse between the blood moving among the lamellae and the air in the lung chamber. Other arachnids possess a series of branched, chitin-lined tubules that deliver air directly to body tissues. These tubule systems, called **tracheae** (sing., trachea), open to the outside through openings called **spiracles** along the ventral or lateral aspects of the abdomen. (Tracheae are also present in insects but had a separate evolutionary origin. Aspects of their physiology are described in chapter 15.)

The circulatory system of arachnids, like that of most arthropods, is an open system in which a dorsal contractile vessel (usually called the dorsal aorta or "heart") pumps blood into tissue spaces of the hemocoel. Blood bathes the tissues and then returns to the dorsal aorta through openings in the aorta called ostia. Arachnid blood contains the dissolved respiratory pigment hemocyanin and has amoeboid cells that aid in clotting and body defenses.

The nervous system of all arthropods is ventral and, in ancestral arthropods, must have been laid out in a pattern similar to

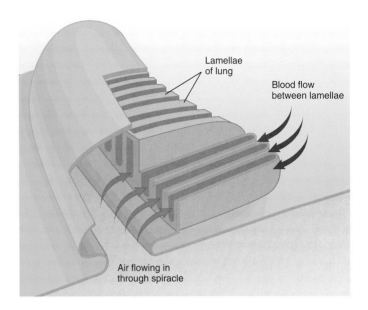

FIGURE 14.9

Arachnid Book Lung.   Air and blood moving on opposite sides of a lamella of the lung exchange respiratory gases by diffusion. Figure 14.12 shows the location of book lungs in spiders.

that of the annelids (*see figure 13.8a*). With the exception of scorpions, the nervous system of arachnids is centralized by the fusion of ganglia.

The body of an arachnid has a variety of sensory structures. Most mechanoreceptors and chemoreceptors are modifications of the exoskeleton, such as projections, pores, and slits, together with sensory and accessory cells. Collectively, these receptors are called **sensilla.** For example, setae are hairlike, cuticular modifications that may be set into membranous sockets. Displacement of a seta initiates a nerve impulse in an associated nerve cell (figure 14.10a). Vibration receptors are very important to some arachnids. Spiders that use webs to capture prey, for example, determine both the size of the insect and its position on the web by the vibrations the insect makes while struggling to free itself. The chemical sense of arachnids is comparable to taste and smell in vertebrates. Small pores in the exoskeleton are frequently associated with peglike, or other, modifications of the exoskeleton, and they allow chemicals to stimulate nerve cells. Arachnids possess one or more pairs of eyes, which they use primarily for detecting movement and changes in light intensity (figure 14.10b). The eyes of some hunting spiders probably form images.

Arachnids are dioecious. Paired genital openings are on the ventral side of the second abdominal segment. Sperm transfer is usually indirect. The male often packages sperm in a spermatophore, which is then transferred to the female. Courtship rituals confirm that individuals are of the same species, attract a female to the spermatophore, and position the female to receive the spermatophore. In some taxa (e.g., spiders), copulation occurs, and sperm is transferred via a modified pedipalp of the male. Development is direct, and the young hatch from eggs as miniature adults. Many arachnids tend their developing eggs and young during and after development.

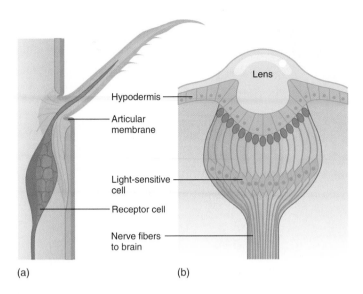

(a)

FIGURE 14.10

Arthropod Seta and Eye (Ocellus). (*a*) A seta is a hairlike modification of the cuticle set in a membranous socket. Displacement of the seta initiates a nerve impulse in a receptor cell (sensillum) associated with the base of the seta. (*b*) The lens of this spider eye is a thickened, transparent modification of the cuticle. Below the lens and hypodermis are light-sensitive sensillae with pigments that convert light energy into nerve impulses.

Order Scorpionida

Members of the order Scorpionida (skor"pe-ah-ni'dah) are the scorpions (figure 14.11*a*). They are common from tropical to warm temperate climates. Scorpions are secretive and nocturnal, hiding during most daylight hours under logs and stones.

Scorpions have a prosoma that is fused into a shield-like carapace, and small chelicerae project anteriorly from the front of the carapace (figure 14.11*b*). A pair of enlarged, chelate pedipalps is posterior to the chelicerae. The opisthosoma is divided. An anterior preabdomen contains the slitlike openings to book lungs, comblike tactile and chemical receptors called pectines, and genital openings. The postabdomen (commonly called the tail) is narrower than the preabdomen and is curved dorsally and anteriorly over the body when aroused. At the tip of the postabdomen is a sting. The sting has a bulbular base that contains venom-producing glands and a hollow, sharp, barbed point. Smooth muscles eject venom during stinging. Only a few scorpions have venom that is highly toxic to humans. Species in the genera *Androctonus* (northern Africa) and *Centuroides* (Mexico, Arizona, and New Mexico) have been responsible for human deaths. Other scorpions from the southern and southwestern areas of North America give stings comparable to wasp stings.

Prior to reproduction, male and female scorpions have a period of courtship that lasts from five minutes to several hours. Male and female scorpions face each other and extend their abdomens high into the air. The male seizes the female with his pedipalps, and they repeatedly walk backward and then forward. The lower portion of the male reproductive tract forms a sper-

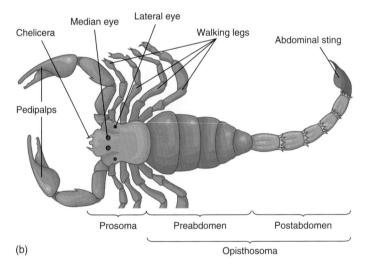

(b)

FIGURE 14.11

Order Scorpionida. (*a*) *Centuroides sculpturatus* is shown here. (*b*) External anatomy of a scorpion.

matophore that is deposited on the ground. During courtship, the male positions the female so that the genital opening on her abdomen is positioned over the spermatophore. Downward pressure of the female's abdomen on a triggerlike structure of the spermatophore releases sperm into the female's genital chamber.

Most arthropods are **oviparous;** females lay eggs that develop outside the body. Many scorpions and some other arthropods are **ovoviviparous;** development is internal, although large, yolky eggs provide all the nourishment for development. Some scorpions, however, are **viviparous,** meaning that the mother provides nutrients to nourish the embryos. Eggs develop in diverticula of

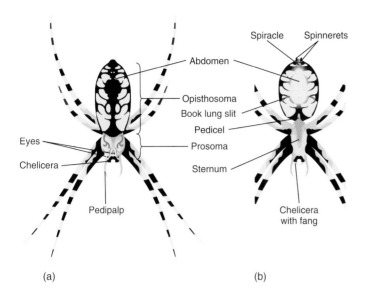

FIGURE 14.12

External Structure of a Spider.   (*a*) Dorsal view. (*b*) Ventral view.
Sources: (a) After Sherman and Sherman. (b) After the Kastons.

FIGURE 14.13

Order Araneae.   Members of the family Araneidae, the orb weavers, produce some of the most beautiful and intricate spider webs. Many species are relatively large, like this garden spider—*Argiope*. A web is not a permanent construction. When webs become wet with rain or dew, or when they age, they lose their stickiness. The entire web, or at least the spiraled portion, is then eaten and replaced.

the ovary that are closely associated with diverticula of the digestive tract. Nutrients pass from the digestive tract diverticula to the developing embryos. Development requires up to 1.5 years, and 20 to 40 young are brooded. After birth, the young crawl onto the mother's back, where they remain for up to a month.

Order Araneae

With about 34,000 species, the order Araneae (ah-ran′a-e) is the largest group of arachnids (figure 14.12). The prosoma of spiders bears chelicerae with poison glands and fangs. Pedipalps are leglike and, in males, are modified for sperm transfer. The dorsal, anterior margin of the carapace usually has six to eight eyes.

A slender, waistlike pedicel attaches the prosoma to the opisthosoma. The abdomen is swollen or elongate and contains openings to the reproductive tract, book lungs, and tracheae. It also has six to eight conical projections, called spinnerets, that are associated with silk glands. The protein that forms silk is emitted as a liquid, but hardens as it is drawn out. Spiders produce several kinds of silk, each with its own use. In addition to webs for capturing prey (figure 14.13), silk is used to line retreats, to lay a safety line that fastens to the substrate to interrupt a fall, and to wrap eggs into a case for development. Air currents catch silk lines that young spiders produce and disperse them. Silk lines have carried spiders at great altitudes for hundreds of kilometers. This is called ballooning.

Most spiders feed on insects and other arthropods that they hunt or capture in webs. A few (e.g., tarantulas or "bird spiders") feed on small vertebrates. Spiders bite their prey to paralyze them and then sometimes wrap prey in silk. They puncture the prey's body wall and inject enzymes. The spider's pumping stomach then sucks predigested prey products into the spider's digestive tract. The venom of most spiders is harmless to humans. Black widow spiders (*Lactrodectus*) and brown recluse spiders (*Loxosceles*) are exceptions, since their venom is toxic to humans (figure 14.14).

Mating of spiders involves complex behaviors that include chemical, tactile, and/or visual signals. Females deposit chemicals called pheromones on their webs or bodies to attract males. (Pheromones are chemicals that one individual releases into the environment to create a behavioral change in another member of the same species.) A male may attract a female by plucking the strands of a female's web. The pattern of plucking is species specific and helps identify and locate a potential mate and prevents the male spider from becoming the female's next meal. The tips of a male's pedipalps possess a bulblike reservoir with an ejaculatory duct and a penislike structure called an embolus. Prior to mating, the male fills the reservoir of his pedipalps by depositing sperm on a small web and then collecting sperm with his pedipalps. During mating, the pedipalp is engorged with blood, the embolus is inserted into the female's reproductive opening, and sperm are discharged. The female deposits up to 3,000 eggs in a silken egg case, which she then seals and attaches to webbing, places in a retreat, or carries with her.

Order Opiliones

Members of the order Opiliones (o′pi-le″on-es) are the harvestmen or daddy longlegs. The prosoma broadly joins to the opisthosoma, and thus, the body appears ovoid. Legs are very long and slender. Many harvestmen are omnivores (they feed on a variety of plant

(a)

(b)

FIGURE 14.14

Two Venomous Spiders.   (*a*) A black widow spider (*Lactrodectus mactans*) is recognized by its shiny black body with a red hourglass pattern on the ventral surface of its opisthosoma. (*b*) A brown recluse spider (*Loxosceles reclusa*) is recognized by the dark brown, violin-shaped mark on the dorsal aspect of its prosoma.

and animal material), and others are strictly predators. They seize prey with their pedipalps and ingest prey as described for other arachnids. Digestion is both external and internal. Sperm transfer is direct, as males have a penislike structure. Females have a tubular ovipositor that projects from a sheath at the time of egg laying. Females deposit hundreds of eggs in damp locations on the ground.

Order Acarina

Members of the order Acarina (ak′ar-i″nah) are the mites and ticks. Many are ectoparasites (parasites on the outside of the body) on humans and domestic animals. Others are free-living in both terrestrial and aquatic habitats. Of all arachnids, acarines have had the greatest impact on human health and welfare.

Mites are 1 mm or less in length. The prosoma and opisthosoma are fused and covered by a single carapace. An anterior projection called the capitulum carries mouthparts. Chelicerae and pedipalps are variously modified for piercing, biting, anchoring, and sucking, and adults have four pairs of walking legs.

Free-living mites may be herbivores or scavengers. Herbivorous mites, such as spider mites, damage ornamental and agricultural plants. Scavenging mites are among the most common animals in soil and in leaf litter. These mites include some pest species that feed on flour, dried fruit, hay, cheese, and animal fur (figure 14.15).

Parasitic mites usually do not permanently attach to their hosts, but feed for a few hours or days and then drop to the ground. One mite, the notorious chigger or red bug (*Trombicula*), is a parasite during one of its larval stages on all groups of terrestrial vertebrates. A larva enzymatically breaks down and sucks host skin, causing local inflammation and intense itching at the site of the bite. The chigger larva drops from the host and then molts to the next immature stage, called a nymph. Nymphs eventually molt to adults, and both nymphs and adults feed on insect eggs.

A few mites are permanent ectoparasites. The follicle mite, *Demodex folliculorum*, is common (but harmless) in hair follicles of most of the readers of this text. Itch mites cause scabies in humans and other animals. *Sarcoptes scabei* is the human itch mite. It tunnels in the epidermis of human skin, where females lay about 20 eggs each day. Secretions of the mites irritate the skin, and infections are acquired by contact with an infected individual.

Ticks are ectoparasites during their entire life history. They may be up to 3 cm in length, but are otherwise similar to mites. Hooked mouthparts are used to attach to their hosts and to feed on blood. The female ticks, whose bodies are less sclerotized than those of males, expand when engorged with blood. Copulation occurs on the host, and after feeding, females drop to the ground to lay eggs. Eggs hatch into six-legged immatures called seed ticks. Immatures feed on host blood and drop to the ground for each molt. Some ticks transmit diseases to humans and domestic animals. For example, *Dermacentor andersoni* transmits the bacteria that cause Rocky

FIGURE 14.15

Order Acarina.   *Dermatophagoides farinae* (×200) is a mite that is common in homes and grain storage areas. It is believed to be a major cause of dust allergies.

(a)

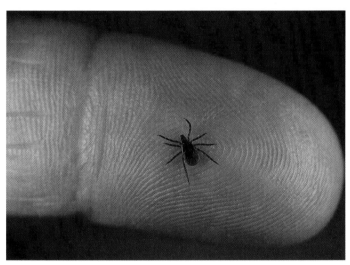

(b)

FIGURE 14.16

Order Acarina. (*a*) *Ixodes scapularis*, the tick that transmits the bacteria that cause Lyme disease. (*b*) The adult (shown here) is about the size of a sesame seed, and the nymph is the size of a poppy seed. People walking in tick-infested regions should examine themselves regularly and remove any ticks found on their skin because ticks can transmit diseases, such as Rocky Mountain spotted fever, tularemia, and Lyme disease.

Mountain spotted fever and tularemia, and *Ixodes scapularis* transmits the bacteria that cause Lyme disease (figure 14.16).

Other orders of arachnids include whip scorpions, whip spiders, pseudoscorpions, and others.

FIGURE 14.17

Class Pycnogonida. Sea spiders are often found in intertidal regions feeding on cnidarian polyps. This male sea spider (*Nymphon gracile*) is carrying eggs on ovigers.

CLASS PYCNOGONIDA

Members of the class Pycnogonida (pik″no-gon′i-dah) are the sea spiders. All are marine and worldwide, but are most common in cold waters (figure 14.17). Pycnogonids live on the ocean floor and frequently feed on cnidarian polyps and ectoprocts. Some sea spiders feed by sucking prey tissues through a proboscis. Others tear at prey with their first pair of appendages, called chelifores.

Pycnogonids are dioecious. Gonads are U-shaped, and branches of the gonads extend into each leg. Gonopores are on one of the pairs of legs. As the female releases eggs, the male fertilizes them, and the fertilized eggs are cemented into spherical masses and attached to a pair of elongate appendages of the male, called ovigers, where they are brooded until hatching.

SUBPHYLUM CRUSTACEA

Some members of the subphylum Crustacea (krus-tās′e-ah) (L. *crustaceus*, hard shelled), such as crayfish, shrimp, lobsters, and crabs, are familiar to nearly everyone. Many others are lesser-known but very common taxa. These include copepods, cladocerans, fairy shrimp, isopods, amphipods, and barnacles. Except for some isopods and crabs, crustaceans are all aquatic.

Crustaceans differ from other living arthropods in two ways. They have two pairs of antennae, whereas all other arthropods have one pair or none. In addition, crustaceans possess biramous appendages, each of which consists of a basal segment, called the **protopodite,** with two rami (distal processes that give the appendage a Y shape) attached. The medial ramus is the **endopodite,** and the lateral ramus is the **exopodite** (figure 14.18). *Trilobites had similar structures, which is evidence that the trilobites were closely related*

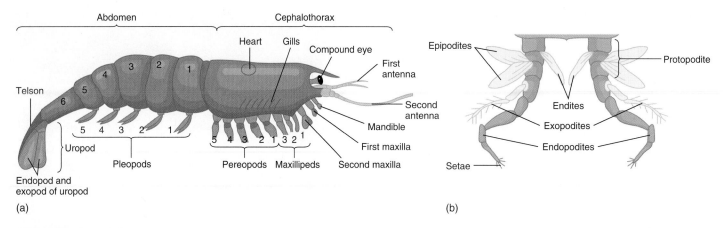

FIGURE 14.18

Crustacean Body Form. (*a*) External anatomy of a crustacean. (*b*) Pair of appendages, showing the generalized biramous structure. A protopodite attaches to the body wall. An exopodite (a lateral ramus) and an endopodite (a medial ramus) attach at the end of the protopodite. In modern crustaceans, both the distribution of appendages along the length of the body and the structure of appendages are modified for specialized functions.

to ancestral crustaceans. There are five classes of crustaceans (*see table 14.1*) and numerous orders. The three most common classes, and selected orders within those classes, are covered next.

CLASS MALACOSTRACA

Malacostraca (mal-ah-kos′trah-kah) (Gr. *malakos*, soft + *ostreion*, shell) is the largest class of crustaceans. It includes crabs, lobsters, crayfish, shrimp, mysids, shrimplike krill, isopods, and amphipods.

The order Decapoda (dek-i-pod′ah) is the largest order of crustaceans and includes shrimp, crayfish, lobsters, and crabs. Shrimp have a laterally compressed, muscular abdomen and pleopods for swimming. Lobsters, crabs, and crayfish are adapted to crawling on the surface of the substrate (figure 14.19). The abdomen of crabs is greatly reduced and is held flexed beneath the cephalothorax.

Crayfish illustrate general crustacean structure and function. They are convenient to study because of their relative

FIGURE 14.19

Order Decapoda. The lobsters, shrimp, crayfish, and crabs comprise the largest crustacean order. The lobster, *Homarus americanus*, is shown here.

abundance and large size. The body of a crayfish is divided into two regions. A cephalothorax is derived from the developmental fusion of a sensory and feeding tagma (the head) with a locomotor tagma (the thorax). The exoskeleton of the cephalothorax extends laterally and ventrally to form a shieldlike carapace. The abdomen is posterior to the cephalothorax, has locomotor and visceral functions, and in crayfish, takes the form of a muscular "tail."

Paired appendages are present in both body regions (figure 14.20). The first two pairs of cephalothoracic appendages are the first and second antennae. The third through fifth pairs of appendages are associated with the mouth. During crustacean evolution, the third pair of appendages became modified into chewing or grinding structures called **mandibles.** The fourth and fifth pairs of appendages, called **maxillae,** are for food handling. The second maxilla bears a gill and a thin, bladelike structure, called a scaphognathite (gill bailer), for circulating water over the gills. The sixth through the eighth cephalothoracic appendages are called maxillipeds and are derived from the thoracic tagma. They are accessory sensory and food-handling appendages. The last two pairs of maxillipeds also bear gills. Appendages 9 to 13 are thoracic appendages called periopods (walking legs). The first periopod, known as the cheliped, is enlarged and chelate (pincherlike) and used in defense and capturing food. All but the last pair of appendages of the abdomen are called pleopods (swimmerets) and are used for swimming. In females, developing eggs attach to pleopods, and the embryos are brooded until after hatching. In males, the first two pairs of pleopods are modified into gonopods (claspers) used for sperm transfer during copulation. The abdomen ends in a median extension called the telson. The telson bears the anus and is flanked on either side by flattened, biramous appendages of the last segment, called uropods. The telson and uropods make an effective flipperlike structure used in swimming and in escape responses.

All crustacean appendages, except the first antennae, have presumably evolved from an ancestral biramous form,

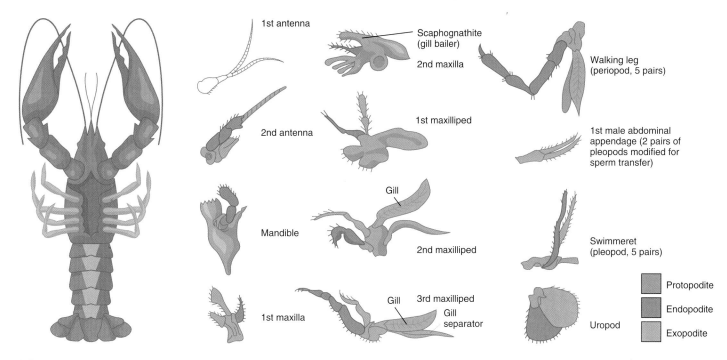

FIGURE 14.20

Crayfish Appendages.   Ventral view of a crayfish. Individual appendages are arranged in sequence. Homologies regarding the structure of appendages are color coded. The origin and homology of the first antennae are uncertain.

as evidenced by their embryological development, in which they arise as simple two-branched structures. (First antennae develop as uniramous appendages and later acquire the branched form. The crayfish and their close relatives are unique in having branched first antennae.) Structures, such as the biramous appendages of a crayfish, whose form is based on a common ancestral pattern and that have similar development in the segments of an animal, are said to be **serially homologous.**

Crayfish prey upon other invertebrates, eat plant matter, and scavenge dead and dying animals. The foregut includes an enlarged stomach, part of which is specialized for grinding. A digestive gland secretes digestive enzymes and absorbs products of digestion. The midgut extends from the stomach and is often called the intestine. A short hindgut ends in an anus and is important in water and salt regulation (figure 14.21*a*).

As previously described, the gills of a crayfish attach to the bases of some cephalothoracic appendages. Gills are in a branchial (gill) chamber, the space between the carapace and the lateral body wall (figure 14.21*b*). The beating of the scaphognathite of the second maxilla drives water anteriorly through the branchial chamber. Oxygen and carbon dioxide are exchanged between blood and water across the gill surfaces, and a respiratory pigment, hemocyanin, carries oxygen in blood plasma.

Circulation in crayfish is similar to that of most arthropods. Dorsal, anterior, and posterior arteries lead away from a muscular heart. Branches of these vessels empty into sinuses of the hemocoel. Blood returning to the heart collects in a ventral sinus and enters the gills before returning to the pericardial sinus, which surrounds the heart (figure 14.21*b*).

Crustacean nervous systems show trends similar to those in annelids and arachnids. Primitively, the ventral nervous system is ladderlike. Higher crustaceans show a tendency toward centralization and cephalization. Crayfish have supraesophageal and subesophageal ganglia that receive sensory input from receptors in the head and control the head appendages. The ventral nerves and segmental ganglia fuse, and giant neurons in the ventral nerve cord function in escape responses (figure 14.21*a*). When nerve impulses are conducted posteriorly along giant nerve fibers of a crayfish, powerful abdominal flexor muscles of the abdomen contract alternately with weaker extensor muscles, causing the abdomen to flex (the propulsive stroke) and then extend (the recovery stroke). The telson and uropods form a paddlelike "tail" that propels the crayfish posteriorly.

In addition to antennae, the sensory structures of crayfish include compound eyes, simple eyes, statocysts, chemoreceptors, proprioceptors, and tactile setae. Chemical receptors are widely distributed over the appendages and the head. Many of the setae covering the mouthparts and antennae are chemoreceptors used in sampling food and detecting pheromones. A single pair of statocysts is at the base of the first antennae. A statocyst is a pitlike invagination of the exoskeleton that contains setae and a group of cemented sand grains called a statolith. Crayfish movements move the statolith and displace setae. Statocysts provide information regarding movement, orientation with respect to the pull of gravity, and vibrations of the substrate. Because the statocyst is cuticular, it is replaced with each molt. Sand is incorporated into the statocyst when the crustacean is buried in sand. Other receptors involved with equilibrium, balance, and position senses are tactile receptors on the appendages and at joints. When a crustacean is

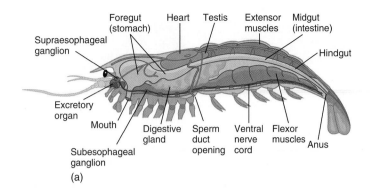

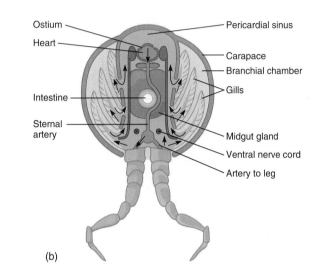

FIGURE 14.21

Internal Structure of a Crayfish. (*a*) Lateral view of a male. In the female, the ovary is in the same place as the testis of the male, but the gonoducts open at the base of the third periopods. (*b*) Cross section of the thorax in the region of the heart. In this diagram, gills are shown attached higher on the body wall than they actually occur to show the path of blood flow (arrows) through them.

crawling or resting, stretch receptors at the joints are stimulated. Crustaceans detect tilting from changing patterns of stimulation. These widely distributed receptors are important to crustaceans that lack statocysts.

Crayfish have compound eyes mounted on movable eyestalks. The lens system consists of 25 to 14,000 individual receptors called ommatidia. Compound eyes also occur in insects, and their structure and function are discussed in chapter 15. Larval crustaceans have a single, median photoreceptor consisting of a few sensilla. These simple eyes, called ocelli, allow larval crustaceans to orient toward or away from the light, but do not form images. Many larvae are planktonic and use their ocelli to orient toward surface waters.

The endocrine system of a crayfish controls functions such as ecdysis, sex determination, and color change. Endocrine glands release chemicals called hormones into the blood, where they circulate and cause responses at certain target tissues. In crustaceans, endocrine functions are closely tied to nervous functions. Nervous

tissues that produce and release hormones are called neurosecretory tissues. X-organs are neurosecretory tissues in the eyestalks of crayfish. Associated with each X-organ is a sinus gland that accumulates and releases the secretions of the X-organ. Other glands, called Y-organs, are not directly associated with nervous tissues. They are near the bases of the maxillae. Both the X-organ and the Y-organ control ecdysis. The X-organ produces molt-inhibiting hormone, and the sinus gland releases it. The target of this hormone is the Y-organ. As long as molt-inhibiting hormone is present, the Y-organ is inactive. Certain conditions prevent the release of molt-inhibiting hormone; when these conditions exist, the Y-organ releases ecdysone hormone, leading to molting. (These "certain conditions" are often complex and species specific. They include factors such as nutritional state, temperature, and photoperiod.) Other hormones that facilitate molting have also been described. These include, among others, a molt-accelerating factor.

Androgenic glands in the cephalothorax of males mediate another endocrine function. (Females possess rudiments of these glands during development, but the glands never mature.) Normally, androgenic hormone(s) promotes the development of testes and male characteristics, such as gonopods. The removal of androgenic glands from males results in the development of female sex characteristics, and if androgenic glands are experimentally implanted into a female, she develops testes and gonopods.

Hormones probably regulate many other crustacean functions. Some that have been investigated include the development of female brooding structures in response to ovarian hormones, the seasonal regulation of ovarian functions, and the regulation of heart rate and body color changes by eyestalk hormones.

The excretory organs of crayfish are called antennal glands (green glands) because they are at the bases of the second antennae and are green in living crayfish. In other crustaceans, they are called maxillary glands because they are at the bases of the second maxillae. They are structurally similar to the coxal glands of arachnids and presumably had a common evolutionary origin. Excretory products form by the filtration of blood. Ions, sugars, and amino acids are reabsorbed in the tubule before the diluted urine is excreted. As with most aquatic animals, ammonia is the primary excretory product. However, crayfish do not rely solely on the antennal glands to excrete ammonia. Ammonia also diffuses across thin parts of the exoskeleton. Even though it is toxic, ammonia is water soluble, and water rapidly dilutes it. All freshwater crustaceans face a continual influx of freshwater and loss of ions. Thus, the elimination of excess water and the reabsorption of ions become extremely important functions. Gill surfaces are also important in ammonia excretion and water and ion regulation (osmoregulation).

Crayfish, and all other crustaceans except the barnacles, are dioecious. Gonads are in the dorsal portion of the thorax, and gonoducts open at the base of the third (females) or fifth (males) periopods. Mating occurs just after a female has molted. The male turns the female onto her back and deposits nonflagellated sperm near the openings of the female's gonoducts. Fertilization occurs after copulation, as the eggs are shed. The eggs are sticky and

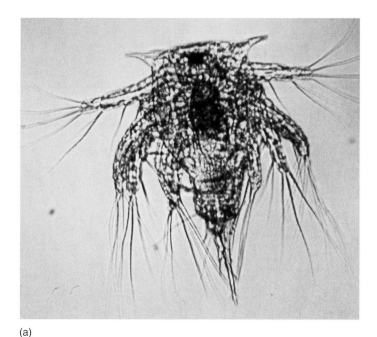

(a)

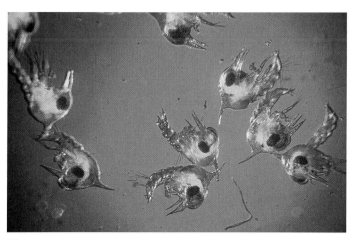

(b)

FIGURE 14.22

Crustacean Larvae.   (*a*) Nauplius larva of a barnacle (0.5 mm).
(*b*) Zoea larvae (1 mm) of a crab (*Pachygrapsus crassipes*).

securely fasten to the female's pleopods. Fanning movements
of the pleopods over the eggs keep the eggs aerated. The development of crayfish embryos is direct, with young hatching as
miniature adults. Many other crustaceans have a planktonic, free-
swimming larva called a nauplius (figure 14.22*a*). In some, the
nauplius develops into a miniature adult. Crabs and their relatives
have a second larval stage called a zoea (figure 14.22*b*). When all
adult features are present, except sexual maturity, the immature
crab is called the postlarva.

Two other orders of malacostracans have members familiar to most humans. Members of the order Isopoda (i″so-
pod′ah) include "pillbugs." Isopods are dorsoventrally flattened,
may be either aquatic or terrestrial, and scavenge decaying
plant and animal material. Some have become modified for
clinging to and feeding on other animals. Terrestrial isopods

(a)

(b)

FIGURE 14.23

Orders Isopoda and Amphipoda.   (*a*) Some isopods roll into a ball
when disturbed or threatened with drying—thus, the name "pillbug."
(*b*) This amphipod (*Orchestoidea californiana*) spends some time out of
the water hopping along beach sands—thus, the name "beachhopper."

live under rocks and logs and in leaf litter (figure 14.23*a*).
Members of the order Amphipoda (am″fi-pod′ah) have a laterally compressed body that gives them a shrimplike appearance.
Amphipods move by crawling or swimming on their sides along
the substrate. Some species are modified for burrowing, climbing, or jumping (figure 14.23*b*). Amphipods are scavengers, and
a few species are parasites.

CLASS BRANCHIOPODA

Members of the class Branchiopoda (brang″ke-o-pod′ah) (Gr.
branchio, gill + *podos*, foot) primarily live in freshwater. All branchiopods possess flattened, leaflike appendages used in respiration,
filter feeding, and locomotion.

Fairy shrimp and brine shrimp comprise the order Anostraca
(an-ost′ra-kah). Fairy shrimp usually live in temporary ponds that
spring thaws and rains form. Eggs are brooded, and when the female
dies, and the temporary pond begins to dry, the embryos become dormant in a resistant capsule. Embryos lie on the forest floor until the
pond fills again the following spring, at which time they hatch into
nauplius larvae. Animals, wind, or water currents may carry the embryos to other locations. Their short and uncertain life cycle is an
adaptation to living in ponds that dry up. The vulnerability of these
slowly swimming and defenseless crustaceans probably explains why

WILDLIFE ALERT
A Cave Crayfish (*Cambarus aculabrum*)

VITAL STATISTICS

Classification: Phylum Arthropoda, class Malacostraca, order Decapoda

Habitat: Two limestone caves of northwest Arkansas

Number Remaining: 33

Status: Endangered

NATURAL HISTORY AND ECOLOGICAL STATUS

So very little is known of this crayfish species that it does not even have a common name that is distinct from other cave crayfish. It lives only in caves, and like many other obligate cave dwellers, it lacks pigmentation and functional eyes, but other sensory organs are greatly enhanced (box figure 14.1). The origins of such morphological adaptations to caves, called "troglomorphy," are the subject of debate by evolutionary biologists. Some cave biologists (biospeleologists) believe that the lack of food resources selects for traits that conserve energy. Others believe that caves lack predators and lack selective pressures that maintain functional organs (like vision to help detect predators), and thus, unneeded structures tend to degenerate due to the accumulation of mutations. This species has been found only in two caves in Benton County, Arkansas (box figure 14.2). In one cave, there is a 2 km stream and an underground lake. Up to 24 crayfish have been observed in one visit to this cave by researchers. The second cave has less available habitat,

BOX FIGURE 14.2 The Distribution of *Cambarus aculabrum.*

and up to nine crayfish have been observed on a single visit. Although underground habitat that is inaccessible to researchers does exist and may harbor other individuals, surveys from 1990 to 2000 of caves in the Ozarks region did not reveal any other populations of this species.

These crayfish are small and are usually observed near the edges of underground streams and pools. Females carrying eggs or young have never been observed. Reproductive males are present between October and January. They are thought to live much longer than surface crayfish, perhaps even as long as humans.

The major threat to this cave crayfish is groundwater pollution. The primary source of this pollution is from agricultural operations and septic systems near caves. Soils in the area are extremely shallow, and pollutants can pass directly into groundwater reservoirs within fractured limestone bedrock. Pesticide use and accidental spills of toxic chemicals are also potential threats. One cave is surrounded by a retirement community development consisting of over 36,000 lots. Septic tank pollution and alteration of the hydrology by extensive pavement are important concerns for crayfish in this cave. There is no documentation that indicates present populations are very small because of prior contamination. The endangered status of this crayfish is warranted because present and future environmental contamination could quickly eliminate such small, local populations. Most work being done on this crayfish is currently being carried out by the Subterranean Biodiversity Project, Department of Biological Sciences, University of Arkansas.

BOX FIGURE 14.1 A Cave Crayfish (*Cambarus aculabrum*). Note the absence of pigmentation and reduced eyes that are often characteristic of cave-dwelling (troglobitic) invertebrates.

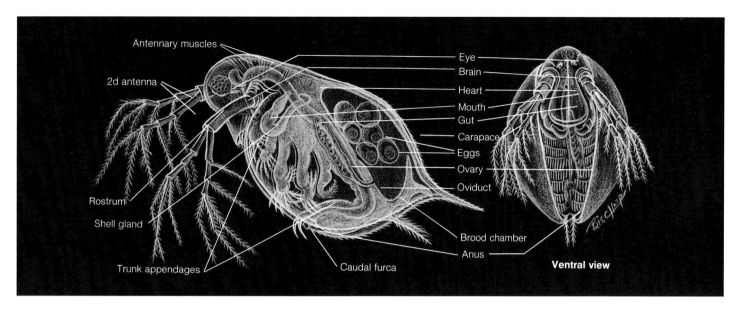

FIGURE 14.24

Class Branchiopoda.   The structure of a cladoceran water flea (*Daphnia*). Lateral and ventral views (2 mm).

they live primarily in temporary ponds, a habitat that contains few larger predators. Brine shrimp also form resistant embryos. They live in salt lakes and ponds (e.g., the Great Salt Lake in Utah).

Members of the order Cladocera (kla-dos′er-ah) are called water fleas (figure 14.24). A large carapace covers their bodies, and they swim by repeatedly thrusting their second antennae downward to create a jerky, upward locomotion. Females reproduce parthenogenetically (without fertilization) in spring and summer, and can rapidly populate a pond or lake. Eggs are brooded in an egg case beneath the carapace. At the next molt, the egg case is released and either floats or sinks to the bottom of the pond or lake. In response to decreasing temperature, changing photoperiod, or decreasing food supply, females produce eggs that develop parthenogenetically into males. Sexual reproduction produces resistant "winter eggs" that overwinter and hatch in the spring.

CLASS MAXILLOPODA

Members of the class Maxillopoda include a variety of small (with the exception of the barnacles) and sometimes bizarre crustaceans that are recognized by their short bodies and the unique combination of five head, six thoracic, and four abdominal segments, plus a telson. Interestingly, one subclass, Pentastomida, is made up of parasites of the respiratory passages of reptiles, birds, and mammals. Until recently, these animals were not recognized as crustaceans and were placed in their own phylum. Their inclusion in the Maxillopoda is the result of molecular studies of DNA coding for ribosomal particles. Two groups of common maxillopods, the copepods and the barnacles, are described next.

Subclass Copepoda

Members of the subclass Copepoda (ko″pe-pod′ah) (Gr. *kope*, oar + *podos*, foot) include some of the most abundant crustaceans (*see*

figure 14.1). There are both marine and freshwater species. Copepods have a cylindrical body and a median ocellus that develops in the nauplius stage and persists into the adult stage. The first antennae (and the thoracic appendages in some) are modified for swimming, and the abdomen is free of appendages. Most copepods are planktonic and use their second maxillae for filter feeding. Their importance in marine food webs was noted in the "Evolutionary Perspective" that opened this chapter. A few copepods live on the substrate, a few are predatory, and others are commensals or parasites of marine invertebrates, fishes, or marine mammals.

Subclass Thecostraca

The barnacles are members of the infraclass Cirripedia (sir″ĭ-ped′e-ah). They are sessile and highly modified as adults (figure 14.25a). They are exclusively marine and include about one thousand species. Most barnacles are monoecious. The planktonic nauplius of barnacles is followed by a planktonic larval stage, called a cypris larva, which has a bivalved carapace. Cypris larvae attach to the substrate by their first antennae and metamorphose to adults. In the process of metamorphosis, the abdomen is reduced, and the gut tract becomes U-shaped. Thoracic appendages are modified for filtering and moving food into the mouth. Calcareous plates cover the larval carapace in the adult stage.

Barnacles attach to a variety of substrates, including rock outcroppings, ship bottoms, whales, and other animals. Some barnacles attach to their substrate by a stalk (figure 14.25b). Others are nonstalked and are called acorn barnacles. Barnacles that colonize ship bottoms reduce both ship speed and fuel efficiency. Much time, effort, and money have been devoted to research on keeping ships free of barnacles.

Some barnacles have become highly modified parasites. *The evolution of parasitism in barnacles is probably a logical consequence of living attached to other animals.*

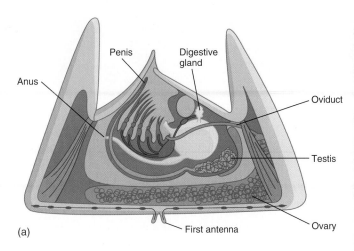

FIGURE 14.25

Class Maxillopoda, Infraclass Cirripedia. (*a*) Internal structure of a stalkless (acorn) barnacle. (*b*) Stalked (gooseneck) barnacles (*Lepas*).

FURTHER PHYLOGENETIC CONSIDERATIONS

As this chapter indicates, the arthropods have been very success-ful. This is evidenced by the diverse body forms and lifestyles of copepods, crabs, lobsters, crayfish, and barnacles, which demon-strate adaptive radiation. Few aquatic environments are without some crustaceans.

The subphylum Chelicerata is a very important group of an-imals from an evolutionary standpoint, even though they are less numerous in terms of numbers of species and individuals than are many of the crustacean groups. *Their arthropod exoskeleton and the evolution of excretory and respiratory systems that minimize water loss resulted in ancestral members of this subphylum be-coming some of the first terrestrial animals.* Chelicerates, how-ever, are not the only terrestrial arthropods. In terms of numbers of species and numbers of individuals, chelicerates are dwarfed in terrestrial environments by the insects and their relatives. These groups and the evolutionary relationships within the entire phy-lum are the subject of chapter 15.

SUMMARY

1. Arthropods and annelids are closely related animals. Living arthropods are divided into four subphyla: Chelicerata, Crustacea, Hexapoda, and Myriapoda. All members of a fifth subphylum, Trilobitomorpha, are extinct.

2. Arthropods have four distinctive characteristics: they are metameric and display tagmatization, they possess an exoskeleton, they have an open circulatory system contained within a hemocoel, and many undergo metamorphosis during development.

3. Members of the extinct subphylum Trilobitomorpha had oval, flat-tened bodies that consisted of three tagmata and three longitudinal lobes. Appendages were biramous.

4. The subphylum Chelicerata has members whose bodies are divided into a prosoma and an opisthosoma. They also possess a pair of feeding appendages called chelicerae.

5. The horseshoe crabs and the giant water scorpions belong to the class Merostomata.

6. The class Arachnida includes spiders, mites, ticks, scorpions, and others. Their exoskeleton partially preadapted the arachnids for their terrestrial habitats.

7. The sea spiders are the only members of the class Pycnogonida.

8. The subphylum Crustacea contains animals characterized by two pairs of antennae and biramous appendages. All crustaceans, except for some isopods, are primarily aquatic.

9. The class Malacostraca includes the crabs, lobsters, crayfish, shrimp, isopods, and amphipods. This is the largest crustacean class in terms of numbers of species and contains the largest crustaceans.

10. Members of the class Branchiopoda have flattened, leaflike ap-pendages. Examples are fairy shrimp, brine shrimp, and water fleas.

11. Members of the class Maxillopoda include the copepods and the barnacles.

SELECTED KEY TERMS

biramous appendages (p. 218)
chelicerae (p. 219)
ecdysis (p. 217)
exoskeleton or cuticle (p. 216)
hemocoel (p. 217)

mandibles (p. 226)
oviparous (p. 222)
ovoviviparous (p. 222)
serially homologous (p. 227)
viviparous (p. 222)

CRITICAL THINKING QUESTIONS

1. What is tagmatization, and why is it advantageous for metameric animals?

2. Explain how, in spite of being an armorlike covering, the exoskeleton permits movement and growth.

3. Why is the arthropod exoskeleton often cited as the major reason for arthropod success?

4. Explain why excretory and respiratory systems of ancestral arachnids probably preadapted these organisms for terrestrial habitats.

5. Barnacles are very successful arthropods. What factors do you think are responsible for the evolution of their highly modified body form?

ONLINE LEARNING CENTER

Visit our Online Learning Center (OLC) at www.mhhe.com/zoology (click on this book's title) to find the following chapter-related materials:

- CHAPTER QUIZZING
- RELATED WEB LINKS
    Phylum Arthropoda
        Subphylum Chelicerata
        Class Arachnida
        Subphylum Crustacea
- BOXED READINGS ON
    A Fearsome Twosome
    Sacculina: A Highly Modified Parasite
    How Insects Fly—Secrets Revealed
    Communication in Honeybees
    "Killer Bees?"
- SUGGESTED READINGS
- LAB CORRELATIONS

Check out the OLC to find specific information on these related lab exercises in the *General Zoology Laboratory Manual,* 5th edition, by Stephen A. Miller:

Exercise 15 *Arthropoda*

CHAPTER 15

THE HEXAPODS AND MYRIAPODS:

TERRESTRIAL TRIUMPHS

Outline

Concepts

1. Flight, along with other arthropod characteristics, has resulted in insects becoming the most abundant and diverse group of terrestrial animals.
2. The members of the subphylum Myriapoda are in the classes Diplopoda, Chilopoda, Pauropoda, and Symphyla.
3. Members of the subphylum Hexapoda are characterized by three pairs of legs and three tagmata. They are divided into two classes, Entognatha and Insecta.
4. Adaptations for living on land are reflected in many aspects of insect structure and function.
5. Insects have important effects on human health and welfare.
6. Arthropoda is a monophyletic grouping; arthropods are among the oldest of all animals; and ancient crustaceans provided the ancestral stock for all modern arthropods.

EVOLUTIONARY PERSPECTIVE

By almost any criterion, the insects have been enormously successful. Zoologists have described approximately 750,000 species, and some estimate that the number of insect species may be as high as 30 million! Most of the undescribed species are in tropical rain forests. Since just the described insect species comprise 75% of all living species, the total number of described and undescribed insect species dwarfs all other kinds of living organisms. Although freshwater and parasitic insect species are numerous, the success of insects has largely been due to their ability to exploit terrestrial habitats (figure 15.1).

During the late Silurian and early Devonian periods (about 400 million years ago), terrestrial environments were largely uninhabited by animals. Low-growing herbaceous plants and the first forests were beginning to flourish, and enough ozone had accumulated in the upper atmosphere to filter ultraviolet radiation from the sun. Animals with adaptations that permitted life on land had a wealth of photosynthetic production available, and unlike in marine habitats, faced little competition from other animals for resources. However, the problems associated with terrestrial life were substantial. Support and movement outside of a watery environment were difficult on land, as were water, ion (electrolyte), and temperature regulation.

A number of factors contributed to insect dominance of terrestrial habitats. *The exoskeleton preadapted the insects for life on land. Not only is the exoskeleton supportive, but the evolution of a waxy epicuticle enhanced the exoskeleton's water-conserving properties. The evolution of flight also played a big role in insect success.*

This chapter contains evolutionary concepts, which are set off in this font.

FIGURE 15.1

Class Insecta. Insects were early inhabitants of terrestrial environments. The exoskeleton and the evolution of flight have contributed to the enormous success of this group of arthropods.

The ability to fly allowed insects to use widely scattered food resources, to invade new habitats, and to escape unfavorable environments. These factors—along with desiccation-resistant eggs, metamorphosis, high reproductive potential, and diversification of mouthparts and feeding habits—permitted insects to become the dominant class of organisms on the earth.

This chapter covers two arthropod subphyla: Myriapoda and Hexapoda (table 15.1). The Myriapoda include the familiar millipedes and centipedes as well as two other less-familiar groups. The Hexapoda include the insects and their relatives. The myriapods and hexapods have been traditionally considered to be closely related based on similar excretory and respiratory structures and uniramous legs. Recent research is causing zoologists to rethink their perceptions of arthropod phylogeny, and this is affecting our view of the relationships between the two subphyla covered in this chapter. Phylogenetic implications of this research are discussed at the end of this chapter.

SUBPHYLUM MYRIAPODA

The subphylum Myriapoda (mir"e-a-pod'ah) (Gr. *myriad*, ten thousand + *podus*, foot) is divided into four classes: Diplopoda (millipedes), Chilopoda (centipedes), Symphyla (symphylans), and Pauropoda (pauropodans) (*see table 15.1*). They are characterized by a body consisting of two tagmata (head and trunk) and uniramous appendages. All modern myriapods are terrestrial.

CLASS DIPLOPODA

The class Diplopoda (dip'lah-pod'ah) (Gr. *diploos*, twofold + *podus*, foot) contains the millipedes. Ancestors of this group appeared on land during the Devonian period and were among the first terrestrial animals. Millipedes have 11 to 100 trunk segments

derived from an embryological and evolutionary fusion of primitive metameres. An obvious result of this fusion is the occurrence of two pairs of appendages on each apparent trunk segment. Each segment is actually the fusion of two segments. Fusion is also reflected internally by two ganglia, two pairs of ostia, and two pairs of tracheal trunks per apparent segment. Most millipedes are round in cross section, although some are more flattened (figure 15.2a).

Millipedes are worldwide in distribution and are nearly always found in or under leaf litter, humus, or decaying logs. Their epicuticle does not contain much wax; therefore, their choice of habitat is important to prevent desiccation. Their many legs, simultaneously pushing against the substrate, help millipedes bulldoze through the habitat. Millipedes feed on decaying plant matter using their mandibles in a chewing or scraping fashion. A few millipedes have mouthparts modified for sucking plant juices.

(a)

(b)

FIGURE 15.2

Myriapods. (*a*) A woodland millipede (*Ophyiulus pilosus*). (*b*) A centipede (*Scolopendra heros*).

TABLE 15.1
CLASSIFICATION OF THE SUBPHYLA MYRIAPODA AND HEXAPODA*

Phylum Arthropoda (ar″thra-po′dah)
Animals with metamerism and tagmatization, a jointed exoskeleton, and a ventral nervous system.

Subphylum Myriapoda (mir″e-a-pod′ah) (Gr. *myriad*, ten thousand + *podus*, foot)
Body divided into head and trunk; four pairs of head appendages; uniramous appendages. Millipedes and centipedes.

Class Diplopoda (dip″lah-pod′ah)
Two pairs of legs per apparent segment; body usually round in cross section. Millipedes.

Class Chilopoda (ki″lah-pod′ah)
One pair of legs per segment; body oval in cross section; poison claws. Centipedes.

Class Pauropoda (por″o-pod′ah)
Small (0.5 to 2 mm), soft-bodied animals; 11 segments; nine pairs of legs; live in leaf mold. Pauropods.

Class Symphyla (sim-fi′lah)
Small (2 to 10 mm); long antennae; centipede-like; 10 to 12 pairs of legs; live in soil and leaf mold. Symphylans.

Subphylum Hexapoda (hex″sah-pod′ah) (Gr. *hexa*, six + *podus*, foot)
Body divided into head, thorax, and abdomen; five pairs of head appendages; three pairs of uniramous appendages on the thorax. Insects and their relatives.

Class Entognatha (en″to-na′tha) (Gr. *entos*, within + *gnathos*, jaw)
Mouth appendages hidden within the head; mandibles with single articulation; legs with one undivided tarsus.

Order Collembola (col-lem′bo-lah)
Antennae with four to six segments; compound eyes absent; abdomen with six segments, most with springing appendage on fourth segment; inhabit soil and leaf litter. Springtails.

Order Protura (pro-tu′rah)
Minute, with cone-shaped head; antennae, compound eyes, and ocelli absent; abdominal appendages on first three segments; inhabit soil and leaf litter. Proturans.

Order Diplura (dip-lu′rah)
Head with many segmented antennae; compound eyes and ocelli absent; cerci multisegmented or forcepslike; inhabit soil and leaf litter. Diplurans.

Class Insecta (in-sekt′ah) (L. *insectum*, to cut up)
Mouth appendages exposed and projecting from head; mandibles usually with two points of articulation; well developed Malpighian tubules.

Subclass Archaeognatha (ar″ke-ona′tha)
Order Archaeognatha
Small, wingless, cylindrical and scaly body; mandibles with single articulation; abdomen 11 segmented with 3 to 8 pairs of styli and 3 caudal filaments; ametabolous metamorphosis. Jumping bristletails.

Subclass Zygentoma (xi-gen″to-mah)
Order Thysanura (thi-sa-nu′rah)
Tapering abdomen; flattened; scales on body; terminal cerci; long antennae; ametabolous metamorphosis. Silverfish.

Subclass Pterogota (ter-i-go′tah)
Wings on second and third thoracic segments; wings may be modified or lost; no pregenital abdominal appendages; direct sperm transfer.

Infraclass Palaeoptera (pa′le-op″ter-ah)
Wings incapable of being folded at rest, held vertically above the body or horizontally out from the body; wings with many veins and cross-veins; antennae reduced or vestigial in adults.

Order Ephemeroptera (e-fem-er-op′ter-ah)
Elongate abdomen with two or three tail filaments; two pairs of membranous wings with many veins; forewings triangular; short, bristlelike antennae; hemimetabolous metamorphosis. Mayflies.

Order Odonata (o-do-nat′ah)
Elongate, membranous wings with netlike venation; abdomen long and slender; compound eyes occupy most of head; hemimetabolous metamorphosis. Dragonflies, damselflies.

*Selected orders of insects are described.

Infraclass Neoptera (ne-op′ter-ah)
Wings folded at rest; venation reduced.

Order Plecoptera (ple-kop′ter-ah)
Adults with reduced mouthparts; elongate antennae; long cerci; nymphs aquatic with gills; hemimetabolous metamorphosis. Stoneflies.

Order Mantodea (man-to′deah)
Prothorax long; prothoracic legs long and armed with strong spines for grasping prey; predators; hemimetabolous metamorphosis. Mantids.

Order Blattaria (blat-tar′eah)
Body oval and flattened; head concealed from above by a shieldlike extension of the prothorax; hemimetabolous metamorphosis. Cockroaches.

Order Isoptera (i-sop′ter-ah)
Workers white and wingless; front and hindwings of reproductives of equal size; reproductives and some soldiers may be sclerotized; abdomen broadly joins thorax; social; hemimetabolous metamorphosis. Termites.

Order Dermaptera (der-map′ter-ah)
Elongate; chewing mouthparts; threadlike antennae; abdomen with unsegmented forcepslike cerci; hemimetabolous metamorphosis. Earwigs.

Order Orthoptera (or-thop′ter-ah)
Forewing long, narrow, and leathery; hindwing broad and membranous; chewing mouthparts; hemimetabolous metamorphosis. Grasshoppers, crickets, katydids.

Order Phasmida (fas′mi-dah)
Body elongate and sticklike; wings reduced or absent; some tropical forms are flattened and leaflike; hemimetabolous metamorphosis. Walking sticks, leaf insects.

Order Phthiraptera (fthi-rap′ter-ah)
Small, wingless ectoparasites of birds and mammals; body dorsoventrally flattened; white; hemimetabolous metamorphosis. Sucking and chewing lice.

Order Hemiptera (hem-ip′ter-ah)
Piercing-sucking mouthparts; mandibles and first maxillae styletlike and lying in grooved labium; wings membranous; hemimetabolous metamorphosis. Bugs, cicadas, leafhoppers, aphids.

Order Thysanoptera (thi-sa-nop′ter-ah)
Small bodied; sucking mouthparts; wings narrow and fringed with long setae; plant pests; hemimetabolous metamorphosis. Thrips.

Order Neuroptera (neu-rop′ter-ah)
Wings membranous; hindwings held rooflike over body at rest; holometabolous metamorphosis. Lacewings, snakeflies, antlions, dobsonflies.

Order Coleoptera (ko-le-op′ter-ah)
Forewings sclerotized, forming covers (elytra) over the abdomen; hindwings membranous; chewing mouthparts; the largest insect order; holometabolous metamorphosis. Beetles.

Order Trichoptera (tri-kop′ter-ah)
Mothlike with setae-covered antennae; chewing mouthparts; wings covered with setae and held rooflike over abdomen at rest; larvae aquatic and often dwell in cases that they construct; holometabolous metamorphosis. Caddis flies.

Order Lepidoptera (lep-i-dop′ter-ah)
Wings broad and covered with scales; mouthparts formed into a sucking tube; holometabolous metamorphosis. Moths, butterflies.

Order Diptera (dip′ter-ah)
Mesothoracic wings well developed; metathoracic wings reduced to knoblike halteres; variously modified but never chewing mouthparts; holometabolous metamorphosis. Flies.

Order Siphonaptera (si-fon-ap′ter-ah)
Laterally flattened, sucking mouthparts; jumping legs; parasites of birds and mammals; holometabolous metamorphosis. Fleas.

Order Hymenoptera (hi-men-op′ter-ah)
Wings membranous with few veins; well-developed ovipositor, sometimes modified into a sting; mouthparts modified for biting and lapping; social and solitary species; holometabolous metamorphosis. Ants, bees, wasps.

Millipedes roll into a ball when faced with desiccation or when disturbed. Many also possess repugnatorial glands that produce hydrogen cyanide, which repels other animals. Hydrogen cyanide is not synthesized and stored as hydrogen cyanide because it is caustic and would destroy millipede tissues. Instead, a precursor compound and an enzyme mix as they are released from separate glandular compartments. Repellants increase the likelihood that the millipede will be dropped unharmed and decrease the chances that the same predator will try to feed on another millipede.

Male millipedes transfer sperm to female millipedes with modified trunk appendages, called gonopods, or in spermatophores. Eggs are fertilized as they are laid and hatch in several weeks. Immatures acquire more legs and segments with each molt until they reach adulthood.

CLASS CHILOPODA

Members of the class Chilopoda (ki″lah-pod′ah) (Gr. *cheilos*, lip + *podus*, foot) are the centipedes. Most centipedes are nocturnal and scurry about the surfaces of logs, rocks, or other forest-floor debris. Like millipedes, most centipedes lack a waxy epicuticle and therefore require moist habitats. Their bodies are flattened in cross section, and they have a single pair of long legs on each of their 15 or more trunk segments. The last pair of legs is usually modified into long sensory appendages.

Centipedes are fast-moving predators (figure 15.2*b*). Food usually consists of small arthropods, earthworms, and snails; however, some centipedes feed on frogs and rodents. Poison claws (modified first-trunk appendages called maxillipeds) kill or immobilize prey. Maxillipeds, along with mouth appendages, hold the prey as mandibles chew and ingest the food. Most centipede venom is essentially harmless to humans, although many centipedes have bites that are comparable to wasp stings; a few human deaths have been reported from large, tropical species.

Centipede reproduction may involve courtship displays in which the male lays down a silk web using glands at the posterior tip of the body. He places a spermatophore in the web, which the female picks up and introduces into her genital opening. Eggs are fertilized as they are laid. A female may brood and guard eggs by wrapping her body around the eggs, or they may be deposited in the soil. Young are similar to adults except that they have fewer legs and segments. Legs and segments are added with each molt.

CLASSES PAUROPODA AND SYMPHYLA

Members of the class Pauropoda (por″o-pod′ah) (Gr. *pauros*, small + *podus*, foot) are soft-bodied animals with 11 segments (figure 15.3*a*). These animals live in forest-floor litter, where they feed on fungi, humus, and other decaying organic matter. Their very small size and thin, moist exoskeleton allow gas exchange across the

(a)

(b)

FIGURE 15.3

Subphylum Myriapoda. (*a*) A member of the class Pauropoda (*Pauropus*). (*b*) A member of the class Symphyla (*Scutigerella*).

body surface and diffusion of nutrients and wastes in the body cavity.

Members of the class Symphyla (sim-fi′lah) (Gr. *sym*, same + *phyllos*, leaf) are small arthropods (2 to 10 mm in length) that occupy soil and leaf mold, superficially resemble centipedes, and are often called garden centipedes (figure 15.3*b*). They lack eyes and have 12 leg-bearing trunk segments. The posterior segment may have one pair of spinnerets or long, sensory bristles. Symphylans normally feed on decaying vegetation; however, some species are pests of vegetables and flowers.

SUBPHYLUM HEXAPODA

The subphylum Hexapoda (hex″sah-pod′ah) (Gr. *hexa*, six + *podus*, foot) includes animals whose bodies are divided into three tagmata, have five pairs of head appendages, and three pairs of legs on the thorax. The subphylum is divided into two classes. Entognatha (en″to-na′tha) (Gr. *entos*, within + *gnathos*, jaw) includes the collembolans, proturans, and diplurans (*see table 15.1*). Members of this class have mouthparts that are hidden inside the head capsule, thus the class name. It is probably not a monophyletic grouping because the entognathous mouthparts of the diplurans are apparently not homologous to the mouthparts of the other two orders. Insecta (L. *insectum*, to cut up) includes the 30 orders of insects. Table 15.1 provides a partial listing of these orders. They are all characterized by mouthparts that project from the head capsule. A new insect order, Mantophasmatodea, is the first new insect order to be described since 1914.

CLASS INSECTA

Members of the class Insecta are, in terms of numbers of species and individuals, the most successful land animals. In spite of obvious diversity, common features make insects easy to recognize. Many insects have wings and one pair of antennae, and virtually all adults have three pairs of legs.

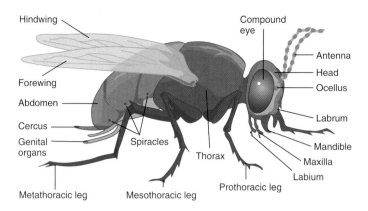

FIGURE 15.4

External Structure of a Generalized Insect.   Insects are characterized by a body divided into head, thorax, and abdomen; three pairs of legs; and two pairs of wings.

External Structure and Locomotion

The body of an insect is divided into three tagmata: head, thorax, and **abdomen** (figure 15.4). The head bears a single pair of antennae, mouthparts, compound eyes, and zero, two, or three ocelli. The thorax consists of three segments. They are, from anterior to posterior, the **prothorax,** the **mesothorax,** and the **metathorax.** One pair of legs attaches along the ventral margin of each thoracic segment, and a pair of wings, when present, attaches at the dorsolateral margin of the mesothorax and metathorax. Wings have thickened, hollow veins for increased strength. The thorax also contains two pairs of spiracles, which are openings to the tracheal system. Most insects have 10 or 11 abdominal segments, each of which has a lateral fold in the exoskeleton that allows the abdomen to expand when the insect has gorged itself or when it is full of mature eggs. Each abdominal segment has a pair of spiracles. Also present are genital structures used during copulation and egg deposition, and sensory structures called cerci. Gills are present on abdominal segments of certain immature aquatic insects.

Insect Flight   Insects move in diverse ways. *From an evolutionary perspective, however, flight is the most important form of insect locomotion. Insects were the first animals to fly.* One of the most popular hypotheses on the origin of flight states that wings may have evolved from rigid, lateral outgrowths of the thorax that probably protected the legs or spiracles. Later, these fixed lobes could have been used in gliding from the tops of tall plants to the forest floor. The ability of the wing to flap, tilt, and fold back over the body probably came later.

Another requirement for flight was the evolution of limited thermoregulatory abilities. Thermoregulation is the ability to maintain body temperatures at a level different from environmental temperatures. Relatively high body temperatures, perhaps 25° C or greater, are needed for flight muscles to contract rapidly enough for flight.

Some insects use a **direct** or **synchronous flight** mechanism, in which muscles acting on the bases of the wings contract to

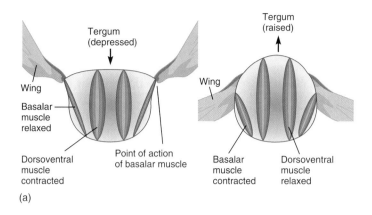

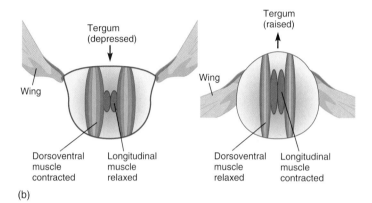

FIGURE 15.5

Insect Flight.   (*a*) Muscle arrangements for the direct or synchronous flight mechanism. Note that muscles responsible for the downstroke attach at the base of the wings. (*b*) Muscle arrangements for the indirect or asynchronous flight mechanism. Muscles changing the shape of the thorax cause wings to move up and down.

produce a downward thrust, and muscles attaching dorsally and ventrally on the exoskeleton contract to produce an upward thrust (figure 15.5*a*). The synchrony of direct flight mechanisms depends on the nerve impulse to the flight muscles that must precede each wingbeat. Butterflies, dragonflies, and grasshoppers are examples of insects with a synchronous flight mechanism.

Other insects use an **indirect** or **asynchronous flight** mechanism. Muscles act to change the shape of the exoskeleton for both upward and downward wing strokes. Dorsoventral muscles pulling the dorsal exoskeleton (tergum) downward produce the upward wing thrust. The downward thrust occurs when longitudinal muscles contract and cause the exoskeleton to arch upward (figure 15.5*b*). The resilient properties of the exoskeleton enhance the power and velocity of these strokes. During a wingbeat, the thorax is deformed, storing energy in the exoskeleton. At a critical point midway into the downstroke, stored energy reaches a maximum, and at the same time, resistance to wing movement suddenly decreases. The wing then "clicks" through the rest of the cycle, using energy stored in the exoskeleton. Asynchrony of this flight mechanism arises from the lack of one-to-one correspondence between nerve impulses and wingbeats. A single nerve impulse can result in approximately 50 cycles of the wing, and frequencies of 1,000 cycles per second (cps) have been recorded in some midges! The asynchrony between wing-beat and nerve impulses is dependent on flight muscles being stretched during the "click" of the thorax. The stretching of longitudinal flight muscles during the upward beat of the wing initiates the subsequent contraction of these muscles. Similarly, stretching during the downward beat of the wing initiates subsequent contraction of dorsoventral flight muscles. Indirect flight muscles are frequently called **fibrillar flight muscles.** Flies and wasps are examples of insects with an asynchronous flight mechanism.

Simple flapping of wings is not enough for flight. The tilt of the wing must be controlled to provide lift and horizontal propulsion. In most insects, muscles that control wing tilting attach to sclerotized plates at the base of the wing.

Other Forms of Locomotion   Insects walk, run, jump, or swim across the ground or other substrates. When they walk, insects have three or more legs on the ground at all times, creating a very stable stance. When they run, fewer than three legs may be in contact with the ground. A fleeing cockroach (order Blattaria) reaches speeds of about 5 km/hour, although it seems much faster when trying to catch one. The apparent speed is the result of their small size and ability to quickly change directions. Jumping insects, such as grasshoppers (order Orthoptera), usually have long, metathoracic legs in which leg musculature is enlarged to generate large, propulsive forces. Energy for a flea's (order Siphonaptera) jump is stored as elastic energy of the exoskeleton. Muscles that flex the legs distort the exoskeleton. A catch mechanism holds the legs in this "cocked" position until special muscles release the catches and allow the stored energy to quickly extend the legs. This action hurls the flea for distances that exceed 100 times its body length. A comparable distance for a human long jumper would be the length of two football fields!

Nutrition and the Digestive System

The diversity of insect feeding habits parallels the diversity of insects themselves. There are many variations on mouthparts of insects, but the mouthparts are based on a common arrangement of structures shown in figure 15.6, which shows the biting-chewing mouthparts of a grasshopper. An upper liplike structure is called the labrum. It is sensory, and unlike the remaining mouthparts, is not derived from segmental, paired appendages. Mandibles are sclerotized chewing mouthparts. They usually bear teeth for grinding and cutting and have a side-to-side movement. The maxillae have cutting surfaces and palps that are sensory and food-holding structures. The **labium** is a sensory lower lip and its palps are also used in food holding. A hypopharynx is a tonguelike sensory structure. The efficiency of these biting-chewing mouthparts is obvious in watching a caterpillar feeding on a leaf or considering a termite feeding on wooden structures.

The structure of mouthparts is modified in insects that suck liquid food, although the basic arrangement of mouthparts is

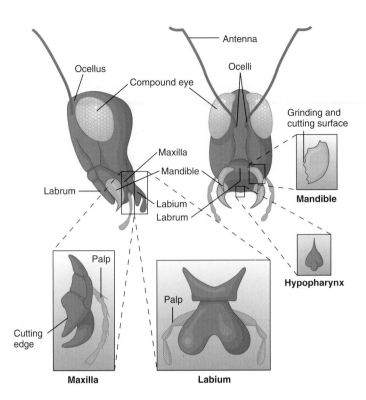

FIGURE 15.6

Head and Mouthparts of a Grasshopper.   All mouthparts except the labrum are derived from segmental appendages. The labrum is a sensory upper lip. The mandibles are heavily sclerotized and used for tearing and chewing. The maxillae have cutting edges and a sensory palp. The labium forms a sensory lower lip. The hypopharynx is a sensory, tongue-like structure.

retained. There are many variations in sucking mouthparts. In mosquitoes, six stylets are formed from the labrum, hypopharynx, mandibles, and maxillae and are used to pierce flesh and suck blood. In the butterflies and moths, the maxillae form a long, coiled tube that is used to suck nectar from flowers (figure 15.7).

The housefly has sponging mouthparts. The labium is expanded into a labellum. Saliva is secreted from the mouth, and minute channels on the labellum provide pathways for liquefied food to move the mouth through capillary action.

The digestive tract, as in all arthropods, is long and straight and consists of three regions: a foregut, a midgut, and a hindgut (figure 15.8). The foregut is often modified into a muscular pharynx. In sucking insects, the pharynx (or in some, the oral cavity) is used for sucking fluids into the digestive tract. Behind the pharynx is a crop that is used in storage. A proventriculus or gizzard regulates movement to the midgut and may function in grinding food. The midgut provides the surfaces for digestion and absorption, and gastric cecae increase the surface area for these functions. The hindgut or intestine is primarily involved with the reabsorption of water.

Gas Exchange

Gas exchange with air requires a large surface area for the diffusion of gases. In terrestrial environments, these surfaces are also

FIGURE 15.7

Specialization of Insect Mouthparts.   The mouthparts of insects are often highly specialized for specific feeding habits. For example, the sucking mouthparts of a butterfly consist of modified maxillae that coil when not in use. Mandibles, labia, and the labrum are reduced in size. A portion of the anterior digestive tract is modified as a muscular pump for drawing liquids through the mouthparts.

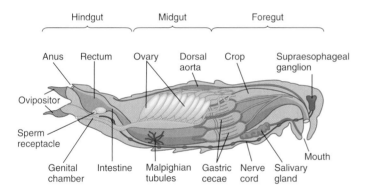

FIGURE 15.8

Internal Structure of a Generalized Insect.   Salivary glands produce enzymes but may be modified for the secretion of silk, anticoagulants, or pheromones. The crop is an enlargement of the foregut and stores food. The proventriculus is a grinding and/or straining structure at the junction of the midgut and hindgut. Gastric cecae secrete digestive enzymes. The intestine and the rectum are modifications of the hindgut that absorb water and the products of digestion.

avenues for water loss. Respiratory water loss in insects, as in some arachnids, is reduced through the invagination of respiratory surfaces to form highly branched systems of chitin-lined tubes, called tracheae.

Tracheae open to the outside of the body through spiracles, which usually have some kind of closure device to prevent excessive water loss. Spiracles lead to tracheal trunks that branch, eventually giving rise to smaller branches, the tracheoles. Taenidia are rings or spiral thickenings of tracheal trunks that keep tracheae from collapsing and allow lengthwise expansion with body movements. Tracheoles end intracellularly and are especially abundant in metabolically active tissues, such as

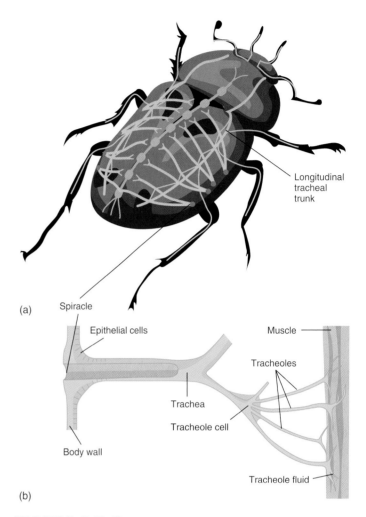

FIGURE 15.9

Tracheal System of an Insect.   (*a*) Major tracheal trunks. (*b*) Tracheoles end in cells, and the terminal portions of tracheoles are fluid filled.

flight muscles. No cells are more than 2 or 3 μm from a tracheole (figure 15.9).

Most insects have ventilating mechanisms that move air into and out of the tracheal system. For example, contracting flight muscles alternately compress and expand the larger tracheal trunks and thereby ventilate the tracheae. In some insects, carbon dioxide that metabolically active cells produce is sequestered in the hemocoel as bicarbonate ions (HCO_3^-). As oxygen diffuses from the tracheae to the body tissues, and is not replaced by carbon dioxide, a vacuum is created that draws more air into the spiracles. This process is called passive suction. Periodically, the sequestered bicarbonate ions are converted back into carbon dioxide, which escapes through the tracheal system. Other insects contract abdominal muscles in a pumplike fashion to move air into and out of their tracheal systems.

In many aquatic insects, spiracles are nonfunctional and gases diffuse across the body wall. In others (some aquatic beetles and hemipterans), a bubble of air covers the spiracles and is carried underwater with the insect, which may periodically surface to refresh the air. Alternatively, gases may diffuse into and out of the bubble directly from the water. Some aquatic insects (immature mayflies and some immature stoneflies) have tracheal gills. Gases diffuse across the gill surface into branches of the tracheal system that extend into the gills.

Circulation and Temperature Regulation

The circulatory system of insects is similar to that described for other arthropods, although the blood vessels are less well developed. Blood distributes nutrients, hormones, and wastes, and amoeboid blood cells participate in body defense and repair mechanisms. Blood is not important in gas transport.

As described earlier, thermoregulation is a requirement for flying insects. Virtually all insects warm themselves by basking in the sun or resting on warm surfaces. Because they use external heat sources in temperature regulation, insects are generally considered ectotherms. Other insects (e.g., some moths, alpine bumblebees, and beetles) can generate heat by rapid contraction of flight muscles, a process called shivering thermogenesis. Metabolic heat generated in this way can raise the temperature of thoracic muscles from near 0 to 30° C. Because some insects rely to a limited extent on metabolic heat sources, they have a variable body temperature and are sometimes called heterotherms. Insects can also cool themselves by seeking cool, moist habitats. Honeybees can cool a hive by beating their wings at the entrance of the hive, thus circulating cooler outside air through the hive.

Nervous and Sensory Functions

The nervous system of insects is similar to the pattern described for annelids and other arthropods (see figure 15.8). The supraesophageal ganglion is associated with sensory structures of the head. Connectives join the supraesophageal ganglion to the subesophageal ganglion, which innervates the mouthparts and salivary glands and has a general excitatory influence on other body parts. Segmental ganglia of the thorax and abdomen fuse to various degrees in different taxa. Insects also possess a well-developed visceral nervous system that innervates the gut, reproductive organs, and heart.

Research has demonstrated that insects are capable of some learning and have a memory. For example, bees (order Hymenoptera) instinctively recognize flowerlike objects by their shape and ability to absorb ultraviolet light, which makes the center of the flower appear dark. If a bee is rewarded with nectar and pollen, it learns the odor of the flower. Bees that feed once at artificially scented feeders choose that odor in 90% of subsequent feeding trials. Odor is a very reliable cue for bees because it is more constant than color and shape. Wind, rain, and herbivores may damage the latter.

Sense organs of insects are similar to those found in other arthropods, although they are usually specialized for functioning on land. Mechanoreceptors perceive physical displacement of the body or of body parts. Setae are distributed over the mouthparts, antennae, and legs (see figure 14.10a). Touch, air movements, and vibrations of the substrate can displace setae. Stretch receptors at the joints, on other parts of the cuticle, and on muscles monitor posture and position.

Hearing is a mechanoreceptive sense in which airborne pressure waves displace certain receptors. All insects can respond to pressure waves with generally distributed setae; others have specialized receptors. For example, **Johnston's organs** are in the base of the antennae of most insects, including mosquitoes and midges (order Diptera). Long setae that vibrate when certain frequencies of sound strike them cover the antennae of these insects. Vibrating setae move the antenna in its socket, stimulating sensory cells. Sound waves in the frequency range of 500 to 550 cycles per second (cps) attract and elicit mating behavior in the male mosquito *Aedes aegypti*. These waves are in the range of sounds that the wings of females produce. **Tympanal (tympanic) organs** are in the legs of crickets and katydids (order Orthoptera), in the abdomen of grasshoppers (order Orthoptera) and some moths (order Lepidoptera), and in the thorax of other moths. Tympanal organs consist of a thin, cuticular membrane covering a large air sac. The air sac acts as a resonating chamber. Just under the membrane are sensory cells that detect pressure waves. Grasshopper tympanal organs can detect sounds in the range of 1,000 to 50,000 cps. (The human ear can detect sounds between 20 and 20,000 cps.) Bilateral placement of tympanal organs allows insects to discriminate the direction and origin of a sound.

Insects use chemoreception in feeding, selection of egglaying sites, mate location, and sometimes, social organization. Chemoreceptors are usually abundant on the mouthparts, antennae, legs, and ovipositors, and take the form of hairs, pegs, pits, and plates that have one or more pores leading to internal nerve endings. Chemicals diffuse through these pores and bind to and excite nerve endings.

All insects are capable of detecting light and may use light in orientation, navigation, feeding, or other functions. **Compound eyes** are well developed in most adult insects. They are similar in structure and function to those of other arthropods, and recent evidence points to their homology (common ancestry) with those of crustaceans. Compound eyes consist of a few to 28,000 receptors, called **ommatidia,** that fuse into a multifaceted eye. The outer surface of each ommatidium is a lens and is one facet of the eye. Below the lens is a crystalline cone. The lens and the crystalline cone are light-gathering structures. Certain cells of an ommatidium, called retinula cells, have a special light-collecting area, called the rhabdom. The rhabdom converts light energy into nerve impulses. Pigment cells surround the crystalline cone, and sometimes the rhabdom, and prevent the light that strikes one rhabdom from reflecting into an adjacent ommatidium (figure 15.10).

Although many insects form an image of sorts, the concept of an image has no real significance for most species. The compound eye is better suited for detecting movement. Movement of a point of light less than 0.1° can be detected as light successively strikes adjacent ommatidia. For this reason, bees are attracted to flowers blowing in the wind, and predatory insects select moving prey. Compound eyes detect wavelengths of light that the human

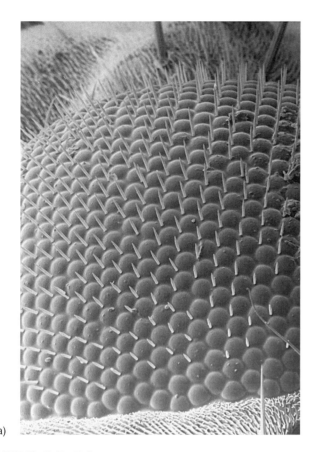

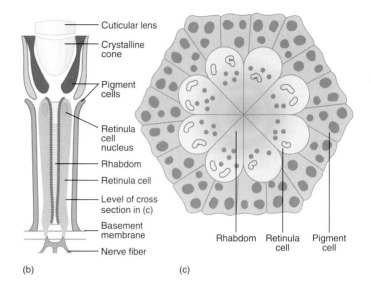

Cuticular lens

Crystalline cone

Pigment cells

Retinula cell nucleus

Rhabdom

Retinula cell

Level of cross section in (c)

Basement membrane

Nerve fiber

(b)

Rhabdom   Retinula cell   Pigment cell

(c)

FIGURE 15.10

Compound Eye of an Insect.   (*a*) Compound eye of *Drosophila* (SEM ×300). Each facet of the eye is the lens of a single sensory unit called an ommatidium. (*b*) Structure of an ommatidium. The lens and the crystalline cone are light-gathering structures. Retinula cells have light-gathering areas, called rhabdoms. Pigment cells prevent light in one ommatidium from reflecting into adjacent ommatidia. In insects that are active at night, the pigment cells are often migratory, and pigment can be concentrated around the crystalline cone. In these insects, low levels of light from widely scattered points can excite an ommatidium. (*c*) Cross section through the rhabdom region of an ommatidium.

eye cannot detect, especially in the ultraviolet end of the spectrum. In some insects, compound eyes also detect polarized light, which may be used for navigation and orientation.

Ocelli consist of 500 to 1,000 receptor cells beneath a single cuticular lens (*see figure 14.10*b). Ocelli are sensitive to changes in light intensity and may be important in the regulation of daily rhythms.

Excretion

The primary insect excretory structures are the Malpighian tubules and the rectum. Malpighian tubules end blindly in the hemocoel and open to the gut tract at the junction of the midgut and the hindgut. Microvilli cover the inner surfaces of their cells. Various ions are actively transported into the tubules, and water passively follows. Uric acid is secreted into the tubules and then into the gut, as are amino acids and ions (figure 15.11). In the rectum, water, certain ions, and other materials are reabsorbed, and the uric acid is eliminated.

As described in chapter 14, the excretion of uric acid is advantageous for terrestrial animals because it minimizes water

loss. There is, however, an evolutionary trade-off. The conversion of primary nitrogenous wastes (ammonia) to uric acid is energetically costly. Nearly half of the food energy a terrestrial insect consumes may be used to process metabolic wastes! In aquatic insects, ammonia simply diffuses out of the body into the surrounding water.

Chemical Regulation

The endocrine system controls many physiological functions of insects, such as cuticular sclerotization (*see chapter 14*), osmoregulation, egg maturation, cellular metabolism, gut peristalsis, and heart rate. As in all arthropods, ecdysis is under neuroendocrine control. In insects, the subesophageal ganglion and two endocrine glands—the corpora allata and the prothoracic glands—control these activities.

Neurosecretory cells of the subesophageal ganglion manufacture ecdysiotropin. This hormone travels in neurosecretory cells to a structure called the corpora cardiaca. The corpora cardiaca then releases thoracotropic hormone, which stimulates the prothoracic gland to secrete ecdysone. Ecdysone initiates the

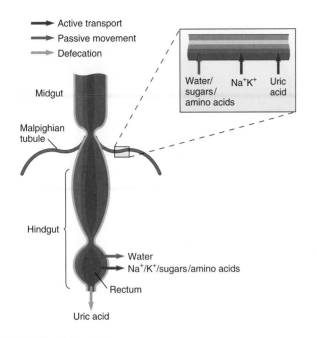

FIGURE 15.11

Insect Excretion. Malpighian tubules remove nitrogenous wastes from the hemocoel. Various ions are actively transported across the outer membranes of the tubules. Water follows these ions into the tubules and carries amino acids, sugars, and some nitrogenous wastes along passively. Some water, ions, and organic compounds are reabsorbed in the basal portion of the Malpighian tubules and the hindgut; the rest are reabsorbed in the rectum. Uric acid moves into the hindgut and is excreted.

reabsorption of the inner portions of the procuticle and the formation of the new exoskeleton. Chapter 25 discusses these events further. Other hormones are also involved in ecdysis. The recycling of materials absorbed from the procuticle, changes in metabolic rates, and pigment deposition are a few of probably many functions that hormones control.

In immature stages, the corpora allata produces and releases small amounts of juvenile hormone. The amount of juvenile hormone circulating in the hemocoel determines the nature of the next molt. Large concentrations of juvenile hormone result in a molt to a second immature stage, intermediate concentrations result in a molt to a third immature stage, and low concentrations result in a molt to the adult stage. Decreases in the level of circulating juvenile hormone also lead to the degeneration of the prothoracic gland so that, in most insects, molts cease once adulthood is reached. Interestingly, after the final molt, the level of juvenile hormone increases again, but now it promotes the development of accessory sexual organs, yolk synthesis, and the egg maturation.

Pheromones are chemicals an animal releases that cause behavioral or physiological changes in another member of the same species. Zoologists have described many different insect uses of pheromones (table 15.2). Pheromones are often so specific that the stereoisomer (chemical mirror image) of the pheromone may be ineffective in initiating a response. Wind or water may carry pheromones several kilometers, and a few pheromone molecules

TABLE 15.2
FUNCTIONS OF INSECT PHEROMONES

Sex pheromones—Excite or attract members of the opposite sex; accelerate or retard sexual maturation. Example: Female moths produce and release pheromones that attract males.

Caste-regulating pheromones—Used by social insects to control the development of individuals in a colony. Example: The amount of "royal jelly" fed a female bee larva determines whether the larva will become a worker or a queen.

Aggregation pheromones—Produced to attract individuals to feeding or mating sites. Example: Certain bark beetles aggregate on pine trees during an attack on a tree.

Alarm pheromones—Released to warn other individuals of danger; may cause orientation toward the pheromone source and elicit a subsequent attack or flight from the source. Example: A sting from one bee alarms other bees in the area, who are are likely to attack.

Trailing pheromones—Laid down by foraging insects to help other members of a colony identify the location and quantity of food found by one member of the colony. Example: Ants often trail on a pheromone path to and from a food source. The pheromone trail is reinforced each time an ant travels over it.

falling on a chemoreceptor of another individual may be enough to elicit a response.

Reproduction and Development

One of the reasons for insects' success is their high reproductive potential. Reproduction in terrestrial environments, however, has its risks. Temperature, moisture, and food supplies vary with the season. Internal fertilization requires highly evolved copulatory structures because gametes dry quickly on exposure to air. In addition, mechanisms are required to bring males and females together at appropriate times.

Complex interactions between internal and external environmental factors regulate sexual maturity. Internal regulation includes interactions between endocrine glands (primarily the corpora allata) and reproductive organs. External regulating factors may include the quantity and quality of food. For example, the eggs of mosquitoes (order Diptera) do not mature until after the female takes a meal of blood, and the number of eggs produced is proportional to the quantity of blood ingested. Many insects use the photoperiod (the relative length of daylight and darkness in a 24-hour period) for timing reproductive activities because it indicates seasonal changes. Population density, temperature, and humidity also influence reproductive activities.

A few insects, including silverfish (order Thysanura) and springtails (order Collembola) have indirect fertilization. The male deposits a spermatophore that the female picks up later. Most insects have complex mating behaviors for locating and recognizing a potential mate, for positioning a mate for copulation, or for pacifying an aggressive mate. Mating behavior may involve pheromones (moths, order Lepidoptera), visual signals (fireflies, order Coleoptera), and auditory signals (cicadas, order Hemiptera; and grasshoppers, crickets, and katydids, order Orthoptera). Once other stimuli have brought the male and female together, tactile stimuli from the antennae and other appendages help position the insects for mating.

Abdominal copulatory appendages of the male usually transfer the sperm to an outpocketing of the female reproductive tract, the sperm receptacle (*see figure 15.8*). Eggs are fertilized as they leave the female and are usually laid near the larval food supply. Females may use an **ovipositor** to deposit eggs in or on some substrate.

Insect Development and Metamorphosis

Insect evolution has resulted in the divergence of immature and adult body forms and habits. For insects in the superorder Endopterygota (*see table 15.1*), immature stages, called **larval instars,** are a time of growth and accumulation of reserves for the transition to adulthood. The adult stage is associated with reproduction and dispersal. In these orders, insects tend to spend a greater part of their lives in juvenile stages. The developmental patterns of insects reflect degrees of divergence between immatures and adults and are classified into three (or sometimes four) categories.

In insects that display **ametabolous** (Gr. *a*, without + *metabolos*, change) **metamorphosis,** the primary differences between adults and larvae are body size and sexual maturity. Both adults and larvae are wingless. The number of molts in the ametabolous development of a species varies, and unlike most other insects, molting continues after sexual maturity. Silverfish (order Thysanura) have ametabolous metamorphosis.

Hemimetabolous (Gr. *hemi*, half) **metamorphosis** involves a species-specific number of molts between egg and adult stages, during which immatures gradually take on the adult form. The external wings develop (except in those insects, such as lice, that have secondarily lost wings), adult body size and proportions are attained, and the genitalia develop during this time. Immatures are called **nymphs.** Grasshoppers (order Orthoptera) and chinch bugs (order Hemiptera) show hemimetabolous metamorphosis (figure 15.12). When immature stages are aquatic, they often have gills (e.g., mayflies, order Ephemeroptera; dragonflies, order Odonata). These immatures are called **naiads** (L. *naiad*, water nymph).

In **holometabolous** (Gr. *holos*, whole) **metamorphosis,** immatures are called larvae because they are very different from the adult in body form, behavior, and habitat (figure 15.13). The number of larval instars is species specific, and the last larval molt forms the **pupa.** The pupa is a time of apparent inactivity but is actually a time of radical cellular change, during which all characteristics of the adult insect develop. A protective case may enclose the pupal stage. The last larval instar (e.g., moths, order Lepidoptera) constructs a **cocoon** partially or entirely from silk. The **chrysalis** (e.g., butterflies, order Lepidoptera) and **puparium** (e.g., flies, order Diptera) are the last larval exoskeletons and are retained through the pupal stage. Other insects (e.g., mosquitoes, order Diptera) have pupae that are unenclosed by a larval exoskeleton, and the pupa may be active. The final molt to the adult stage usually occurs within the cocoon, chrysalis, or puparium,

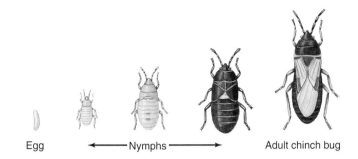

FIGURE 15.12

Hemimetabolous Development of the Chinch Bug, *Blissus leucopterus* (Order Hemiptera).   Eggs hatch into nymphs. Note the gradual increase in nymph size and the development of external wing pads. In the adult stage, the wings are fully developed, and the insect is sexually mature.

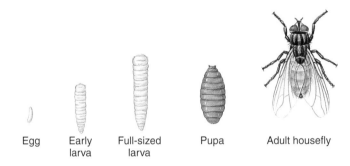

FIGURE 15.13

Holometabolous Development of the Housefly, *Musca domestica* (Order Diptera).   The egg hatches into a larva that is different in form and habitat from the adult. After a certain number of larval instars, the insect pupates. During the pupal stage, all adult characteristics form.

and the adult then exits, frequently using its mandibles to open the cocoon or other enclosure. This final process is called emergence or eclosion.

Insect Behavior

Insects have many complex behavior patterns. Most of these are innate (genetically programmed). For example, a newly emerged queen in a honeybee hive will search out and try to destroy other queen larvae and pupae in the hive. This behavior is innate because no experiences taught the potential queen that her survival in the hive required the death or dispersal of all other potential queens. Similarly, no experience taught her how queen-rearing cells differ from the cells containing worker larvae and pupae. Some insects are capable of learning and remembering, and these abilities play important roles in insect behavior.

Social Insects

Social behavior has evolved in many insects and is particularly evident in those insects that live in colonies. Usually, different members of the colony are specialized, often structurally as well as behaviorally, for performing different tasks.

(a)   (b)   (c)

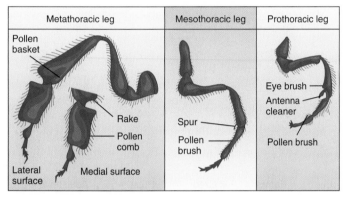

(d)

FIGURE 15.14

Honeybees (Order Hymenoptera).   Honeybees have a social organization consisting of three castes. Eye size and overall body size distinguish the castes. (*a*) A worker bee. (*b*) A drone bee. (*c*) A queen bee marked with blue to identify her. (*d*) The inner surface of metathoracic legs have setae, called the pollen comb, that remove pollen from the mesothoracic legs and the abdomen. Pollen is then compressed into a solid mass by being squeezed in a pollen press and moved to a pollen basket on the outer surface of the leg, where the pollen is carried. The mesothoracic legs gather pollen from body parts. The prothoracic legs of a worker bee clean pollen from the antennae and body.

Social behavior is most highly evolved in the bees, wasps, and ants (order Hymenoptera) and in termites (order Isoptera). Each kind of individual in an insect colony is called a **caste.** Often, three or four castes are present in a colony. Reproductive females are referred to as queens. Workers may be sterile males and females (termites) or sterile females (order Hymenoptera), and they support and protect the colony. Their reproductive organs are often degenerate. Reproductive males inseminate the queen(s) and are called kings or drones. Soldiers are usually sterile and may possess large mandibles to defend the colony.

Honeybees (order Hymenoptera) have three of these castes in their colonies (figure 15.14). A single queen lays all the eggs. Workers are female, and they construct the comb out of wax that they produce. They also gather nectar and pollen, feed the queen and drones, care for the larvae, and guard and clean the hive.

These tasks are divided among workers according to age. Younger workers take care of jobs around the hive, and older workers forage for nectar and pollen. Except for those that overwinter, workers live for about one month. Drones develop from unfertilized eggs, do not work, and are fed by workers until they leave the hive to attempt mating with a queen.

A pheromone that the queen releases controls the honeybee caste system. Workers lick and groom the queen and other workers. In so doing, they pick up and pass to other workers a caste-regulating pheromone. This pheromone inhibits the workers from rearing new queens. As the queen ages, or if she dies, the amount of caste-regulating pheromone in the hive decreases. As the pheromone decreases, workers begin to feed the food for queens ("royal jelly") to several female larvae developing in the hive. This food contains chemicals that promote the development of queen characteristics. The larvae that receive royal jelly develop into queens, and as they emerge, the new queens begin to eliminate each other until only one remains. The queen that remains goes on a mating flight and returns to the colony, where she lives for several years.

The evolution of social behavior involving many individuals leaving no offspring and sacrificing individuals for the perpetuation of the colony has puzzled evolutionists for many years. It may be explained by the concepts of kin selection and altruism.

Insects and Humans

Only about 0.5% of insect species adversely affect human health and welfare. Many others have provided valuable services and commercially valuable products, such as wax, honey, and silk, for thousands of years. Insects are responsible for the pollination of approximately 65% of all plant species. Insects and flowering plants have coevolutionary relationships that directly benefit humans. The annual value of insect-pollinated crops is estimated at $19 billion per year in the United States.

Insects are also agents of biological control. The classic example of one insect regulating another is the vedalia (lady bird) beetles' control of cottony-cushion scale. The scale insect, *Icerya purchasi,* was introduced into California in the 1860s. Within 20 years, the citrus industry in California was virtually destroyed. The vedalia beetle (*Vedalia cardinalis*) was brought to the United States in 1888 and 1889 and cultured on trees infested with scale. In just a few years, the scale was under control, and the citrus industry began to recover.

Many other insects are also beneficial. Soil-dwelling insects play important roles in aeration, drainage, and turnover of soil, and they promote decay processes. Other insects serve important roles in food webs. Insects are used in teaching and research, and have contributed to advances in genetics, population ecology, and physiology. Insects have also given endless hours of pleasure to those who collect them and enjoy their beauty.

Some insects, however, are parasites and vectors of disease. Parasitic insects include head, body, and pubic lice (order Anoplura); bedbugs (order Hemiptera); and fleas (order Siphonaptera). Other insects transmit disease-causing microorganisms, nematodes, and flatworms. Insect-transmitted diseases, such as

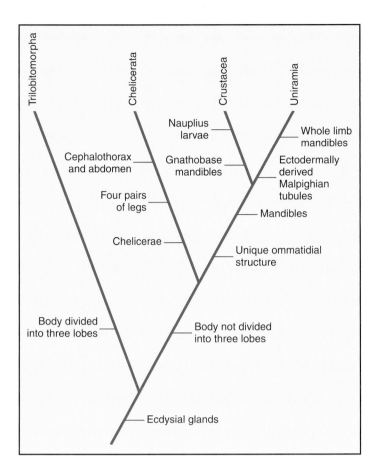

FIGURE 15.15

A Traditional Interpretation of Arthropod Phylogeny.   Systematists have considered the trilobites to be the ancestors all arthropods. The Myriapoda and Hexapoda have been considered closely related and grouped here into a single subphylum, Uniramia. Crustaceans are represented as having close ties to the Myriapoda and Hexapoda (Uniramia). Alternate traditional phylogenies have been proposed, but virtually all represented the trilobites as ancestral and the crustaceans as a true monophyletic group.

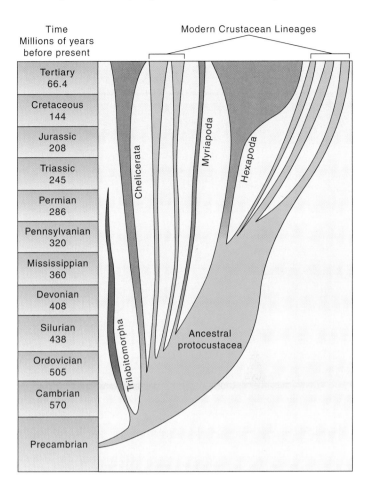

FIGURE 15.16

A Highly Speculative Interpretation of Arthropod Phylogeny.   Recent research is interpreted to suggest that ancient crustaceans are the ancestors of all arthropods. Crustacea is a paraphyletic grouping with modern crustaceans diverging from the ancestral stock at various points. Trilobites were the first noncrustacean lineage to diverge from ancient crustaceans. The uniramous appendages, Malpighian tubules, and tracheal systems of myriapods and hexapods are probably convergent, and these groups are represented as having separate ancestry.

malaria, yellow fever, bubonic plague, encephalitis, leishmaniasis, and typhus, have changed the course of history.

Other insects are pests of domestic animals and plants. Some reduce the health of domestic animals and the quality of animal products. Insects feed on crops and transmit plant diseases, such as Dutch elm disease, potato virus, and asters yellow. Annual lost revenue from insect damage to crops or insect-transmitted diseases in the United States is approximately $5 billion.

FURTHER PHYLOGENETIC CONSIDERATIONS

A fundamental question regarding arthropod evolution has been whether or not the Arthropoda is a monophyletic taxon, that is, whether or not the arthropods originated from a single ancestral species. That question seems to be settled in the minds of virtually all arthropod systematists. *Arthropoda is a monophyletic grouping.*

Until recently, zoologists thought they had a good idea of the relationships of taxa within the phylum—at least there were only two major competing views regarding relationships among the arthropod subphyla. These ideas were based primarily on paleontological and morphological data. One example is shown in figure 15.15. Recently, data from developmental biology, molecular biology, and new paleontological studies are causing a major rethinking of these relationships. This revision is in its infancy, and so any representation of arthropod relationships, such as that shown in figure 15.16, is highly speculative. A few of the major points that are emerging are described in the following paragraphs.

Arthropods were some of the earliest animals to evolve. They are represented in the fossils of the Ediacaran fauna (*see chapter 7*) and, therefore, date to Precambrian times about 600 million years ago (*see table 4.1*). Interestingly, these earliest fossils are not of trilobites but are similar to crustaceans. This evidence,

WILDLIFE ALERT
The Karner Blue Butterfly (*Lycaeides melissa samuelis*)

VITAL STATISTICS

Classification: Phylum Arthropoda, class Hexapoda, order Lepidoptera

Range: New England to the Great Lakes Region (historical), Western Great Lakes Region (current)

Habitat: Sand plains, oak savannas, or pine barrens in association with wild lupine

Number remaining: Less than 1% of its population in 1900

Status: Endangered

NATURAL HISTORY AND ECOLOGICAL STATUS

The male Karner blue butterfly is silvery or dark blue with a narrow black margin on the upper surface of the wing (box figure 15.1). The female has grayish brown wings with a dark border and orange bands.

The Karner blue butterfly lives in sand plains, oak savannas, or pine barrens. These habitats are patchily distributed across the northeastern and midwestern parts of the United States (box figure 15.2). They consist of grassland areas with scattered trees and are maintained in their natural state by periodic disturbances from fire. Without fire disturbance, shrub and tree vegetation soon overruns these habitats. These grassy islands are the home of a plant called wild lupine (*Lupinus perennis*), which is the sole food source for caterpillars of the Karner blue butterfly.

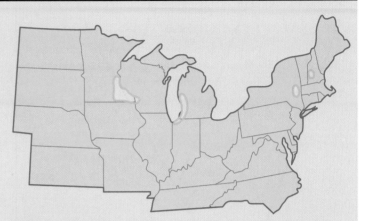

BOX FIGURE 15.2 Approximate Distribution of Karner Blue Butterfly (*Lycaeides melissa samuelis*).

In the spring, Karner blue eggs, which have overwintered, hatch, and the larvae feed on wild lupine. By the end of May, the larvae have grown and pupated. Adults emerge and mate, and the females lay eggs on or near wild lupine. Eggs quickly hatch, and the larvae feed, grow, and pupate. By the end of July, the second generation of adults emerges and mates. This generation of females lays eggs among the plant litter near wild lupine. These eggs remain dormant until the following April. After August, no adults or larvae of the Karner blue butterfly are found.

The endangered status of the Karner blue butterfly is a result of habitat loss. As humans develop land, they quickly bring fires under control. Fire control allows a wild lupine habitat to overgrow with shrubs and trees, making it no longer suitable for wild lupine or the Karner blue butterfly. Habitat loss is devastating for a species, especially if habitat distribution is patchy. Patchy distribution does not provide corridors for movement and dispersal to new areas. When a patch of habitat is lost, the species present in that patch usually have no place to go. Even when patches of habitat are not entirely lost, but simply broken up by human development, dispersal within the habitat may be nearly impossible. The construction of roads, buildings, and off-road vehicle trails presents formidable obstacles for a species as fragile and specialized as the Karner blue butterfly.

This example vividly illustrates an organism's dependence on habitat preservation. The struggle of the Karner blue butterfly is a subtle reminder that something is wrong in our treatment of the land. Protecting pine barrens and oak savannas will improve the chances for the survival of wild lupine, the Karner blue butterfly, and other species specialized for life in these fragile habitats.

BOX FIGURE 15.1 Male Karner Blue Butterfly (*Lycaeides melissa samuelis*).

along with data from molecular biology, is causing zoologists to reject the prior assumption that the first arthropods were trilobites. *We now view ancestral crustaceans, or protocrustaceans, as the stem group from which other arthropods evolved.*

The trilobites appeared in the early Cambrian and were probably the first major lineage to diverge from ancestral crustaceans. They were the most abundant arthropods until their extinction at the Permian-Triassic boundary 240 million years ago.

The Chelicerata were common in the Ordovician period about 500 million years ago, and by the Silurian period (450 million years ago) chelicerates had become the first, or some of the first, land animals. The Myriapoda is the next lineage to appear in the fossil record. Even though all modern myriapods are terrestrial, the first myriapods were marine animals. *Terrestrial myriapods—millipedes and centipedes—quickly joined the chelicerates on land during the Silurian period.*

The Hexapoda is the last arthropod lineage to appear. Collembolans and archaeognathans are found in the Devonian fossil record and date to about 400 million years ago. The rise of flowering plants about 130 million years ago, along with the evolution of flight, probably promoted the rapid diversification of insects during the Cretaceous period.

Recent research is challenging former ideas regarding the relationships of the Myriapoda and Hexapoda. Shared characteristics like their tracheal systems, uniramous legs, and Malpighian tubules have been interpreted as supporting a close relationship of these taxa. In figure 15.15, these two groups are united under the subphylum Uniramia. *New evidence from molecular, developmental, and morphological investigations suggests that the Hexapoda are more closely related to the Crustacea than to the Myriapoda.* This means that the shared myriapod/hexapod characters are convergent, not homologous. On the other hand, shared crustacean/hexapod characters like the compound eye may be homologous.

Modern crustacean lineages probably arose at various times from the ancestral crustaceans. In figure 15.16, these lineages are not labeled because of the highly speculative nature of the tree. What is emerging from all of these studies is that Arthropoda is a monophyletic grouping; arthropods are among the oldest of all animals; and that ancient crustaceans provided the ancestral stock for all modern arthropods. The subphylum Crustacea is thus paraphyletic

SUMMARY

1. During the Devonian period, insects began to exploit terrestrial environments. Flight, the exoskeleton, and metamorphosis are probably keys to insect success.

2. The subphylum Myriapoda includes four classes of arthropods. Members of the class Diplopoda (the millipedes) are characterized by apparent segments bearing two pairs of legs. Members of the class Chilopoda (the centipedes) are characterized by a single pair of legs on each of their 15 segments and a body that is flattened in cross section. The class Pauropoda contains soft-bodied animals that feed on fungi and decaying organic matter in forest-floor litter. Members of the class Symphyla are centipede-like arthropods that live in soil and leaf mold, where they feed on decaying vegetation.

3. Animals in the subphylum Hexapoda are characterized by bodies divided into three tagmata, five pairs of head appendages, and three pairs of legs. It includes two classes, Entognatha and Insecta.

4. Insect flight involves either a direct (synchronous) flight mechanism or an indirect (asynchronous) flight mechanism.

5. Mouthparts of insects are adapted for chewing, piercing, and/or sucking, and the gut tract may be modified for pumping, storage, digestion, and water conservation.

6. In insects, gas exchange occurs through a tracheal system.

7. The insect nervous system is similar to that of other arthropods. Sensory structures include tympanal organs, compound eyes, and ocelli.

8. Malpighian tubules transport uric acid to the digestive tract. Conversion of nitrogenous wastes to uric acid conserves water but is energetically expensive.

9. Hormones regulate many insect functions, including ecdysis and metamorphosis. Pheromones are chemicals emitted by one individual that alter the behavior of another member of the same species.

10. Insect adaptations for reproduction on land include resistant eggs, external genitalia, and behavioral mechanisms that bring males and females together at appropriate times.

11. Metamorphosis of an insect may be ametabolous, hemimetabolous, or holometabolous. Neuroendocrine and endocrine secretions control metamorphosis.

12. Insects show both innate and learned behavior.

13. Many insects are beneficial to humans, and a few are parasites and/or transmit diseases to humans or agricultural products. Others attack cultivated plants and stored products.

14. New research is causing zoologists to rethink arthropod phylogeny. Arthropods are considered monophyletic, and ancestral crustaceans are probably the stock from which other arthropods evolved. Crustacea is a paraphyletic grouping.

SELECTED KEY TERMS

ametabolous metamorphosis (p. 245)
caste (p. 246)
direct or synchronous flight (p. 239)
hemimetabolous metamorphosis (p. 245)
holometabolous metamorphosis (p. 245)

indirect or asynchronous flight (p. 239)
larval instars (p. 245)
nymphs (p. 245)
pupa (p. 245)

CRITICAL THINKING QUESTIONS

1. What problems are associated with living and reproducing in terrestrial environments? Explain how insects overcome these problems.

2. List as many examples as you can of how insects communicate with each other. In each case, what is the form and purpose of the communication?

3. In what way does holometabolous metamorphosis reduce competition between immature and adult stages? Give specific examples.

4. What role does each stage play in the life history of holometabolous insects?

5. Discuss the phylogenetic implications of recent research that has been interpreted to suggest that Crustacea is a paraphyletic group. How might future analyses deal with this group?

ONLINE LEARNING CENTER

Visit our Online Learning Center www.mhhe.com/zoology (click on this book's title) to find the following chapter-related materials:

- CHAPTER QUIZZING
- RELATED WEB LINKS
    Subphylum Hexapoda
    Minor Classes of Myriapods
    Class Insecta
- BOXED READINGS ON
    A Fearsome Twosome
    Sacculina: A Highly Modified Parasite
    How Insects Fly—Secrets Revealed
    Communication in Honeybees
    "Killer Bees?"
- READINGS ON LESSER-KNOWN INVERTEBRATES
    Possible Arthropod Relatives
- SUGGESTED READINGS
- LAB CORRELATIONS

Check out the OLC to find specific information on these related lab exercises in the *General Zoology Laboratory Manual*, 5th edition, by Stephen A. Miller:

Exercise 15   *Arthropoda*

CHAPTER 16

THE ECHINODERMS

Outline

Concepts

1. Echinoderms are a part of the deuterostome evolutionary lineage. They are characterized by pentaradial symmetry, a calcium carbonate internal skeleton, and a water-vascular system.
2. Although many classes of echinoderms are extinct, living echinoderms are divided into five classes: (a) Asteroidea, sea stars; (b) Ophiuroidea, brittle stars and basket stars; (c) Echinoidea, sea urchins and sand dollars; (d) Holothuroidea, sea cucumbers; and (e) Crinoidea, sea lilies and feather stars.
3. Pentaradial symmetry of echinoderms probably developed during the evolution of sedentary lifestyles, in which echinoderms used the water-vascular system in suspension feeding. Later, evolution resulted in some echinoderms becoming more mobile, and the water-vascular system came to be used primarily in locomotion.

EVOLUTIONARY PERSPECTIVE

If you could visit 400-million-year-old Paleozoic seas, you would see representatives of nearly every phylum studied in the previous eight chapters of this text. In addition, you would observe many representatives of the phylum Echinodermata (i-ki"na-dur′ma-tah) (Gr. *echinos*, spiny + *derma*, skin + *ata*, to bear). Many ancient echinoderms attached to their substrates and probably lived as suspension feeders—a feature found in only one class of modern echinoderms (figure 16.1). Today, the relatively common sea stars, sea urchins, sand dollars, and sea cucumbers represent this phylum. In terms of numbers of species, echinoderms may seem to be a declining phylum. Fossil records indicate that about 12 of 18 classes of echinoderms have become extinct. That does not mean, however, that living echinoderms are of minor importance. Members of three classes of echinoderms have flourished and often make up a major component of the biota of marine ecosystems (table 16.1).

Characteristics of the phylum Echinodermata include:

1. Calcareous endoskeleton in the form of ossicles that arise from mesodermal tissue
2. Adults with pentaradial symmetry and larvae with bilateral symmetry
3. Water-vascular system composed of water-filled canals used in locomotion, attachment, and/or feeding
4. Complete digestive tract that may be secondarily reduced
5. Hemal system derived from coelomic cavities
6. Nervous system consisting of a nerve net, nerve ring, and radial nerves

This chapter contains evolutionary concepts, which are set off in this font.

FIGURE 16.1

Phylum Echinodermata.   This feather star (*Comanthina*) uses its highly branched arms in suspension feeding. Although this probably reflects the original use of echinoderm appendages, most modern echinoderms use arms for locomotion, capturing prey, and scavenging the substrate for food. Feather stars can detach from the substrate and also use their arms in swimming and crawling.

RELATIONSHIPS TO OTHER ANIMALS

*M*ost zoologists believe that echinoderms share a common an-cestry with hemichordates and chordates because of the deuterostome characteristics that they share (see figure 7.12): an anus that develops in the region of the blastopore, a coelom that forms from outpockets of the embryonic gut tract (verte-brate chordates are an exception), and radial, indeterminate cleavage. Unfortunately, no known fossils document a common ancestor for these phyla or demonstrate how the deuterostome lin-eage was derived from ancestral diploblastic or triploblastic stocks (figure 16.2).

*A*lthough echinoderm adults are radially symmetrical, most zoologists believe that echinoderms evolved from bilater-ally symmetrical ancestors. Evidence for this relationship in-cludes bilaterally symmetrical echinoderm larval stages and extinct forms that were not radially symmetrical.

ECHINODERM CHARACTERISTICS

The approximately seven thousand species of living echinoderms are exclusively marine and occur at all depths in all oceans. Mod-ern echinoderms have a form of radial symmetry, called **pentara-dial symmetry,** in which body parts are arranged in fives, or a multiple of five, around an oral-aboral axis (figure 16.3*a*). Radial symmetry is adaptive for sedentary or slowly moving animals be-cause it allows a uniform distribution of sensory, feeding, and

TABLE 16.1
CLASSIFICATION OF THE PHYLUM ECHINODERMATA

Phylum Echinodermata (i-ki″na-dur′ma-tah)
The phylum of triploblastic, coelomate animals whose members are pentaradially symmetrical as adults and possess a water-vascular system and an endoskeleton covered by epithelium. Pedicellaria often present.

   Class Crinoidea (krin-oi′de-ah)
   Free-living or attached by an aboral stalk of ossicles; flourished in the Paleozoic era. Approximately 230 living species. Sea lilies, feather stars.

   Class Asteroidea (as″te-roi′de-ah)
   Rays not sharply set off from central disk; ambulacral grooves with tube feet; suction disks on tube feet; pedicellariae present. Sea stars. About 1,500 species.

   Class Ophiuroidea (o-fe-u-roi′de-ah)
   Arms sharply marked off from the central disk; tube feet without suction disks. Brittle stars. Over 2,000 species.

   Class Echinoidea (ek″i-noi′de-ah)
   Globular or disk shaped; no rays; movable spines; skeleton (test) of closely fitting plates. Sea urchins, sand dollars. Approximately 1,000 species.

   Class Holothuroidea (hol″o-thu-roi′de-ah)
   No rays; elongate along the oral-aboral axis; microscopic ossicles embedded in a muscular body wall; circumoral tentacles. Sea cucumbers. Approximately 1,500 species.

This listing reflects a phylogenetic sequence; however, the discussion that follows begins with the echinoderms that are familiar to most students.

other structures around the animal. Some modern mobile echino-derms, however, have secondarily returned to a basically bilateral form.

The echinoderm skeleton consists of a series of calcium carbonate plates called ossicles. These plates are derived from mesoderm, held in place by connective tissues, and covered by an epidermal layer. If the epidermal layer is abraded away, the skele-ton may be exposed in some body regions. The skeleton is fre-quently modified into fixed or articulated spines that project from the body surface.

*T*he evolution of the skeleton may be responsible for the pentaradial body form of echinoderms. The joints between two skeletal plates represent a weak point in the skeleton (figure 16.3b). By not having weak joints directly opposite one an-other, the skeleton is made stronger than if the joints were arranged opposite each other.

The **water-vascular system** of echinoderms is a series of water-filled canals, and their extensions are called tube feet. It orig-inates embryologically as a modification of the coelom and is cili-ated internally. The water-vascular system includes a ring canal that surrounds the mouth (figure 16.4). The ring canal usually opens to the outside or to the body cavity through a stone canal and an opening called the madreporite. In sea stars, the madreporite is

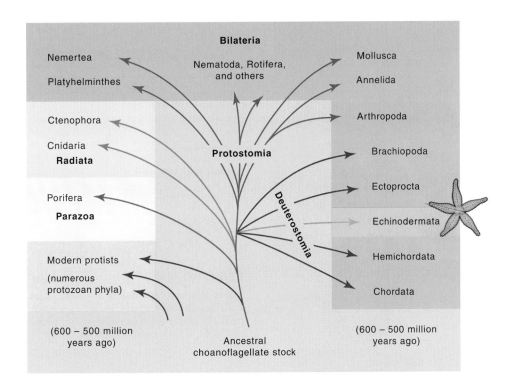

FIGURE 16.2

Evolutionary Relationships of the Echinoderms.   The echinoderms (shaded in orange) diverged from the deuterostomate lineage at least 600 million years ago. Although modern echinoderms are pentaradially symmetrical, the earliest echinoderms were probably bilaterally symmetrical.

a sievelike plate. In others it is a simple opening. The madreporite may serve as an inlet to replace water lost from the water-vascular system and may help equalize pressure differences between the water-vascular system and the outside. Tiedemann bodies are swellings often associated with the ring canal. They are believed to be sites for the production of phagocytic cells, called coelomocytes, whose functions are described later in this chapter. Polian vesicles are sacs that are also associated with the ring canal and function in fluid storage for the water-vascular system.

Five (or a multiple of five) radial canals branch from the ring canal. Radial canals are associated with arms of star-shaped echinoderms. In other echinoderms, they may be associated with the body wall and arch toward the aboral pole. Many lateral canals branch off each radial canal and end at the tube feet.

Tube feet are extensions of the canal system and usually emerge through openings in skeletal ossicles (*see figure 16.3a*). Internally, tube feet usually terminate in a bulblike, muscular ampulla. When an ampulla contracts, it forces water into the tube foot, which then extends. Valves prevent the backflow of water from the tube foot into the lateral canal. A tube foot often has a suction cup at its distal end. When the foot extends and contacts solid substrate, muscles of the suction cup contract and create a vacuum. In some taxa, tube feet have a pointed or blunt distal end. These echinoderms may extend their tube feet into a soft substrate to secure contact during locomotion or to sift sediment during feeding.

The water-vascular system has other functions in addition to locomotion. As is discussed at the end of this chapter, the original function of water-vascular systems was probably feeding, not locomotion. In addition, the soft membranes of the tube feet permit the exchange of respiratory gases and nitrogenous wastes with the environment. Tube feet also have sensory functions.

A **hemal system** consists of strands of tissue that encircle an echinoderm near the ring canal of the water-vascular system and run into each arm near the radial canals (*see figure 16.4*). The hemal system is derived from the coelom and circulates fluid using cilia that line its channels. The function of the hemal system is largely unknown, but it probably helps distribute nutrients absorbed from the digestive tract. It may aid in the transport of large molecules, hormones, or coelomocytes, which are cells that engulf and transport waste particles within the body.

CLASS ASTEROIDEA

The sea stars make up the class Asteroidea (as″te-roi′de-ah) (Gr. *aster,* star + *oeides,* in the form of) and include about 1,500 species. They often live on hard substrates in marine environments, although some species also live in sandy or muddy substrates. Sea stars may be brightly colored with red, orange, blue, or gray. *Asterias* is an orange sea star common along the Atlantic coast of North America and is frequently studied in introductory zoology laboratories.

Sea stars usually have five arms that radiate from a central disk. The oral opening, or mouth, is in the middle of one side of the central disk. It is normally oriented downward, and movable oral spines surround it. Movable and fixed spines project from the

(a)

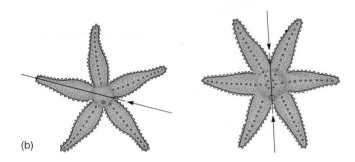

(b)

FIGURE 16.3

Pentaradial Symmetry. (*a*) Echinoderms exhibit pentaradial symme-try, in which body parts are arranged in fives around an oral-aboral axis. Note the madreporite between the bases of the two arms in the fore-ground and the tube feet on the tips of the upturned arm. (*b*) Compari-son of hypothetical penta- and hexaradial echinoderms. The five-part organization may be advantageous because joints between skeletal ossi-cles are never directly opposite one another, as they would be with an even number of parts. Having joints on opposite sides of the body in line with each other (arrows) could make the skeleton weaker.

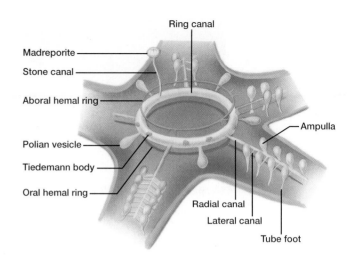

FIGURE 16.4

Water-Vascular System of a Sea Star. The ring canal gives rise to radial canals that lead into each arm. It opens to the outside or to the body cavity through a stone canal that ends at a madreporite on the ab-oral surface. Polian vesicles and Tiedemann bodies are often associated with the ring canal.

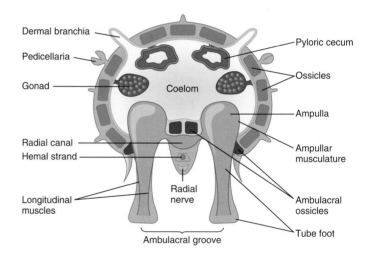

FIGURE 16.5

Body Wall and Internal Anatomy of a Sea Star. A cross section through one arm of a sea star shows the structures of the water-vascular system and the tube feet extending through the ambulacral groove.

skeleton and roughen the aboral surface. Thin folds of the body wall, called **dermal branchiae** or **papulae,** extend between ossicles and function in gas exchange (figure 16.5). In some sea stars, the aboral surface has numerous pincherlike structures called **pedicel-lariae,** which clean the body surface of debris and have protective functions. Pedicellariae may be attached on a movable spine, or they may be immovably fused to skeletal ossicles.

A series of ossicles in the arm form an **ambulacral groove** that runs the length of the oral surface of each arm. The ambulacral groove houses the radial canal, and paired rows of tube feet protrude through the body wall on either side of the ambulacral groove. Tube feet of sea stars move in a stepping motion. Alternate extension, at-tachment, and contraction of tube feet move sea stars across their substrate. The nervous system coordinates the tube feet so that all feet move the sea star in the same direction; however, the tube feet do not move in unison. The suction disks of tube feet are effective attachment structures, allowing sea stars to maintain their position, or move from place to place, in spite of strong wave action.

MAINTENANCE FUNCTIONS

Sea stars feed on snails, bivalves, crustaceans, polychaetes, corals, detritus, and a variety of other food items. The mouth opens to a short esophagus and then to a large stomach that fills most of the coelom of the central disk. The stomach is divided into two re-gions. The larger, oral stomach, sometimes called the cardiac stomach, receives ingested food (figure 16.6). It joins the smaller, aboral stomach, sometimes called the pyloric stomach. The aboral (pyloric) stomach gives rise to ducts that connect to secretory and absorptive structures called pyloric cecae. Two pyloric cecae extend into each arm. A short intestine leads to rectal cecae

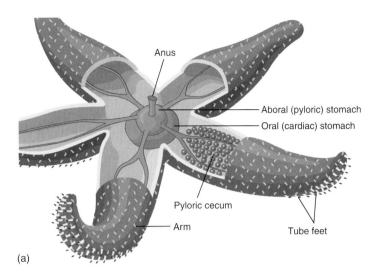

(a)

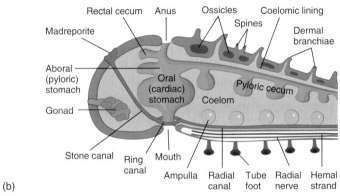

(b)

FIGURE 16.6

Digestive Structures in a Sea Star. A mouth leads to a large oral (cardiac) stomach and an aboral (pyloric) stomach. Pyloric cecae extend into each arm. (*a*) Aboral view. (*b*) Lateral view through central disk and one arm.

(uncertain functions) and to a nearly nonfunctional anus, which opens on the aboral surface of the central disk.

Some sea stars ingest whole prey, which are digested extracellularly within the stomach. Undigested material is expelled through the mouth. Many sea stars feed on bivalves by forcing the valves apart. (Anyone who has tried to pull apart the valves of a bivalve shell can appreciate that this is a remarkable accomplishment.) When a sea star feeds on a bivalve, it wraps itself around the bivalve's ventral margin. Tube feet attach to the outside of the shell, and the body-wall musculature forces the valves apart. (This is possible because the sea star changes tube feet when the muscles of engaged tube feet begin to tire.) When the valves are opened about 0.1 mm, increased coelomic pressure everts the oral (cardiac) portion of the sea star's stomach into the bivalve shell. Digestive enzymes are released, and partial digestion occurs in the bivalve shell. This digestion further weakens the bivalve's adductor muscles, and the shell eventually opens completely. Partially digested tissues are taken into the aboral (pyloric) portion of the stomach, and into the pyloric cecae for further digestion and absorption. After feeding and

initial digestion, the sea star retracts the stomach, using stomach retractor muscles.

Gases, nutrients, and metabolic wastes are transported in the coelom by diffusion and by the action of ciliated cells lining the body cavity. Gas exchange and excretion of metabolic wastes (principally ammonia) occur by diffusion across dermal branchiae, tube feet, and other membranous structures. A sea star's hemal system consists of strands of tissue that encircle the mouth near the ring canal, extend aborally near the stone canal, and run into the arms near radial canals (*see figure 16.4*).

The nervous system of sea stars consists of a nerve ring that encircles the mouth and radial nerves that extend into each arm. Radial nerves lie within the ambulacral groove, just oral to the radial canal of the water-vascular system and the radial strands of the hemal system (*see figure 16.5*). Radial nerves coordinate the functions of tube feet. Other nervous elements are in the form of a nerve net associated with the body wall.

Most sensory receptors are distributed over the surface of the body and tube feet. Sea stars respond to light, chemicals, and various mechanical stimuli. They often have specialized photoreceptors at the tips of their arms. These are actually tube feet that lack suction cups but have a pigment spot surrounding a group of photoreceptors called ocelli.

REGENERATION, REPRODUCTION, AND DEVELOPMENT

Sea stars are well known for their powers of regeneration. They can regenerate any part of a broken arm. In a few species, an entire sea star can be regenerated from a broken arm if the arm contains a portion of the central disk. Regeneration is a slow process, taking up to a year for complete regeneration. Asexual reproduction occurs in some asteroids and involves division of the central disk, followed by regeneration of each half.

Most sea stars are dioecious, but sexes are indistinguishable externally. Two gonads are present in each arm, and these enlarge to nearly fill an arm during the reproductive periods. Gonopores open between the bases of each arm.

The embryology of echinoderms has been studied extensively because of the relative ease of inducing spawning and maintaining embryos in the laboratory. External fertilization is the rule. Because gametes cannot survive long in the ocean, maturation of gametes and spawning must be coordinated if fertilization is to take place. The photoperiod (the relative length of light and dark in a 24-hour period) and temperature are environmental factors used to coordinate sexual activity. In addition, gamete release by one individual is accompanied by the release of spawning pheromones, which induce other sea stars in the area to spawn, increasing the likelihood of fertilization.

Embryos are planktonic, and cilia are used in swimming (figure 16.7). After gastrulation, bands of cilia differentiate, and a bilaterally symmetrical larva, called a bipinnaria larva, forms. The larva usually feeds on planktonic protists. The development of larval arms results in a brachiolaria larva, which settles to the substrate, attaches, and metamorphoses into a juvenile sea star.

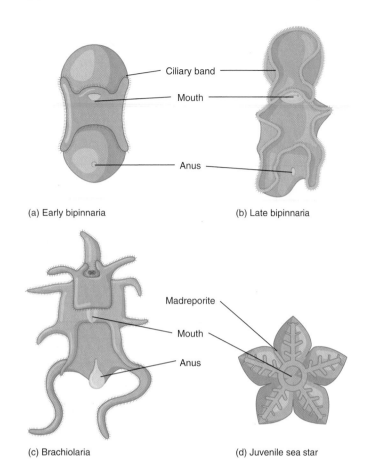

(a) Early bipinnaria

(b) Late bipinnaria

(c) Brachiolaria

(d) Juvenile sea star

FIGURE 16.7

Development of a Sea Star. Later embryonic stages are bilaterally symmetrical, ciliated, and swim and feed in the plankton. In a few species, embryos develop from yolk stored in the egg during gamete formation. Following blastula and gastrula stages, larvae develop. (*a*) Early bipinnaria larva (0.5 mm). (*b*) Late bipinnaria larva (1 mm). (*c*) Brachiolaria larva (1 mm). (*d*) Juvenile sea star (1 to 2 mm).

SEA DAISIES

One group of very unusual echinoderms has previously been assigned to its own class, Concentricycloidea. The current consensus is that they are highly modified members of the class Asteroidea. Two species of sea daisies have been described (figure 16.8). They lack arms and are less than 1 cm in diameter. The most distinctive features of this group are the two circular water-vascular rings that encircle the disklike body. The inner of the two rings probably corresponds to the ring canal of other asteroids. The outer ring contains tube feet and ampulae and probably corresponds to radial canals of other asteroids. Sea daisies lack an internal digestive system. Instead, a thin membrane, called a velum, covers the surface of the animal that is applied to the substrate (e.g., decomposing organic matter) and digests and absorbs nutrients. Internally, five pairs of brood pouches hold embryos during development. No free-swimming larval stages are known.

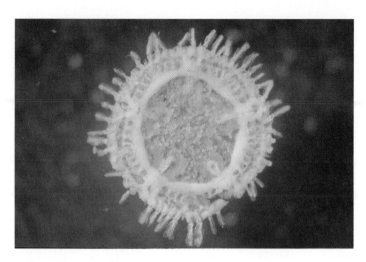

FIGURE 16.8

A Sea Daisy. A preserved sea daisy (*Xyloplax medusiformis*). This specimen is 3 mm in diameter.

CLASS OPHIUROIDEA

The class Ophiuroidea (o-fe-u-roi′de-ah) (Gr. *ophis*, snake + *oura*, tail + *oeides*, in the form of) includes the basket stars and the brittle stars or serpent stars. With over two thousand species, this is the most diverse group of echinoderms. Ophiuroids, however, are often overlooked because of their small size and their tendency to occupy crevices in rocks and coral or to cling to algae.

The arms of ophiuroids are long and, unlike those of asteroids, are sharply set off from the central disk, giving the central disk a pentagonal shape. Brittle stars have unbranched arms, and most have a central disk that ranges in size from 1 to 3 cm (figure 16.9*a*). Basket stars have arms that branch repeatedly (figure 16.9*b*). Neither dermal branchiae nor pedicellariae are present in ophiuroids. The tube feet of ophiuroids lack suction disks and ampullae, and the contraction of muscles associated with the base of a tube foot extends the tube foot. Unlike the sea stars, the madreporite of ophiuroids is on the oral surface.

The water-vascular system of ophiuroids is not used for locomotion. Instead, the skeleton is modified to permit a unique form of grasping and movement. Superficial ossicles, which originate on the aboral surface, cover the lateral and oral surfaces of each arm. The ambulacral groove—containing the radial nerve, hemal strand, and radial canal—is thus said to be "closed." Ambulacral ossicles are in the arm, forming a central supportive axis. Successive ambulacral ossicles articulate with one another and are acted upon by relatively large muscles to produce snakelike movements (hence the derivation of the class name) that allow the arms to curl around a stalk of algae or to hook into a coral crevice. During locomotion, the central disk is held above the substrate, and two arms pull the animal along, while other arms extend forward and/or trail behind the animal.

(a)

(b)

FIGURE 16.9

Class Ophiuroidea.   (*a*) This brittle star (*Ophiopholis aculeata*) uses its long, snakelike arms for crawling along its substrate and curling around objects in its environment. (*b*) Basket stars have five highly branched arms. They wave the arms in the water and with the mucus-covered tube feet capture planktonic organisms.

MAINTENANCE FUNCTIONS

Ophiuroids are predators and scavengers. They use their arms and tube feet in sweeping motions to collect prey and particulate matter, which are then transferred to the mouth. Basket stars are suspension feeders that wave their arms and trap plankton on mucus-covered tube feet. Trapped plankton is passed from tube foot to tube foot along the length of an arm until it reaches the mouth.

The mouth of ophiuroids is in the center of the central disk, and five triangular jaws form a chewing apparatus. The mouth leads to a saclike stomach. There is no intestine, and no part of the digestive tract extends into the arms.

The coelom of ophiuroids is reduced and is mainly confined to the central disk, but it still serves as the primary means for the

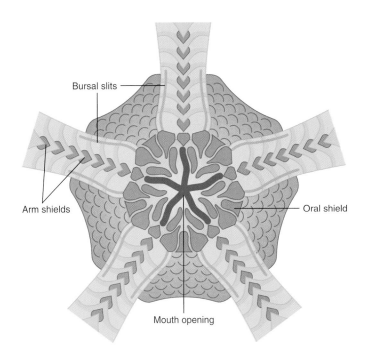

FIGURE 16.10

Class Ophiuroidea.   Oral view of the disk of the brittle star, *Ophiomusium*.   *Redrawn from L. Hyman, The Invertebrates, Volume IV. Copyright © 1959 McGraw-Hill, Inc. Used by permission.*

distribution of nutrients, wastes, and gases. Coelomocytes aid in the distribution of nutrients and the expulsion of particulate wastes. Ammonia is the primary nitrogenous waste product, and it is lost by diffusion across tube feet and membranous sacs, called **bursae,** that invaginate from the oral surface of the central disk. Slits in the oral disk, near the base of each arm, allow cilia to move water into and out of the bursae (figure 16.10).

REGENERATION, REPRODUCTION, AND DEVELOPMENT

Like sea stars, ophiuroids can regenerate lost arms. If a brittle star is grasped by an arm, the contraction of certain muscles may sever and cast off the arm—hence the common name brittle star. This process, called autotomy (Gr. *autos*, self + *tomos*, to cut), is used in escape reactions. The ophiuroid later regenerates the arm. Some species also have a fission line across their central disk. When an ophiuroid splits into halves along this line, two ophiuroids regenerate.

Ophiuroids are dioecious. Males are usually smaller than females, who often carry the males. The gonads are associated with each bursa, and gametes are released into the bursa. Eggs may be shed to the outside or retained in the bursa, where they are fertilized and held through early development. Embryos are protected in the bursa and are sometimes nourished by the parent. A larval stage, called an ophiopluteus, is planktonic. Its long arms bear ciliary bands used to feed on plankton, and it undergoes metamorphosis before sinking to the substrate.

(a)

(b)

FIGURE 16.11

Class Echinoidea. (*a*) A sea urchin (*Strongylocentrotus*) (*b*) Sand dollars are specialized for living in soft substrates, where they are often partially buried.

CLASS ECHINOIDEA

The sea urchins, sand dollars, and heart urchins make up the class Echinoidea (ek″i-noi′de-ah) (Gr. *echinos*, spiny + *oeides*, in the form of). The approximately one thousand species are widely distributed in nearly all marine environments. Sea urchins are specialized for living on hard substrates, often wedging themselves into crevices and holes in rock or coral (figure 16.11*a*). Sand dollars and heart urchins usually live in sand or mud, and burrow just below the surface (figure 16.11*b*). They use tube feet to catch organic matter settling on them or passing over them. Sand dollars often live in dense beds, which favors efficient reproduction and feeding.

Sea urchins are rounded, and their oral end is oriented toward the substrate. Their skeleton, called a test, consists of 10 sets of closely fitting plates that arch between oral and aboral

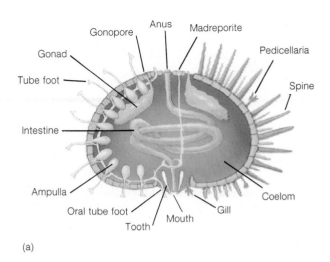

(a)

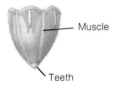

(b)

FIGURE 16.12

Internal Anatomy of a Sea Urchin. (*a*) Sectional view. (*b*) Aristotle's lantern is a chewing structure consisting of about 35 ossicles and associated muscles.

ends. Five rows of ambulacral plates have openings for tube feet, and alternate with five interambulacral plates, which have tubercles for the articulation of spines. The base of each spine is a concave socket, and muscles at its base move the spine. Spines are often sharp and sometimes hollow, and they may contain venom dangerous to swimmers. The pedicellariae of sea urchins have either two or three jaws and connect to the body wall by a relatively long stalk (figure 16.12*a*). They clean the body of debris and capture planktonic larvae, which provide an extra source of food. Pedicellariae of some sea urchins contain venom sacs and are grooved or hollow to inject venom into a predator, such as a sea star.

The water-vascular system is similar to that of other echinoderms. Radial canals run along the inner body wall between the oral and the aboral poles. Tube feet possess ampullae and suction cups, and the water-vascular system opens to the outside through many pores in one aboral ossicle that serves as a madreporite.

Echinoids move by using spines for pushing against the substrate and tube feet for pulling. Sand dollars and heart urchins use spines to help them to burrow in soft substrates. Some sea urchins burrow into rock and coral to escape the action of waves and strong currents. They form cup-shaped depressions and deeper burrows, using the action of their chewing Aristotle's lantern, which is described next.

MAINTENANCE FUNCTIONS

Echinoids feed on algae, bryozoans, coral polyps, and dead animal remains. Oral tube feet surrounding the mouth manipulate food. A chewing apparatus, called **Aristotle's lantern,** can be projected from the mouth (figure 16.12*b*). It consists of about 35 ossicles and attached muscles and cuts food into small pieces for ingestion. The mouth cavity leads to a pharynx, an esophagus, and a long, coiled intestine that ends aborally at the anus.

Echinoids have a large coelom, and coelomic fluids are the primary circulatory medium. Small gills, found in a thin membrane surrounding the mouth, are outpockets of the body wall and are lined by ciliated epithelium. Gas exchange occurs by diffusion across this epithelium and across the tube feet. Ciliary currents, changes in coelomic pressure, and the contraction of muscles associated with Aristotle's lantern move coelomic fluids into and out of gills. Excretory and nervous functions are similar to those described for asteroids.

REPRODUCTION AND DEVELOPMENT

Echinoids are dioecious. Gonads are on the internal body wall of the interambulacral plates. During breeding season, they nearly fill the spacious coelom. One gonopore is in each of five ossicles, called genital plates, located at the aboral end of the echinoid. The sand dollars are an exception, usually having only four gonads and gonopores. Gametes are shed into the water, and fertilization is external. Development eventually results in a pluteus larva that spends several months in the plankton and eventually undergoes metamorphosis to the adult.

CLASS HOLOTHUROIDEA

The class Holothuroidea (hol″o-thu-roi′de-ah) (Gr. *holothourion,* sea cucumber + *oeides,* in the form of) has approximately 1,500 species, whose members are commonly called sea cucumbers. Sea cucumbers are found at all depths in all oceans, where they crawl over hard substrates or burrow through soft substrates (figure 16.13).

Sea cucumbers have no arms, and they are elongate along the oral-aboral axis. They lie on one side, which is usually flattened as a permanent ventral side, giving them a secondary bilateral symmetry. Tube feet surrounding the mouth are enlarged, highly modified, and referred to as tentacles. Most adults range in length between 10 and 30 cm. Their body wall is thick and muscular, and it lacks protruding skeletal spines and pedicellariae. Beneath the epidermis is the dermis, a thick layer of connective tissue with embedded ossicles. Sea cucumber ossicles are microscopic and do not function in determining body shape. Larger ossicles form a calcareous ring that encircles the oral end of the digestive tract, serving as a point of attachment for body wall muscles (figure 16.14). Beneath the dermis is a layer of circular muscles overlying longitudinal muscles. The body wall of sea cucumbers, when boiled and dried, is known as trepang in Asian countries. It

FIGURE 16.13

Class Holothuroidea. A sea cucumber (*Parastichopus californicus*).

may be eaten as a main-course item or added to soups as flavoring and a source of protein.

The madreporite of sea cucumbers is internal, and the water-vascular system is filled with coelomic fluid. The ring canal encircles the oral end of the digestive tract and gives rise to one to ten Polian vesicles. Five radial canals and the canals to the tentacles branch from the ring canal. Radial canals and tube feet, with suction cups and ampullae, run between the oral and aboral poles. The side of a sea cucumber resting on the substrate contains three of the five rows of tube feet, which are primarily used for attachment. The two rows of tube feet on the upper surface may be reduced in size or may be absent.

Sea cucumbers are mostly sluggish burrowers and creepers, although some swim by undulating their bodies from side to side. Locomotion using tube feet is inefficient, because the tube feet are not anchored by body wall ossicles. Locomotion more commonly results from contractions of body-wall muscles that produce wormlike, locomotor waves that pass along the length of the body.

MAINTENANCE FUNCTIONS

Most sea cucumbers ingest particulate organic matter using their tentacles. Mucus covering the tentacles traps food as the tentacles sweep across the substrate or are held out in seawater. The digestive tract consists of a stomach; a long, looped intestine; a rectum; and an anus (figure 16.14). Sea cucumbers thrust tentacles into their mouths to wipe off trapped food. During digestion, coelomocytes move across the intestinal wall, secrete enzymes to aid in digestion, and engulf and distribute the products of digestion.

The coelom of sea cucumbers is large, and the cilia of the coelomic lining circulate fluids throughout the body cavity, distributing respiratory gases, wastes, and nutrients. The hemal system of sea cucumbers is well developed, with relatively large sinuses and a network of channels containing coelomic fluids. Its primary role is food distribution.

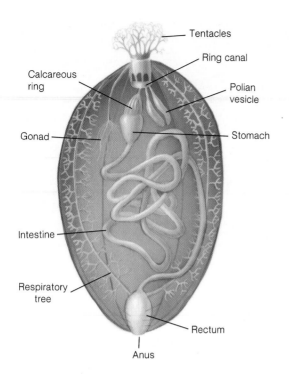

FIGURE 16.14

Internal Structure of a Sea Cucumber, *Thyone*. The mouth leads to a stomach supported by a calcareous ring. The calcareous ring is also the attachment site for longitudinal retractor muscles of the body. Contractions of these muscles pull the tentacles into the anterior end of the body. The stomach leads to a looped intestine. The intestine continues to the rectum and anus. (The anterior portion of the digestive tract is displaced aborally in this illustration.)

A pair of tubes called **respiratory trees** attach at the rectum and branch throughout the body cavity of sea cucumbers. The pumping action of the rectum circulates water into these tubes. When the rectum dilates, water moves through the anus into the rectum. Contraction of the rectum, along with contraction of an anal sphincter, forces water into the respiratory tree. Water exits the respiratory tree when tubules of the tree contract. Respiratory gases and nitrogenous wastes move between the coelom and seawater across these tubules.

The nervous system of sea cucumbers is similar to that of other echinoderms but has additional nerves supplying the tentacles and pharynx. Some sea cucumbers have statocysts, and others have relatively complex photoreceptors.

Casual examination suggests that sea cucumbers are defenseless against predators. Many sea cucumbers, however, produce toxins in their body walls that discourage predators. Other sea cucumbers can evert tubules of the respiratory tree, called Cuverian tubules, through the anus. These tubules contain sticky secretions and toxins capable of entangling and immobilizing predators. In addition, contractions of the body wall may result in the expulsion of one or both respiratory trees, the digestive tract, and the gonads through the anus. This process, called evisceration, occurs in response to chemical and physical stress and may be a defensive adaptation that discourages predators. Regeneration of lost parts follows.

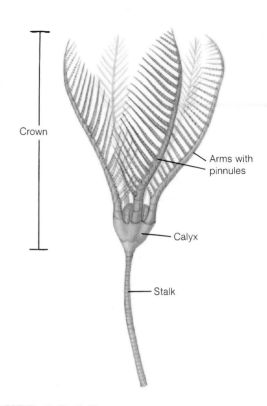

FIGURE 16.15

Class Crinoidea. A sea lily (*Ptilocrinus*).

REPRODUCTION AND DEVELOPMENT

Most sea cucumbers are dioecious. They possess a single gonad, located anteriorly in the coelom, and a single gonopore near the base of the tentacles. Fertilization is usually external, and embryos develop into planktonic larvae. Metamorphosis precedes settling to the substrate. In some species, a female's tentacles trap eggs as the eggs are released. After fertilization, eggs are transferred to the body surface, where they are brooded. Although rare, coelomic brooding also occurs. Eggs are released into the body cavity, where fertilization (by an unknown mechanism) and early development occur. The young leave through a rupture in the body wall. Sea cucumbers can also reproduce by transverse fission, followed by regeneration of lost parts.

CLASS CRINOIDEA

Members of the class Crinoidea (krin-oi′de-ah) (Gr. *krinon*, lily + *oeides*, in the form of) include the sea lilies and the feather stars. They are the most primitive of all living echinoderms and are very different from any covered thus far. Approximately 630 species are living today; however, an extensive fossil record indicates that many more were present during the Paleozoic era, 200 to 600 million years ago.

Sea lilies attach permanently to their substrate by a stalk (figure 16.15). The attached end of the stalk bears a flattened disk or rootlike extensions that are fixed to the substrate. Disklike ossicles of the stalk appear to be stacked on top of one another and are held

WILDLIFE ALERT
Imperiled Sea Cucumbers (*Isotichopus fuscus*)

VITAL STATISTICS

Classification: Phylum Echinodermata, class Holothuroidea

Range: Galápagos Archipelago

Habitat: The seafloor around the Galápagos Islands

Number remaining: Unknown

Status: Endangered

NATURAL HISTORY AND ECOLOGICAL STATUS

Isotichopus fuscus is one of 14 species of sea cucumbers native to the Galápagos Archipelago (box figure 16.1). Sea cucumbers have been called "earthworms of the sea" because they feed on detritus and turn over the seafloor, much like earthworms do on land. Sea cucumbers grow slowly, so a population that is decimated will require decades to recover.

There is an increasing demand for sea cucumbers in Asian markets. Muscles of *Isotichopus fuscus* are served in sushi bars, dried cucumbers are added to soup and sautéed vegetables and meats, or they are served separately with rice as a part of a larger meal. The intestine is used to prepare a gourmet Japanese dish called konowata. Alleged medicinal uses for sea cucumbers include treatments for ulcers, cuts, and arthritis and use as an aphrodisiac.

(a)

(b)

BOX FIGURE 16.2   Sea Cucumber Harvesting.   (*a*) Sea cucumber harvesting by "pepineros." (*b*) *Isotichopus* eviscerated and ready for drying.

When the demand for sea cucumbers resulted in the decimation of populations in the western Pacific, demand spread to the eastern Pacific. Since the 1980s, sea cucumber harvesting has been a lucrative profession along the coasts of North and South America. Harvesting occurs by either dragging the sea bottom from boats or by diving (box figure 16.2).

Sea cucumber harvesting began in 1988 along Ecuador's Pacific coast and spread to the Galápagos Islands by early 1992. In response to a fishing frenzy by "pepineros" (sea cucumber harvesters), Ecuador's government quickly imposed a ban on cucumber fishing. Before the ban was in place, between 12 and 30 million sea cucumbers had been harvested in the Galápagos. The ban was lifted in 1993, and within two months seven million more cucumbers were taken—in spite of a three-month quota of 550,000. The ban was reestablished in December 1994, but the ban is nearly impossible to enforce because of the lack of enforcement resources and the expanse of ocean that needs to be patrolled.

BOX FIGURE 16.1   The Distribution of *Isotichopus fuscus* in the Galápagos Archipelago.

Sea cucumber harvesting has had devastating effects on the Galápagos Islands and their inhabitants. The islands are host to hundreds of unique species that are of inestimable economic and scientific value. They attract thousands of tourists to the islands each year and are the focus of research efforts of scientists from around the world. Sea cucumber fishing threatens not only sea cucumbers, but also the existence of many other species. Pepineros cut large mangrove trees as fuel for drying sea cucumbers. This cutting threatens the mangrove swamps, which host hundreds of the islands' unique species. Pepinero camps introduce nonnative species such as feral pigs, dogs, brown rats, and fire ants into the very fragile Galápagos ecosystems. Social unrest has also resulted from cucumber fishing. Conflicts between pepineros and conservationists and the Ecuadoran government have even resulted in violence. One park warden was killed while investigating illegal poaching by pepineros. The islands' unique tortoises have been killed in protest of fishing bans. The protesters think that the government and conservationists consider the survival of sea cucumbers more important than the economic survival of Ecuadoran people.

These problems illustrate the complex difficulties associated with saving endangered animals. They involve demands for animal products in distant countries, the economic and survival interests of native human populations, and scientific conservation efforts. Balancing all of these interests requires multinational and multidisciplinary approaches to conservation. Wildlife conservation involves much more than understanding the biology of a threatened animal!

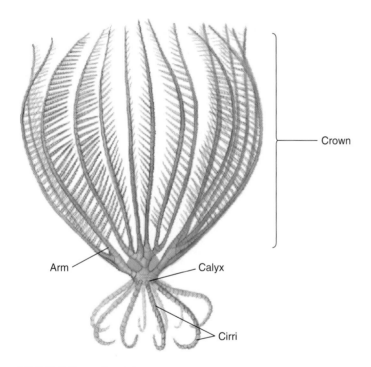

FIGURE 16.16

Class Crinoidea. A feather star (*Neometra*).

together by connective tissues, giving a jointed appearance. The stalk usually bears projections, or cirri, arranged in whorls. The unattached end of a sea lily is called the crown. The aboral end of the crown attaches to the stalk and is supported by a set of ossicles, called the **calyx.** Five arms also attach at the calyx. They are branched, supported by ossicles, and bear smaller branches (pinnules)—giving them a featherlike appearance. Tube feet are in a double row along each arm. Ambulacral grooves on the arms lead toward the mouth. The mouth and anus open onto the upper (oral) surface.

Feather stars are similar to sea lilies, except they lack a stalk and are swimming and crawling animals (figure 16.16). The aboral end of the crown bears a ring of rootlike cirri, which cling when the animal is resting on a substrate. Feather stars swim by raising and lowering the arms, and they crawl over substrate by pulling with the tips of the arms.

MAINTENANCE FUNCTIONS

Circulation, gas exchange, and excretion in crinoids are similar to these functions in other echinoderms. In feeding, however, crinoids use outstretched arms for suspension feeding. A planktonic organism that contacts a tube foot is trapped, and cilia in ambulacral grooves carry it to the mouth. *Although this method of feeding is different from how other modern echinoderms feed, it probably reflects the original function of the water-vascular system.*

Crinoids lack the nerve ring found in most echinoderms. Instead, a cup-shaped nerve mass below the calyx gives rise to radial nerves that extend through each arm and control the tube feet and arm musculature.

REPRODUCTION AND DEVELOPMENT

Many crinoids, like other echinoderms, are dioecious. Others are monoecious, with male gametes developing before female gametes. This sequence of gamete development is called protandry and ensures that cross-fertilization will occur. Gametes form from germinal epithelium in the coelom and are released through ruptures in the walls of the arms. Some species spawn in seawater, where fertilization and development occur. Other species brood embryos on the outer surface of the arms. Metamorphosis occurs after larvae attach to the substrate. Like other echinoderms, crinoids can regenerate lost parts.

FURTHER PHYLOGENETIC CONSIDERATIONS

As mentioned earlier, most zoologists believe that echinoderms evolved from bilaterally symmetrical ancestors. Radial symmetry probably evolved during the transition from active to more

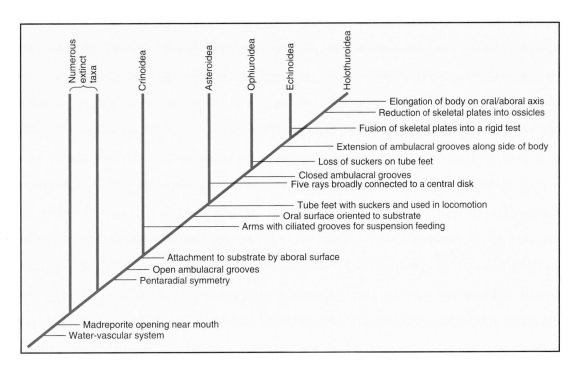

FIGURE 16.17

Echinoderm Phylogeny. The evolutionary relationships among echinoderms are not clear. This interpretation shows a relatively distant relationship between the Asteroidea and Ophiuroidea. Some taxonomists interpret the five-rayed body form as a synapomorphy that links these two groups to a single ancestral lineage.

sedentary lifestyles; however, the oldest echinoderm fossils, about 600 million years old, give little direct evidence of how this transition occurred.

Ancient fossils do give clues regarding the origin of the water-vascular system and the calcareous endoskeleton. Of all living echinoderms, the crinoids most closely resemble the oldest fossils. *Because crinoids use their water-vascular system primarily for suspension feeding rather than locomotion, the former was probably the original function of the water-vascular system.* As do crinoids, early echinoderms probably assumed a mouth-up position and attached aborally. They may have used arms and tube feet to capture food and move it to the mouth. *The calcium carbonate endoskeleton may have evolved to support extended filtering arms and to protect these sessile animals.*

Many modern echinoderms are more mobile. This free-living lifestyle is probably secondarily derived, as is the mouth-down orientation of most echinoderms. The mouth-down position is advantageous for predatory and scavenging lifestyles.

Similarly, changes in the water-vascular system, such as the evolution of ampullae, suction disks, and feeding tentacles, can be interpreted as adaptations for locomotion and feeding in a more mobile lifestyle. The idea that the free-living lifestyle is secondary is reinforced by the observation that some echinoderms, such as the irregular echinoids and the holothuroids, have bilateral symmetry imposed upon a pentaradial body form.

The evolutionary relationships among the echinoderms are not clear. Numerous fossils date into the Cambrian period, but no interpretation of the evolutionary relationships among living and extinct echinoderms is definitive. Figure 16.17 shows one interpretation of the evolutionary relationships among extant (with living members) echinoderm classes. *Most taxonomists agree that the echinoids and holothuroids are closely related. Whether the ophiuroids are more closely related to the echinoid/holothuroid lineage or the asteroid lineage is debated.*

SUMMARY

1. Echinoderms, chordates, and other deuterostomes share a common, but remote, ancestry. Modern echinoderms were probably derived from bilaterally symmetrical ancestors.

2. Echinoderms are pentaradially symmetrical, have an endoskeleton of interlocking calcium carbonate ossicles, and have a water-vascular system that is used for locomotion, food gathering, attachment, and exchanges with the environment.

3. Members of the class Asteroidea are the sea stars. They are predators and scavengers, and their arms are broadly joined to the central disk. Sea stars are dioecious, and external fertilization results in the formation of planktonic bipinnaria and brachiolaria larvae. Sea stars also have remarkable powers of regeneration.

4. The brittle stars and basket stars make up the class Ophiuroidea. Arms are sharply set off from the central disk. Ophiuroids are dioecious. Externally fertilized eggs may either develop in the plankton or they may be brooded.

5. The class Echinoidea includes the sea urchins, heart urchins, and sand dollars. They have a specialized chewing structure, called Aristotle's lantern. External fertilization results in a planktonic pluteus larva.

6. Members of the class Holothuroidea include the sea cucumbers. They rest on one side and are elongate along their oral-aboral axis. Their body wall contains microscopic ossicles. Many sea cucumbers eviscerate themselves when disturbed. Sea cucumbers are dioecious, and fertilization and development are external.

7. The class Crinoidea contains the sea lilies and feather stars. They are oriented oral side up and use arms and tube feet in suspension feeding. Crinoids are dioecious, and fertilization and development are external.

8. Radial symmetry of echinoderms probably evolved during a transition to a sedentary, filter-feeding lifestyle. The water-vascular system and the calcareous endoskeleton are probably adaptations for that lifestyle. The evolution of a more mobile lifestyle has resulted in the use of the water-vascular system for locomotion and in the assumption of a mouth-down position.

SELECTED KEY TERMS

ambulacral groove (p. 254)   pentaradial symmetry (p. 252)
Aristotle's lantern (p. 259)   respiratory trees (p. 260)
dermal branchiae (p. 254)   tube feet (p. 253)
pedicellariae (p. 254)   water-vascular system (p. 252)

CRITICAL THINKING QUESTIONS

1. What is pentaradial symmetry, and why is it adaptive for echinoderms?

2. Why do zoologists think that pentaradial symmetry was not present in the ancestors of echinoderms?

3. Compare and contrast the structure and function of the water-vascular systems of asteroids, ophiuroids, echinoids, holothuroids, and crinoids.

4. In which of the groups in question 3 is the water-vascular system probably most similar in form and function to an ancestral condition? Explain your answer.

5. What physical process is responsible for gas exchange and excretion in all echinoderms? What structures facilitate these exchanges in each echinoderm class?

ONLINE LEARNING CENTER

Visit our Online Learning Center (OLC) at www.mhhe.com/zoology (click on this book's title) to find the following chapter-related materials:

- CHAPTER QUIZZING
- RELATED WEB LINKS
  Phylum Echinodermata
    Class Asteroidea
    Class Ophiuroidea
    Class Echinoidea
    Class Holothuroidea
    Class Crinoidea

- BOXED READINGS ON
  A Thorny Problem for Australia's Barrier Reef
  Suspension Feeding in Invertebrates and Nonvertebrate Chordates

- READINGS ON LESSER-KNOWN INVERTEBRATES
  The Lophophorates, Entoprocts, Cycliophores, and Chaetognaths

- SUGGESTED READINGS
- LAB CORRELATIONS

Check out the OLC to find specific information on these related lab exercises in the *General Zoology Laboratory Manual*, 5th edition, by Stephen A. Miller:

   Exercise 16   *Echinodermata*

CHAPTER 17

HEMICHORDATA AND INVERTEBRATE CHORDATES

Outline

Concepts

1. Members of the phyla Echinodermata, Hemichordata, and Chordata are probably derived from a common diploblastic or triploblastic ancestor.
2. The phylum Hemichordata includes the acorn worms (class Enteropneusta) and the pterobranchs (class Pterobranchia). Hemichordates live in or on marine substrates and feed on sediment or suspended organic matter.
3. Animals in the phylum Chordata are characterized by a notochord, pharyngeal slits or pouches, a dorsal tubular nerve cord, and a postanal tail.
4. The urochordates are marine and are called tunicates. They are attached or planktonic, and solitary or colonial as adults. All are filter feeders.
5. Members of the subphylum Cephalochordata are called lancelets. They are filter feeders that spend most of their time partly buried in marine substrates.
6. Motile, fishlike chordates may have evolved from sedentary, filter-feeding ancestors as a result of paedomorphosis in a motile larval stage.

EVOLUTIONARY PERSPECTIVE

Some members of one of the phyla discussed in this chapter are more familiar to you than members of any other group of animals. This familiarity is not without good reason, for you yourself are a member of one of these phyla—Chordata. Other members of these phyla, however, are much less familiar. During a walk along a seashore at low tide, you may observe coiled castings (sand, mud, and excrement) at the openings of U-shaped burrows. Excavating these burrows reveals a wormlike animal that is one of the members of a small phylum—Hemichordata. Other members of this phylum include equally unfamiliar filter feeders called pterobranchs.

While at the seashore, you could also see animals clinging to rocks exposed by low tide. At first glance, you might describe them as jellylike masses with two openings at their unattached end. Some live as solitary individuals; others live in colonies. If you handle these animals, you may be rewarded with a stream of water squirted from their openings. Casual observations provide little evidence that these small filter feeders, called sea squirts or tunicates, are chordates. However, detailed studies have made that conclusion certain. Tunicates and a small group of fishlike cephalochordates are often called the invertebrate chordates because they lack a vertebral column (figure 17.1).

This chapter contains evolutionary concepts, which are set off in this font.

PHYLOGENETIC RELATIONSHIPS

Animals in the phyla Hemichordata and Chordata share deuterostome characteristics with echinoderms (figure 17.2). Most zoologists, therefore, believe that ancestral representatives of these phyla were derived from a common, as yet undiscovered, diploblastic or triploblastic ancestor. The chordates are characterized by a dorsal tubular nerve cord, a notochord, pharyngeal slits or pouches, and a postanal tail. The only characteristics they share with the hemichordates are pharyngeal slits and, in some species, a dorsal tubular nerve cord. **T**herefore, most zoologists agree that the evolutionary ties between the chordates and hemichordates are closer than those between echinoderms and either phylum. Chordates and hemichordates, however, probably diverged from widely separated points along the deuterostome lineage. The diverse body forms and lifestyles present in these phyla support this generalization.

PHYLUM HEMICHORDATA

The phylum Hemichordata (hem″i-kor-da′tah) (Gr. *hemi*, half + L. *chorda*, cord) includes the acorn worms (class Enteropneusta) and the pterobranchs (class Pterobranchia) (table 17.1). Members of both classes live in or on marine sediments.

Characteristics of the phylum Hemichordata include:

1. Marine, deuterostomate animals with a body divided into three regions: proboscis, collar, and trunk; coelom divided into three cavities

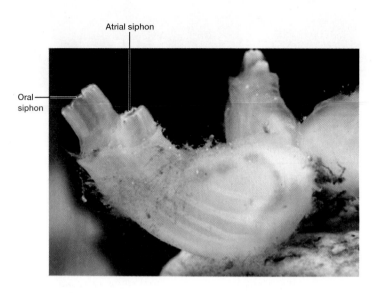

FIGURE 17.1

Phylum Chordata.   This tunicate, or sea squirt (*Ciona intestinalis*), is an invertebrate chordate that attaches to substrates in marine environments. Note the two siphons for circulating water through a filter-feeding apparatus.

2. Ciliated pharyngeal slits
3. Open circulatory system
4. Complete digestive tract
5. Dorsal, sometimes tubular, nerve cord

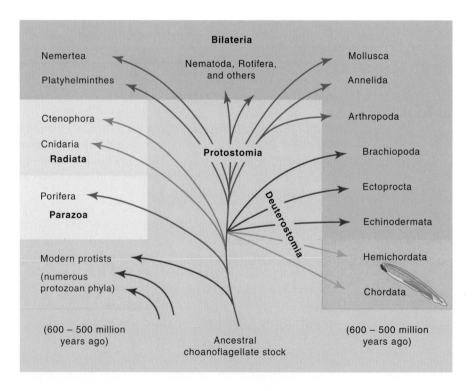

FIGURE 17.2

Phylogenetic Relationships among the Hemichordata and Chordata.   Hemichordates and chordates (shaded in orange) are distantly related deuterostomes derived from a common, as yet undiscovered, diploblastic or triploblastic ancestor.

TABLE 17.1
CLASSIFICATION OF THE HEMICHORDATA AND CHORDATA

Phylum Hemichordata (hem″i-kor-da′tah)
Widely distributed in shallow, marine, tropical waters and deep, cold waters; softbodied and wormlike; epidermal nervous system; most with pharyngeal slits.

    Class Enteropneusta (ent″er-op-nus′tah)
    Shallow-water, wormlike animals; inhabit burrows on sandy shorelines; body divided into three regions: proboscis, collar, and trunk. Acorn worms (*Balanoglossus, Saccoglossus*). About 75 species.

    Class Pterobranchia (ter″o-brang′ke-ah)
    With or without pharyngeal slits; two or more arms; often colonial, living in an externally secreted encasement. *Rhabdopleura*. About 20 species.

    Class Planctosphaeroidea (plank″to-sfer-roi′de-ah)
    Spherical body with ciliary bands covering the surface; U-shaped digestive tract; coelom poorly developed; planktonic. Only one species is known to exist (*Planctosphaera pelagica*).

Phylum Chordata (kor-dat′ah)
Occupy a wide variety of marine, freshwater, and terrestrial habitats. A notochord, pharyngeal slits, a dorsal tubular nerve cord, and a postanal tail are all present at some time in chordate life histories. About 45,000 species.

    Subphylum Urochordata (u″ro-kor-da′tah)
    Notochord, nerve cord, and postanal tail present only in free-swimming larvae; adults sessile, or occasionally planktonic, and enclosed in a tunic that contains some cellulose; marine. Sea squirts or tunicates.

        Class Ascidiacea (as-id″e-as′e-ah)
        All sessile as adults; solitary or colonial; colony members interconnected by stolons. Sea squirts (*Ascidia, Ciona*).

        Class Appendicularia (a-pen″di-ku-lar′e-ah)
        (Larvacea) (lar-vas′e-ah)
        Planktonic; adults retain tail and notochord; lack a cellulose tunic; epithelium secretes a gelatinous covering of the body. Appendicularians (*Fritillaria*).

        Class Sorberacea (sor″ber-as′e-ah)
        Ascidian-like urochordates possessing dorsal nerve cords as adults; deep water, benthic; carnivorous. *Octacnemus*.

        Class Thaliacea (tal″e-as′e-ah)
        Planktonic; adults are tailless and barrel shaped; oral and atrial openings are at opposite ends of the tunicate; muscular contractions of the body wall produce water currents. Salps (*Salpa, Thetys*).

Subphylum Cephalochordata (sef″a-lo-kor-dat′ah)
Body laterally compressed and transparent; fishlike; all four chordate characteristics persist throughout life. Amphioxus (*Branchiostoma*). About 45 species.

Craniata (kra″ne-ah′tah)
Skull surrounds the brain, olfactory organs, eyes, and inner ear. Unique embryonic tissue, neural crest, contributes to a variety of adult structures including sensory nerve cells, and some skeletal and other connective tissue structures.

Subphylum Hyperotreti (hi″per-ot′re′te)
Fishlike; skull consisting of cartilaginous bars; jawless; no paired appendages; mouth with four pairs of tentacles; olfactory sacs open to mouth cavity; 5 to 15 pairs of pharyngeal slits; ventrolateral slime glands. Hagfishes.

Subphylum Vertebrata (ver″te-bra′tah)
Notochord, nerve cord, postanal tail, and pharyngeal slits present at least in embryonic stages; vertebrae surround nerve cord and serve as primary axial support.

    Class Cephalaspidomorphi (sef″ah-las″pe-do-morf′e)
    Fishlike; jawless; no paired appendages; cartilaginous skeleton; sucking mouth with teeth and rasping tongue. Lampreys.

    Class Chondrichthyes (kon-drik′thi-es)
    Fishlike; jawed; paired appendages and cartilaginous skeleton; no swim bladder. Skates, rays, sharks.

    Class Osteichthyes (os″te-ik′the-es)
    Bony skeleton; swim bladder and operculum present. Bony fishes.

    Class Amphibia (am-fib′e-ah)
    Skin with mucoid secretions; possess lungs and/or gills; moist skin serves as respiratory organ; aquatic developmental stages usually followed by metamorphosis to an adult. Frogs, toads, salamanders.

    Class Reptilia (rep-til′e-ah)
    Dry skin with epidermal scales; amniotic eggs; terrestrial embryonic development. Snakes, lizards, alligators.

    Class Aves (a′vez)
    Scales modified into feathers for flight; efficiently regulate body temperature (endothermic); amniotic eggs. Birds.

    Class Mammalia (mah-ma′le-ah)
    Bodies at least partially covered by hair; endothermic; young nursed from mammary glands; amniotic eggs. Mammals.

CLASS ENTEROPNEUSTA

Members of the class Enteropneusta (ent″er-op-nus′tah) (Gr. *entero*, intestine + *pneustikos,* for breathing) are marine worms that usually range in size between 10 and 40 cm, although some can be as long as 2 m. Zoologists have described about 70 species, and most occupy U-shaped burrows in sandy and muddy substrates between the limits of high and low tides. The common name of the enteropneusts—acorn worms—is derived from the appearance of the proboscis, which is a short, conical projection at the worm's anterior end. A ringlike collar is posterior to the proboscis, and an elongate trunk is the third division of the body (figure 17.3). A ciliated epidermis and gland cells cover acorn worms. The mouth is located ventrally between the proboscis and the collar.

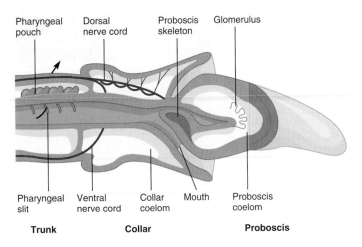

FIGURE 17.3

Class Enteropneusta. Longitudinal section showing the proboscis, collar, pharyngeal region, and internal structures. The black arrow shows the path of water through a pharyngeal slit.

A variable number of pharyngeal slits, from a few to several hundred, are positioned laterally on the trunk. Pharyngeal slits are openings between the anterior region of the digestive tract, called the pharynx, and the outside of the body.

Maintenance Functions

Cilia and mucus assist acorn worms in feeding. Detritus and other particles adhere to the mucus-covered proboscis. Tracts of cilia transport food and mucus posteriorly and ventrally. Ciliary tracts converge near the mouth and form a mucoid string that enters the mouth. Acorn worms may reject some substances trapped in the mucoid string by pulling the proboscis against the collar. Ciliary tracts of the collar and trunk transport rejected material and discard it posteriorly.

The digestive tract of enteropneusts is a simple tube. Food is digested as diverticula of the gut, called hepatic sacs, release enzymes. The worm extends its posterior end out of the burrow during defecation. At low tide, coils of fecal material, called castings, lie on the substrate at burrow openings.

The nervous system of enteropneusts is ectodermal in origin and lies at the base of the ciliated epidermis. It consists of dorsal and ventral nerve tracts and a network of epidermal nerve cells, called a nerve plexus. In some species, the dorsal nerve is tubular and usually contains giant nerve fibers that rapidly transmit impulses. There are no major ganglia. Sensory receptors are unspecialized and widely distributed over the body.

Because acorn worms are small, respiratory gases and metabolic waste products (principally ammonia) probably are exchanged by diffusion across the body wall. In addition, respiratory gases are exchanged at the pharyngeal slits. Cilia associated with pharyngeal slits circulate water into the mouth and out of the body through the pharyngeal slits. As water passes through the pharyngeal slits, gases are exchanged by diffusion between water and blood sinuses surrounding the pharynx.

The circulatory system of acorn worms consists of one dorsal and one ventral contractile vessel. Blood moves anteriorly in the dorsal vessel and posteriorly in the ventral vessel. Branches from these vessels lead to open sinuses. All blood flowing anteriorly passes into a series of blood sinuses, called the glomerulus, at the base of the proboscis. Excretory wastes may be filtered through the glomerulus, into the coelom of the proboscis, and released to the outside through one or two pores in the wall of the proboscis. The blood of acorn worms is colorless, lacks cellular elements, and distributes nutrients and wastes.

Reproduction and Development

Enteropneusts are dioecious. Two rows of gonads lie in the body wall in the anterior region of the trunk, and each gonad opens separately to the outside. Fertilization is external. Spawning by one worm induces others in the area to spawn—behavior that suggests the presence of spawning pheromones. Ciliated larvae, called **tornaria,** swim in the plankton for several days to a few weeks. The larvae settle to the substrate and gradually transform into the adult form (figure 17.4).

CLASS PTEROBRANCHIA

Pterobranchia (ter″o-brang′ke-ah) (Gk. *pteron,* wing or feather + *branchia,* gills) is a small class of hemichordates found mostly in deep, oceanic waters of the Southern Hemisphere. A few live in European coastal waters and in shallow waters near Bermuda. Zoologists have described approximately 20 species of pterobranchs.

Pterobranchs are small, ranging in size from 0.1 to 5 mm. Most live in secreted tubes in asexually produced colonies. As in enteropneusts, the pterobranch body is divided into three regions. The proboscis is expanded and shieldlike (figure 17.5). It secretes the tube and aids in movement in the tube. The collar possesses two to nine arms with numerous ciliated tentacles. The trunk is U-shaped.

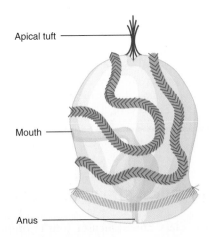

FIGURE 17.4

Tornaria Larva of an Enteropneust (*Balanoglossus*). When larval development is complete, a tornaria locates a suitable substrate, settles, and begins to burrow and elongate (1 mm).

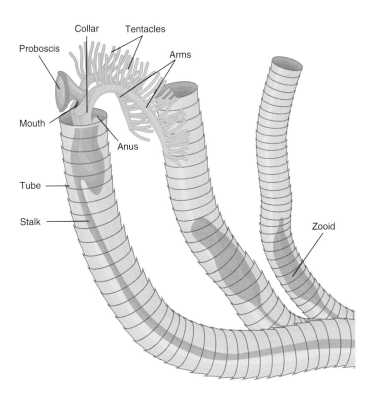

FIGURE 17.5

External Structure of the Pterobranch, *Rhabdopleura.* Ciliated tracts on tentacles and arms direct food particles toward the mouth (5 mm).

Maintenance Functions

Pterobranchs use water currents that cilia on their arms and tentacles generate to filter feed. Cilia trap and transport food particles toward the mouth. Although one genus has a single pair of pharyngeal slits, respiratory and excretory structures are unnecessary in animals as small as pterobranchs because gases and wastes exchange by diffusion.

Reproduction and Development

Asexual budding is common in pterobranchs and is responsible for colony formation. Pterobranchs also possess one or two gonads in the anterior trunk. Most species are dioecious, and external fertilization results in the development of a planula-like larva that lives for a time in the tube of the female. This nonfeeding larva eventually leaves the female's tube, settles to the substrate, forms a cocoon, and metamorphoses into an adult.

PHYLUM CHORDATA

Although the phylum Chordata (kor-dat′ah) (L. *chorda*, cord) does not have an inordinately large number of species (about 45,000), its members have been very successful at adapting to aquatic and terrestrial environments throughout the world. Sea squirts, members of the subphylum Urochordata, are briefly described in the "Evolutionary Perspective" that opens this chapter. Other chordates include lancelets (subphylum Cephalochordata)

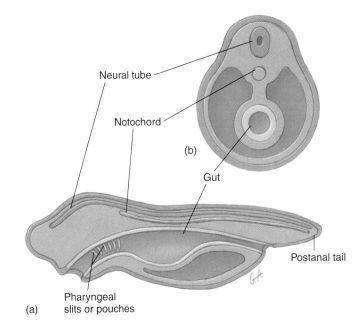

FIGURE 17.6

Chordate Body Plan. The development of all chordates involves the formation of a neural tube, the notochord, pharyngeal slits or pouches, and a postanal tail. Derivatives of all three primary germ layers are present. (*a*) Lateral view. (*b*) Cross section.

and the vertebrates (subphylum Vertebrata) (*see table 17.1*). Characteristics of the phylum Chordata include:

1. Bilaterally symmetrical, deuterostomate animals
2. A unique combination of four characteristics present at some stage in development: notochord, pharyngeal slits or pouches, dorsal tubular nerve cord, and postanal tail
3. Presence of an endostyle or thyroid gland
4. Complete digestive tract
5. Ventral, contractile blood vessel (heart)

The four characteristics listed in number 2 are unique to chordates and are discussed further in the paragraphs that follow (figure 17.6).

The phylum is named after the **notochord** (Gr. *noton*, the back + L. *chorda*, cord), a supportive rod that extends most of the length of the animal dorsal to the body cavity and into the tail. It consists of a connective-tissue sheath that encloses cells, each of which contains a large, fluid-filled vacuole. This arrangement gives the notochord some turgidity, which prevents compression along the anteroposterior axis. At the same time, the notochord is flexible enough to allow lateral bending, as in the lateral undulations of a fish during swimming. In most adult vertebrates, cartilage or bone partly or entirely replaces the notochord.

Pharyngeal slits are a series of openings in the pharyngeal region between the digestive tract and the outside of the body. In some chordates, diverticula from the gut in the pharyngeal region never break through to form an open passageway to the outside. These diverticula are then called pharyngeal pouches. The earliest chordates used the slits for filter feeding; some living chordates still use them for feeding. Other chordates have developed gills in the pharyngeal pouches for gas exchange. The pharyngeal slits of terrestrial vertebrates are mainly embryonic features and may be incomplete.

The **tubular nerve cord** and its associated structures are largely responsible for chordate success. The nerve cord runs along the longitudinal axis of the body, just dorsal to the notochord, and usually expands anteriorly as a brain. This central nervous system is associated with the development of complex systems for sensory perception, integration, and motor responses.

The fourth chordate characteristic is a **postanal tail.** (A postanal tail extends posteriorly beyond the anal opening.) Either the notochord or vertebral column supports the tail.

SUBPHYLUM UROCHORDATA

Members of the subphylum Urochordata (u″ro-kor-dah′tah) (Gr. *uro*, tail + L. *chorda*, cord) are the tunicates or sea squirts. The ascidians comprise the largest class of tunicates (*see table 17.1*). They are sessile as adults and are either solitary or colonial. The appendicularians and thaliaceans are planktonic as adults (figure 17.7). In some localities, tunicates occur in large enough numbers to be considered a dominant life-form.

Sessile urochordates attach their saclike bodies to rocks, pilings, ship hulls, and other solid substrates. The unattached end of urochordates contains two siphons that permit seawater to circulate through the body. One siphon is the oral siphon, which is the inlet for water circulating through the body and is usually directly opposite the attached end of the ascidian (figure 17.8). It also serves as the mouth opening. The second siphon, the atrial siphon, is the opening for excurrent water.

The body wall of most tunicates (L. *tunicatus*, to wear a tunic or gown) is a connective-tissue-like covering, called the tunic, that appears gel-like but is often quite tough. Secreted by the epidermis, it is composed of proteins, various salts, and cellulose. Some mesodermally derived tissues, including blood vessels and blood cells, are incorporated into the tunic. Rootlike extensions of the tunic, called stolons, help anchor a tunicate to the substrate and may connect individuals of a colony.

Maintenance Functions

Longitudinal and circular muscles below the body wall epithelium help to change the shape of the adult tunicate. They act against the elasticity of the tunic and the hydrostatic skeleton that seawater confined to internal chambers creates.

The nervous system of tunicates is largely confined to the body wall. It forms a nerve plexus with a single ganglion located on the wall of the pharynx between the oral and atrial openings (figure 19.9*a*). This ganglion is not vital for coordinating bodily functions. Tunicates are sensitive to many kinds of mechanical and chemical stimuli, and receptors for these senses are distributed over the body wall, especially around the siphons. There are no complex sensory organs.

The most obvious internal structures of the urochordates are a very large pharynx and a cavity, called the atrium, that surrounds the pharynx laterally and dorsally (figure 17.9*b*). The pharynx of tunicates originates at the oral siphon and is continuous with the

(a)

(b)

FIGURE 17.7

Subphylum Urochordata. (*a*) Members of the class Appendicularia are planktonic and have a tail and notochord that persist into the adult stage. (*b*) The thaliaceans, or salps, are barrel-shaped, planktonic urochordates. Oral and atrial siphons are at opposite ends of the body, and muscles of the body wall contract to create a form of weak jet propulsion.

remainder of the digestive tract. The oral margin of the pharynx has tentacles that prevent large objects from entering the pharynx. Numerous pharyngeal slits called stigmas perforate the pharynx. Cilia associated with the stigmas cause water to circulate into the pharynx, through the stigmas, and into the surrounding atrium. Water leaves the tunicate through the atrial siphon.

The digestive tract of adult tunicates continues from the pharynx and ends at the anus near the atrial siphon. During feeding, cells of a ventral, ciliated groove, called the **endostyle,** form a mucous sheet (figure 17.9*b*). Cilia move the mucous sheet dorsally across the pharynx. Food particles, brought into the oral siphon

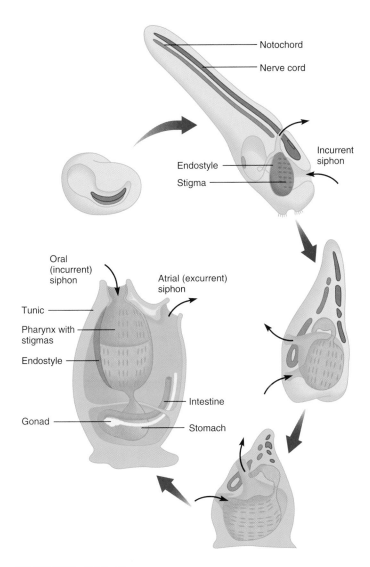

FIGURE 17.8

Tunicate Metamorphosis. Small black arrows show the path of water through the body.

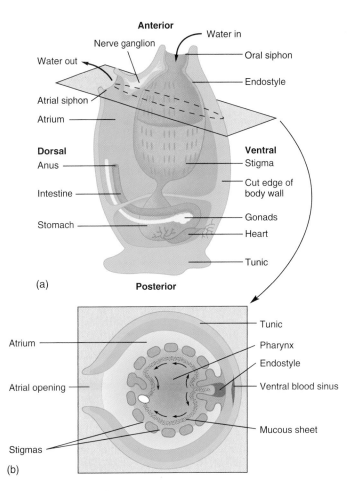

FIGURE 17.9

Internal Structure of a Tunicate. (*a*) Longitudinal section. Black arrows show the path of water. (*b*) Cross section at the level of the atrial siphon. Small black arrows show movement of food trapped in mucus that the endostyle produces.

with incurrent water, are trapped in the mucous sheet and passed dorsally. Food is incorporated into a string of mucus that ciliary action moves into the next region of the gut tract. Digestive enzymes are secreted in the stomach, and most absorption occurs across the walls of the intestine. Excurrent water carries digestive wastes from the anus out of the atrial siphon.

In addition to its role in feeding, the pharynx also functions in gas exchange. Gases are exchanged as water circulates through the tunicate.

The tunicate heart lies at the base of the pharynx. One vessel from the heart runs anteriorly under the endostyle, and another runs posteriorly to the digestive organs and gonads. Blood flow through the heart is not unidirectional. Peristaltic contractions of the heart may propel blood in one direction for a few beats; then the direction is reversed. The significance of this reversal is not understood. Tunicate blood plasma is colorless and contains various kinds of amoeboid cells.

Ammonia diffuses into water that passes through the pharynx and is excreted. In addition, amoeboid cells of the circulatory system accumulate uric acid and sequester it in the intestinal loop. Pyloric glands on the outside of the intestine are also thought to have excretory functions.

Reproduction and Development

Urochordates are monoecious. Gonads are located near the loop of the intestine, and genital ducts open near the atrial siphon. Gametes may be shed through the atrial siphon for external fertilization, or eggs may be retained in the atrium for fertilization and early development. Although self-fertilization occurs in some species, cross-fertilization is the rule. Development results in the formation of a tadpolelike larva with all four chordate characteristics. Metamorphosis begins after a brief free-swimming larval existence, during which the larva does not feed. The larva settles to a firm substrate and attaches by adhesive papillae located below the mouth. During metamorphosis, the outer epidermis shrinks and pulls the notochord and other tail structures internally for reorganization into adult tissues. The internal structures rotate 180°, positioning the oral siphon opposite the adhesive papillae and bending the digestive tract into a **U** shape (*see figure 17.8*).

SUBPHYLUM CEPHALOCHORDATA

Members of the subphylum Cephalochordata (sef"a-lo-kor-dah'tah) (Gr. *kephalo,* head + L. *chorda,* cord) are called lancelets. Lancelets clearly demonstrate the four chordate characteristics, and for that reason they are often studied in introductory zoology courses.

The cephalochordates consist of two genera, *Branchiostoma* (amphioxus) and *Asymmetron,* and about 45 species. They are distributed throughout the world's oceans in shallow waters that have clean sand substrates.

Cephalochordates are small (up to 5 cm long), tadpolelike animals. They are elongate, laterally flattened, and nearly transparent. In spite of their streamlined shape, cephalochordates are relatively weak swimmers and spend most of their time in a filter feeding position—partly to mostly buried with their anterior end sticking out of the sand (figure 17.10).

The notochord of cephalochordates extends from the tail to the head, giving them their name. Unlike the notochord of other chordates, most of the cells are muscle cells, making the notochord somewhat contractile. Both of these characteristics are probably adaptations to burrowing. Contraction of the muscle cells increases the rigidity of the notochord by compressing the fluids within, giving additional support when pushing into sandy substrates. Relaxation of these muscle cells increases flexibility for swimming.

Muscle cells on either side of the notochord cause undulations that propel the cephalochordate through the water. Longitudinal, ventrolateral folds of the body wall help stabilize cephalochordates during swimming, and a median dorsal fin and a caudal fin also aid in swimming.

An oral hood projects from the anterior end of cephalochordates. Ciliated, fingerlike projections, called cirri, hang from the ventral aspect of the oral hood and are used in feeding. The posterior wall of the oral hood bears the mouth opening that leads to a large pharynx. Numerous pairs of pharyngeal slits perforate the pharynx and are supported by cartilaginous gill bars. Large folds of the body wall extend ventrally around the pharynx and fuse at the ventral midline of the body, creating the atrium, a chamber that surrounds the pharyngeal region of the body. It may protect the delicate, filtering surfaces of the pharynx from bottom sediments. The opening from the atrium to the outside is called the atriopore (figure 17.10).

Maintenance Functions

Cephalochordates are filter feeders. During feeding, they are partially or mostly buried in sandy substrates with their mouths pointed upward. Cilia on the lateral surfaces of gill bars sweep water into the mouth. Water passes from the pharynx, through pharyngeal slits to the atrium, and out of the body through the atriopore. Food is initially sorted at the cirri. Larger materials catch on cilia of the cirri. As these larger particles accumulate, contractions of the cirri throw them off. Smaller, edible particles are pulled into the mouth with water and are collected by cilia on the gill bars and in mucus secreted by the endostyle. As in tunicates, the endostyle is a ciliated groove that extends longitudinally along the midventral aspect of the pharynx. Cilia move food and mucus dorsally, forming a food cord to the gut. A ring of cilia rotates the food cord, dislodging food. Digestion is both extracellular and intracellular. A diverticulum off the gut, called the midgut cecum, extends anteriorly. It ends blindly along the right side of the pharynx and secretes digestive enzymes. An anus is on the left side of the ventral fin.

Cephalochordates do not possess a true heart. Contractile waves in the walls of major vessels propel blood. Blood contains amoeboid cells and bathes tissues in open spaces.

Excretory tubules are modified coelomic cells closely associated with blood vessels. This arrangement suggests active transport of materials between the blood and excretory tubules.

The coelom of cephalochordates is reduced, compared to that of most other chordates. It is restricted to canals near the gill bars, the endostyle, and the gonads.

Reproduction and Development

Cephalochordates are dioecious. Gonads bulge into the atrium from the lateral body wall. Gametes are shed into the atrium and leave the body through the atriopore. External fertilization leads to a bilaterally symmetrical larva. Larvae are free-swimming, but they eventually settle to the substrate before metamorphosing into adults.

FURTHER PHYLOGENETIC CONSIDERATIONS

The evolutionary relationships between the hemichordates and chordates are difficult to document with certainty. *The dorsal tubular nerve cord and pharyngeal slits of hemichordates are*

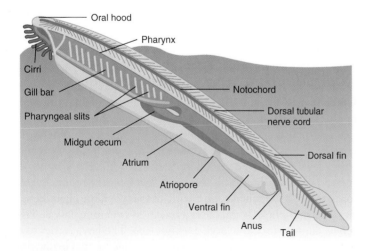

FIGURE 17.10

Subphylum Cephalochordata.   Internal structure of *Branchiostoma* (amphioxus) shown in its partially buried feeding position.

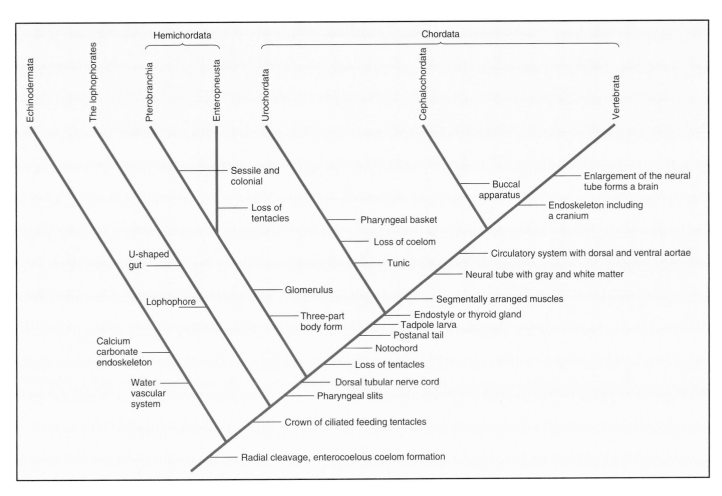

FIGURE 17.11

One Interpretation of Deuterostomate Phylogeny.   The dorsal tubular nerve cord and pharyngeal slits are possible synapomorphies that link the Hemichordata and the Chordata. The notochord, postanal tail, and endostyle or thyroid gland are important characteristics that distinguish the Hemichordata and Chordata. Some of the synapomorphies that distinguish the chordate subphyla are shown.

evidence of evolutionary ties between these phyla (figure 17.11). Questions regarding the homologies of these structures, however, remain. Synapomorphies that distinguish chordates from hemichordates include tadpole larvae, a notochord, a postanal tail, and an endostyle.

Figure 17.11 also shows evolutionary relationships between members of the three chordate subphyla. As discussed in chapter 16, the earliest echinoderms were probably sessile filter feeders. The lifestyle of adult urochordates suggests a similar ancestry (perhaps from a common ancestor with echinoderms) for chordates. The evolution of motile chordates from attached ancestors may have involved the development of a tadpolelike larva. Increased larval mobility is often adaptive for species with sedentary adults because it promotes dispersal. In 1928, W. Garstang suggested that the evolution of motile adults could have resulted from paedomorphosis, which is the development of sexual maturity in the larval body form. (Paedomorphosis is well documented in the animal kingdom, especially among amphibians.) Paedomorphosis could have led to a small, sexually

reproducing, fishlike chordate that could have been the ancestor of higher chordates.

The largest and most successful chordates belong to the subphylum Vertebrata. Bony or cartilaginous vertebrae that completely or partially replace the notochord characterize the vertebrates. The development of the anterior end of the nerve cord into a brain and the development of specialized sense organs on the head are evidence of a high degree of cephalization. The skeleton is modified anteriorly into a skull or cranium. There are seven classes of vertebrates (see table 17.1). Because of their cartilaginous and bony endoskeletons, vertebrates have left an abundant fossil record. Ancient jawless fishes were common in the Ordovician period, approximately 500 million years ago. Over a period of approximately 100 million years, fishes became the dominant vertebrates. Near the end of the Devonian period, approximately 400 million years ago, terrestrial vertebrates made their appearance. Since that time, vertebrates have radiated into most of the earth's habitats. Chapters 18 through 22 give an account of these events.

EVOLUTIONARY INSIGHTS
Early Deuterostome Evolution

The origin of the deuterostome lineage is an unresolved question. The fact that mesoderm forms while the coelom is forming (*see figure 7.12*) suggests to some zoologists that this lineage was derived from diploblastic ancestors. (In the protostomes, the coelom forms from preformed mesoderm, suggesting triploblastic ancestry.) A hypothetical, bilateral, diploblastic animal similar to the planula larva of cnidarians has been suggested as a common ancestor of all deuterostomes. The echinoderms originated very early in this lineage—the earliest fossils that are truly echinoderm are found in early Cambrian fossil beds.

W. Garstang suggested that the origin of the primitive chordates may have occurred when ciliated bands present on most deuterostome larvae (*see figure 7.12*) shifted dorsally to form the neural tube. Some modern ultrastructural evidence suggests homology between embryonic ciliary bands and certain cells in the nervous system of amphioxus. The origin of pharyngeal slits would have promoted water movement and thus filter feeding and gas exchange. Garstang's paedomorphosis hypothesis was described earlier and is accepted as being a reasonable explanation of the transition from an attached filter-feeder to a free-swimming, fishlike animal. The oldest chordate fossils are from fossil beds in China and date to about 530 million years before the present. Slightly younger (520 million years before the present) chordate fossils have been found in the Burgess Shale of British Columbia (box figure 17.1).

Recent molecular evidence is providing cause for rethinking the relationships between protostomes and deuterostomes. A nineteenth-century zoologist, Geoffroy Saint-Hilaire, studied the anatomy of a lobster from a ventral view. He noticed that, in this orientation, the ventral-to-dorsal sequence of nerve cord, segmental muscles, and digestive tract was similar to the dorsal-to-ventral sequence of these same structures in chordates. He hypothesized that the dorsal/ventral axis in arthropods was homologous to, although inverted from, that of chordates.

This old hypothesis is being revived. Molecular biologists have discovered genes of fruit flies (*Drosophila melanogaster*) that control the development of structures on the dorsal aspect of the fly and other genes that control the development of structures on the ventral aspect

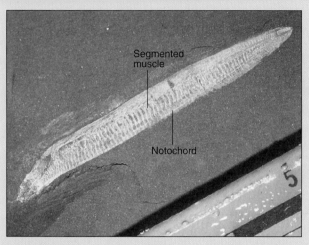

BOX FIGURE 17.1   An Early Chordate. This photograph shows a fossil of a 520-million-year-old cephalochordate, *Pikaia gracilens.*

of the fly. Similarly, there are genes that control the development of dorsal and ventral structures in a frog (*Xenopus laevis*) and a fish (*Danio rerio*). Interestingly, the genes that control ventral development of the two chordates are very similar to the genes that control the dorsal development of the fly and vice versa. Injection into a frog embryo of the messenger RNA or protein products of the fly gene that controls dorsal development promotes the development of ventral features in the frog. The implication of this work is that the dorsal surface of deuterostomes may be homologous to the ventral surface of protostomes. Once again, molecular biology is providing a way of peering back into the evolutionary history of animals. In this case, it may be showing us one of the important changes that occurred at the divergence of protostomate and deuterostomate lineages.

SUMMARY

1. Echinoderms, hemichordates, and chordates share deuterostome characteristics and are believed to have evolved from a common diploblastic or triploblastic ancestor.

2. Members of the phylum Hemichordata include the acorn worms and the pterobranchs. Acorn worms are burrowing marine worms, and pterobranchs are marine hemichordates whose collar has arms with many ciliated tentacles.

3. Chordates have four unique characteristics. A notochord is a supportive rod that extends most of the length of the animal. Pharyngeal slits are a series of openings between the digestive tract and the outside of the body. The tubular nerve cord lies just above the notochord and expands anteriorly into a brain. A postanal tail extends posteriorly to the anus and is supported by the notochord or the vertebral column.

4. Members of the subphylum Urochordata are the tunicates or sea squirts. Urochordates are sessile or planktonic filter feeders. Their development involves a tadpolelike larva.

5. The subphylum Cephalochordata includes small, tadpolelike filter feeders that live in shallow marine waters with clean, sandy substrates. Their notochord extends from the tail into the head and is somewhat contractile.

6. Pharyngeal slits and a tubular nerve cord link hemichordates and chordates to the same evolutionary lineage.

7. Chordates probably evolved from a sessile, filter-feeding ancestor. A larval stage of this sedentary ancestor may have undergone paedomorphosis to produce a small, sexually reproducing, fishlike chordate.

SELECTED KEY TERMS

endostyle (p. 270)

notochord (p. 269)

pharyngeal slits (p. 269)

postanal tail (p. 270)

tubular nerve cord (p. 270)

CRITICAL THINKING QUESTIONS

1. What evidence links hemichordates and chordates to the same evolutionary lineage?

2. What evidence of chordate affinities is present in adult tunicates? In larval tunicates?

3. What is paedomorphosis? What is a possible role for paedomorphosis in chordate evolution?

4. Discuss the possible influence of filter-feeding lifestyles on early chordate evolution.

5. What selection pressures could have favored a foraging or predatory lifestyle for later chordates?

ONLINE LEARNING CENTER

Visit our Online Learning Center (OLC) at www.mhhe.com/zoology (click on this book's title) to find the following chapter-related materials:

- CHAPTER QUIZZING
- RELATED WEB LINKS
    Phylum Chaetognatha
    Phylum Hemichordata
    Phylum Chordata
        General Chordata References
        Subphylum Urochordata
        Subphylum Cephalochordata

- BOXED READINGS ON
    Planktonic Tunicates
    The Endostyle and the Vertebrate Thyroid Gland
    Suspension Feeding in Invertebrates and Nonvertebrate Chordates

- SUGGESTED READINGS
- LAB CORRELATIONS

Check out the OLC to find specific information on these related lab exercises in the *General Zoology Laboratory Manual*, 5th edition, by Stephen A. Miller:

Exercise 17   *Chordata*

CHAPTER 18

THE FISHES:

VERTEBRATE SUCCESS IN WATER

Outline

Concepts

1. Cladistic analysis suggests that hagfishes are the most primitive craniates known. Craniate fossils can be traced back 530 million years.
2. Members of the subphylum Hyperotreti include the hagfishes. Hagfishes live in marine environments where they prey on invertebrates and scavenge dead and dying fish.
3. Members of the subphylum Vertebrata include the extinct ostracoderms and the lampreys and gnathostomes. The gnathostomes include the jawed fishes and tetrapods. Members of the class Chondrichthyes include the cartilaginous fishes—sharks, skates, rays, and ratfishes. Members of the class Osteichthyes include the bony fishes—sarcopterygians and actinopterygians.
4. Aquatic environments have selected for certain adaptations in fishes. These include the ability to move in a relatively dense medium, to exchange gases with water or air, to regulate buoyancy, to detect environmental changes, to regulate electrolytes (ions) and water in tissues, and to successfully reproduce.
5. Adaptive radiation resulted in the large variety of fishes present today. Evolution of some fishes led to the terrestrial vertebrates.

EVOLUTIONARY PERSPECTIVE

Water, a buoyant medium that resists rapid fluctuations in temperature, covers over 70% of the earth's surface. Because life began in water, and living tissues are made mostly of water, it might seem that nowhere else would life be easier to sustain. This chapter describes why that is not entirely true.

You do not need to wear scuba gear to appreciate that fishes are adapted to aquatic environments in a fashion that no other group of animals can surpass. If you spend recreational hours with hook and line, visit a marine theme park, or simply glance into a pet store when walking through a shopping mall, you can attest to the variety and beauty of fishes. *This variety is evidence of adaptive radiation that began more than 500 million years ago and shows no sign of ceasing. Fishes dominate many watery environments and are also the ancestors of all other members of the subphylum Vertebrata.*

This chapter contains evolutionary concepts, which are set off in this font.

FIGURE 18.1

Fishes.   Five hundred million years of evolution have resulted in unsurpassed diversity in fishes. The spines of this beautiful marine lionfish (*Pterois*) are extremely venomous.

PHYLOGENETIC RELATIONSHIPS

Fishes are included, along with the chordates covered in the following chapters, in a group called the Craniata (kra″ne-ah′tah) (*see table 17.1*; figure 18.1). This name is descriptive of a skull that surrounds the brain, olfactory organs, eyes, and inner ear. These animals are divided into two subphyla. The subphylum Hyperotreti (hi″per-ot′re-te) includes the hagfishes, and the subphylum Vertebrata (ver″te-bra′tah) includes the lampreys, cartilaginous fishes, and bony fishes (table 18.1). Traditional taxonomic groupings have combined the hagfishes, lampreys, and an extinct assemblage called ostracoderms into a superclass called Agnatha based on the absence of jaws in these groups. Modern cladistic analysis has revealed, however, that the lampreys share more characteristics with cartilaginous and bony fishes than with the hagfishes. The term "agnatha" no longer has taxonomic significance, but it is often used as a convenient nontaxonomic reference to jawless fishes.

Zoologists do not know what animals were the first craniates. Molecular evidence gathered by comparing gene similarities of cephalochordates and vertebrates suggests that the vertebrate lineage may go back about 750 million years. This date has not been confirmed by fossil evidence. Recent cladistic analysis of vertebrate evolution indicates that a group of fishes called hagfishes are the most primitive vertebrates known (living or extinct). Making a connection between this lineage and other vertebrates depends on the analysis of two key vertebrate characteristics, the brain and bone.

Chinese researchers have recently unearthed the oldest alleged craniate fossils—a small, lancelet-shaped animal that has characteristics that suggest an active, predatory lifestyle. A brain is present that would have processed sensory information from the pair of eyes that are seen in fossils. Muscle blocks along the body wall suggest an active swimming existence. This evidence means that these 530-million-year-old animals located prey by sight and then pursued it through prehistoric seas.

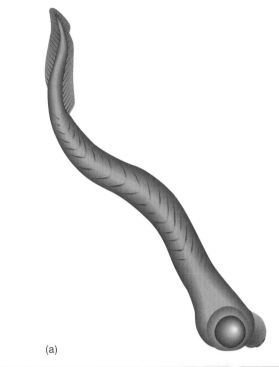

(a)

(b)

FIGURE 18.2

A Conodont (*Clydagnathus*) Reconstruction.   (*a*) A wealth of conodont fossils have been found that date to 510 million years ago. These animals have been assigned to a variety of phyla, but recent information has led many zoologists to accept them as some of the very early vertebrates. Their two large eyes, eel-like body, and toothlike denticles suggest that they lived as predators in prehistoric seas (1 cm). (*b*) A photomicrograph showing fossilized teeth of a conodont. These teeth were probably used for crushing food.

The origin of bone in craniates is equally intriguing. A group of ancient eel-like animals, the conodonts, are known from fossils that date back about 510 million years (figure 18.2). They have been assigned to a variety of phyla, but recent evidence

TABLE 18.1
CLASSIFICATION OF LIVING FISHES

Craniata (kra″ne-ah′tah)
Skull surrounds the brain, olfactory organs, eyes, and inner ear. Unique embryonic tissue, neural crest, contributes to a variety of adult structures, including sensory nerve cells and some skeletal and other connective tissue structures.

Subphylum Hyperotreti (hi″per-ot′re-te)
Fishlike; skull consisting of cartilaginous bars; jawless; no paired appendages; mouth with four pairs of tentacles; olfactory sacs open to mouth cavity; 5 to 15 pairs of pharyngeal slits; ventrolateral slime glands. Hagfishes.

Subphylum Vertebrata (ver″te-bra′tah)
Vertebrae surround nerve cord and serve as primary axial support.

Class Cephalaspidomorphi (sef″-ah-las″pe-do-morf′e)
Sucking mouth with teeth and rasping tongue; seven pairs of pharyngeal slits; blind olfactory sacs. Lampreys.

Gnathostomata (na′tho-sto′ma-tah)
Hinged jaws and paired appendages; vertebral column may have replaced notochord; three semicircular canals.

Class Chondrichthyes (kon-drik′thi-es)
Tail fin with large upper lobe (heterocercal tail); cartilaginous skeleton; lack opercula and a swim bladder or lungs. Sharks, skates, rays, ratfishes.

Subclass Elasmobranchii (e-laz″mo-bran′ke-i)
Cartilaginous skeleton may be partially ossified; placoid scales or no scales. Sharks, skates, rays.

Subclass Holocephali (hol″o-sef′a-li)
Operculum covers pharyngeal slits; lack scales; teeth modified into crushing plates; lateral-line receptors in an open groove. Ratfishes.

Class Osteichthyes (os″-te-ik′-the-es)
Most with bony skeleton; operculum covers single gill opening; pneumatic sacs function as lungs or swim bladders. Bony fishes.

Subclass Sarcopterygii (sar-kop-te-rij′e-i)
Paired fins with muscular lobes; pneumatic sacs function as lungs. Lungfishes and coelacanths (lobe-finned fishes).

Subclass Actinopterygii (ak″tin-op′te-rig-e-i)
Paired fins supported by dermal rays; basal portions of paired fins not especially muscular; tail fin with approximately equal upper and lower lobes (homocercal tail); blind olfactory sacs. Ray-finned fishes.

is bringing a growing number of zoologists to accept them as full-fledged craniates. They had two large eyes and a mouth filled with toothlike structures made of dentine—a component found in the craniate skeleton. These structures may represent the earliest occurrence of bone in vertebrates. Other hypotheses on the origin of bone suggest that it may have arisen as denticles in the skin (similar to those of sharks), in association with certain sensory receptors, or as structures for mineral (especially calcium phosphate) storage.

Regardless of its origin, bone was well developed by 500 million years ago. It was present in the bony armor of a group of fishes called ostracoderms. Ostracoderms were relatively inactive filter feeders that lived on the bottom of prehistoric lakes and seas. *They possessed neither jaws nor paired appendages; however, the evolution of fishes resulted in both jaws and paired appendages as well as many other structures.* The results of this adaptive radiation are described in this chapter.

Did ancestral fishes live in freshwater or in the sea? The answer to this question is not simple. The first vertebrates were probably marine, because ancient stocks of other deuterostome phyla were all marine. *Vertebrates, however, adapted to freshwater very early, and much of the evolution of fishes occurred there. Apparently, early vertebrate evolution involved the movement of fishes back and forth between marine and freshwater environments. The majority of the evolutionary history of some fishes took place in ancient seas, and most of the evolutionary history of others occurred in freshwater. The importance of freshwater in the evolution of fishes is evidenced by the fact that*

over 41% of all fish species are found in freshwater, even though freshwater habitats represent only a small percentage (0.0093% by volume) of the earth's water resources.

SURVEY OF FISHES

The taxonomy of fishes has been the subject of debate for many years. Modern cladistic analysis has resulted in complex revisions in the taxonomy of this group of vertebrates. Figure 18.3 includes the tetrapods as members of the superclass Gnathostomata. Inclusion of the tetrapods in this cladogram emphasizes the relationship of the fishes to land vertebrates and makes the lineage monophyletic. The tetrapods are discussed in chapters 19 through 22.

SUBPHYLUM HYPEROTRETI— HAGFISHES

Hagfishes are members of the class Myxini (mik′si-ne) (Gr. *myxa*, slime). There are about 20 species divided into four genera. Their heads are supported by cartilaginous bars and their brains are enclosed in a fibrous sheath. They lack vertebrae and retain the notochord as the axial supportive structure. They have four pairs of sensory tentacles surrounding their mouths and ventrolateral slime glands that produce copious amounts of slime (figure 18.4). Hagfishes are found in cold-water marine

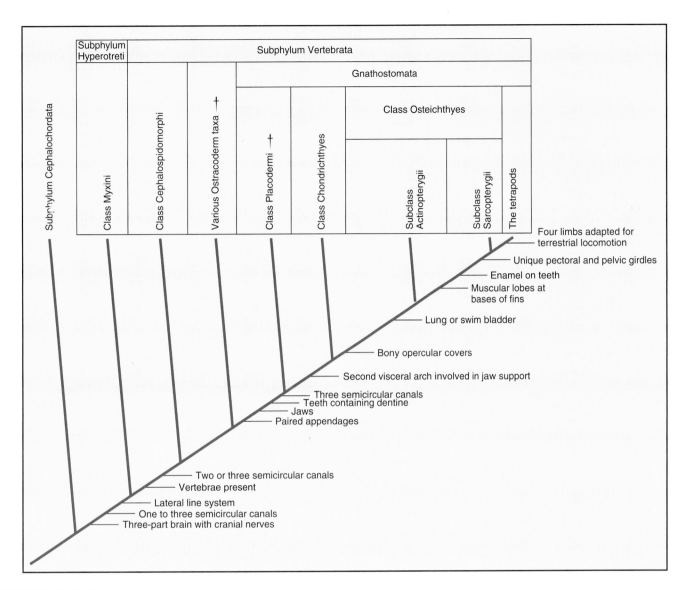

FIGURE 18.3

Or⏸ **Interpretation of the Phylogeny of the Craniata with Emphasis on the Fishes.** The evolutionary relationships among fishes are unsettled. This cladogram shows a few selected ancestral and derived characters. Each lower taxon has numerous synapomorphies that are not shown. Most zoologists consider the ostracoderms a paraphyletic group (having multiple lineages). Their representation as a monophyletic group is an attempt to simplify this presentation. Daggers (†) indicate groups whose members are extinct.

habitats of both the Northern and Southern Hemispheres. Most zoologists now consider the hagfishes to be the most primitive group of craniates.

Hagfishes live buried in the sand and mud of marine environments, where they feed on soft-bodied invertebrates and scavenge dead and dying fish. When hagfishes find a suitable fish, they enter the fish through the mouth and eat the contents of the body, leaving only a sack of skin and bones. Anglers must contend with hagfishes because they will bite at a baited hook. Hagfishes have the annoying habit of swallowing a hook so deeply that the hook is frequently lodged near the anus. The excessively slimy bodies of hagfishes make all but the grittiest fishermen cut their lines and tie on a new hook. Their tough skin is sold as "eel skin."

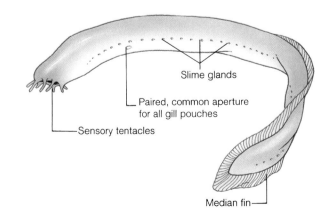

FIGURE 18.4

Class Myxini. Hagfish external structure.

FIGURE 18.5

Artist's Rendering of an Ancient Silurian Seafloor.   Two ostracoderms, *Pteraspis* and *Anglaspis*, are in the background.

SUBPHYLUM VERTEBRATA— OSTRACODERMS, LAMPREYS, AND GNATHOSTOME FISHES

The vertebrates are characterized by vertebrae that surround a
nerve cord and serve as primary axial support. Today, most vertebrates are members of the superclass Gnathostomata. They include the jawed fishes and the tetrapods. About 400 million years
ago, that was not the case—the jawless ostracoderms were a very
early and successful group of vertebrates. A third group of vertebrates, the lampreys, is also jawless and lives in both marine and
freshwater environments.

Ostracoderms

Ostracoderms are extinct agnathans that belonged to several
classes. The fossils of predatory water scorpions (phylum Arthropoda [*see figure 14.7*]) are often found with fossil ostracoderms. As
sluggish as ostracoderms apparently were, bony armor was probably
their only defense. Ostracoderms were bottom dwellers, often about
15 cm long (figure 18.5). Most were probably filter feeders, either
filtering suspended organic matter from the water or extracting annelids and other animals from muddy sediments. Some ostracoderms may have used bony plates around the mouth in a jawlike
fashion to crack gastropod shells or the exoskeletons of arthropods.

FIGURE 18.6

Class Cephalaspidomorphi.   A lamprey (*Petromyzon marinus*). Note
the sucking mouth and teeth used to feed on other fish.

Class Cephalaspidomorphi

Lampreys are agnathans in the class Cephalaspidomorphi (sef"ah-
las"pe-do-morf'e) (Gr. *kephale*, head + *aspidos*, shield + *morphe*,
form). They are common inhabitants of marine and freshwater environments in temperate regions. Most adult lampreys prey on other
fishes, and the larvae are filter feeders. The mouth of an adult is
suckerlike and surrounded by lips that have sensory and attachment
functions. Numerous epidermal teeth line the mouth and cover a
movable tonguelike structure (figure 18.6). Adults attach to prey
with their lips and teeth and use their tongues to rasp away scales.
Lampreys have salivary glands with anticoagulant secretions and
feed mainly on the blood of their prey. Some lampreys, however, are
not predatory. For example, some members of the genus *Lampetra*
are called brook lampreys. The larval stages of brook lampreys last
for about three years, and the adults neither feed nor leave their
stream. They reproduce soon after metamorphosis and then die.

Adult sea lampreys (*Petromyzon marinus*) live in the ocean
or the Great Lakes. Near the end of their lives, they migrate—
sometimes hundreds of miles—to a spawning bed in a freshwater
stream. Once lampreys reach their spawning site, usually in relatively shallow water with swift currents, they begin building a nest
by making small depressions in the substrate. When the nest is
prepared, a female usually attaches to a stone with her mouth. A
male uses his mouth to attach to the female's head and wraps his
body around the female (figure 18.7). Eggs are shed in small

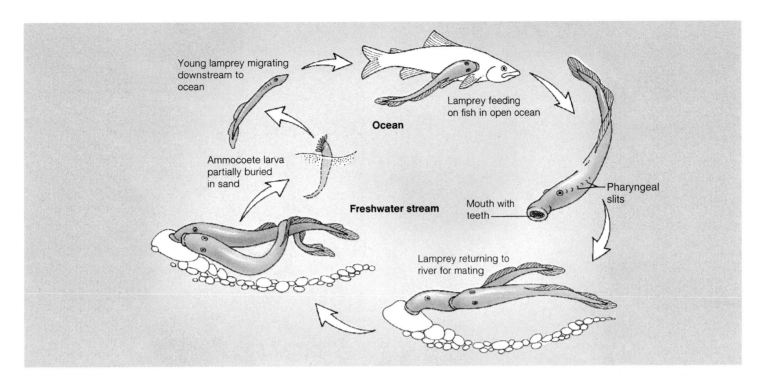

FIGURE 18.7

External Structure and Life History of a Sea Lamprey.   Sea lampreys feed in the open ocean and, near the end of their lives, migrate into freshwater streams, where they mate. They deposit eggs in nests on the stream bottom, and young ammocoete larvae hatch in about three weeks. Ammocoete larvae live as filter feeders until they attain sexual maturity.

batches over a period of several hours, and fertilization is external. The relatively sticky eggs are then covered with sand.

Eggs hatch in approximately three weeks into ammocoete larvae. The larvae drift downstream to softer substrates, where they bury themselves in sand and mud and filter feed in a fashion similar to amphioxus (*see figure 17.10*).

Ammocoete larvae grow from 7 mm to about 17 cm over three to seven years. During later developmental stages, the larvae metamorphose to the adult over a period of several months. The mouth becomes suckerlike, and the teeth, tongue, and feeding musculature develop. Lampreys eventually leave the mud permanently and begin a journey to the sea to begin life as predators. Adults return only once to the headwaters of their stream to spawn and die.

Gnathostomata

Jaws of vertebrates evolved from the most anterior pair of pharyngeal arches (the skeletal supports for the pharyngeal slits). This extremely important event in vertebrate evolution permitted more efficient gill ventilation and the capture and ingestion of a variety of food sources. Similarly, paired appendages were extremely important evolutionary developments. Their origin has been debated for many years, and the debate remains unresolved. In the absence of paired appendages, fishes must have been relatively sluggish bottom dwellers. Increased activity without paired appendages would have led to instability. Paired appendages can be used to counter the tendency to roll during locomotion. They can be used to control the tilt or pitch of the swimming fish and can be used in lateral steering. Pectoral fins of fishes are appendages usually just behind the head, and pelvic fins are usually located ventrally and more posteriorly. In modern bony fishes the pelvic fins are usually

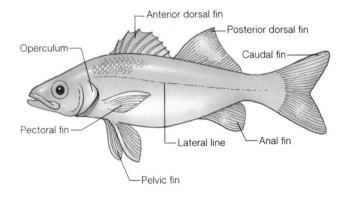

FIGURE 18.8

Paired Pectoral and Pelvic Appendages.   Appendages of a member of the gnathostomes. These appendages are secondarily reduced in some species.

positioned just behind the pectoral fins (figure 18.8). *The evolution of jaws and paired appendages must have revolutionized the life of early fishes. Both contributed to the evolution of the predatory lifestyles of many fishes. The ability to feed efficiently allowed fishes to produce more offspring and exploit new habitats. These new habitats, in turn, fostered the adaptive radiation of fishes.* The results of this adaptive radiation are described in this chapter.

Two classes of gnathostomes still have living members: the cartilaginous fishes (class Chondrichthyes) and the bony fishes (class Osteichthyes). Another class, the armored fishes, or placoderms, contained the earliest jawed fishes (*see figure 18.3*). They are now extinct and apparently left no descendants. A fourth group of ancient, extinct fishes, the acanthodians, may be more closely related to the bony fishes.

(a)

(b)

(c)

FIGURE 18.9

Class Chondrichthyes. (*a*) A gray reef shark (*Carcharhinus*). (*b*) A manta ray (*Manta hamiltoni*) with two remoras (*Remora remora*) attached to its ventral surface. (*c*) A bullseye stingray (*Urolophus concentricus*).

Class Chondrichthyes Members of the class Chondrichthyes (kon-drik′thi-es) (Gr. *chondros*, cartilage + *ichthyos*, fish) include the sharks, skates, rays, and ratfishes (*see table 18.1*). Most chondrichthians are carnivores or scavengers, and most are marine. In addition to their biting mouthparts and paired appendages, chondrichthians possess placoid scales and a cartilaginous endoskeleton.

The subclass Elasmobranchii (e-laz′mo-bran′ke-i) (Gr. *elasmos*, plate metal + *branchia*, gills), which includes the sharks,

(a)

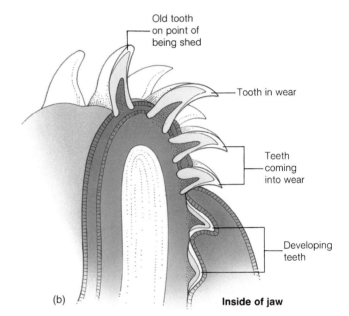

(b)

Inside of jaw

FIGURE 18.10

Scales and Teeth of Sharks. (*a*) Section of shark skin magnified to show posteriorly pointing placoid scales (SEM ×500). (*b*) The teeth of sharks develop as modified placoid scales. Newer teeth that move from the inside to the outside of the jaw continuously replace older teeth.

skates, and rays, has about 820 species (figure 18.9). Sharks arose from early jawed fishes midway through the Devonian period, about 375 million years ago. The absence of certain features characteristic of bony fishes (e.g., a swim bladder to regulate buoyancy, a gill cover, and a bony skeleton) is sometimes interpreted as evidence of the primitiveness of elasmobranchs. This interpretation is mistaken, as these characteristics simply resulted from different adaptations in the two groups to similar selection pressures. Some of these adaptations are described later in this chapter.

Tough skin with dermal, placoid scales covers sharks (figure 18.10*a*). These scales project posteriorly and give the skin a tough, sandpaper texture. (In fact, dried shark skin has been used for sandpaper.) Posteriorly pointed scales also reduce friction with the water as a shark swims.

Shark teeth are actually modified placoid scales. The row of teeth on the outer edge of the jaw is backed up by rows of teeth attached to a ligamentous band that covers the jaw cartilage inside the mouth. As the outer teeth wear and become useless, newer teeth move into position from inside the jaw and replace them. In young sharks, this replacement is rapid, with a new row of teeth developing every seven or eight days (figure 18.10b). Crowns of teeth in different species may be adapted for shearing prey or for crushing the shells of molluscs.

Sharks range in size from less than 1 m (e.g., *Squalus*, the laboratory dissection specimen) to greater than 10 m (e.g., basking sharks and whale sharks). The largest living sharks are not predatory but are filter feeders. They have pharyngeal-arch modifications that strain plankton. The fiercest and most feared sharks are the great white shark (*Carcharodon*) and the mako (*Isurus*). Extinct specimens may have reached lengths of 15 m or more.

Skates and rays are specialized for life on the ocean floor. They usually inhabit shallow water, where they use their blunt teeth to feed on invertebrates. Their most obvious modification for life on the ocean floor is a lateral expansion of the pectoral fins into winglike appendages. Locomotion results from dorsoventral muscular waves that pass posteriorly along the fins. Frequently, elaborate color patterns on the dorsal surface of these animals provide effective camouflage (*see figure 18.9c*). The sting ray (*Dasyatis*) has a tail modified into a defensive lash; a group of modified placoid scales persists as a venomous spine. Also included in this group are the electric rays (*Narcine* and *Torpedo*) and manta rays (*Manta*) (*see figure 18.9b*).

A second major group of chondrichthians, in the subclass Holocephali (hol″o-sef′a-li) (Gr. *holos*, whole + *kephalidos*, head), contains about 30 species. A frequently studied example, *Chimaera*, has a large head with a small mouth surrounded by large lips. A narrow, tapering tail has resulted in the common name "ratfish." Holocephalans diverged from other chondrichthians nearly 300 million years ago. Since that time, specializations not found in other elasmobranchs have evolved, including a gill cover, called an **operculum,** and teeth modified into large plates for crushing the shells of molluscs. Holocephalans lack scales.

Class Osteichthyes

Members of the class Osteichthyes (os″te-ik′the-es) (Gr. *osteon*, bone + *ichthyos*, fish) are characterized by having at least some bone in their skeleton and/or scales, bony operculum covering the gill openings, and lungs or a swim bladder. *Any group that has at least 24,000 species and is a major life-form in most of the earth's vast aquatic habitats must be judged very successful from an evolutionary perspective.*

The first fossils of bony fishes are from late Silurian deposits (approximately 405 million years old). By the Devonian period (350 million years ago), the two subclasses were in the midst of their adaptive radiations (*see table 18.1; see also figure 18.3*).

Members of the subclass Sarcopterygii (sar-kop-te-rij′e-i) (Gr. *sark*, flesh + *pteryx*, fin) have muscular lobes associated with their fins and usually use lungs in gas exchange. One group of sarcopterygians are the lungfishes. Only three genera survive today, and all live in regions where seasonal droughts are common.

FIGURE 18.11

Subclass Sarcopterygii.   The lungfish, *Lepidosiren paradoxa*, has lungs that allow it to withstand stagnation of its habitat.

When freshwater lakes and rivers begin to stagnate and dry, these fishes use lungs to breathe air (figure 18.11). Some (*Neoceratodus*) inhabit the freshwaters of Queensland, Australia. They survive stagnation by breathing air, but they normally use gills and cannot withstand total drying. Others are found in freshwater rivers and lakes in tropical Africa (*Protopterus*) and tropical South America (*Lepidosiren*). They can survive when rivers or lakes are dry by burrowing into the mud. They keep a narrow air pathway open by bubbling air to the surface. After the substrate dries, the only evidence of a lungfish burrow is a small opening in the earth. Lungfishes may remain in aestivation for six months or more. (Aestivation is a dormant state that helps an animal withstand hot, dry periods.) When rain again fills the lake or riverbed, lungfishes emerge from their burrows to feed and reproduce.

A second group of sarcopterygians are the coelacanths. The most recent coelacanth fossils are over 70 million years old. In 1938, however, people fishing in deep water off the coast of South Africa brought up one fish that was identified as a coelacanth (figure 18.12). Since then, numerous other specimens have been caught in deep water around the Comoro Islands off Madagascar. The discovery of this fish, *Latimeria chalumnae*, was a milestone event because coelacanths had previously been known only from the fossil record. It is large—up to 80 kg—and has heavy scales. A second species of coelacanth, *Latimeria menadoensis*, was discovered in 1977 off the coast of Indonesia. Ancient coelacanths lived in freshwater lakes and rivers; thus, the ancestors of *Latimeria* must have moved from freshwater habitats to the deep sea.

A third group of sarcopterygians, called osteolepiforms, became extinct before the close of the Paleozoic period. They are believed to have been the ancestors of ancient amphibians. They are discussed further at the end of this chapter.

The subclass Actinopterygii (ak″tin-op′te-rig-e-i) (Gr. *aktis*, ray + *pteryx*, fin) contains fishes that are sometimes called the ray-finned fishes because their fins lack muscular lobes. They usually possess **swim bladders,** gas-filled sacs along the dorsal wall of the body cavity that regulate buoyancy. *Zoologists now realize that there have been many points of divergence in the evolution of the Actinopterygii. One modern classification system divides the Actinopterygii into two infraclasses.*

FIGURE 18.12

A Sarcopterygian, the Coelacanth. *Latimeria* is the only known surviving coelacanth.

One group of actinopterygians, the chondrosteans, contains many species that lived during the Permian, Triassic, and Jurassic periods (215 to 120 million years ago), but only 25 species remain today. Ancestral chondrosteans had a bony skeleton, but living members, the sturgeons and paddlefishes, have cartilaginous skeletons. Chondrosteans also have a tail with a large upper lobe.

Most sturgeons live in the sea and migrate into rivers to breed (figure 18.13a). (Some sturgeons live in freshwater but maintain the migratory habits of their marine relatives.) They are large (up to 1,000 kg), and bony plates cover the anterior portion of the body. Heavy scales occur along the lateral surface. The sturgeon mouth is small, and jaws are weak. Sturgeons feed on invertebrates that they stir up from the sea or riverbed using their snouts. Because sturgeons are valued for their eggs (caviar), they have been severely overfished.

Paddlefishes are large, freshwater chondrosteans. They have a large, paddlelike rostrum that is innervated with sensory organs believed to detect weak electrical fields (figure 18.13b). They swim through the water with their large mouths open, filtering crustaceans and small fishes. They are found mainly in lakes and large rivers of the Mississippi River basin and are also known from China.

The second group of actinopterygians (Neopterygii) flourished in the Jurassic period and succeeded most chondrosteans. Two very primitive genera occur in temperate to warm freshwaters of North America. *Lepisosteus*, the garpike, has thick scales and long jaws that it uses to catch fishes. *Amia* is commonly referred to as the dogfish or bowfin. Most living fishes are members of this group and are referred to as teleosts or modern bony fishes. *After their divergence from ancient marine actinopterygians in the late Triassic period, teleosts experienced a remarkable evolutionary diversification and adapted to nearly every avail-*

(a)

(b)

FIGURE 18.13

Subclass Actinopterygii, the Chondrosteans. (a) Shovelnose sturgeons (*Scaphirhynchus platorynchus*). Sturgeons are covered anteriorly by heavy bony plates and posteriorly by scales. (b) The distinctive rostrum of a paddlefish (*Polydon spathula*) is densely innervated with sensory structures that are probably used to detect minute electrical fields. Note the mouth in its open, filter-feeding position.

able aquatic habitat (figure 18.14). The number of teleost species exceeds 24,000.

The next section in this chapter, explores some of the keys to the evolutionary success of this largest vertebrate group. A part of the answer to why the modern bony fishes have become so diverse and plentiful lies in the fact that 73% of the earth's surface is covered by water and an abundance of aquatic habitats are available. That, however, cannot be the whole answer because many other aquatic animals have been much less successful. The keys to teleost success lie in their ability to adapt to a demanding environment. Highly efficient respiratory systems allow fishes to extract oxygen from an environment that holds little oxygen per unit volume; efficient locomotor structures allow fishes to move through a buoyant, but viscous medium; highly efficient sensory systems include typical vertebrate systems, but also a lateral-line system that detects low-pressure waves and electroreception; and efficient reproductive mechanisms have the potential to produce overwhelming numbers of offspring.

(a)

(b)

(c)

FIGURE 18.14

Subclass Actinopterygii, the Teleosts. (*a*) Flat fish, such as this winter flounder (*Pseudopleuronectes americanus*), have both eyes on the right side of the head, and they often rest on their side fully or partially buried in the substrate. (*b*) The yellowtail snapper (*Ocyurus chrysurus*) is a popular sport fish and food item found in offshore tropical water. It reaches a length of 75 cm and a mass of 2.5 kg. (*c*) The sarcastic fringehead (*Neoclinus blanchardi*) retreats to holes on the mud bottom of the ocean. It is an aggressive predator that will charge and bite any intruder.

EVOLUTIONARY PRESSURES

Why is a fish fishlike? This apparently redundant question is unanswerable in some respects because some traits of animals are selectively neutral and, thus, neither improve nor detract from overall fitness. On the other hand, aquatic environments have physical characteristics that are important selective forces for aquatic animals. Although animals have adapted to aquatic environments in different ways, you can understand many aspects of the structure and function of a fish by studying the fish's habitat. This section will help you appreciate the many ways that a fish is adapted for life in water.

LOCOMOTION

Picture a young child running full speed down the beach and into the ocean. She hits the water and begins to splash. At first, she lifts her feet high in the air between steps, but as she goes deeper, her legs encounter more and more resistance. The momentum of her upper body causes her to fall forward, and she resorts to labored and awkward swimming strokes. The density of the water makes movement through it difficult and costly. For a fish, however, swimming is less energetically costly than running is for a terrestrial organism. The streamlined shape of a fish and the mucoid secretions that lubricate its body surface reduce friction between the fish and the water. Water's buoyant properties also contribute to the efficiency of a fish's movement through the water. A fish expends little energy in support against the pull of gravity.

Fishes move through the water using their fins and body wall to push against the incompressible surrounding water. Anyone who has eaten a fish filet probably realizes that muscle bundles of most fishes are arranged in a ≷ pattern. Because these muscles extend posteriorly and anteriorly in a zigzag fashion, contraction of each muscle bundle can affect a relatively large portion of the body wall. Very efficient, fast-swimming fishes, such as tuna and mackerel, supplement body movements with a vertical caudal (tail) fin that is tall and forked. The forked shape of the caudal fin reduces surface area that could cause turbulence and interfere with forward movement.

NUTRITION AND THE DIGESTIVE SYSTEM

The earliest fishes were probably filter feeders and scavengers that sifted through the mud of ancient seafloors for decaying organic matter, annelids, molluscs, or other bottom-dwelling invertebrates. Fish nutrition dramatically changed when the evolution of jaws transformed early fishes into efficient predators.

Most modern fishes are predators and spend much of their lives searching for food. Their prey vary tremendously. Some fishes feed on invertebrate animals floating or swimming in the plankton or living in or on the substrate. Many feed on other vertebrates. Similarly, the kinds of food that one fish eats at different times in its life varies. For example, as a larva, a fish may feed on

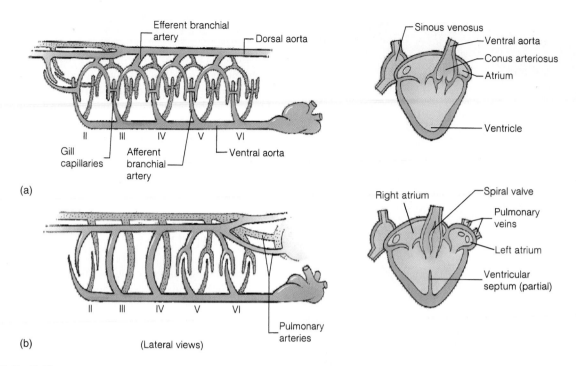

FIGURE 18.15

Circulatory System of Fishes. Diagrammatic representation of the circulatory systems of (*a*) bony fishes and (*b*) lungfishes. Hearts are drawn from a ventral view. Major branches of arteries carrying blood to and from the gills are called branchial arteries (or embryologically, aortic arches) and are numbered with Roman numerals. They begin with II because aortic arch I is lost during embryonic development.

plankton; as an adult, it may switch to larger prey, such as annelids or smaller fish. Fishes usually swallow prey whole. Teeth capture and hold prey, and some fishes have teeth that are modified for crushing the shells of molluscs or the exoskeletons of arthropods. To capture prey, fishes often use the suction that closing the opercula and rapidly opening the mouth creates, which develops a negative pressure that sweeps water and prey inside the mouth.

Other feeding strategies have also evolved in fishes. Herring, paddlefishes, and whale sharks are filter feeders. Long gill processes, called **gill rakers,** trap plankton while the fish is swimming through the water with its mouth open (*see figure 18.13b*). Other fishes, such as carp, feed on a variety of plants and small animals. A few, such as the lamprey, are external parasites for at least a portion of their lives. A few are primarily herbivores, feeding on plants.

The fish digestive tract is similar to that of other vertebrates. An enlargement, called the stomach, stores large, often infrequent, meals. The small intestine, however, is the primary site for enzyme secretion and food digestion. Sharks and other elasmobranchs have a spiral valve in their intestine, and bony fishes possess outpockets of the intestine, called pyloric ceca, which increase absorptive and secretory surfaces.

CIRCULATION AND GAS EXCHANGE

All vertebrates have a closed circulatory system in which a heart pumps blood, with red blood cells containing hemoglobin, through a series of arteries, capillaries, and veins. *The evolution of lungs in fishes was paralleled by changes in vertebrate circulatory systems. These changes are associated with the loss of gills, delivery of blood to the lungs, and separation of oxygenated and unoxygenated blood in the heart.*

The vertebrate heart develops from four embryological enlargements of a ventral aorta. In fishes, blood flows from the venous system through the thin-walled sinus venosus into the thin-walled, muscular atrium. From the atrium, blood flows into a larger, more muscular ventricle. The ventricle is the primary pumping structure. Anterior to the ventricle is the conus arteriosus, which connects to the ventral aorta. In teleosts, the conus arteriosus is replaced by an expansion of the ventral aorta called the bulbus arteriosus (figure 18.15*a*). Blood is carried by the ventral aorta to afferent vessels leading to the gills. These vessels break into capillaries and blood is oxygenated. Blood is then collected by efferent vessels, delivered to the dorsal aorta, and distributed to the body where it enters a second set of capillaries. Blood then returns to the heart through the venous system.

In most fishes, blood passes through the heart once with every circuit around the body. A few fishes (for example, the lungfishes) have lungs, and the pattern of circulation is altered. Understanding the pattern of circulation in these fishes is important because it was an important preadaptation for terrestrial life. In the lungfish, circulation to gills continues, but a vessel to the lungs has developed as a branch off aortic arch VI (figure 18.15*b*). This vessel is now called the pulmonary artery. Blood from the lungs returns to the heart through pulmonary veins and enters the left side of the heart. The atrium and ventricle of the lungfish heart are partially divided. These partial divisions help keep less oxygenated blood from the body separate from the oxygenated blood from the lungs. A spiral valve in the conus arteriosus helps direct blood from the right side of the heart to the pulmonary artery and blood from the left side of the heart to the remaining aortic arches. *Thus, the lungfishes show a distinction between a pulmonary circuit and a systemic circuit.*

Gas Exchange

Fishes live in an environment that contains less than 2.5% of the oxygen present in air. To maintain adequate levels of oxygen in their bloodstream, fishes must pass large quantities of water across gill surfaces and extract the small amount of oxygen present in the water.

Most fishes have a muscular pumping mechanism to move the water into the mouth and pharynx, over the gills, and out of the fish through gill openings. Muscles surrounding the pharynx and the opercular cavity, which is between the gills and the operculum, power this pump.

Some elasmobranchs and open-ocean bony fishes, such as the tuna, maintain water flow by holding their mouths open while swimming. This method is called **ram ventilation.** Elasmobranchs do not have opercula to help pump water, and therefore, some sharks must keep moving to survive. Others move water over their gills with a pumping mechanism similar to that just described. Rather than using an operculum in the pumping process, however, these fishes have gill bars with external flaps that close and form a cavity functionally similar to the opercular cavity of other fishes. Spiracles are modified first pharyngeal slits that open just behind the eyes of elasmobranchs and are used as an alternate route for water entering the pharynx.

Gas exchange across gill surfaces is very efficient. **Gill (visceral) arches** support gills. **Gill filaments** extend from each gill arch and include vascular folds of epithelium, called **pharyngeal lamellae** (figure 18.16*a,b*). Branchial arteries carry blood to the gills and into gill filaments. The arteries break into capillary beds in pharyngeal lamellae. Gas exchange occurs as blood and water move in opposite directions on either side of the lamellar epithelium. This **countercurrent exchange mechanism** provides very efficient gas exchange by maintaining a concentration gradient between the blood and the water over the entire length of the capillary bed (figure 18.16*c,d*).

Swim Bladders and Lungs

The Indian climbing perch spends its life almost entirely on land. These fishes, like most bony fishes, have gas chambers called **pneumatic sacs.** In nonteleost fishes and some teleosts, a pneumatic duct connects the pneumatic sacs to the esophagus or another part of the digestive tract. Swallowed air enters these sacs, and gas exchange occurs across vascular surfaces. Thus, in the Indian climbing perch, lungfishes, and ancient rhipidistians, pneumatic sacs function(ed) as lungs. In other bony fishes, pneumatic sacs act as swim bladders.

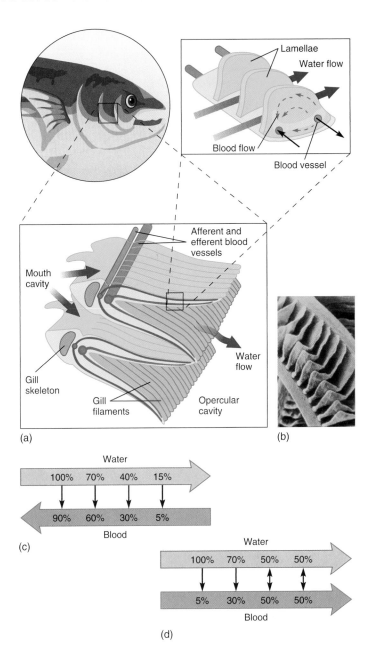

FIGURE 18.16

Gas Exchange at the Pharyngeal Lamellae. (*a*) The gill arches under the operculum support two rows of gill filaments. Blood flows into gill filaments through afferent branchial arteries, and these arteries break into capillary beds in the pharyngeal lamellae. Water and blood flow in opposite directions on either side of the lamellae. (*b*) Electron micrograph of the tip of a trout gill filament, showing numerous lamellae. (*c, d*) A comparison of countercurrent and parallel exchanges. Water entering the spaces between pharyngeal lamellae is saturated with oxygen in both cases. In countercurrent exchange (*c*), this water encounters blood that is almost completely oxygenated, but a diffusion gradient still favors the movement of more oxygen from the water to the blood. As water continues to move between lamellae, it loses oxygen to the blood because it is continually encountering blood with a lower oxygen concentration. Thus, a diffusion gradient is maintained along the length of the lamellae. If blood and water moved in parallel fashion (*d*), oxygen would diffuse from water to blood only until the oxygen concentration in blood equaled the oxygen concentration in water, and the exchange would be much less efficient.

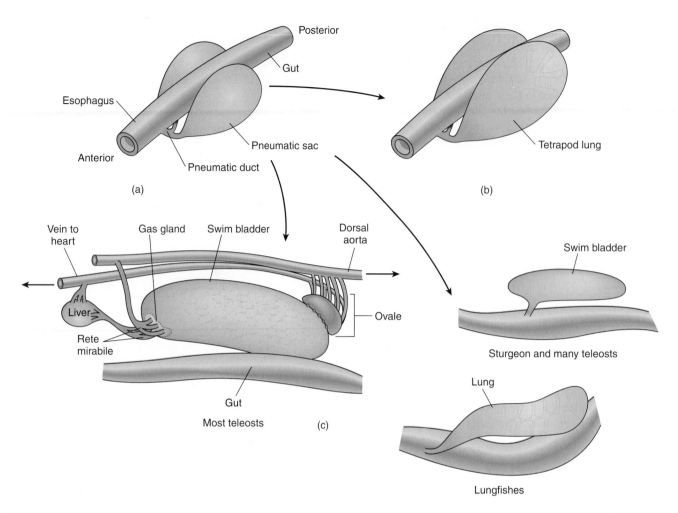

FIGURE 18.17

Possible Sequence in the Evolution of Pneumatic Sacs. (*a*) Pneumatic sacs may have originally developed from ventral outgrowths of the esophagus. Many ancient fishes probably used pneumatic sacs as lungs. (*b*) Primitive lungs developed further during the evolution of vertebrates. Internal compartmentalization increases surface area for gas exchange in land vertebrates. (*c*) In most bony fishes, pneumatic sacs are called swim bladders, and they are modified for buoyancy regulation. Swim bladders are dorsal in position to prevent a tendency for the fish to "belly up" in the water. Pneumatic duct connections to the esophagus are frequently lost, and gases transfer from the blood to the swim bladder through a countercurrent exchange mechanism called a rete mirabile and the gas gland. The ovale, at the posterior end of the swim bladder, returns gases to the bloodstream.

Most zoologists believe that lungs are more primitive than swim bladders. Much of the early evolution of bony fishes occurred in warm, freshwater lakes and streams during the Devonian period. These bodies of water frequently became stagnant and periodically dried. Having lungs in these habitats could have meant the difference between life and death. On the other hand, the later evolution of modern bony fishes occurred in marine and freshwater environments, where stagnation was not a problem. In these environments, the use of pneumatic sacs in buoyancy regulation would have been adaptive (figure 18.17).

Buoyancy Regulation

Did you ever consider why you can float in water? Water is a supportive medium, but that is not sufficient to prevent you from sinking. Even though you are made mostly of water, other constituents of tissues are more dense than water. Bone, for example, has a specific gravity twice that of water. Why, then, can you float? You can float because of two large, air-filled organs called lungs.

Fishes maintain their vertical position in a column of water in one or more of four ways. One way is to incorporate low-density compounds into their tissues. Fishes (especially their livers) are saturated with buoyant oils. A second way fishes maintain vertical position is to use fins to provide lift. The pectoral fins of a shark are planing devices that help to create lift as the shark moves through the water. Also, the large upper lobe of a shark's caudal fin provides upward thrust for the posterior end of the body (*see figure 18.9a*). A third adaptation is the reduction of heavy tissues in fishes. The bones of fishes are generally less dense than those of terrestrial vertebrates. One of the adaptive features of the elasmobranch cartilaginous skeleton probably results from cartilage being only slightly heavier than water. The fourth adaptation is the swim bladder. A fish regulates buoyancy by precisely controlling the volume of gas in its swim bladder. (You can mimic this adaptation while floating in water. How well do you float after forcefully exhaling as much air as possible?)

The pneumatic duct connects the swim bladders of garpike, sturgeons, and other primitive bony fishes to the esophagus or another part of the digestive tract. These fishes gulp air at the surface to force air into their swim bladders.

Most teleosts have swim bladders that have lost a functional connection to the digestive tract. The blood secretes gases (various mixtures of nitrogen and oxygen) into the swim bladder using a countercurrent exchange mechanism in a vascular network called the rete mirabile ("miraculous net"). Gases are secreted from the rete mirabile into the swim bladder through a gas gland. Gases may be reabsorbed into the blood at the posterior end of the bladder, the ovale (see figure 18.17c).

NERVOUS AND SENSORY FUNCTIONS

The central nervous system of fishes, as in other vertebrates, consists of a brain and a spinal cord. Sensory receptors are widely distributed over the body. In addition to generally distributed receptors for touch and temperature, fishes possess specialized receptors for olfaction, vision, hearing, equilibrium and balance, and for detecting water movements.

Openings, called external nares, in the snouts of fishes lead to olfactory receptors. In most fishes, receptors are in blind-ending olfactory sacs. In a few fishes, the external nares open to nasal passages that lead to the mouth cavity. Recent research has revealed that some fishes rely heavily on their sense of smell. For example, salmon and lampreys return to spawn in the streams in which they hatched years earlier. Their migrations to these streams often involve distances of hundreds of kilometers, and the fishes' perception of the characteristic odors of their spawning stream guide them.

The eyes of fishes are similar in most aspects of structure to those in other vertebrates. They are lidless, however, and the lenses are round. Focusing requires moving the lens forward or backward in the eye. (Most other vertebrates focus by changing the shape of the lens.)

Receptors for equilibrium, balance, and hearing are in the inner ears of fishes, and their functions are similar to those of other vertebrates. Semicircular canals detect rotational movements, and other sensory patches help with equilibrium and balance by detecting the direction of the gravitational pull. Fishes lack the outer and/or middle ear, which conducts sound waves to the inner ear in other vertebrates. Anyone who enjoys fishing knows, however, that most fishes can hear. Vibrations may pass from the water through the bones of the skull to the middle ear, and a few fishes have chains of bony ossicles (modifications of vertebrae) that connect the swim bladder to the back of the skull. Vibrations strike the fish, are amplified by the swim bladder, and are sent through the ossicles to the skull.

Running along each side and branching over the head of most fishes is a lateral-line system. The **lateral-line system** consists of sensory pits in the epidermis of the skin that connect to canals that run just below the epidermis. In these pits are receptors that are stimulated by water moving against them (see figure 24.19). Lateral lines are used to detect either water currents, or a predator or a prey that may be causing water movements, in the vicinity of the fish. Fishes may also detect low-frequency sounds with these receptors.

Electroreception and Electric Fishes

A U.S. Navy pilot has just ejected from his troubled aircraft over shark-infested water! What measures can the pilot take to ensure survival under these hostile conditions? The Navy has considered this scenario. One of the solutions to the problem is a polyvinyl bag suspended from an inflatable collar. The polyvinyl bag helps conceal the downed flyer from a shark's vision and keen sense of smell. But is that all that is required to ensure protection?

All organisms produce weak electrical fields from the activities of nerves and muscles. **Electroreception** is the detection of electrical fields that the fish or another organism in the environment generates. Electroreception and/or electrogeneration has been demonstrated in over five hundred species of fishes in seven families of Chondrichthyes and Osteichthyes. These fishes use their electroreceptive sense for detecting prey and for orienting toward or away from objects in the environment.

Prey detection with this sense is highly developed in the rays and sharks. Spiny dogfish sharks, the common laboratory specimens, locate prey by electroreception. A shark can find and eat a flounder that is buried in sand, and it will try to find and eat electrodes that are creating electrical signals similar to those that the flounder emits. On the other hand, a shark cannot find a dead flounder buried in the sand or a live flounder covered by an insulating polyvinyl sheet.

Some fishes are not only capable of electroreception, but can also generate electrical currents. An electric fish (*Gymnarchus niloticus*) lives in freshwater systems of Africa. Muscles near its caudal fin are modified into organs that produce a continuous electrical discharge. This current spreads between the tail and the head. Porelike perforations near the head contain electroreceptors. The electrical waves circulating between the tail and the head are distorted by objects in their field. This distortion is detected in changing patterns of receptor stimulation (figure 18.18). The electrical sense of *Gymnarchus* is an adaptation to living in murky freshwater habitats where eyes are of limited value.

The fishes best known for producing strong electrical currents are the electric eel (a bony fish) and the electric ray (an elasmobranch). The electric eel (*Electrophorus*) occurs in rivers of the Amazon basin in South America. The organs for producing electrical currents are in the trunk of the electric eel and can deliver shocks in excess of 500 volts. The electric ray (*Narcine*) has electric organs in its fins that are capable of producing pulses of 50 amperes at about 50 volts (figure 18.19). Shocks that these fishes produce are sufficiently strong to stun or kill prey, discourage large predators, and teach unwary humans a lesson that will never need to be repeated.

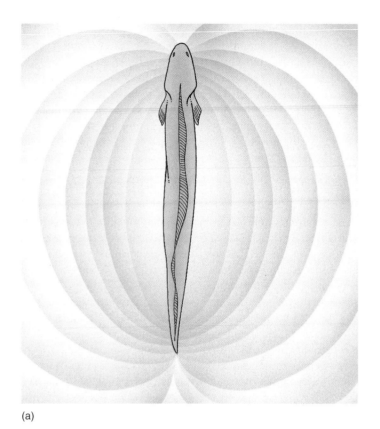

(a)

(b)

FIGURE 18.18

Electric Fishes. (*a*) The electrical field of a fish detects the presence of prey and other objects in the fish's murky environment. Currents circulate from electrical organs in the fish's tail to electroreceptors near its head. An object in this electrical field changes the pattern of stimulation of electroreceptors. (*b*) The electric fish (*Gymnarchus niloticus*).

EXCRETION AND OSMOREGULATION

Fishes, like all animals, must maintain a proper balance of electrolytes (ions) and water in their tissues. This osmoregulation is a major function of the kidneys and gills of fishes. Kidneys are located near the midline of the body, just dorsal to the peritoneal

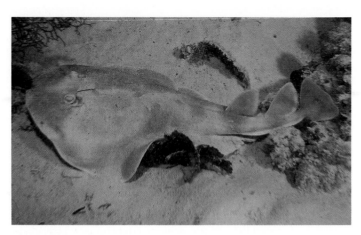

FIGURE 18.19

Electric Fishes. A lesser electric ray (*Narcine brasiliensis*).

membrane that lines the body cavity. As with all vertebrates, the excretory structures in the kidneys are called **nephrons.** Nephrons filter bloodborne nitrogenous wastes, ions, water, and small organic compounds across a network of capillaries called a **glomerulus.** The filtrate then passes into a tubule system, where essential components may be reabsorbed into the blood. The filtrate remaining in the tubule system is then excreted.

Freshwater fishes live in an environment containing few dissolved substances. Osmotic uptake of water across gill, oral, and intestinal surfaces and the loss of essential ions by excretion and defecation are constant. To control excess water buildup and ion loss, freshwater fishes never drink and only take in water when feeding. Also, the numerous nephrons of freshwater fishes often have large glomeruli and relatively short tubule systems. Reabsorption of some ions and organic compounds follows filtration. Because the tubule system is relatively short, little water is reabsorbed. Thus, freshwater fishes produce large quantities of very dilute urine. Ions are still lost, however, through the urine and by diffusion across gill and oral surfaces. Active transport of ions into the blood at the gills compensates for this ion loss. Freshwater fishes also get some salts in their food (figure 18.20*a*).

Marine fishes face the opposite problems. Their environment contains 3.5% ions, and their tissues contain approximately 0.65% ions. Marine fishes, therefore, must combat water loss and accumulation of excess ions. They drink water and eliminate excess ions by excretion, defecation, and active transport across gill surfaces. The nephrons of marine fishes often possess small glomeruli and long tubule systems. Much less blood is filtered than in freshwater fishes, and water is efficiently, although not entirely, reabsorbed from the nephron (figure 18.20*b*).

Elasmobranchs have a unique osmoregulatory mechanism. They convert some of their nitrogenous wastes into urea in the liver. This in itself is somewhat unusual, because most fishes excrete ammonia rather than urea. Even more unusual, however, is that urea is sequestered in tissues all over the body. Enough urea is stored to make body tissues slightly hyperosmotic to seawater. (That is, the concentration of solutes in a shark's tissues is essentially the same as the concentration of ions in seawater.)

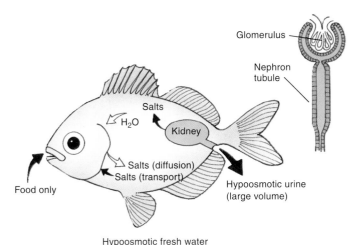

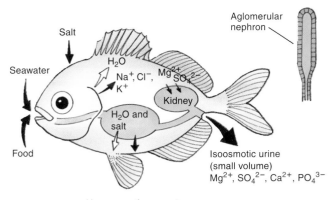

FIGURE 18.20

Osmoregulation by (a) Freshwater and (b) Marine Fishes. Large arrows indicate passive uptake or loss of water or electrolytes (ions) through ingestion and excretion. Small, solid arrows indicate active transport processes at gill membranes and kidney tubules. Small, open arrows indicate passive uptake or loss by diffusion through permeable surfaces. Insets of kidney nephrons depict adaptations within the kidney. Water, ions, and small organic molecules are filtered from the blood at the glomerulus of the nephron. Essential components of the filtrate can be reabsorbed within the tubule system of the nephron. Marine fishes conserve water by reducing the size of the glomerulus of the nephron, and thus reducing the quantity of water and ions filtered from the blood. Other ions can be secreted from the blood into the kidney tubules. Marine fishes can produce urine that is isoosmotic with the blood. Freshwater fishes have enlarged glomeruli and short tubule systems. They filter large quantities of water from the blood, and tubules reabsorb some ions from the filtrate. Freshwater fishes produce a hypoosmotic urine.

Therefore, the problem most marine fishes have of losing water to their environment is much less severe for elasmobranchs. Energy that does not have to be devoted to water conservation can now be used in other ways. This adaptation required the development of tolerance to high levels of urea, because urea disrupts important enzyme systems in the tissues of most other animals.

In spite of this unique adaptation, elasmobranchs must still regulate the ion concentrations in their tissues. In addition to having ion-absorbing and secreting tissues in their gills and kidneys, elasmobranchs possess a rectal gland that removes excess sodium chloride from the blood and excretes it into the cloaca. (A **cloaca** is a common opening for excretory, digestive, and reproductive products.)

Diadromous fishes migrate between freshwater and marine environments. Salmon (*e.g., Oncorhynchus*) and marine lampreys (*Petromyzon*) migrate from the sea to freshwater to spawn, and the freshwater eel (*Anguilla*) migrates from freshwater to marine environments to spawn. Diadromous migrations require gills capable of coping with both uptake and secretion of ions. Osmoregulatory powers needed for migration between marine and freshwater environments may not be developed in all life-history stages. Young salmon, for example, cannot enter the sea until certain cells on the gills develop ion-secreting powers.

Fishes have few problems getting rid of the nitrogenous byproducts of protein metabolism. Up to 90% of nitrogenous wastes are eliminated as ammonia by diffusion across gill surfaces. Even though ammonia is toxic, aquatic organisms can have it as an excretory product because ammonia diffuses in the surrounding water. The remaining 10% of nitrogenous wastes are excreted as urea, creatine, or creatinine. These wastes are produced in the liver and are excreted via the kidneys.

REPRODUCTION AND DEVELOPMENT

Imagine, 45 kg of caviar from a single, 450 kg sturgeon! Admittedly, a 450 kg sturgeon is a very large fish (even for a sturgeon), but a fish producing millions of eggs in a single season is not unusual. These numbers simply reflect the hazards of developing in aquatic habitats unattended by a parent. The vast majority of these millions of potential adults will never survive to reproduce. Many eggs will never be fertilized, many fertilized eggs may wash ashore and dry, currents and tides will smash many eggs and embryos, and others will fall victim to predation. In spite of all of these hazards, if only four of the millions of embryos of each breeding pair survive and reproduce, the population will double.

Producing overwhelming numbers of eggs, however, is not the only way that fishes increase the chances that a few of their offspring will survive. Some fishes show mating behavior that helps ensure fertilization, or nesting behavior that protects eggs from predation, sedimentation, and fouling.

Mating may occur in large schools, and one individual releasing eggs or sperm often releases spawning pheromones that induce many other adults to spawn. Huge masses of eggs and sperm released into the open ocean ensure the fertilization of many eggs.

The vast majority of fishes are oviparous, meaning that eggs develop outside the female from stored yolk. Some elasmobranchs are ovoviviparous, and their embryos develop in a modified oviduct of the female. Nutrients are supplied from yolk stored in the egg. Other elasmobranchs, including gray reef sharks and hammerheads, are viviparous. A placenta-like outgrowth of a modified oviduct diverts nutrients from the female to the yolk sacs of developing embryos. Internal development of viviparous bony fishes usually occurs in ovarian follicles, rather than in the

FIGURE 18.21

Male Garibaldi (*Hypsypops rubicundus*). The male cultivates a nest of filamentous red algae and then entices a female to lay eggs in the nest. Males also defend the nest against potential predators.

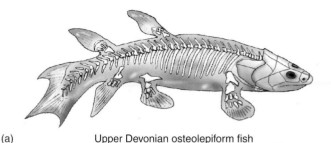

(a)   Upper Devonian osteolepiform fish

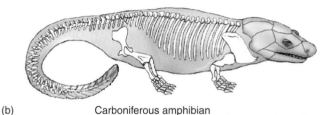

(b)   Carboniferous amphibian

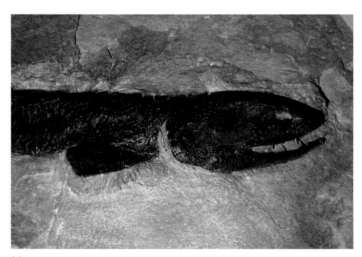

(c)

FIGURE 18.22

Osteolepiform Fishes and a Primitive Amphibian. The osteolepiform fish (*a*), *Eusthenopteron*, shares many skeletal features with an ancient amphibian (*b*), *Diplovertebron*. Differences include the loss of median fins in the amphibian and the development of stout ribs and a pelvic girdle in the amphibian. A fossil *Eusthenopteron* is shown in (*c*).

oviduct. In guppies (*Lebistes*), eggs are retained in the ovary, and fertilization and early development occur there. Embryos are then released into a cavity within the ovary and development continues, with nourishment coming partly from yolk and partly from ovarian secretions.

Some fishes have specialized structures that aid in sperm transfer. Male elasmobranchs, for example, have modified pelvic fins called claspers. During copulation, the male inserts a clasper into the cloaca of a female. Sperm travel along grooves of the clasper. Fertilization occurs in the female's reproductive tract and usually results in a higher proportion of eggs being fertilized than in external fertilization. Thus, fishes with internal fertilization usually produce fewer eggs.

In many fishes, care of the embryos is limited or nonexistent. Some fishes, however, construct and tend nests (figure 18.21), and some carry embryos during development. Clusters of embryos may be brooded in special pouches attached to some part of the body, or they may be brooded in the mouth. Some of the best-known brooders include the seahorses (*Hippocampus*) and pipefishes (e.g., *Syngnathus*). Males of these closely related fishes carry embryos throughout development in ventral pouches. The male Brazilian catfish (*Loricaria typhys*) broods embryos in an enlarged lower lip.

Sunfishes and sticklebacks provide short-term care of posthatching young. Male sticklebacks assemble fresh plant material into a mass in which the young take refuge. If one offspring wanders too far from the nest, the male snaps it up in its mouth and spits it back into the nest. Sunfish males do the same for young that wander from schools of recently hatched fish. The Cichlidae engage in longer-term care (*see figures 1.1 and 1.2*). In some species, young are mouth brooded, and other species tend young in a nest. After hatching, the young venture from the parent's mouth or nest but return quickly when the parent signals danger with a flicking of the pelvic fins.

FURTHER PHYLOGENETIC CONSIDERATIONS

*T*wo important series of evolutionary events occurred during the evolution of the Osteichthyes. One of these was an evolutionary explosion that began about 150 million years ago and resulted in the vast diversity of teleosts living today. The last half of this chapter should have helped you appreciate some of these events.

The second series of events involves the evolution of terrestrialism. The presence of functional lungs in modern lungfishes

WILDLIFE ALERT
The Pallid Sturgeon (*Scaphirhynchus albus*)

VITAL STATISTICS

Classification: Phylum Chordata, class Osteichthyes, order Acipenseriformes, family Acipenseridae

Range: Headwaters of the Missouri River to the Mississippi River south of St. Louis, Missouri

Habitat: Areas of strong current and firm sand bottom in the main channels of the Missouri and Mississippi Rivers

Number remaining: Unknown

Status: Endangered

NATURAL HISTORY AND ECOLOGICAL STATUS

The pallid sturgeon's native range is in the Missouri and Mississippi Rivers from headwaters near Great Falls, Montana, to New Orleans (box figure 18.1). It is a large (76–152 cm, 40 kg), freshwater fish that is light brown dorsally and white ventrally. This sturgeon has a flattened, shovel-shaped snout with fleshy barbels between its mouth and snout (box figure 18.2). It feeds on aquatic insects and small fish, which it sucks from the substrate with its toothless mouth. Little is known of its behavior, movements, and habitat requirements.

BOX FIGURE 18.2   The Pallid Sturgeon (*Scaphirhynchus albus*).

The pallid sturgeon is rare throughout its range and was placed on the Federal Endangered Species List in 1990. The decline in population numbers of this sturgeon since it was described by early explorers is probably due to habitat destruction. Originally the Missouri and Mississippi Rivers had wide, meandering channels and extensive backwater areas. They provided diverse substrates, water depths, and current characteristics. Only about one-third of the length of these rivers remain similar to their natural state. They have been impounded by dams for flood control and channelized for river navigation. The diversity of habitats has been severely reduced. Municipal, industrial, and agricultural pollution have further degraded water quality.

Recovery efforts have been focused in three directions. First, three captive populations have been established that are being used to restock these rivers. In 1997, the Missouri Department of Conservation stocked 412 25-cm sturgeons into the Platte River, a Missouri River tributary. Second, these stocked fish were tagged with plastic and ultrasonic tags to study their long-range movements, survival, and preferred habitat areas. Third, six recovery-management areas have been designated along the Montana–Louisiana range. New information on habitat preferences is being used in habitat restoration efforts, which will promote the recovery of this fish's population.

One hopeful sign regarding the fate of this species was the collection of a single young-of-the-year pallid sturgeon in 1998 from the Mississippi River near Cape Girardeau, MO. Until this collection, there had been no recent documentation of successful wild sturgeon reproduction.

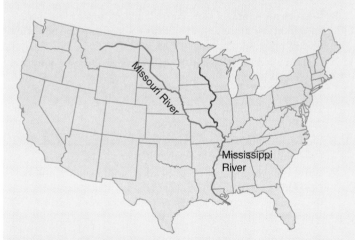

BOX FIGURE 18.1   The Original Range of the Pallid Sturgeon (*Scaphirhynchus albus*).

has led to the suggestion that the lungfish lineage may have been ancestral to modern terrestrial vertebrates. Most cladistic and anatomical evidence indicates that the lungfish lineage gave rise to no other vertebrate taxa. *Recent molecular studies, however, suggest a close relationship between lungfishes and modern amphibians. This evidence is causing some to look again at the lungfish lineage as a possible route to terrestrial vertebrates.*

The traditional explanation for the origin of terrestrial vertebrates involves osteolepiform sarcopterygians. They possessed several unique characteristics in common with early amphibians. These included structures of the jaw, teeth, vertebrae, and limbs, among others (figure 18.22). Studies of limb morphology suggest that muscular limbs may have been used for crawling along the bottom of shallow ponds and swamps. Shallow-water habitats would have allowed them to avoid falling prey to larger predatory fishes. Young osteolepiforms may have inhabited very shallow water. In the warm, shallow-water environments, depleted oxygen would have favored the evolution of sarcopterygian lungs. Abundant food, in the form of arthropods and other invertebrates, at the water's edge would have favored shallow-water and terrestrial feeding, appendages that allowed the fish to wallow at the land/water interface, and eye lenses that could function in air. While millions of years would elapse before any vertebrate could be considered terrestrial, brief excursions onto land allowed some vertebrates to exploit resources that for the previous 50 million years had only been available to terrestrial arthropods.

SUMMARY

1. Zoologists do not know what animals were the first vertebrates. The oldest alleged vertebrate fossils are of 530-million-year-old predators. A group of ancient eel-like animals, the conodonts, are known from fossils that date back about 510 million years. The hagfishes are the most ancient living fishes.

2. Members of the subphylum Hyperotreti include the hagfishes. Hagfishes lack vertebrae, have a cranium consisting of cartilaginous bars, four pairs of sensory tentacles surrounding their mouths, and slime glands along their bodies. They are predators and scavengers and are considered the most primitive of all living craniates.

3. Members of the subphylum Vertebrata include the extinct ostracoderms and the lampreys and gnathostomes. They possess vertebrae. Ostracoderms included several classes of jawless vertebrates that were very common bottom dwellers about 400 million years ago. Lampreys (class Cephalaspidomorphi) are modern jawless vertebrates that inhabit both marine and freshwater. Members of the superclass Gnathostomata are the jawed vertebrates. In addition to the tetrapods, this superclass includes two classes of fishes. Members of the class Chondrichthyes include the cartilaginous fishes, and members of the class Osteichthyes include the bony fishes.

4. The class Osteichthyes has two subclasses. The subclass Sarcopterygii includes the lungfishes, the coelacanths, and the rhipidistians; and the subclass Actinopterygii includes the ray-finned fishes. In the Actinopterygii, the teleosts are the modern bony fishes. Members of this very large group have adapted to virtually every available aquatic habitat.

5. Fishes show numerous adaptations to living in aquatic environments. These adaptations include an arrangement of body-wall muscles that creates locomotor waves in the body wall; mechanisms that constantly move water across gill surfaces; a countercurrent exchange mechanism to promote efficient gas exchange; buoyancy regulation; well-developed sensory receptors, including eyes, inner ears, and lateral-line receptors; mechanisms of osmoregulation; and mechanisms that help ensure successful reproduction.

6. Two evolutionary lineages in the Actinopterygii are very important. One of these resulted in the adaptive radiation of modern bony fishes, the teleosts. The second evolutionary line probably diverged from the Sarcopterygii. Adaptations that favored osteolepiform survival in early Devonian streams preadapted some osteolepiforms for terrestrial habitats.

SELECTED KEY TERMS

cloaca (p. 291)
countercurrent exchange
  mechanism (p. 287)
gill (visceral) arches (p. 287)
gill filaments (p. 287)
lateral-line system (p. 289)

operculum (p. 283)
pharyngeal lamellae (p. 287)
pneumatic sacs (p. 287)
ram ventilation (p. 287)
swim bladders (p. 283)

CRITICAL THINKING QUESTIONS

1. What characteristic of water makes it difficult to move through, but also makes support against gravity a minor consideration? How is a fish adapted for moving through water?

2. Could a fish drown? Explain. Would it make a difference if the fish were an open-ocean fish, such as a tuna, or a fish such as a freshwater perch?

3. Why is it a mistake to consider the cartilaginous skeleton of chondrichthians a primitive characteristic?

4. Would swim bladders with functional pneumatic ducts work well for a fish that lives at great depths? Why or why not?

5. What would happen to a deep-sea fish brought rapidly to the surface? Explain your answer in light of the fact that gas pressure in the swim bladders of some deep-sea fishes increases up to about 300 atmospheres.

ONLINE LEARNING CENTER

Visit our Online Learning Center (OLC) at www.mhhe.com/zoology (click on this book's title) to find the following chapter-related materials:

- CHAPTER QUIZZING
- ANIMATION
    Evolution of Fish Jaws
- RELATED WEB LINKS
    Subphylum Vertebrata
        Systematics and Characteristics of the Craniates
        Early Vertebrates
        Subphylum Hyperotreti
        Class Myxini

CHAPTER 19

AMPHIBIANS:

THE FIRST TERRESTRIAL VERTEBRATES

Outline

Concepts

1. Adaptations that favored the survival of fishes during periodic droughts preadapted vertebrates to life on land. There were two lineages of ancient amphibians: one gave rise to modern amphibians, and the other lineage resulted in amniote vertebrates.
2. Modern amphibians belong to three orders. Caudata contains the salamanders, Gymnophiona contains the caecilians, and Anura contains the frogs and toads.
3. Although amphibians are restricted to moist habitats, most spend much of their adult life on land. Virtually all amphibian body systems show adaptations for living on land.
4. Eggs and developmental stages that were resistant to drying probably evolved in some ancient amphibians. This development was a major step in vertebrate evolution, because it weakened vertebrate ties to moist environments.

EVOLUTIONARY PERSPECTIVE

Who, while walking along the edge of a pond or stream, has not been startled by the "plop" of an equally startled frog jumping to the safety of its watery retreat? Or who has not marveled at the sounds of a chorus of frogs breaking through an otherwise silent spring evening? These experiences and others like them have led some to spend their lives studying members of the class Amphibia (am-fib′e-ah) (L. *amphibia*, living a double life): frogs, toads, salamanders, and caecilians (figure 19.1). The class name implies that amphibians either move back and forth between water and land, or live one stage of their life in water and another on land. One or both of these descriptions is accurate for most amphibians.

Amphibians are **tetrapods** (Gr. *tetra*, four + *podos*, foot). The name is derived from the presence of four muscular limbs and feet with toes and fingers (digits). Currently, the trend is to reserve the term tetrapod for "crown group animals." The tetrapod crown group includes the extant (living) tetrapods plus their most recent common ancestor. Tetrapod, in this sense, refers to living amphibians (often called **Lissamphibia** [lis′am-fib′e-ah]), the reptiles, birds, and mammals, and the common ancestor of these groups. A host of extinct amphibians is excluded from the tetrapod designation. Stegocephalia is a more inclusive taxonomic designation that includes the tetrapods and their extinct four-legged relatives.

PHYLOGENETIC RELATIONSHIPS

Chapter 18 describes ideas regarding the origin of amphibians from ancient sarcopterygians. The fossil record provides evidence of many extinct taxa, and no one knows what

> *T*his chapter contains evolutionary concepts, which are set off in this font.

FIGURE 19.1

Class Amphibia.   Amphibians, like this tree frog (*Hyla andersoni*), are common vertebrates in most terrestrial and freshwater habitats. Their ancestors were the first terrestrial vertebrates.

TABLE 19.1
CLASSIFICATION OF LIVING AMPHIBIANS

Class Amphibia (am-fib′e-ah)
Skin with mucoid secretions and lacking epidermal scales, feathers, or hair; larvae usually aquatic and undergo metamorphosis to the adult; two atrial chambers in the heart; one cervical and one sacral vertebra.

   Order Caudata (kaw′dat-ah)
   Long tail, two pairs of limbs; lack middle ear. Salamanders, newts.
   Order Gymnophiona (jim″no-fi′o-nah)
   Elongate, limbless; segmented by annular grooves; specialized for burrowing; tail short and pointed; rudimentary left lung. Caecilians.
   Order Anura (ah-noor′ah) or **Salientia** (sā′le-en′tia)
   Tailless; elongate hindlimbs modified for jumping and swimming; five to nine presacral vertebrae with transverse processes (except the first); postsacral vertebrae fused into rodlike urostyle; tympanum and larynx well developed. Frogs, toads.

SURVEY OF AMPHIBIANS

Amphibians occur on all continents except Antarctica, but they are absent from many oceanic islands. The three thousand modern species are a mere remnant of this once-diverse group. Modern amphibians belong to three orders: Caudata, the salamanders; Anura, the frogs and toads; and Gymnophiona, the caecilians (table 19.1).

ORDER CAUDATA

Members of the order Caudata (kaw′dat-ah) (L. *cauda*, tail + Gr. *ata*, to bear) are the salamanders. They possess a tail throughout life, and both pairs of legs, when present, are relatively unspecialized (figure 19.4).

Approximately 115 of the 350 described species of salamanders live in North America. Most terrestrial salamanders live in moist forest-floor litter and have aquatic larvae. A number of families live in caves, where constant temperature and moisture conditions create a nearly ideal environment. Salamanders in the family Plethodontidae are the most fully terrestrial salamanders in that their eggs are laid on land, and the young hatch as miniatures of the adult. Members of the family Salamandridae are commonly called newts. They spend most of their lives in water and frequently retain caudal fins. Salamanders range in length from only a few centimeters to 1.5 m (the Japanese giant salamander, *Andrias japonicus*). The largest North American salamander is the hellbender (*Cryptobranchus alleganiensis*), which reaches lengths of about 65 cm.

Most salamanders have internal fertilization. Males produce a pyramidal, gelatinous spermatophore that is capped with sperm and deposited on the substrate. Females pick up the sperm cap with the cloaca and store the sperm in a special pouch, the

---

animal was the first stegocephalian. The best-known and some of the earliest fossils were first discovered in Greenland in 1932. They are of a 400-million-year-old group called Ichthyostegalia. *Ichthyostega* (figure 19.2) was not the ancestor of all stegocephalians, but it has been influential in formulating ideas regarding what the ancestral animals must have been like. Important characteristics evident in these fossils include the loss of some cranial bones and the appearance of a mobile neck, the loss of opercular bones, a reduction of the notochord, the formation of a more rigid vertebral column, four muscular limbs with discrete digits, loss of fin rays, and the presence of a sacral vertebra that fuses the vertebral column and the pelvis (*see figure 18.22*).

Adaptive radiation of the stegocephalian lineage resulted in a variety of taxa. Later convergent evolution and widespread extinctions have clouded evolutionary pathways. The phylogenetic relationships among these groups are very controversial. Virtually all taxonomists do agree, however, that the extant amphibians are monophyletic and closely related to reptiles, birds, and mammals. The latter three groups are called the amniotes, a name derived from the presence of an amniotic egg that resists drying and allows development to occur in a terrestrial environment. These groups are covered in chapters 20 through 22. Figure 19.3 shows one interpretation of the evolutionary relationships in the stegocephalian lineage. It is important to reemphasize the fact that these relationships are highly controversial. This figure shows just one of a number of interpretations that are being debated.

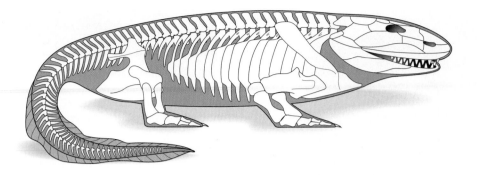

FIGURE 19.2

***Ichthyostega*: An Early Amphibian.** Fossils of this early amphibian were discovered in eastern Greenland in late Devonian deposits. The total length of the restored specimen is about 65 cm. Terrestrial adaptations are heavy pectoral and pelvic girdles and sturdy limbs that probably helped lift the body off the ground. Strong jaws suggest that it was a predator in shallow water, perhaps venturing onto shore. Other features include a skull that is similar in structure to ancient sarcopterygian fishes and a finlike tail. Note that bony rays dorsal to the spines of the vertebrae support the tail fin. This pattern is similar to the structure of the dorsal fins of fishes and is unknown in any other tetrapod.

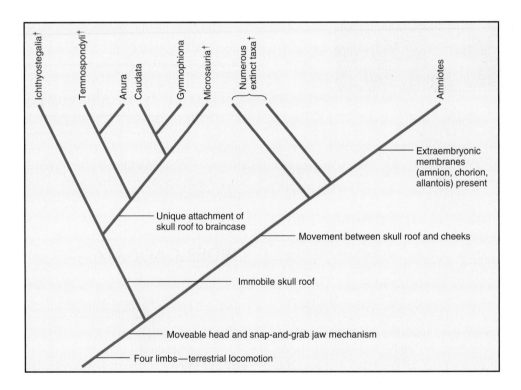

FIGURE 19.3

One Hypothesis of Evolutionary Relationships among the Stegocephalia. The earliest amphibians arose during the Devonian period. There are three classes of modern amphibians and numerous extinct taxa. The amniotic lineage of early tetrapods gave rise to reptiles, birds, mammals, and other extinct taxa. Daggers (†) indicate extinct taxa. Synapomorphic characters for lower taxonomic groups are not indicated.

spermatheca. Eggs are fertilized as they pass through the cloaca and are usually deposited singly, in clumps, or in strings (figure 19.5*a*). Larvae are similar to adults but smaller. They often possess external gills, a tail fin, larval dentition, and a rudimentary tongue (figure 19.5*b*). The aquatic larval stage usually metamorphoses into a terrestrial adult (figure 19.5*c*). Many other salamanders undergo incomplete metamorphosis and are paedomorphic (e.g., *Necturus*); that is, they become sexually mature while still showing larval characteristics.

ORDER GYMNOPHIONA

Members of the order Gymnophiona (jim″no-fi′o-nah) (Gr. *gymnos*, naked + *ophineos*, like a snake) are the caecilians (figure 19.6). Zoologists have described about 160 species confined to tropical regions. Caecilians are wormlike burrowers that feed on worms and other invertebrates in the soil. Caecilians appear segmented because of folds in the skin that overlie separations between muscle bundles. A retractile tentacle between their eyes

FIGURE 19.4
Order Caudata.   The Blue Ridge Spring salamander (*Gyrinophilus danielsi*).

and nostrils may transport chemicals from the environment to olfactory cells in the roof of the mouth. Skin covers the eyes; thus, caecilians are probably nearly blind.

Fertilization is internal in caecilians. Larval stages are often passed within the oviducts, where they scrape the inner lining of the oviducts with fetal teeth to feed. The young emerge from the female as miniature adults. Other caecilians lay eggs that develop into either aquatic larvae or embryos that develop on land.

ORDER ANURA (SALIENTIA)

The order Anura (ah-noor'ah) (Gr. *a*, without + *oura*, tail) or Salientia (sā″le-en'tia) includes about 3,500 species of frogs and toads. Anurans live in most moist environments, except in high latitudes and on some oceanic islands. A few even occur in very dry deserts. Adults lack tails, and caudal (tail) vertebrae fuse into a rodlike structure called the urostyle. Hindlimbs are long and muscular and end in webbed feet.

Anurans have diverse life histories. Fertilization is almost always external, and eggs and larvae are typically aquatic. Larval stages, called tadpoles, have well-developed tails. Their plump bodies lack limbs until near the end of their larval existence. Unlike adults, the larvae are herbivores and possess a proteinaceous, beaklike structure used in feeding. Anuran larvae undergo a drastic and rapid metamorphosis from the larval to the adult body form.

The distinction between "frog" and "toad" is more vernacular than scientific. "Toad" usually refers to anurans with relatively dry and warty skin that are more terrestrial than other members of the order. A number of distantly related taxa have these characteristics. True toads belong to the family Bufonidae (figure 19.7).

EVOLUTIONARY PRESSURES

Most amphibians divide their lives between freshwater and land. This divided life is reflected in body systems that show adaptations to both environments. In the water, amphibians are supported by water's buoyant properties, they exchange gases with the water, and they face the same osmoregulatory problems as freshwater fishes. On land, amphibians support themselves against gravity, exchange gases with the air, and tend to lose water to the air.

EXTERNAL STRUCTURE AND LOCOMOTION

Vertebrate skin protects against infective microorganisms, ultraviolet light, desiccation, and mechanical injury. As discussed later in this chapter, the skin of amphibians also functions in gas exchange, temperature regulation, and absorption and storage of water.

Amphibian skin lacks a covering of scales, feathers, or hair. It is, however, highly glandular, and its secretions aid in protection. These glands keep the skin moist to prevent drying. They also produce sticky secretions that help a male cling to a female during mating and produce toxic chemicals that discourage potential predators. The skin of many amphibians is smooth, although epidermal thickenings may produce warts, claws, or sandpapery textures, which are usually the result of keratin deposits or the formation of hard, bony areas.

Chromatophores are specialized cells in the epidermis and dermis of the skin that are responsible for skin color and color changes. Cryptic coloration, aposematic coloration, and mimicry are all common in amphibians.

Support and Movement

Water buoys and supports aquatic animals. The skeletons of fishes function primarily in protecting internal organs, providing points of attachment for muscles, and keeping the body from collapsing during movement. In terrestrial vertebrates, however, the skeleton is modified to provide support against gravity, and it must be strong enough to support the relatively powerful muscles that propel terrestrial vertebrates across land. The amphibian skull is flattened, is relatively smaller, and has fewer bony elements than the skull of fishes. These changes lighten the skull so it can be supported out of the water. Changes in jaw structure and musculature allow terrestrial vertebrates to crush prey held in the mouth.

The vertebral column of amphibians is modified to provide support and flexibility on land (figure 19.8). It acts somewhat like the arch of a suspension bridge by supporting the weight of the body between anterior and posterior paired appendages. Supportive processes called zygapophyses on each vertebra prevent twisting. Unlike fishes, amphibians have a neck. The first vertebra is a cervical vertebra, which moves against the back of the skull and allows the head to nod vertically. The last trunk vertebra is a sacral vertebra. This vertebra anchors the pelvic girdle to the vertebral column to provide increased support. A ventral plate of bone, called the sternum, is present in the anterior ventral trunk region

(a)

(b)

(c)

FIGURE 19.5

Order Caudata.   (*a*) Eggs, (*b*) larva, and (*c*) adult of the spotted salamander, *Ambystoma maculatum*. Adults feed on worms and small arthropods. Larvae are omnivores.

FIGURE 19.6

Order Gymnophiona.   A caecilian (*Ichthyophis glutinosus*).

FIGURE 19.7

Order Anura.   An American toad (*Bufo americanus*).

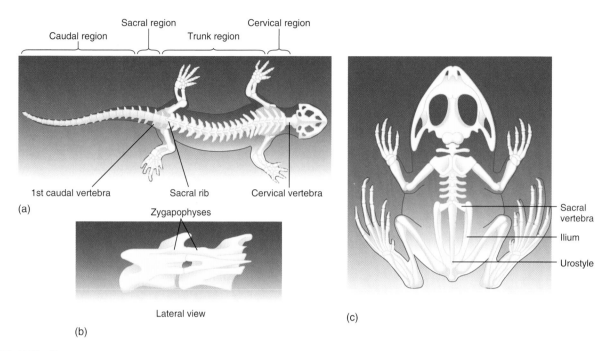

FIGURE 19.8

Skeletons of Amphibians.   (*a*) The salamander skeleton is divided into four regions: cervical, trunk, sacral, and caudal. (*b*) Interlocking processes, called zygapophyses, prevent twisting between vertebrae. (*c*) A frog skeleton shows adaptations for jumping. Note the long back legs and the firm attachment of the back legs to the vertebral column through the ilium and urostyle.

and supports the forelimbs and protects internal organs. It is reduced or absent in the Anura.

The origin of the bones of vertebrate appendages is not precisely known; however, similarities in the structures of the bones of the amphibian appendages and the bones of the fins of ancient sarcopterygian fishes suggest possible homologies (figure 19.9). Joints at the shoulder, hip, elbow, knee, wrist, and ankle allow freedom of movement and better contact with the substrate. The pelvic girdle of amphibians consists of three bones (the ilium, ischium, and pubis) that firmly attach pelvic appendages to the vertebral column. These bones, which are present in all tetrapods, but not fishes, are important for support on land.

Tetrapods depend more on appendages than on the body wall for locomotion. Thus, body-wall musculature is reduced, and appendicular musculature predominates. (Contrast, for example, what you eat in a fish dinner as compared to a plate of frog legs.)

Salamanders employ a relatively unspecialized form of locomotion that is reminiscent of the undulatory waves that pass along the body of a fish. Terrestrial salamanders also move by a pattern of limb and body movements in which the alternate movement of appendages results from muscle contractions that throw the body into a curve to advance the stride of a limb (figure 19.10). Caecilians have an accordion-like movement in which adjacent body parts push or pull forward at the same time. The long hindlimbs and the pelvic girdle of anurans are modified for jumping. The dorsal bone of the pelvis (the ilium) extends anteriorly and securely attaches to the vertebral column, and the urostyle extends posteriorly and attaches to the pelvis (*see figure 19.8*). These skeletal modifications stiffen the posterior half of the anuran. Long hindlimbs and powerful muscles form an efficient lever system for jumping. Elastic connective tissues and muscles attach the

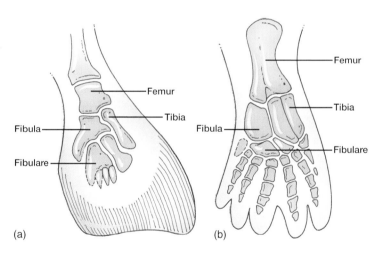

FIGURE 19.9

Origin of Tetrapod Appendages.   A comparison of (*a*) the fin bones of an osteolepiform and (*b*) the limb bones of a tetrapod suggests that the basic bone arrangements in tetrapod limbs were already present in primitive fishes.

pectoral girdle to the skull and vertebral column, and function as shock absorbers for landing on the forelimbs.

NUTRITION AND THE DIGESTIVE SYSTEM

Most adult amphibians are carnivores that feed on a wide variety of invertebrates. The diets of some anurans, however, are more diverse. For example, a bullfrog will prey on small mammals, birds,

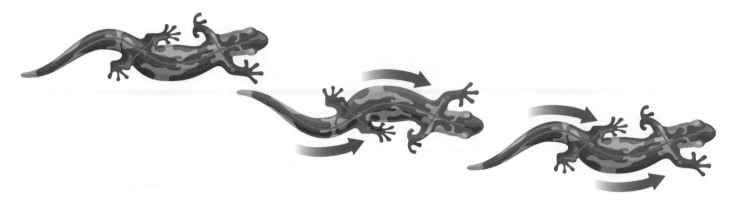

FIGURE 19.10

Salamander Locomotion. Pattern of leg movement in salamander locomotion. Blue arrows show leg movements.

and other anurans. The main factors that determine what amphibians will eat are prey size and availability. Most larvae are herbivorous and feed on algae and other plant matter. Most amphibians locate their prey by sight and simply wait for prey to pass by. Olfaction plays an important role in prey detection by aquatic salamanders and caecilians.

Many salamanders are relatively unspecialized in their feeding methods, using only their jaws to capture prey. Anurans and plethodontid salamanders, however, use their tongue and jaws in a flip-and-grab feeding mechanism (figure 19.11). A true tongue is first seen in amphibians. (The "tongue" of fishes is simply a fleshy fold on the floor of the mouth. Fish food is swallowed whole and not manipulated by the "tongue.") The amphibian tongue attaches at the anterior margin of the jaw and folds back over the floor of the mouth. Mucous and buccal glands on the tip of the tongue exude sticky secretions. When prey comes within range, an amphibian lunges forward and flicks out its tongue. The tongue turns over, and the lower jaw is depressed. The head tilts on its single cervical vertebra, which helps aim the strike. The tip of the tongue entraps the prey, and the tongue and prey are flicked back inside the mouth. All of this may happen in 0.05 to 0.15 second! The amphibian holds the prey by pressing it against teeth on the roof of the mouth, and the tongue and other muscles of the mouth push food toward the esophagus. The eyes sink downward during swallowing and help force food toward the esophagus.

CIRCULATION, GAS EXCHANGE, AND TEMPERATURE REGULATION

The circulatory system of amphibians shows remarkable adaptations for a life divided between aquatic and terrestrial habitats. The separation of pulmonary and systemic circuits is less efficient in amphibians than in lungfishes (figure 19.12; *see also figure 18.15b*). The atrium is partially divided in urodeles and completely divided in anurans. The ventricle has no septum. A spiral valve in the conus arteriosus or ventral aorta helps direct blood into pulmonary and systemic circuits. As discussed later, gas exchange occurs across the skin of amphibians, as well

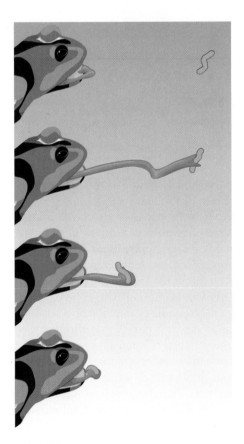

FIGURE 19.11

Flip-and-Grab Feeding in a Toad. The tongue attaches at the anterior margin of the toad's jaw and flips out to capture a prey item on its sticky secretions.

as in the lungs. Therefore, blood entering the right side of the heart is nearly as well oxygenated as blood entering the heart from the lungs. When an amphibian is completely submerged, all gas exchange occurs across the skin and other moist surfaces; therefore, blood coming into the right atrium has a higher oxygen concentration than blood returning to the left atrium from the lungs. Under these circumstances, blood vessels leading to the lungs constrict, reducing blood flow to the lungs and

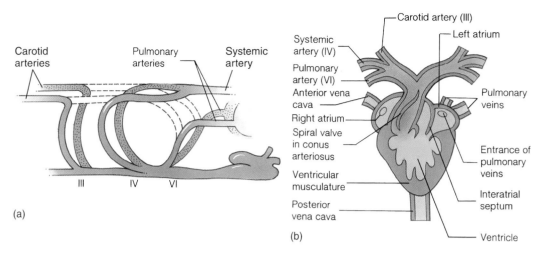

FIGURE 19.12

Diagrammatic Representation of an Anuran Circulatory System. (*a*) The Roman numerals indicate the various aortic arches. Vessels shown in dashed outline are lost during embryological development. (*b*) A ventral view of the heart.

conserving energy. *This adaptation is especially valuable for those frogs and salamanders that overwinter in the mud at the bottom of a pond.*

Adult amphibians have fewer aortic arches than fishes. After leaving the conus arteriosus, blood may enter the carotid artery (aortic arch III), which takes blood to the head; the systemic artery (aortic arch IV), which takes blood to the body; or the pulmonary artery (aortic arch VI).

In addition to a vascular system that circulates blood, amphibians have a well-developed lymphatic system of blind-ending vessels that returns fluids, proteins, and ions filtered from capillary beds in tissue spaces to the circulatory system. The lymphatic system also transports water absorbed across the skin. Unlike other vertebrates, amphibians have contractile vessels, called lymphatic hearts, that pump fluid through the lymphatic system. Lymphatic spaces between body-wall muscles and the skin transport and store water absorbed across the skin.

Gas Exchange

Terrestrial animals expend much less energy moving air across gas-exchange surfaces than do aquatic organisms because air contains 20 times more oxygen per unit volume than does water. On the other hand, exchanges of oxygen and carbon dioxide require moist surfaces, and exposure of respiratory surfaces to air may result in rapid water loss.

Anyone who has searched pond and stream banks for frogs knows that the skin of amphibians is moist. Amphibian skin is also richly supplied with capillary beds. These two factors permit the skin to function as a respiratory organ. Gas exchange across the skin is called **cutaneous respiration** and can occur either in water or on land. This ability allows a frog to spend the winter in the mud at the bottom of a pond. In salamanders, 30 to 90% of gas exchange occurs across the skin. Gas exchange also occurs across the moist surfaces of the mouth and pharynx. This **buccopharyngeal respiration** accounts for 1 to 7% of total gas exchange.

Most amphibians, except for plethodontid salamanders, possess lungs (figure 19.13*a*). The lungs of salamanders are relatively simple sacs. The lungs of anurans are subdivided, increasing the surface area for gas exchange. Pulmonary (lung) ventilation occurs by a **buccal pump** mechanism. Muscles of the mouth and pharynx create a positive pressure to force air into the lungs (figure 19.13*b–e*).

Cutaneous and buccopharyngeal respiration have a disadvantage in that their contribution to total gas exchange is relatively constant. The quantity of gas exchanged across these surfaces cannot be increased when the metabolic rate increases. Lungs, however, compensate for this shortcoming. As environmental temperature and activity increase, lungs contribute more to total gas exchange. At 5° C, approximately 70% of gas exchange occurs across the skin and mouth lining of a frog. At 25° C, the absolute quantity of oxygen exchanged across external body surfaces does not change significantly, but because pulmonary respiration increases, exchange across skin and mouth surfaces accounts for only about 30% of total oxygen exchange.

Amphibian larvae and some adults respire using external gills. Cartilaginous rods that form between embryonic pharyngeal slits support three pairs of gills. During metamorphosis, the gills are usually reabsorbed, pharyngeal slits close, and lungs become functional.

Temperature Regulation

Amphibians are ectothermic. (They depend on external heat sources to maintain body temperature [*see chapter 28.*]) Any poorly insulated aquatic animal, regardless of how much metabolic heat it produces, loses heat as quickly as it is produced because of powerful heat-absorbing properties of the water. Therefore, when amphibians are in water, they take on the temperature of their environment. On land, however, their body temperatures can differ from that of the environment.

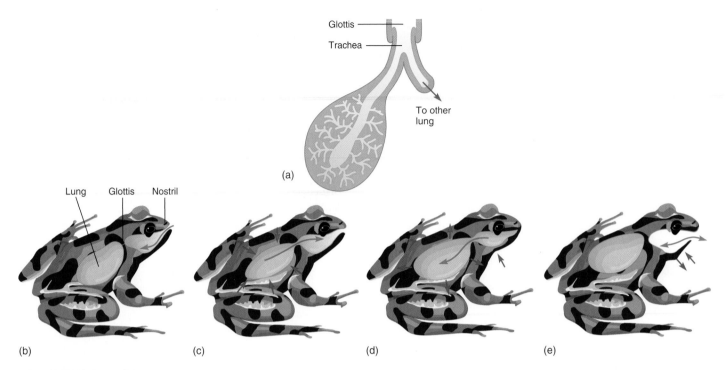

FIGURE 19.13

Amphibian Lung Structure, Buccal Pump, and Buccopharyngeal Ventilation.   (*a*) Lung structure of a frog. (*b*) With the opening of the respiratory tract (the glottis) closed, the floor of the mouth lowers, and air enters the mouth cavity. (*c*) The glottis opens, and the elasticity of the lungs and contraction of the body wall force air out of the lungs, over the top of air just brought into the mouth. (*d*) The mouth and nares close, and the floor of the mouth raises, forcing air into the lungs. (*e*) With the glottis closed, oscillations of the floor of the mouth exchange air in the mouth cavity to facilitate buccopharyngeal respiration. Blue arrows show air movements. Red arrows show movements of the lungs and body wall.

Temperature regulation is mainly behavioral. Some cooling results from evaporative heat loss. In addition, many amphibians are nocturnal and remain in cooler burrows or under moist leaf litter during the hottest part of the day. Amphibians may warm themselves by basking in the sun or on warm surfaces. Body temperatures may rise 10° C above the air temperature. Basking after a meal is common because increased body temperature increases the rate of all metabolic reactions—including digestive functions, growth, and the fat deposition necessary to survive periods of dormancy.

Amphibians' daily and seasonal environmental temperatures often fluctuate widely, and therefore, amphibians have correspondingly wide temperature tolerances. Critical temperature extremes for some salamanders lie between −2 and 27° C, and for some anurans between 3 and 41° C.

NERVOUS AND SENSORY FUNCTIONS

The nervous system of amphibians is similar to that of other vertebrates. The brain of adult vertebrates develops from three embryological subdivisions. In amphibians, the forebrain contains olfactory centers and regions that regulate color change and visceral functions. The midbrain contains a region called the optic tectum that assimilates sensory information and initiates motor responses. The midbrain also processes visual sensory information.

The hindbrain functions in motor coordination and in regulating heart rate and the mechanics of respiration.

Many amphibian sensory receptors are widely distributed over the skin. Some of these are simply bare nerve endings that respond to heat, cold, and pain. The lateral-line system is similar in structure to that found in fishes, and it is present in all aquatic larvae, aquatic adult salamanders, and some adult anurans. Lateral-line organs are distributed singly or in small groups along the lateral and dorsolateral surfaces of the body, especially the head. These receptors respond to low-frequency vibrations in the water and movements of the water relative to the animal. On land, however, lateral-line receptors are less important.

Chemoreception is an important sense for many amphibians. Chemoreceptors are in the nasal epithelium and the lining of the mouth, on the tongue, and over the skin. Olfaction is used in mate recognition, as well as in detecting noxious chemicals and in locating food.

Vision is one of the most important senses in amphibians because they are primarily sight feeders, often responding to the movements of their prey. (Caecilians are an obvious exception.) A number of adaptations allow the eyes of amphibians to function in terrestrial environments (figure 19.14). The eyes of some amphibians (i.e., anurans and some salamanders) are on the front of the head, providing the binocular vision and well-developed depth perception necessary for capturing prey. Other amphibians with smaller lateral eyes (some salamanders) lack binocular vision. The lower eyelid is movable, and it cleans and protects the

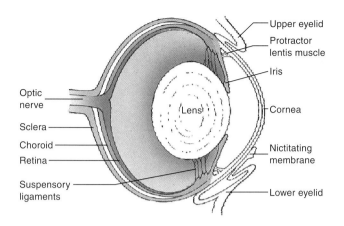

FIGURE 19.14

Amphibian Eye.   Longitudinal section of the eye of the leopard frog, *Rana pipiens*.

eye. Much of it is transparent and is called the **nictitating membrane.** When the eyeball retracts into the orbit of the skull, the nictitating membrane is drawn up over the cornea. In addition, orbital glands lubricate and wash the eye. Together, eyelids and glands keep the eye free of dust and other debris. The lens is large and nearly round. It is set back from the cornea, and a fold of epithelium called the iris surrounds it. The iris can dilate or constrict to control the size of the pupil.

Focusing, or accommodation, involves bending (refracting) light rays to a focal point on the retina. Light waves moving from air across the cornea are refracted because of the change in density between the two media. The lens provides further refraction. Like the eyes of most tetrapods, the amphibian eye focuses on distant objects when the eye is at rest. To focus on near objects, the protractor lentis muscle must move the lens forward (figure 19.14). Receptors called rods and cones are in the retina. Because cones are associated with color vision in some other vertebrates, their occurrence suggests that amphibians can distinguish between some wavelengths of light. The extent to which color vision is developed is unknown. The neuronal interconnections in the retina are complex and allow an amphibian to distinguish between flying insect prey, shadows that may warn of an approaching predator, and background movements, such as blades of grass moving with the wind.

The auditory system of amphibians is clearly an evolutionary adaptation to life on land. It transmits both substrate-borne vibrations and, in anurans, airborne vibrations. The ears of anurans consist of a tympanic membrane, a middle ear, and an inner ear. The tympanic membrane is a piece of integument stretched over a cartilaginous ring that receives airborne vibrations and transmits them to the middle ear, which is a chamber beneath the tympanic membrane. Abutting the tympanic membrane is a middle-ear ossicle (bone) called the stapes (columella), which transmits vibrations of the tympanic membrane into the inner ear (*see figure 24.20*). High-frequency (1,000 to 5,000 Hz) airborne vibrations are transmitted to the inner ear through the tympanic membrane. Low-frequency (100 to 1,000 Hz) substrate-borne vibrations are transmitted through the front appendages

and the pectoral girdle to the inner ear through a second ossicle, called the operculum.

Muscles attached to the operculum and stapes can lock either or both of these ossicles, allowing an anuran to screen out either high- or low-frequency sounds. *This mechanism is adaptive because anurans use low- and high-frequency sounds in different situations. Mating calls are high-frequency sounds that are of primary importance for only a part of the year (the breeding season). At other times, low-frequency sounds may warn of approaching predators.*

Salamanders lack a tympanic membrane and middle ear. They live in streams, ponds, caves, and beneath leaf litter. They have no mating calls, and the only sounds they hear are probably low-frequency vibrations transmitted through the substrate and skull to the stapes and inner ear.

The sense of equilibrium and balance is similar to that described for fishes in chapter 18. The inner ear of amphibians has semicircular canals that help detect rotational movements and other sensory patches that respond to gravity and detect linear acceleration and deceleration.

EXCRETION AND OSMOREGULATION

The kidneys of amphibians lie on either side of the dorsal aorta on the dorsal wall of the body cavity. A duct leads to the cloaca, and a storage structure, the urinary bladder, is a ventral outgrowth of the cloaca.

The nitrogenous waste product that amphibians excrete is either ammonia or urea. Amphibians that live in freshwater excrete ammonia. It is the immediate end product of protein metabolism; therefore, no energy is expended converting it into other products. The toxic effects of ammonia are avoided because it rapidly diffuses into the surrounding water. Amphibians that spend more time on land excrete urea that is produced from ammonia in the liver. Although urea is less toxic than ammonia, it still requires relatively large quantities of water for its excretion. Unlike ammonia, urea can be stored in the urinary bladder. Some amphibians excrete ammonia when in water and urea when on land.

One of the biggest problems that amphibians face is osmoregulation. In water, amphibians face the same osmoregulatory problems as freshwater fishes. They must rid the body of excess water and conserve essential ions. Amphibian kidneys produce large quantities of hypotonic urine, and the skin and walls of the urinary bladder transport Na^+, Cl^- and other ions into the blood.

On land, amphibians must conserve water. Adult amphibians do not replace water by intentional drinking, nor do they have the impermeable skin characteristic of other tetrapods or kidneys capable of producing a hypertonic urine. Instead, amphibians limit water loss by behavior that reduces exposure to desiccating conditions. Many terrestrial amphibians are nocturnal. During daylight hours, they retreat to areas of high humidity, such as under stones, or in logs, leaf mulch, or burrows. Water lost on nighttime foraging trips must be replaced by water uptake across the skin while in the retreat. Diurnal amphibians usually live in areas of high humidity and rehydrate themselves by entering the water. Many amphibians reduce evaporative water loss by reducing the amount of body surface exposed to air. They may curl their bodies and tails into tight coils and tuck their limbs close to their bodies (figure 19.15a). Individuals may form closely packed aggregations to reduce overall surface area.

Some amphibians have protective coverings that reduce water loss. Hardened regions of skin are resistant to water loss and may be used to plug entrances to burrows or other retreat openings to maintain high humidity in the retreat. Other amphibians prevent water loss by forming cocoons that encase the body during long periods of dormancy. Cocoons are made from outer layers of the skin that detach and become parchmentlike. These cocoons open only at the nares or the mouth and, in experimental situations, reduce water loss 20 to 50% over noncocooned individuals (figure 19.15b).

Paradoxically, the skin—the most important source of water loss—is also the most important structure for rehydration. When an amphibian flattens its body on moist surfaces, the skin, especially in the ventral pelvic region, absorbs water. The skin's permeability, vascularization, and epidermal sculpturing all promote water reabsorption. Minute channels increase surface area and spread water over surfaces not necessarily in direct contact with water.

Amphibians can also temporarily store water. Water accumulated in the urinary bladder and lymph sacs can be selectively reabsorbed to replace evaporative water loss. Amphibians living in very dry environments can store volumes of water equivalent to 35% of their total body weight.

REPRODUCTION, DEVELOPMENT, AND METAMORPHOSIS

Amphibians are dioecious, and ovaries and testes are located near the dorsal body wall. Fertilization is usually external, and because the developing eggs lack any resistant coverings, development is tied to moist habitats, usually water. A few anurans have terrestrial nests that are kept moist by being enveloped in foam or by being located near the water and subjected to flooding. In a few species, larval stages are passed in the egg membranes, and the immatures hatch into an adultlike body. The main exceptions to external fertilization in amphibians are the salamanders. Only about 10% of all salamanders have external fertilization. All others produce spermatophores, and fertilization is internal. Eggs may be deposited in soil or water or retained in the oviduct during development. All caecilians have internal fertilization, and 75% have internal development. Amphibian development, which zoologists have studied extensively, usually includes larval stages called tadpoles. Amphibian tadpoles often differ from the adults in mode of respiration, form of locomotion, and diet. These differences reduce competition between adults and larvae.

Interactions between internal (largely hormonal) controls and extrinsic factors determine the timing of reproductive activities. In temperate regions, temperature seems to be the most important environmental factor that induces physiological changes associated with breeding, and breeding periods are seasonal, occurring in spring two summer. In tropical regions, amphibian breeding correlates with rainy seasons.

(a)

(b)

FIGURE 19.15

Water Conservation by Anurans. (a) Daytime sleeping posture of the green tree frog, *Hyla cinerea*. The closely tucked appendages reduce exposed surface area. (b) The Australian burrowing frog, *Cyclorana alboguttatus*, in its burrow and water-retaining skin.

FIGURE 19.16

Amplexus. In frogs, eggs are released and fertilized when the male (smaller frog) mounts and grasps the female. This positioning is called amplexus.

Courtship behavior helps individuals locate breeding sites and identify potential mates. It also prepares individuals for reproduction and ensures that eggs are fertilized and deposited in locations that promote successful development.

Salamanders rely primarily on olfactory and visual cues in courtship and mating, whereas male vocalizations and tactile cues are important for anurans. Many species congregate in one location during times of intense breeding activity. Male vocalizations are usually species specific and function in the initial attraction and contact between mates. After that, tactile cues become more important. The male grasps the female—his forelimbs around her waist—so that they are oriented in the same direction, and the male is dorsal to the female. This positioning is called **amplexus** and may last from 1 to 24 hours (figure 19.16). During amplexus, the male releases sperm as the female releases eggs.

Little is known of caecilian breeding behavior. Males have an intromittent organ that is a modification of the cloacal wall, and fertilization is internal.

Vocalization

Sound production is primarily a reproductive function of male anurans. Advertisement calls attract females to breeding areas and announce to other males that a given territory is occupied. Advertisement calls are species specific, and the repertoire of calls for any one species is limited. The calls may also help induce psychological and physiological readiness to breed. Females respond by making reciprocation calls to indicate receptiveness. Release calls inform a partner that a frog is incapable of reproducing. Unresponsive females give release calls if a male attempts amplexus, as do males that have been mistakenly identified as female by another male. Distress calls are not associated with reproduction; either sex produces these calls in response to pain or being seized by a predator. The calls may be loud enough to cause a predator to release the frog. The distress call of the South American jungle frog, *Leptodactylus pentadactylus*, is a loud scream similar to the call of a cat in distress.

The sound-production apparatus of frogs consists of the larynx and its vocal cords. This laryngeal apparatus is well developed in males, who also possess a vocal sac. In the majority of frogs, vocal sacs develop as a diverticulum from the lining of the buccal cavity (figure 19.17). Air from the lungs is forced over the vocal cords and cartilages of the larynx, causing them to vibrate. Muscles control the tension of the vocal cords and regulate the frequency of the sound. Vocal sacs act as resonating structures and increase the volume of the sound.

The use of sound to attract mates is especially useful in organisms that occupy widely dispersed habitats and must come together for breeding. Because many species of frogs often converge at the same pond for breeding, finding a mate of the proper species could be chaotic. Vocalizations help to reduce the chaos.

Parental Care

Parental care increases the chances of any one egg developing, but it requires large energy expenditures on the part of the parent. The most common form of parental care in amphibians is atten-

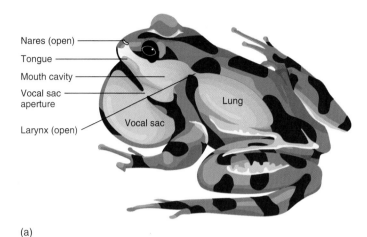

Nares (open)
Tongue
Mouth cavity
Vocal sac aperture
Larynx (open)
Vocal sac
Lung

(a)

(b)

FIGURE 19.17

Anuran Vocalization. (*a*) Generalized vocal apparatus of an anuran. (*b*) Inflated vocal sac of the Great Plains toad, *Bufo cognatus*.

dance of the egg clutch by either parent. Maternal care occurs in species with internal fertilization (predominantly salamanders and caecilians), and paternal care may occur in species with external fertilization (predominantly anurans). It may involve aeration of aquatic eggs, cleaning and/or moistening of terrestrial eggs, protection of eggs from predators, or removal of dead and infected eggs.

Eggs may be transported when development occurs on land. Females of the genus *Pipa* carry eggs on their backs. Two species of *Rheobatrachus* were discovered in Australia within the last 30 years. They have not been observed in the wild since the 1980s and are presumed extinct. *Rheobatrachus* females brooded tadpoles in their stomachs, and the young emerged from the females' mouths (figure 19.18)! Unfortunately, we will never know whether the female swallowed fertilized eggs and all development occurred in her stomach or whether she swallowed tadpoles. During brooding, the female's stomach expanded to fill most of her body cavity, and the stomach stopped producing digestive secretions. Viviparity and ovoviviparity occur primarily in salamanders and caecilians.

FIGURE 19.18

Parental Care of Young. Female *Rheobatrachus* with young emerging from her mouth. This Australian frog species has not been observed in the wild since the 1980s and is now presumed extinct.

Metamorphosis

Metamorphosis is a series of abrupt structural, physiological, and behavioral changes that transform a larva into an adult. A variety of environmental conditions, including crowding and food availability, influence the time required for metamorphosis. Most directly, however, metamorphosis is under the control of neurosecretions of the hypothalamus, hormones of the anterior lobe of the pituitary gland (the adenohypophysis), and the thyroid gland.

Morphological changes associated with the metamorphosis of caecilians and salamanders are relatively minor. Reproductive structures develop, gills are lost, and a caudal fin (when present) is lost. In the Anura, however, changes from the tadpole into the small frog are more dramatic (figure 19.19). Limbs and lungs develop, the tail is reabsorbed, the skin thickens, and marked changes in the head and digestive tract (associated with a new mode of nutrition) occur.

The mechanics of metamorphosis explain paedomorphosis in amphibians. Some salamanders are paedomorphic because cells fail to respond to thyroid hormones, whereas others are paedomorphic because they fail to produce the hormones associated with metamorphosis. In some salamander families, paedomorphosis is the rule. In other families, the occurrence of paedomorphosis is variable and influenced by environmental conditions.

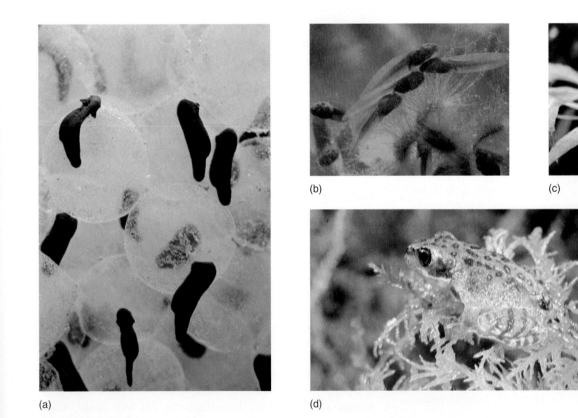

(a)　　　(b)　　　(c)　　　(d)

FIGURE 19.19

Events of Metamorphosis in the Frog, *Rana temporaria.* (*a*) Before metamorphosis. Prolactin secretion, controlled by the hypothalamus and the adenohypophysis, promotes the growth of larval structures. (*b–d*) Metamorphosis. The median eminence of the hypothalamus develops and initiates the secretion of thyroid-stimulating hormone (TSH). TSH begins to inhibit prolactin release. TSH causes the release of large quantities of T_4 and T_3, which promote the growth of limbs, reabsorption of the tail, and other changes of metamorphosis, resulting eventually in a young, adult frog.

WILDLIFE ALERT
Red Hills Salamander (*Phaeognathus hubrichti*)

VITAL STATISTICS

Classification: Phylum Chordata, class Amphibia, order Caudata, family Plethodontidae

Range: Southern Alabama

Habitat: Burrows on the slopes of moist, cool ravines shaded by an overstory of hardwood trees

Number remaining: Unknown

Status: Endangered throughout its range

NATURAL HISTORY AND ECOLOGICAL STATUS

The Red Hills salamander is a fairly large, terrestrial salamander that grows to about 255 mm in length (box figure 19.1). Its body color is gray to brownish without distinct markings. Sexual activity peaks in the spring, and eggs are deposited within the burrow system. There is no aquatic larval stage. Adults reach sexual maturity in two to three years and have a sexual longevity of two to five years.

BOX FIGURE 19.2 Range of Red Hills Salamander in Alabama.

The range of the Red Hills salamander is restricted to a narrow belt of two geological formations in southern Alabama (box figure 19.2). This salamander inhabits burrows on the slopes of moist, cool ravines shaded by an overstory of hardwood trees. These areas are underlain by a subsurface siltstone stratum containing many crevices, root tracings, and solution channels that the salamander uses. The topsoil habitat is typically sandy loam.

Primary threats to this species include its restricted range, loss of habitat, a low reproductive rate, and a limited capability for dispersal. Of the approximately 63,000 acres of remaining habitat, about 60% is currently owned or leased by paper companies that primarily use a clear-cutting system of forest management. This technique, along with the mechanical preparation of the site for logging, appears to completely destroy the habitat for the Red Hills salamander.

Needed management activities include: acquisition of major habitats, conservation agreements and easements to maintain the viability of remaining populations, continued research on biological and management-related questions, and care in the planning of forestry operations.

BOX FIGURE 19.1 Red Hills Salamander (*Phaeognathus hubrichti*).

AMPHIBIANS IN PERIL

Frogs and salamanders are disappearing at an alarming rate—and no one knows exactly why. Certainly, local events can result in the decimation of amphibian populations. Clear-cutting forests allows sunlight to reach forest floors and dry the moist habitats that amphibians require. Mining, drilling, industrial and agricultural operations, and urban sprawl also destroy habitat. Amphibian populations are disappearing, however, from vast areas of the earth, often in regions where local damage has not occurred.

Two additional culprits seem to be partly responsible for the decline of amphibian populations—acid deposition and ultraviolet radiation (*see chapter 6*). Amphibian embryos are especially susceptible to changes in the pH of their watery environment. A pH of 5 or less will kill most embryos. In the northern hemisphere, it is estimated that environments are 100 times more acidic than they were before the Industrial Revolution. Ultraviolet radiation, especially in the 280 to 320 nm range (UV-B), also kills amphibian eggs and embryos. Pollutants that contribute to acid deposition and deplete the ozone shield of the earth may be partly responsible for the alarming reductions in amphibian populations.

None of these explanations account for all problems with amphibian populations. Even in remote tropical regions, where none of these problems exist, amphibian populations are experiencing similar declines. The causes are unknown. It is imperative that the world's governments and funding agencies make funds available to help scientists discover the causes and reverse this alarming trend.

FURTHER PHYLOGENETIC CONSIDERATIONS

In the past, there has been some debate as to whether or not the three modern orders of amphibians (Lissamphibia) represent a monophyletic grouping. This debate is essentially over. Common characteristics such as the stapes/operculum complex, the importance of the skin in gas exchange, aspects of the structure of the skull and teeth, and molecular evidence have convinced most zoologists of the close relationships within this group. The exact nature of these relationships, however, remains controversial. The relationships depicted in figure 19.3 represent one of a number of hypotheses.

Chapters 20 through 22 focus on descendants of the amniote lineage (*see figure 19.3*). *Three sets of evolutionary changes in amphibian lineages allowed movement onto land. Two of these occurred early enough that they are found in all amphibians. One was the set of changes in the skeleton and muscles that allowed greater mobility on land. A second change involved a jaw mechanism and movable head that permitted effective exploitation of insect resources on land. A jaw-muscle arrangement that permitted rhipidistian fishes to snap, grab, and hold prey was adaptive when early tetrapods began feeding on insects in terrestrial environments. The third set of changes occurred in the amniote lineage—the development of an egg that was resistant to drying.* Although the **amniotic egg** is not completely independent of water, the extraembryonic membranes that form during development protect the embryo from desiccation, store wastes, and promote gas exchange. In addition, this egg has a leathery or calcified shell that is protective, yet porous enough to allow gas exchange with the environment. *These evolutionary events eventually resulted in the remaining three vertebrate groups: reptiles, birds, and mammals.*

SUMMARY

1. Terrestrial vertebrates are called tetrapods and probably arose from sarcopterygians. Most taxonomists agree that amphibians are monophyletic, but the exact relationships among living and extinct groups are controversial.

2. Members of the order Caudata are the salamanders. Salamanders are widely distributed, usually have internal fertilization, and may have aquatic larvae or direct development.

3. The order Gymnophiona contains the caecilians. Caecilians are tropical, wormlike burrowers. They have internal fertilization and many are viviparous.

4. Frogs and toads comprise the order Anura. Anurans lack tails and possess adaptations for jumping and swimming. External fertilization results in tadpole larvae, which metamorphose to adults.

5. The skin of amphibians is moist and functions in gas exchange, water regulation, and protection.

6. Skeletal and muscular systems of amphibians are adapted for movement on land.

7. Amphibians are carnivores that capture prey in their jaws or by using their tongues.

8. The circulatory system of amphibians is modified to accommodate the presence of lungs, gas exchange at the skin, and loss of gills in most adults.

9. Gas exchange is cutaneous, buccopharyngeal, and pulmonary. A buccal pump accomplishes pulmonary ventilation. A few amphibians retain gills as adults.

10. Sensory receptors of amphibians, especially the eye and ear, are adapted for functioning on land.

11. Amphibians excrete ammonia or urea. Ridding the body of excess water when in water and conserving water when on land are functions of the kidneys, the skin, and amphibian behavior.

12. The reproductive habits of amphibians are diverse. Many have external fertilization and development. Others have internal fertilization and development. Courtship, vocalizations, and parental care are common in some amphibians. The nervous and endocrine systems control metamorphosis.

13. An egg that is resistant to drying evolved in the amniote lineage, which is represented today by reptiles, birds, and mammals.

14. Local habitat destruction, global acid deposition and ozone depletion, and other unknown causes are resulting in an alarming reduction in amphibian populations around the world.

SELECTED KEY TERMS

amniotic egg (p. 310)
amplexus (p. 307)
buccal pump (p. 303)
buccopharyngeal respiration (p. 303)

cutaneous respiration (p. 303)
Lissamphibia (p. 296)
nictitating membrane (p. 305)
tetrapods (p. 296)

CRITICAL THINKING QUESTIONS

1. How are the skeletal and muscular systems of amphibians adapted for life on land?

2. Would the buccal pump be more important for an active amphibian or for one that is becoming inactive for the winter? Explain your answer.

3. Why is the separation of oxygenated and nonoxygenated blood in the heart not as important for amphibians as it is for other terrestrial vertebrates?

4. Explain how the skin of amphibians is used in temperature regulation, protection, gas exchange, and water regulation. Under what circumstances might cooling interfere with water regulation?

5. In what ways could anuran vocalizations have influenced the evolution of that order?

ONLINE LEARNING CENTER

Visit our Online Learning Center (OLC) at www.mhhe.com/zoology (click on this book's title) to find the following chapter-related materials:

- CHAPTER QUIZZING
- ANIMATION
     Transition to Land

- RELATED WEB LINKS
     Class Amphibia
          Order Gymnophiona
          Order Caudata
          Order Salienta
          Dissection Guides for Amphibians
          Conservation Issues Concerning Amphibians
- BOXED READINGS ON
     Poison Frogs of South America
     Neurotoxins
- SUGGESTED READINGS

CHAPTER 20

REPTILES:

THE FIRST AMNIOTES

Concepts

1. Adaptive radiation of primitive amniotes resulted in the three or four lineages of reptiles. These lineages have given rise to four orders of modern reptiles, the birds, and the mammals.
2. The class Reptilia has four orders: Testudines includes the turtles; Rhynchocephalia includes a single genus, *Sphenodon;* Squamata includes the lizards, the snakes, and the worm lizards; and Crocodilia includes the alligators and crocodiles.
3. Many reptiles have adaptations that allow them to spend most of their lives apart from standing or flowing water. These include adaptations for support and movement, feeding, gas exchange, temperature regulation, excretion, osmoregulation, and reproduction.
4. Two reptilian evolutionary lineages gave rise to two other vertebrate classes: Aves and Mammalia.

EVOLUTIONARY PERSPECTIVE

The earliest members of the class Reptilia (rep-til'e-ah) (L. *reptus*, to creep) were the first vertebrates to possess **amniotic eggs** (figure 20.1). Amniotic eggs have extraembryonic membranes that protect the embryo from desiccation, cushion the embryo, promote gas transfer, and store waste materials (figure 20.2). The amniotic eggs of reptiles and birds also have leathery or hard shells that protect the developing embryo, albumen that cushions and provides moisture and nutrients for the embryo, and yolk that supplies food to the embryo. All of these features are adaptations for development on land. (The amniotic egg is not, however, the only kind of land egg: some arthropods, amphibians, and even a few fishes have eggs that develop on land.) *The amniotic egg is the major synapomorphy that distinguishes the reptiles, birds, and mammals from other vertebrates. Even though the amniotic egg has played an important role in vertebrates' successful invasion of terrestrial habitats, it is just one of many reptilian adaptations that have allowed members of this class to flourish on land. Other adaptations for terrestrialism that will be described in this chapter include an impervious skin, horny nails for digging and locomotion, water-conserving kidneys, and enlarged lungs. Reptiles have also lost the lateral-line system of fishes and amphibians. Living representatives of the class Reptilia include the turtles, lizards, snakes, worm lizards, crocodilians, and the tuatarans (table 20.1).*

Even though fossil records of many reptiles are abundant, much remains to be learned of reptilian origins. *As indicated by the circle at the base of the cladogram in figure 20.3, the ancestral amniote has not yet been discovered. The adaptive radiation*

This chapter contains evolutionary concepts, which are set off in this font.

FIGURE 20.1

Class Reptilia. Members of the class Reptilia were the first vertebrates to possess amniotic eggs, which develop free from standing or flowing water. Numerous other adaptations have allowed members of this class to flourish on land. A Nile crocodile (*Crocodylus niloticus*) is shown here.

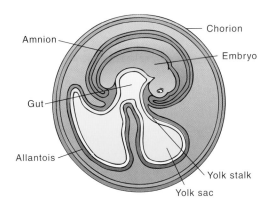

FIGURE 20.2

The Amniotic Egg. The amniotic egg provides a series of extraembryonic membranes that protect the embryo from desiccation. In reptiles, birds, and one group of mammals, the egg is enclosed within a shell (not shown). The embryo develops at the surface of a mass of yolk. The amnion encloses the embryo in a fluid-filled sac and protects against shock and desiccation. The chorion is nearer the shell and becomes highly vascular and aids in gas exchange. The allantois is a ventral outgrowth of the gut and stores nitrogenous wastes (e.g., uric acid).

of the early amniotes began in the late Carboniferous and early Permian periods. This time coincided with the adaptive radiation of terrestrial insects, the major prey of early amniotes. The adaptive radiation of the amniotes resulted in the lineages described in the paragraphs that follow. Skull structure, particularly the modifications in jaw muscle attachment, is one way these lineages are distinguished (figure 20.4).

Reptiles in the subclass Anapsida (Gr. *an*, without + *hapsis*, arch) lack openings or fenestrae in the temporal (posterolateral) region of the skull. The turtles represent this lineage today. *Recent evidence suggests that the anapsid lineage probably does not have close evolutionary ties to other reptiles. Changes have occurred in their long evolutionary history, but the fundamental form of their skull and shell is recognizable in 200-million-year-*

TABLE 20.1
CLASSIFICATION OF LIVING REPTILES

Class Reptilia* (rep-til′e-ah)
Dry skin with epidermal scales; skull with one point of articulation with the vertebral column (occipital condyle); respiration via lungs; metanephric kidneys; internal fertilization; amniotic eggs.

 Order Testudines (tes-tu′din-ez) or **Chelonia** (ki-lo′ne-ah)
 Teeth absent in adults and replaced by a horny beak; short, broad body; shell consisting of a dorsal carapace and ventral plastron. Turtles.

 Order Rhynchocephalia (rin″ko-se-fa′le-ah)
 Contains very primitive; lizardlike reptiles; well-developed parietal eye. Two species survive in New Zealand. Tuataras.

 Order Squamata (skwa-ma′tah)
 Recognized by specific characteristics of the skull and jaws (temporal arch reduced or absent and quadrate movable or secondarily fixed); the most successful and diverse group of living reptiles. Snakes, lizards, worm lizards.

 Order Crocodilia (krok″o-dil′e-ah)
 Elongate, muscular, and laterally compressed; tongue not protrusible; complete ventricular septum. Crocodiles, alligators, caimans, gavials.

*The class Reptilia as shown here is a paraphyletic grouping. A monophyletic representation would include the entire amniote lineage.

old fossils. Evidence of the anapsid lineage has been found in 245-million-year-old rocks from South Africa.

A second subclass of reptiles is Diapsida (Gr. *di*, two). Members of this group have upper and lower openings in the temporal region of the skull. *One group, the Lepidosauromorpha, includes modern snakes, lizards, and tuataras. A second group, Archosauromorpha, underwent extensive evolutionary radiation in the Mesozoic era and includes the dinosaurs. Most archosaurs are now extinct. Living archosaurs include the crocodilians and the dinosaurs' closest living relatives—the birds.*

Another subclass of reptiles are the Synapsida (Gr. *syn*, with). They possess a single dorsal opening in the temporal region of the skull. *Although no reptilian descendants of this group survive today, they are important because a group of synapsids, called therapsids, gave rise to the mammals.*

CLADISTIC INTERPRETATION OF THE AMNIOTIC LINEAGE

Cladistic taxonomic methods have resulted in reexamination and reinterpretation of the amniotic lineage. As shown in figure 19.3, the amniotic lineage is monophyletic. Figure 20.3 shows that the birds (traditionally the class Aves) and the mammals (traditionally the class Mammalia) share a common ancestor with the reptiles. The rules of cladistic analysis state that all descendants of a

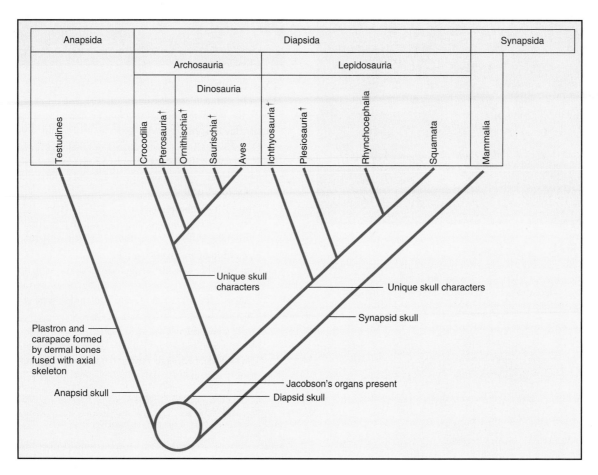

FIGURE 20.3

Amniote Phylogeny.   This cladogram shows one interpretation of amniote phylogeny. The circle at the base of the cladogram indicates that the ancestral amniotes are not described. Some researchers believe that the Diapsida is not a single lineage and that the Archosauria and Lepidosauria should be elevated to subclass status. Synapomorphies used to distinguish lower taxa are not shown. Daggers (†) indicate some extinct taxa. Other numerous extinct taxa are not shown.

most recent common ancestor must be included in a particular taxon. Clearly, that is not the case with the traditional class Reptilia; the birds and mammals are excluded, even though they share a common ancestry with reptiles. According to cladistic interpretations, birds should be classified as "reptiles" with their closest relatives, the dinosaurs. Similarly, cladistic interpretations take into account the close relationships of the mammals and ancient synapsid reptiles.

Evolutionary systematists disagree with cladists' interpretations. They contend that both the birds and the mammals have important morphological, behavioral, and ecological characteristics (e.g., feathers and endothermy in the birds; hair, mammary glands, and endothermy in mammals) that warrant their assignment to separate classes. In effect, evolutionary systematists weigh these characters and conclude that they are of overriding importance in the taxonomy of these groups.

This text presents the traditional interpretation of amniote classification. This may change in future editions, however, since the disagreements between cladists and evolutionary systematists will probably continue. Further challenges to traditional amniote classification are coming from the analysis of molecular data.

SURVEY OF THE REPTILES

Reptiles are characterized by a skull with one surface (condyle) for articulation with the first neck vertebra, respiration by lungs, metanephric kidneys, internal fertilization, and amniotic eggs. Reptiles also have dry skin with keratinized epidermal scales. **Keratin** is a resistant protein found in epidermally derived structures of amniotes. It is protective, and when chemically bonded to phospholipids, prevents water loss across body surfaces. Members of three of the four orders described here live on all continents except Antarctica. However, reptiles are a dominant part of any major ecosystem only in tropical and subtropical environments. There are 17 orders of reptiles, but members of most orders are extinct. The four orders containing living representatives are described next (*see table 20.1*).

ORDER TESTUDINES (CHELONIA)

Members of the order Testudines (tes-tu′din-ez) (L. *testudo*, tortoise), or Chelonia (ki-lo′ne-ah) (Gr. *chelone*, tortoise), are the turtles. The approximately 225 species of turtles are characterized

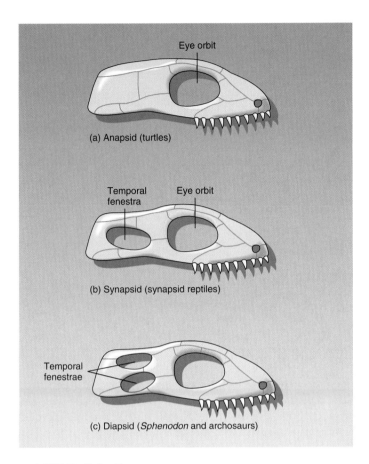

FIGURE 20.4

Amniote Skull Characteristics.   Amniotes are classified according to skull characteristics and jaw muscle attachment. (*a*) Anapsid skulls lack openings (fenestrae) in the temporal region. This kind of skull is characteristic of turtles. (*b*) Synapsid skulls have a single temporal opening and are characteristic of the lineage of amniotes leading to mammals. (*c*) Diapsid skulls have two temporal openings. This kind of skull is characteristic of lizards, snakes, worm lizards, the tuatara, and birds.

by a bony shell, limbs articulating internally to the ribs, and a keratinized beak rather than teeth. The dorsal portion of the shell is the **carapace,** which forms from a fusion of vertebrae, expanded ribs, and bones in the dermis of the skin. Keratin covers the bone of the carapace. The ventral portion of the shell is the **plastron.** It forms from bones of the pectoral girdle and dermal bone, and keratin also covers it (figure 20.5). In some turtles, such as the North American box turtle (*Terrapene*), the shell has flexible areas, or hinges, that allow the anterior and posterior edges of the plastron to be raised. The hinge allows the shell openings to close when the turtle withdraws into the shell. Turtles have eight cervical vertebrae that can be articulated into an S-shaped configuration, which allows the head to be drawn into the shell.

Turtles have long life spans. Most reach sexual maturity after seven or eight years, and live 14 or more years. Large tortoises, like those of the Galápagos Islands, may live in excess of 100 years (*see figure 4.3*). (Tortoises are entirely terrestrial and lack webbing in their feet.) All turtles are oviparous. Females use their hindlimbs to excavate nests in the soil. There they lay and

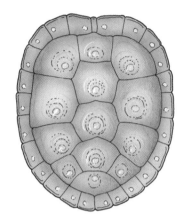

(a)

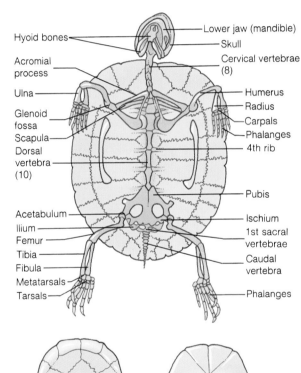

(b)

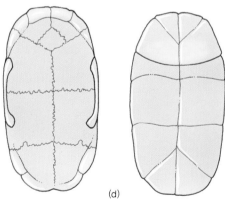

(c)   (d)

FIGURE 20.5

Skeleton of a Turtle.   (*a*) Dorsal view of the carapace. (*b*) Ventral view of the carapace and appendicular skeleton. Keratin covers the carapace, which is composed of fused vertebrae, expanded ribs, and dermal bone. (*c*) Dorsal view of the plastron. (*d*) Ventral view of the plastron. The plastron forms from dermal bone and bone of the pectoral girdle. It is also covered by keratin.

FIGURE 20.6
Order Testudines. Green sea turtles (*Chelonia mydas*) nest every two to four years and migrate many miles to nesting beaches in the Caribbean and South Atlantic Oceans.

FIGURE 20.7
Order Rhynchocephalia. The tuatara (*Sphenodon punctatus*).

cover with soil clutches of 5 to 100 eggs. Development takes from four weeks to one year, and the parent does not attend to the eggs during this time. The young are independent of the parent at hatching.

In recent years, turtle conservation programs have been enacted. Slow rates of growth and long juvenile periods make turtles vulnerable to extinction in the face of high mortality rates. Turtle hunting and predation on young turtles and turtle nests by dogs and other animals have severely threatened some species—in particular, sea turtles that nest on certain beaches year after year. Conservation of sea turtles is complicated by their having ranges of thousands of square kilometers of ocean, so that protective areas must include waters under the jurisdiction of many different nations (figure 20.6).

ORDER RHYNCHOCEPHALIA

The two surviving species of the order Rhynchocephalia (rin"ko-se-fa'le-ah) (Gr. *rhynchos*, snout + *kephale*, head) are the tuataras (*Sphenodon punctatus* and *S. guntheri*) (figure 20.7). These superficially lizardlike reptiles are virtually unchanged from extinct relatives that were present at the beginning of the Mesozoic era, nearly 200 million years ago. Tooth attachment and structure distinguish the tuataras from other reptiles. Two rows of teeth on the upper jaw and a single row of teeth in the lower jaw produce a shearing bite that can decapitate a small bird. Formerly more widely distributed in New Zealand, the tuataras fell prey to human influences and domestic animals. They are now present only on remote offshore islands and are protected by New Zealand law. They are oviparous and share underground burrows with ground-nesting seabirds. Tuataras venture out of their burrows at dusk and dawn to feed on insects or, occasionally, small vertebrates.

ORDER SQUAMATA

The order Squamata (skwa-ma'tah) (L. *squama*, scale + *ata*, to bear) is divided into three suborders. Ancestral members of these suborders originated in the lepidosaur lineage about 150 million years ago and diverged into numerous modern forms.

Suborder Sauria—The Lizards

About 3,300 species of lizards are in the suborder Sauria (sawr'e-ah) (Gr. *sauro*, lizard). In contrast to snakes, lizards usually have two pairs of legs, and their upper and lower jaws unite anteriorly. The few lizards that are legless retain remnants of a pectoral girdle and sternum. Lizards vary in length from only a few centimeters to as large as 3 m. Many lizards live on surface substrates and retreat under rocks or logs when necessary. Others are burrowers or tree dwellers. Most lizards are oviparous; some are ovoviviparous or viviparous. They usually deposit eggs under rocks or debris or in burrows.

Geckos, commonly found on the walls of human dwellings in semitropical areas, are short and stout. They are nocturnal, and unlike most lizards, are capable of clicking vocalizations. Their large eyes, with pupils that contract to a narrow slit during the day and dilate widely at night, are adapted for night vision. Adhesive disks on their digits aid in clinging to trees and walls.

Iguanas have robust bodies, short necks, and distinct heads. This group includes the marine iguanas of the Galápagos Islands and the flying dragons (*Draco*) of Southeast Asia. The latter have lateral folds of skin supported by ribs. Like the ribs of an umbrella, the ribs of *Draco* can expand to form a gliding surface. When this lizard launches itself from a tree, it can glide 30 m or more!

Another group of iguanas, the chameleons, lives mainly in Africa and India. Chameleons are adapted to arboreal lifestyles and use a long, sticky tongue to capture insects. *Anolis*, or the "pet-store chameleon," is also an iguanid, but is not a true chameleon. Chameleons and *Anolis* are well known for their ability to change

FIGURE 20.8
Order Squamata.   The gila monster (*Heloderma suspectum*) is a poisonous lizard of southwestern North America.

FIGURE 20.9
Order Squamata.   An amphisbaenian "worm lizard" (*Amphisbaenia alba*), sometimes called a two-headed snake.

color in response to illumination, temperature, or their behavioral state.

The only venomous lizards are the gila monster (*Heloderma suspectum*) (figure 20.8) and the Mexican beaded lizard (*Heloderma horridum*). These heavy-bodied lizards are native to southwestern North America. Venom is released into grooves on the surface of teeth and introduced into prey as the lizard chews. Lizard bites are seldom fatal to humans.

Suborder Serpentes—The Snakes

About 2,300 species are in the suborder Serpentes (ser-pen'tez) (L. *serpere*, to crawl). Although the vast majority of snakes are not dangerous to humans, about three hundred species are venomous. Worldwide, 30,000 to 40,000 people die from snake bites each year. Most of these deaths are in Southeast Asia. In the United States, fewer than one hundred people die each year from snake bites.

Snakes are elongate and lack limbs, although vestigial pelvic girdles and appendages are present in pythons and boas. The skeleton may contain more than two hundred vertebrae and pairs of ribs. Joints between vertebrae make the body very flexible. Snakes possess skull adaptations that facilitate swallowing large prey. These adaptations include upper jaws that are movable on the skull, and upper and lower jaws that are loosely joined so that each half of the jaw can move independently. Other differences between lizards and snakes include the mechanism for focusing the eyes and the morphology of the retina. Elongation and narrowing of the body has resulted in the reduction or loss of the left lung and displacement of the gallbladder, the right kidney, and often, the gonads. Most snakes are oviparous, although a few, such as the New World boas and garter snakes, give birth to live young.

Zoologists debate the evolutionary origin of the snakes. The earliest fossils are from 135-million-year-old Cretaceous deposits. Some zoologists believe that the earliest snakes were

burrowers. Loss of appendages and changes in eye structure could be adaptations similar to those seen in caecilians (see figure 19.6). Others believe that the loss of legs could be adaptive if early snakes were aquatic or lived in densely tangled vegetation. Recently discovered fossils of a primitive (90-million-year-old) snake from Australia support this second hypothesis. These snakes were large and lacked burrowing adaptations.

Suborder Amphisbaenia—Worm Lizards

About 135 species are in the suborder Amphisbaenia (am″ fis-be′ne-ah) (Gr. *amphi*, double + *baen*, to walk). They are specialized burrowers that live in soils of Africa, South America, the Caribbean, and the Mideast (figure 20.9). Most are legless, and their skulls are wedge or shovel shaped. A single median tooth in the upper jaw distinguishes amphisbaenians from all other vertebrates. The skin of amphisbaenians has ringlike folds called annuli and loosely attaches to the body wall. Muscles of the skin cause it to telescope and bulge outward, forming an anchor against a burrow wall. Amphisbaenians move easily forward or backward—thus, the suborder name. They feed on worms and small insects and are oviparous.

ORDER CROCODILIA

The order Crocodilia (krok″o-dil′e-ah) (Gr. *krokodeilos*, lizard) has 21 species. Along with dinosaurs, crocodilians are derived from the archosaurs and distinguished from other reptiles by certain skull characteristics: openings in the skull in front of the eye, triangular rather than circular eye orbits, and laterally compressed teeth. Living crocodilians include the alligators, crocodiles, gavials, and caimans.

Crocodilians have not changed much over their 170-million-year history. The snout is elongate and often used to capture

FIGURE 20.10

Chuckwalla (*Sauromalus obesus*). Many reptiles, like this chuck-walla, possess adaptations that make life apart from standing or running water possible.

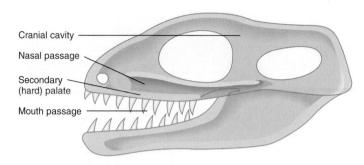

FIGURE 20.11

Secondary Palate. Sagittal section of the skull of a synapsid reptile, showing the secondary palate that separates the nasal and mouth cavities. Extension of the bones of the anterior skull forms the anterior portion of the secondary palate (the hard palate), and skin and soft connective tissues form the posterior portion of the secondary palate (the soft palate).

food by a sideways sweep of the head. The nostrils are at the tip of the snout, so the animal can breathe while mostly submerged. Air passageways of the head lead to the rear of the mouth and throat, and a flap of tissue near the back of the tongue forms a watertight seal that allows breathing without inhaling water in the mouth. A plate of bone, called the secondary palate, evolved in the archosaurs and separates the nasal and mouth passageways. The muscular, elongate, and laterally compressed tail is used for swimming, offensive and defensive maneuvers, and attacking prey. Teeth are used only for seizing prey. Food is swallowed whole, but if a prey item is too large, crocodilians tear it apart by holding onto a limb and rotating their bodies wildly until the prey is dismembered. The stomach is gizzardlike, and crocodilians swallow rocks and other objects as abrasives for breaking apart ingested food. Crocodilians are oviparous and display parental care of hatchlings that parallels that of birds. Nesting behavior and parental care may be traced back to the common ancestor of both groups.

EVOLUTIONARY PRESSURES

The lifestyles of most reptiles reveal striking adaptations for terrestrialism. For example, a lizard common to deserts of the southwestern United States—the chuckwalla (*Sauromalus obesus*)—survives during late summer when temperatures exceed 40° C (104° F) and when arid conditions wither plants and blossoms upon which chuckwallas browse (figure 20.10). To withstand these hot and dry conditions, chuckwallas disappear below ground and aestivate. Temperatures moderate during the winter, but little rain falls, so life on the desert surface is still not possible for the chuckwalla. The summer's sleep, therefore, merges into a winter's sleep. The chuckwalla does not emerge until March, when rain falls, and the desert explodes with greenery and flowers. The chuckwalla browses and drinks, storing water in large reservoirs under its skin. Chuckwallas are not easy prey. If threatened, a chuckwalla takes refuge in the nearest rock crevice. There, it inflates its lungs with air, increasing its girth and wedging itself against the rock walls of its refuge. Friction of its body scales against the rocks makes the chuckwalla nearly impossible to dislodge.

The adaptations that chuckwallas display are not exceptional for reptiles. This section discusses some of these adaptations that make life apart from an abundant water supply possible.

EXTERNAL STRUCTURE AND LOCOMOTION

Unlike that of amphibians, the skin of reptiles has no respiratory functions. Reptilian skin is thick, dry, and keratinized. Scales may be modified for various functions. For example, the large belly scales of snakes provide contact with the substrate during locomotion. Although reptilian skin is much less glandular than that of amphibians, secretions include pheromones that function in sex recognition and defense.

All reptiles periodically shed the outer, epidermal layers of the skin in a process called ecdysis. (The term ecdysis is also used for a similar, though unrelated, process in arthropods [*see figure 14.5*].) Because the blood supply to the skin does not extend into the epidermis, the outer epidermal cells lose contact with the blood supply and die. Movement of lymph between the inner and outer epidermal layers loosens the outer epidermis. Ecdysis generally begins in the head region, and in snakes and many lizards, the epidermal layers come off in one piece. In other lizards, smaller pieces of skin flake off. The frequency of ecdysis varies from one species to another, and it is greater in juveniles than adults.

The chromatophores of reptiles are primarily dermal in origin and function much like those of amphibians. Cryptic coloration, mimicry, and aposematic coloration occur in reptiles. Color and color change also function in sex recognition and thermoregulation.

Support and Movement

The skeletons of snakes, amphisbaenians, and turtles show modifications; however, in its general form, the reptilian skeleton is based on one inherited from ancient amphibians. The skeleton is highly ossified to provide greater support. The skull is longer than that of amphibians, and a plate of bone, the **secondary palate,** partially separates the nasal passages from the mouth cavity (figure 20.11). **As described earlier, the secondary palate evolved in**

FIGURE 20.12
Order Squamata.   A chameleon (*Chameleo chameleon*) using its tongue to capture prey. Note the prehensile tail.

archosaurs, where it was most likely an adaptation for breathing when the mouth was full of water or food. It is also present in other reptiles, although developed to a lesser extent. Longer snouts also permit greater development of olfactory epithelium and increased reliance on the sense of smell.

Reptiles have more cervical vertebrae than do amphibians. The first two cervical vertebrae (atlas and axis) provide greater freedom of movement for the head. An atlas articulates with a single condyle on the skull and facilitates nodding. An axis is modified for rotational movements. A variable number of other cervical vertebrae provide additional neck flexibility.

The ribs of reptiles may be highly modified. Those of turtles and the flying dragon were described previously. The ribs of snakes have muscular connections to large belly scales to aid locomotion. The cervical vertebrae of cobras have ribs that may be flared in aggressive displays.

Two or more sacral vertebrae attach the pelvic girdle to the vertebral column. The caudal vertebrae of many lizards possess a vertical fracture plane. When a lizard is grasped by the tail, caudal vertebrae can be broken, and a portion of the tail is lost. Tail loss, or **autotomy,** is an adaptation that allows a lizard to escape from a predator's grasp, or the disconnected, wiggling piece of tail may distract a predator from the lizard. The lizard later regenerates the lost portion of the tail.

Locomotion in primitive reptiles is similar to that of salamanders. The body is slung low between paired, stocky appendages, which extend laterally and move in the horizontal plane. The limbs of other reptiles are more elongate and slender and are held closer to the body. The knee and elbow joints rotate posteriorly; thus, the body is higher off the ground, and weight is supported vertically. Many prehistoric reptiles were bipedal, meaning that they walked on hindlimbs. They had a narrow pelvis

and a heavy, outstretched tail for balance. Bipedal locomotion freed the front appendages, which became adapted for prey capture or flight in some animals.

NUTRITION AND THE DIGESTIVE SYSTEM

Most reptiles are carnivores, although turtles will eat almost anything organic. The tongues of turtles and crocodilians are nonprotrusible and aid in swallowing. Like some anurans, some lizards and the tuatara have sticky tongues for capturing prey. The tongue extension of chameleons exceeds their body length (figure 20.12).

Probably the most remarkable adaptations of snakes involve modifications of the skull for feeding. The bones of the skull and jaws loosely join and may spread apart to ingest prey much larger than a snake's normal head size (figure 20.13a). The bones of the upper jaw are movable on the skull, and ligaments loosely join the halves of both of the upper and lower jaws anteriorly. Therefore, each half of the upper and lower jaws can move independently of one another. After a prey is captured, opposite sides of the upper and lower jaws are alternately thrust forward and retracted. Posteriorly pointing teeth prevent prey escape and help force the food into the esophagus. The glottis, the respiratory opening, is far forward so that the snake can breathe while slowly swallowing its prey.

Vipers (family Viperidae) possess hollow fangs on the maxillary bone at the anterior margin of the upper jaw (figure 20.13b). These fangs connect to venom glands that inject venom when the viper bites. The maxillary bone (upper jaw bone) of vipers is hinged so that when the snake's mouth is closed, the fangs fold back and lie along the upper jaw. When the mouth opens, the maxillary bone rotates and causes the fangs to swing down (figure 20.13c). Because the fangs project outward from the mouth, vipers

(a)

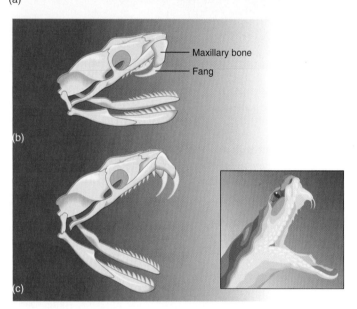

Maxillary bone

Fang

(b)

(c)

FIGURE 20.13

Feeding Adaptations of Snakes.   (*a*) A copperhead (*Ankistrodon*) ingesting a prey. Flexible joints allow the bones of the skull to separate during feeding. Note the pit organ just anterior to the eye. (*b*) The skull of a viper. The hinge mechanism of the jaw allows upper and lower bones on one side of the jaw to slide forward and backward alternately with bones of the other side. Posteriorly curved teeth hold prey as it is worked toward the esophagus. (*c*) Note that the maxillary bone, into which the fang is embedded, swings forward when the mouth opens.

may strike at objects of any size. Rear-fanged snakes (family Colubridae) possess grooved rear teeth. In those that are venomous, venom is channeled along these grooves and worked into prey to quiet them during swallowing. These snakes usually do not strike, and most are harmless to humans; however, the African boomslang (*Dispholidus typus*) has caused human fatalities. Coral snakes, sea snakes, and cobras have fangs that rigidly attach to the upper jaw in an erect position. When the mouth is closed, the fangs fit into a pocket in the outer gum of the lower jaw. Fangs are grooved or hollow, and contraction of muscles associated with venom glands injects venom into the fangs. Some cobras can "spit"

venom at their prey; if not washed from the eyes, the venom may cause blindness.

Venom glands are modified salivary glands. Most snake venoms are mixtures of neurotoxins and hemotoxins. The venoms of coral snakes, cobras, and sea snakes are primarily neurotoxins that attack nerve centers and cause respiratory paralysis. The venoms of vipers are primarily hemotoxins. They break up blood cells and attack blood vessel linings.

CIRCULATION, GAS EXCHANGE, AND TEMPERATURE REGULATION

The circulatory system of reptiles is based on that of amphibians. Because reptiles are, on average, larger than amphibians, their blood must travel under higher pressures to reach distant body parts. To take an extreme example, the blood of the dinosaur *Brachiosaurus* had to be pumped a distance of about 6 m from the heart to the head—mostly uphill! (The blood pressure of a giraffe is about double that of a human to move blood the 2 m from heart to head.)

Like amphibians, reptiles possess two atria that are completely separated in the adult and have veins from the body and lungs emptying into them. Except for turtles, the sinus venosus is no longer a chamber but has become a patch of cells that acts as a pacemaker. The ventricle of most reptiles is incompletely divided (figure 20.14). (Only in crocodilians is the ventricular septum complete.) The ventral aorta and the conus arteriosus divide during development and become three major arteries that leave the heart. A pulmonary artery leaves the ventral side of the ventricle and takes blood to the lungs. Two systemic arteries, one from the ventral side of the heart and the other from the dorsal side of the heart, take blood to the lower body and the head.

Blood low in oxygen enters the ventricle from the right atrium and leaves the heart through the pulmonary artery and moves to the lungs. Blood high in oxygen enters the ventricle from the lungs via pulmonary veins and the left atrium, and leaves the heart through left and right systemic arteries. The incomplete separation of the ventricle permits shunting of some blood away from the pulmonary circuit to the systemic circuit by the constriction of muscles associated with the pulmonary artery. This is advantageous because virtually all reptiles breathe intermittently. When turtles withdraw into their shells, their method of lung ventilation cannot function. They also stop breathing during diving. During periods of apnea ("no breathing"), blood flow to the lungs is limited, which conserves energy and permits more efficient use of the pulmonary oxygen supply.

Gas Exchange

Reptiles exchange repiratory gases across internal respiratory surfaces to avoid losing large quantities of water. A larynx is present; however, vocal cords are usually absent. Cartilages support the respiratory passages of reptiles, and lungs are partitioned into spongelike, interconnected chambers. Lung chambers provide a large surface area for gas exchange.

In most reptiles, a negative-pressure mechanism is responsible for lung ventilation. A posterior movement of the ribs and the

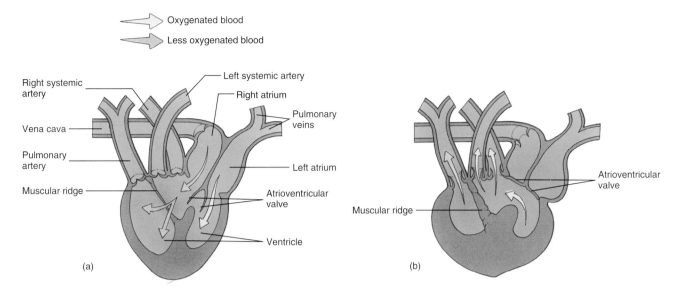

FIGURE 20.14

Heart and Major Arteries of a Lizard. (*a*) When the atria contract, blood enters the ventricle. An atrioventricular valve prevents the mixing of oxygenated and less oxygenated blood across the incompletely separated ventricle. (*b*) When the ventricle contracts, a muscular ridge closes to direct oxygenated blood to the systemic arteries and less oxygenated blood to the pulmonary artery.

body wall expands the body cavity, decreasing pressure in the lungs and drawing air into the lungs. Air is expelled by elastic recoil of the lungs and forward movements of the ribs and body wall, which compress the lungs. The ribs of turtles are a part of their shell; thus, movements of the body wall to which the ribs attach are impossible. Turtles exhale by contracting muscles that force the viscera upward, compressing the lungs. They inhale by contracting muscles that increase the volume of the visceral cavity, creating negative pressure to draw air into the lungs.

Temperature Regulation

Unlike aquatic animals, terrestrial animals may face temperature extremes (−65 to 70° C) that are incompatible with life. Temperature regulation, therefore, is important for animals that spend their entire lives out of water. Most reptiles use external heat sources for thermoregulation and are, therefore, ectotherms. Brooding Indian pythons, however, can use metabolic heat to increase body temperature. Female pythons coil around their eggs and elevate their body temperature as much as 7.3° C above the air temperature using metabolic heat sources.

Some reptiles can survive wide temperature fluctuations (e.g., −2 to 41° C for some turtles). To sustain activity, however, body temperatures are regulated within a narrow range, between 25 and 37° C. If that is not possible, the reptile usually seeks a retreat where body temperatures are likely to remain within the range compatible with life.

Most thermoregulatory activities of reptiles are behavioral, and they are best known in the lizards. To warm itself, a lizard orients itself at right angles to the sun's rays, often on a surface inclined toward the sun, and presses its body tightly to a warm surface to absorb heat by conduction. To cool itself, a lizard orients its body parallel to the sun's rays, seeks shade or burrows, or assumes an erect posture (legs extended and tail arched) to reduce conduction from warm surfaces. In hot climates, many reptiles are nocturnal.

Various physiological mechanisms also regulate body temperature. As temperatures rise, some reptiles begin panting, which releases heat through evaporative cooling. (Little evaporative cooling occurs across the dry skin of reptiles.) Marine iguanas divert blood to the skin while basking in the sun and warm up quickly. On diving into the cool ocean, however, marine iguanas reduce heart rate and blood flow to the skin, which slows heat loss. Chromatophores also aid in temperature regulation. Dispersed chromatophores (thus, a darker body) increase the rate of heat absorption.

In temperate regions, many reptiles withstand cold winter temperatures by entering torpor when body temperatures and metabolic rates decrease. Individuals that are usually solitary may migrate to a common site, called a hibernaculum, to spend the winter. Heat loss from individuals in hibernacula is reduced because the total surface area of many individuals clumped together is reduced compared to widely separated animals. Unlike true hibernators, the body temperatures of reptiles in torpor are not regulated, and if the winter is too cold or the retreat is too exposed, the animals can freeze and die. Death from freezing is an important cause of mortality for temperate reptiles.

NERVOUS AND SENSORY FUNCTIONS

The brain of reptiles is similar to the brains of other vertebrates. The cerebral hemispheres are somewhat larger than those of amphibians. This increased size is associated with an improved sense of smell. The optic lobes and the cerebellum are also enlarged, which reflects increased reliance on vision and more refined coordination of muscle functions.

The complexity of reptilian sensory systems is evidenced by a chameleon's method of feeding. Its protruding eyes swivel independently, and each has a different field of view. Initially, the brain keeps both images separate, but when an insect is spotted, both eyes converge on the prey. Binocular vision then provides the depth perception for determining whether or not the insect is within range of the chameleon's tongue (see figure 20.12).

Vision is the dominant sense in most reptiles, and their eyes are similar to those of amphibians (see figure 19.14). Snakes focus on nearby objects by moving the lens forward. Contraction of the iris places pressure on the gel-like vitreous body in the posterior region of the eye, and displacement of this gel pushes the lens forward. All other reptiles focus on nearby objects when the normally elliptical lens is made more spherical, as a result of ciliary muscles pressing the ciliary body against the lens. Reptiles have a greater number of cones than do amphibians and probably have well-developed color vision.

Upper and lower eyelids, a nictitating membrane, and a blood sinus protect and cleanse the surface of the eye. In snakes and some lizards, the upper and lower eyelids fuse in the embryo to form a protective window of clear skin, called the spectacle. (During ecdysis, the outer layers of the spectacle become clouded and impair the vision of snakes.) The blood sinus, which is at the base of the nictitating membrane, swells with blood to help force debris to the corner of the eye, where it may be rubbed out. Horned lizards squirt blood from their eyes by rupturing this sinus in a defensive maneuver to startle predators.

Some reptiles possess a **median (parietal) eye** that develops from outgrowths of the roof of the forebrain (see figure 24.30). In the tuatara, it is an eye with a lens, a nerve, and a retina. In other reptiles, the parietal eye is less developed. Parietal eyes are covered by skin and probably cannot form images. They can, however, differentiate light and dark periods and are used in orientation to the sun.

The structure of reptilian ears varies. The ears of snakes detect substrate vibrations. They lack a middle-ear cavity, an auditory tube, and a tympanic membrane. A bone of the jaw articulates with the stapes and receives substrate vibrations. Snakes can also detect airborne vibrations. In other reptiles, a tympanic membrane may be on the surface or in a small depression in the head. The inner ear of reptiles is similar to that of amphibians.

Olfactory senses are better developed in reptiles than in amphibians. In addition to the partial secondary palate providing more surface for olfactory epithelium, many reptiles possess blind-ending pouches that open through the secondary palate into the mouth cavity. These pouches, called **Jacobson's (vomeronasal) organs,** are in diapsid reptiles; however, they are best developed in the squamates. Jacobson's organs develop in embryonic crocodilians but are not present in adults of this group. Anapsids (turtles) lack these olfactory organs. The protrusible, forked tongues of snakes and lizards are accessory olfactory organs for sampling airborne chemicals. A snake's tongue flicks out and then moves to the Jacobson's organs, which perceive odor molecules. Tuataras use Jacobson's organs to taste objects held in the mouth.

Rattlesnakes and other pit vipers have heat-sensitive **pit organs** on each side of the face between the eye and nostril (see figure 20.13a). These depressions are lined with sensory epithelium and are used to detect objects with temperatures different from the snake's surroundings. Pit vipers are usually nocturnal, and their pits help them to locate small, warm-blooded prey.

EXCRETION AND OSMOREGULATION

The kidneys of embryonic reptiles are similar to those of fishes and amphibians. Life on land, increased body size, and higher metabolic rates, however, require kidneys capable of processing wastes with little water loss. A kidney with many more blood-filtering units, called nephrons, replaces the reptilian embryonic kidney during development. The functional kidneys of adult reptiles are called metanephric kidneys. Their function depends on a circulatory system that delivers more blood at greater pressures to filter large quantities of blood.

Most reptiles excrete uric acid. It is nontoxic, and being relatively insoluble in water, it precipitates in the excretory system. The urinary bladder or the cloacal walls reabsorb water, and the uric acid can be stored in a pastelike form. Utilization of uric acid as an excretory product also made possible the development of embryos in terrestrial environments, because nontoxic uric acid can be concentrated in egg membranes.

In addition to the excretory system's reabsorption of water, internal respiratory surfaces and relatively impermeable exposed surfaces reduce evaporative water loss. The behaviors that help regulate temperature also help conserve water. Nocturnal habits and avoiding hot surface temperatures during the day by burrowing reduce water loss. When water is available, many reptiles (e.g., chuckwallas) store large quantities of water in lymphatic spaces under the skin or in the urinary bladder. Many lizards possess salt glands below the eyes for ridding the body of excess salt.

REPRODUCTION AND DEVELOPMENT

Vertebrates could never be truly terrestrial until their reproduction and embryonic development became separate from standing or running water. For vertebrates, internal fertilization and the amniotic egg (see figure 20.2) made complete movement to land possible. The amniotic egg, however, is not completely independent of water. Pores in the eggshell permit gas exchange but also allow water to evaporate. Amniotic eggs require significant energy expenditures by parents. Parental care occurs in some reptiles and may involve maintaining relatively high humidity around the eggs. These eggs are often supplied with large quantities of yolk for long developmental periods, and parental energy and time are sometimes invested in the posthatching care of dependent young.

Accompanying the development of amniotic eggs is the necessity for internal fertilization. Fertilization must occur in the reproductive tract of the female before protective egg membranes are laid down around an egg. All male reptiles, except tuataras, possess an intromittent organ for introducing sperm into the female reproductive tract. Lizards and snakes possess paired hemipenes at the base of the tail that are erected by being turned inside out, like the finger of a glove.

Gonads lie in the abdominal cavity. In males, a pair of ducts delivers sperm to the cloaca. After copulation, sperm may be stored in a seminal receptacle in the female reproductive tract. Secretions of the seminal receptacle nourish the sperm and arrest their activity. Sperm may be stored for up to four years in some turtles, and up to six years in some snakes! In temperate latitudes, sperm can be stored over winter. Copulation may take place in the fall, when individuals congregate in hibernacula, and fertilization and development may occur in the spring, when temperatures favor successful development. Fertilization occurs in the upper regions of the oviduct, which leads from the ovary to the cloaca. Glandular regions of the oviduct secrete albumen and the eggshell. The shell is usually tough yet flexible. In some crocodilians, the eggshell is calcareous and rigid, like the eggshells of birds.

Parthenogenesis has been described in six families of lizards and one species of snakes. In these species, no males have been found. Populations of parthenogenetic females have higher reproductive potential than bisexual populations. A population that suffers high mortality over a cold winter can repopulate its habitat rapidly because all surviving individuals can produce offspring. This apparently offsets disadvantages of genetic uniformity resulting from parthenogenesis.

Reptiles often have complex reproductive behaviors that may involve males actively seeking out females. As in other animals, courtship functions in sexual recognition and behavioral and physiological preparation for reproduction. Head-bobbing displays by some male lizards reveal bright patches of color on the throat and enlarged folds of skin. Courtship in snakes is based primarily on tactile stimulation. Tail-waving displays are followed by the male running his chin along the female, entwining his body around her, and creating wavelike contractions that pass posteriorly to anteriorly along his body. Recent research indicates that lizards and snakes also use sex pheromones to assess the reproductive condition of a potential mate. Vocalizations are important only in crocodilians. During the breeding season, males are hostile and may bark or cough as territorial warnings to other males. Roaring vocalizations also attract females, and mating occurs in the water.

After they are laid, reptilian eggs are usually abandoned (figure 20.15). Virtually all turtles bury their eggs in the ground or in plant debris. Other reptiles lay their eggs under rocks, in debris, or in burrows. About one hundred species of reptiles have some degree of parental care of eggs. One example is the American alligator, *Alligator mississippiensis* (figure 20.16). The female builds a mound of mud and vegetation about 1 m high and 2 m in diameter. She hollows out the center of the mound, partially fills it with mud and debris, deposits her eggs in the cavity, and then covers the eggs. Temperature within the nest influences the sex of the hatchlings. Temperatures at or below 31.5° C result in female offspring. Temperatures between 32.5 and 33° C result in male offspring. Temperatures around 32° C result in both male and female offspring. (Similar temperature effects on sex determination are known in some lizards and many turtles.) The female remains in the vicinity of the nest throughout development to protect the eggs from predation. She frees hatchlings from the nest in response to their high-pitched calls and picks them up in her mouth to transport them to water. She may scoop shallow pools for the young and remain with them for up to two years. Young feed on scraps of food the female drops when she feeds and on small vertebrates and invertebrates that they catch on their own.

FIGURE 20.15
Reptile Eggs and Young.   This young giant Madagascar day gecko (*Phelsuma madagascariensis*) is hatching from its egg.

FIGURE 20.16
Parental Care in Reptiles.   A female American alligator (*Alligator mississippiensis*) tending to her nest.

FURTHER PHYLOGENETIC CONSIDERATIONS

The archosaur and synapsid lineages of ancient reptiles diverged from ancient amniotes about 280 million years ago and are ancestral to animals described in chapters 21 and 22 (see figure 20.3). The archosaur lineage not only included the dinosaurs and gave rise to crocodilians, but it also gave rise to two groups of fliers. The pterosaurs (Gr. *pteros*, wing + *sauros*, lizard) ranged from sparrow size to animals with wingspans of 13 m. An elongation of the fourth finger supported their membranous wings, their sternum was adapted for the attachment of flight muscles, and their bones were hollow to lighten the skeleton for flight. As presented in chapter 21, these adaptations are paralleled by,

WILDLIFE ALERT
Kemp's Ridley Sea Turtle (*Lepidochelys kempii*)

VITAL STATISTICS

Classification: Phylum Chordata, class Reptilia, order Testudines

Range: Gulf of Mexico and Atlantic Ocean

Habitat: Shallow coastal and estuarine waters, often associated with red mangrove swamps

Number remaining: Unknown

Status: Endangered

NATURAL HISTORY AND ECOLOGICAL STATUS

Kemp's Ridley sea turtle (*Lepidochelys kempii*) is a relatively small sea turtle. Adults range in size from 52 to 75 cm (carapace length) and 32 to 48 kg. The carapace is nearly round, and the turtle is olive gray dorsally and white or yellow ventrally (box figure 20.1). Its range includes all of the Gulf of Mexico and both coasts of the Atlantic Ocean (box figure 20.2). Adults are primarily found in the gulf. Most Kemp's Ridley sea turtles nest on 40 km of beach at Rancho Nuevo, Mexico. A few nests have been observed along the Texas coast. In 1998, 3,845 nests were counted in Mexico. Since females lay an average of 2.3 clutches of eggs, this number of nests corresponds to approximately 1,671 females. Since females do not lay eggs every year, the population is probably somewhat larger than these numbers suggest.

Kemp's Ridley sea turtles inhabit shallow coastal and estuarine waters. They feed primarily on crabs, shrimp, snails, sea urchins, and some fishes.

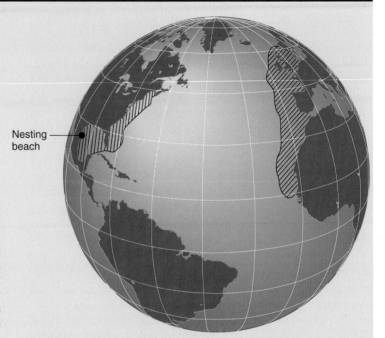

BOX FIGURE 20.2 Distribution of Kemp's Ridley Sea Turtle.

Nesting occurs from April to June when females come ashore in large numbers to deposit an average of 110 eggs in the sand along the shore. Eggs incubate in the absence of the adult turtle for 45 to 58 days and reach maturity in 10 to 12 years (five years in captivity). Adult Kemp's Ridley sea turtles live mostly in the Gulf of Mexico, but young turtles range throughout the gulf and migrate along both coasts of the Atlantic Ocean.

The Kemp's Ridley sea turtle was placed on the federal endangered species list in 1970. By 1985, the population of nesting females had dropped from an estimated 40,000 in 1947 to fewer than 200. This decline is the result of collecting eggs for human consumption, harvesting adults for food and their skin, drowning when accidentally caught in shrimp nets, and the development of nesting beaches. In recent years the number of Kemp's Ridley sea turtles has increased as a result of international conservation efforts. These efforts include the establishment of a marine preserve that protects nesting beaches, the use of turtle excluder devices that prevent turtles from being trapped in shrimp nets, turtle rearing projects, and public awareness projects. The increased numbers of nests since 1985 is an encouraging sign that these efforts are beginning to succeed. There is still much work to be done to achieve the recovery plan's goal of at least 10,000 females nesting in a season.

BOX FIGURE 20.1 Kemp's Ridley Sea Turtle (*Lepidochelys kempii*).

though not identical to, adaptations in the birds—the descendants of the second lineage of flying archosaurs.

The synapsid lineage eventually gave rise to the mammals. The legs of synapsids were relatively long and held their bodies off the ground. Teeth and jaws were adapted for effec- *tive chewing and tearing. Additional bones were incorporated into the middle ear. These and other mammal-like characteristics developed between the Carboniferous and Triassic periods.* The "Evolutionary Perspective" section of chapter 22 describes more about the nature of this transition.

SUMMARY

1. The earliest amniotes are classified as reptiles. The evolution of the amniotes resulted in lineages leading to the turtles; birds and dinosaurs; squamates (snakes, lizards, and worm lizards) and tuataras; and mammals.
2. The order Testudines contains the turtles. Turtles have a bony shell and lack teeth. All are oviparous.
3. The order Rhynchocephalia contains two species of tuataras. They are found only on remote islands of New Zealand.
4. The order Squamata contains the lizards, snakes, and worm lizards. Lizards usually have two pairs of legs, and most are oviparous. Snakes lack developed limbs and have skull adaptations for swallowing large prey. Worm lizards are specialized burrowers. They have a single median tooth in the upper jaw, and most are oviparous.
5. The order Crocodilia contains alligators, crocodiles, caimans, and gavials. These groups have a well-developed secondary palate and display nesting behaviors and parental care.
6. The skin of reptiles is dry and keratinized, and it provides a barrier to water loss. It also has epidermal scales and chromatophores.
7. The reptilian skeleton is modified for support and movement on land. Loss of appendages in snakes is accompanied by greater use of the body wall in locomotion.
8. Reptiles have a tongue that may be used in feeding. Bones of the skulls of snakes are loosely joined and spread apart during feeding.
9. The circulatory system of reptiles is divided into pulmonary and systemic circuits and functions under relatively high blood pressures. Blood may be shunted away from the pulmonary circuit during periods of apnea.
10. Gas exchange occurs across convoluted lung surfaces. Ventilation of lungs occurs by a negative-pressure mechanism.
11. Reptiles are ectotherms and mainly use behavioral mechanisms to thermoregulate.
12. Vision is the dominant sense in most reptiles. Median (parietal) eyes, ears, Jacobson's organs, and pit organs are important receptors in some reptiles.
13. Because uric acid is nontoxic and relatively insoluble in water, reptiles can store and excrete it as a semisolid. Internal respiratory surfaces and dry skin also promote water conservation.
14. The amniotic egg and internal fertilization permit development on land. They require significant parental energy expenditure.
15. Some reptiles use visual, olfactory, and auditory cues for reproduction. Parental care is important in crocodilians.
16. Descendants of the diapsid evolutionary lineage include the birds. Descendants of the synapsid lineage are the mammals.

SELECTED KEY TERMS

amniotic eggs (p. 312)
Jacobson's organs (p. 322)
keratin (p. 314)
median (parietal) eye (p. 322)
pit organs (p. 322)
secondary palate (p. 318)

CRITICAL THINKING QUESTIONS

1. Explain the nature of the controversy between cladists and evolutionary systematists regarding the higher taxonomy of the amniotes. Do you think that the Reptilia should be retained as a formal class designation? If so, what groups of animals should it contain?
2. What characteristics of the life history of turtles make them vulnerable to extinction? What steps do you think should be taken to protect endangered turtle species?
3. What might explain why parental care is common in crocodilians and birds?
4. List the adaptations that make life on land possible for a reptile. Explain why each is adaptive.
5. The incompletely divided ventricle of reptiles is sometimes portrayed as an evolutionary transition between the heart of primitive amphibians and the completely divided ventricles of birds and mammals. Do you agree with this portrayal? Why or why not?
6. What effect could significant global warming have on the sex ratios of crocodilians? Speculate on what long-term effects might be seen in populations of crocodilians as a result of global warming.

ONLINE LEARNING CENTER

Visit our Online Learning Center (OLC) at www.mhhe.com/zoology (click on this book's title) to find the following chapter-related materials:

- CHAPTER QUIZZING
- RELATED WEB LINKS
    Class Reptilia
        Order Testudines
        Marine Turtles
        Order Squamata
        Suborder Lacertilia
        Suborder Serpentes
        Order Sphenodonta
        Superorder Archosauria and Related Mesozoic Reptiles
        Dissection Guides for Reptiles
        Conservation Issues Concerning Reptiles
- BOXED READINGS ON
    Collision or Coincidence?
    Bones and Scales
    Changing Perceptions of Ancient Lifestyles
    Extrarenal Secretion by Avian and Reptilian Salt Glands
- SUGGESTED READINGS

CHAPTER 21

BIRDS:

FEATHERS, FLIGHT, AND ENDOTHERMY

Outline

Concepts

1. Fossils of the earliest birds clearly show reptilian features. Fossils of *Archaeopteryx*, *Sinornis*, and *Eoalulavis* give clues to the origin of flight in birds.
2. Integumentary, skeletal, muscular, and gas exchange systems of birds are adapted for flight and endothermic temperature regulation.
3. Large regions of the brains of birds are devoted to integrating sensory information.
4. Birds' complex mating systems and behavioral patterns increase the chances of offspring survival.
5. Migration and navigation allow birds to live, feed, and reproduce in environments favorable to the survival of adults and young.

EVOLUTIONARY PERSPECTIVE

Drawings of birds on the walls of caves in southern France and Spain, bird images of ancient Egyptian and American cultures, and the bird images in Biblical writings are evidence that humans have marveled at birds and bird flight for thousands of years. From Leonardo da Vinci's early drawings of flying machines (1490) to Orville Wright's first successful powered flight on 17 December 1903, humans have tried to take to the sky and experience soaring like a bird.

Birds' ability to navigate long distances between breeding and wintering grounds is just as impressive as flight. For example, Arctic terns have a migratory route that takes them from the Arctic to the Antarctic and back again each year, a distance of approximately 35,000 km (figure 21.1). Their rather circuitous route takes them across the northern Atlantic Ocean, to the coast of Europe and Africa, and then across vast stretches of the southern Atlantic Ocean before they reach their wintering grounds.

PHYLOGENETIC RELATIONSHIPS

Birds are traditionally classified as members of the class Aves (a'ves) (L. *avis*, bird). The major characteristics of this class are adaptations for flight, including appendages modified as wings, feathers, endothermy, a high metabolic rate, a vertebral column modified for flight, and bones lightened by numerous air spaces. In addition, modern birds possess a horny bill and lack teeth.

The similarities between birds and reptiles are so striking that birds are often referred to as "glorified reptiles." These similarities include features such as a single occipital

This chapter contains evolutionary concepts, which are set off in this font.

(a)

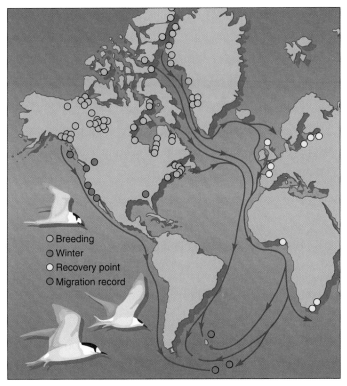

○ Breeding
○ Winter
○ Recovery point
○ Migration record

(b)

FIGURE 21.1

Class Aves. (*a*) The birds were derived from the archosaur lineage of ancient reptiles. Adaptations for flight include appendages modified as wings, feathers, endothermy, a high metabolic rate, a vertebral column modified for flight, and bones lightened by numerous air spaces. Flight has given birds, like this Arctic tern (*Sterna arctica*), the ability to exploit resources unavailable to other vertebrates. (*b*) Migration route of the Arctic tern. Arctic terns breed in northern North America, Greenland, and the Arctic. Migrating birds cross the Atlantic Ocean on their trip to wintering grounds in Antarctica. In the process, they fly about 35,000 km (22,000 mi) each year.

condyle on the skull (the point of articulation between the skull and the first cervical vertebra), a single ear ossicle, lower jaw structure, and dozens of other technical skeletal characteristics. Physiological characteristics, such as the presence of nucleated red blood cells and aspects of liver and kidney function, are shared by reptiles and birds. Some birds and reptiles even share behavioral characteristics, for example, those related to nesting and care of young.

Birds descended from ancient archosaurs—a lineage shared by the dinosaurs and crocodilians (*see figure 20.3*). There has been debate for many years regarding bird affinities: are they more closely related to the dinosaurs or to the crocodilians? While the question is not completely settled, most zoologists would probably favor the Saurischian lineage of dinosaurs as bird ancestors. This is the lineage that also included bipedal dinosaurs like *Tyrannosaurus*. This conclusion has received support from recent discoveries of fossils of feathered dinosaurs from fossil beds in China. Three species have been described that all lived between 120 and 135 million years ago. They were about the size of turkeys, and none could fly. *Sinosauropteryx* had a coat of downy, featherlike fibers and *Protarchaeopteryx* and *Caudipteryx* had long, symmetrical feathers on their foreappendages and tails. These feathers may have served as insulation to aid in temperature regulation, and the foreappendages may have been used as balancing devices while running along the ground. The discovery of dinosaur fossils showing a furcula, or wishbone, further supports the dinosaur ancestry hypothesis. (The furcula is derived from a fusion of the clavicles and, as described later, is an adaptation for flight in birds.)

ANCIENT BIRDS AND THE EVOLUTION OF FLIGHT

In 1861, one of the most important vertebrate fossils was found in a slate quarry in Bavaria, Germany (figure 21.2). It was a fossil of a pigeon-sized animal that lived during the Jurassic period, about 150 million years ago. It had a long, reptilian tail and clawed fingers. The complete head of this specimen was not preserved, but imprints of feathers on the tail and on short, rounded wings were the main evidence that this was the fossil of an ancient bird. It was named *Archaeopteryx* (Gr. *archaios*, ancient + *pteron*, wing). Sixteen years later, a more complete fossil was discovered, revealing teeth in beaklike jaws. Four later discoveries of *Archaeopteryx* fossils have reinforced the ideas of reptilian ancestry for birds. Most zoologists consider *Archaeopteryx* to be the oldest bird yet discovered and very close to the main line of evolution between the reptiles and birds.

Interpretations of the lifestyle of Archaeopteryx *have been important in the development of hypotheses on the origin of flight.* The clavicles (wishbone) of *Archaeopteryx* were well developed and probably provided points of attachment for wing muscles. The sternum, wing bones, and other sites for flight muscle attachment were less developed than in modern birds. These observations indicate that *Archaeopteryx* may have primarily been a glider that was capable of flapping flight over short distances.

Some zoologists think that the clawed digits of the wings may have been used to climb trees and cling to branches. *A sequence in the evolution of flight may have involved jumping*

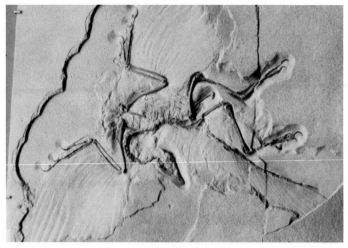

(a)

(b)

FIGURE 21.2

Archaeopteryx, an Ancient Bird.   (*a*) *Archaeopteryx* fossil. (*b*) Artist's representation. Some zoologists think that *Archaeopteryx* was a ground dweller rather than the tree dweller depicted here.

from branch to branch, or branch to ground. At some later point, gliding evolved. Still later, weak flapping supplemented gliding, and finally, wing-powered flight evolved.

Other zoologists note that the hindlimb structure of the earliest birds suggests that they may have been bipedal, running and hopping along the ground. Their wings may have functioned in batting flying insects out of the air or in trapping insects and other prey against the ground. The teeth and claws, which resemble talons of modern predatory birds, may have been used to grasp prey. *Wings would have been useful in providing stability during horizontal jumps when pursuing prey, and they would also have allowed flight over short distances. The benefits of such flight may have led eventually to wing-powered flight.*

A second ancient bird (*Sinornis*), discovered in China, is consistent with the view that *Archaeopteryx* was closely related to

ancestral bird stocks. *Sinornis* fossils are 135 million years old—only 15 million years younger than *Archaeopteryx.* In addition to some primitive dinosaur-like characteristics, *Sinornis* had characterisitics similar to those of modern birds, including a shortened body and tail, and a sternum with a large surface area for flight muscles. The claws were reduced, and the forelimbs were modified to permit the folding of wings at rest. These characteristics all indicate that powered flight was well developed in birds 135 million years ago.

A very recent discovery of a fossil bird from early Cretaceous deposits in Spain provides additional important information on the origin of flight. This bird, *Eoalulavis,* was found in 115-million-year-old deposits and had a wingspan of 17 cm (roughly the same as a goldfinch). This fossil showed that *Eoalulavis* had a wing structure called an alula. As described later in this chapter, the alula is present in many modern birds that engage in slow, hovering flight. Its presence in this fossil indicates that complex flight mechanisms associated with slow, hovering, and highly maneuverable flight evolved at least 115 million years ago.

DIVERSITY OF MODERN BIRDS

Archaeopteryx, Sinornis, and Eoalulavis provide evidence of the transition between reptiles and birds. Zoologists do not know, however, whether or not one of these birds is the direct ancestor of modern birds. A variety of fossil birds for the period between 100 million and 70 million years ago has been found. Some of these birds were large, flightless birds; others were adapted for swimming and diving; and some were fliers. Most, like *Archaeopteryx,* had reptilelike teeth. Most of the lineages that these fossils represent became extinct, along with the dinosaurs, at the end of the Mesozoic era.

Some of the few birds that survived into the Tertiary period were the ancestors of modern, toothless birds. The phylogeny of modern birds is controversial. It is sufficient to say that adaptive radiation has resulted in about 9,100 species of living birds, which are divided into about 27 orders (table 21.1). (The number of orders varies, depending on the classification system used.) Characteristic behaviors, songs, anatomical differences, and ecological niches distinguish the orders.

EVOLUTIONARY PRESSURES

Virtually every body system of a bird shows some adaptation for flight. Endothermy, feathers, acute senses, long, flexible necks, and lightweight bones are a few of the many adaptations described in this section.

EXTERNAL STRUCTURE AND LOCOMOTION

The covering of feathers on a bird is called the plumage. Feathers have two primary functions essential for flight. They form the flight surfaces that provide lift and aid steering, and they prevent

TABLE 21.1
CLASSIFICATION OF THE BIRDS

Class Aves (a′ves) (L. *avis*, bird)*

Adaptations for flight include: foreappendages modified as feathered wings, endothermic, high metabolic rate, flexible neck, fused posterior vertebrae, and bones lightened by numerous air spaces. The skull is lightened by a reduction in bone and the presence of a horny bill that lacks teeth. The birds.

Order Sphenisciformes (sfe-nis″i-for′mez)
Heavy bodied; flightless, flipperlike wings for swimming; well insulated with fat. Penguins.

Order Struthioniformes (stroo″the-on-i-for′mez)
Large, flightless birds; wings with numerous fluffy plumes. Ostriches.

Order Gaviiformes (ga″ve-i-for′mez)
Strong, straight bill; diving adaptations include legs far back on body, bladelike tarsus, webbed feet, and heavy bones. Loons.

Order Podicipediformes (pod″i-si-ped″i-for′mez)
Short wings; soft and dense plumage; feet webbed with flattened nails. Grebes.

Order Procellariiformes (pro-sel-lar-e-i-for′mez)
Tubular nostrils, large nasal glands; long and narrow wings. Albatrosses, shearwaters, petrels.

Order Pelecaniformes (pel″e-can-i-for′mez)
Four toes joined in common web; nostrils rudimentary or absent; large gular sac. Pelicans, boobies, cormorants, anhingas, frigate-birds.

Order Ciconiiformes (si-ko″ne-i-for′mez)
Long neck, often folded in flight; long-legged waders. Herons, egrets, storks, wood ibises, flamingos.

Order Anseriformes (an″ser-i-for′mez)
South American screamers, ducks, geese, and swans; the latter three groups possess a wide, flat bill and an undercoat of dense down; webbed feet.

Order Falconiformes (fal″ko-ni-for′mez)
Strong, hooked beak; large wings; raptorial feet. Vultures, secretarybirds, hawks, eagles, ospreys, falcons.

Order Galliformes (gal″li-for′mez)
Short beak; short, concave wings; strong feet and claws. Curassows, grouse, quail, pheasants, turkeys.

Order Gruiformes (gru″i-for′mez)
Order characteristics variable and not diagnostic. Marsh birds, including cranes, limpkins, rails, coots.

Order Charadriiformes (ka-rad″re-i-for′mez)
Order characteristics variable. Shorebirds, gulls, terns, auks.

Order Columbiformes (co-lum″bi-for′mez)
Dense feathers loosely set in skin; well-developed crop. Pigeons, doves, sandgrouse.

Order Psittaciformes (sit″ta-si-for′mez)
Maxilla hinged to skull; thick tongue; reversible fourth toe; usually brightly colored. Parrots, lories, macaws.

Order Cuculiformes (ku-koo″li-for′mez)
Reversible fourth toe; soft, tender skin. Plantaineaters, roadrunners, cuckoos.

Order Strigiformes (strij″i-for′mez)
Large head with fixed eyes directed forward; raptorial foot. Owls.

Order Caprimulgiformes (kap″ri-mul″ji-for′mez)
Owl-like head and plumage, but weak bill and feet; beak with wide gape; insectivorous. Whippoorwills, other goatsuckers.

Order Apodiformes (a-pod″i-for′mez)
Long wings; weak feet. Swifts, hummingbirds.

Order Coraciiformes (kor″ah-si″ah-for′mez)
Large head; large beak; metallic plumage. Kingfishers, todies, bee eaters, rollers.

Order Piciformes (pis″i-for′mez)
Usually long strong beak; strong legs and feet with fourth toe permanently reversed in woodpeckers. Woodpeckers, toucans, honeyguides, barbets.

Order Passeriformes (pas″er-i-for′mez)
Largest avian order; 69 families of perching birds; perching foot; variable external features. Swallows, larks, crows, titmice, nuthatches, and many others.

*Selected bird orders are described.

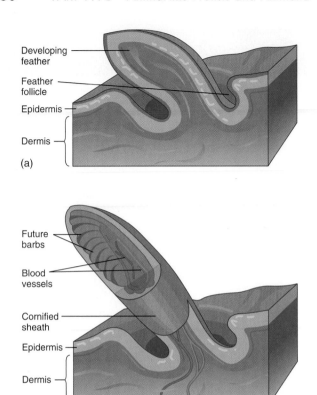

FIGURE 21.3

Formation of Bird Feathers during Embryonic Development.
(*a*) Feathers form from epidermal evaginations. (*b*) Blood flow to the feather supports initial feather development. Later in development, the blood supply to the feather is cut off and the feather becomes a dead, keratinized, epidermal structure seated in a feather follicle.

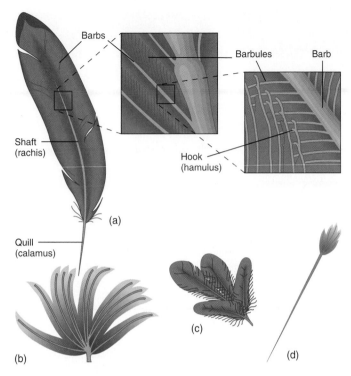

FIGURE 21.4

Anatomy of Selected Feather Types. (*a*) Anatomy of a contour feather, showing enlargements of barbs and barbules. (*b*) Down feather. Various types of down feathers provide insulation for adult and immature birds. (*c*) Contour feather. (*d*) Filoplume. Filoplume feathers are usually covered by contour feathers and are associated with nerve endings in the skin, thus serving as sensory structures.

excessive heat loss, permitting the endothermic maintenance of high metabolic rates. Feathers also have roles in courtship, incubation, and waterproofing.

Feathers develop in a fashion similar to the epidermal scales of reptiles, and this similarity is one source of evidence that demonstrates the evolutionary ties between reptiles and birds (figure 21.3). Only the inner pulp of feathers contains dermal elements, such as blood vessels, which supply nutrients and pigments for the growing feather. As feathers mature, their blood supply is cut off, and the feathers become dead, keratinized, epidermal structures seated in epidermal invaginations of the skin called feather follicles.

The most obvious feathers are **contour feathers,** which cover the body, wings, and tail (Figure 21.4*a, c*). Contour feathers consist of a vane with its inner and outer webs, and a supportive shaft. Feather barbs branch off the shaft, and barbules branch off the barbs. Barbules of adjacent barbs overlap one another. The ends of barbules lock with hooklike hamuli (sing., hamulus). Interlocking barbs keep contour feathers firm and smooth. Other types of feathers include **down feathers,** which function as insulating feathers, and **filoplume feathers** (pinfeathers), which have sensory functions (figure 21.4*b, d*).

Birds maintain a clean plumage to rid the feathers and skin of parasites. Preening, which is done by rubbing the bill over the feathers, keeps the feathers smooth, clean, and in place. Hamuli that become dislodged can be rehooked by running a feather through the bill. Secretions from an oil gland (uropygial gland) at the base of the tail of many birds are spread over the feathers during preening to keep the plumage water repellent and supple. The secretions also lubricate the bill and legs to prevent chafing. Anting is a maintenance behavior common to many songbirds and involves picking up ants in the bill and rubbing them over the feathers. The formic acid that ants secrete is apparently toxic to feather mites.

Feather pigments deposited during feather formation produce most colors in a bird's plumage. Other colors, termed structural colors, arise from irregularities on the surface of the feather that diffract white light. For example, blue feathers are never blue because of the presence of blue pigment. A porous, nonpigmented outer layer on a barb reflects blue wavelengths of light. The other wavelengths pass into the barb and are absorbed by the dark pigment melanin. Iridescence results from the interference of light waves caused by a flattening and twisting of barbules. An example of iridescence is the perception of interchanging colors on the neck and back of hummingbirds and grackles. Color patterns are involved in cryptic coloration, species and sex recognition, and sexual attraction.

Mature feathers receive constant wear; thus, all birds periodically shed and replace their feathers in a process called

molting. The timing of molt periods varies in different taxa. The following is a typical molting pattern for songbirds. After hatching, a chick is covered with down. Juvenile feathers replace the down at the juvenile molt. A postjuvenile molt in the fall results in plumage similar to that of the adult. Once sexual maturity is attained, a prenuptial molt occurs in late winter or early spring, prior to the breeding season. A postnuptial molt usually occurs between July and October. Flight feathers are frequently lost in a particular sequence so that birds are not wholly deprived of flight during molt periods. However, many ducks, coots, and rails cannot fly during molt periods and hide in thick marsh grasses.

The Skeleton

The bones of most birds are lightweight yet strong. Some bones, such as the humerus (forearm bone), have large air spaces and internal strutting (reinforcing bony bars), which increase strength (figure 21.5c). (Engineers take advantage of this same principle. They have discovered that a strutted girder is stronger than a solid girder of the same weight.) Birds also have a reduced number of skull bones, and a lighter, keratinized sheath called a bill replaces the teeth. The demand for lightweight bones for flight is countered in some birds with other requirements. For example, some aquatic birds (e.g., loons) have dense bones, which help reduce buoyancy during diving.

The appendages involved in flight cannot manipulate nesting materials or feed young. The bill and very flexible neck and feet make these activities possible. The cervical vertebrae have saddle-shaped articular surfaces that permit great freedom of movement. In addition, the first cervical vertebra (the atlas) has a single point of articulation with the skull (the occipital condyle), which permits a high degree of rotational movement between the skull and the neck. (*The single occipital condyle is another characteristic that birds share with reptiles.*) This flexibility allows the bill and neck to function as a fifth appendage.

The pelvic girdle, vertebral column, and ribs are strengthened for flight. The thoracic region of the vertebral column contains ribs, which attach to thoracic vertebrae. Most ribs have posteriorly directed uncinate processes that overlap the next rib to strengthen the rib cage (figure 21.5a). (*Uncinate processes are also present on the ribs of most reptiles and are additional evidence of their common ancestry.*) Posterior to the thoracic region is the lumbar region. The **synsacrum** forms by the fusion of the posterior thoracic vertebrae, all the lumbar and sacral vertebrae, and the anterior caudal vertebrae. Fusion of these bones helps to maintain the proper flight posture and supports the hind appendages during landing, hopping, and walking. The posterior caudal vertebrae are fused into a **pygostyle,** which helps support the tail feathers that are important in steering.

The sternum of most birds bears a large, median keel for the attachment of flight muscles. (Exceptions to this include some flightless birds, such as ostriches.) The keel attaches firmly to the rest of the axial skeleton by the ribs. Paired clavicles fuse medially and ventrally into a **furcula** (wishbone). The furcula braces the pectoral girdle against the sternum and serves as an additional site for the attachment of flight muscles.

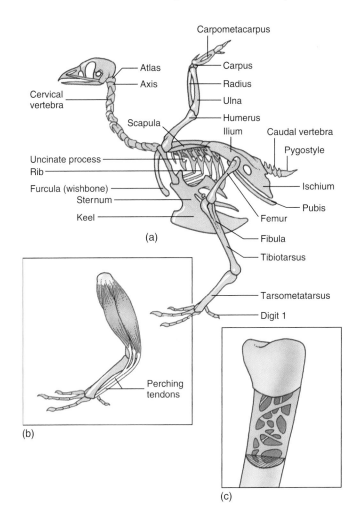

FIGURE 21.5

Bird Skeleton. (*a*) Skeleton of a pigeon. (*b*) Perching tendons run from the toes across the back of the ankle joint, which cause the foot to grip a perch. (*c*) Internal structure of the humerus. Note the air spaces in this pneumatic bone.

The appendages of birds have also been modified. Some bones of the front appendages have been lost or fused, and they are points of attachment for flight feathers. The rear appendages are used for hopping, walking, running, and perching. Perching tendons run from the toes across the back of the ankle joint to muscles of the lower leg. When the ankle joint is flexed, as in landing on a perch, tension on the perching tendons increases, and the foot grips the perch (figure 21.5b). This automatic grasp helps a bird perch even while sleeping. The muscles of the lower leg can increase the tension on these tendons, for example, when an eagle grasps a fish in its talons.

Muscles

The largest, strongest muscles of most birds are the flight muscles. They attach to the sternum and clavicles and run to the humerus. The muscles of most birds are adapted physiologically for flight. Flight muscles must contract quickly and fatigue very slowly. These muscles have many mitochondria and produce large

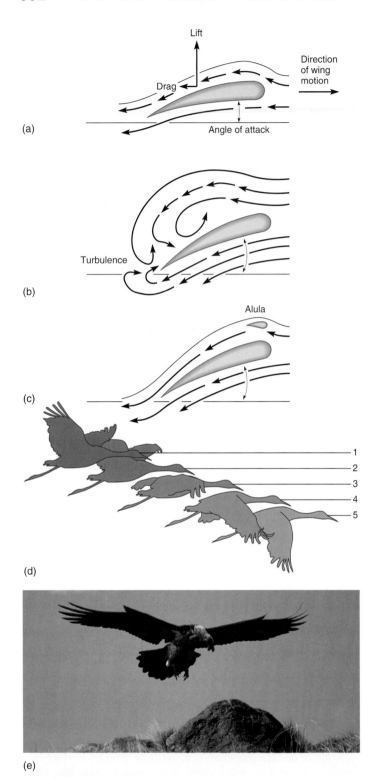

(a)

(b)

(c)

(d)

(e)

FIGURE 21.6

Mechanics of Bird Flight. (*a*) A bird's wing acts as an airfoil. Air passing over the top of the wing travels farther and faster than air passing under the wing, creating lift. (*b*) Increasing the angle of attack increases lift but also increases turbulence. (*c*) The alula reduces turbulence. (*d*) Wing orientation during a downstroke. (*e*) Note the alula on the wings of this bearded vulture (*Gypaetus barbatus*). This species was widely distributed in the Alps. Excessive hunting nearly eliminated this species around 1900. It is now making a comeback, inhabiting large territories above the Alps forest line.

quantities of ATP to provide the energy required for flight, especially long-distance migrations. Domestic fowl have been selectively bred for massive amounts of muscle ("white meat") that humans like as food but that is poorly adapted for flight because it contains fibers that can rapidly contract but contain few mitochondria and have poor vascularization.

Flight

The wings of birds are adapted for different kinds of flight. However, regardless of whether a bird soars, glides, or has a rapid flapping flight, the mechanics of staying aloft are similar. Bird wings form an **airfoil**. The anterior margin of the wing is thicker than the posterior margin. The upper surface of the wing is slightly convex, and the lower surface is flat or slightly concave. Air passing over the wing travels farther and faster than air passing under the wing, decreasing air pressure on the upper surface of the wing and creating lift (figure 21.6*a*). The lift the wings create must overcome the bird's weight, and the forces that propel the bird forward must overcome the drag that the bird moving through the air creates. Increasing the angle that the leading edge of the wing makes with the oncoming air (the angle of attack) increases lift. As the angle of attack increases, however, the flow of air over the upper surface becomes turbulent, reducing lift (figure 21.6*b*). Turbulence is reduced if air can flow rapidly through slots at the leading edge of the wing. Slotting the feathers at the wing tips and the presence of an alula on the anterior margin of the wing reduce turbulence. The **alula** is a group of small feathers that bones of the medial digit support. During takeoff, landing, and hovering flight, the angle of attack increases, and the alula is elevated (figure 21.6*c,e*). During soaring and fast flight, the angle of attack decreases, and slotting is reduced.

The distal part of the wing generates most of the propulsive force of flight. Because it is farther from the shoulder joint, the distal part of the wing moves farther and faster than the proximal part of the wing. During the downstroke (the powerstroke), the leading edge of the distal part of the wing is oriented slightly downward and creates a thrust somewhat analogous to the thrust that an airplane propeller creates (figure 21.6*d*). During the upstroke (the recovery stroke), the distal part of the wing is oriented upward to decrease resistance. Feathers on a wing overlap so that, on the downstroke, air presses the feathers at the wing margins together, allowing little air to pass between them, enhancing both lift and propulsive forces. On the upstroke, wings fold inward slightly. Also, feathers part slightly, allowing air to pass between them, which reduces resistance during this recovery stroke.

The tail of a bird serves a variety of balancing, steering, and braking functions during flight. It also enhances lift that the wings produce during low-speed flight. During horizontal flight, spreading the tail feathers increases lift at the rear of the bird and causes the head to dip for descent. Closing the tail feathers has the opposite effect. Tilting the tail sideways turns the bird. When a bird lands, its tail deflects downward, serving as an air brake. In the males of some species—for example, sunbirds (*Nectarinia*) and widow birds (*Euplectes*)—tails possess dramatic ornamentation that attracts females and improves reproductive success.

Different birds, or the same bird at different times, use different kinds of flight. During gliding flight, the wing is stationary, and a bird loses altitude. Waterfowl coming in for a landing use gliding flight. Flapping flight generates the power for flight and is the most common type of flying. Many variations in wing shape and flapping patterns result in species-specific speed and maneuverability. Soaring flight allows some birds to remain airborne with little energy expenditure. During soaring, wings are essentially stationary, and the bird utilizes updrafts and air currents to gain altitude. Hawks, vultures, and other soaring birds are frequently observed circling along mountain valleys, soaring downwind to pick up speed and then turning upwind to gain altitude. As the bird slows and begins to lose altitude, it turns downwind again. The wings of many soarers are wide and slotted to provide maximum maneuverability at relatively low speeds. Oceanic soarers, such as albatrosses and frigate birds, have long, narrow wings that provide maximum lift at high speeds, but they compromise maneuverability and ease of take-off and landing. Hummingbirds perform hovering flight. They hover in still air by fanning their wings back and forth (50 to 80 beats per second) to remain suspended in front of a flower or feeding station. The wings move in a figure-eight pattern. As they move, the wings are flipped from right-side-up to upside-down, and lift is generated from both the top and bottom sides of the wings.

NUTRITION AND THE DIGESTIVE SYSTEM

Most birds have ravenous appetites! This appetite supports a high metabolic rate that makes endothermy and flight possible. For example, hummingbirds feed almost constantly during the day. In spite of high rates of food consumption, they often cannot sustain their rapid metabolism overnight, and they may become torpid, with reduced body temperature and respiratory rate, until they can feed again in the morning.

Bird bills and tongues are modified for a variety of feeding habits and food sources (figures 21.7 and 21.8). For example, a woodpecker's tongue is barbed for extracting grubs from the bark of trees (see figure 27.6d). Sapsuckers excavate holes in trees and use a brushlike tongue for licking the sap that accumulates in these holes. The tongues of hummingbirds and other nectar feeders roll into a tube for extracting nectar from flowers. Birds' bills are used in feeding, preening, nest building, courtship displays, and defense. The neck, head, and bill combination functions as a fifth appendage. Modifications of the bill reflect specific functions in various species. For example, the bill of an eagle is modified for tearing prey, the bill of a cardinal is specialized for cracking seeds, and the bill of a flamingo is used to strain food from the water.

In many birds, a diverticulum of the esophagus, called the crop, is a storage structure that allows birds to quickly ingest large quantities of locally abundant food. They can then seek safety while digesting their meal. The crop of pigeons produces "pigeon's milk," a cheesy secretion formed by the proliferation and slough-

FIGURE 21.7

Bird Flight and Feeding Adaptations. This ruby-throated hummingbird (*Archilochus colubris*) hovers while feeding on flower nectar. Hummingbird beaks often match the length and curvature of the flower from which the birds extract nectar.

ing of cells lining the crop. Young pigeons (squabs) feed on pigeon's milk until they are able to eat grain. Cedar waxwings, vultures, and birds of prey use their esophagus for similar storage functions. Crops are less well developed in insect-eating birds because insectivorous birds feed throughout the day on sparsely distributed food.

The stomach of birds is modified into two regions. The proventriculus secretes gastric juices that initiate digestion (figure 21.9). The ventriculus (gizzard) has muscular walls to abrade and crush seeds or other hard materials. Birds may swallow sand and other abrasives to aid digestion. The bulk of enzymatic digestion and absorption occurs in the small intestine, aided by secretions from the pancreas and liver. Paired ceca may be located at the union of the large and small intestine. These blind-ending sacs contain bacteria that aid in cellulose digestion. Birds usually eliminate undigested food through the cloaca; however, owls form pellets of bone, fur, and feathers that are ejected from the ventriculus through the mouth. Owl pellets accumulate in and around owl nests and are useful in studying their food habits.

Birds are often grouped by their feeding habits. These groupings are somewhat artificial, however, because birds may eat different kinds of food at different stages in their life history, or they may change diets simply because of changes in food availability. Robins, for example, feed largely on worms and other invertebrates when these foods are available. In the winter, however, robins may feed on berries.

In some of their feeding habits, birds directly conflict with human interests. Bird damage to orchard and grain crops is tallied in the millions of dollars each year. Flocking and roosting habits of some birds, such as European starlings and redwing blackbirds, concentrate millions of birds in local habitats, where they devastate fields of grain. Recent monocultural practices tend to aggravate problems with grain-feeding birds by encouraging the formation of very large flocks.

(a)

(b)

(c)

FIGURE 21.8

Some Specializations of Bird Bills.   (*a*) The bill of a bald eagle (*Haliacetus leucocephalus*) is specialized for tearing prey. (*b*) The thick, powerful bill of this cardinal (*Cardinalis cardinalis*) cracks tough seeds. (*c*) The bill of a flamingo (*Phoenicopterus ruber*) strains food from the water in a head-down feeding posture. Large bristles fringe the upper and lower mandibles. As water is sucked into the bill, larger particles are filtered and left outside. Inside the bill, tiny inner bristles filter smaller algae and animals. The tongue removes food from the bristles.

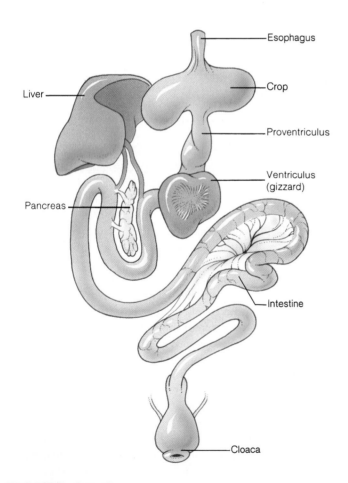

Esophagus

Liver

Crop

Proventriculus

Ventriculus
(gizzard)

Pancreas

Intestine

Cloaca

FIGURE 21.9

Digestive System of a Pigeon.   Birds have high metabolic rates that require a nearly constant nutrient supply.

Birds of prey have minimal impact on populations of poultry and game birds, and on commercial fisheries. Unfortunately, the mistaken impression that they are responsible for significant losses has led humans to poison and shoot them.

CIRCULATION, GAS EXCHANGE, AND TEMPERATURE REGULATION

The circulatory system of birds is similar to that of reptiles, except that the heart has completely separated atria and ventricles, resulting in separate pulmonary and systemic circuits. This separation prevents any mixing of highly oxygenated blood with less oxygenated blood. *In vertebrate evolution, the sinus venosus has gradually decreased in size.* It is a separate chamber in fishes, amphibians, and turtles and receives blood from the venous system. In other reptiles, it is a group of cells in the right atrium that serves as the pacemaker for the heart. In birds, the sinus venosus also persists only as a patch of pacemaker tissue in the right atrium. The bird heart is relatively large (up to 2.4% of total body weight), and it beats rapidly. Rates in excess of 1,000 beats per minute have been recorded for hummingbirds under stress. Larger birds have relatively smaller hearts and slower heart rates. The heart rate of an ostrich, for example, varies between 38 and 176 beats per minute. *A large heart, rapid heart rate, and complete separation of highly oxygenated blood from less oxygenated blood are important adaptations for delivering the large quantities of blood required for endothermy and flight.*

GAS EXCHANGE

The respiratory system of birds is extremely complex and efficient. It consists of external nares, which lead to nasal passageways and the pharynx. Bone and cartilage support the trachea. A special voice box, called the **syrinx,** is located where the trachea divides into bronchi. The muscles of the syrinx and bronchi, as well as characteristics of the trachea, produce bird vocalizations. The bronchi lead to a complex system of air sacs that occupy much of the body and extend into some of the bones of the skeletal system (figure 21.10*a*). The air sacs and bronchi connect to the lungs. The lungs of birds are made of small air tubes called **parabronchi.** Air capillaries about 10 mm in diameter branch from the parabronchi and are associated with capillary beds for gas exchange.

Inspiration and expiration result from increasing and decreasing the volume of the thorax and from alternate expansion and compression of air sacs during flight and other activities. During breathing, the movement of the sternum and the posterior ribs compresses the thoracic air sacs. X-ray movies of European starlings in a wind tunnel show that the contraction of flight muscles distorts the furcula. Alternate distortion and recoiling helps compress and expand air sacs between the bone's two shafts.

It takes two ventilatory cycles to move a particular volume of air through the respiratory system of a bird. During the first inspiration, air moves into the abdominal air sacs. At the same time, air already in the lungs moves through parabronchi into the thoracic air sacs (figure 21.10*b*, cycle 1). During expiration, the air in the thoracic air sacs moves out of the respiratory system, and the air in the abdominal air sacs moves into parabronchi (figure 21.10*c*). At the second inspiration (figure 21.10*b*, cycle 2), the air moves into the thoracic air sacs, and it is expelled during the next expiration.

Because of high metabolic rates associated with flight, birds have a greater rate of oxygen consumption than any other vertebrate. When other tetrapods inspire and expire, air passes into and out of respiratory passageways in a simple back-and-forth cycle. Ventilation is interrupted during expiration, and much "dead air" (air not forced out during expiration) remains in the lungs. Because of the unique system of air sacs and parabronchi, bird lungs have a nearly continuous movement of oxygen-rich air over respiratory surfaces during both inspiration and expiration. The quantity of "dead air" in the lungs is sharply reduced, compared with other vertebrates. *This avian system of gas exchange is more efficient than that of any other tetrapod. In addition to supporting high metabolic rates, this efficient gas exchange system probably also explains how birds can live and fly at high altitudes, where oxygen tensions are low.* During their migrations, bar-headed geese fly over the peaks of the Himalayas at altitudes of 9,200 m. A human mountain climber begins to feel the symptoms of altitude sickness and must carry auxiliary oxygen supplies above 2,500 m.

Thermoregulation

Birds maintain body temperatures between 38 and 45° C. Lethal extremes are lower than 32 and higher than 47° C. On a cold day, a resting bird fluffs its feathers to increase their insulating properties, and to increase the dead air space within them. It also tucks its bill into its feathers to reduce heat loss from the respiratory tract. The most exposed parts of a bird are the feet and tarsi, which have neither fleshy muscles nor a rich blood supply. Temperatures in these extremities are allowed to drop near freezing to prevent heat loss. Countercurrent heat exchange between the warm blood flowing to the legs and feet, and the cooler blood flowing to the

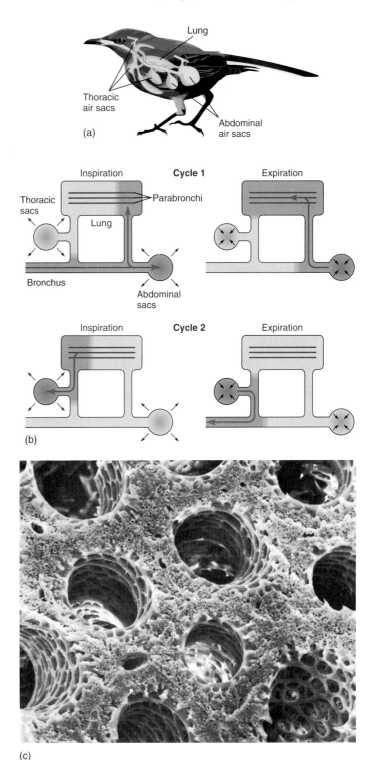

FIGURE 21.10

Respiratory System of a Bird.   (*a*) Air sacs branch from the respiratory tree. (*b*) Air flow during inspiration and expiration. Air flows through the parabronchi during both inspiration and expiration. The shading represents the movement through the lungs of one inspiration. (*c*) Scanning electron micrograph of parabronchi.

body core from the legs and feet, prevents excessive heat loss at the feet. Heat is returned to the body core before it goes to the extremities and is lost to the environment (*see figure 28.5b*). Shivering also generates heat in extreme cold. Increases in metabolism during winter months require additional food.

Some birds become torpid and allow their body temperatures to drop on cool nights. For example, whippoorwills allow their body temperatures to drop from about 40° C to near 16° C, and respiratory rates become very slow.

Muscular activity during flight produces large quantities of heat, which birds can dissipate by panting. Evaporative heat loss from the floor of the mouth is enhanced by fluttering vascular membranes of this region.

NERVOUS AND SENSORY SYSTEMS

A mouse, enveloped in the darkness of night, skitters across the floor of a barn. An owl in the loft overhead turns in the direction of the faint sounds the tiny feet make. As the sounds made by hurrying feet change to a scratchy gnawing of teeth on a sack of feed, the barn owl dives for its prey (figure 21.11). Fluted tips of flight feathers make the owl's approach imperceptible to the mouse, and the owl's ears, not its eyes, guide it to its prey. Barn owls successfully locate and capture prey in over 75% of attempts! This ability is just one example of the many sensory adaptations of birds.

The forebrain of birds is much larger than that of reptiles due to the enlargement of the cerebral hemispheres, including a region of gray matter, the corpus striatum. The corpus striatum functions in visual learning, feeding, courtship, and nesting. A pineal body on the roof of the forebrain appears to stimulate ovarian development and to regulate other functions influenced by light and dark periods. The optic tectum (the roof of the midbrain), along with the corpus striatum, plays an important role in integrating sensory functions. The midbrain also receives sensory input from the eyes. As in reptiles, the hindbrain includes the cerebellum and the medulla oblongata, which coordinate motor activities and regulate heart and respiratory rates, respectively.

Vision is an important sense for most birds. The structures of bird eyes are similar to those of other vertebrates, but bird eyes are much larger relative to body size than those of other vertebrates (*see figure 19.14*). The eyes are usually somewhat flattened in an anteroposterior direction; however, the eyes of birds of prey protrude anteriorly because of a bulging cornea. Birds have a unique double-focusing mechanism. Padlike structures (similar to those of reptiles) control the curvature of the lens, and ciliary muscles change the curvature of the cornea. Double, nearly instantaneous focusing allows an osprey or other bird of prey to remain focused on a fish throughout a brief, but breathtakingly fast, descent.

The retina of a bird's eye is thick and contains both rods and cones. Rods are active under low light intensities, and cones are active under high light intensities. Cones are especially concentrated (1,000,000/mm^2) at a focal point called the fovea. Unlike other vertebrates, some birds have two foveae per eye. The one at the center of the retina is sometimes called the "search fovea" because it gives the bird a wide angle of monocular vision. The other

FIGURE 21.11

Barn Owl *(Tyto alba).* A keen sense of hearing allows barn owls to find prey in spite of the darkness of night.

fovea is at the posterior margin of the retina. It functions with the posterior fovea of the other eye to allow binocular vision. The posterior fovea is called the "pursuit fovea" because binocular vision produces depth perception, which is necessary to capture prey. The words "search" and "pursuit" are not meant to imply that only predatory birds have these two foveae. Other birds use the "search fovea" to observe the landscape below them during flight and the "pursuit" fovea when depth perception is needed, as in landing on a branch of a tree.

The position of the eyes on the head also influences the degree of binocular vision (figure 21.12). Pigeons have eyes located well back on the sides of their head, giving them a nearly 360° monocular field, but a narrow binocular field. They do not have to pursue their food (grain), and a wide monocular field of view helps them stay alert to predators while feeding on the ground. Hawks and owls have eyes farther forward on the head. This increases their binocular field of view and correspondingly decreases their monocular field of view.

Like reptiles, birds have a nictitating membrane that is drawn over the surface of the eye to cleanse and protect the eye.

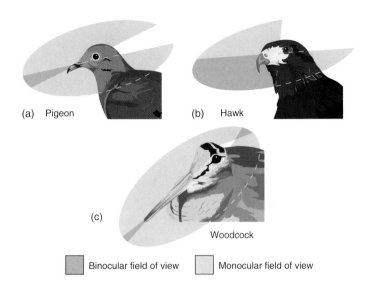

(a) Pigeon

(b) Hawk

(c)

Woodcock

☐ Binocular field of view   ☐ Monocular field of view

FIGURE 21.12

Avian Vision.   The fields of view of (*a*) a pigeon, (*b*) a hawk, and (*c*) a woodcock. Woodcocks have eyes located far posteriorly and have a narrow field of binocular vision in front and behind. They can focus on predators circling above them while probing mud with their long beaks.

Olfaction apparently plays a minor role in the lives of most birds. External nares open near the base of the beak, but the olfactory epithelium is poorly developed. Exceptions include turkey vultures, which locate their dead and dying prey largely by smell.

In contrast, most birds have well-developed hearing. Loose, delicate feathers called auriculars cover the external ear opening. Middle- and inner-ear structures are similar to those of reptiles. The sensitivity of the avian ear (100 to 15,000 Hz) is similar to that of the human ear (16 to 20,000 Hz).

EXCRETION AND OSMOREGULATION

Birds and reptiles face essentially identical excretory and osmoregulatory demands. Like reptiles, birds excrete uric acid, which is temporarily stored in the cloaca. Water is also reabsorbed in the cloaca. As with reptiles, the excretion of uric acid conserves water and promotes embryo development in terrestrial environments. In addition, some birds have supraorbital salt glands that drain excess sodium chloride through the nasal openings to the outside of the body. These are especially important in marine birds that drink seawater and feed on invertebrates containing large quantities of salt in their tissues. Salt glands can secrete salt in a solution that is two to three times more concentrated than other body fluids. Salt glands, therefore, compensate for the kidney's inability to concentrate salts in the urine.

REPRODUCTION AND DEVELOPMENT

The sexual activities of birds have been observed more closely than those of any other group of animals. These activities include establishing territories, finding mates, constructing nests, incubating eggs, and feeding young.

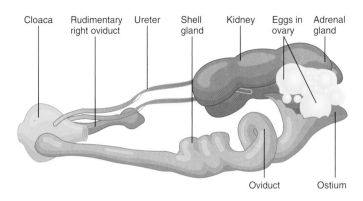

FIGURE 21.13

Urogenital System of a Female Pigeon.   The right ovary and oviduct are rudimentary in most female birds.

All birds are oviparous. Gonads are in the dorsal abdominal region, next to the kidneys. Testes are paired, and coiled tubules (vasa deferentia) conduct sperm to the cloaca. An enlargement of the vasa deferentia, the seminal vesicle, is a site for temporary storage and maturation of sperm prior to mating. Testes enlarge during the breeding season. Except for certain waterfowl and ostriches, birds have no intromittent organ, and sperm are transferred by cloacal contact when the male briefly mounts the female.

In females, two ovaries form during development, but usually only the left ovary fully develops (figure 21.13). A large, funnel-shaped opening (the ostium) of the oviduct envelops the ovary and receives eggs after ovulation. The egg is fertilized in the upper portions of the oviduct, and albumen that glandular regions of the oviduct wall secrete gradually surrounds the zygote as it completes its passage. A shell gland in the lower region of the oviduct adds a shell. The oviduct opens into the cloaca.

Many birds establish territories prior to mating. Although size and function vary greatly among species, territories generally allow birds to mate without interference. They provide nest locations and sometimes food resources for adults and offspring. Breeding birds defend their territories and expel intruders of the same sex and species. Threats are common, but actual fighting is minimal.

Mating may follow the attraction of a mate to a territory. For example, male woodpeckers drum on trees to attract females. Male ruffed grouse fan their wings on logs and create sounds that can be heard for many miles. Cranes have a courtship dance that includes stepping, bowing, stretching, and jumping displays. Mating occurs when a mate's call or posture signals readiness. It happens quickly but repeatedly to assure fertilization of all the eggs that will be laid.

Most birds are **monogamous.** A single male pairs with a single female during the breeding season. Some birds (swans, geese, eagles) pair for life. Frequent mating apparently strengthens the pair bonds that develop. Monogamy is common when resources are widely and evenly distributed, and one bird cannot control the access to resources. Monogamy is also advantageous because both parents usually participate in nest building and care of the young. One parent incubates and protects the eggs or chicks while the other searches for food.

FIGURE 21.14

Courtship Displays. A male greater prairie chicken (*Tympanuchus cupido*) displaying in a lek.

Some birds are **polygynous.** Males mate with more than one female, and the females care for the eggs and chicks. Polygyny tends to occur in species whose young are less dependent at hatching and in situations where patchy resource distribution may attract many females to a relatively small breeding area. Prairie chickens are polygynous, and males display in groups called leks. In prairie chicken leks, the males in the center positions are preferred and attract the majority of females (figure 21.14).

A few bird species are **polyandrous,** and the females mate with more than one male. For example, female spotted sandpipers are larger than males, and they establish and defend their territories from other females. They lay eggs for each male that is attracted to and builds a nest in their territory. If a male loses his eggs to a predator, the female replaces them. Polyandry results in the production of more eggs than in monogamous matings. It is thought to be advantageous when food is plentiful but, because of predation or other threats, the chances of successfully rearing young are low.

Nest construction usually begins after pair formation. The female usually initiates this instinctive behavior. A few birds do not make nests. Emperor penguins, for example, breed on the snow and ice of Antarctica, where no nest materials are available. Their single egg is incubated on the web of the foot (mostly the male's foot), tucked within a fold of abdominal skin.

Nesting Activities

The nesting behavior of birds is often species specific. Some birds choose nest sites away from other members of their species, and other birds nest in large flocks. Unfortunately, predictable nesting behavior has led to the extinction of some species of birds.

The group of eggs laid and chicks produced by a female is called a **clutch.** Clutch size usually varies. Most birds incubate their eggs, and some birds have a featherless, vascularized incubation or brood patch (*see figure 25.13*) that helps keep the eggs at temperatures between 33 and 37° C. Birds turn the eggs to prevent egg membranes from adhering in the egg and deforming the

(a)

(b)

FIGURE 21.15

Altricial and Precocial Chicks. (*a*) An American robin (*Turdus migratorius*) feeding nestlings. Robins have altricial chicks that are helpless at hatching. (*b*) Killdeer (*Charadrius vociferus*) have precocial chicks that are down covered and can move about.

embryo. Adults of some species sprinkle the eggs with water to cool and humidify them. The Egyptian plover carries water from distant sites in the breast feathers. The incubation period lasts between 10 and 80 days and correlates with egg size and degree of development at hatching. One or two days before hatching, the young bird penetrates an air sac at the blunt end of its egg, inflates its lungs, and begins breathing. Hatching occurs as the young bird pecks the shell with a keratinized egg tooth on the tip of the upper jaw and struggles to free itself.

Some birds are helpless at hatching; others are more independent. Those that are entirely dependent on their parents are said to be **altricial** (L. *altricialis*, to nourish), and they are often naked at hatching (figure 21.15*a*). Altricial young must be brooded constantly at first because endothermy is not developed. They grow rapidly, and when they leave the nest, they are nearly as large as their parents. For example, American robins weigh 4 to 6 g at hatching and leave the nest 13 days later weighing 57 g. **Precocial** (L. *pracoci*, early ripe) young are alert and lively at hatching (figure 21.15*b*). They are usually covered with down and can walk, run, swim, and feed themselves—although one parent is usually present to lead the young to food and shelter.

Young altricial birds have huge appetites and keep one or both parents continually searching for food. They may consume a mass of food that equals their own weight each day. Adults bring food to the nest or regurgitate food stored in the crop or esophagus. Vocal signals or color patterns on the bills or throats of adults initiate feeding responses in the young. Parents instinctively feed gaping mouths, and many hatchlings have brightly colored mouth linings or spots that attract a parent's attention. The first-hatched young is fed first—most often because it is usually the largest and can stretch its neck higher than can its nestmates.

Life is usually brief for birds. Approximately 50% of eggs laid yield birds that leave the nest. Most birds, if kept in captivity, have a potential life span of 10 to 20 years. Natural longevity is much shorter. The average American robin lives 1.3 years, and the average black-capped chickadee lives less than one year. Mortality is high in the first year from predators and inclement weather.

MIGRATION AND NAVIGATION

Over 20 centuries ago, Aristotle described birds migrating to escape the winter cold and summer heat. He had the mistaken impression that some birds disappear during winter because they hibernate and that others transmutate to another species. It is now known that some birds migrate long distances. Modern zoologists study the timing of migration, stimuli for migration, and physiological changes during migration, as well as migration routes and how birds navigate over huge expanses of land or water.

Migration (as used here) refers to periodic round trips between breeding and nonbreeding areas. Most migrations are annual, with nesting areas in northern regions and wintering grounds in the south. (Migration is more pronounced for species found in the Northern Hemisphere because about 70% of the earth's land is in the Northern Hemisphere.) Migrations occasionally involve east/west movements or altitude changes. Migration allows birds to avoid climatic extremes and to secure adequate food, shelter, and space throughout the year.

Birds migrate in response to species-specific physiological conditions. Innate (genetic) clocks and environmental factors influence their preparation for migration. The photoperiod is an important migratory cue for many birds, particularly for birds in temperate zones. The changing photoperiod initiates seasonal changes in gonadal development that often serve as migratory stimuli. Increasing day length in the spring promotes gonadal development, and decreasing day length in the fall initiates gonadal regression. In many birds, the changing photoperiod also appears to promote fat deposition, which acts as an energy reserve for migration. The anterior lobe of the pituitary gland and the pineal body have been implicated in mediating photoperiod responses.

The mechanics of migration are species specific. Some long-distance migrants may store fat equal to 50% of their body weight and make nonstop journeys. Other species that take a more leisurely approach to migration begin their journeys early and stop frequently to feed and rest. In clear weather, many birds fly at altitudes greater than 1,000 m, which reduces the likelihood of hitting tall obstacles. Many birds have very specific migration routes (*see figure 21.1*).

Navigation

Homing pigeons have served for many years as a pigeon postal service. In ancient Egyptian times and as recently as World War II, pigeons returned messages from the battlefield.

Birds use two forms of navigation. Route-based navigation involves keeping track of landmarks (visual or auditory) on an outward journey so that those landmarks can be used in a reverse sequence on the return trip. Location-based navigation is based on establishing the direction of the destination from information available at the journey's site of origin. It involves the use of sun compasses, other celestial cues, and/or the earth's magnetic field.

Birds' lenses are transparent to ultraviolet light, and their photoreceptors respond to it, allowing them to orient using the sun, even on cloudy days. This orientation cue is called a sun compass. Because the sun moves through the sky between sunrise and sunset, birds use internal clocks to perceive that the sun rises in the east, is approximately overhead at noon, and sets in the west. The biological clocks of migratory birds can be altered. For example, birds ready for northward migration can be held in a laboratory in which the "laboratory sunrise" occurs later than the natural sunrise. When released to natural light conditions, they fly in a direction they perceive to be north, but which is really northwest. Night migrators can also orient using the sun by flying in the proper direction from the sunset.

Celestial cues other than the sun can be used to navigate. Humans recognize that in the Northern Hemisphere, the North Star lines up with the axis of rotation of the earth. The angle between the North Star and the horizon decreases as you move toward the equator. Birds may use similar information to determine latitude. Experimental rotations of the night sky in a planetarium have altered the orientation of birds in test cages.

Some zoologists have long speculated that birds employ magnetic compasses to detect the earth's magnetic field, and thus, determine direction. Direct evidence of the existence of magnetic compasses now has been uncovered. Magnets strapped to the heads of pigeons severely disorient the birds. European robins and a night migrator, the garden warbler, orient using the earth's magnetic field. However, no discrete magnetic receptors have been found in either birds or other animals. Early reports of finding a magnetic iron, magnetite, in the head and necks of pigeons did not lead to a greater understanding of magnetic compasses. Further experiments failed to demonstrate magnetic properties in these regions. Magnetic iron has been found in bacteria and a variety of animal tissues. None is clearly associated with a magnetic sense, although the pineal body of pigeons has been implicated in the use of a sun compass and in responses to magnetic fields.

There is redundancy in bird navigational mechanisms, which suggests that under different circumstances, birds probably use different sources of information.

WILDLIFE ALERT
Red-Cockaded Woodpecker (*Picoides borealis*)

VITAL STATISTICS

Classification: Phylum Chordata, class Aves, order Piciformes, family Picidae

Range: Fragmented, isolated populations where southern pines exist in the United States, (Alabama, Arkansas, Georgia, Kentucky, Louisiana, Mississippi, North Carolina, South Carolina, and Tennessee)

Habitat: Open stands of pines with a minimum age of 80 to 120 years

Number remaining: Approximately 4,500 groups, or 10,000 to 12,000 birds

Status: Endangered throughout its range

NATURAL HISTORY AND ECOLOGICAL STATUS

Red-cockaded woodpeckers are 18 to 20 cm long with a wingspan of 35 to 38 cm. They have black-and-white horizontal stripes on the back, and their cheeks and underparts are white (box figure 21.1). Males have a small, red spot on each side of their black cap. After the first postfledgling molt, fledgling males have a red crown patch. The diet of these woodpeckers consists mostly of insects and wild fruit.

BOX FIGURE 21.2   Distribution of the Red-Cockaded Woodpecker (*Picoides borealis*).

Eggs are laid from April through June, with females utilizing their mate's roosting cavity as a nest. The average clutch size is three to five eggs. Most often, the parent birds and one or more male offspring from previous nests form a family unit called a group. A group may include one breeding pair and as many as seven other birds. Rearing the young becomes a shared responsibility of the group.

The range of red-cockaded woodpeckers is closely tied to the distribution of southern pines, with open stands of trees in the 80- to 120-year-old group being the favored nesting habitat (box figure 21.2). Dense stands or hardwoods are usually avoided. The woodpeckers excavate roosting cavities in living pines, usually those infected with a fungus that produces what is known as red-heart disease. The aggregate of cavity trees is called a cluster and may include 1 to 20 or more cavity trees on 3 to 60 acres. Completed cavities in active use have numerous small resin wells, which exude sap. The birds keep the sap flowing, apparently as a cavity defense mechanism against rat snakes and other predators. The territory for a group averages about 200 acres.

The decline of red-cockaded woodpecker populations is due primarily to the cutting of pine forests with trees that are 80 years or more old and the encroachment of hardwood understories. Recommendations for management and protection include: survey, monitor, and assess the status of individual populations and the species; protect and manage nesting and foraging habitats on federal lands; encourage protection and management on private lands; and inform and involve the public.

BOX FIGURE 21.1   Red-Cockaded Woodpecker (*Picoides borealis*).

SUMMARY

1. Birds are members of the archosaur lineage. Fossils of ancient feathered dinosaurs and birds—*Archaeopteryx, Sinornis,* and *Eoalularis*—show reptilian affinities and give clues to the origin of flight.

2. Feathers evolved from reptilian scales and function in flight, insulation, sex recognition, and waterproofing. Feathers are maintained and periodically molted.

3. The bird skeleton is light and made more rigid by the fusion of bones. Birds use the neck and bill as a fifth appendage.

4. Bird wings form airfoils that provide lift. Tilting the wing during flapping generates propulsive force. Gliding, flapping, soaring, and hovering flight are used by different birds or by the same bird at different times.

5. Birds feed on a variety of foods, as reflected in the structure of the bill and other parts of the digestive tract.

6. The heart of birds consists of two atria and two ventricles. Rapid heart rate and blood flow support the high metabolic rate of birds.

7. The respiratory system of birds provides one-way, nearly constant air movement across respiratory surfaces.

8. Birds are able to maintain high body temperatures endothermically because of insulating fat deposits and feathers.

9. The development of the corpus striatum enlarged the cerebral hemispheres of birds. Vision is the most important avian sense.

10. Birds are oviparous. Reproductive activities include the establishment and defense of territories, courtship, and nest building.

11. Either or both bird parents incubate the eggs, and one or both parents feed the young. Altricial chicks are helpless at hatching, and precocial chicks are alert and lively shortly after hatching.

12. Migration allows some birds to avoid climatic extremes and to secure adequate food, shelter, and space throughout the year. The photoperiod is the most important migratory cue for birds.

13. Birds use both route-based navigation and location-based navigation.

SELECTED KEY TERMS

airfoil (p. 332)
altricial (p. 338)
molting (p. 331)
monogamous (p. 337)

polyandrous (p. 338)
polygynous (p. 338)
precocial (p. 339)

CRITICAL THINKING QUESTIONS

1. Birds are sometimes called "glorified reptiles." Discuss why this description is appropriate.

2. What adaptations of birds promote endothermy and flight? Why is endothermy important for birds?

3. Birds are, without exception, oviparous. Why do you think that is true?

4. What are the advantages that offset the great energy expenditure that migration requires?

5. Compare and contrast the advantages of monogamy, polygyny, and polyandry for birds. In what ways are the advantages and disadvantages of each related to the abundance and utilization of food and other resources?

ONLINE LEARNING CENTER

Visit our Online Learning Center (OLC) at www.mhhe.com/zoology (click on this book's title) to find the following chapter-related materials:

- CHAPTER QUIZZING
- RELATED WEB LINKS
- CLASS AVES
    Class Aves
    Marine Birds
    Dissection Guides for Birds
    Conservation Issues Concerning Birds
- BOXED READINGS ON
    Bones and Scales
    Sailors' Curse—Gliders' Envy
    Bright Skies and Silent Thunder
    Extrarenal Secretion by Avian and Reptilian Salt Glands
- SUGGESTED READINGS

CHAPTER 22

MAMMALS:

SPECIALIZED TEETH, HAIR, ENDOTHERMY, AND VIVIPARITY

Outline

Concepts

1. Mammalian characteristics evolved gradually over a 200-million-year period in the synapsid lineage.
2. Two subclasses of mammals evolved during the Mesozoic era—the Prototheria and the Theria. Modern mammals include monotremes, marsupial mammals, and placental mammals.
3. The skin of mammals is thick and protective and has an insulating covering of hair.
4. Adaptations of teeth and the digestive tract allow mammals to exploit a wide variety of food resources.
5. Efficient systems for circulation and gas exchange support the high metabolic rate associated with mammalian endothermy.
6. The brain of mammals has an expanded cerebral cortex that processes information from various sensory structures.
7. Metanephric kidneys permit mammals to excrete urea without excessive water loss.
8. Complex behavior patterns enhance mammalian survival.
9. Most mammals are viviparous and have reproductive cycles that help ensure internal fertilization and successful development.

EVOLUTIONARY PERSPECTIVE

The fossil record that documents the origin of the mammals from ancient reptilian ancestors is very complete and relatively uncontroversial. It is being used to test, and has confirmed, many macroevolutionary theories (*see chapter 4*). The beginning of the Tertiary period, about 70 million years ago, was the start of the "age of mammals." It coincided with the extinction of many reptilian lineages, which led to the adaptive radiation of the mammals. Tracing the roots of the mammals, however, requires returning to the Carboniferous period, when the synapsid lineage diverged from other amniote lineages (*see figure 20.3*).

Mammalian characteristics evolved gradually over a period of 200 million years (figure 22.1). Most of what we know about early synapsids is based on skeletal characteristics. The early synapsids were the pelycosaurs. Some were herbivores; others showed skeletal adaptations that reflect increased effectiveness as predators (figure 22.2a). The anterior teeth of the upper jaw were large and were separated from the posterior teeth by a gap that accommodated the enlarged anterior teeth of the lower jaw when the jaw closed. The palate was

This chapter contains evolutionary concepts, which are set off in this font.

FIGURE 22.1

Class Mammalia.   The decline of the ruling reptiles about 70 million years ago permitted mammals to radiate into diurnal habitats previously occupied by dinosaurs and other reptiles. Hair, endothermy, and mammary glands characterize mammals. The lowland gorilla (*Gorilla gorilla graueri*, order Primates) is shown here.

arched, which strengthened the upper jaw and allowed air to pass over prey held in the mouth. Early synapsids had a reptile-like sprawling gate. In later synapsids, the legs were longer and slimmer, and the body was held higher from the ground.

By the middle of the Permian period, other successful mammal-like reptiles had arisen from the pelycosaurs. They were a diverse group known as the therapsids. Some were predators, and others were herbivores. In the predatory therapsids, teeth were concentrated at the front of the mouth and enlarged for holding and tearing prey. The posterior teeth were reduced in size and number. The jaws of some therapsids were elongate and generated a large biting force when snapped closed. The teeth of the herbivorous therapsids were also mammal-like. Some had a large space, called the diastema, separating the anterior and posterior teeth. The posterior teeth had ridges (cusps) and cutting edges that were probably used to shred plant material. Unlike other reptiles, therapsids held hindlimbs directly beneath the body and moved them parallel to the long axis of the body. Changes in the size and shape of the ribs suggest the separation of the trunk into thoracic and abdominal regions and a breathing mechanism similar to that of mammals. The last therapsids were a group called the cynodonts (figure 22.2b). Some of these were as large as a big dog, but most were small and little different from the earliest mammals.

The first mammals were small (less than 10 cm long) with delicate skeletons. Most knowledge of early mammalian phylogeny comes from the study of their fossilized teeth and skull fragments. These studies suggest that the mammals of the Jurassic and Cretaceous periods were mostly predators that fed on other vertebrates and arthropods. A few were herbivores, and others combined predatory and herbivorous feeding habits. Changes in the structure of the middle ear and the regions of the brain devoted to hearing and olfaction indicate that these senses were important during the early evolution of mammals.

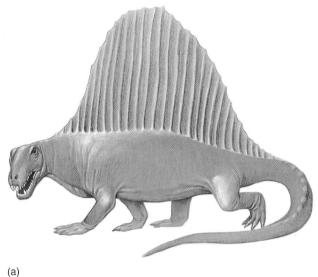

(a)

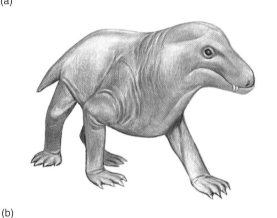

(b)

FIGURE 22.2

Members of the Subclass Synapsida.   (*a*) *Dimetrodon* was a 3 m long pelycosaur. It probably fed on other reptiles and amphibians. The large sail may have been a recognition signal and a thermoregulatory device. (*b*) *Cynognathus* was a mammal-like reptile that probably foraged for small animals, much like a badger does today. The badger-sized animal was a cynodont within the order Therapsida, the stock from which mammals arose during the mid-Triassic period.

*S*ome zoologists speculate that the small size, well-developed olfactory and auditory functions, and lack of color vision in most mammals suggest that early mammals were nocturnal. This habit may have allowed them to avoid competition with the much larger dinosaurs and the smaller diurnal (day-active [L. diurnalis, daily]) reptiles living at the same time. Again, it is speculative, but nocturnal habits could have led to endothermy. Endothermy would have allowed small mammals to maintain body temperatures above that of their surroundings after the sun had set and the air temperature began to fall.

DIVERSITY OF MAMMALS

Hair, mammary glands, specialized teeth, three middle-ear ossicles, endothermy, and other characteristics listed in table 22.1 characterize modern members of the class Mammalia (mah-ma'le-ah)

TABLE 22.1
CLASSIFICATION OF LIVING MAMMALS

Class Mammalia (mah-ma'le-ah)
Mammary glands; hair; diaphragm; three middle-ear ossicles; heterodont dentition; sweat, sebaceous, and scent glands; four-chambered heart; large cerebral cortex.

Subclass Prototheria (pro"to-ther'e-ah)
This subclass formerly contained the monotremes. Monotremes have recently been reclassified, and this subclass now contains only extinct species.

Subclass Theria (ther'e-ah)
Technical characteristics of the skull distinguish members of this subclass.

Infraclass Ornithodelphia (or"ne-tho-del'fe-ah)
Technical characteristics of the skull distinguish members of this infraclass, Monotremes.

Infraclass Metatheria (met"ah-ther'e-ah)
Viviparous; primitive placenta; young are born early and often are carried in a marsupial pouch on the female's belly. Marsupials.

Infraclass Eutheria (u-ther'e-ah)*
Complex placenta; young develop to advanced stage prior to birth. Placentals.

Order Insectivora (in-sec-tiv'or-ah)
Diverse group of small, primitive mammals; third largest mammalian order. Hedgehogs, tenrecs, moles, shrews.

Order Chiroptera (ki-rop'ter-ah)
Cosmopolitan, but especially abundant in the tropics; bones of the arm and hand are elongate and slender; flight membranes extend from the body, between digits of forelimbs, to the hindlimbs; most are insectivorous, but some are fruit eaters, fish eaters, and blood feeders; second largest mammalian order. Bats.

Order Primates (pri-ma'tez)
Adaptations of primates reflect adaptations for increased agility in arboreal (tree-dwelling) habitats; omnivorous diets; unspecialized teeth; grasping digits; freely movable limbs; nails on digits; reduced nasal cavity; enlarged stereoscopic eyes and cerebral hemispheres. Lemurs (Madagascar and the Comoro Islands), tarsiers (jungles of Sumatra and the East Indies), monkeys, gibbons, great apes (apes and humans).

Order Xenarthra (ze'nar-thra) or **Edentata** (e"den-ta'tah)
Incisors and canines absent; cheek teeth, when present, lack enamel; braincase is long and cylindrical; hindfoot is four toed; forefoot has two or three prominent toes with large claws; limbs are specialized for climbing or digging; xenarthrous lumbar vertebrae. Anteaters, tree sloths, armadillos.

Order Lagomorpha (lag"o-mor'fah)
Two pairs of upper incisors; one pair of lower incisors; incisors are ever-growing and slowly worn down by feeding on vegetation. Rabbits, pikas.

Order Rodentia (ro-den'che-ah)
Largest mammalian order; upper and lower jaws bear a single pair of ever-growing incisors. Squirrels, chipmunks, rats, mice, beavers, porcupines, woodchucks, lemmings.

Order Cetacea (se-ta'she-ah)
Streamlined, nearly hairless, and insulated by thick layers of fat (blubber); no sebaceous glands; forelimbs modified into paddlelike flippers for swimming; hindlimbs reduced and not visible externally; tail fins (flukes) flattened horizontally; external naris (blowhole) on top of skull. Toothed whales (beaked whales, narwhals, sperm whales, dolphins, porpoises, killer whales); toothless, filter-feeding baleen whales (right whales, gray whales, blue whales, and humpback whales).

Order Carnivora (kar-niv'o-rah)
Predatory mammals; usually have a highly developed sense of smell and a large braincase; premolars and molars modified into carnassial apparatus; three pairs of upper and lower incisors usually present, and canines are well developed. Dogs, cats, bears, raccoons, minks, sea lions, seals, walruses, otters.

Order Proboscidea (pro"bah-sid'e-ah)
Long, muscular proboscis (trunk) with one or two fingerlike processes at the tip; short skull with the second incisor on each side of the upper jaw modified into tusks; six cheek teeth are present in each half of each jaw; teeth erupt (grow into place) in sequence from front to rear, so that one tooth in each jaw is functional. African and Indian elephants.

Order Sirenia (si-re'ne-ah)
Large, aquatic herbivores that weigh in excess of 600 kg; nearly hairless, with thick, wrinkled skin; heavy skeleton; forelimb is flipperlike, and hindlimb is vestigial; horizontal tail fluke is present; horizontally oriented diaphragm; teeth lack enamel. Manatees (coastal rivers of the Americas and Africa), dugongs (western Pacific and Indian Oceans).

Order Perissodactyla (pe-ris"so-dak'ti-lah).
Hoofed; axis of support passes through the third digit. Skull usually elongate, large molars and premolars; primarily grazers. (The Artiodactyla also have hoofs. Artiodactyls and perissodactyls are, therefore, called ungulates [L. *ungula*, hoof]). Odd-toed ungulates: horses, rhinoceroses, zebras, tapirs.

Order Artiodactyla (ar"te-o-dak'ti-lah)
Hoofed; axis of support passes between third and fourth digits; digits one, two, and five reduced or lost; primarily grazing and browsing animals (pigs are an obvious exception). Even-toed ungulates: pigs, hippopotamuses, camels, antelope, deer, sheep, giraffes, cattle.

*Selected eutherian orders are described.

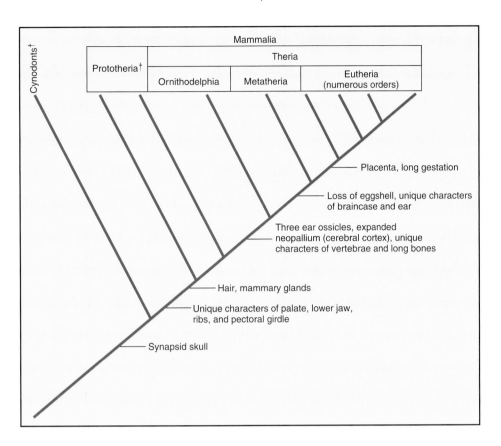

FIGURE 22.3
Mammalian Phylogeny.   A cladogram showing the evolutionary relationships among mammals. Selected characters are shown. Daggers (†) indicate some extinct taxa. However, numerous extinct groups have been omitted from the cladogram. The names of the various orders of eutherians have also been omitted (*see table 22.1*).

(L. *mamma*, breast). *The extent to which all of these characteristics were developed in the earliest mammals cannot currently be determined because hair, mammary glands, and endothermy are hard to detect in the fossil record. The presence of large sails on some early synapsids suggests that they relied on the environment for behavioral warming and cooling. The first evidence of hair in the fossil record dates to about 60 million years ago. The presence of hair in all groups of extant mammals, however, suggests that that characteristic was present in the common ancestor of modern mammals about 130 million years ago. Although there is some debate, most zoologists recognize two early lineages that are usually designated as subclasses: the Prototheria and the Theria (figure 22.3).* Until recently, monotremes (the duck-billed platypus and the echidna) were classified in the subclass Prototheria (Gr. *protos*, first + *therion*, wild beast). Recent fossil evidence showing monotreme dentition that is characteristic of the subclass Theria has resulted in this group being reassigned to the latter subclass. The Prototheria, therefore, contains only extinct forms.

The subclass Theria diverged into three infraclasses by the late Cretaceous period. The infraclass Ornithodelphia (Gr. *ornis*, bird + *delphia*, birthplace) contains the monotremes (Gr. *monos*,

one + *trema*, opening). This name refers to the fact that monotremes, unlike other mammals, possess a cloaca. Unlike other mammals, monotremes are oviparous (figure 22.4*a,b*). The six species of monotremes are found in Australia and New Guinea.

The infraclass Metatheria (Gr. *meta*, after) contains the marsupial mammals. They are viviparous but have very short gestation periods. A protective pouch, called the marsupium, covers the female mammary glands. The young crawl into the marsupium after birth, where they feed and complete development. About 250 species of marsupials live in the Australian region and the Americas (figure 22.4*c*; *see also figure 22.17*).

The other therian infraclass, Eutheria (Gr. *eu*, true), contains the placental mammals. They are usually born at an advanced stage of development, having been nourished within the uterus. Exchanges between maternal and fetal circulatory systems occur by diffusion across an organ called the **placenta,** which is composed of both maternal and fetal tissue. The approximately 3,800 species of eutherians are classified into 17 orders (figures 22.5 and 22.6; *see also figures 22.11 and 22.15–22.17*). Evolutionary relationships within these orders are the subject of intensive research and debate.

(a)

(b)

(c)

FIGURE 22.4

Representatives of the Mammalian Infraclasses Ornithodelphia and Metatheria.   The infraclass Ornithodelphia: (*a*) A duck-billed platypus (*Ornithorhychus anatinus*). (*b*) An echidna or spiny anteater (*Tachyglossus aculeatus*). The infraclass Metatheria: (*c*) The koala (*Phascolarctos cinereus*) feeds on *Eucalyptus* leaves in Australia.

(a)

(b)

FIGURE 22.5

Order Xenarthra.   (*a*) A giant anteater (*Myrmecophaga tridactyla*). Anteaters lack teeth. They use powerful forelimbs to tear into an insect nest and a long tongue covered with sticky saliva to capture prey. (*b*) An armadillo (*Dasypus novemcinctus*).

FIGURE 22.6
Order Carnivora. An Arctic fox (*Alopex lagopus*) with its winter pelage. With its spring molt, the Arctic fox acquires a gray-and-yellow-colored coat.

EVOLUTIONARY PRESSURES

Mammals are naturally distributed on all continents except Antarctica, and they live in all oceans. This section discusses the many adaptations that have accompanied their adaptive radiation.

EXTERNAL STRUCTURE AND LOCOMOTION

The skin of a mammal, like that of other vertebrates, consists of epidermal and dermal layers. It protects from mechanical injury, invasion by microorganisms, and the sun's ultraviolet light. Skin is also important in temperature regulation, sensory perception, excretion, and water regulation (*see figure 23.9*).

Hair is a keratinized derivative of the epidermis of the skin and is uniquely mammalian. It is seated in an invagination of the epidermis called a hair follicle. A coat of hair, called pelage, usually consists of two kinds of hair. Long guard hairs protect a dense coat of shorter, insulating underhairs.

Because hair is composed largely of dead cells, it must be periodically molted. In some mammals (e.g., humans), molting occurs gradually and may not be noticed. In others, hair loss occurs rapidly and may result in altered pelage characteristics. In the fall, many mammals acquire a thick coat of insulating underhair, and the pelage color may change. For example, the Arctic fox takes on a white or cream color with its autumn molt, which helps conceal the fox in a snowy environment. With its spring molt, the Arctic fox acquires a gray and yellow pelage (*see figure 22.6*).

Hair is also important for the sense of touch. Mechanical displacement of a hair stimulates nerve cells associated with the hair root. Guard hairs may sometimes be modified into thick-shafted hairs called vibrissae. Vibrissae occur around the legs, nose, mouth, and eyes of many mammals. Their roots are richly innervated and very sensitive to displacement.

Air spaces in the hair shaft and air trapped between hair and the skin provide an effective insulating layer. A band of smooth muscle, called the arrector pili muscle, runs between the hair follicle and the lower epidermis. When the muscle contracts, the hairs stand upright, increasing the amount of air trapped in the pelage and improving its insulating properties. Arrector pili muscles are under the control of the autonomic nervous system, which also controls a mammal's "fight-or-flight" response. In threatening situations, the hair (especially on the neck and tail) stands on end and may give the perception of increased size and strength.

Hair color depends on the amount of pigment (melanin) deposited in it and the quantity of air in the hair shaft. The pelage of most mammals is dark above and lighter underneath. This pattern makes them less conspicuous under most conditions. Some mammals advertise their defenses using aposematic (warning) coloration. The contrasting markings of a skunk are a familiar example.

Pelage is reduced in large mammals from hot climates (e.g., elephants and hippopotamuses) and in some aquatic mammals (e.g., whales) that often have fatty insulation. A few mammals (e.g., naked mole rats) have almost no pelage.

Claws are present in all amniote classes. They are used for locomotion and offensive and defensive behavior. Claws form from accumulations of keratin that cover the terminal phalanx (bone) of the digits. In some mammals, they are specialized to form nails or hooves (figure 22.7).

Glands develop from the epidermis of the skin. **Sebaceous (oil) glands** are associated with hair follicles, and their oily secretion lubricates and waterproofs the skin and hair. Most mammals also possess **sudoriferous (sweat) glands.** Small sudoriferous glands (eccrine glands) release watery secretions used in evaporative cooling. Larger sudoriferous glands (apocrine glands) secrete a mixture of salt, urea, and water, which microorganisms on the skin convert to odorous products.

Scent or **musk glands** are around the face, feet, or anus of many mammals. These glands secrete pheromones, which may be involved with defense, species and sex recognition, and territorial behavior.

Mammary glands are functional in female mammals and are present, but nonfunctional, in males. The milk that mammary glands secrete contains water, carbohydrates (especially the sugar lactose), fat, protein, minerals, and antibodies. *Mammary glands are probably derived evolutionarily from apocrine glands and usually contain substantial fatty deposits.*

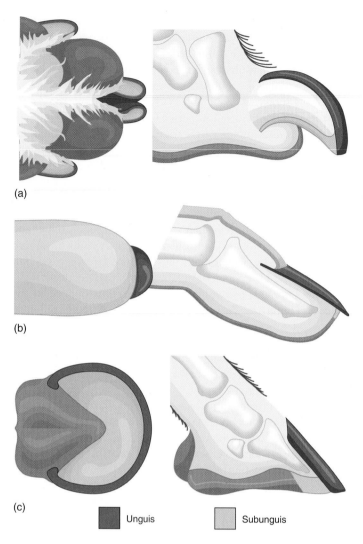

(a)

(b)

(c)

Unguis   Subunguis

FIGURE 22.7

Structure of Claws, Nails, and Hooves.   (*a*) Claws. (*b*) Nails are flat, broad claws found on the hands and feet of primates and are an adaptation for arboreal habits, where grasping is essential. (*c*) Hooves are characteristic of ungulate mammals. The number of toes is reduced, and the animals walk or run on the tips of the remaining digits. The unguis is a hard, keratinized dorsal plate, and the subunguis is a softer ventral plate.

Monotremes have mammary glands that lack nipples. The glands discharge milk into depressions on the belly, where the young lap it up. In other mammals, mammary glands open via nipples or teats, and the young suckle for their nourishment (figure 22.8).

The Skull and Teeth

The skulls of mammals show important modifications of the reptilian pattern. One feature that zoologists use to distinguish reptilian from mammalian skulls is the method of jaw articulation. In reptiles, the jaw articulates at two small bones at the rear of the jaw. In mammals, these bones have moved into the middle ear, and along with the stapes, form the middle-ear ossicles. A single bone of the lower jaw articulates the mammalian jaw.

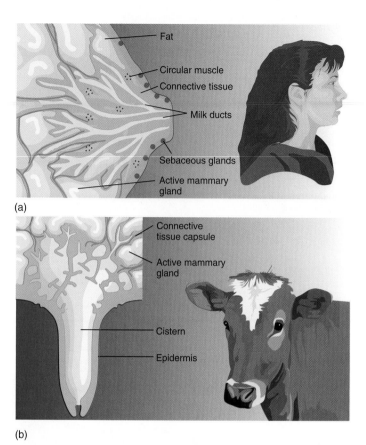

(a)

(b)

FIGURE 22.8

Mammary Glands.   Mammary glands are specialized to secrete milk following the birth of young. (*a*) Many ducts lead from the glands to a nipple. Parts of the duct system are enlarged to store milk. Suckling by an infant initiates a hormonal response that causes the mammary glands to release milk. (*b*) Some mammals (e.g., cattle) have teats that form by the extension of a collar of skin around the opening of mammary ducts. Milk collects in a large cistern prior to its release. The number of nipples or teats varies with the number of young produced.

A secondary palate evolved twice in vertebrates—in the archosaur lineage (*see figure 20.3*) and in the synapsid lineage. In some therapsids, small, shelflike extensions of bone (the hard palate) partially separated the nasal and oral passageways (*see figure 20.11*). In mammals, the secondary palate extends posteriorly by a fold of skin, called the soft palate, which almost completely separates the nasal passages from the mouth cavity. Unlike other vertebrates that swallow food whole or in small pieces, some mammals chew their food. The more extensive secondary palate allows mammals to breathe while chewing. Breathing needs to stop only briefly during swallowing (figure 22.9).

The structure and arrangement of teeth are important indicators of mammalian lifestyles. In reptiles, the teeth are uniformly conical, a condition referred to as **homodont.** In mammals, the teeth are often specialized for different functions, a condition called **heterodont.** Reptilian teeth attach along the top or inside of the jaw, whereas in mammals, the teeth are set into sockets of the jaw. Most mammals have two sets of teeth during their lives. The first teeth emerge before or shortly after birth and are called deciduous or milk teeth. These teeth are lost, and permanent teeth replace them.

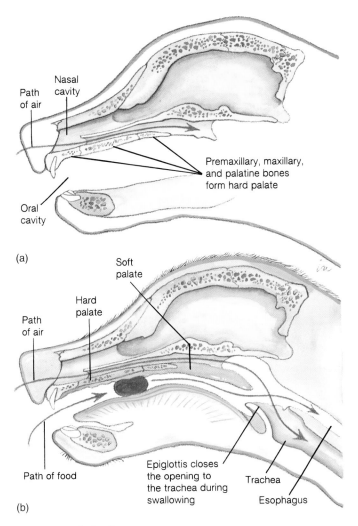

FIGURE 22.9

Secondary Palate. (*a*) The secondary palate (consisting of the hard and soft palates) of a mammal almost completely separates the nasal and oral cavities. (*b*) Breathing stops only momentarily during swallowing.

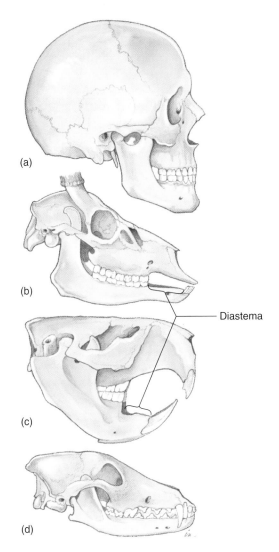

FIGURE 22.10

Specializations of Teeth. (*a*) An omnivore (*Homo sapiens*). (*b*) An herbivore, the male fallow deer (*Dama dama*). (*c*) A rodent, the beaver (*Castor canadensis*). (*d*) A carnivore, the coyote (*Canis latrans*).

Adult mammals have up to four kinds of teeth. Incisors are the most anterior teeth in the jaw. They are usually chisel-like and used for gnawing or nipping. Canines are often long, stout, and conical, and are usually used for catching, killing, and tearing prey. Canines and incisors have single roots. Premolars are positioned next to canines, and have one or two roots and truncated surfaces for chewing. Molars have broad chewing surfaces and two (upper molars) or three (lower molars) roots.

Mammalian species have characteristic numbers of each kind of adult tooth. Zoologists use a **dental formula** to characterize taxa. It is an expression of the number of teeth of each kind in one-half of the upper and lower jaws. The teeth of the upper jaw are listed above those of the lower jaw and in the following order: incisors, canine, premolars, and molars. For example:

Human	Beaver
2•1•2•3	1•0•1•3
2•1•2•3	1•0•1•3

Mammalian teeth (dentition) may be specialized for particular diets. In some mammals, the dentition is reduced, sometimes to the point of having no teeth. For example, armadillos and the giant anteater (order Edentata) feed on termites and ants, and their teeth are reduced.

Some mammals (e.g., humans, order Primates; and pigs, order Artiodactyla) are omnivorous; they feed on a variety of plant and animal materials. They have anterior teeth with sharp ripping and piercing surfaces, and posterior teeth with flattened grinding surfaces for rupturing plant cell walls (figure 22.10*a*).

Mammals that eat plant material often have flat, grinding posterior teeth and incisors, and sometimes have canines modified for nipping plant matter (e.g., horses, order Perissodactyla; deer, order Artiodactyla) or gnawing (e.g., rabbits, order Lagomorpha; beavers, order Rodentia) (figure 22.10*b,c*). In rodents, the incisors grow throughout life. Although most mammals have enamel covering the entire tooth, rodents have enamel only on the front surfaces of their incisors. The teeth are kept sharp by slower wear in front than in back. A gap called the diastema separates the anterior food-procuring teeth from the posterior grinding teeth. The diastema results from an elongation of the snout that allows the anterior teeth to reach close to the ground or into narrow openings to procure food. The posterior teeth have a high, exposed surface (crown) and continuous growth, which allows these teeth to withstand years of grinding tough vegetation.

Predatory mammals use canines and incisors for catching, killing, and tearing prey. In members of the order Carnivora (e.g., coyotes, dogs, and cats), the fourth upper premolars and first lower molars form a scissorlike shearing surface, called the carnassial apparatus, that is used for cutting flesh from prey (figure 22.10d).

The Vertebral Column and Appendicular Skeleton

The vertebral column of mammals is divided into five regions. As with reptiles and birds, the first two cervical vertebrae are the atlas and axis. Five other cervical vertebrae usually follow. Even the giraffe and the whale have seven neck vertebrae, which are greatly elongate or compressed, respectively. In contrast, tree sloths have either six or nine cervical vertebrae, and the manatee has six cervical vertebrae.

The trunk is divided into thoracic and lumbar regions, as is the case for birds. In mammals, the division is correlated with their method of breathing. The thoracic region contains the ribs. Most ribs connect to the thoracic vertebrae and to the sternum via costal cartilages. Other ribs attach only to the thoracic vertebrae or to other ribs through cartilages. All ribs protect the heart and lungs. The articulation between the thoracic vertebrae provides the flexibility needed in turning, climbing, and lying on the side to suckle young. Lumbar vertebrae have interlocking processes that give support, but little freedom of movement.

The appendicular skeleton of mammals rotates under the body so that the appendages are usually directly beneath the body. Joints usually limit the movement of appendages to a single anteroposterior plane, causing the tips of the appendages to move in long arcs. The bones of the pelvic girdle are fused in the adult, a condition that is advantageous for locomotion but presents problems during the birth of offspring. In a pregnant female, the ventral joint between the halves of the pelvis—the pubic symphysis—loosens before birth, allowing the pelvis to spread during birth.

Muscles

Because the appendages are directly beneath the body of most mammals, the skeleton bears the weight of the body. Muscle mass is concentrated in the upper appendages and girdles. Many running mammals (e.g., deer, order Artiodactyla) have little muscle in their lower leg that would slow leg movement. Instead, tendons run from muscles high in the leg to cause movement at the lower joints.

NUTRITION AND THE DIGESTIVE SYSTEM

The digestive tract of mammals is similar to that of other vertebrates but has many specializations for different feeding habits. Some specializations of teeth have already been described.

FIGURE 22.11

Order Perissodactyla. This plains zebra (*Equus burchelli*) is native to the savannas of eastern Africa.

The feeding habits of mammals are difficult to generalize. Feeding habits reflect the ecological specializations that have evolved. For example, most members of the order Carnivora feed on animal flesh and are, therefore, carnivores. Other members of the order, such as bears, feed on a variety of plant and animal products and are omnivores. Some carnivorous mammals are specialized for feeding on arthropods or soft-bodied invertebrates, and are often referred to as (rather loosely) insectivores. These include animals in the orders Insectivora (e.g., shrews), Chiroptera (bats), and Edentata (anteaters) (*see figure 22.5a*). Herbivores such as deer (order Artiodactyla) and zebras (order Perissodactyla) (figure 22.11) feed mostly on vegetation, but their diet also includes invertebrates inadvertently ingested while feeding.

Specializations in the digestive tract of most herbivores reflect the difficulty of digesting food rich in cellulose. Horses, rabbits, and many rodents have an enlarged **cecum** at the junction of the large and small intestines. A cecum is a fermentation pouch where microorganisms aid in cellulose digestion. Sheep, cattle, and deer are called ruminants (L. *ruminare*, to chew the cud). Their stomachs are modified into four chambers. The first three chambers are storage and fermentation chambers and contain microorganisms that synthesize a cellulose-digesting enzyme (cellulase). Gases that fermentation produces are periodically belched, and some plant matter (cud) is regurgitated and rechewed. Other microorganisms convert nitrogenous compounds in the food into new proteins.

CIRCULATION, GAS EXCHANGE, AND TEMPERATURE REGULATION

The hearts of birds and mammals are superficially similar. Both are four-chambered pumps that keep blood in the systemic and pulmonary circuits separate, and both evolved from the hearts of ancient reptiles. *Their similarities, however, are a result of adaptations to active lifestyles. The evolution of similar structures in different lineages is called convergent evolution. The*

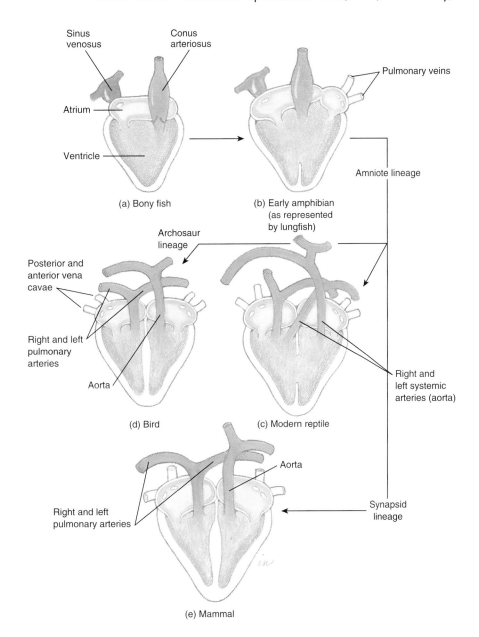

FIGURE 22.12

Possible Sequence in the Evolution of the Vertebrate Heart.   (*a*) Diagrammatic representation of a bony fish heart. (*b*) In lungfish, partially divided atria and ventricles separate pulmonary and systemic circuits. This heart was probably similar to that in primitive amphibians and early amniotes. (*c*) The hearts of modern reptiles were derived from the pattern in (*b*). (*d*) The archosaur and (*e*) synapsid lineages resulted in completely separated, four-chambered hearts.

mammalian heart evolved from the synapsid reptilian lineage, whereas the avian heart evolved from the archosaur lineage (figure 22.12).

One of the most important adaptations in the circulatory system of eutherian mammals concerns the distribution of respiratory gases and nutrients in the fetus (figure 22.13*a*). Exchanges between maternal and fetal blood occur across the placenta. Although maternal and fetal blood vessels are intimately associated, no blood actually mixes. Nutrients, gases, and wastes simply diffuse between fetal and maternal blood supplies.

Blood entering the right atrium of the fetus is returning from the placenta and is highly oxygenated. Because fetal lungs are not inflated, resistance to blood flow through the pulmonary arteries is high. Therefore, most of the blood entering the right atrium bypasses the right ventricle and passes instead into the left atrium through a valved opening between the atria (the foramen ovale). Some blood from the right atrium, however, does enter the right ventricle and the pulmonary artery. Because of the resistance at the uninflated lungs, most of this blood is shunted to the aorta through a vessel connecting the aorta and the left pulmonary artery (the ductus arteriosus). At birth, the placenta is lost, and the lungs are inflated. Resistance to blood flow through the lungs is reduced, and blood flow to them increases. Flow through the ductus arteriosus decreases, and the vessel is gradually reduced to a ligament. Blood flow back to the left atrium from the lungs correspondingly increases, and the valve of the foramen ovale closes

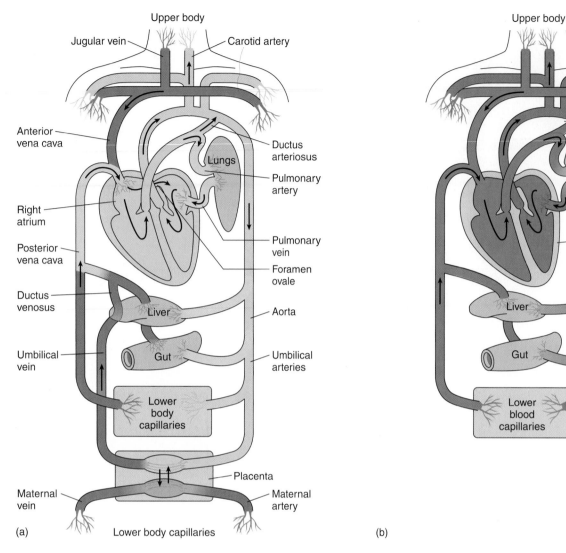

FIGURE 22.13

Mammalian Circulatory Systems. The circulatory patterns of (a) fetal and (b) adult mammals. Highly oxygenated blood is shown in red, and less oxygenated blood is shown in blue. In fetal circulation, highly oxygenated blood from the placenta mixes with less oxygenated blood prior to entering the right atrium. Thus, most arterial blood of the fetus is moderately oxygenated. The rose color in (a) symbolizes this state of oxygenation.

and gradually fuses with the tissue separating the right and left atria (figure 22.13b).

Gas Exchange

High metabolic rates require adaptations for efficient gas exchange. Most mammals have separate nasal and oral cavities and longer snouts, which provide an increased surface area for warming and moistening inspired air. Respiratory passageways are highly branched, and large surface areas exist for gas exchange. Mammalian lungs resemble a highly vascular sponge, rather than the saclike structures of amphibians and a few reptiles

Mammalian lungs, like those of reptiles, inflate using a negative-pressure mechanism. Unlike reptiles and birds, however, mammals possess a muscular **diaphragm** that separates the thoracic and abdominal cavities. Inspiration results from contraction of the diaphragm and expansion of the rib cage, both of which

decrease the intrathoracic pressure and allow air to enter the lungs. Expiration is normally by elastic recoil of the lungs and relaxation of inspiratory muscles, which decreases the volume of the thoracic cavity. The contraction of other thoracic and abdominal muscles can produce forceful exhalation.

Temperature Regulation

Mammals are widely distributed over the earth, and some face harsh environmental temperatures. Nearly all face temperatures that require them to dissipate excess heat at some times and to conserve and generate heat at other times.

The heat-producing mechanisms of mammals are divided into two categories. Shivering thermogenesis is muscular activity that generates large amounts of heat but little movement. Nonshivering thermogenesis involves heat production by general cellular metabolism and the metabolism of special fat deposits called brown fat. Chapter 28 discusses these heat-generating processes in more detail.

Heat production is effective in thermoregulation because mammals are insulated by their pelage and/or fat deposits. Fat deposits are also sources of energy to sustain high metabolic rates.

Mammals without a pelage can conserve heat by allowing the temperature of surface tissues to drop. A walrus in cold, arctic waters has a surface temperature near 0° C; however, a few centimeters below the skin surface, body temperatures are about 35° C. Upon emerging from the icy water, the walrus quickly warms its skin by increasing peripheral blood flow. Most tissues cannot tolerate such rapid and extreme temperature fluctuations. Further investigations are likely to reveal some unique biochemical characteristics of these skin tissues.

Even though most of the body of an arctic mammal is unusually well insulated, appendages often have thin coverings of fur as an adaptation to changing thermoregulatory needs. Even in winter, an active mammal sometimes produces more heat than is required to maintain body temperature. Patches of poorly insulated skin allow excess heat to be dissipated. During periods of inactivity or extreme cold, however, arctic mammals must reduce heat loss from these exposed areas, often by assuming heat-conserving postures. Mammals sleeping in cold environments conserve heat by tucking poorly insulated appendages and their faces under well-insulated body parts.

Countercurrent heat-exchange systems may help regulate heat loss from exposed areas (figure 22.14). Arteries passing peripherally through the core of an appendage are surrounded by veins that carry blood back toward the body. When blood returns to the body through these veins, heat transfers from arterial blood to venous blood and returns to the body rather than being lost to the environment. When excess heat is produced, blood is shunted away from the countercurrent veins toward peripheral vessels, and excess heat is radiated to the environment.

Mammals have few problems getting rid of excess heat in cool, moist environments. Heat can be radiated into the air from vessels near the surface of the skin or lost by evaporative cooling from either sweat glands or respiratory surfaces during panting.

Hot, dry environments present far greater problems, because evaporative cooling may upset water balances. Jackrabbits and elephants use their long ears to radiate heat. Small mammals often avoid the heat by remaining in burrows during the day and foraging for food at night. Other mammals seek shade or watering holes for cooling.

Winter Sleep and Hibernation

Mammals react in various ways to environmental extremes. Caribou migrate to avoid extremes of temperature, and wildebeest migrate to avoid seasonal droughts. Other mammals retreat to burrows under the snow, where they become less active but are still relatively alert and easily aroused—a condition called **winter sleep.** For example, bears and raccoons retreat to dens in winter. Their body temperatures and metabolic rates decrease somewhat, but they do not necessarily remain inactive all winter.

Hibernation is a period of winter inactivity in which the hypothalamus of the brain slows the metabolic, heart, and respiratory

(a)

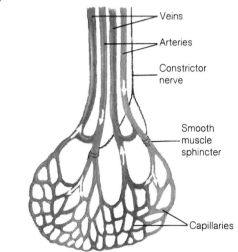

(b)

Countercurrent vessels

FIGURE 22.14

Countercurrent Heat Exchange. Countercurrent heat exchangers conserve body heat in animals adapted to cold environments. (*a*) Systems similar to the one depicted here are found in the legs of reindeer (*Rangifer tarandus*) and in the flippers of dolphins. (*b*) Heat transferred from blood moving peripherally in arteries warms venous blood returning from an extremity. During winter, the lower part of a reindeer's leg may be at 10° C, while body temperature is about 40° C. White arrows indicate direction of blood flow.

rates. True hibernators include the monotremes (echidna and duck-billed platypus) and many members of the Insectivora (e.g., moles and shrews), Rodentia (e.g., chipmunks and woodchucks), and Chiroptera (bats). In preparation for hibernation, mammals usually accumulate large quantities of body fat. After a hibernating mammal retreats to a burrow or a nest, the hypothalamus sets the body's thermostat to about 2° C. The respiratory rate of a hibernating ground squirrel falls from 100 to 200 breaths per minute to about four breaths per minute. The heart rate falls from 200 to 300 beats per minute to about 20 beats per minute. During hibernation, a mammal may lose a third to half of its body weight. Arousal from hibernation occurs by metabolic heating, frequently using brown fat deposits, and it takes several hours to raise body temperature to near 37° C.

NERVOUS AND SENSORY FUNCTIONS

The basic structure of the vertebrate nervous system is retained in mammals. The development of complex nervous and sensory functions goes hand-in-hand with active lifestyles and is most evident in the enlargement of the cerebral hemispheres and the cerebellum of mammals. Most integrative functions shift to the enlarged cerebral cortex (neocortex).

In mammals, the sense of touch is well developed. Receptors are associated with the bases of hair follicles and are stimulated when a hair is displaced.

Olfaction was apparently an important sense in early mammals, because fossil skull fragments show elongate snouts, which would have contained olfactory epithelium. Cranial casts of fossil skulls show enlarged olfactory regions. Olfaction is still an important sense for many mammals. Mammals can perceive olfactory stimuli over long distances during either the day or night to locate food, recognize members of the same species, and avoid predators.

Auditory senses were similarly important to early mammals. More recent adaptations include an ear flap (the pinna) and the external ear canal leading to the tympanum that directs sound to the middle ear. The middle ear contains three ear ossicles that conduct vibrations to the inner ear. The sensory patch of the inner ear that contains the sound receptors is long and coiled and is called the cochlea. This structure provides more surface area for receptor cells and gives mammals greater sensitivity to pitch and volume than is present in reptiles. Cranial casts of early mammals show well-developed auditory regions.

Vision is an important sense in many mammals, and eye structure is similar to that described for other vertebrates. Accommodation occurs by changing the shape of the lens (see figure 24.29). Color vision is less well developed in mammals than in reptiles and birds. Rods dominate the retinas of most mammals, which supports the hypothesis that early mammals were nocturnal. Primates, squirrels, and a few other mammals have well-developed color vision.

EXCRETION AND OSMOREGULATION

Mammals, like all amniotes, have a metanephric kidney. Unlike reptiles and birds, which excrete mainly uric acid, mammals excrete urea. Urea is less toxic than ammonia and does not require large quantities of water in its excretion. Unlike uric acid, however, urea is highly water soluble and cannot be excreted in a semisolid form; thus, some water is lost. Excretion in mammals is always a major route for water loss.

In the nephron of the kidney, fluids and small solutes are filtered from the blood through the walls of a group of capillary-like vessels, called the glomerulus. The remainder of the nephron consists of tubules that reabsorb water and essential solutes and secrete particular ions into the filtrate.

The primary adaptation of the mammalian nephron is a portion of the tubule system called the loop of the nephron. The transport processes in this loop and the remainder of the tubule system allow mammals to produce urine that is more concentrated than blood. For example, beavers produce urine that is twice as concentrated as blood, while Australian hopping mice produce urine that is 22 times more concentrated than blood. This accomplishes the same function that nasal and orbital salt glands do in reptiles and birds.

Water loss varies greatly, depending on activity, physiological state, and environmental temperature. Water is lost in urine and feces, in evaporation from sweat glands and respiratory surfaces, and during nursing. Mammals in very dry environments have many behavioral and physiological mechanisms to reduce water loss. The kangaroo rat, named for its habit of hopping on large hind legs, is capable of extreme water conservation (figure 22.15). It is native to the southwestern deserts of the United States and Mexico, and it survives without drinking water. Its feces are almost dry, and its nocturnal habits reduce evaporative water loss. Condensation as warm air in the respiratory passages encounters the cooler nasal passages minimizes respiratory water loss. A low-protein diet, which reduces urea production, minimizes excretory water loss. The nearly dry seeds that the kangaroo rat eats are rich sources of carbohydrates and fats. Metabolic oxidation of carbohydrates produces water as a by-product.

BEHAVIOR

Mammals have complex behaviors that enhance survival. Visual cues are often used in communication. The bristled fur, arched back, and open mouth of a cat communicate a clear message to curious dogs or other potential threats. A tail-wagging display of a dog has a similarly clear message. A wolf defeated in a fight with other wolves lies on its back and exposes its vulnerable throat and belly. Similar displays may allow a male already recognized as being subordinate to another male to avoid conflict within a social group.

Pheromones are used to recognize members of the same species, members of the opposite sex, and the reproductive state of a member of the opposite sex. Pheromones may also induce sexual behavior, help establish and recognize territories, and ward off predators. The young of many mammalian species recognize their parents, and parents recognize their young, by smell. Bull elk smell the rumps of females during the breeding season to recognize those in their brief receptive period. They also urinate on their own bellies and underhair to advertise their reproductive status to females and other males. Male mammals urinate on objects in the environment to establish territories and to allow females to become accustomed to their odors. Rabbits and rodents spray urine on a member of the opposite sex to inform the second individual of the first's readiness to mate. Skunks use chemicals to ward off predators.

(a)

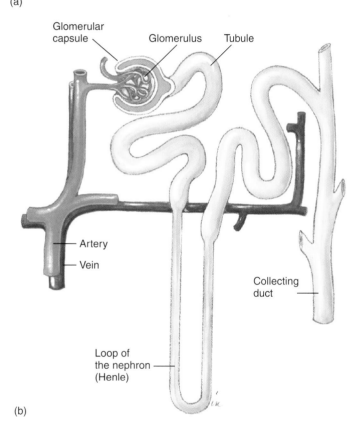

(b)

FIGURE 22.15

Order Rodentia.   (*a*) The kangaroo rat (*Dipodomys ordii*). (*b*) The long loop of the nephron of this desert animal conserves water, preventing dehydration.

Auditory and tactile communication are also important in the lives of mammals. Herd animals stay together and remain calm as long as familiar sounds (e.g., bellowing, hooves walking over dry grasses and twigs, and rumblings from ruminating stomachs) are uninterrupted. Unfamiliar sounds may trigger alarm and flight.

Vocalizations and tactile communication are important in primate social interactions. Tactile communication ranges from precopulatory "nosing" that occurs in many mammals to grooming. Grooming helps maintain a healthy skin and pelage, but also reinforces important social relationships within primate groups.

FIGURE 22.16

Order Carnivora.   California sea lions (*Zalophus californianus*) on a beach during the breeding season. The adult males in the foreground are vocalizing and posturing.

Territoriality

Many mammals mark and defend certain areas from intrusion by other members of the same species. When cats rub their face and neck on humans or on furniture, the behavior is often interpreted as affection. Cats, however, are really staking claim to their territory, using odors from facial scent glands. Some territorial behavior attracts females to, and excludes other males from, favorable sites for mating and rearing young.

Male California sea lions (*Zalophus californianus*) establish territories on shorelines where females come to give birth to young. For about two weeks, males engage in vocalizations, displays, and sometimes serious fighting to stake claim to favorable territories (figure 22.16). Older, dominant bulls are usually most successful in establishing territories, and young bulls generally swim and feed just offshore. When they arrive at the beaches, females select a site for giving birth. Selection of the birth site also selects the bull that will father next year's offspring. Mating occurs approximately two weeks after the birth of the previous year's offspring. Development is arrested for the three months during which the recently born young do most of their nursing. This mechanism is called embryonic diapause. Thus, even though actual development takes about nine months, the female carries the embryo and fetus for a period of one year.

REPRODUCTION AND DEVELOPMENT

In no other group of animals has viviparity developed to the extent it has in mammals. Mammalian viviparity requires a large expenditure of energy on the part of the female during development and on the part of one or both parents caring for young after they are born. Viviparity is advantageous because females are not necessarily tied to a single nest site but can roam or migrate to find food or a proper climate. Viviparity is accompanied by the evolution of

WILDLIFE ALERT
The Southern (California) Sea Otter (*Enhydra lutris nereis*)

VITAL STATISTICS

Classification: Phylum Chordata, class Mammalia, order Carnivora

Range: Southern California coast

Habitat: Kelp beds in near-shore waters

Number remaining: 2,000

Status: Threatened

NATURAL HISTORY AND ECOLOGICAL STATUS

Sea otters (*Enhydra lutris*) are divided into three subspecies based upon morphological and molecular characteristics. Their historic range includes most of the northern Pacific rim from Hokkaido, Japan, to Baja California (box figure 22.1). Prior to the 1700s, the sea otter population probably numbered between 150,000 and 300,000 individuals. Of the three subspecies, the southern (California) sea otter (*E. lutris nereis*) has been in the greatest danger of extinction.

Sea otters are the smallest marine mammals (box figure 22.2). Mature males average 29 kg and mature females average 20 kg. They feed on molluscs, sea urchins, and crabs. They use shells and rocks to pry their prey from the substrate and to crack shells and tests of their food items. Unlike other marine mammals, they have no blubber for insula-

BOX FIGURE 22.2   The Southern (California) Sea Otter (*Enhydra lutris nereis*).

tion from cold water. Their very thick fur, with about 150,000 hairs per cm^2, is their insulation. (The human head has about 42,000 hairs per cm^2.) Sea otters are considered a keystone predator. By preying on a variety of kelp herbivores, they enhance the productivity of kelp beds and increase the diversity of the kelp ecosystem. (The kelp ecosystem is one of the most diverse ecosystems in temperate regions of the earth.)

Southern sea otters have faced, and continue to face, pressures that threaten their survival. In the 1700s, they were hunted extensively for their thick fur. They are sensitive to contaminants in the ecosystem. Poisons such as pesticides, PCBs, and tributylin (a component of antifouling agents used on boat hulls) accumulate in their tissues and weaken the animals. When oil from tanker spills becomes trapped in an otter's thick fur, it destroys its insulating qualities and quickly kills the otter. All of these pressures devastated southern sea otter populations. Historically, there were about 13,000 to 20,000 southern sea otters along their range, which extended along what is now the California coast. In the early 1900s, they were thought to be extinct until a small group of otters was observed on California's Big Sur coast.

Southern sea otters are now protected by the International Convention for the Preservation and Protection of Fur Seals, the Marine Mammal Protection Act, and the Endangered Species Act. This protection and other recovery efforts have protected the otters and sheltered other species in the kelp ecosystem. The population has slowly grown from the 1935 low to 2,400 in 1995. Unfortunately, and for unknown reasons, the population has declined 3% each year since 1995.

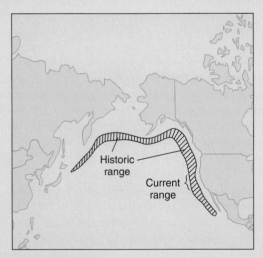

BOX FIGURE 22.1   **Range of Sea Otters.**   The historic range of sea otters (shaded) probably consisted of a cline (a gradual transition between) of the three subspecies. Southern sea otters (*Enhydra lutris nereis*) now occupy a portion of the California coast between Half Moon Bay and Gaviota.

a portion of the reproductive tract where the young are nourished and develop. In viviparous mammals, the oviducts are modified into one or two uteri (sing., uterus).

Reproductive Cycles

Most mammals have a definite time or times during the year in which ova (eggs) mature and are capable of being fertilized.

Reproduction usually occurs when climatic conditions and resource characteristics favor successful development. Mammals living in environments with few seasonal changes and those that exert considerable control over immediate environmental conditions (e.g., humans) may reproduce at any time of the year. However, they are still tied to physiological cycles of the female that determine when ova can be fertilized.

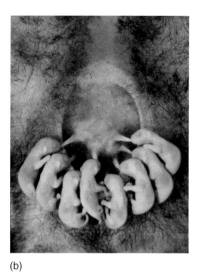

(a)   (b)

FIGURE 22.17

Order Marsupialia.   (*a*) An opossum (*Didelphis marsupialis*) with young. (*b*) Opossum young nursing in a marsupial pouch.

Most female mammals undergo an **estrus** (Gr. *oistros*, a vehement desire) **cycle,** which includes a time during which the female is behaviorally and physiologically receptive to the male. During the estrus cycle, hormonal changes stimulate the maturation of ova in the ovary and induce ovulation (release of one or more mature ova from an ovarian follicle). A few mammals (e.g., rabbits, ferrets, and mink) are induced ovulators; coitus (copulation) induces ovulation.

Hormones also mediate changes in the uterus and vagina. As the ova are maturing, the inner lining of the uterus proliferates and becomes more vascular in preparation for receiving developing embryos. External swelling in the vaginal area and increased glandular discharge accompany the proliferation of vaginal mucosa. During this time, males show heightened interest in females, and females are receptive to males. If fertilization does not occur, the changes in the uterus and vagina are reversed until the next cycle begins. No bleeding or sloughing of uterine lining usually occurs.

Many mammals are monestrus and have only a single yearly estrus cycle that is sharply seasonal. Wild dogs, bears, and sea lions are monestrus; domestic dogs are diestrus. Other mammals are polyestrus. Rats and mice have estrus cycles that repeat every four to six days.

The menstrual cycle of female humans, apes, and monkeys is similar to the estrus cycle in that it results in a periodic proliferation of the inner lining of the uterus and correlates with the maturation of an ovum. If fertilization does not occur before the end of the cycle, menses—the sloughing of the uterine lining—occurs. Chapter 29 describes human menstrual and ovarian cycles.

Fertilization usually occurs in the upper third of the oviduct within hours of copulation. In a few mammals, fertilization may be delayed. In some bats, for example, coitus occurs in autumn, but fertilization is delayed until spring. Females store sperm in the uterus for periods in excess of two months. This **delayed fertilization** is apparently an adaptation to winter dormancy. Fertilization

can occur immediately after females emerge from dormancy rather than waiting until males attain their breeding state.

In many other mammals, fertilization occurs right after coitus, but development is arrested after the first week or two. This **embryonic diapause,** which was described previously for sea lions, also occurs in some bats, bears, martens, and marsupials. The adaptive significance of embryonic diapause varies with species. In the sea lion, embryonic diapause allows the mother to give birth and mate within a short interval, but not have her resources drained by both nursing and pregnancy. It also allows young to be born at a time when resources favor their survival. In some bats, fertilization occurs in the fall before hibernation, but birth is delayed until resources become abundant in the spring.

Modes of Development

Monotremes are oviparous. The ovaries release ova with large quantities of yolk. After fertilization, shell glands in the oviduct deposit a shell around the ovum, forming an egg. Female echidnas incubate eggs in a ventral pouch. Platypus eggs are laid in their burrows.

All other mammals have a placenta through which young are nourished for at least a portion of their development. The maternal bloodstream, not yolk, supplies nutrients.

In marsupials, most nourishment for the fetus comes from "uterine milk" that uterine cells secrete. Some nutrients diffuse from maternal blood into a highly vascular yolk sac that makes contact with the uterus. This connection in marsupials is a primitive placenta. The marsupial **gestation period** (the length of time young develop within the female reproductive tract) varies between 8 and 40 days in different species. The gestation period is short because of marsupials' inability to sustain the production of hormones that maintain the uterine lining. After birth, tiny young crawl into the marsupium and attach to a nipple, where they suckle for an additional 60 to 270 days (figure 22.17).

In eutherian mammals, the embryo implants deeply into the uterine wall. Embryonic and uterine tissues grow rapidly and become highly folded and vascular, forming the placenta. Although maternal and fetal blood do not mix, nutrients, gases, and wastes diffuse between the two bloodstreams. Gestation periods of eutherian mammals vary from 20 days (some rodents) to 19 months (the African elephant). Following birth, the placenta and other tissues that surrounded the fetus in the uterus are expelled as "afterbirth." The newborns of many species (e.g., humans) are helpless at birth (altricial); others (e.g., deer and horses) can walk and run shortly after birth (precocial).

SUMMARY

1. Mammalian characteristics evolved from the synapsid lineage over a period of about 200 million years. Mammals evolved from a group of synapsids called therapsids.
2. Modern mammals include the monotremes, marsupials, and placental mammals.
3. Hair is uniquely mammalian. It functions in sensory perception, temperature regulation, and communication.
4. Mammals have sebaceous, sudoriferous, scent, and mammary glands.
5. The teeth and digestive tracts of mammals are adapted for different feeding habits. Flat, grinding teeth and fermentation structures for digesting cellulose characterize herbivores. Predatory mammals have sharp teeth for killing and tearing prey.
6. The mammalian heart has four chambers, and circulatory patterns are adapted for viviparous development.
7. Mammals possess a diaphragm that alters intrathoracic pressure, which helps ventilate the lungs.
8. Mammalian thermoregulation involves metabolic heat production, insulating pelage, and behavior.
9. Mammals react to unfavorable environments by migration, winter sleep, and hibernation.
10. The nervous system of mammals is similar to that of other vertebrates. Olfaction and hearing were important for early mammals. Vision, hearing, and smell are the dominant senses in many modern mammals.
11. The nitrogenous waste of mammals is urea, and the kidney is adapted for excreting a concentrated urine.
12. Mammals have complex behavior to enhance survival. Visual cues, pheromones, and auditory and tactile cues are important in mammalian communication.
13. Most mammals have specific times during the year when reproduction occurs. Female mammals have estrus or menstrual cycles. Monotremes are oviparous. All other mammals nourish young by a placenta.

SELECTED KEY TERMS

cecum (p. 350)
delayed fertilization (p. 357)
dental formula (p. 349)
diaphragm (p. 352)
embryonic diapause (p. 357)

estrus cycle (p. 357)
gestation period (p. 357)
heterodont (p. 348)
homodont (p. 348)

CRITICAL THINKING QUESTIONS

1. Why is tooth structure important in the study of mammals?
2. What does the evolution of secondary palates have in common with the evolution of completely separated, four-chambered hearts?
3. Why is classifying mammals by feeding habits not particularly useful to phylogenetic studies?
4. Under what circumstances is endothermy disadvantageous for a mammal?
5. Discuss the possible advantages of embryonic diapause for marsupials that live in climatically unpredictable regions of Australia.
6. What is induced ovulation? Why might it be adaptive for a mammal?

ONLINE LEARNING CENTER

Visit our Online Learning Center (OLC) at www.mhhe.com/zoology (click on this book's title) to find the following chapter-related materials:

- CHAPTER QUIZZING
- RELATED WEB LINKS
- CLASS MAMMALIA
  Marine Mammals
  Cat Dissections
  Rat Dissections
  Fetal Pig Dissections
  Conservation Issues Concerning Mammals
- BOXED READINGS ON
  Bones and Scales
  Horns and Antlers
  Mammalian Echolocation
  How the One-Humped Dromedary (Camel) of the Arabian and
    African Deserts Thrives in Some of the Hottest and Driest
    Climates on Earth
  The Participation of the Fetus During Childbirth
- SUGGESTED READINGS

PART THREE

FORM AND FUNCTION: A COMPARATIVE PERSPECTIVE

Is the whole of an animal equal to the sum of its parts? A superficial answer would be "yes." However, the structure and function of an animal are never as simple as this answer implies. A body is composed of many parts (e.g., cells, tissues, organs, and organ systems), yet rarely are any of these parts independent of one another. Simple additive relationships fail to describe adequately the interactions between the body's parts. Instead, an animal is the product of many complex interactions. In understanding the structure and function of any system, you only begin to understand the whole animal.

Cells, tissues, organs, and organ systems all interact to maintain a steady homeostatic state compatible with life. Ultimately, you need to look inward to see the genetic potential of the animal, outward to see how environmental constraints limit the fulfillment of that potential, and backward to see the evolutionary pressures that shaped the particular species.

Parts One and Two of this text examine animal life at molecular, cellular, genetic, developmental, behavioral, and taxonomic levels. Throughout, the evolutionary forces and pressures that influenced the development of a vast array of animal life-forms are presented, concluding with five chapters on the vertebrates. Part Three (chapters 23 through 29) continues this coverage of animal life by presenting an overview of the various organ systems: integumentary, skeletal, muscular, nervous and sensory, endocrine, circulatory, lymphatic, respiratory, digestive, urinary, and reproductive.

The major theme in Part Three is that all organ systems are specialized and coordinate with each other and that they constantly adjust to changes inside and outside the animal. Although each system has its own specialized function, none operates without assistance from the others. As you will see, the structure of each system determines its particular function.

Photo (top): This Goldentail Moray (Gymnothorax miliaris), an endangered species, lives in coral crevices like that created by these Star Corals (Montastrea cavernosa). Although snake-like in appearance, it is actually a fish that emerges from coral at night to find prey.

359

PROTECTION, SUPPORT, AND MOVEMENT

Outline

Concepts

1. The integumentary system of animals includes an outer protective body covering called the integument and associated structures and secretions. The integument of most multicellular invertebrates consists of a single layer of cells. The vertebrate integument is multilayered and is called skin. Skin contains nerves and blood vessels, as well as derivatives such as glands, hair, and nails.

2. Skeletal systems move primarily by the actions of antagonistic muscles. Animals have three types of skeletons: fluid hydrostatic skeletons, rigid exoskeletons, and rigid endoskeletons. Many invertebrates have hydrostatic skeletons consisting of a core of liquid wrapped in a tension-resistant sheath containing muscles. Rigid exoskeletons completely surround an animal and are sites for muscle attachment and counterforces for muscle movements. They also offer protection and support. The vertebrate skeletal system is an endoskeleton. It consists mainly of supportive tissue composed of cartilage and bone.

3. Contractile proteins and muscles provide the force for movement in animals from cnidarians to vertebrates. Vertebrates use their endoskeletons in conjunction with muscles to move. In vertebrates, striated skeletal muscles move the body, smooth muscles move material through tubular organs and change the size of tubular openings, and cardiac muscle produces the beating of the heart.

In animals, structure and function have evolved together. Several results of this evolution are protection, support, and movement. The integumentary, skeletal, and muscular systems are primarily responsible for these functions.

PROTECTION: INTEGUMENTARY SYSTEMS

The **integument** (L. *integumentum*, cover) is the external covering of an animal. It protects the animal from mechanical and chemical injury and invasion by microorganisms. Many other diverse functions of the integument have evolved in different animal groups. These functions include regulation of body temperature; excretion of waste materials; conversion of sunlight into vitamin D; reception of environmental stimuli, such as pain, temperature, and pressure; locomotion; and movement of nutrients and gases.

This chapter contains evolutionary concepts, which are set off in this font.

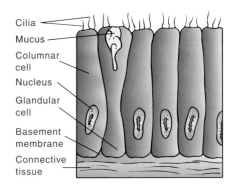

Cilia
Mucus
Columnar cell
Nucleus
Glandular cell
Basement membrane
Connective tissue

FIGURE 23.1

Integument of Invertebrates. The integument of many invertebrates consists of a simple layer of columnar epithelial cells (epidermis) resting on a basement membrane. A thin layer of connective tissue lies under the basement membrane. Cilia and glandular cells may or may not be present.

THE INTEGUMENTARY SYSTEM OF INVERTEBRATES

Some single-celled protozoa have only a **plasma membrane** for an external covering. This membrane is structurally and chemically identical to the plasma membrane of multicellular organisms (*see figure 2.4*). In protozoa, the plasma membrane has a large surface area relative to body volume, so gas exchange and the removal of soluble wastes occur by simple diffusion. This large surface area also facilitates the uptake of dissolved nutrients from surrounding fluids. Other protozoa, such as *Paramecium*, have a thick protein coat called a **pellicle** (L. *pellicula*, thin skin) outside the plasma membrane. This pellicle offers further environmental protection and is a semirigid structure that transmits the force of cilia or flagella to the entire body of the protozoan as it moves.

Most multicellular invertebrates have an integument consisting of a single layer of columnar epithelial cells (figure 23.1). This outer layer, the **epidermis** (Gr. *epi*, upon + *derm*, skin), rests on a basement membrane. Beneath the basement membrane is a thin layer of connective tissue fibers and cells. Epidermal cells exposed at the surface of the animal may possess cilia. The epidermis of some invertebrates also contains glandular cells, which secrete an overlying, noncellular material that encases part or most of the animal.

Some invertebrates possess **cuticles** (L. *cuticula, cutis*, skin) that are highly variable in structure (figure 23.2). For example, in some animals (rotifers), cuticles are thin and elastic, whereas in others (crustaceans, arachnids, insects), cuticles are thick and rigid and support the body. Such cuticles consist of chitin and proteins in rigid plates that a flexible membrane links together. A disadvantage of cuticles is that animals have difficulty growing within them. As a result, most of these invertebrates (e.g., arthropods) periodically shed the old, outgrown cuticle in a process called molting or ecdysis (*see figure 14.5*).

In cnidarians, such as *Hydra*, the epidermis is only one cell layer thick. Other cnidarians (e.g., the corals) have mucous glands that secrete a calcium carbonate ($CaCO_3$) **shell.** The outer cover-

ing of parasitic flukes and tapeworms is a complex syncytium called a **tegument** (L. *tegumentum, tegere*, to cover). Its main functions are nutrient ingestion and protection against digestion by host enzymes. Nematodes and annelids have an epidermis that is one cell thick and secretes a cuticle that has many layers. The integument of echinoderms consists of a thin, usually ciliated epidermis and an underlying connective-tissue dermis containing $CaCO_3$. Arthropods have the most complex of invertebrate integuments, in part because their integument is a specialized exoskeleton.

THE INTEGUMENTARY SYSTEM OF VERTEBRATES

Skin is the vertebrate integument. It is the largest organ (with respect to surface area) of the vertebrate body and grows with the animal. Skin has two main layers. As in invertebrates, the epidermis is the outermost layer of epithelial tissue and is one to several cells thick. The **dermis** (Gr. *derma*, hide, skin) is the connective tissue meshwork of collagenous, reticular, and elastic fibers beneath the epidermis. A **hypodermis** ("below the skin"), consisting of loose connective tissue, adipose tissue, and nerve endings, separates the skin from deeper tissues.

The Skin of Jawless Fishes

Jawless fishes, such as lampreys and hagfishes, have relatively thick skin (figure 23.3). Of the several types of epidermal glandular cells that may be present, one secretes a protective cuticle. In hagfishes, multicellular slime glands produce large amounts of mucous slime that covers the body surface. This slime protects the animals from external parasites and has earned hagfishes the descriptive name "slime eels."

The Skin of Cartilaginous Fishes

The skin of cartilaginous fishes (e.g., sharks) is multilayered and contains mucous and sensory cells (figure 23.4). The dermis contains bone in the form of small placoid scales called **denticles** (L. *denticulus*, little teeth). Denticles contain blood vessels and nerves and are similar to vertebrate teeth. Because cartilaginous fishes grow throughout life, the skin area also increases. New denticles are produced to maintain enough of these protective structures at the skin surface. Like teeth, once denticles reach maturity, they do not grow; thus, they continually wear down and are lost. Since denticles project above the surface of the skin, they give cartilaginous fishes a sandpaper texture.

The Skin of Bony Fishes

The skin of bony fishes (teleosts) contains **scales** (Fr. *escale*, shell, husk) composed of dermal bone. A thin layer of dermal tissue overlaid by the superficial epidermis normally covers scales (figure 23.5). Because scales are not shed, they grow at the margins and over the lower surface. In many bony fishes, growth lines, which are useful in determining the age of a fish, can often be detected. The skin of bony fishes is permeable and functions in gas

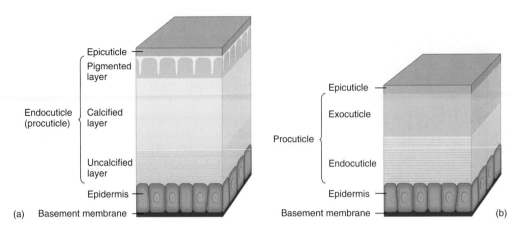

FIGURE 23.2

Cuticles.   The cuticles of (*a*) a crustacean and (*b*) an insect. The underlying epidermis secretes the cuticles of both groups of animals.   *From: "A LIFE OF INVERTEBRATES" © 1979 W. D. Russell-Hunter.*

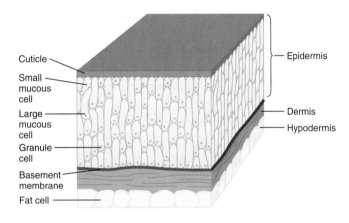

FIGURE 23.3

Skin of Jawless Fishes.   The skin of an adult lamprey has a multilayered epidermis with glandular cells and fat storage cells and an underlying hypodermis.

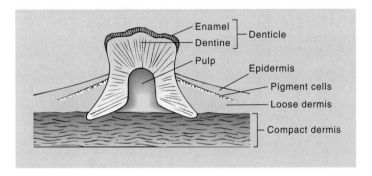

FIGURE 23.4

Skin of Cartilaginous Fishes.   Shark skin contains toothlike denticles that become exposed through the loss of the epidermal covering. The skin is otherwise fishlike in structure.

exchange, particularly in the smaller fishes that have a large skin surface area relative to body volume. The dermis is richly supplied with capillary beds to facilitate its use in respiration. The epidermis also contains many mucous glands. Mucus production helps prevent bacterial and fungal infections, and it reduces friction as the fish swims. Some species have granular glands that secrete an irritating—or to some species, poisonous—alkaloid. Many teleosts that live in deep aquatic habitats have photophores that facilitate species recognition or act like lures and warning signals.

The Skin of Amphibians

Amphibian skin consists of a stratified epidermis and a dermis containing mucous and serous glands plus pigmentation cells (figure 23.6). *Phylogenetically, amphibians are transitional between aquatic and terrestrial vertebrates. The earliest amphibians were covered by dermal bone scales like their fish ancestors.* Three problems associated with terrestrial environments

are desiccation, the damaging effects of ultraviolet light, and physical abrasion. During amphibian evolution, keratin production increased in the outer layer of skin cells. (Keratin is a tough, impermeable protein that protects the skin in the physically abrasive, rigorous terrestrial environment.) The increased keratin in the skin also protects the cells, especially their nuclear material, from ultraviolet light. The mucus that mucous glands produce helps prevent desiccation, facilitates gas exchange when the skin is used as a respiratory organ, and makes the body slimy, which facilitates escape from predators.

Within the dermis of some amphibians are poison glands that produce an unpleasant-tasting or toxic fluid that acts as a predator deterrent. Sensory nerves penetrate the epidermis as free nerve endings. Interestingly, the "warts" of toads seem to be specialized sensory structures, since they contain many sensory cells.

The Skin of Reptiles

The skin of reptiles reflects their greater commitment to a terrestrial existence. The outer layer of the epidermis (stratum corneum) is thick (figure 23.7), lacks glands, and is modified into

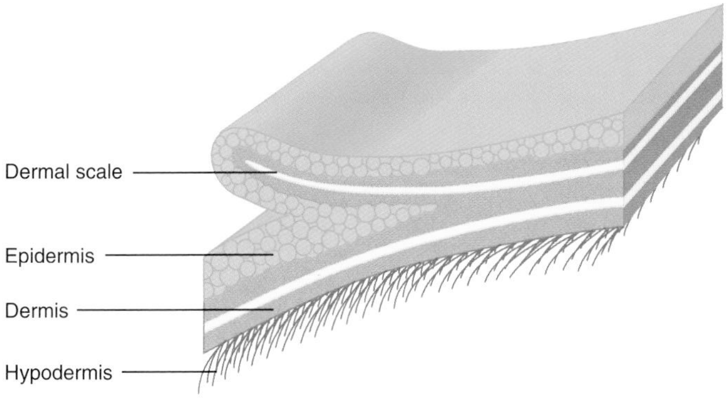

FIGURE 23.5

Skin of Bony Fishes. The skin of a typical bony fish has overlapping scales (two are shown here). The scales are layers of collagenous fibers covered by a thin, flexible layer of bone.

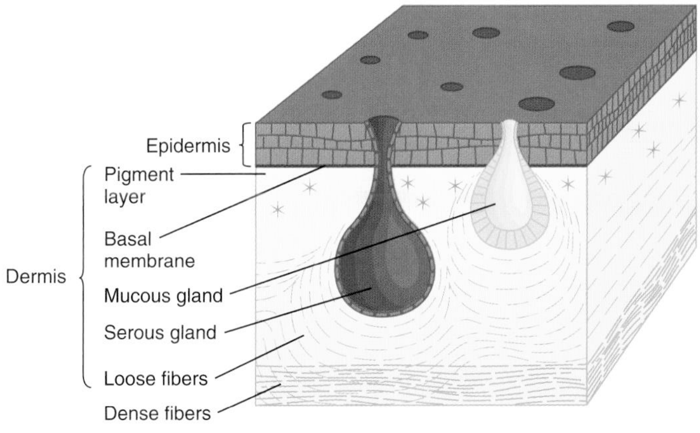

FIGURE 23.6

Skin of Amphibians. Frog skin has a stratified epidermis and several types of glands in the dermis. Notice the pigment layer in the upper part of the dermis.

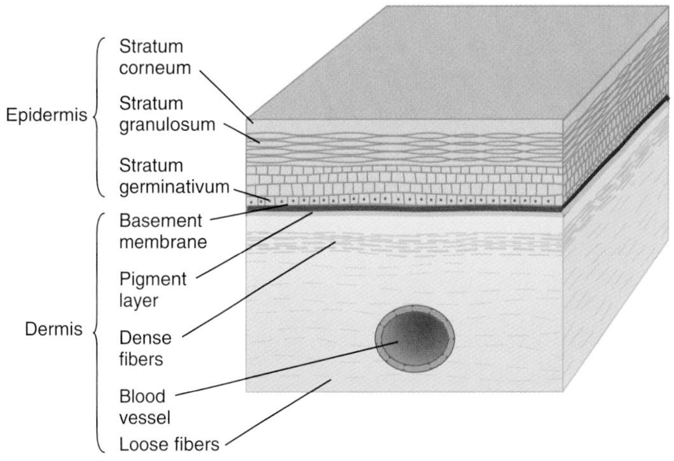

FIGURE 23.7

Skin of Reptiles. Lizard skin has the heavily keratinized outer epidermis (scales) characteristic of reptiles. Notice the absence of integumentary glands, making reptilian skin exceptionally dry.

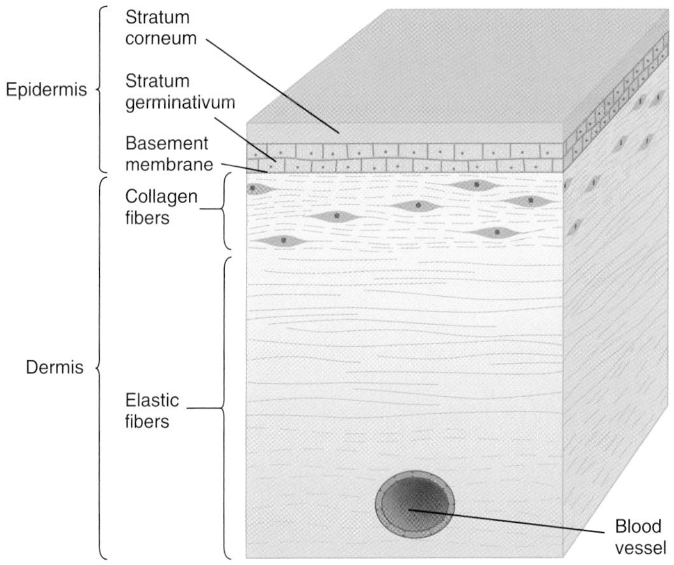

FIGURE 23.8

Skin of Birds. Bird skin has a relatively soft and thin epidermis with no epidermal glands.

keratinized scales, scutes (thick scales) in snakes and turtles, beaks in turtles, rattles on snakes, and claws, plaques, and spiny crests on most other reptiles. This thick, keratinized layer resists abrasion, inhibits dehydration, and protects like a suit of armor. During shedding or molting of the skin of many reptiles (e.g., snakes and lizards), the old outer layer separates from newly formed epidermis. Diffusion of fluid between the layers aids this separation.

The Skin of Birds

The skin of birds shows many typically reptilian features with no epidermal glands (the only epidermal gland of birds is the uropygial or preen gland). Over most of the bird's body, the epidermis is usually thin and only two or three cell layers thick (figure 23.8). Indeed, the term "thin skinned," sometimes applied figuratively to humans, is literal when applied to birds. The outer keratinized layer is often quite soft. The most prominent parts of the epidermis are the feathers. *Feathers are derived from the scales of reptilian ancestors and are the most complex of all the derivatives of the vertebrate stratum corneum* (see figure 21.4).

The dermis of birds is similar in structure to that of reptiles and contains blood and lymphatic vessels, nerves, and epidermally derived sensory bodies. Air spaces that are part of the avian respiratory system extend into the dermis. These air spaces are involved in thermal regulation. Associated with the feathers and their normal functioning is an array of dermal smooth-muscle fibers that control the position of the feathers. Feather position is important in thermal regulation, flying, and behavior. Aquatic birds may also have fat deposits in the hypodermal layer that store energy and help insulate the body.

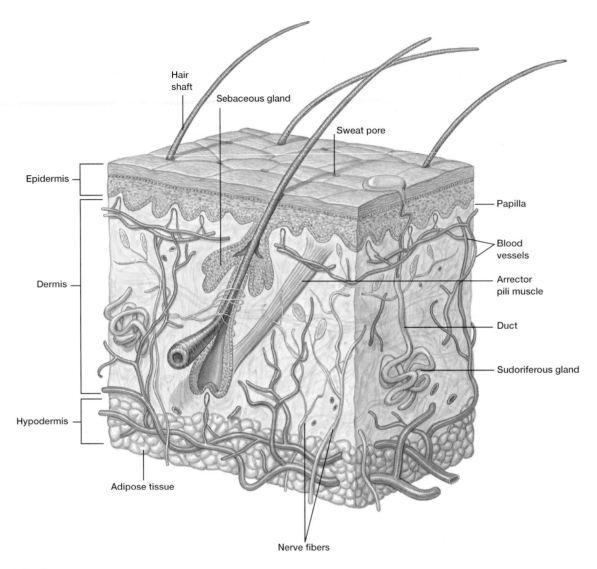

FIGURE 23.9

Skin of Mammals. Notice the various structures in the dermis of human skin.

The Skin of Mammals

The notable features of mammalian skin are: (1) hair; (2) a greater variety of epidermal glands than in any other vertebrate class; (3) a highly stratified, cornified epidermis; and (4) a dermis many times thicker than the epidermis.

The epidermis of mammalian skin is composed of stratified squamous epithelium and consists of several layers of a variety of cells. Rapid cell divisions in the deepest layer of the epidermis push cells toward the surface of the skin. As cells progress toward the surface, they die and become keratinized (contain the protein keratin). Keratinized cells make up the outer skin layer, called the stratum corneum. Because keratin is virtually insoluble in water, the stratum corneum prevents dehydration and is a first line of defense against many toxic substances and microorganisms. *The prevention of dehydration is one of the evolutionary reasons mammals and other animals have been able to colonize terrestrial environments.*

The thickest portion of mammalian skin is composed of dermis, which contains blood vessels, lymphatic vessels, nerve endings, hair follicles, small muscles, and glands (figure 23.9). A special tanning process makes leather from the dermal layer of mammalian skin.

The hypodermis underneath the dermis is different from that of other vertebrate classes in that it consists of loose connective tissue, adipose tissue, and skeletal muscles. Adipose tissue stores energy in the form of fat and provides insulation in cold environments. Skeletal muscle allows the skin above it to move somewhat independently of underlying tissues. Blood vessels thread from the hypodermis to the dermis and are absent from the epidermis.

In humans and a few other animals (e.g., horses), the skin regulates body temperature by opening and closing sweat pores and perspiring or sweating. The skin screens out excessive harmful ultraviolet rays from the sun, but it also lets in some necessary rays that convert a chemical in the skin into vitamin D. The skin is also an important sense organ, containing sensory receptors for heat, cold, touch, pressure, and pain. Its many nerve endings keep the animal responsive to potentially harmful factors in the environment.

The skin of humans and other mammals contains several types of glands. **Sudoriferous glands** (L. *sudor*, sweat), also called sweat glands, are distributed over most of the human body surface (figure 23.9). These glands secrete sweat by a process called **perspiration** (L. *per*, through + *spirare*, to breathe). Perspiration helps to regulate body temperature and maintain homeostasis, largely by the cooling effect of evaporation. In some mammals, certain sweat glands also produce pheromones. (A pheromone is a chemical that an animal secretes and that communicates with other members of the same species to elicit certain behavioral responses.) **Sebaceous (oil) glands** (L. *sebum*, tallow or fat) are simple glands connected to hair follicles in the dermis (figure 23.9). They lubricate and protect by secreting **sebum.** Sebum is a permeability barrier, an emollient (skin-softening agent), and a protective agent against microorganisms. Sebum can also act as a pheromone.

Mammalian skin color is due either to pigments or to anatomical structures that absorb or reflect light. Pigments (e.g., melanin in human skin) are within the cells of the epidermal layer, in hair, or in specialized cells called chromatophores. Some skin color is due to the color of blood in superficial blood vessels reflected through the epidermis. Bright skin colors in venomous, toxic, or bad-tasting animals may deter potential predators. Other skin colors may camouflage the animal. In addition, colors serve in social communication, helping members of the same species to identify each other, their sex, reproductive status, or social rank.

Hair is composed of keratin-filled cells that develop from the epidermis. The portion of hair that protrudes from the skin is the hair shaft, and the portion embedded beneath the skin is the root (figure 23.9). An arrector pili muscle (smooth muscle; involuntary muscle) attaches to the connective-tissue sheath of a hair follicle surrounding the bulb of the hair root. When this muscle contracts, it pulls the follicle and its hair to an erect position. In humans, this is referred to as a "goose bump." In other mammals, this action helps warm the animal by producing an insulating layer of warm air between the erect hair and skin. If hair is erect because the animal is frightened instead of cold, the erect hair also makes the animal look larger and less vulnerable to attack.

Nails, like hair, are modifications of the epidermis. Nails are flat, horny plates on the dorsal surface of the distal segments of the digits (e.g., fingers and toes of primates). Other mammals have **claws** and hooves (*see figure 22.7*). Other keratinized derivatives of mammalian skin are **horns** (not to be confused with bony antlers) and the **baleen plates** of the toothless whales.

MOVEMENT AND SUPPORT: SKELETAL SYSTEMS

As organisms evolved from the ancestral protists to the multicellular animals, body size increased dramatically. Systems involved in movement and support evolved simultaneously with the increase in body size.

Four cell types contribute to movement: (1) amoeboid cells, (2) flagellated cells, (3) ciliated cells, and (4) muscle cells. With respect to support, organisms have three kinds of skeletons: (1) fluid hydrostatic skeletons, (2) rigid exoskeletons, and (3) rigid endoskeletons. These skeletal systems also function in animal movement that requires muscles working in opposition (antagonism) to each other.

THE SKELETAL SYSTEM OF INVERTEBRATES

Many invertebrates use their body fluids for internal support. For example, sea anemones (figure 23.10*a*) and earthworms have a form of internal support called the hydrostatic skeleton.

Hydrostatic Skeletons

The **hydrostatic** (Gr. *hydro*, water + *statikos*, to stand) **skeleton** is a core of liquid (water or a body fluid such as blood) surrounded by a tension-resistant sheath of longitudinal and/or circular muscles. It is similar to a water-filled balloon because the force exerted against the incompressible fluid in one region can be transmitted to other regions. Contracting muscles push against a hydrostatic skeleton, and the transmitted force generates body movements, as the movement of a sea anemone illustrates (figure 23.10*b, c*). Another example is the earthworm, *Lumbricus terrestris*. It contracts its longitudinal and circular muscles alternately, creating a rhythm that moves the earthworm through the soil. In both of these examples, the hydrostatic skeleton keeps the body from collapsing when its muscles contract.

The invertebrate hydrostatic skeleton can take many forms and shapes, such as the gastrovascular cavity of acoelomates, a rhynchocoel in nemertines, a pseudocoelom in aschelminths, a coelom in annelids, or a hemocoel in molluscs. *Overall, the hydrostatic skeleton of invertebrates is an excellent example of adaptation of major body functions to this simple but efficient principle of hydrodynamics—use of the internal pressure of body fluids.*

Exoskeletons

Rigid **exoskeletons** (Gr. *exo*, outside + skeleton) also have locomotor functions because they provide sites for muscle attachment and counterforces for muscle movements. Exoskeletons also support and protect the body, but these are secondary functions.

In arthropods, the epidermis of the body wall secretes a thick, hard cuticle that waterproofs the body (*see figure 14.3*). The cuticle also protects and supports the animal's soft internal organs. In crustaceans (e.g., crabs, lobsters, shrimp), the exoskeleton contains calcium carbonate crystals that make it hard and inflexible—except at the joints. Besides providing shieldlike protection from enemies and resistance to general wear and tear, the exoskeleton also prevents internal tissues from drying out. *This important evolutionary adaptation contributed to arthropods' successful colonization of land. Exoskeletons, however,*

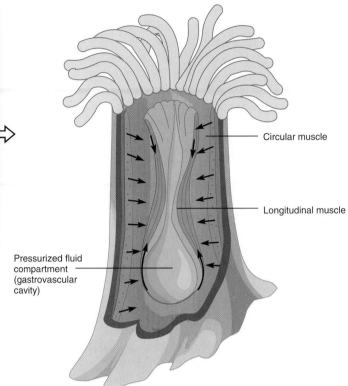

FIGURE 23.10

Hydrostatic Skeletons. (*a*) The hydrostatic skeleton of sea anemones (*Corynactis californic*) allows them to shorten or close when longitudinal muscles contract, or to lengthen or open when circular muscles contract. (*b,c*) How a hydrostatic skeleton changes an invertebrate's shape with only longitudinal muscles. Because the fluid volume is constant, a change (increase) in width must accompany a change (decrease) in length.

limit an animal's growth. Most animals shed the exoskeleton periodically, as arthropods do when they molt (figure 23.11*a*).

Certain regions of the arthropod body have a thin, flexible cuticle, and joints (articulations) (figure 23.11*b*). It is in these areas that pairs of antagonistic muscles function through a system of levers to produce coordinated movement. Interestingly, some arthropod joints (e.g., the wing joints of flying beetles and the joints of fleas involved in jumping) possess a highly elastic protein called "animal rubber," or resilin. Resilin stores energy on compression and then releases the energy to produce movement (*see figure 23.23*). *From an evolutionary perspective, the development of a jointed, flexible exoskeleton that permitted flight is one of the reasons for the success of arthropods.*

Endoskeletons

Like the term implies, other body tissues enclose **endoskeletons** (Gr. *endo*, within + skeleton). For example, the endoskeletons of sponges consist of mineral spicules and fibers of spongin that keep the body from collapsing (*see figure 9.5*). Since adult sponges attach to the substrate, they have no need for muscles attached to the endoskeleton. Similarly, the endoskeletons of echinoderms (sea stars, sea urchins) consist of small, calcareous plates called ossicles. The most familiar endoskeletons, however, are in vertebrates and are discussed under "The Skeletal System of Vertebrates."

Mineralized Tissues and the Invertebrates

Hard, mineralized tissues are not unique to the vertebrates. In fact, over two-thirds of the living species of animals that contain mineralized tissues are invertebrates. Most invertebrates have inorganic calcium carbonate crystals embedded in a collagen matrix. (Vertebrates have calcium phosphate crystals.) Bone, dentin, cartilage, and enamel were all present in Ordovician ostracoderms (*see figure 18.3*).

Cartilage is the supportive tissue that makes up the major skeletal component of some gastropods, invertebrate chordates (amphioxus), jawless fishes such as hagfishes and lampreys, and sharks and rays. Since cartilage is lighter than bone, it gives these predatory fishes the speed and agility to catch prey. It also provides buoyancy without the need for a swim bladder.

(a)

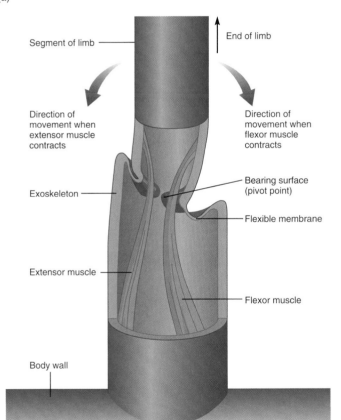

Segment of limb

End of limb

Direction of movement when extensor muscle contracts

Direction of movement when flexor muscle contracts

Exoskeleton

Bearing surface (pivot point)

Flexible membrane

Extensor muscle

Flexor muscle

Body wall

(b)

FIGURE 23.11

Exoskeletons. (*a*) A cicada nymph (*Platypedia*) leaves its old exoskeleton as it molts. This exoskeleton provides external support for the body and attachment sites for muscles. (*b*) In an arthropod, muscles attach to the interior of the exoskeleton. In this articulation of an arthropod limb, the cuticle is hardened everywhere except at the joint, where the membrane is flexible. Notice that the extensor muscle is antagonistic to (works in an opposite direction from) the flexor muscle.   (*b*) *Source: After Russell-Hunter.*

THE SKELETAL SYSTEM OF VERTEBRATES

The vertebrate skeletal system is an endoskeleton enclosed by other body tissues. This endoskeleton consists of two main types of supportive tissue: cartilage and bone.

Cartilage

Cartilage is a specialized type of connective tissue that provides a site for muscle attachment, aids in movement at joints, provides support (*see figure 2.24h–j*), and transmits the force of muscular contraction from one part of the body to another during movement. Like other connective tissues, it consists of cells (chondrocytes), fibers, and a cellular matrix.

Bone or Osseous Tissue

Bone (osseous) tissue is a specialized connective tissue that provides a point of attachment for muscles and transmits the force of muscular contraction from one part of the body to another during movement (figure 23.12*a*). In addition, bones of the skeleton support the internal organs of many animals, store reserve calcium and phosphate, and manufacture red blood cells and some white blood cells.

Bone tissue is more rigid than other connective tissues because its homogeneous, organic ground substance also contains inorganic salts—mainly calcium phosphate and calcium carbonate. When an animal needs the calcium or phosphate stored within bones, metabolic reactions (under endocrine control) release the required amounts.

Bone cells (osteocytes) are in minute chambers called lacunae (sing., lacuna), which are arranged in concentric rings around osteonic canals (formerly called Haversian systems) (figure 23.12*b*). These cells communicate with nearby cells by means of cellular processes passing through small channels called canaliculi (sing., canaliculus).

The Skeleton of Fishes

Both cartilaginous and bony endoskeletons first appeared in the vertebrates. Since water has a buoyant effect on the fish body, the requirement for skeletal support is not as demanding in these vertebrates as it is in terrestrial vertebrates. Although most vertebrates have a well-defined vertebral column (the reason they are called "vertebrates"), the jawless vertebrates do not. For example, lampreys only have isolated cartilaginous blocks along the notochord, and hagfishes do not even have these.

Most jawed fishes have an axial skeleton (so named because it forms the longitudinal axis of the body) that includes a notochord, ribs, and cartilaginous or bony vertebrae (figure 23.13). Muscles used in locomotion attach to the axial skeleton.

The Skeleton of Tetrapods

Tetrapods must lift themselves to walk on land. The first amphibians needed support to replace the buoyancy of water. *For the earliest terrestrial animals, support and locomotion were difficult*

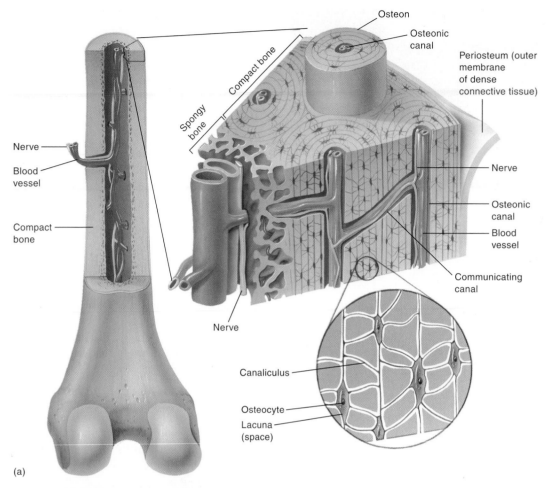

(a)

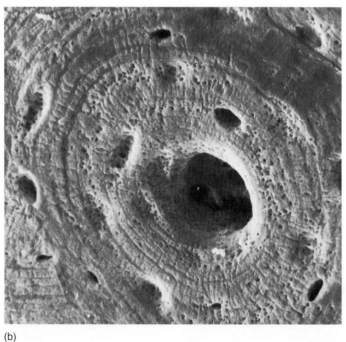

(b)

FIGURE 23.12

Bone. (*a*) Structural organization of a long bone (femur) of mammals. Compact bone is composed of osteons connected together. Spongy bone is latticelike rather than dense. (*b*) Single osteon in compact bone (SEM ×450).

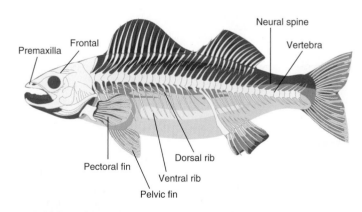

FIGURE 23.13

Fish Endoskeleton. Lateral view of the perch skeleton.

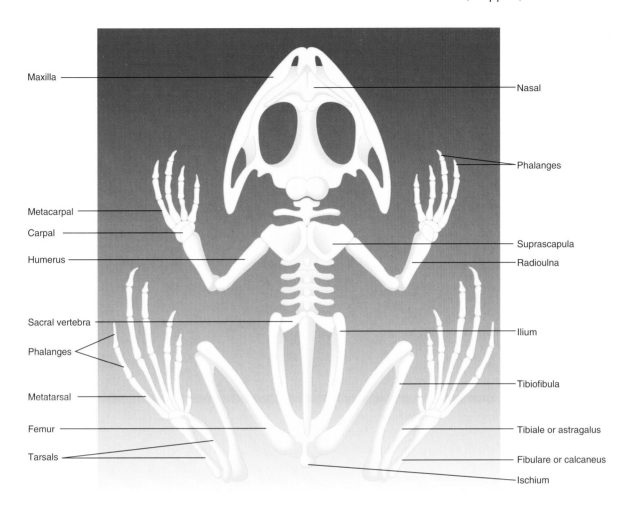

Maxilla

Nasal

Phalanges

Metacarpal

Carpal

Humerus

Suprascapula

Radioulna

Sacral vertebra

Phalanges

Ilium

Metatarsal

Femur

Tibiofibula

Tarsals

Tibiale or astragalus

Fibulare or calcaneus

Ischium

FIGURE 23.14

Tetrapod Endoskeleton.   Dorsal view of the frog skeleton.

and complicated processes. Adaptations for support and movement on land occurred over a period of approximately 200 million years. During this evolution, the tetrapod endoskeleton became modified for support on land (figure 23.14). This added support resulted from the specializations of the intervertebral disks that articulate with adjoining vertebrae. The intervertebral disks help hold the vertebral column together, and they also absorb shock and provide joint mobility. Bone replaced cartilage in the ribs, which became more rigid. The various types of connective tissue that connect to the axial skeleton helped keep elevated portions from sagging. Appendages became elongated for support on a hard surface, and changes in the shoulder enabled the neck to move more freely.

The Human Endoskeleton

The human endoskeleton has two major parts: the axial skeleton and the appendicular skeleton. The **axial skeleton** is made up of the skull, vertebral column, sternum, and ribs. The **appendicular skeleton** is composed of the appendages, the pectoral girdle, and the pelvic girdles. These girdles attach the upper and lower appendages to the axial skeleton.

MOVEMENT: NONMUSCULAR MOVEMENT AND MUSCULAR SYSTEMS

Movement is a characteristic of certain cells, protists, and animals. For example, certain white blood cells, coelomic cells, and protists such as *Amoeba* utilize nonmuscular amoeboid movement. Amoeboid movement also occurs in embryonic tissue movements, in wound healing, and in many cell types growing in tissue culture. Other protists and some invertebrates utilize cilia or flagella for movement. Muscles and muscle systems are found in various invertebrate groups from the primitive cnidarians to the arthropods (e.g., insect flight muscles). In more complex animals, the muscles attach to exo- and endoskeletal systems to form a motor system, which allows complex movements.

NONMUSCULAR MOVEMENT

Nearly all cells have some capacity to move and change shape due to their cytoskeleton (*see figure 2.19*). *It is from this basic framework of the cell that specialized contractile mechanisms*

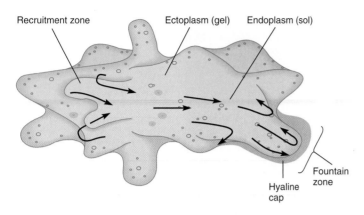

FIGURE 23.15

Mechanism of Amoeboid Movement.   Endoplasm (sol) flows into an advancing pseudopodium. At the tip (fountain zone) of the pseudopodium, endoplasm changes into ectoplasm (gel). At the opposite end (recruitment zone) of the amoeba, ectoplasm changes into endoplasm and begins flowing in the direction of movement.

emerged. For example, protozoan protists move by means of specific nonmuscular structures (pseudopodia, flagella, or cilia) that involve the contractile proteins actin and myosin. *Interactions between these proteins are also responsible for muscle contraction in animals, and the presence of actin and myosin in protozoa and animals is evidence of evolutionary ties between the two groups.*

Amoeboid Movement

As the name suggests, **amoeboid movement** was first observed in *Amoeba.* The plasma membrane of an amoeba has adhesive properties since new **pseudopodia** (sing., pseudopodium) (Gr. *pseudes,* false + *podion,* little foot) attach to the substrate as they form. The plasma membrane also seems to slide over the underlying layer of cytoplasm when an amoeba moves. The plasma membrane may be "rolling" in a way that is (roughly) analogous to a bulldozer track rolling over its wheels. A thin fluid layer between the plasma membrane and the ectoplasm may facilitate this rolling.

As an amoeba moves, the fluid endoplasm flows forward into the fountain zone of an advancing pseudopodium. As it reaches the tip of a pseudopodium, endoplasm changes into ectoplasm. At the same time, ectoplasm near the opposite end in the recruitment zone changes into endoplasm and begins flowing forward (figure 23.15).

Ciliary and Flagellar Movement

With the exception of the arthropods, locomotor cilia and flagella occur in every animal phylum. Structurally, **cilia** (sing., cilium) (L. "eyelashes") and **flagella** (sing., flagellum) (L. "small whips") are similar, but cilia are shorter and more numerous, whereas flagella are long and generally occur singly or in pairs.

Ciliary movements are coordinated. For example, in some ciliated protozoa, pairs of cilia occur in rows. Rows of cilia beat slightly out of phase with one another so that ciliary waves

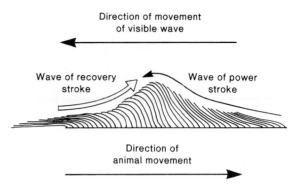

FIGURE 23.16

Ciliary Movement.   A metachronal (coordinated) wave passing along a row of cilia.

periodically pass over the surface of the protozoan (figure 23.16). In fact, many ciliates can rapidly reverse the direction of ciliary beating, which changes the direction of the ciliary waves and the direction of movement.

The epidermis of free-living flatworms (e.g., turbellarians) and nemertines is abundantly ciliated. The smallest specimens (about 1 mm long) lie at the upper end of the size range for efficient locomotion using cilia. Larger flatworms (e.g., triclads and polyclads) have retained **ciliary creeping** as the principal means of locomotion, and the largest animals to move by ciliary creeping are the nemertines. The muscular activities of the flatworms and nemertines are varied and involve pedal locomotion, peristalsis, or looping movements with anterior and posterior adhesion. *Since ciliary and muscular means of movement (locomotion) coexist in some free-living flatworms and nemertines, the transition from ciliary to muscular locomotion is likely to have taken place among the flatworm-like ancestors.*

AN INTRODUCTION TO ANIMAL MUSCLES

Muscular tissue is the driving force, the power behind movement in most invertebrates and vertebrates. The basic physiological property of muscle tissue is contractility, the ability to contract or shorten. In addition, muscle tissue has three other important properties: (1) excitability (or irritability), the capacity to receive and respond to a stimulus; (2) extensibility, the ability to be stretched; and (3) elasticity, the ability to return to its original shape after being stretched or contracted.

Animals may have one or more of the following types of muscle tissue: smooth, cardiac, and skeletal. The contractile cells of these tissues are called **muscle fibers.**

Smooth muscle is also called involuntary muscle because higher brain centers do not control its contractions. Smooth-muscle fibers have a single nucleus, are spindle shaped, and are arranged in a parallel pattern to form sheets (*see figure 2.25p*). Smooth muscle maintains good tone (a normal degree of vigor and tension) even without nervous stimulation. It contracts slowly, but it can sustain prolonged contractions and does not fatigue (tire) easily.

Smooth muscle is the predominant muscle type in many invertebrates. For example, it forms part of the adductor ("catch") muscles that close the valves of clams and other bivalve molluscs. These smooth muscles give bivalves the ability to "clam up" against predators for days with little or no energy expenditure.

Striated muscle fibers (cells) with single nuclei are common in invertebrates, but they occur in adult vertebrates only in the heart, where they are called cardiac muscle. **Cardiac muscle** fibers are involuntary, have a single nucleus, are striated (have dark and light bands), and are branched (*see figure 2.25*q). This branching allows the fibers to interlock for greater strength during contraction. Hearts do not fatigue because cardiac fibers relax completely between contractions.

Skeletal muscle, also a striated muscle, is a voluntary muscle because the nervous system consciously controls its contractions. Skeletal muscle fibers are multinucleated and striated (*see figure 2.25*o). Skeletal muscles attach to skeletons (both endo- and exoskeletons). When skeletal muscles contract, they shorten. Thus, muscles can only pull; they cannot push. Therefore, skeletal muscles work in antagonistic pairs. For example, one muscle of a pair bends (flexes) a joint and brings a limb close to the body. The other member of the pair straightens (extends) the joint and extends the limb away from the body (*see figure 23.11*b).

THE MUSCULAR SYSTEM OF INVERTEBRATES

A few functional differences among invertebrate muscles indicate some of the differences from the vertebrate skeletal muscles (discussed next). In arthropods, at least two motor nerves innervate a typical muscle fiber. One motor nerve fiber causes a fast contraction and the other a slow contraction. Another variation occurs in certain insect (bees, wasps, flies, beetles) flight muscles. These muscles are called asynchronous muscles because the upward wing movement (rather than a nerve impulse) activates the muscles that produce the downstroke. In the midge (a dipteran related to the fly/mosquito), for example, this can happen a thousand times a second.

An understanding of the structure and function of invertebrate locomotion (movement) is crucial to an understanding of the evolutionary origins of the various invertebrate groups. Discussion of several types of invertebrate locomotion that involve muscular systems follows.

The Locomotion of Soft-Bodied Invertebrates

Many soft-bodied invertebrates can move over a firm substratum. For example, flatworms, some cnidarians, and the gastropod molluscs move by means of waves of activity in the muscular system that are applied to the substrate. This type of movement is called **pedal locomotion.** Pedal locomotion can be easily seen

by examining the undersurface of a planarian or a snail while it crawls along a glass plate. In the land snail *Helix,* several waves cross the length of the foot simultaneously, each moving in the same direction as the locomotion of the snail, but at a greater rate.

Many large flatworms and most nemertine worms exhibit a muscular component to their locomotion. In this type of movement, alternating waves of contraction of circular and longitudinal muscles generate peristaltic waves, which enhance the locomotion that the surface cilia also provide. This system is most highly developed in the septate coelomate worms, especially earthworms (figure 23.17).

Leeches and some insect larvae exhibit **looping movements.** Leeches have anterior and posterior suckers that provide alternating temporary points of attachment (figure 23.18*a*). Lepidopteran caterpillars exhibit similar locomotion, in which arching movements are equivalent to the contraction of longitudinal muscles (figure 23.18*b*).

Polychaete worms move by the alternate movement of multiple limbs (parapodia), the tips of which move backward relative to the body; however, since the tips attach to the ground, the body of the worm moves forward (figure 23.19).

The **water-vascular system** of echinoderms provides a unique means of locomotion. For example, sea stars typically have five arms, with a water-vascular canal in each. Along each canal are reservoir ampullae and tube feet (figure 23.20*a, b*). Contraction of the muscles comprising the ampullae drives water into the tube feet, whereas contraction of the tube feet moves water into the ampullae. Thus, the tube feet extend by hydraulic pressure and can perform simple steplike motions (figure 23.20*c*).

Terrestrial Locomotion in Invertebrates: Walking

Invertebrates (terrestrial arthropods) living in/on terrestrial environments are much denser than the air in which they live. As a result, they require structural support, and those that move quickly make use of rigid skeletal elements that interact with the ground. These elements include flexible joints, tendons, and muscles that attach to a rigid cuticle and form limbs. The walking limbs of the most highly evolved arthropods (Crustacea, Chelicerata, and Uniramia) are remarkably uniform in structure. The limbs are composed of a series of jointed elements that become progressively less massive toward the tip (figure 23.21*a*). Each joint is articulated to allow movement in only one plane. These limb joints allow extension (a motion that increases the angle of a joint) and flexion (a motion that decreases the angle of a joint) of the limb. The limb plane at the basal joint with the body can also rotate, and this rotation is responsible for forward movement. The body is typically carried slung between the laterally projected limbs (figure 23.21*b*), and walking movements do not involve raising or lowering the body. Depending on the arthropod, the trajectory of each limb is different and nonoverlapping (figure 23.22). Most arthropods walk forward, rotating the basal joint of the limb relative to the body, but crabs walk in a sideways fashion.

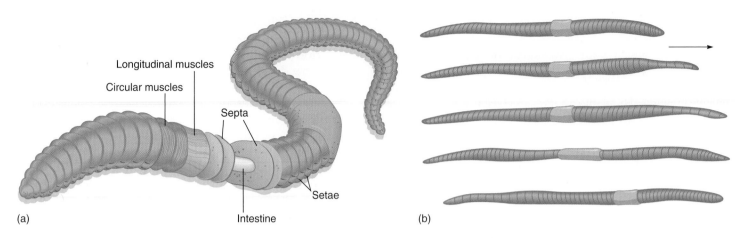

FIGURE 23.17

Successive Stages in Earthworm Movement. (*a*) When the longitudinal muscles contract and the circular muscles relax, the segments of the earthworm bulge and are stationary with respect to the ground. (*b*) In front of each region of longitudinal muscle contraction, circular muscles contract, causing the segments to elongate and push forward. Contraction of longitudinal muscles in segments behind a bulging region cause those segments to be pulled forward. For reasons of simplification, setae movements are not shown.

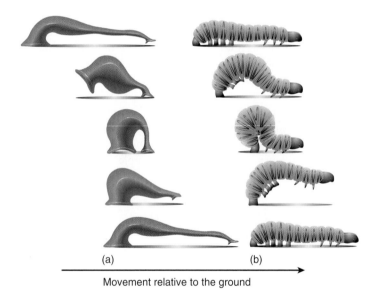

(a) (b)

Movement relative to the ground

FIGURE 23.18

Looping Movements. (*a*) Leeches have anterior and posterior suckers, which they alternately attach to the substrate in looping movements to move forward. (*b*) Some insect larvae, such as lepidopteran caterpillars, exhibit similar movements. The caterpillar uses arching movements to move forward.

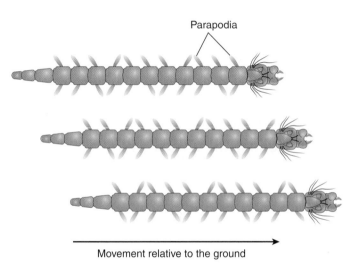

Parapodia

Movement relative to the ground

FIGURE 23.19

Locomotion in a Polychaete. When a polychaete (e.g., *Nereis*) crawls slowly, the tips of the multiple limbs (parapodia) move backward relative to the body. Since the tips of the parapodia touch the ground, this moves the body forward. In addition, a coordinated wave of activity in the parapodia passes forward from the tail to the head, with the left and right parapodia being exactly one-half wavelength out of phase. This ensures that each parapodium executes its power stroke without interfering with the parapodium immediately posterior. For simplification, setae movements are not shown.

Terrestrial Locomotion in Invertebrates: Flight

The physical properties (e.g., strengthening so that the exoskeleton does not deform under muscle contraction) of an arthropod cuticle are such that true flight evolved for the pterygote insects some 200 million years ago. Since then, the basic mechanism of flight has been modified. Consequently, present-day insects exhibit a wide range of structural adaptations and mechanisms for flight (*see figure 15.5*).

Terrestrial Locomotion in Invertebrates: Jumping

Some insects (fleas, grasshoppers, leafhoppers) can jump. Most often, this is an escape reaction. To jump, an insect must exert a force against the ground sufficient to impart a takeoff velocity greater than its weight (figure 23.23). Long legs increase the mechanical advantage of the leg extensor muscles. This is why insects that jump have relatively long legs. The mechanical

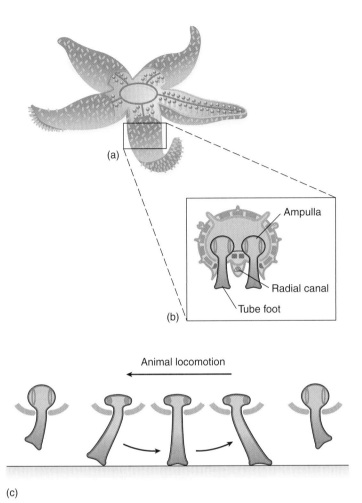

Animal locomotion

(c)

FIGURE 23.20

Water-Vascular System of Echinoderms. (*a*) General arrangement of the water-vascular system. (*b*) Cross section of an arm, showing the radial canal, ampullae, and tube feet of the water-vascular system. (*c*) Stepping cycle of a single tube foot. For simplification, the retractor muscles in the tube foot are not shown.

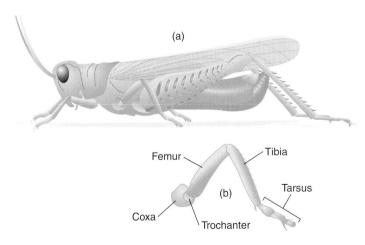

FIGURE 23.21

Typical Arthropod Limb. (*a*) Notice that most of the muscles are in the basal section. (*b*) Characteristic projection of the arthropod limb.

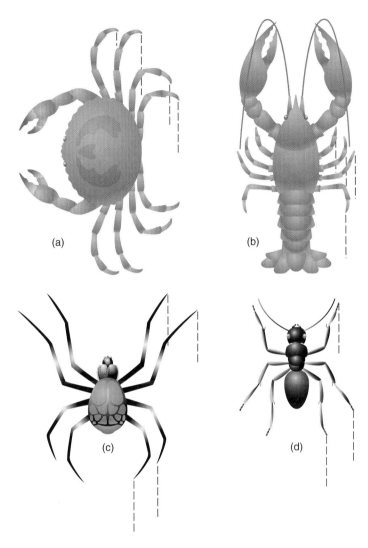

FIGURE 23.22

Walking: Limb Trajectories of Several Arthropods. (*a*) Crabs walk in a sideways fashion, a movement achieved by extension and retraction of the lower limb joints. Other arthropods, such as (*b*) the lobster, (*c*) the spider, and (*d*) an insect, have nonoverlapping limb trajectories and move forward by rotating the basal joint of the limb relative to their body.

strength of the insect cuticle acting as the lever in this system probably determines the limit to this line of evolution.

THE MUSCULAR SYSTEM OF VERTEBRATES

The vertebrate endoskeleton provides sites for skeletal muscles to attach. **Tendons,** which are tough, fibrous bands or cords, attach skeletal muscles to the skeleton.

Most of the musculature of fishes consists of segmental **myomeres** (Gr. *myo*, muscle + *meros*, part) (figure 23.24*a*). Myomere segments cause the lateral undulations of the trunk and tail that produce fish locomotion (figure 23.24*b*).

The transition from water to land entailed changes in the body musculature. As previously noted, the appendages became

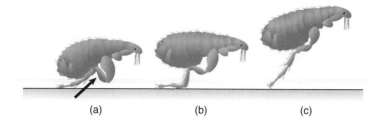

(a)   (b)   (c)

FIGURE 23.23

Jump of Flea.   A flea has a jointed exoskeleton. (*a*) When a flea is resting, the femur (black arrow) of the leg (for simplicity, only one leg is shown) is raised, the joints are locked, and energy is stored in the deformed elastic protein ("animal rubber" or resilin) of the cuticle. (*b*) As a flea begins to jump, the relaxation of muscles unlocks the joints. (*c*) The force exerted against the ground by the tibia gives the flea a specific velocity that determines the height of the jump. The jump is the result of the explosive release of the energy stored in the resilin of the cuticle.

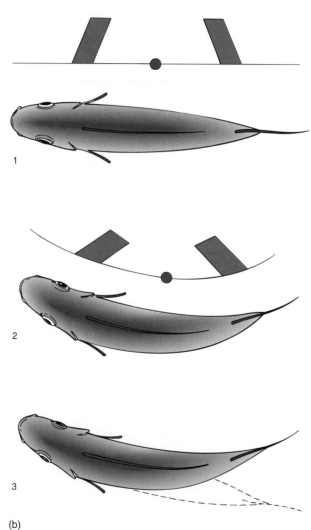

(b)

(a)

FIGURE 23.24

Fish Musculature.   (*a*) Skeletal muscles of a bony fish (perch), showing mainly the large muscles of the trunk and tail. These muscles occur in blocks called myomeres separated by connective tissue sheaths. Notice that the myomeres are flexed so that they resemble the letter W tipped at a 90° angle. The different colors (red, orange, blue) represent different myomeres. (*b*) Fish movements based on myomere contractions. (*1*) Muscular forces cause the myomere segments to rotate rather than constrict. (*2*) The rotation of cranial and caudal myomere segments bends the fish's body about a point midway between the two segments. (*3*) Alternate bends of the caudal end of the body propel the fish forward.

increasingly important in locomotion, and movements of the trunk became less important. The segmental nature of the myomeres in the trunk muscles was lost. Back muscles became more numerous and powerful. These evolutionary adaptations are well illustrated by comparing what is eaten in a fish dinner to a plate of frog legs.

Skeletal Muscle Contraction

When observed with the light microscope, each skeletal muscle fiber (cell) has a pattern of alternate dark and light bands (*see figure 2.25o*). This striation of whole fibers arises from the alternating dark and light bands of the many smaller, threadlike **myofibrils** in each muscle fiber (figure 23.25*a–c*). Electron

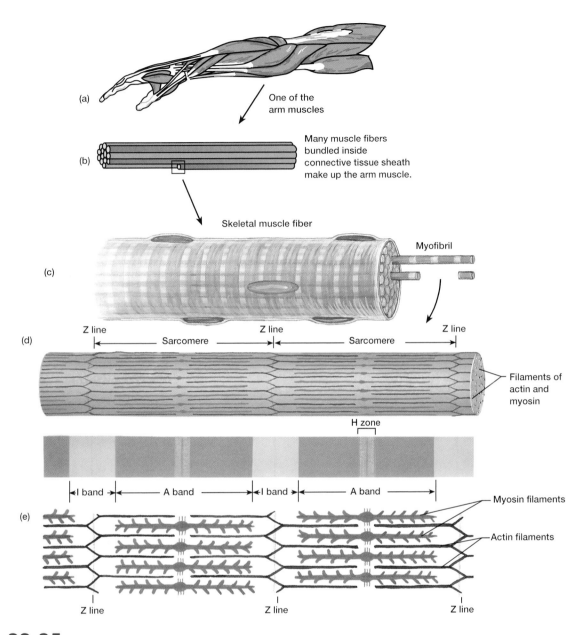

FIGURE 23.25

Structure of Skeletal Muscle Tissue.   (*a*) A skeletal muscle in the forearm consists of many muscle fibers (cells) (*b*) bundled inside a connective tissue sheath. (*c*) A skeletal muscle fiber contains many myofibrils, each consisting of (*d*) functional units called sarcomeres. (*e*) The characteristic striations of a sarcomere are due to the arrangement of actin and myosin filaments.

microscopy and biochemical analysis show that these bands are due to the placement of the muscle proteins **actin** and **myosin** within the myofibrils. Myosin occurs as thick filaments and actin as thin filaments. As figure 23.25*c–e* illustrates, the lightest region of a myofibril (the I band) contains only actin, whereas the darkest region (the A band) contains both actin and myosin.

The functional (contractile) unit of a myofibril is the **sarcomere,** each of which extends from one Z line to another Z line. Notice that the actin filaments attach to the Z lines, whereas myosin filaments do not (figure 23.25*e*). When a sar-

comere contracts, the actin filaments slide past the myosin filaments as they approach one another. This process shortens the sarcomere. The combined decreases in length of the individual sarcomeres account for contraction of the whole muscle fiber, and in turn, the whole muscle. This movement of actin in relation to myosin is called the sliding-filament model of muscle contraction.

A ratchet mechanism between the two filament types produces the actual contraction. Myosin contains globular projections that attach to actin at specific active binding sites, forming attachments called cross-bridges (figure 23.26). Once

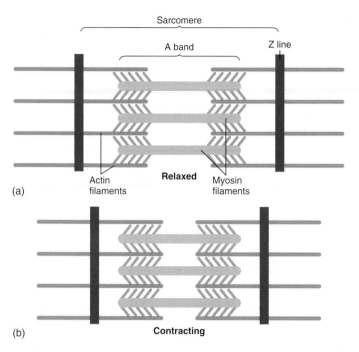

(a) **Relaxed**

Sarcomere

A band

Z line

Actin filaments

Myosin filaments

(b) **Contracting**

FIGURE 23.26

Sliding-Filament Model of Muscle Contraction. (*a*) A sarcomere in a relaxed position. (*b*) As the sarcomere contracts, the myosin filaments form attachments called cross-bridges to the actin filaments and pull the actin filaments so that they slide past the myosin filaments. Compare the length of the sarcomere in (*a*) to that in (*b*).

cross-bridges form, they exert a force on the thin actin filament and cause it to move.

Control of Skeletal Muscle Contraction

When a motor nerve conducts nerve impulses to skeletal muscle fibers, the fibers are stimulated to contract via a motor unit. A **motor unit** consists of one motor nerve fiber and all the muscle fibers with which it communicates. A space (cleft) separates the specialized end of the motor nerve fiber from the membrane (**sarcolemma**) of the muscle fiber. The motor end plate is the specialized portion of the sarcolemma of a muscle fiber surrounding the terminal end of the nerve. This arrangement of structures is called a **neuromuscular junction** (figure 23.27).

When nerve impulses reach the ends of the nerve fiber branches, synaptic vesicles in the nerve ending release a chemical called acetylcholine. Acetylcholine diffuses across the neuromuscular cleft between the nerve ending and the muscle-fiber sarcolemma and binds with acetylcholine receptors on the sarcolemma. The sarcolemma is normally polarized; the outside is positive, and the inside is negative. When acetylcholine binds to the receptors, ions are redistributed on both sides of the membrane, and the polarity is altered. This altered polarity flows in a wavelike progression into the muscle fiber by conducting paths called transverse tubules. Associated with the transverse tubules is the endoplasmic reticulum (*see figure 2.15*) of muscle cells, called sarcoplasmic reticulum. The altered polarity of the transverse tubules causes the sarcoplasmic reticulum to release

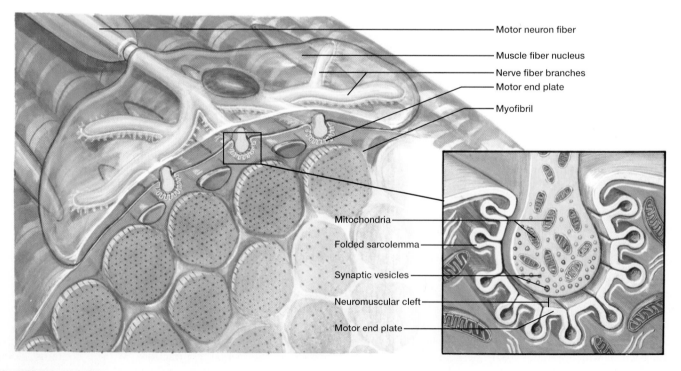

Motor neuron fiber

Muscle fiber nucleus

Nerve fiber branches

Motor end plate

Myofibril

Mitochondria

Folded sarcolemma

Synaptic vesicles

Neuromuscular cleft

Motor end plate

FIGURE 23.27

Nerve-Muscle Motor Unit. A motor unit consists of one motor nerve and all the muscle fibers that it innervates. A neuromuscular junction, or cleft, is the place where the nerve fiber and muscle fiber meet.

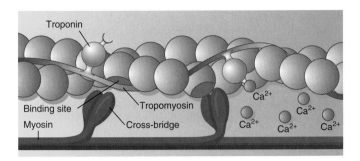

FIGURE 23.28

Model of the Calcium-Induced Changes in Troponin that Allow Cross-Bridges to Form between Actin and Myosin.   The attachment of Ca^{2+} to troponin moves the troponin-tropomyosin complex, which exposes a binding site on the actin. The myosin cross-bridge can then attach to actin and undergo a power stroke.

calcium ions (Ca^{2+}), which diffuse into the cytoplasm. The calcium then binds with a regulatory protein called troponin that is on another protein called tropomyosin. This binding exposes the myosin binding sites on the actin molecule that tropomyosin had blocked (figure 23.28). Once the binding sites are open, the myosin filament can form cross-bridges with actin, and power strokes of cross-bridges result in filament sliding and muscular contraction.

Relaxation follows contraction. During relaxation, an active transport system pumps calcium back into the sarcoplasmic reticulum for storage. By controlling the nerve impulses that reach the sarcoplasmic reticulum, the nervous system controls Ca^{2+} levels in skeletal muscle tissue, thereby exerting control over contraction.

SUMMARY

1. The integumentary system is the external covering of an animal. It primarily protects against mechanical injury and invasion by microorganisms.

2. Some single-celled protozoa have only a plasma membrane for an external covering. Other protozoa have a thick protein coat, called a pellicle, outside the plasma membrane. Most invertebrates have an integument consisting of a single layer of columnar epithelial cells called an epidermis. Specializations outside of this epithelial layer may be in the form of cuticles, shells, or teguments.

3. Skin is the vertebrate integument. It has two main layers: the epidermis and the dermis. Skin structure varies considerably among vertebrates. Some of these variable structures include scales, hairs, feathers, claws, nails, and baleen plates.

4. The skin of jawless fishes (lampreys and hagfishes) is thick. The skin of cartilaginous fishes (sharks) is multilayered and contains bone in the form of denticles. The skin of bony fishes (teleosts) contains scales. The skin of amphibians is stratified and contains mucous and serous glands plus pigmentation. The skin of reptiles is thick and modified into keratinized scales. The skin of birds is thin and soft and contains feathers. Mammalian skin consists of several layers of a variety of cells.

5. Animals have three types of skeletons: hydrostatic skeletons, exoskeletons, and endoskeletons. These skeletons function in animal movement that requires muscles working in opposition (antagonism) to each other.

6. The hydrostatic skeleton is a core of liquid (water or a body fluid such as blood) surrounded by a tension-resistant sheath of longitudinal and/or circular muscles. Hydrostatic skeletons are found in invertebrates and can take many forms and shapes, such as the gastrovascular cavity of acoelomates, the rhynchocoel in nemertines, a pseudocoelom in aschelminths, a coelom in annelids, or a hemocoel in molluscs.

7. Rigid exoskeletons also have locomotor functions because they provide sites for muscle attachment and counterforces for muscle movements. Exoskeletons also support and protect the body, but these are secondary functions. In arthropods, the epidermis of the body wall secretes a thick, hard cuticle. In crustaceans (crabs, lobsters, shrimp), the exoskeleton contains calcium carbonate crystals that make it hard and inflexible, except at the joints.

8. Rigid endoskeletons are enclosed by other body tissues. For example, the endoskeletons of sponges consist of mineral spicules, and the endoskeletons of echinoderms (sea stars, sea urchins) are made of calcareous plates called ossicles.

9. The most familiar endoskeletons, both cartilaginous and bony, first appeared in the vertebrates. Endoskeletons consist of two main types of supportive connective tissue: cartilage and bone. Cartilage provides a site for muscle attachment, aids in movement at joints, and provides support. Bone provides a point of attachment for muscles and transmits the force of muscular contraction from one part of the body to another.

10. Movement (locomotion) is characteristic of certain cells, protists, and animals. Amoeboid movement and movement by cilia and flagella are examples of locomotion that does not involve muscles.

11. The power behind muscular movement in both invertebrates and vertebrates is muscular tissue. The three types of muscular tissue are smooth, cardiac, and skeletal. Muscle tissue exhibits contractility, excitability, extensibility, and elasticity.

12. The functional (contractile) unit of a muscle myofibril is the sarcomere. Nerves control skeletal muscle contraction.

SELECTED KEY TERMS

amoeboid movement (p. 370)
ciliary creeping (p. 370)
denticle (p. 361)
dermis (p. 361)
endoskeleton (p. 366)
epidermis (p. 361)

exoskeleton (p. 365)
hydrostatic skeleton (p. 365)
integument (p. 360)
neuromuscular junction (p. 376)
skin (p. 361)

CRITICAL THINKING QUESTIONS

1. How does the structure of skin relate to its functions of protection, temperature control, waste removal, water conservation, radiation protection, vitamin production, and environmental responsiveness?

2. How does the epidermis of an invertebrate differ from that of a vertebrate?

3. Give an example of an animal with each type of skeleton (hydro-exo-, endoskeleton), and explain how the contractions of its muscles produce locomotion.

4. Give one similarity and one difference between vertebrate skeletal muscle and the following: asynchronous insect muscle; the "catch" muscle of molluscs; the movement of cilia; amoeboid movement.

ONLINE LEARNING CENTER

Visit our Online Learning Center (OLC) at www.mhhe.com/zoology (click on this book's title) to find the following chapter-related materials:

- CHAPTER QUIZZING
- ANIMATIONS
  Striated Muscle Contraction
  Muscle Contraction Action Potential
  Detailed Striated Muscle Contraction
  Actin-Myosin Cross-bridges
  Walking

- RELATED WEB LINKS
  Support, Protection, and Movement
    Integumentary Systems
    Vertebrates: Macroscopic Anatomy of the Integument
    Vertebrates: Microscopic Anatomy of the Integument
    Human Integumentary System Topics
  Skeletal Systems
    Vertebrates: Macroscopic Anatomy of the Skeleton
    Axial Skeleton
    Appendicular Skeleton: Gills, Fins, and Limbs
    Vertebrates: Microscopic Anatomy of the Skeleton
    Bone and Cartilage
    Ligaments and Joints
    Human Skeletal System Topics
  Animal Movement and Musculature
- SUGGESTED READINGS
- LAB CORRELATIONS

Check out the OLC to find specific information on these related lab exercises in the *General Zoology Laboratory Manual*, 5th edition, by Stephen A. Miller:

Exercise 18   *Vertebrate Musculoskeletal Systems*

CHAPTER 24

COMMUNICATION I:

NERVOUS AND SENSORY SYSTEMS

Outline

Concepts

1. The nervous system helps to communicate, integrate, and coordinate the functions of the various organs and organ systems in the animal body.
2. Information flow through the nervous system has three main steps: (1) the collection of information from outside and inside the body (sensory activities), (2) the processing of this information in the nervous system, and (3) the initiation of appropriate responses.
3. Information is transmitted between neurons directly (electrically) or by means of chemicals called neurotransmitters.
4. The evolution of the nervous system in invertebrates has led to the elaboration of organized nerve cords and the centralization of responses in the anterior portion of the animal.
5. The vertebrate nervous system consists of the central nervous system, made up of the brain and spinal cord, and the peripheral nervous system, composed of the nerves in the rest of the body.
6. Nervous systems evolved through the gradual layering of additional nervous tissue over reflex pathways of more ancient origin.
7. Sensory receptors or organs permit an animal to detect changes in its body, as well as in objects and events in the world around it. Sensory receptors collect information that is then passed to the nervous system, which determines, evaluates, and initiates an appropriate response.
8. Sensory receptors initiate nerve impulses by opening channels in sensory neuron plasma membranes, depolarizing the membranes, and causing a generator potential. Receptors differ in the nature of the environmental stimulus that triggers an eventual nerve impulse.
9. Many kinds of receptors have evolved among invertebrates and vertebrates, and each receptor is sensitive to a specific type of stimulus.
10. The nature of its sensory receptors gives each animal species a unique perception of its body and environment.

The two forms of communication in an animal that integrate body functions to maintain homeostasis are: (1) neurons, which transmit electrical signals that report information or initiate a quick response in a specific tissue; and (2) hormones, which are slower, chemical signals that initiate a widespread, prolonged response, often in a variety of tissues. This chapter focuses on the function of the neuron, the anatomical organization of the nervous system in animals, and the ways in which the senses collect information and transmit it along nerves to the central nervous system. To conclude the study of communication, chapter 25 examines how hormones affect long-term changes in an animal's body.

This chapter contains evolutionary concepts, which are set off in this font.

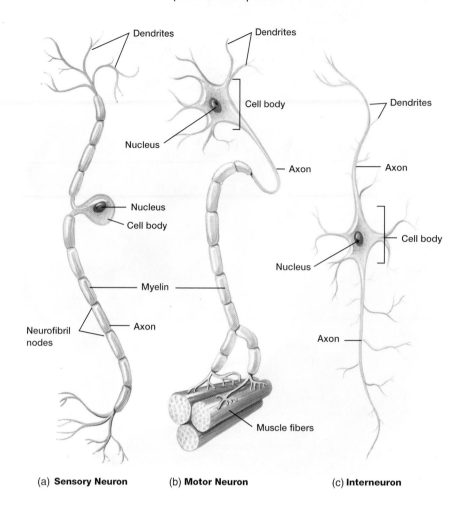

(a) **Sensory Neuron**   (b) **Motor Neuron**   (c) **Interneuron**

FIGURE 24.1

Types of Vertebrate Neurons.   (*a*) Sensory neurons transmit information from the environment to the central nervous system. (*b*) Motor neurons transmit information from the central nervous system to muscles or glands; they tend to have short dendrites and long axons. (*c*) Interneurons connect other neurons, permitting the integration of information.

NEURONS: THE BASIC FUNCTIONAL UNITS OF THE NERVOUS SYSTEM

The functional unit of the nervous system is a highly specialized cell called the **neuron** (Gr. "nerve"). Neurons are specialized to produce signals that can be communicated over short to relatively long distances, from one part of an animal's body to another. Neurons have two important properties: (1) excitability, the ability to respond to stimuli; and (2) conductivity, the ability to conduct a signal.

The three functional types of neurons are sensory neurons, interneurons, and motor neurons. **Sensory** (**receptor** or **afferent**) **neurons** either act as receptors of stimuli themselves or are activated by receptors (figure 24.1*a*). Changes in the internal or external environments stimulate sensory neurons, which respond by sending signals to the major integrating centers where information is processed. **Interneurons** (figure 24.1*c*) receive signals from the sensory neurons and transmit them to motor neurons. **Motor** (**effector** or **efferent**) **neurons** (figure 24.1*b*) send the processed

information via a signal to the body's effectors (e.g., muscles), causing them to contract, or to glands, causing them to secrete. Figure 24.2 summarizes the flow of information in the nervous system.

NEURON STRUCTURE: THE KEY TO FUNCTION

Most neurons contain three principal parts: a cell body, dendrites, and an axon (*see figure 24.1*). The **cell body** has a large, central nucleus. The motor neuron in figure 24.1*b* has many short, threadlike branches called **dendrites** (Gr. *dendron*, tree), which are actually extensions of the cell body and conduct signals toward the cell body. The **axon** is a relatively long, cylindrical process that conducts signals (information) away from the cell body.

The neurons of hydras and sea anemones do not have a sheath covering the axon of the neuron. Other invertebrates and all vertebrates have sheathed neurons. When present, the laminated lipid sheath is called **myelin.** In some neurons, a **neurolemmocyte** (formerly known as a Schwann cell) wraps the myelin

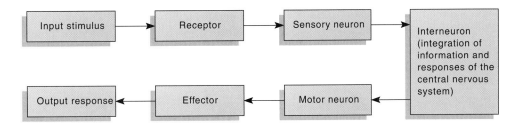

FIGURE 24.2

Generalized Pathway for the Flow of Information within the Nervous System.   An input stimulus initiates impulses within some sensory structure (the receptor); the impulses are then transferred via sensory neurons to interneurons. After response selection, nerve impulses are generated and transferred along motor neurons to an effector (e.g., a muscle or gland), which elicits the appropriate output response.

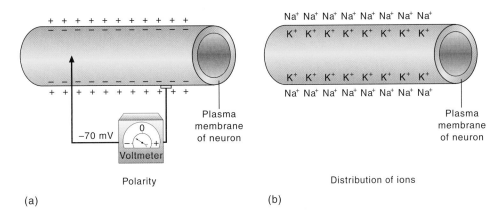

FIGURE 24.3

Resting Membrane Potential.   (a) A voltmeter measures the difference in electrical potential between two electrodes. When one microelectrode is placed inside a neuron at rest, and one is placed outside, the electrical potential inside the cell is −70 mV relative to the outside. (b) In a neuron at rest, sodium is more concentrated outside and potassium is more concentrated inside the cell. A neuron in this resting condition is said to be polarized.

sheath in layers. In these neurons, gaps called **neurofibril nodes** (formerly **nodes of Ranvier**) segment the myelin sheath at regular intervals. The neurolemmocyte also assists in the regeneration of injured myelinated neurons.

The nervous system receives data (input stimulus), integrates it, and effects a change (output response) in the animal's physiology. In a given neuron, the dendrites are the receptors, the cell body is the integrator, and the ends of the axon are the effectors.

NEURON COMMUNICATION

The language (signal) of a neuron is the nerve impulse or action potential. The key to this nerve impulse is the neuron's plasma membrane and its properties. Changes in membrane permeability and the subsequent movement of ions produce a nerve impulse that travels along the plasma membrane of the dendrites, cell body, and axon of each neuron.

RESTING MEMBRANE POTENTIAL

A "resting" neuron is not conducting a nerve impulse. The plasma membrane of a resting neuron is polarized; the fluid on the inner side of the membrane is negatively charged with respect to the

positively charged fluid outside the membrane (figure 24.3). The difference in electrical charge between the inside and the outside of the membrane at any given point is due to the relative numbers of positive and negative ions in the fluids on either side of the membrane, and to the permeability of the plasma membrane to these ions. The difference in charge is called the **resting membrane potential.** All cells have such a resting potential, but neurons and muscle cells are specialized to transmit and recycle it rapidly.

The resting potential is measured in millivolts (mV). A millivolt is 1/1,000 of a volt. Normally, the resting membrane potential is about −70 mV, due to the unequal distribution of various electrically charged ions. Sodium (Na^+) ions are more highly concentrated in the fluid outside the plasma membrane, and potassium (K^+) and negative protein ions are more highly concentrated inside.

The Na^+ and K^+ ions constantly diffuse through ion channels in the plasma membrane, moving from regions of higher concentrations to regions of lower concentrations. (There are also larger Cl^- ions and huge negative protein ions, which cannot move easily from the inside of the neuron to the outside.) However, the concentrations of Na^+ and K^+ ions on the two sides of the membrane remain constant due to the action of the **sodium-potassium ATPase pump,** which is

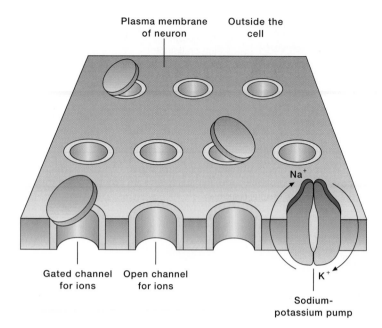

Plasma membrane
of neuron

Outside the
cell

Na$^+$

K$^+$

Gated channel
for ions

Open channel
for ions

Sodium-
potassium pump

FIGURE 24.4

Ion Channels and the Sodium-Potassium Pump.   These mechanisms maintain a balance between the sodium ions and potassium ions on both sides of the membrane and create a membrane potential. Some channels are always open, but others open or close by the position of gates, which are proteins that change shape to block or clear the channel. Whether a gate opens or closes a channel depends on the membrane potential. Such gates are said to be voltage regulated. Some of these membrane channels are specific for sodium ions, and others are specific for potassium ions.

powered by ATP (figure 24.4). The pump actively moves Na$^+$ ions to the outside of the cell and K$^+$ ions to the inside of the cell. Because it moves three Na$^+$ molecules out for each two K$^+$ molecules that it moves in, the pump works to establish the resting potential across the membrane. Both ions leak back across the membrane—down their concentration gradients. K$^+$ ions, however, move more easily back to the outside, adding to the positive charge there and contributing to the membrane potential of −70 mV.

MECHANISM OF NEURON ACTION: CHANGING THE RESTING MEMBRANE POTENTIAL INTO THE ACTION POTENTIAL (NERVE IMPULSE)

Changing the resting electrical potential across the plasma membrane is the key factor in the creation and subsequent conduction of a nerve impulse. A stimulus that is strong enough to initiate an impulse is called a threshold stimulus. When such a stimulus is applied to a point along the resting plasma membrane, the permeability to Na$^+$ ions increases at that point. The inflow of positively charged Na$^+$ ions causes the membrane potential to go from −70 mV toward 0. This loss in membrane polarity is called **depolarization** (figure 24.5). When depolarization reaches

a certain level, special Na$^+$ channels (voltage-gated) that are sensitive to changes in membrane potential quickly open, and more Na$^+$ ions rush to the inside of the neuron. Shortly after the Na$^+$ ions move into the cell, the Na$^+$ gates close, but now voltage-gated K$^+$ channels open, and K$^+$ ions rapidly diffuse outward. The movement of the K$^+$ ions out of the cell builds up the positive charge outside the cell again, and the membrane becomes **repolarized.** This series of membrane changes triggers a similar cycle in an adjacent region of the membrane, and the wave of depolarization moves down the axon as an **action potential.** Overall, the transmission of an action potential along the neuron plasma membrane is a wave of depolarization and repolarization.

After each action potential, there is an interval of time when it is more difficult for another action potential to occur because the membrane has become hyperpolarized (more negative than −70 mV) due to the large number of K$^+$ ions that rushed out. This brief period is called the **refractory period.** During this period, the resting potential is being restored at the part of the membrane where the impulse has just passed. Afterward, the neuron is repolarized and ready to transmit another impulse.

A minimum stimulus (threshold) is necessary to initiate an action potential, but an increase in stimulus intensity does not increase the strength of the action potential. The principle that states that an axon will "fire" at full power or not at all is the **all-or-none law.**

Increasing the axon diameter and/or adding a myelin sheath increases the speed of conduction of a nerve impulse. Axons with a large diameter transmit impulses faster than smaller ones. Large-diameter axons are common among many invertebrates (e.g., crayfishes, earthworms). The largest are those of the squid (*Loligo*), where axon diameter may be over 1 mm, and the axons have a conduction velocity greater than 36 m/second! (The giant squid axons provide a simple, rapid triggering mechanism for quick escape from predators. A single action potential elicits a maximal contraction of the mantle muscle that it innervates. Mantle contraction rapidly expels water, "jetting" the squid away from the predator.) Most vertebrate axons have a diameter of less than 10 µm; however, some fishes and amphibians have evolved large, unmyelinated axons 50 µm in diameter. These extend from the brain, down the spinal cord, and they activate skeletal muscles for rapid escapes.

Regardless of an axon's diameter, the myelin sheath greatly increases conduction velocity. The reason for this velocity increase is that myelin is an excellent insulator and effectively stops the movement of ions across it. Action potentials are generated only at the neurofibril nodes. In fact, the action potential "jumps" from one node to the next node. For this reason, conduction along myelinated fibers is known as **saltatory conduction** (L. *saltare,* to jump). It takes less time for an impulse to jump from node to node along a myelinated fiber than to travel smoothly along an unmyelinated fiber. *Myelination allows rapid conduction in small neurons and thus provides for the evolution of nervous systems that do not occupy much space within the animal.*

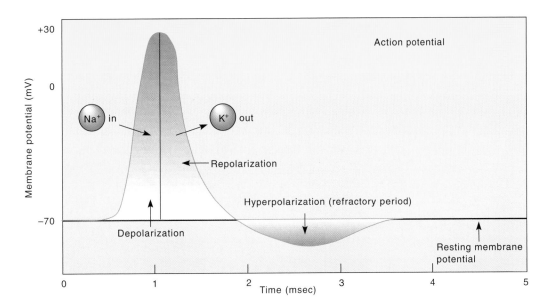

FIGURE 24.5

Action Potential as Recorded on an Oscilloscope. During the depolarization phase of the action potential, sodium (Na^+) ions rush to the inside of a neuron. The repolarization phase is characterized by a rapid increase in potassium (K^+) ions on the outside of the neuron. The action potential is sometimes called a "spike" because of its shape on an oscilloscope screen.

TRANSMISSION OF THE ACTION POTENTIAL BETWEEN CELLS

After an action potential travels along an axon, it reaches the end of a branching axon terminal called the **end bulb.** The **synapse** (Gr. *synapsis*, connection) is the junction between the axon of one neuron and the dendrite of another neuron or effector cell. The space (junction) between the end bulb and the dendrite of the next neuron is the **synaptic cleft.** The neuron carrying the action potential toward a synapse is the presynaptic ("before the synapse") neuron. It initiates a response in the receptive segment of a postsynaptic ("after the synapse") neuron leading away from the synapse. The presynaptic cell is always a neuron, but the postsynaptic cell can be a neuron, muscle cell, or gland cell.

Synapses can be electrical or chemical. In an **electrical synapse,** nerve impulses transmit directly from neuron to neuron when positively charged ions move from one neuron to the next. These ions depolarize the postsynaptic membrane, as though the two neurons were electrically coupled. An electrical synapse can rapidly transmit impulses in both directions. Electrical synapses are common in fishes and partially account for their ability to dart swiftly away from a threatening predator.

In a **chemical synapse,** two cells communicate by means of a chemical agent called a **neurotransmitter,** which the presynaptic neuron releases. A neurotransmitter changes the resting potential in the plasma membrane of the receptive segment of the postsynaptic cell, creating an action potential in that cell, which continues the transmission of the impulse.

When a nerve impulse reaches an end bulb, it causes storage vesicles (containing the chemical neurotransmitter) to fuse with the plasma membrane. The vesicles release the neurotransmitter by exocytosis into the synaptic cleft (figure 24.6). One common neurotransmitter is the chemical **acetylcholine;** another is **norepinephrine.** (More than 50 other possible transmitters are known.)

When the released neurotransmitter (e.g., acetylcholine) binds with receptor protein sites in the postsynaptic membrane, it causes a depolarization similar to that of the presynaptic cell. As a result, the impulse continues its path to an eventual effector. Once acetylcholine has crossed the synaptic cleft, the enzyme acetylcholinesterase quickly inactivates it. Without this breakdown, acetylcholine would remain and would continually stimulate the postsynaptic cell, leading to a diseased state. You have probably created a similar diseased state at the synapses of the fleas on your dog or cat. The active ingredient in most flea sprays and powders is parathion. It prevents the breakdown of acetylcholine in the fleas, as well as pets and people. However, because fleas are so small, the low dose that immobilizes the fleas does not affect pets or humans.

INVERTEBRATE NERVOUS SYSTEMS

All cells respond to some stimuli and relay information both internally and externally. Thus, even when no real nervous system is present, such as in the protozoa and sponges, coordination and reaction to external and internal stimuli do occur. For example, the regular beating of protozoan cilia (*see figure 23.16*) or the response of flagellates to varying light intensities requires intracellular coordination. Only animals that have achieved the tissue level of

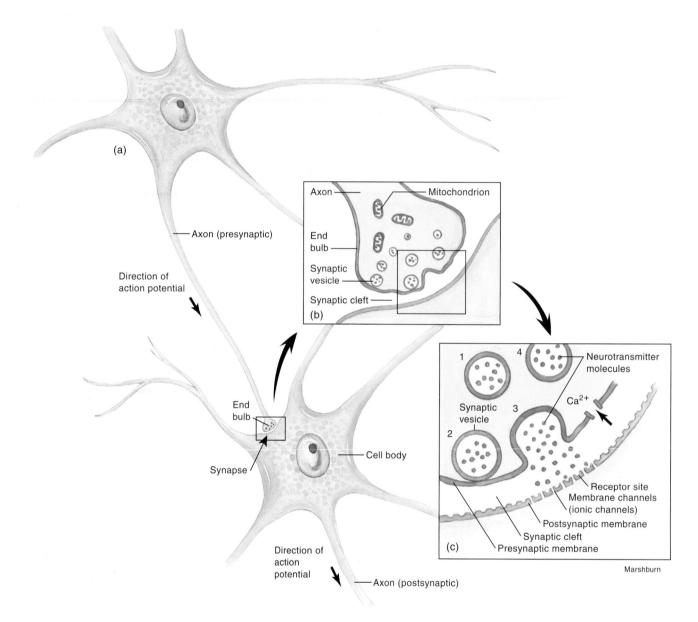

FIGURE 24.6

Chemical Transmission across a Synapse.   (*a*) Pre- and postsynaptic neurons with synaptic end bulb. (*b*) Enlarged view of the end bulb containing synaptic vesicles. (*c*) Enlargement of a portion of the end bulb showing exocytosis. The sequence of events in neurotransmitter release is: (1) a synaptic vesicle containing neurotransmitter approaches the plasma membrane; (2) due to the influx of calcium ions, the vesicle fuses with the membrane; (3) exocytosis occurs; and (4) the vesicle reforms and begins to fill with more neurotransmitter.

organization (e.g., the diploblastic and triploblastic animals) have true nervous systems, however. This clearly excludes the protozoa and sponges.

Among animals more complex than sponges, five general evolutionary trends in nervous system development are apparent. *The first has been integrated throughout Part Two of this text. More complex animals possess more detailed nervous systems.*

Of all animals, the cnidarians (hydras, jellyfishes, and sea anemones) have the simplest form of nervous organization. These animals have a **nerve net,** a latticework that conducts impulses from one area to another (figure 24.7*a*). In nerve nets,

impulse conduction by neurons is bidirectional. Cnidarians lack brains and even local clusters of neurons. Instead, a nerve stimulus anywhere on the body initiates a nerve impulse that spreads across the nerve net to other body regions. In jellyfishes, this type of nervous system is involved in slow swimming movements and in keeping the body right-side up. At the cellular level, the neurons function in the way discussed earlier in this chapter.

Echinoderms (e.g., sea stars, sea urchins, sea cucumbers) still have nerve nets, but of increasing complexity. For example, sea stars have three distinct nerve nets. The one that lies just under the skin has a circumoral ring and five sets of nerve cords

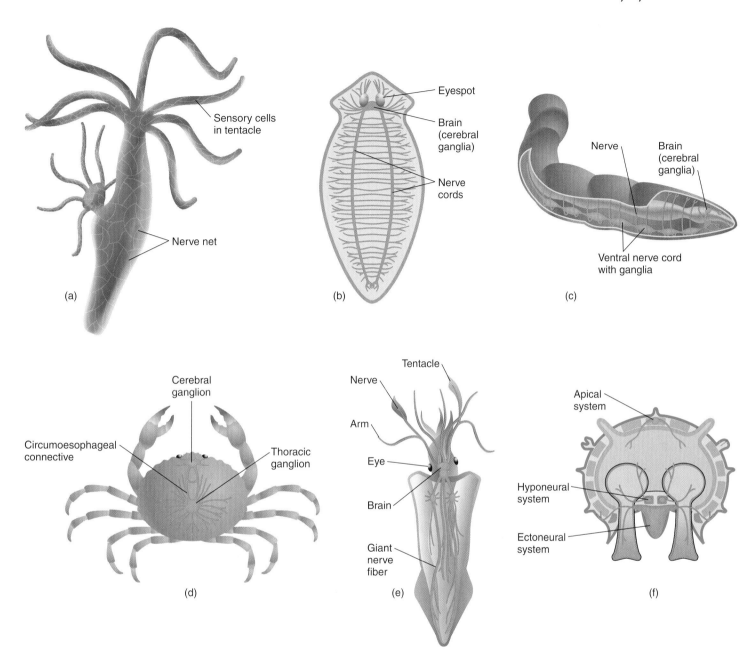

FIGURE 24.7

Some Invertebrate Nervous Systems. (*a*) The nerve net of *Hydra*, a cnidarian, represents the simplest neural organization. (*b*) Brain and paired nerve cords of a planarian flatworm. This is the first nervous system showing differentiation into a peripheral nervous system and a central nervous system. (*c*) Brain, ventral nerve cord, ganglia, and peripheral nerves of the earthworm, an annelid worm. (*d*) A crustacean, showing the principal ganglia and visceral connective nerves. The most primitive crustaceans have nervous systems similar to those of the platyhelminths, whereas (*e*) some cephalopods (such as the squid) have brains and behavior as complex as those of fishes. (*f*) Cross section of a starfish arm. Nerves from the ectoneural system terminate on the surface of the hyponeural system, but the two systems have no contact.

running out to the animal's arms. Another net serves the muscles between the skin plates, called ossicles. The third net connects to the tube feet. This degree of nerve net complexity permits locomotion, a variety of useful reflexes, and some degree of "central" coordination. For example, when a sea star is flipped over, it can right itself.

Animals, such as flatworms and roundworms, that move in a forward direction have sense organs concentrated in the body region that first encounters new environmental stimuli. *Thus, the second trend in nervous system evolution involves cephalization, which is a concentration of receptors and nervous tissue in the animal's anterior end.* For example, a flatworm's nervous system contains **ganglia** (sing., ganglion), which are distinct aggregations of nerve cells in the head region. Ganglia function as a primitive "brain" (figure 24.7*b*). Distinct lateral nerve cords (collections of neurons) on either side of the body carry sensory information from the periphery to the head ganglia and carry motor impulses from the head ganglia back to muscles, allowing the animal to react to environmental stimuli.

These lateral nerve cords reveal that flatworms also exhibit the third trend in nervous system evolution: bilateral symmetry. Bilateral symmetry (a body plan with roughly equivalent right and left halves) could have led to paired neurons, muscles, sensory structures, and brain centers. This pairing facilitates coordinated movements, such as climbing, crawling, flying, or walking.

In other invertebrates, such as crustaceans, segmented worms, and arthropods, the organization of the nervous system shows further advances. In these invertebrates, axons join into nerve cords, and in addition to a small, centralized brain, smaller peripheral ganglia help coordinate outlying regions of the animal's body. Ganglia can occur in each body segment or can be scattered throughout the body close to the organs they regulate (figure 24.7 c, d, e). *Regardless of the arrangement, these ganglia represent the fourth evolutionary trend. The more complex an animal, the more interneurons it has. Because interneurons in ganglia do much of the integrating that takes place in nervous systems, the more interneurons, the more complex behavior patterns an animal can perform.*

In echinoderms, such as starfishes, the nervous system is divided into several parts (figure 24.7f). The ectoneural system retains a primitive epidermal position and combines sensory and motor functions. A radial nerve extends down the lower surface of each arm. A deeper hyponeural system has a motor function, and the apical system may have some sensory functions.

The fifth trend in the evolution of invertebrate nervous systems is a consequence of the increasing number of interneurons. The brain contains the largest number of neurons, and the more complex the animal, and the more complicated its behavior, the more neurons (especially interneurons) are concentrated in an anterior brain and bilaterally organized ganglia. Vertebrate brains are an excellent example of this trend.

VERTEBRATE NERVOUS SYSTEMS

The basic organization of the nervous system is similar in all vertebrates. *Bilateral symmetry, a notochord, and a tubular nerve cord characterize the evolution of vertebrate nervous systems.*

The **notochord** is a rod of mesodermally derived tissue encased in a firm sheath that lies ventral to the neural tube. It first appeared in marine chordates and is present in all vertebrate embryos, but is greatly reduced or absent in adults. During embryological development in most vertebrate species, vertebrae serially arranged into a vertebral column replace the notochord. *This vertebral column led to the development of strong muscles, allowing vertebrates to become fast-moving, predatory animals. Some of the other bones developed into*

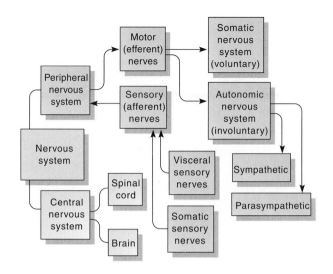

FIGURE 24.8

The Basic Organization of the Nervous System Is Similar in All Vertebrates.   This flowchart shows the divisions and nerves of the vertebrate nervous system. Arrows indicate the directional flow of nerve impulses (information).

powerful jaws, which facilitated the predatory nature of these animals.

A related character in vertebrate evolution was the development of a single, tubular nerve cord above the notochord. During early evolution, the nerve cord underwent expansion, regional modification, and specialization into a spinal cord and brain. Over time, the anterior end thickened variably with nervous tissue and functionally divided into the hindbrain, midbrain, and forebrain. In the sensory world of the fast-moving and powerful vertebrates, the anterior sensory receptors became more complex and bilaterally symmetrical. For example, paired structures, such as eyes and ears, developed to better gather information from the outside environment.

The nervous system of vertebrates has two main divisions (figure 24.8). The **central nervous system** is composed of the brain and spinal cord and is the site of information processing. The **peripheral nervous system** is composed of all the nerves of the body outside the brain and spinal cord. These nerves are commonly divided into two groups: **sensory (afferent) nerves,** which transmit information to the central nervous system; and **motor (efferent) nerves,** which carry commands away from the central nervous system. The motor nerves divide into the **voluntary (somatic) nervous system,** which relays commands to skeletal muscles, and the **involuntary (visceral** or **autonomic) nervous system,** which stimulates other muscles (smooth and cardiac) and glands of the body. The nerves of the autonomic nervous system divide into **sympathetic** and **parasympathetic** systems.

Nervous system pathways are composed of individual neuronal axons bundled like the strands of a telephone cable. In the central nervous system, these bundles of nerve fibers are called **tracts.** In the peripheral nervous system, they are called **nerves.** The cell bodies from which the axons extend often cluster into groups. These groups are called nuclei if they are in the central

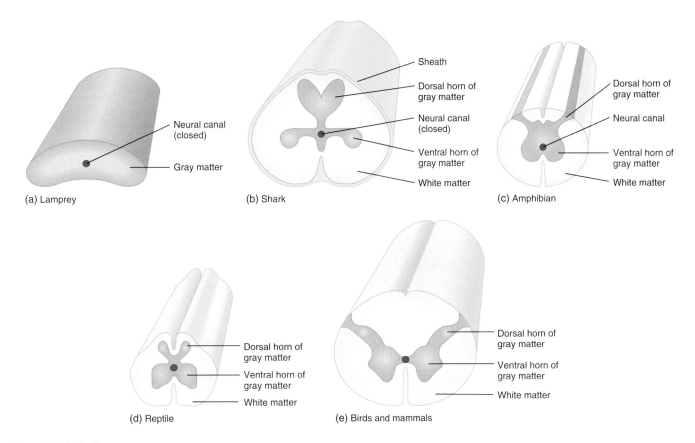

FIGURE 24.9

Spinal Cords of Vertebrates. (*a*) The spinal cord of a typical agnathan (lamprey) is flattened and possesses no myelinated axons. Its shape facilitates the diffusion of gases, nutrients, and other products. (*b, c*) In fishes and amphibians, the spinal cord is larger, well vascularized, and rounded. With more white matter, the spinal cord bulges outward. The gray matter in the spinal cord of (*d*) a reptile and (*e*) birds and mammals has a characteristic butterfly shape.

nervous system and ganglia if they are part of the peripheral nervous system.

THE SPINAL CORD

The spinal cord serves two important functions in an animal; it is the connecting link between the brain and most of the body, and it is involved in spinal reflex actions. A reflex is a predictable, involuntary response to a stimulus. Thus, both voluntary and involuntary limb movements, as well as certain organ functions, depend on this link.

The spinal cord is the part of the central nervous system that extends from the brain to near or into the tail (figure 24.9). A cross section shows a neural canal that contains cerebrospinal fluid. The gray matter consists of cell bodies and dendrites, and is concerned mainly with reflex connections at various levels of the spinal cord. Extending from the spinal cord are the ventral and dorsal roots of the spinal nerves. These roots contain the main motor and sensory fibers (axons and/or dendrites), respectively, that contribute to the major spinal nerves. The white matter of the spinal cord gets its name from the whitish myelin that covers the axons.

Three layers of protective membranes called **meninges** (pl. of meninx, membrane) surround the spinal cord. They are continuous with similar layers that cover the brain. The outer layer, the **dura mater,** is a tough, fibrous membrane. The middle layer, the **arachnoid,** is delicate and connects to the innermost layer, the **pia mater.** The pia mater contains small blood vessels that nourish the spinal cord.

SPINAL NERVES

Generally, the number of spinal nerves is directly related to the number of segments in the trunk and tail of a vertebrate. For example, a frog has evolved strong hind legs for swimming or jumping, a reduced trunk, and no tail in the adult. It has only 10 pairs of spinal nerves. By contrast, a snake, which moves by lateral undulations of its long trunk and tail, has several hundred pairs of spinal nerves.

THE BRAIN

Anatomically, the vertebrate brain develops at the anterior end of the spinal cord. During embryonic development, the brain undergoes regional expansion as a hollow tube of nervous tissue forms and develops into the hindbrain, midbrain, and forebrain (figure 24.10).

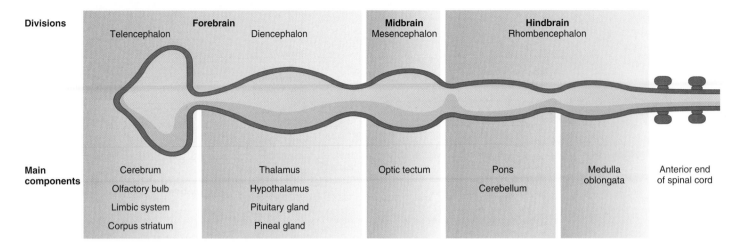

Divisions	Forebrain		Midbrain	Hindbrain
	Telencephalon	Diencephalon	Mesencephalon	Rhombencephalon

Main components	Cerebrum	Thalamus	Optic tectum	Pons	Medulla oblongata	Anterior end of spinal cord
	Olfactory bulb	Hypothalamus		Cerebellum		
	Limbic system	Pituitary gland				
	Corpus striatum	Pineal gland				

FIGURE 24.10

Development of the Vertebrate Brain.   Summary of the three major subdivisions and some of the structures they contain. This drawing is highly simplified and flattened.

The central canal of the spinal cord extends up into the brain and expands into chambers called ventricles. The ventricles are filled with cerebrospinal fluid.

Hindbrain

The **hindbrain** is continuous with the spinal cord and includes the medulla oblongata, cerebellum, and pons. The **medulla oblongata** is the enlargement where the spinal cord enters the brain. It contains reflex centers for breathing, swallowing, cardiovascular function, and gastric secretion. The medulla oblongata is well developed in all jawed vertebrates, reflecting its ability to control visceral functions and to serve as a screen for information that leaves or enters the brain.

The **cerebellum** is an outgrowth of the medulla oblongata. It coordinates motor activity associated with limb movement, maintaining posture, and spatial orientation. The cerebellum in cartilaginous fishes has distinct anterior and posterior lobes. In teleosts, the cerebellum is large in active swimmers and small in relatively inactive fishes. Amphibians often have a rudimentary cerebellum, reflecting their relatively simple locomotor patterns (figure 24.11). In tetrapods, the cerebellum is laterally expanded. These expanded lateral lobes provide locomotor control of muscles of the appendages. *The cerebellum is much larger in birds and mammals—a reflection of complex locomotor patterns and a common evolutionary history of limb development and phylogeny as terrestrial vertebrates.*

The **pons** is a bridge of transverse nerve tracts from the cerebrum of the forebrain to both sides of the cerebellum. It also contains tracts that connect the forebrain and spinal cord in all vertebrates.

Midbrain

The **midbrain** was originally a center for coordinating reflex responses to visual input. As the brain evolved, it took on added functions relating to tactile (touch) and auditory (hearing) input,

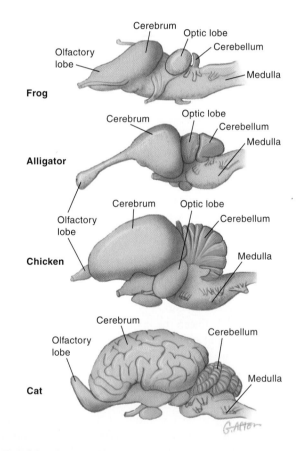

FIGURE 24.11

Vertebrate Brains.   Comparison of several vertebrate brains, as viewed from the side. The drawings are not drawn to the same scale. Notice the increase in relative size of the cerebrum from amphibian (frog) to mammal (cat).

but it did not change in size. The roof of the midbrain, called the optic tectum, is a thickened region of gray matter that integrates visual and auditory signals.

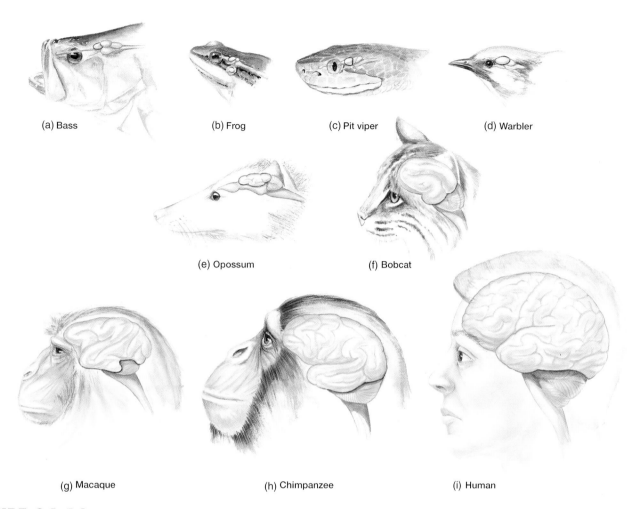

(a) Bass   (b) Frog   (c) Pit viper   (d) Warbler

(e) Opossum   (f) Bobcat

(g) Macaque   (h) Chimpanzee   (i) Human

FIGURE 24.12

Cerebrum in Different Vertebrate Species.   The cerebrum increases in both size and complexity of its neural connections in more advanced groups. (*a*) Fishes and (*b*) amphibians lack cerebral cortices, whereas (*c*) reptiles and (*d*) birds have a small amount of gray matter covering their cerebrums. Most primitive mammals, such as (*e*) the opossum, have smooth cortices. Carnivores, such as (*f*) the bobcat, have larger cerebrums, and the cortex has a few convolutions. (*g*, *h*) In the primates, the cerebrum is much increased relative to other brain structures, and the cortex is highly convoluted. (*i*) The human cerebrum dominates in brain evolution and is highly convoluted.

Forebrain

The vertebrate forebrain has changed a great deal during vertebrate evolution. The **forebrain** has two main parts: the diencephalon and telencephalon (*see figure 24.10*). The diencephalon lies just in front of the midbrain and contains the pineal gland, pituitary gland, hypothalamus, and thalamus. The **thalamus** relays all sensory information to higher brain centers. The **hypothalamus** lies below the thalamus and regulates many functions, such as body temperature, sexual drive, carbohydrate metabolism, hunger, and thirst. The pineal gland controls some body rhythms. The pituitary is a major endocrine gland, and chapter 25 discusses it in detail.

In fishes and amphibians, the diencephalon processes sensory information. In reptiles and birds, the most important part of the brain is the corpus striatum, which plays a role in their complex behavior patterns (*see chapter 21*).

As the diencephalon slowly expanded during evolution and handled more and more sensory functions, the telencephalon (the front part of the forebrain) expanded rapidly in both size and complexity.

External to the corpus striatum is the **cerebrum,** which a large groove divides into right and left cerebral hemispheres. The parts of the brain related to sensory and motor integration changed greatly as vertebrates became more agile and inquisitive (figure 24.12). Many functions shifted from the optic tectum to the expanding cerebral hemispheres. The increasing importance of the cerebrum affected many other brain regions, especially the thalamus and cerebellum. *In mammals, the outermost part of the cerebrum, called the cerebral cortex, progressively increased in size and complexity. This layer folds back on itself to a remarkable extent, suggesting that the evolution of the mammalian cerebrum outpaced the enlargement of the skullbones housing it.*

Different parts of the cerebrum have specific functions. For example, the cerebral cortex contains primary sensory areas and primary motor areas. Other areas of the cortex are involved in the perception of visual or auditory signals from the environment. In humans, this includes the ability to use language—both written and spoken.

TABLE 24.1
FUNCTIONS OF THE CRANIAL NERVES OF REPTILES, BIRDS, AND MAMMALS

NERVE		TYPE	INNERVATION AND FUNCTION
I	Olfactory	Sensory	Smell
II	Optic	Sensory	Vision
III	Oculomotor	Primarily motor	Eyelids, eyes, adjustments of light entering eyes, lens focusing (motor)
IV	Trochlear	Primarily motor	Condition of muscles (sensory)
			Eye muscles (motor)
V	Trigeminal	Mixed	Condition of muscles (sensory)
	Ophthalmic division		Eyes, tear glands, scalp, forehead, and upper eyelids (sensory)
	Maxillary division		Upper teeth, upper gum, upper lip, lining of the palate, and skin of the face (sensory)
	Mandibular division		Scalp, skin of the jaw, lower teeth, lower gum, and lower lip (sensory)
VI	Abducens	Primarily motor	Jaws, floor of the mouth (motor)
			Eye muscles (motor)
VII	Facial	Mixed	Condition of muscles (sensory)
			Taste receptors of the anterior tongue (sensory)
			Facial expression, tear glands, and salivary glands (motor)
VIII	Vestibulocochlear	Sensory	
	Vestibular branch		Equilibrium; vestibule
	Cochlear branch		Hearing; cochlea
IX	Glossopharyngeal	Mixed	Pharynx, tonsils, posterior tongue, and carotid arteries (sensory)
			Pharynx and salivary glands (motor)
X	Vagus	Mixed	Speech and swallowing, heart, and visceral organs in the thorax and abdomen (motor)
			Pharynx, larynx, esophagus, and visceral organs of the thorax and abdomen (sensory)
XI	Accessory	Motor	
	Cranial branch		Soft palate, pharynx, and larynx
	Spinal branch		Neck and back
XII	Hypoglossal	Motor	Tongue muscles

CRANIAL NERVES

In addition to the paired spinal nerves, the peripheral nervous system of vertebrates includes paired cranial nerves (table 24.1). Reptiles, birds, and mammals have 12 pairs of cranial nerves. Fishes and amphibians have only the first 10 pairs. Some of the nerves (e.g., optic nerve) contain only sensory axons, which carry signals to the brain. Others contain sensory and motor axons, and are termed mixed nerves. For example, the vagus nerve has sensory axons leading to the brain as well as motor axons leading to the heart and smooth muscles of the visceral organs in the thorax and abdomen.

THE AUTONOMIC NERVOUS SYSTEM

The vertebrate autonomic nervous system is composed of two divisions that act antagonistically (in opposition to each other) to control the body's involuntary muscles (smooth and cardiac)

and glands. The **parasympathetic nervous system** functions during relaxation. It contains nerves that arise from the brain and sacral region of the spinal cord. It consists of a network of long efferent nerve fibers that synapse at ganglia in the immediate vicinity of organs, and short efferent neurons that extend from the ganglia to the organs. The **sympathetic nervous system** is responsible for the "fight-or-flight" response. It contains nerves that arise from the thoracic and lumbar regions of the spinal cord. It is a network of short efferent central nervous system fibers that extend to ganglia near the spine, and long efferent neurons extending from the ganglia directly to each organ.

Many organs receive input from both the parasympathetic and sympathetic systems. For example, parasympathetic input stimulates salivary gland secretions and intestinal movements, contracts pupillary muscles in the eyes, and relaxes sphincter muscles. Sympathetic input controls antagonistic actions: the inhibition of salivary gland secretions and intestinal movements, the relaxation of pupillary muscles, and the contraction of sphincters.

SENSORY RECEPTION

About two thousand years ago, Aristotle identified five human senses—sight, hearing, smell, taste, and touch—commonly referred to as the "five senses." Today, zoologists know that animals also have other senses. For example, invertebrates possess an impressive array of sensory receptors through which they receive information about their environment. Common examples include tactile receptors that sense touch; georeceptors that sense the pull of gravity; hygroreceptors that detect the water content of air; proprioceptors that respond to mechanically induced changes caused by stretching, compression, bending, and tension; phonoreceptors that are sensitive to sound; baroreceptors that respond to pressure changes; chemoreceptors that respond to air- and waterborne molecules; photoreceptors that sense light; and thermoreceptors that are influenced by temperature changes.

Most vertebrates have a sense of equilibrium (balance) and a sense of body movement, and they are also sensitive to fine touch, touch-pressure, heat, taste, vision, olfaction, audition, cold, pain, and various other tactile stimuli. In addition, receptors in the circulatory system register changes in blood pressure and blood levels of carbon dioxide and hydrogen ions, and receptors in the digestive system are involved in the perception of hunger and thirst.

Overall, an animal's senses limit and define its impression of the environment. In fact, all awareness depends on the reception and decoding of stimuli from the external environment and from within an animal's body. The rest of this chapter examines how animals use sensory information to help maintain homeostasis.

Sensory receptors consist of cells that can convert environmental information (stimuli) into nerve impulses. A **stimulus** (pl., stimuli) is any form of energy an animal can detect with its receptors. All receptors are **transducers** ("to change over"); that is, they convert one form of energy into another. Because all nerve impulses are the same, different types of receptors convert different kinds of stimuli, such as light or heat, into a local electrical potential called the **generator potential.** If the generator potential reaches the sensory neuron's threshold potential, it causes channels in the plasma membrane to open and creates an action potential. The impulse then travels along the cell's axon toward a synaptic junction and becomes information going to the central nervous system or brain.

As presented in the beginning of this chapter, all action potentials are alike. Furthermore, an action potential is an all-or-none phenomenon; it either occurs or it doesn't. How, then, does a common action potential give rise to different sensations, such as taste, color, or sound, or different degrees of sensation? In those animals that have brains, some nerve signals from specific receptors always end up in a specific part of the brain for interpretation; therefore, a stimulus that goes to the optic center is interpreted as a visual stimulus. Another factor that characterizes a particular stimulus is its intensity. When the stimulus strength increases, the number of action potentials per unit of time also increases. Thus, the brain can perceive the intensity and type of stimulus from the timing of the impulses and the "wiring" of neurons.

Sensory receptors have the following basic features:

1. They contain sensitive receptor cells or finely branched peripheral endings of sensory neurons that respond to a stimulus by creating a generator potential.
2. Their structure is designed to receive a specific stimulus.
3. Their receptor cells synapse with afferent nerve fibers that travel to the central nervous system along specific neural pathways.
4. In the central nervous system, the nerve impulse is translated into a recognizable sensation, such as sound.

INVERTEBRATE SENSORY RECEPTORS

An animal's behavior is largely a function of its responses to environmental information. Invertebrates possess an impressive array of receptor structures through which they receive information about their environment. Some common examples are now discussed from a structural and functional perspective.

BARORECEPTORS

Baroreceptors (Gr. *baros*, weight + receptor) sense changes in pressure. However, zoologists have not identified any specific structures for baroreception in invertebrates. Nevertheless, responses to pressure changes have been identified in ocean-dwelling copepod crustaceans, ctenophores, jellyfish medusae, and squids. Some intertidal crustaceans coordinate migratory activity with daily tidal movements, possibly in response to pressure changes accompanying water depth changes.

CHEMORECEPTORS

Chemoreceptors (Gr. *chemeia*, pertaining to chemistry) respond to chemicals. Chemoreception is a direct sense in that molecules act specifically to stimulate a response. Chemoreception is the oldest and most universal sense in the animal kingdom. For example, protozoa have a chemical sense; they respond with avoidance behavior to acid, alkali, and salt stimuli. Specific chemicals attract predatory ciliates to their prey. The chemoreceptors of many aquatic invertebrates are located in pits or depressions, through which water carrying the specific chemicals may be circulated. In arthropods, the chemoreceptors are usually on the antennae, mouthparts, and legs in the form of hollow hairs (**sensilla;** sing., sensillum) containing chemosensory neurons (figure 24.13).

The types of chemicals to which invertebrates respond are closely associated with their lifestyles. Examples include chemoreceptors that provide information that the animal uses to perform tasks, such as humidity detection, pH assessment, prey tracking, food recognition, and mate location. With respect to mate location, the antennae of male silkworm moths (*Bombyx mori*) can detect one bombykol molecule in over a trillion molecules of air.

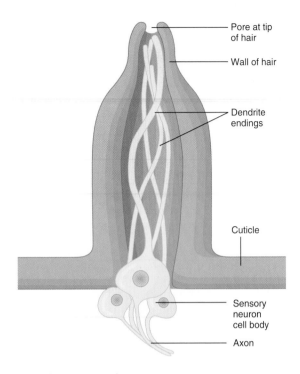

FIGURE 24.13

Invertebrate Chemoreceptor.   Cross section through an insect sensillum. The receptor is a projection of the cuticle with a pore at the tip. Each chemoreceptor generally contains four to five dendrites, which lead to sensory neuron cell bodies underneath the cuticle. Each sensory cell has its own spectrum of chemical responses. Thus, a single sensillum with four or five dendrites and cell bodies may be capable of discriminating between many different chemicals.

Female silk moths secrete bombykol as a sex attractant, which enables a male to find a female at night from several miles downwind, an ability that confers obvious reproductive advantage in a widely dispersed species.

GEORECEPTORS

Georeceptors (Gr. *geo*, earth + receptor) respond to the force of gravity. This gives an animal information about its orientation relative to "up" and "down." Most georeceptors are **statocysts** (Gr. *statos*, standing + *kystis*, bladder) (figure 24.14). Statocysts consist of a fluid-filled chamber lined with cilia-bearing sensory epithelium; within the chamber is a solid granule called a **statolith** (Gr. *lithos*, stone). Any movement of the animal changes the position of the statolith and moves the fluid, thus altering the intensity and pattern of information arising from the sensory epithelium. For example, when an animal moves, both the movement of the statolith and the flow of fluid over the sensory epithelium provide information about the animal's linear and rotational acceleration relative to the environment.

Statocysts are found in various gastropods, cephalopods, crustaceans, nemertines, polychaetes, and scyphozoans. These

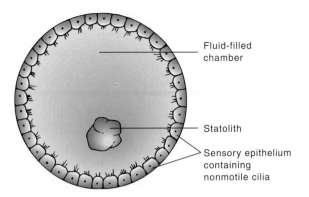

FIGURE 24.14

Invertebrate Georeceptor.   A statocyst (cross section) consists of a fluid-filled chamber containing a solid granule called the statolith. The inner lining of the chamber contains tactile epithelium from which cilia associated with underlying neurons project.

animals use information from statocysts in different ways. For example, burrowing invertebrates cannot rely on photoreceptors for orientation; instead, they rely on georeceptors for orientation within the substratum. Planktonic animals orient in their three-dimensional aquatic environment using statocysts. This is especially important at night and in deep water where there is little light.

In addition to having statocysts, a number of aquatic insects detect gravity from air bubbles trapped in certain passageways (e.g., tracheal tubes). Analogous to the air bubble in a carpenter's level, these air bubbles move according to their orientation to gravity. The air bubbles stimulate sensory bristles that line the tubes.

HYGRORECEPTORS

Hygroreceptors (Gr. *hygros*, moist) detect the water content of air. For example, some insects have hygroreceptors that can detect small changes in the ambient relative humidity. This sense enables them to seek environments with a specific humidity or to modify their physiology or behavior with respect to the ambient humidity (e.g., to control the opening or closing of spiracles). Zoologists have identified a variety of hygrosensory structures on the antennae, palps, underside of the body, and near the spiracles of insects. However, how a hygroreceptor transduces humidity into an action potential is not known.

PHONORECEPTORS

True **phonoreceptors** (Gr. *phone*, voice + receptor) that respond to sound have been demonstrated only in insects, arachnids, and centipedes, although other invertebrates seem to respond to sound-induced vibrations of the substratum. For example, crickets, grasshoppers, and cicadas possess phonoreceptors called **tympanic** or **tympanal organs** (figure 24.15). This organ consists of a

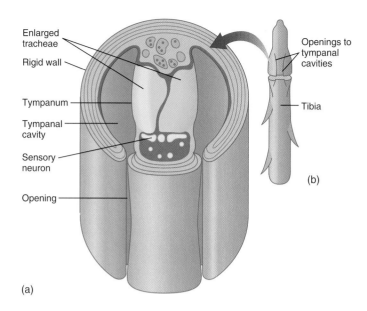

(a)

FIGURE 24.15

Invertebrate Phonoreceptor (Tympanal Organ). (a) This organ functions on the drumhead principle. The flattened outer wall (tympanum) of each trachea functions as a "drumhead." As the tympanum vibrates in response to sound waves, pressure changes within the tracheae affect the sensory neuron, causing a generator potential. (b) The slit openings on the leg (tibia) of a cricket lead to the tympanal cavities.

tough, flexible tympanum that covers an internal sac that allows the tympanum to vibrate when sound waves strike it. Sensory neurons attached to the tympanum are stimulated and produce a generator potential.

Most arachnids possess phonoreceptors in their cuticle called slit sense organs that can sense sound-induced vibrations. Centipedes have organs of Tomosvary, which some zoologists believe may be sensitive to sound. However, the physiology of both slit sense organs and organs of Tomosvary is poorly understood.

PHOTORECEPTORS

Photoreceptors (Gr. *photos*, light + receptor) are sensitive to light. All photoreceptors possess light-sensitive pigments (e.g., carotenoids, rhodopsin). These pigments absorb photons of light energy and then produce a generator potential. Beyond this basic commonality, the complexity and arrangement of photoreceptors within various animals vary incredibly.

Certain flagellated protozoa (*Euglena*) that contain chlorophyll possess a mass of bright red photoreceptor granules called the **stigma** (pl., stigmata) (figure 24.16a). The granules are carotenoid pigments. The actual photoreceptor is the swelling at the base of the flagellum. The stigma probably serves as a shield, which is essential if the photoreceptor is to detect light coming from certain directions but not from others. Thus, the photoreceptor plus the stigma enable *Euglena* to orient itself so that its

photoreceptor is exposed to light. This helps the protozoan maintain itself in the region of the water column where sufficient light is available for photosynthesis.

Some animals, such as the earthworm *Lumbricus*, have simple unicellular photoreceptor cells scattered over the epidermis or concentrated in particular areas of the body. Others possess multicellular photoreceptors that can be classified into three basic types: ocelli, compound eyes, and complex eyes.

An **ocellus** (L. dim. of *oculus*, eye) (pl., ocelli) is simply a small cup lined with light-sensitive receptors and backed by light-absorbing pigment (figure 24.16b). The light-sensitive cells are called retinular cells and contain a photosensitive pigment. Stimulation by light causes a chemical change in the pigment, leading to a generator potential, which causes an action potential that sensory neurons carry for interpretation elsewhere in the animal's body. This type of visual system gives an animal information about light direction and intensity, but not image formation. Ocelli are common in many phyla (e.g., Annelida, Mollusca, and Arthropoda).

Compound eyes consist of a few to many distinct units called **ommatidia** (Gr. *ommato*, eye + *ium*, little) (sing., ommatidium) (figure 24.16c). Although compound eyes occur in some annelids and bivalve molluscs, they are best developed and understood in arthropods. A compound eye may contain thousands of ommatidia, each oriented in a slightly different direction from the others as a result of the eye's overall convex shape. The visual field of a compound eye is very wide, as anyone who has tried to catch a fly knows. Each ommatidium has its own nerve tract leading to a large optic nerve. The visual fields of adjacent ommatidia overlap to some degree. Thus, if an object within the total visual field shifts position, the level of stimulation of several ommatidia changes. As a result of this physiology, as well as a sufficiently sophisticated central nervous system, compound eyes are very effective in detecting movements and are probably capable of forming an image. In addition, most compound eyes can adapt to changes in light intensities, and some provide for color vision. Color vision is particularly important in active, day-flying, nectar-drinking insects, such as honeybees. Honeybees learn to recognize particular flowers by color, scent, and shape.

The **complex camera eyes** of squids and octopuses are the best image-forming eyes among the invertebrates. In fact, the giant squid's eye is the largest of any animal's, exceeding 38 cm in diameter. Cephalopod eyes are often compared to those of vertebrates because they contain a thin, transparent cornea and a lens that focuses light on the retina and is suspended by, and controlled by, ciliary muscles (figure 24.16d). However, the complex eyes of squids are different from the vertebrate eye in that the receptor sites on the retinal layer face in the direction of light entering the eye. In the vertebrate eye, the retinal layer is inverted, and the receptors are the deepest cells in the retina. Both eyes are focusing and image-forming, although the process differs in detail. In terrestrial vertebrates, muscles that alter the shape (thickness) of the lens focus light. In fishes and cephalopods, light is focused by muscles that move the lens toward or away from the retina (like

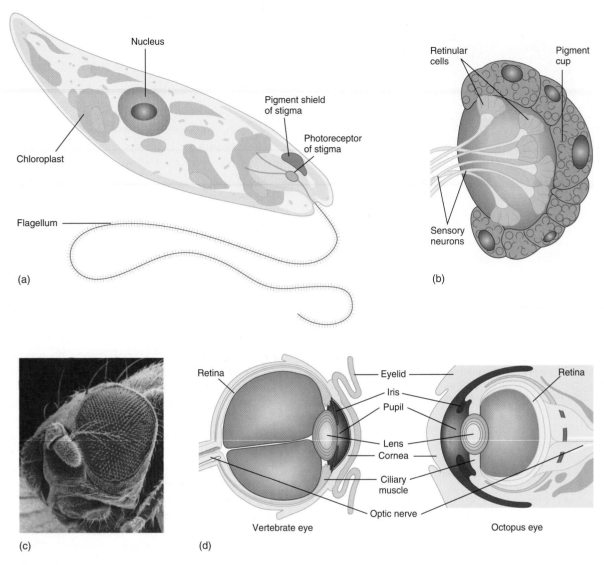

FIGURE 24.16

Invertebrate Photoreceptors. (*a*) Stigma. The protozoan, *Euglena,* contains a mass of bright red granules called the stigma. The actual photoreceptor is the swelling at the base of the flagellum. (*b*) Ocellus. The inverted pigment cup ocellus of a flatworm. (*c*) Compound eye. The compound eye of a fly contains hundreds of ommatidia. Note the eye's convex shape; no two ommatidia are oriented in precisely the same direction (SEM). (*d*) Complex camera eyes. Comparison of a vertebrate eye and an octopus eye (vertical sections).

moving a magnifying glass back and forth to achieve proper focus), and by altering the shape of the eyeball.

PROPRIOCEPTORS

Proprioceptors (L. *proprius,* one's self + receptor), commonly called "stretch receptors," are internal sense organs that respond to mechanically induced changes caused by stretching, compression, bending, or tension. These receptors give an animal information about the movement of its body parts and their positions relative to each other. Proprioceptors have been most thoroughly studied in arthropods, where they are associated with appendage joints and body extensor muscles (figure 24.17). In these animals, the sensory neurons involved in proprioception are

associated with and attached to some part of the body that is stretched. These parts may be specialized muscle cells, elastic connective-tissue fibers, or various membranes that span joints. As these structures change shape, sensory nerve endings of the attached nerves distort accordingly and initiate a generator potential.

TACTILE RECEPTORS

Tactile (touch) receptors are generally derived from modifications of epithelial cells associated with sensory neurons. Most tactile receptors of animals involve projections from the body surface. Examples include various bristles, spines, setae, and tubercles. When an animal contacts an object in the environment, these

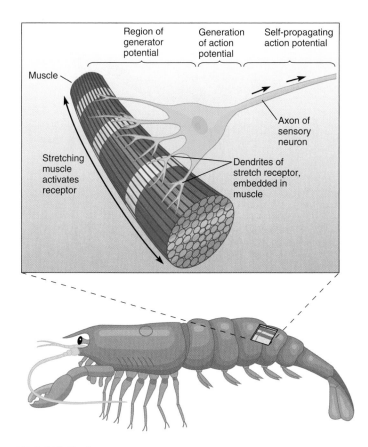

FIGURE 24.17

Invertebrate Proprioceptor.   Crayfish stretch receptors are neurons attached to muscles. In this example, when the crayfish arches its abdomen while swimming, the stretch receptor detects the change in muscle length. When the muscle is stretched, so is the receptor. The stretch increases the sodium permeability of the receptor cell plasma membrane by mechanically opening sodium channels. The inflow of sodium ions produces depolarization and a generator potential that evokes an action potential. The axon of the sensory neuron then transmits the action potential to the central nervous system, where it is interpreted.

receptors are mechanically deformed. These deformations activate the receptor, which, in turn, activates underlying sensory neurons, initiating a generator potential.

Most tactile receptors are also sensitive to mechanically induced vibrations propagated through water or a solid substrate. For example, tube-dwelling polychaetes bear receptors that allow them to retract quickly into their tubes in response to movements in their surroundings. Web-building spiders have tactile receptors that can sense struggling prey in webs through vibrations of the web threads.

THERMORECEPTORS

Thermoreceptors (Gr. *therme*, heat + receptors) respond to temperature changes. Some invertebrates can directly sense differences in environmental temperatures. For example, the protozoan *Paramecium* collects in areas where water temperature is moderate, and it avoids temperature extremes. Somehow, a heat-sensing mechanism draws leeches and ticks to warm-blooded hosts.

Certain insects, some crustaceans, and the horseshoe crab (*Limulus*) can also sense thermal variations. In all of these cases, however, specific receptor structures have not been identified.

VERTEBRATE SENSORY RECEPTORS

V*ertebrate sensory receptors reflect adaptations to the nature of sensory stimuli in different external and internal environments. Each environment has chemical and physical characteristics that affect the kinds of energy and molecules that carry sensory information.* For example, your external environment consists of the media that surrounds you: the earth that you stand on and the air that you breathe. Other animals may have different external environments: a trout may be immersed in the cool, clear water of a mountain stream; a turtle may be submerged in the turbid water of a swamp; and a salmon may be swimming in the salty water of the sea.

Each of the previous media contains only certain environmental stimuli. For example, air transmits light very well and conducts sound waves rather efficiently. But air can carry only a limited assortment of small molecules detectable using the sense of smell and can pass little or no electrical energy. In water, however, sound travels both faster and farther than in air, and water dissolves and carries a wide range of chemicals. Water, especially seawater, is also an excellent conductor of electricity, but it absorbs (and hence fails to transmit) many wavelengths of light. As these examples indicate, *vertebrate sensory receptors (organs), like invertebrate sensory receptors, have evolved in ways that relate to the environment in which they must function.*

Many underlying similarities unite all vertebrate senses. For each sense, there is a fascinating story of environmental information, the evolutionary adaptation of receptor cells to detect that information, and the processing in the central nervous system of the information so that the animal can use it. What follows is a discussion of particular vertebrate receptors (e.g., lateral-line systems, ears, eyes, skin sensors) that detect changes in the external environment, and of several receptors (e.g., pain, proprioception) that detect changes in the internal environment of some familiar vertebrate animals.

LATERAL-LINE SYSTEM AND ELECTRICAL SENSING

S*pecialized organs for equilibrium and gravity detection, audition, and magnetoreception have evolved from the lateral-line system of fishes.* The **lateral-line system** for electrical sensing is in the head and body areas of most fishes, some amphibians, and the platypus (figure 24.18*a*). It consists of sensory pores in the epidermis of the skin that connect to canals leading into **electroreceptors** called **ampullary organs** (figure 24.18*b*). These organs can sense electrical currents in the surrounding water. Most living organisms generate weak electrical fields. The ability to detect these fields helps a fish to find mates, capture prey, or avoid predators. This is

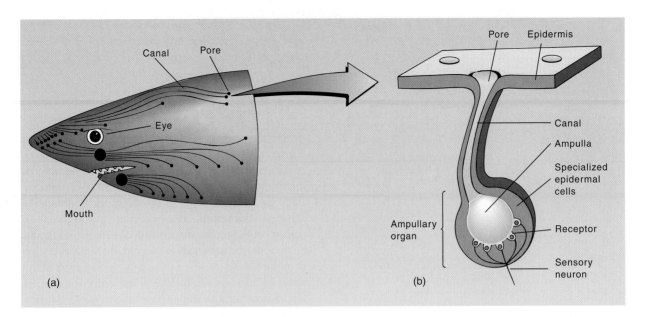

FIGURE 24.18

Lateral-Line System and Electrical Sensing.   (*a*) In jawless fishes, jawed fishes, and amphibians, electroreceptors are in the epidermis along the sides of the head and body. (*b*) Pores of the lateral-line system lead into canals that connect to an ampullary organ that functions in electroreception and the production of a generator potential.

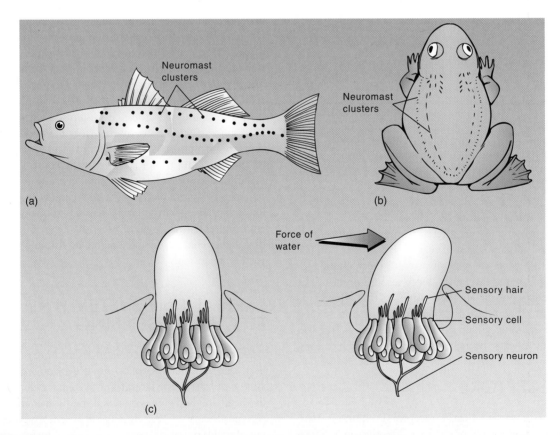

FIGURE 24.19

Lateral-Line System and Mechanoreception.   The lateral-line system of (*a*) a bony fish and (*b*) a frog, showing the various neuromast clusters. (*c*) Action of neuromast stimulation. The water movement (blue arrow) forces the cap-like structure covering a group of neuromast cells to bend or distort, thereby distorting the small sensory hairs of the neuromast cells, producing a generator potential. The generator potential causes an action potential in the sensory neuron.

an especially valuable sense in deep, turbulent, or murky water, where vision is of little use. In fact, some fishes actually generate electrical fields and then use their electroreceptors (electrocommunication) to detect how surrounding objects distort the field. This allows these fishes to navigate in murky or turbulent waters.

LATERAL-LINE SYSTEM AND MECHANORECEPTION

A **mechanoreceptor** is excited by mechanical pressures or distortions (e.g., sound, touch, and muscular contractions). The lateral-line system of cyclostomes, sharks, some of the more advanced fishes, and aquatic amphibians includes several different kinds of hair-cell mechanoreceptors called **neuromasts.** Neuromasts are in pits along the body, but not in the head region (figure 24.19*a, b*). All neuromasts are responsive to local water displacement or disturbance. When the water near the lateral line moves, it moves the water in the pits and distorts the hair cells, causing a generator potential in the associated sensory neurons (figure 24.19*c*). Thus, the animal can detect the direction and force of water currents and the movement of other animals or prey in the water. For example, this sense enables a trout to orient with its head upstream.

HEARING AND EQUILIBRIUM IN AIR

Hearing may initially have been important to vertebrates as a mechanism to alert them to either nearby or faraway potentially dangerous activity. It also became important in the search for food and mates, and in communication. Hearing (audition) and equilibrium (balance) are considered together because both sensations are received in the same vertebrate organ—the ear. The vertebrate ear has two functional units: (1) the auditory apparatus is concerned with hearing, and (2) the vestibular apparatus is concerned with posture and equilibrium.

Sound results when pressure waves transmit energy through some medium, such as air or water. Hearing in air poses serious problems for vertebrates, since middle-ear transformers are sound pressure sensors, but in air, sound produces less than 0.1% of the pressure it produces in water. *Adaptation to hearing in air resulted from the evolution of an acoustic transformer that incorporates a thin, stretched membrane, called either an eardrum, tympanic membrane, or tympanum, that is exposed to the air.* *The tympanum first evolved in the amphibians.* The ears of anurans (frogs) consist of a tympanum, a middle ear, and an inner ear (figure 24.20). The tympanum is modified integument stretched over a cartilaginous ring. It vibrates in response to sounds and transmits these movements to the middle ear, a chamber behind the tympanum. Touching the tympanum is an ossicle (a small bone or bony structure) called the columella or stapes. The opposite end of the columella (stapes) touches the membrane of the oval window, which stretches between the middle and inner ears. High-frequency (1,000 to 5,000 Hz) sounds strike the tympanum and are transmitted through the middle ear via the columella and cause pressure waves in the fluid of the semicircular canals. These pressure waves in the inner ear fluid stimulate receptor

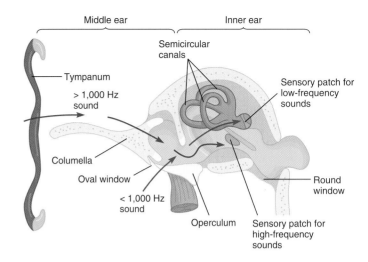

FIGURE 24.20

Ear of an Anuran (Posterior View). Red arrows show the pathway of low-frequency sounds, via the operculum. Dark-blue arrows show the pathway of high-frequency sounds, via the columella (stapes).

cells. A second small ossicle, the operculum, also touches the oval window. Substrate-borne vibrations transmitted through the front appendages and the pectoral girdle cause this ossicle to vibrate. The resulting pressure waves in the inner ear stimulate a second patch of sensory receptor cells that is sensitive to low-frequency (100 to 1,000 Hz) sounds. Muscles attached to the operculum and columella can lock either or both of these ossicles, allowing a frog to screen out either high- or low-frequency sounds. *This mechanism is adaptive because frogs use low- and high-frequency sounds in different situations. For example, mating calls are high-frequency sounds that are of primary importance for only part of the year (breeding season). At other times, low-frequency sounds may warn of approaching predators.*

Salamanders lack a tympanum and middle ear. They live in streams, ponds, and caves, and beneath leaf litter. They have no mating calls, and the only sounds they hear are probably transmitted through the substratum and skull to the inner ear.

The sense of equilibrium and balance in amphibians involves the semicircular canals. These canals help detect rotational movements and gravity. Since the semicircular canals have a similar function in all vertebrates, they are discussed later in this section in information about the human ear.

The structures of reptilian ears vary. For example, the ears of snakes lack a middle-ear cavity and a tympanum. A bone of the jaw articulates with the stapes and receives vibrations of the substratum. In other reptiles, a tympanum may be on the surface or in a small depression in the head. The inner ear of reptiles is similar to that of amphibians.

Hearing is well developed in most birds. Loose, delicate feathers cover the external ear opening. Middle- and inner-ear structures are similar to those of mammals.

Auditory senses were also important to the early mammals. Adaptations include an ear flap (the auricle) and the auditory tube (external auditory canal) leading to the tympanum that

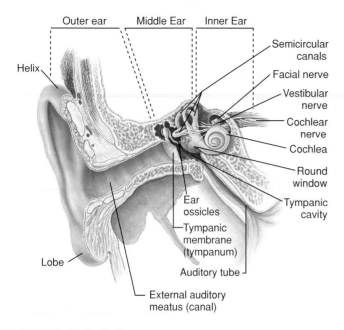

FIGURE 24.21

Anatomy of the Human Ear.   Note the outer, middle, and inner regions. The inner ear includes the semicircular canals, which are involved with equilibrium, and the cochlea, which is involved with hearing.

directs sounds to the middle ear. In mammals, the long, coiled, sensory structure of the inner ear that contains receptors for sound is the cochlea. *This structure provides more surface area for receptor cells and gives mammals greater sensitivity to pitch and volume than is present in other animals.* Since the structure and function of all mammal ears are basically the same, the familiar human ear is a good example.

The human ear has three divisions: the outer, middle, and inner ear. The outer ear consists of the auricle and external auditory canal (figure 24.21). The middle ear begins at the tympanic membrane (tympanum or eardrum) and ends inside the skull, where two small membranous openings, the oval and round windows, are located. Three small ossicles are between the tympanic membrane and the oval window. They include the malleus (hammer), incus (anvil), and stapes (stirrup), so named for their shapes. The malleus adheres to the tympanic membrane and connects to the incus. The incus connects to the stapes, which adheres to the oval window. The auditory (eustachian) tube extends from the middle ear to the nasopharynx and equalizes air pressure between the middle ear and the throat.

The inner ear has three components. The first two, the vestibule and the semicircular canals, are concerned with equilibrium, and the third, the cochlea, is involved with hearing. The semicircular canals are arranged so that one is in each dimension of space. The process of hearing can be summarized as follows:

1. Sound waves enter the outer ear and create pressure waves that reach the tympanic membrane.
2. Air molecules under pressure vibrate the tympanic membrane. The vibrations move the malleus on the other side of the membrane.

3. The handle of the malleus articulates with the incus, vibrating it.
4. The vibrating incus moves the stapes back and forth against the oval window.
5. The movements of the oval window set up pressure changes that vibrate the fluid in the inner ear. These vibrations are transmitted to the basilar membrane, causing it to ripple.
6. Receptor hair cells of the organ of Corti that are in contact with the overlying tectorial membrane are bent, causing a generator potential, which leads to an action potential that travels along the vestibulocochlear nerve to the brain for interpretation.
7. Vibrations in the cochlear fluid dissipate as a result of movements of the round window.

Humans are not able to hear low-pitched sounds, below 20 cycles per second, although some other vertebrates can. Young children can hear high-pitched sounds up to 20,000 cycles per second, but this ability decreases with age. Other vertebrates can hear sounds at much higher frequencies. For example, dogs can easily detect sounds of 40,000 cycles per second. Thus, dogs can hear sounds from a high-pitched dog whistle that seems silent to humans.

The sense of equilibrium (balance) can be divided into two separate senses. Static equilibrium refers to sensing movement in one plane (either vertical or horizontal), and dynamic equilibrium refers to sensing angular and/or rotational movement.

When the body is still, the otoliths in the semicircular canals rest on hair cells (figure 24.22a). When the head or body moves horizontally, or vertically, the granules are displaced, causing the gelatinous material to sag (figure 24.22b). This displacement bends the hairs slightly so that hair cells initiate a generator potential and then an action potential. Continuous movement of the fluid in the semicircular canals may cause motion sickness or seasickness in humans.

HEARING AND EQUILIBRIUM IN WATER

In bony fishes, receptors for equilibrium, balance, and hearing are in the inner ear, and their functions are similar to those of other vertebrates (figure 24.23). For example, semicircular canals detect rotational movements, and other sensory patches help with equilibrium and balance by detecting the direction of gravitational pull. Since fishes lack the outer and/or middle ear found in other vertebrates, vibrations pass from the water through the bones of the skull to the inner ear. A few fishes have chains of bony ossicles (modifications of vertebrae) that pass between the swim bladder and the back of the skull. Vibrations that strike the fishes are thus amplified by the swim bladder and sent through the ossicles to the skull.

SKIN SENSORS OF DAMAGING STIMULI

Pain receptors are bare sensory nerve endings that are present throughout the body of mammals, except for the brain and intestines. These nerve endings are also called **nociceptors** (L. *no-cere*,

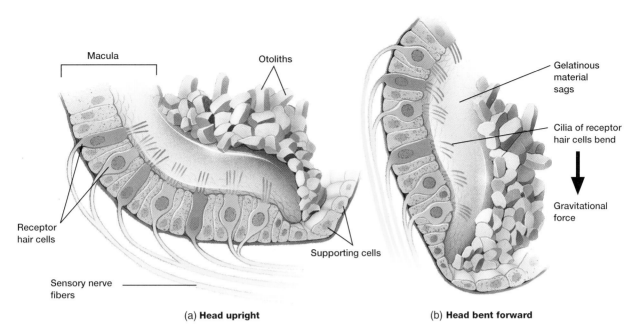

(a) **Head upright**   (b) **Head bent forward**

FIGURE 24.22

Static Equilibrium (Balance).   Receptor hair cells in the utricle and saccule respond to sideways or up or down movement. (*a*) When the head is upright, otoliths are balanced directly over the cilia of receptor hair cells. (*b*) When the head bends forward, the otoliths shift, and the cilia of hair cells bend. This bending of hairs initiates a generator potential.

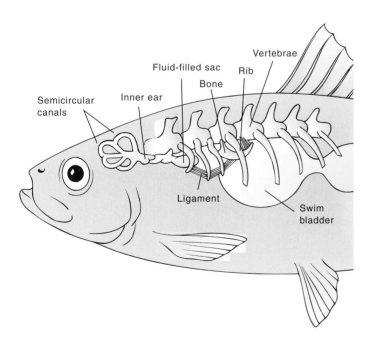

FIGURE 24.23

Inner Ear of a Bony Fish.   Sound waves that enter the pharynx are transmitted to gas in the swim bladder, causing it to expand and contract at frequencies and amplitudes corresponding to the incoming sound waves. Contacting the swim bladder is a bone that is suspended by ligaments and vibrates at the same frequency. The vibrations pass forward along a chain of bones (ossicles) and then to a fluid-filled sac connected directly to the inner ear.

to injure + receptor). Severe heat, cold, irritating chemicals, and strong mechanical stimuli (e.g., penetration) may elicit a response from nociceptors that the brain interprets as pain or itching. Details of the structure and physiology of pain receptors, however, are unknown.

SKIN SENSORS OF HEAT AND COLD

Sensors of temperature (thermoreceptors) are also bare sensory nerve endings. Thermoreceptors may be present in either the epidermis or dermis. Mammals have a distinctly different distribution of areas sensitive to either cold or warm. These areas are called cold or warm spots. A spot refers to a small area of the skin that, when stimulated, yields a temperature sensation of warmth or cold. Cold receptors in the skin respond to temperatures below skin temperature, and heat receptors respond to temperatures above skin temperature. Materials coming into contact with the skin need not be warm or cold to produce temperature sensations. For example, when metal is placed on the skin, it absorbs heat and you feel a sense of coldness. Wood placed on the skin absorbs less heat and, therefore, feels warmer than metal.

   The ability to detect changes in temperature has become well developed in a number of animals. For example, rattlesnakes and other pit vipers have heat-sensitive **pit organs** on each side of the face between the eye and nostril (figure 24.24). These depressions are lined with sensory epithelium containing receptor cells that respond to temperatures (infrared thermal regulation) different from the snakes' surroundings. Snakes use these pit organs to locate warm-blooded prey.

Pit organ

FIGURE 24.24

Thermoreception.   A rattlesnake (*Crotalus vergrandis*) has a pit organ between each eye and nostril that detects heat and allows the snake to locate warm prey in the dark.

SKIN SENSORS OF MECHANICAL STIMULI

Many animals rely on tactile (pertaining to touch) stimuli to obtain information about their environment. Mechanical sensory receptors in vertebrate skin detect stimuli that the brain interprets as light touch, touch-pressure, and vibration.

Light touch is perceived when the skin is touched, but not strongly deformed. Receptors of light touch include **bare sensory nerve endings** and **tactile (Meissner's) corpuscles** (figure 24.25). Bare sensory nerve endings are the most widely distributed receptors in the vertebrate body, and are involved with pain and thermal stimuli, as well as light touch. The **bulbs of Krause** are mechanoreceptors, found in the dermis in certain parts of the body, that respond to some physical stimuli, such as position changes. Other receptors for touch-pressure are **Pacinian corpuscles** and the **organs of Ruffini.**

Many mammals have specially adapted sensory hairs called **vibrissae** (sing., vibrissa) on their wrists, snouts, and eyebrows (e.g., cat whiskers). Around the base of each vibrissa is a blood sinus. Nerves that border the sinus carry impulses from several kinds of mechanoreceptors to the brain for interpretation.

SONAR

Bats, shrews, several cave-dwelling birds (oilbird, cave swiftlet), whales, and dolphins can determine distance and depth by a form of echolocation called **sonar (biosonar).** These animals emit high-frequency sounds and then determine how long it takes for the sounds to return after bouncing off objects in the environment. For example, some bats emit clicks that last from 2 to 3 milliseconds and are repeated several hundred times per second. The returning echo created when a moth or other insect flies past the bat can provide enough information for the bat to locate and catch its prey. Overall, the three-dimensional imaging achieved with this auditory sonar system is quite sophisticated.

SMELL

The sense of smell, or **olfaction** (L. *olere*, to smell + *facere*, to make), is due to olfactory neurons (receptor cells) in the roof of the vertebrate nasal cavity (figure 24.26). These cells, which are specialized endings of the fibers that make up the olfactory nerve, lie among supporting epithelial cells. They are densely packed; for example, a dog has up to 40 million olfactory receptor cells per square centimeter. Each olfactory cell ends in a tuft of cilia containing receptor sites for various chemicals.

Several theories have been proposed to explain how odors are perceived. The most likely one is that odor molecules physically interact with protein receptors on the receptor-plasma membrane. Such an interaction somehow alters membrane permeability and leads to a generator potential.

In most fishes, openings (external nares) in the snout lead to the olfactory receptors. Recent research has revealed that some fishes rely heavily on their sense of smell. For example, salmon and lampreys return to spawn in the same streams in which they hatched years earlier. Their migrations to these streams often involve distances of hundreds of miles and are guided by the fishes' perception of characteristic odors of their spawning stream.

Olfaction is an important sense for many amphibians. It is used in mate recognition, as well as in detecting noxious chemicals and locating food.

Olfactory senses are better developed in reptiles than in amphibians. In addition to having more olfactory epithelium, most reptiles (except crocodilians) possess blind-ending pouches that open into the mouth. These pouches, called **Jacobson's (vomeronasal) organs,** are best developed in snakes and lizards (figure 24.27). The protrusible, forked tongues of snakes and lizards are accessory olfactory organs used to sample airborne chemicals. A snake's tongue flicks out and then moves to the Jacobson's organs, which perceive odor molecules. Turtles and the tuatara use Jacobson's organs to taste objects held in the mouth.

Olfaction apparently plays a minor role in the lives of most birds. External nares open near the base of the beak, but the olfactory epithelium is poorly developed. Vultures are exceptions, in that they locate dead and dying prey largely by smell.

Many mammals can perceive olfactory stimuli over long distances during either the day or night. They use the stimuli to locate food, recognize members of the same species, and avoid predators.

TASTE

The receptors for taste, or **gustation** (L. *gustus*, taste), are chemoreceptors. They may be on the body surface of an animal or in the mouth and throat. For example, the surface of the

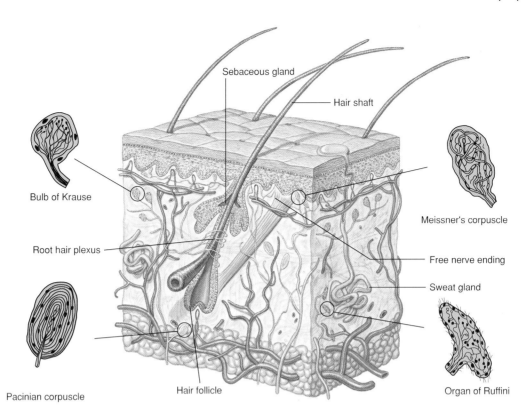

FIGURE 24.25

Different Sensory Receptors to Mechanical Stimuli.   Sensory receptors in the skin for light touch (Meissner's corpuscles), touch-pressure (organs of Ruffini and Pacinian corpuscles), position (bulbs of Krause), and pain (free nerve endings).

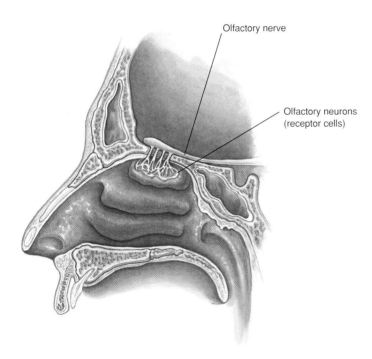

FIGURE 24.26

Smell.   Position of olfactory receptors in a human nasal passageway. Columnar epithelial cells support the receptor cells, which have hair-like processes (analogous to dendrites) projecting into the nasal cavity. When chemicals in the air stimulate these receptor cells, the olfactory nerves conduct nerve impulses to the brain.

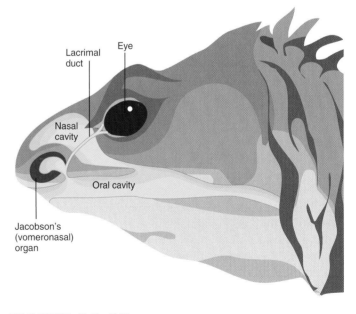

FIGURE 24.27

Smell.   Anatomic relationships of Jacobson's (vomeronasal) organ in a generalized lizard. Only the left organ alongside the nasal cavity is shown. Jacobson's organ is a spherical structure, with the ventral side invaginated into a sphere the shape of a mushroom. A narrow duct connects the interior of Jacobson's organ to the oral cavity. In many lizards, fluid draining from the eye via the lacrimal duct may bring odoriferous molecules into contact with the sensory epithelium of Jacobson's organ.

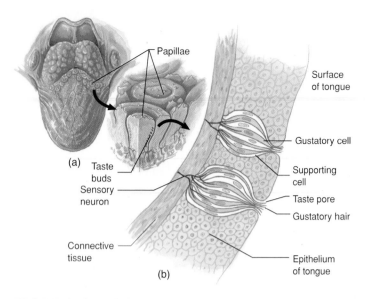

Papillae

Surface
of tongue

Gustatory cell

Supporting
cell

Taste pore

Gustatory hair

(a)

Taste
buds
Sensory
neuron

Connective
tissue

Epithelium
of tongue

(b)

FIGURE 24.28

Taste. (*a*) Surface view of the human tongue, showing the many papillae and the numerous taste buds between papillae. (*b*) Supporting cells encapsulate the gustatory cell and its associated gustatory hair.

mammalian tongue is covered with many small protuberances called papillae (sing., papilla). Papillae give the tongue its "bumpy" appearance (figure 24.28*a*). In the crevices between the papillae are thousands of specialized receptors called **taste buds** (figure 24.28*b*). Taste buds are barrel-shaped clusters of chemoreceptor cells called gustatory cells and supporting cells arranged like alternating segments of an orange. Extending from each receptor cell are gustatory hairs that project through a tiny opening called the taste pore. Sensory neurons are associated with the basal ends of the gustatory cells.

The four generally recognized taste sensations are sweet (sugars), sour (acids), bitter (alkaloids), and salty (electrolytes). The exact mechanism(s) that stimulate a chemoreceptor taste cell are not known. One theory is that different types of gustatory stimuli cause proteins on the surface of the receptor-cell plasma membrane to change the permeability of the membrane—in effect, "opening and closing gates" to chemical stimuli and causing a generator potential.

Vertebrates other than mammals may have taste buds on other parts of the body. For example, reptiles and birds do not usually have taste buds on the tongue; instead, most taste buds are in the pharynx. In fishes and amphibians, taste buds may also be found in the skin. For example, a sturgeon's taste buds are abundant on its head projection, which is called the rostrum. As the sturgeon glides over the bottom, it can obtain a foretaste of potential food before the mouth reaches the food. In other fishes, taste buds are widely distributed in the roof, side walls, and floor of the pharynx, where they monitor the incoming flow of water. In fishes that feed on the bottom (catfish, carp, suckers), taste buds are distributed over the entire surface of the head and body to the tip of the tail. They are also abundant on the barbels ("whiskers") of catfish.

VISION

Vision (photoreception) is the primary sense that vertebrates in a light-filled environment use, and consequently, their photoreceptive structures are well developed. Most vertebrates have eyes capable of forming visual images. As figure 24.29 indicates, the eyeball has a lens, a sclera (the tough outer coat), a choroid layer (a thin middle layer), and an inner retina containing many light-sensitive receptor cells (photoreceptors). The transparent cornea is continuous with the sclera and covers the front of the eyeball. Choroid tissue also extends to the front of the eyeball to form the iris, ciliary body, and suspensory ligaments. The colored iris is heavily endowed with light-screening pigments, and it has radial and circular smooth muscles for regulating the amount of light entering the pupil. A clear fluid (aqueous humor) fills the anterior and posterior chambers, which lie between the lens and the cornea. The lens is behind the iris, and a jellylike vitreous body fills the vitreous chamber behind the lens. The moist mucous membrane that covers the eyeball is the conjunctiva.

Vertebrates can adjust their vision for light coming from either close-up or distant objects. This process of focusing light rays precisely on the retina is called **accommodation.** Vertebrates rely on the coordinated stretching and relaxation of the eye muscles and fibers (the ciliary body and suspensory ligaments) that attach to the lens for accommodation.

The eyes of fishes are similar in most aspects of structure and function to those in other vertebrates. However, fish eyes are lidless, and the lens is rounded and close to the cornea. Focusing requires moving the lens forward or backward.

Vision is one of the most important senses in amphibians because they are primarily sight feeders. ▲ *number of adaptations allow the eyes of amphibians to function in terrestrial environments.* For example, the eyes of some amphibians (e.g., anurans, salamanders) are close together on the front of the head and provide the binocular vision and well-developed depth perception necessary for capturing prey. Other amphibians with smaller and more lateral eyes (e.g., some salamanders) lack binocular vision. However, their more laterally placed eyes permit these animals to see well off to their sides. The transparent **nictitating membrane** (an "inner eyelid") is movable and cleans and protects the eye.

Vision is the dominant sense in most reptiles, and their eyes are similar to those of amphibians. Upper and lower eyelids, a nictitating membrane, and a blood sinus protect and cleanse the surface of the eye. In snakes and some lizards, the upper and lower eyelids fuse in the embryo to form a protective window of clear skin called the spectacle. Some reptiles possess a **median (parietal) eye** that develops from outgrowths of the roof of the optic tectum (midbrain) (figure 24.30). In the tuatara, the median eye is complete with a lens, nerve, and retina. In other reptiles, the median eye is less developed. Skin covers median eyes, which probably cannot form images. They can, however, differentiate light and dark periods and are used in orientation to the sun.

Vision is an important sense for most birds. The structure of the bird eye is similar to that of other vertebrates (*see figure 24.29*). Birds have a unique, double-focusing mechanism. Padlike structures

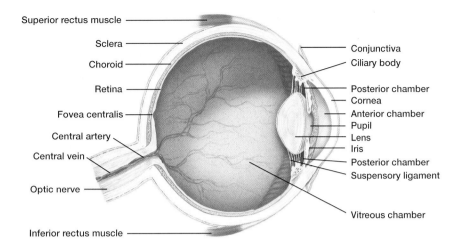

FIGURE 24.29

Internal Anatomy of the Human Eyeball.   Light passes through the transparent cornea. The lens focuses the light on the rear surface of the eye, the retina, at the fovea centralis. The retina is rich in rods and cones.

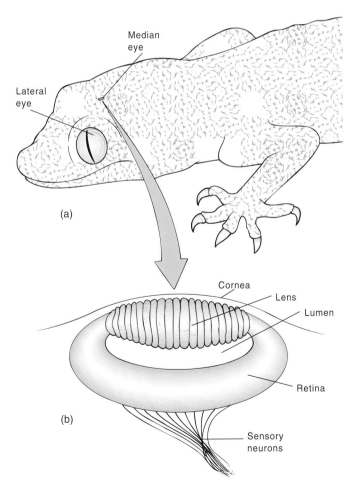

FIGURE 24.30

Median Eye of Reptiles.   (*a*) Median eye in a reptile and its relationship to the lateral eyes (dorsal view). (*b*) Sagittal section of the median eye.

control the curvature of the lens, and ciliary muscles change the curvature of the cornea. Double, nearly instantaneous focusing allows an osprey, or other bird of prey, to focus on a fish throughout a brief, but breathtakingly fast, descent. Like reptiles, birds have a nictitating membrane that is drawn over the eyeball surface to cleanse and protect it.

In all vertebrates, the retina is well developed. Its basement layer is composed of pigmented epithelium that covers the choroid layer. Nervous tissue that contains photoreceptors lies on this basement layer. The photoreceptors are called **rod** and **cone cells** because of their shape. Rods are sensitive to dim light, whereas cones respond to high-intensity light and are involved in color perception.

When a pigment (**rhodopsin**) in a rod cell absorbs light energy, the energy that this reaction releases triggers the generator potential in an axon and then an action potential that leaves the eyeball via the optic nerve. When the photoreceptor cells are not being stimulated (i.e., in the dark), vitamin A and energy from ATP convert rhodopsin back to its light-sensitive form.

Nineteenth-century poet Leigh Hunt said, "Colors are the smiles of Nature." How does an animal distinguish one smile from another? The answer lies to a great extent in the three types of cone-shaped, color-sensitive cells in the retinas of the eyes of primates, birds, reptiles, and fishes. Each type of cone cell responds differently to light reflected from a colored object, depending on whether the cells have red-, green-, or blue-absorbing pigments. The pigments are light-absorbing proteins that are particularly sensitive to either the long-wavelength (red), intermediate-wavelength (green), or short-wavelength (blue) region of the visible spectrum. The retinal nerves translate the relative amounts of light that each type of cone absorbs into generator potentials that are then transmitted as a nerve impulse to the brain, where the overall pattern evokes the perception of a specific hue.

SUMMARY

1. The functional unit of the nervous system is the neuron. Neurons are specialized to produce signals that can be communicated from one part of an animal's body to another. Neurons have two important properties: excitability and conductivity.

2. A typical neuron has three anatomical parts: dendrites, a cell body, and an axon.

3. The language (signal) of a neuron is the nerve impulse or action potential.

4. The plasma membrane of a resting neuron is polarized, meaning that the fluid on the inner side of the membrane is negatively charged with respect to the positively charged fluid outside the membrane. The sodium-potassium ATPase pump and diffusion of ions across membrane channels maintain this polarization.

5. When a threshold stimulus is applied to a resting neuron, the neuron depolarizes, causing an action potential. In myelinated neurons, the action potential jumps from one neurofibril node to the next in a process known as saltatory conduction.

6. Neuronal activity is transmitted between cells at the synapse. Although some animals have electrical synapses that transmit neuronal activity from one neuron to the next, most advanced animals use chemical neurotransmitter molecules.

7. Among animals more complex than sponges, five general evolutionary trends in the nervous system are apparent. That is, the more complex an animal,
   a. the more detailed its nervous system.
   b. the more cephalization concentrates receptors and nervous tissue in an animal's anterior end.
   c. the more bilateral symmetry, which has led to paired nerves, muscles, sensory structures, and brain centers. This pairing facilitates ambulatory movements, such as climbing, crawling, flying, or walking.
   d. the more interneurons it has.
   e. the more complicated its behavior, and the more neurons (especially interneurons) concentrated in an anterior brain and bilaterally organized ganglia.

8. The vertebrate nervous system has two main divisions. The central nervous system is composed of the brain and spinal cord, and the peripheral nervous system is composed of all the nerves (bundles of axons and/or dendrites) outside the brain and spinal cord.

9. The spinal cord of a vertebrate serves two important functions. It is the connecting link between the brain of an animal and most of the body, and it is involved in spinal reflex actions. A reflex is a predictable, involuntary response to a stimulus.

10. The number of spinal nerves is directly related to the number of segments in the trunk and tail of a vertebrate.

11. The vertebrate brain divides into the hindbrain, midbrain, and forebrain. The hindbrain is continuous with the spinal cord and includes the medulla oblongata, cerebellum, and pons.

12. The midbrain is a thickened region of gray matter that integrates visual and auditory signals. The forebrain contains the pineal gland, pituitary gland, hypothalamus, and thalamus. The anterior part of the forebrain expanded during evolution to give rise to the cerebral cortex.

13. In addition to the paired spinal nerves, the vertebrate peripheral nervous system includes 12 pairs of cranial nerves in reptiles, birds, and mammals. Fishes and amphibians have only the first 10 pairs.

14. The autonomic nervous system consists of two antagonistic parts: the sympathetic and parasympathetic divisions.

15. A stimulus is any form of energy an animal can detect with its receptors. Receptors are nerve endings of sensory neurons or specialized cells that respond to stimuli, such as chemical energy, mechanical energy, light energy, or radiant energy.

16. Receptors transduce energy from one form to another. Stimulation of a receptor initiates a generator potential, which creates an action potential that travels along a nerve pathway to another part of the nervous system, where it is perceived.

17. Invertebrates possess an impressive array of receptor structures through which they receive information about their environment. Examples include chemoreceptors that respond to chemicals in the environment; georeceptors, called statocysts, that respond to the force of gravity; hygroreceptors that detect the water content of air; phonoreceptors, such as tympanic organs, that respond to sound; photoreceptors, such as stigmata, ocelli, compound eyes, and complex camera eyes, that respond to light; proprioceptors that respond to mechanically induced changes caused by stretching; tactile receptors, such as bristles, sensilla, spines, setae, and tubercles, that sense touch; and thermoreceptors that respond to temperature changes.

18. Invertebrate and vertebrate sensory receptors (organs) have evolved in ways that relate to the environment in which they must function.

19. The lateral-line system for electrical sensing is in the head area of most fishes, some amphibians, and the platypus. This system can sense electrical currents in the surrounding water. The lateral-line system of fishes and amphibians also contains neuromasts, which are responsive to local water displacement or disturbances.

20. The vertebrate ear has two functional units: the auditory apparatus is concerned with hearing, and the vestibular apparatus is concerned with posture and equilibrium.

21. Pain receptors (nociceptors) are bare sensory nerve endings that produce a painful or itching sensation.

22. Sensors of temperature (thermoreceptors) are bare sensory nerve endings and the simplest vertebrate receptors. Some snakes have heat-sensitive pit organs on each side of the face.

23. Many vertebrates rely on tactile (pertaining to touch) stimuli to respond to their environment. Receptors include bare sensory nerve endings, tactile (Meissner's) corpuscles, bulbs of Krause, Pacinian corpuscles, organs of Ruffini, and vibrissae.

24. Bats, shrews, whales, and dolphins can determine distance and depth by sonar.

25. The receptors for taste (gustation) are chemoreceptors on the body surface of an animal or in the mouth and throat.

26. The sense of smell is due to olfactory neurons in the roof of the vertebrate nasal cavity.

27. Vision (photoreception) is the primary sense that vertebrates in a light-filled environment use. Consequently, their photoreceptive structures are well developed. Most vertebrates have eyes capable of forming visual images.

SELECTED KEY TERMS

accommodation (p. 402)
action potential (p. 382)
axon (p. 380)
cell body (p. 380)
cerebellum (p. 388)
cerebrum (p. 389)
dendrites (p. 380)
generator potential (p. 391)
median (parietal) eye (p. 402)
medulla oblongata (p. 388)

neuron (p. 380)
ocellus (p. 393)
pit organs (p. 399)
rhodopsin (p. 403)
saltatory conduction (p. 382)
sonar (p. 400)
statocysts (p. 392)
stigma (p. 393)
synapse (p. 383)
taste buds (p. 402)

CRITICAL THINKING QUESTIONS

1. What are several ways in which drugs that are stimulants could increase the activity of the human nervous system by acting at the synapse?

2. How can a neuron integrate information?

3. Surveying the functions of the evolutionarily oldest parts of the vertebrate brain gives us some idea of the original functions of the brain. Explain this statement.

4. What are the possible advantages and disadvantages in the evolutionary trend toward cephalization of the nervous system?

5. How does an action potential cross a synapse?

6. Why is it correct to compare a vertebrate eye to a camera?

7. Most animals lack cones in their retinas. How then does such an animal view the visual world?

8. Why is the sense of gravity considered a sense of equilibrium?

9. How does vitamin A deficiency result in night blindness?

10. How would you expect your inner ear to behave in zero gravity?

ONLINE LEARNING CENTER

Visit our Online Learning Center (OLC) at www.mhhe.com/zoology (click on this book's title) to find the following chapter-related materials:

- CHAPTER QUIZZING
- ANIMATIONS
   Action Potential
   Development of Membrane Potential
   Reflex Arc
   Muscle Contraction Action Potential
   Smell
   Taste
   Stroke
   Hearing
   Sense of Balance
   Sense of Rotational Acceleration
   Vision
- RELATED WEB LINKS
   Nervous Systems and Sense Organs
      Neurons
      The Brain and Spinal Cord
      Cranial Nerves
      Sense Organs; Vision, Hearing, and Other Senses
      Human Nervous System Topics
- BOXED READINGS ON
   Neurotoxins
   Hunting with Sound
- SUGGESTED READINGS
- LAB CORRELATIONS

Check out the OLC to find specific information on these related lab exercises in the *General Zoology Laboratory Manual*, 5th edition, by Stephen A. Miller:

Exercise 19   *Vertebrate Nervous Regulation*

CHAPTER 25

COMMUNICATION II:
THE ENDOCRINE SYSTEM
AND CHEMICAL MESSENGERS

Outline

Concepts

1. Chemical messengers are involved in communication, in maintaining homeostasis in an animal's body, and in the body's response to various stimuli. One type of chemical messenger is a hormone. Only those cells that have specific receptors for a hormone can respond to that hormone.
2. Hormones work with nerves to communicate, coordinate, and integrate activities within the body of an animal.
3. Almost every invertebrate produces hormones. However, the physiology of invertebrate hormones is often quite different from that of vertebrate hormones.
4. The major endocrine glands of vertebrates include the hypothalamus, pituitary, thyroid, parathyroids, adrenals, pineal, thymus, pancreas, and gonads. Various other tissues, however, such as the kidneys, heart, digestive system, and placenta also secrete hormones.

Chapter 24 discussed ways that the nervous and sensory systems work together to rapidly communicate information and maintain homeostasis in an animal's body. In addition, many animals have a second, slower form of communication and coordination—the endocrine system with its chemical messengers.

Some scientists suggest that chemical messengers may initially have evolved in single-celled organisms to coordinate feeding or reproduction. As multicellularity evolved, more complex organs also evolved to govern the many individual coordination tasks, but control centers relied on the same kinds of messengers that were present in the simpler organisms. Some of the messengers worked fairly slowly but had long-lasting effects on distant cells; these became the modern hormones. Others worked more quickly but influenced only adjacent cells for short periods; these became the neurotransmitters and local chemical messengers. Clearly, chemical messengers have an ancient origin and must have been conserved for hundreds of millions of years.

Evolutionarily, new messengers are uncommon. Instead, "old" messengers are adapted to new purposes. For example, some ancient protein hormones are in species ranging from bacteria to humans.

One key to the survival of any group of animals is proper timing of activity so that growth, maturation, and reproduction coincide with the times of year when climate and food supply favor survival. It seems likely that the chemical messengers regulating growth and reproduction were among the first to appear. These messengers were

This chapter contains evolutionary concepts, which are set off in this font.

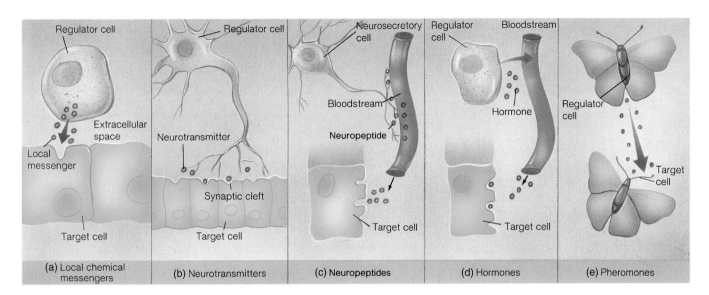

FIGURE 25.1

Chemical Messengers: Targets and Transport.   (*a*) Short-distance local messengers act on an adjacent cell. (*b*) Individual nerve cells secrete neurotransmitters that cross the synaptic cleft to act on target cells. (*c*) Individual nerve cells can also secrete neuropeptides (neurohormones) that travel some distance in the bloodstream to reach a target cell. (*d*) Regulatory cells, usually in an endocrine gland, secrete hormones, which enter the bloodstream and travel to target cells. (*e*) Regulatory cells in exocrine glands secrete pheromones. They leave the body and stimulate target cells in another animal.

probably secretions of neurons. Later, specific hormones developed to play important regulatory roles in molting, growth, metamorphosis, and reproduction in various invertebrates. Chemical messengers and their associated secretory structures became even more complex with the appearance of vertebrates.

CHEMICAL MESSENGERS

The development of most animals commences with fertilization and the subsequent division of the zygote. Further development then depends on continued cell proliferation, growth, and differentiation. The integration of these events, as well as the communication and coordination of physiological processes, such as metabolism, respiration, excretion, movement, and reproduction, depend on chemical messengers—molecules that specialized cells synthesize and secrete. Chemical messengers can be categorized as follows:

1. **Local chemical messengers.** Many cells secrete chemicals that alter physiological conditions in the immediate vicinity (figure 25.1*a*). Most of these chemicals act on adjacent cells and do not accumulate in the blood. Vertebrate examples include some of the chemicals called lumones that the gut produces and that help regulate digestion. In a wound, mast cells secrete a substance called histamine that participates in the inflammatory response.

2. **Neurotransmitters.** As presented in chapter 24, neurons secrete chemicals called neurotransmitters (e.g., nitric oxide and acetylcholine) that act on immediately adjacent target cells (figure 25.1*b*). These chemical messengers reach high concentrations in the synaptic cleft, act quickly, and are actively degraded and recycled.

3. **Neuropeptides.** Some specialized neurons (called neurosecretory cells) secrete neuropeptides (**neurohormones**). The blood or other body fluids transport neuropeptides to nonadjacent target cells, where neuropeptides exert their effects (figure 25.1*c*). In mammals, for example, certain nerve cells in the hypothalamus release a neuropeptide that causes the pituitary gland to release the hormone oxytocin, which induces powerful uterine contractions during the delivery of offspring.

4. **Hormones.** Endocrine glands or cells secrete hormones that the bloodstream transports to nonadjacent target cells (figure 25.1*d*). Many examples are given in the rest of this chapter.

5. **Pheromones.** Pheromones are chemical messengers released to the exterior of one animal that affect the behavior of another individual of the same species (figure 25.1*e*; *see also table 15.2*).

Overall, scientists now recognize that the nervous and endocrine systems work together as an all-encompassing communicative and integrative network called the **neuroendocrine system.** In this system, feedback systems regulate chemical messengers in their short- and long-term coordination of animal body function to maintain homeostasis.

HORMONES AND THEIR FEEDBACK SYSTEMS

A **hormone** (Gr. *hormaein*, to set in motion or to spur on) is a specialized chemical messenger that an endocrine gland or tissue produces and secretes. The study of endocrine glands and their

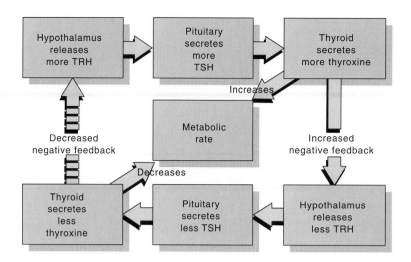

FIGURE 25.2

Hormonal Feedback.   Negative feedback system that helps control metabolic rate in a vertebrate such as a dog. (TRH = thyrotropin-releasing hormone; TSH = thyroid-stimulating hormone.)

hormones is called **endocrinology.** Hormones circulate through body fluids and affect the metabolic activity of a target cell or tissue in a specific way. By definition, a **target cell** has receptors to which chemical messengers either selectively bind or on which they have an effect. Only rarely does a hormone operate independently. More typically, one hormone influences, depends on, and balances another hormone in a controlled feedback network.

BIOCHEMISTRY OF HORMONES

Most hormones are proteins (polypeptides), derivatives of amino acids (amines), or steroids. A few are fatty acid derivatives. For example, most invertebrate neurosecretory cells produce polypeptides called neuropeptides. Hormones that the vertebrate pancreas secretes are proteins; those that the thyroid gland secretes are amines. The ovaries, testes, and cortex of the adrenal glands secrete steroids.

Hormones are effective in extremely small amounts. Only a few molecules of a hormone may be enough to produce a dramatic response in a target cell. In the target cell, hormones help control biochemical reactions in three ways: (1) a hormone can increase the rate at which other substances enter or leave the cell; (2) it can stimulate a target cell to synthesize enzymes, proteins, or other substances; or (3) it can prompt a target cell to activate or suppress existing cellular enzymes. As is the case for enzymes, hormones are not changed by the reaction they regulate.

FEEDBACK CONTROL SYSTEM OF HORMONE SECRETION

Although hormones are always present in some amount in endocrine cells or glands, they are not secreted continuously. Instead, the glands secrete the amount of hormone that the animal needs to maintain homeostasis. A feedback control system monitors changes in the animal or in the external environment and sends information to a central control unit (such as the central nervous system), which makes adjustments. A feedback system that produces a response that counteracts the initiating stimulus is called a negative feedback system. In contrast, a positive feedback system reinforces the initial stimulus. Positive feedback systems are relatively rare in animals because they usually lead to instability or pathological states.

Negative feedback systems monitor the amount of hormone secreted, altering the amount of cellular activity as needed to maintain homeostasis. For example, suppose that the rate of chemical activity (metabolic rate) in the body cells of a dog slows (figure 25.2). The hypothalamus responds to this slow rate by releasing more thyrotropin-releasing hormone (TRH), which causes the pituitary gland to secrete more thyrotropin, or thyroid-stimulating hormone (TSH). This hormone, in turn, causes the thyroid gland to secrete a hormone called thyroxine. Thyroxine increases the metabolic rate, restoring homeostasis. Conversely, if the metabolic rate speeds up, the hypothalamus releases less TRH, the pituitary secretes less TSH, the thyroid secretes less thyroxine, and the metabolic rate decreases once again, restoring homeostasis.

MECHANISMS OF HORMONE ACTION

Hormones modify the biochemical activity of a target cell or tissue. Two basic mechanisms are involved. The first, the fixed-membrane-receptor mechanism, applies to hormones that are proteins or amines. Because they are water-soluble and cannot diffuse across the plasma membrane, these hormones initiate their response by means of specialized receptors on the plasma membrane

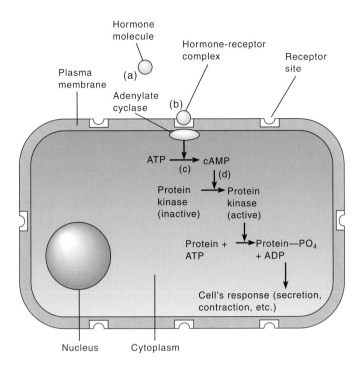

FIGURE 25.3

Steps of the Fixed-Membrane-Receptor Mechanism of Hormonal Action. (*a*) A protein hormone molecule (such as epinephrine) diffuses from the blood to a target cell. (*b*) The binding of the hormone to a specific plasma membrane receptor activates adenylate cyclase (a membrane-bound enzyme system). (*c*) This enzyme system catalyzes cyclic AMP formation (the second messenger) inside the cell. (*d*) Cyclic AMP (cAMP) diffuses throughout the cytoplasm and activates an enzyme called protein kinase, which then phosphorylates specific proteins in the cell, thereby triggering the biochemical reaction, leading ultimately to the cell's response.

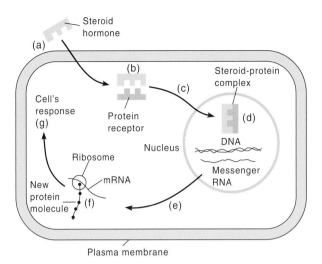

FIGURE 25.4

Steps of the Mobile-Receptor Mechanism. (*a*) A steroid hormone molecule (e.g., testosterone) diffuses from the blood to a target cell and then across the plasma membrane of the target cell. (*b*) Once in the cytoplasm, the hormone binds to a receptor that (*c*) carries it into the nucleus. (*d*) This steroid-protein complex triggers the transcription of specific gene regions of DNA. (*e*) The messenger RNA transcript is then translated into a gene product via (*f*) protein synthesis in the cytoplasm. (*g*) The new protein then mediates the cell's response.

esterase inactivates cyclic AMP. In the meantime, the receptor on the plasma membrane loses the first messenger and now becomes available for a new reaction.

MOBILE-RECEPTOR MECHANISM

Because steroid hormones pass easily through the plasma membrane, their receptors are inside the target cells. The mobile-receptor mechanism involves the stimulation of protein synthesis. After being released from a carrier protein in the bloodstream, the steroid hormone enters the target cell by diffusion and binds to a specific protein receptor in the cytoplasm (figure 25.4*a,b*). This newly formed steroid-protein complex acquires an affinity for DNA that causes it to enter the nucleus of the cell, where it binds to DNA and regulates the transcription of specific genes to form messenger RNA (figure 25.4*c,d*). The newly transcribed mRNA leaves the nucleus and moves to the rough endoplasmic reticulum, where it initiates protein synthesis (figure 25.4*e,f*). Some of the newly synthesized proteins may be enzymes whose effects on cellular metabolism constitute the cellular response attributable to the specific steroid hormone (figure 25.4*g*).

of the target cell. The second, the mobile-receptor mechanism, applies to steroid hormones. These hormones are lipid-soluble and diffuse easily into the cytoplasm, where they initiate their response by binding to cytoplasmic receptors.

FIXED-MEMBRANE-RECEPTOR MECHANISM

With the fixed-membrane-receptor mechanism, an endocrine cell secretes a water-soluble hormone that circulates through the bloodstream (figure 25.3*a*). At the cells of the target organ, the hormone acts as a "first or extracellular messenger," binding to a specific receptor site for that hormone on the plasma membrane (figure 25.3*b*). The hormone-receptor complex activates the enzyme adenylate cyclase in the membrane (figure 25.3*c*). The activated enzyme converts ATP into a nucleotide called cyclic AMP, which becomes the "second (or intracellular) messenger." Cyclic AMP diffuses throughout the cytoplasm and activates an enzyme called protein kinase, which causes the cell to respond with its distinctive physiological activity (figure 25.3*d*). After inducing the target cell to perform its specific function, the enzyme phosphodi-

SOME HORMONES OF INVERTEBRATES

The survival of any group of animals depends on growth, maturation, and reproduction coinciding with the most favorable seasons of the year so that climate and food supply are optimal.

Thus, chemicals regulating growth, maturation, and reproduction probably were among the first hormones to appear during the course of animal evolution.

The first hormones were probably neurosecretions. As discussed next, most of the chemicals functioning as hormones in invertebrate animals are neurosecretions called neuropeptides. Only a few of the more complex invertebrates (e.g., molluscs, arthropods, echinoderms) have hormones other than neurosecretions. What follows is a brief overview of some invertebrate neuropeptides and hormones.

PORIFERA

The porifera (sponges) do not have classical endocrine glands. Since sponges do not have neurons, they also do not have neurosecretory cells.

CNIDARIANS

The nerve cells of *Hydra* contain a growth-promoting hormone that stimulates budding, regeneration, and growth. For example, when the hormone is present in the medium in which fragments of *Hydra* are incubated, "head" regeneration is accelerated. This so-called "head activator" also stimulates mitosis in *Hydra*.

PLATYHELMINTHS

Zoologists identified neurosecretory cells in various flatworms over 30 years ago. These cells are in the cerebral ganglion and along major nerve cords. The neuropeptides that the cells produce function in regeneration, asexual reproduction, and gonad maturation. For example, neurosecretory cells in the scolex of some tapeworms control shedding of the proglottids or the initiation of strobilization.

NEMERTEANS

Nemerteans have more cephalization than platyhelminths and a larger brain, composed of a dorsal and ventral pair of ganglia connected by a nerve ring. The neuropeptide that these ganglia produce appears to control gonadal development and to regulate water balance.

NEMATODES

Although no classical endocrine glands have been identified in nematodes, they do have neurosecretory cells associated with the central nervous system. The neuropeptide that this nervous tissue produces apparently controls ecdysis of the old cuticle. The neuropeptide is released after a new cuticle is produced and stimulates the excretory gland to secrete an enzyme (leucine aminopeptidase) into the space between the old and new cuticle. The accumulation of fluid in this space causes the old cuticle to split and be shed.

MOLLUSCS

The ring of ganglia that constitutes the central nervous system of molluscs is richly endowed with neurosecretory cells. The neuropeptides that these cells produce help regulate heart rate, kidney function, and energy metabolism.

In certain gastropods, such as the common land snail *Helix*, a specific hormone stimulates spermatogenesis; another hormone, termed egg-laying hormone, stimulates egg development; and hormones from the ovary and testis stimulate accessory sex organs. In all snails, a growth hormone controls shell growth.

In cephalopods, such as the octopus and squid, the optic gland in the eye stalk produces one or more hormones that stimulate egg development, proliferation of spermatogonia, and the development of secondary sexual characteristics.

ANNELIDS

Since annelids have a well-developed and cephalized nervous system, a well-developed circulatory system, and a large coelom, their correspondingly well-developed endocrine control of physiological functions is not surprising. The various endocrine systems of annelids are generally involved with morphogenesis, development, growth, regeneration, and gonadal maturation. For example, in polychaetes, juvenile hormone inhibits the gonads and stimulates growth and regeneration. Another hormone, gonadotropin, stimulates the development of eggs. In leeches, a neuropeptide stimulates gamete development and triggers color changes. Osmoregulatory hormones have been reported in oligochaetes, and a hyperglycemic hormone that maintains a high concentration of blood glucose has been reported for the oligochaete *Lumbricus*.

ARTHROPODS

The endocrine systems of advanced invertebrates (crustaceans and insects) are excellent examples of how hormones regulate growth, maturation, and reproduction. Much is known about hormones and their functioning in these animals.

The endocrine system of a crustacean, such as a crayfish, controls functions such as ecdysis (molting), sex determination, and color changes. Only ecdysis is discussed here.

X-organs are neurosecretory tissues in the crayfish eye stalks (figure 25.5*a*). Associated with each X-organ is a sinus gland that accumulates and releases the secretions of the X-organ. Other glands, called Y-organs, are at the base of the maxillae. X-organs and Y-organs control ecdysis as follows. In the absence of an appropriate stimulus, the X-organ produces molt-inhibiting hormone (MIH), and the sinus gland releases it (figure 25.5*b*). The target of this hormone is the Y-organ. When MIH is present in high concentrations, the Y-organ is inactive. Under appropriate internal and external stimuli, MIH release is prevented, and the Y-organ releases the hormone ecdysone, which leads to molting (figure 25.5*c*).

The sequence of events in insects is similar to that of crustaceans, but it does not involve a molt-inhibiting hormone. The

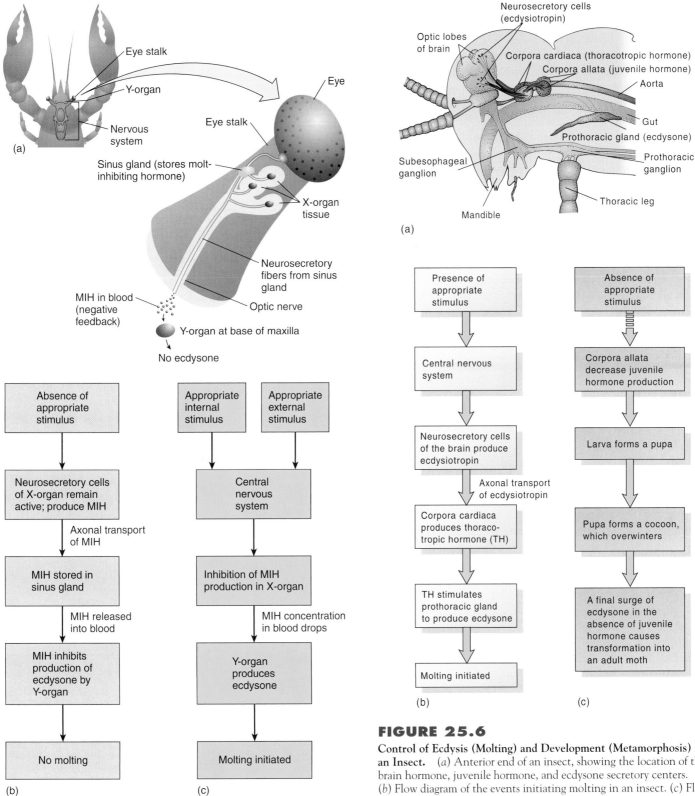

FIGURE 25.5

Hormonal Control of Ecdysis (Molting) in Crustaceans.
(*a*) Neurosecretory apparatus in a crustacean eye stalk. (*b*) Flow diagram of the events inhibiting molting and (*c*) causing molting. (MIH = molt-inhibiting hormone.)

FIGURE 25.6

Control of Ecdysis (Molting) and Development (Metamorphosis) in an Insect.   (*a*) Anterior end of an insect, showing the location of the brain hormone, juvenile hormone, and ecdysone secretory centers. (*b*) Flow diagram of the events initiating molting in an insect. (*c*) Flow diagram of the events of insect metamorphosis as regulated by a decrease in the production of juvenile hormone.

presence of an appropriate stimulus to the central nervous system
activates certain neurosecretory cells (pars intercerebralis) in the
optic lobes of the brain (figure 25.6a). These cells secrete the
hormone ecdysiotropin, which axons transport to the corpora cardiaca (a mass of neurons associated with the brain). The corpora
cardiaca produces thoracotropic hormone, which is carried to the
prothoracic glands, stimulating them to produce and release
ecdysone, which induces molting (figure 25.6b)—in particular,
the reabsorption of some of the old cuticle and the development
of a new cuticle.

Other neurosecretory cells in the brain and nerve cords produce the hormone bursicon. Bursicon influences certain aspects of
epidermal development, such as tanning (i.e., hardening and
darkening of the chitinous outer cuticle layer). Tanning is completed several hours after each molt.

Another hormone, juvenile hormone (JH), is also involved in the morphological differentiation that occurs during
the molting of insects. Just behind the insect brain are the paired
corpora allata (figure 25.6a). These structures produce JH. High
concentrations of JH in the blood of an insect inhibit differentiation. In the absence of an appropriate environmental stimulus,
the corpora allata decrease JH production, which causes the insect larva to differentiate into a pupa (figure 25.6c). The pupa
then forms a cocoon to overwinter. In the spring, a final surge of
ecdysone, in the absence of JH, transforms the pupa into an
adult moth.

ECHINODERMS

Since echinoderms are deuterostomes, they are more closely
allied with chordates than are the protostome invertebrates.
However, the endocrine systems of echinoderms provide few
insights into the evolution of chordate endocrine systems, because echinoderm hormones and endocrine glands are very different from those of chordates. Zoologists do know, however,
that the radial nerves of sea stars contain a neuropeptide called
gonad-stimulating substance. When this neuropeptide is injected into a mature sea star, it induces immediate shedding of
the gametes, spawning behavior, and meiosis in the oocytes. The
neuropeptide also causes the release of a hormone called matu-
ration-inducing substance, which has various effects on the
reproductive system.

AN OVERVIEW OF THE VERTEBRATE ENDOCRINE SYSTEM

Since vertebrates have been studied more than invertebrates, vertebrates have the best understood system of hormonal control. *As
the earliest vertebrates evolved, hormone-producing cells and
tissues developed, and they came to be controlled in several
ways. Sets of nerve cells in the brain direct some endocrine tissues, such as the medullary areas of the adrenal glands. The*

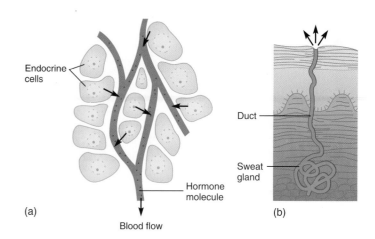

FIGURE 25.7

Vertebrate Glands with and without Ducts.   (a) An endocrine gland,
such as the thyroid, secretes hormones into the extracellular fluid. From
there, the hormones pass into blood vessels and travel throughout the
body. (b) An exocrine gland, such as a sudoriferous (sweat) gland,
secretes material (sweat) into a duct that leads to a body surface.

*hypothalamus of the brain and the pituitary gland (hypophysis)
control others. Still others function independently of either
nerves or the pituitary gland.*

Vertebrates possess two types of glands (figure 25.7). One
type, **exocrine** (Gr. *exo*, outside + *krinein*, to separate) **glands,**
secrete chemicals into ducts that, in turn, empty into body cavities or onto body surfaces (e.g., mammary, salivary, and sweat
glands). The second type, **endocrine** (Gr. *endo*, within + *krinein*,
to separate) **glands,** have no ducts, and instead secrete chemical
messengers, called hormones, directly into the tissue space next
to each endocrine cell. The hormones then diffuse into the
bloodstream, which carries them throughout the body to their
target cells.

ENDOCRINE SYSTEMS OF VERTEBRATES OTHER THAN BIRDS OR MAMMALS

Since this text uses a phylogenetic approach to describe form and
function in animals, endocrine regulation in vertebrates other
than birds or mammals is now discussed. Birds and mammals are
the subject of the last part of this chapter.

Vertebrates other than birds or mammals have somewhat
similar endocrine systems, but differences do exist. Recent research has revealed the following three aspects of endocrinology
that relate to species differences among these vertebrates:

1. Hormones (or neuropeptides) with the same function in different species may not be chemically identical.
2. Certain hormones are species-specific with respect to their
   function; conversely, some hormones produced in one
   species may be completely functional in another species.

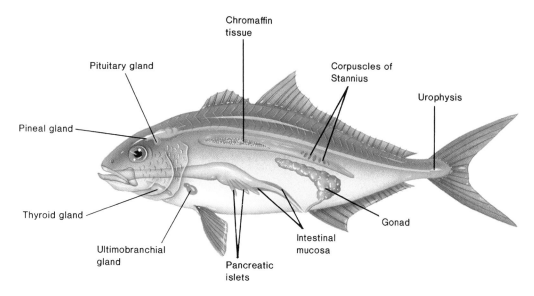

FIGURE 25.8

Approximate Locations of Endocrine Tissues (Glands) in a Bony Fish.

3. A hormone from one species may elicit a different response in the same target cell or tissue of a different species.

The examples that follow illustrate these three principles and also present a comparative survey of endocrine function in selected vertebrates.

When more ancient groups of vertebrates are compared with more recent ones, one general tendency surfaces: older groups seem to have simpler endocrine systems. For example, many of the hormones present in mammals are absent in fishes. In fishes, the brain and spinal cord are the most important producers of hormones, with other glands being rudimentary (figure 25.8). In jawed fishes, three major regions secrete neuropeptides. The two in the brain are the **pineal gland** of the epithalamus and the **preoptic nuclei** of the hypothalamus. The pineal gland produces neuropeptides that affect pigmentation and apparently inhibit reproductive development, both of which are stimulated by light. One specific hormone that the pineal gland produces, **melatonin,** has broad effects on body metabolism by synchronizing activity patterns with light intensity and day length. The preoptic nuclei produce various other neuropeptides that control different functions in fishes (e.g., growth, sleep, locomotion). The third major region of fishes that has neuropeptide function is the urophysis. The **urophysis** (Gr. *oura*, tail + *physis*, growth) is a discrete structure in the spinal cord of the tail. The urophysis produces neuropeptides that help control water and ion balance, blood pressure, and smooth muscle contractions. Other than these functions, little else is known of its functions or the significance of its absence in more complex vertebrates.

In many fishes, amphibians, and reptiles, hormones (e.g., melatonin) from the pineal gland control variations in skin color. When this hormone produced by one species is injected into another species, it can induce dramatic color changes (figure 25.9). This type of experiment indicates that some hormones have a

FIGURE 25.9

Hormonal Control of Frog Skin Color. The light-colored frog on the left was immersed in water containing the hormone melatonin. The dark-colored frog on the right received an injection of melanocyte-stimulating hormone.

close chemical similarity (point 2 at the beginning of this section), despite the distant evolutionary relationships among the animals producing them. Another example is the hormone prolactin (produced by the **pituitary gland**). Prolactin stimulates reproductive migrations in many animals (e.g., the movement of salamanders to water). Prolactin causes brooding behavior in some fishes. It also helps control water and salt balances, and is

TABLE 25.1
MAJOR VERTEBRATE ENDOCRINE TISSUES AND HORMONES

SOURCE	HORMONES	TARGET CELLS AND PRINCIPAL ACTIONS
Anterior lobe of pituitary (adenohypophysis)	Somatotropin (STH, or growth hormone [GH])	Stimulates growth of bone and muscle; promotes protein synthesis; affects lipid and carbohydrate metabolism; increases cell division
	Adrenocorticotropic hormone (ACTH)	Stimulates secretion of adrenocortical steroids; is involved in stress response
	Thyrotropin (TSH) or thyroid-stimulating hormone	Stimulates thyroid gland to synthesize and release thyroid hormones concerned with growth, development, metabolic rate
	Endorphins	Decrease pain
	Gonadotropins:	
	Luteinizing or interstitial cell–stimulating hormone (LH or ICSH)	In ovary: Forms corpora lutea; secretes progesterone; probably acts in conjunction with FSH
		In testis: Stimulates the interstitial cells, thus promoting the secretion of testosterone
	Follicle-stimulating hormone (FSH)	In ovary: Stimulates growth of follicles; functions with LH to cause estrogen secretion and ovulation
		In testis: Acts on seminiferous tubules to promote spermatogenesis
	Prolactin (PRL)	Initiates milk production by mammary glands; acts on crop sacs of some birds; stimulates maternal behavior in birds
Intermediate or posterior lobe of pituitary	Melanocyte-stimulating hormone (MSH)	Expands amphibian melanophores; contracts iridophores and xanthophores; promotes melanin synthesis; darkens the skin; responds to external stimuli
Posterior lobe of pituitary (neurohypophysis) releases these hormones produced by the hypothalamus	Antidiuretic hormone (ADH or vasopressin)	Elevates blood pressure by acting on arterioles; promotes reabsorption of water by kidney tubules
	Oxytocin	Affects postpartum mammary gland, causing ejection of milk; promotes contraction of uterus; has possible action in parturition and in sperm transport in female reproductive tract

essential for certain saltwater fishes to enter freshwater during spawning runs.

From an evolutionary perspective, evidence indicates that the thyroid gland in the earliest vertebrates evolved from a pouchlike structure (the endostyle) that carried food particles in the front end of the digestive tract. This explains why the thyroid gland is in the neck on the ventral side of the pharynx in all vertebrates. How did this feeding mechanism turn into an endocrine gland? One possible hypothesis is that as the developing pouch gradually lost all connection with the pharynx, it became independent of the digestive system both functionally and structurally. As a result, a functionally novel structure arose from an ancestral structure with an unrelated function. The shape of the thyroid varies among vertebrates. It may be a single structure (e.g., many fishes, reptiles, and some mammals), or it may have several to many lobes. The major hormones that this gland produces are thyroxine (T_4) and triiodothyronine (T_3), which control the rate of metabolism, growth, and tissue differentiation in vertebrates.

As noted in point 3 at the beginning of this section, the same hormone(s) in different vertebrates may regulate related but different processes. The hormones thyroxine and triiodothyronine are excellent examples of this point. For example, in most

animals, thyroxine and triiodothyronine regulate overall metabolism. In amphibians, they play an additional role in metamorphosis (figure 25.10). Specifically timed changes in the concentrations of three hormones—prolactin, thyroxine, and triiodothyronine—control metamorphosis in the frog. Low thyroxine and triiodothyronine concentrations and high prolactin concentrations in young tadpoles stimulate larval growth and prevent metamorphosis. As the hypothalamus and pituitary glands develop in the growing tadpole, the hypothalamus releases thyroid-stimulating hormone and prolactin-inhibiting hormone. Their release causes the pituitary gland to release thyroid-stimulating hormone and to cease production of prolactin. As a result, the concentrations of thyroxine and triiodothyronine rise, triggering the onset of metamorphosis. Tail resorption and other metamorphic changes follow.

In jawed fishes and primitive tetrapods, several small **ultimobranchial glands** form ventral to the esophagus (*see figure 25.8*). These glands produce the hormone calcitonin that helps regulate the concentration of blood calcium.

Specialized endocrine cells (**chromaffin tissue**) or glands (**adrenal glands**) near the kidneys prepare some vertebrates for stressful emergency situations (figure 25.11). These tissues and glands produce two hormones (epinephrine or adrenaline, and

SOURCE	HORMONES	TARGET CELLS AND PRINCIPAL ACTIONS
Hypothalamus	Thyroid-stimulating hormone (TSH)	Stimulates release of TSH by anterior pituitary
	Adrenocorticotropin-releasing hormone (CRH)	Stimulates release of ACTH by anterior pituitary
	Gonadotropin-releasing hormone (GnRH)	Stimulates gonadotropin release by anterior pituitary
	Prolactin-inhibiting factor (PIF)	Inhibits prolactin release by anterior pituitary
	Somatostatin	Inhibits release of STH by anterior pituitary
Thyroid gland	Thyroxine, triiodothyronine	Affect growth, amphibian metamorphosis, molting, metabolic rate in birds and mammals, development
	Calcitonin	Lowers blood calcium level by inhibiting calcium reabsorption from bone
Parathyroid glands	Parathormone	Regulates calcium concentration
Pancreas, islet cells	Insulin (from beta cells)	Promotes glycogen synthesis and glucose utilization and uptake from blood
	Glucagon (from alpha cells)	Raises blood glucose concentration
Adrenal cortex	Glucocorticoids (e.g., cortisol)	Promote synthesis of carbohydrates and breakdown of proteins; initiate antiinflammatory and antiallergic actions; mediate response to stress
	Mineralocorticoids (e.g., aldosterone)	Regulate sodium retention and potassium loss through kidneys, and water balance
Adrenal medulla	Epinephrine (adrenaline)	Mobilizes glucose; increases blood flow through skeletal muscle; increases oxygen consumption; increases heart rate
	Norepinephrine	Elevates blood pressure; constricts arterioles and venules
Testes	Androgens (e.g., testosterone)	Maintain male sexual characteristics; promote spermatogenesis
Ovaries	Estrogens (e.g., estradiol)	Maintain female sexual characteristics; promote oogenesis
Corpus luteum	Progesterone	Maintains pregnancy; stimulates development of mammary glands

norepinephrine or noradrenaline) that cause vasoconstriction, increased blood pressure, changes in the heart rate, and increased blood glucose levels. These hormones are involved in the "fight-or-flight" reactions.

ENDOCRINE SYSTEMS OF BIRDS AND MAMMALS

With some minor exceptions, birds and mammals have a similar complement of endocrine glands (figure 25.12). Table 25.1 summarizes the major hormones these vertebrates produce.

BIRDS

The endocrine glands in birds include the ovary, testes, adrenals, pituitary, thyroid, pancreas, parathyroids, pineal, hypothalamus, thymus, ultimobranchial, and bursa of Fabricius (figure 25.12a). Since the hormones that most of these glands produce and their effects on target tissues are nearly the same as in mammals, they are discussed in the next section on mammals. A discussion of some unique hormones and their functions in birds follows.

In some birds (e.g., pigeons and doves), the pituitary gland secretes the hormone prolactin. Prolactin stimulates the production of "pigeon's milk" by desquamation (sloughing off cells) in the pigeon's crop. Prolactin also stimulates and regulates broodiness and certain other kinds of parental behavior, and along with estrogen, stimulates full development of the **brood (incubation) patch** (figure 25.13). The brood patch helps keep the eggs at a temperature between 33 and 37° C.

The bird's thyroid gland produces the hormone thyroxine. In addition to the major vertebrate functions listed in table 25.1, thyroxine regulates the normal development of feathers and the molt cycle, and it plays a role in the onset of migratory behavior.

In male birds, the testes produce the hormone testosterone. Testosterone controls the secondary sexual characteristics of the male, such as bright plumage color, comb (when present), and spurs—all of which strongly influence sexual behavior.

The ultimobranchial glands are small, paired structures in the neck just below the parathyroid glands. They secrete the hormone **calcitonin,** which is involved in regulating blood calcium concentrations.

The **bursa of Fabricius** is a sac that lies just dorsal to the cloaca and empties into it. Although well developed during the bird's embryological development, it begins to shrink soon after

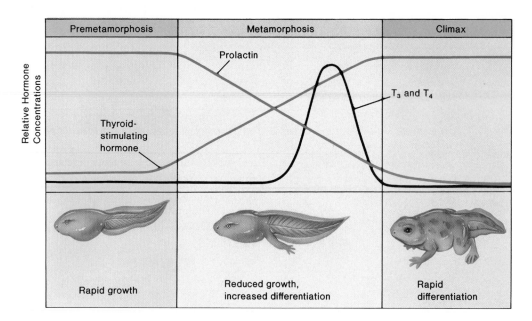

FIGURE 25.10

Frog Tadpole Metamorphosis. The thyroid hormones triiodothyronine (T_3) and thyroxine (T_4) regulate the metamorphosis of an aquatic frog tadpole into a semiterrestrial or terrestrial adult. The anterior pituitary secretes thyroid-stimulating hormone, which regulates thyroid gland activity. During the premetamorphosis (tadpole) stage, the pituitary and thyroid glands are relatively inactive. This keeps the concentration of thyroid-stimulating hormones, T_3 and T_4, at low concentrations. The high prolactin concentration in tadpoles stimulates larval growth and prevents metamorphosis. During metamorphosis, the concentrations of the thyroid hormones markedly increase, and prolactin decreases. These hormonal fluctuations induce rapid differentiation, climaxing in the adult frog.

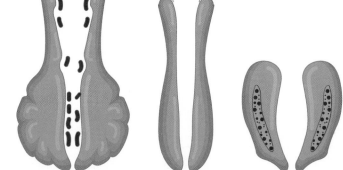

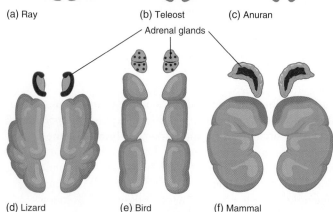

FIGURE 25.11

Chromaffin Tissue and Adrenal Glands in Selected Vertebrates. The chromaffin tissue (steroidogenic) produces steroid hormones and is shown in gray. The aminogenic tissue that produces norepinephrine and epinephrine is shown in black. The kidneys are shown in brown. Note the reversed location of the two components in lizards and mammals. (*a*) In jawless and cartilaginous fishes (elasmobranchs), aminogenic tissue develops as clusters near the kidneys. (*b*) In teleosts, the chromaffin tissue is generally at the anterior end of the kidney (pronephric region). (*c*) In anurans, the chromaffin tissue is interspersed in a diffuse gland on the ventral surface of each kidney. (*d*) In lizards, the chromaffin tissue forms a capsule around the steroidogenic-producing tissue. (*e*) In birds, the chromaffin tissue is interspersed within an adrenal capsule. (*f*) In most mammals, the chromaffin tissue forms an adrenal medulla, and the steroidogenic tissue forms the cortex.

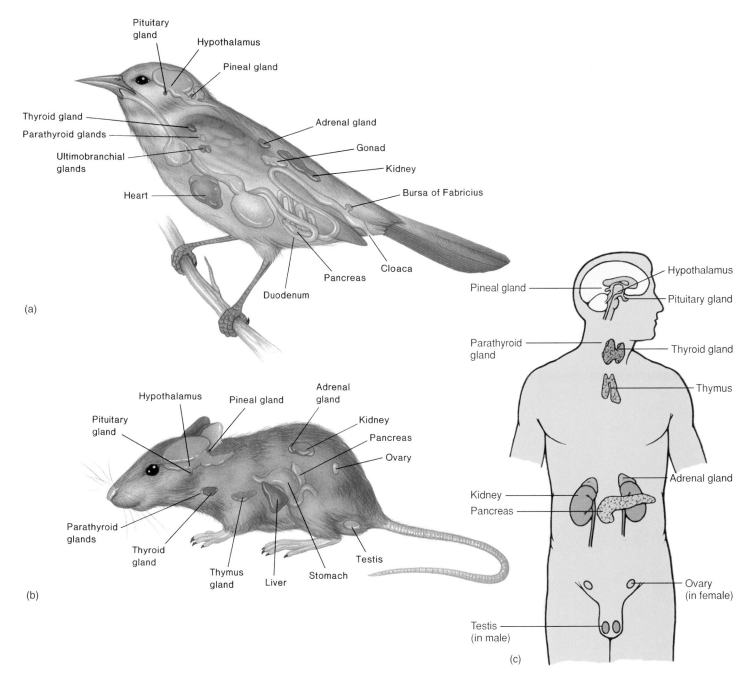

FIGURE 25.12
Endocrine Glands of Birds and Mammals.   Locations of the major endocrine glands of (*a*) a bird, (*b*) a rat, and (*c*) a human.

hatching. Its tissues produce secretions that are responsible for the maturation of white blood cells (B lymphocytes), which play an important role in immunological reactions.

MAMMALS

Zoologists know more about the endocrine organs, hormones, and target tissues of mammals than of any other animal group. This is especially true for the human body. A brief overview of mammalian endocrinology follows.

Pituitary Gland (Hypophysis)

The pituitary gland (also known as the hypophysis) is directly below the hypothalamus (*see figure 25.12c*). The pituitary has two distinct lobes: the anterior lobe (adenohypophysis) and the posterior lobe (neurohypophysis) (figure 25.14). The two lobes differ in several ways: (1) the adenohypophysis is larger than the neurohypophysis; (2) secretory cells called pituicytes are in the adenohypophysis, but not in the neurohypophysis; and (3) the neurohypophysis has a greater supply of nerve endings. Pituicytes produce and secrete hormones directly from the adenohypophysis,

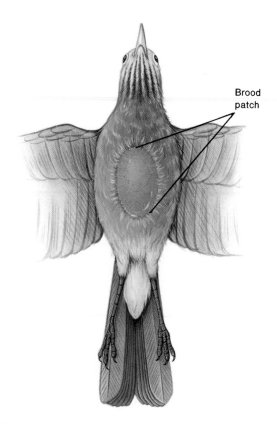

Brood patch

FIGURE 25.13

A Bird's Brood Patch.   In this example, a robin's single brood patch appears (due to the effect of the hormone prolactin) a few days before eggs are laid. Prolactin causes the down feathers to drop from the abdomen of the incubating robin, and the bare patch becomes swollen and richly supplied with blood vessels. After laying the eggs, the robin settles on its nest and brings this warm patch in contact with its eggs, thereby transferring heat to the developing embryos.

whereas the neurohypophysis obtains its hormones from the neurosecretory cells in the hypothalamus, storing and releasing them when they are needed. These modified hypothalamic nerve cells project their axons down a stalk of nerve cells and blood vessels, called the infundibulum, into the pituitary gland, directly linking the nervous and endocrine systems.

The pituitary of many vertebrates (but not in humans, birds, and cetaceans) also has a functional **intermediate lobe (pars intermedia)** of mostly glandular tissue. Its secretions (e.g., melanophore-stimulating hormone) in response to external stimuli induce changes in the coloration of the body surface of many animals.

Hormones of the Neurohypophysis   The neurohypophysis does not manufacture any hormones. Instead, the neurosecretory cells of the hypothalamus synthesize and secrete two hormones, antidiuretic hormone and oxytocin, which move down nerve axons into the neurohypophysis, where they are stored in the axon terminals until released.

Diuretics stimulate urine excretion, whereas antidiuretics decrease urine secretion. When a mammal begins to lose water and

becomes dehydrated, antidiuretic hormone (ADH, or vasopressin) is released and increases water absorption in the kidneys so that less urine is secreted. Because less urine is secreted, water is retained. This negative feedback system thus restores water and solute homeostasis.

Oxytocin plays a role in mammalian reproduction by its effect on smooth muscle. It stimulates contraction of the uterus or uteri to aid in the expulsion of the offspring and promotes the ejection of milk from the mammary glands to provide nourishment for the newborn.

Both ADH and oxytocin are thought to have evolved from a similar ancestral chemical messenger that helped control water loss and, indirectly, solute concentrations. For example, the neurohypophysis is notably larger in animals that live in arid parts of the world, where water conservation is crucial. Also, the structure of the two hormones is similar except for a difference in two of the amino acids.

Hormones of the Adenohypophysis   The true endocrine portion of the pituitary is the adenohypophysis, which synthesizes six different hormones (figure 25.14). All of these hormones are polypeptides, and all but two are true tropic hormones, hormones whose primary target is another endocrine gland. The two nontropic hormones are growth hormone and prolactin.

Growth hormone (GH), or somatotropin (STH), does not influence a particular target tissue; rather, it affects all parts of the body that are concerned with growth. It directly induces the cell division necessary for growth and protein synthesis in most types of cells by stimulating the uptake of amino acids, RNA synthesis, and ribosome activity.

Prolactin (PRL) has the widest range of actions of the adenohypophyseal hormones. It plays an essential role in many aspects of reproduction. For example, it stimulates reproductive migrations in many mammals, such as elk and caribou. Prolactin also enhances mammary gland development and milk production in female mammals. (Oxytocin stimulates milk ejection from the mammary glands, but not its production.)

Thyrotropin, or thyroid-stimulating hormone (TSH), stimulates the thyroid gland's synthesis and secretion of thyroxine, the main thyroid hormone.

Adrenocorticotropic hormone (ACTH) stimulates the adrenal gland to produce and secrete steroid hormones called glucocorticoids (cortisol). The secretion of ACTH is regulated by the secretion of corticotropin-releasing factor from the hypothalamus, which, in turn, is regulated by a feedback system that involves such factors as stress, insulin, ADH, and other hormones.

The adenohypophysis produces two gonadotropins (hormones that stimulate the gonads): luteinizing hormone and follicle-stimulating hormone. Luteinizing hormone (LH) receives its name from the corpus luteum, a temporary endocrine tissue in the ovaries that secretes the female sex hormones estrogen and progesterone. In the female, an increase of LH in the blood stimulates ovulation, the release of a mature egg(s) from an ovary. In the male, the target cells of LH are cells in the testes that secrete the male hormone testosterone. In the female, follicle-stimulating

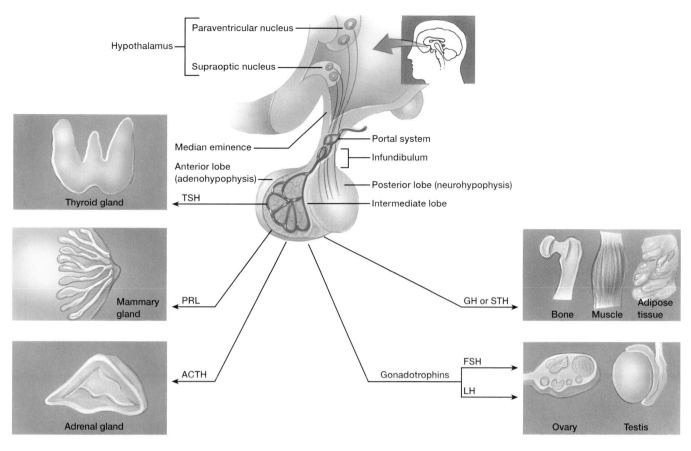

FIGURE 25.14

Functional Links between the Pituitary Gland and the Hypothalamus.   Target areas for each hormone are shown in the corresponding box. The blood vessels that make up the hypothalamic-hypophyseal portal system provide the functional link between the hypothalamus and the adenohypophysis, and the axons of the hypothalamic neurosecretory cells provide the link between the hypothalamus and the neurohypophysis. (TSH = thyroid-stimulating hormone; PRL = prolactin; ACTH = adrenocorticotropic hormone; GH = growth hormone; STH = somatotropin; FSH = follicle-stimulating hormone; LH = luteinizing hormone.)

hormone (FSH) stimulates the follicular cells in the ovaries to develop into mature eggs and to produce estrogen. In the male, FSH stimulates the cells of the testes to produce sperm.

The pineal gland (or pineal body) is so named because it is shaped like a pine cone. Its distinctive cells evolved from the photoreceptors of lower vertebrates; they synthesize melatonin and are most active in the dark. Light inhibits the enzymes needed for melatonin synthesis. Because of its cyclical production, melatonin can affect many physiological processes and adjust them to diurnal and seasonal cycles. *The use of melatonin by mammals is an evolutionary adaptation to help ensure that periodic activities of mammals occur at a time of the year when environmental conditions are optimal for those activities.* In humans, decreased melatonin secretion may help trigger the onset of puberty, the age at which reproductive structures start to mature.

Thyroid Gland

The **thyroid gland** is in the neck, anterior to the trachea (*see figure 25.12*). Two of its secretions are thyroxine and triiodothyronine, both of which influence overall growth, development, and

metabolic rates. Another thyroid hormone, calcitonin, helps control extracellular levels of calcium ions (Ca^{2+}) by promoting the deposition of these ions into bone tissue when their concentrations rise. Once calcium returns to its homeostatic concentration, thyroid cells decrease their secretion of calcitonin.

Parathyroid Glands

The **parathyroid glands** are tiny, pea-sized glands embedded in the thyroid lobes, usually two glands in each lobe (*see figure 25.12*). The parathyroids secrete parathormone (PTH), which regulates the concentrations of calcium (Ca^{2+}) and phosphate (HPO_2^{-4}) ions in the blood.

When the calcium concentration in the blood bathing the parathyroid glands is low, PTH secretion increases and has the following effects: It stimulates bone cells to break down bone tissue and release calcium ions into the blood. It also enhances calcium absorption from the small intestine into the blood. Finally, PTH promotes calcium reabsorption by the kidney tubules to decrease the amount of calcium excreted in the urine. Figure 25.15 shows the negative feedback system for parathormone.

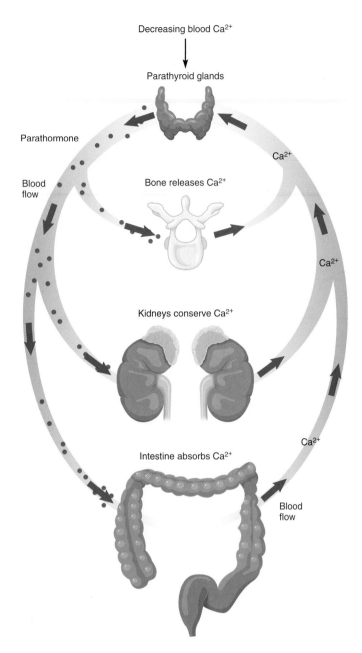

FIGURE 25.15

Hormonal Feedback.   The negative feedback mechanism of the parathyroid glands (parathormone). Parathormone stimulates bones to release calcium and the kidneys to conserve calcium. It indirectly stimulates the intestine to absorb calcium. The result increases blood calcium, which then inhibits parathormone secretion.

Adrenal Glands

In mammals, two adrenal glands rest on top of the kidneys. Each gland consists of two separate glandular tissues. The inner portion is the medulla, and the outer portion, which surrounds the medulla, is the cortex (figure 25.16).

Adrenal Cortex   The adrenal cortex secretes three classes of steroid hormones: glucocorticoids (cortisol), mineralocorticoids (aldosterone), and sex hormones (androgens, estrogens). The

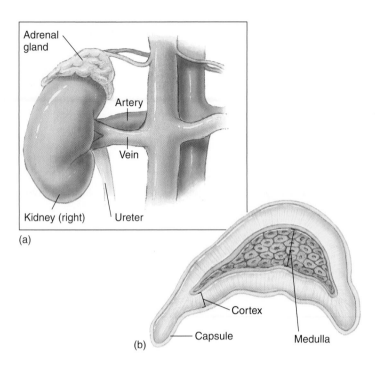

FIGURE 25.16

Adrenal Gland of a Mammal.   (*a*) An adrenal gland sits on top of each kidney. (*b*) Each gland contains two structurally, functionally, and developmentally distinct regions. The outer cortex is endocrine and produces glucocorticoids (cortisol), mineralocorticoids (aldosterone), and androgens (sex hormones). The inner medulla is nervous tissue that produces epinephrine (adrenaline) and norepinephrine (noradrenaline).

glucocorticoids, such as **cortisol,** help regulate overall metabolism and the concentration of blood sugar. They also function in defense responses to infection or tissue injury. Aldosterone helps maintain concentrations of solutes (such as sodium) in the extracellular fluid when either food intake or metabolic activity changes the amount of solutes entering the bloodstream. Aldosterone also promotes sodium reabsorption in the kidneys and, thus, water reabsorption; hence, it plays a major role in maintaining the homeostasis of extracellular fluid. Normally, the sex hormones that the adrenal cortex secretes have only a slight effect on male and female gonads. These sex hormones consist mainly of weak male hormones called androgens and lesser amounts of female hormones called estrogens.

Adrenal Medulla   The adrenal medulla is under neural control. It contains neurosecretory cells that secrete epinephrine (adrenaline) and norepinephrine (noradrenaline), both of which help control heart rate and carbohydrate metabolism. Brain centers and the hypothalamus govern the secretions via sympathetic nerves.

During times of excitement, emergency, or stress, the adrenal medulla contributes to the overall mobilization of the body through the sympathetic nervous system. In response to epinephrine and norepinephrine, the heart rate increases, blood flow increases to many vital organs, the airways in the lungs dilate, and more oxygen is delivered to all cells of the body.

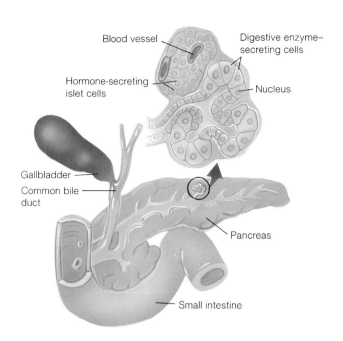

FIGURE 25.17

Pancreas. The hormone-secreting cells of the pancreas are arranged in clusters or islets closely associated with blood vessels. Other pancreatic cells secrete digestive enzymes into ducts.

This group of events is sometimes called the "fight-or-flight" response and permits the body to react strongly and quickly to emergencies.

Pancreas

The **pancreas** is an elongated, fleshy organ posterior to the stomach (figure 25.17). It functions both as an exocrine (with ducts) gland to secrete digestive enzymes and as an endocrine (ductless) gland. The endocrine portion of the pancreas makes up only about 1% of the gland. This portion synthesizes, stores, and secretes hormones from clusters of cells called pancreatic islets.

The pancreas contains 200,000 to 2,000,000 **pancreatic islets** scattered throughout the gland. Each islet contains four special groups of cells, called alpha (α), beta (β), delta (δ), and F cells. The alpha cells produce the hormone glucagon, and beta cells produce insulin. The delta cells secrete somatostatin, the hypothalamic growth-hormone-inhibiting factor that also inhibits glucagon and insulin secretion. F cells secrete a pancreatic polypeptide that is released into the bloodstream after a meal and inhibits somatostatin secretion, gallbladder contraction, and the secretion of pancreatic digestive enzymes.

When glucose concentrations in the blood are high, such as after a meal, beta cells secrete insulin. Insulin promotes the uptake of glucose by the body's cells, including liver cells, where excess glucose can be converted to glycogen (a storage polysaccharide). Insulin and glucagon are crucial to the regulation of blood glucose concentrations. When the blood glucose concentration is low, alpha cells secrete glucagon. Glucagon stimulates

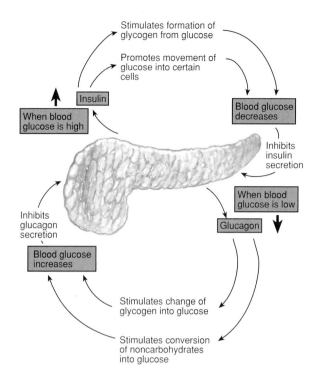

FIGURE 25.18

Two Pancreatic Hormones (Insulin and Glucagon) Regulate the Concentration of Blood Glucose. The negative feedback mechanism for regulating glucagon and insulin secretion helps maintain a relatively stable blood glucose concentration.

the breakdown of glycogen into glucose units, which are released into the bloodstream to raise the blood glucose concentration to the homeostatic level. Figure 25.18 illustrates the negative feedback system that regulates the secretion of glucogon and insulin and the maintenance of appropriate blood glucose concentrations.

Gonads

The **gonads** (ovaries and testes) secrete hormones that help regulate reproductive functions. In the male, the testes secrete testosterone, which acts with luteinizing and follicle-stimulating hormones that the adenohypophysis produces to stimulate spermatogenesis. Testosterone is also necessary for the growth and maintenance of the male sex organs, promotes the development and maintenance of sexual behavior, and in humans, stimulates the growth of facial and pubic hair, as well as the enlargement of the larynx, which deepens the voice. The testes also produce inhibin, which inhibits the secretion of FSH.

Four major classes of ovarian hormones help to regulate female reproductive functions. Estrogens (estrin, estrone, and estradiol) help regulate the menstrual and estrus cycles and the development of the mammary glands and other female secondary sexual characteristics. The progestins (primarily progesterone) also regulate the menstrual and estrus cycles and the development of the mammary glands, and they aid in placenta formation during pregnancy. Relaxin, which is produced in small

TABLE 25.2
SOME OTHER MAJOR SOURCES OF VERTEBRATE HORMONES

GLAND/ORGAN	HORMONE	FUNCTION	TARGET AREA
Placenta	Estrogens, progesterone, human chorionic gonadotropin (hCG), human chorionic somatomammotropin (hCS)	Maintain pregnancy	Ovaries, mammary glands, uterus
Digestive tract	Secretin	Stimulates release of pancreatic juice to neutralize stomach acid	Cells of pancreas
	Gastrin	Stimulates digestive enzymes and HCl in stomach	Stomach mucosa
	Cholecystokinin (CCK)	Stimulates release of pancreatic enzymes and bile from gallbladder	Pancreas, gallbladder
Heart	Atriopeptin	Lowers blood pressure, maintains fluid balance	Blood vessels, kidneys
Kidneys	Erythropoietin	Stimulates red blood cell production	Bone marrow
	Urotensin	Stimulates constriction of arteries	Major arteries
	Calcitrol	Aids in the absorption of dietary calcium and phosphorus	Small intestine
Adipose tissue	Leptin	Suppresses appetite	Brain
Liver	Insulin-like growth factors	Cell division and growth	Various cells
Thymus	Thymopoietin	T lymphocyte function	T cells

quantities, softens the opening of the uterus (cervix) at the time of delivery. The ovaries also produce inhibin, which inhibits the secretion of FSH.

Thymus

The **thymus gland** is near the heart (*see figure 25.12*). It is large and conspicuous in young birds and mammals, but diminishes in size throughout adulthood. The major hormonal product of the thymus is a family of peptide hormones, including thymopoietin (TP) and alpha$_1$ and beta$_4$ thymosin, that appear to be essential for the normal development of the immune system.

Other Sources of Hormones

In addition to the major endocrine glands, other glands and organs carry on hormonal activity. Table 25.2 summarizes some of these.

SUMMARY

1. For metabolic activity to proceed smoothly in an animal, the chemical environment of each cell must be maintained within fairly narrow limits (homeostasis). This is accomplished using negative feedback systems that involve integrating, communicating, and coordinating molecules called messengers.

2. Specialized cells secrete chemical messenger molecules. These chemical messengers can be categorized as local chemical messengers (lumones), neurotransmitters (e.g., acetylcholine), neuropeptides, hormones, and pheromones (e.g., sex attractants).

3. A hormone is a specialized chemical messenger that an endocrine gland or tissue produces and secretes. Hormones are usually steroids, amines, proteins, or fatty acid derivatives. Negative feedback systems often regulate hormone secretion.

4. Hormones modify the biochemical activity of a target cell or tissue (so called because it has receptors to which hormone molecules can bind). Mechanisms of hormone action are the fixed-membrane-receptor mechanism (water-soluble hormones) or the mobile-receptor mechanism (steroid hormones).

5. Most of the chemicals functioning as hormones in invertebrate animals are neurosecretions called neuropeptides. Only a few of the

more advanced invertebrates (e.g., molluscs, arthropods, echinoderms) have nonneurosecretory hormones.

6. In all vertebrates, a neuroendocrine control center coordinates communication and integrative activities for the entire body. This center consists of the hypothalamus and pituitary gland.

7. The vertebrate endocrine system consists of several major glands, the hypothalamus, pituitary gland, pineal gland, thyroid gland, parathyroid glands, adrenal glands, pancreas, gonads, and thymus. In addition to these major glands, other glands and organs, including the placenta, digestive tract, heart, and kidneys, carry on hormonal activity.

SELECTED KEY TERMS

endocrinology (p. 408)
hormones (p. 407)
local chemical messengers (p. 407)
neuroendocrine system (p. 407)

neuropeptides (p. 407)
neurotransmitters (p. 407)
pheromones (p. 407)
target cell (p. 408)

CRITICAL THINKING QUESTIONS

1. How do hormones encode information? How do cells "know what to do" in response to hormonal information?

2. Summarize your knowledge of how endocrine systems work by describing the "life" of a hormone molecule from the time it is secreted until it is degraded or used up.

3. All cells secrete or excrete molecules, and all cells respond to certain biochemical factors in their external environments. Could the origin of endocrine control systems lie in such ordinary cellular events? How might the earliest multicellular organisms have evolved some sort of endocrine coordination?

4. Mental states strongly affect the function of many endocrine glands. This mind-body link occurs through the hypothalamus. Can you describe how thoughts are transformed into physiological responses in the hypothalamus?

5. Compared to enzymes and genes, hormones are remarkably small molecules. Would larger molecules be able to carry more information? Explain.

ONLINE LEARNING CENTER

Visit our Online Learning Center (OLC) at www.mhhe.com/zoology (click on this book's title) to find the following chapter-related materials:

- CHAPTER QUIZZING
- ANIMATIONS
  Endocrine System Regulation
  Peptide Hormone Action
  Parathyroid Hormone
  Glucose Regulation
- RELATED WEB LINKS
  The Endocrine System
    Vertebrates: Macroscopic Anatomy of the Endocrine System
    Vertebrates: Microscopic Anatomy of the Endocrine System
    Human Endocrine Topics
- BOXED READINGS ON
  Diabetes
- SUGGESTED READINGS

CHAPTER 26

CIRCULATION AND GAS EXCHANGE

Outline

Concepts

1. Animal transport and circulatory systems move substances from one part of the body to another, and between the animal's external environment and extracellular fluid.
2. Some invertebrates have specific transport systems, such as gastrovascular cavities. The circulatory system of more complex animals consists of a central pumping heart, blood vessels, blood, and an ancillary lymphatic system.
3. Some invertebrates depend solely on gases, nutrients, and wastes diffusing between body surfaces and individual cells. Others have either open or closed circulatory systems for transporting gases, wastes, and nutrients.
4. Animals have five main types of respiratory systems: simple diffusion across plasma membranes, tracheae, cutaneous (integument or body surface) exchange, gills, and lungs.
5. Gas diffuses between the environment and cells of an animal's body from areas of higher concentration to areas of lower concentration. In large and active animals, respiratory pigments and ventilation—the active movement of air into and out of a respiratory system—increase gas exchange.

INTERNAL TRANSPORT AND CIRCULATORY SYSTEMS

All animals must maintain a homeostatic balance in their bodies. This need requires that nutrients, metabolic wastes, and respiratory gases be circulated through the animal's body. Any system of moving fluids that reduces the functional diffusion distance that nutrients, wastes, and gases must traverse is an internal transport or circulatory system. The nature of the system directly relates to the size, complexity, and lifestyle of the animal in question. The first part of this chapter discusses some of these transport and circulatory systems.

TRANSPORT SYSTEMS IN INVERTEBRATES

Because protozoa are small, with high surface-area-to-volume ratios (*see figure 2.3*), all they need for gas, nutrient, and waste exchange is simple diffusion. In protozoa, the plasma membrane and cytoplasm are the media through which materials diffuse to various parts of the organism, or between the organism and the environment (*see figure 26.11a*).

    Some invertebrates have evolved specific transport systems. For example, sponges circulate water from the external environment through their bodies, instead of circulating an internal fluid (figure 26.1a). Cnidarians, such as *Hydra*, have a fluid-filled internal

This chapter contains evolutionary concepts, which are set off in this font.

gastrovascular cavity (figure 26.1*b*). This cavity supplies nutrients for all body cells lining the cavity, provides oxygen from the water in the cavity, and is a reservoir for carbon dioxide and other wastes. Simple body movement moves the fluid.

The gastrovascular cavity of flatworms, such as the planarian *Dugesia*, is more complex than that of *Hydra*. In the planarian, branches penetrate to all parts of the body (figure 26.1*c*). Because this branched gastrovascular cavity runs close to all body cells, diffusion distances for nutrients, gases, and wastes are short. Body movement helps distribute materials to various parts of the body. One disadvantage of this system is that it limits these animals to relatively small sizes or to shapes that maintain small diffusion distances.

Pseudocoelomate invertebrates, such as rotifers, gastrotrichs, and nematodes, use the coelomic fluid of their body cavity for transport (figure 26.1*d*). Most of these animals are small, and movements of the body against the coelomic fluids, which are in direct contact with the internal tissues and organs, produce adequate transport. A few other invertebrates (e.g., ectoprocts, sipunculans, echinoderms) also depend largely on the body cavity as a coelomic transport chamber.

Beginning with the molluscs, transport functions occur with a separate circulatory system. A **circulatory** or **cardiovascular system** (Gr. *kardia*, heart + L. *vascular*, vessel) is a specialized system in which a muscular, pumping heart moves the fluid medium called either hemolymph or blood in a specific direction determined by the presence of unidirectional blood vessels.

The animal kingdom has two basic types of circulatory systems: open and closed. In an **open circulatory system,** the heart pumps hemolymph out into the body cavity or at least through parts of the cavity, where the hemolymph bathes the cells, tissues, and organs. In a **closed circulatory system,** blood circulates in the confines of tubular vessels. The coelomic fluid of some invertebrates also has a circulatory role either in concert with, or instead of, the hemolymph or blood.

The annelids, such as the earthworm, have a closed circulatory system in which blood travels through vessels delivering nutrients to cells and removing wastes (figure 26.1*e*).

Most molluscs and arthropods have open circulatory systems in which hemolymph directly bathes the cells and tissues rather than being carried only in vessels (figure 26.1*f*). For example, an insect's heart pumps hemolymph through vessels that open into a body cavity (hemocoel).

CHARACTERISTICS OF INVERTEBRATE COELOMIC FLUID, HEMOLYMPH, AND BLOOD CELLS

As previously noted, some animals (e.g., echinoderms, annelids, sipunculans) use coelomic fluid as a supplementary or sole circulatory system. Coelomic fluid may be identical in composition to interstitial fluids or may differ, particularly with respect to specific proteins and cells. Coelomic fluid transports gases, nutrients, and waste products. It also may function in certain invertebrates (annelids) as a hydrostatic skeleton (*see figure 23.10*).

Hemolymph (Gr. *haima*, blood + *lympha*, water) is the circulating fluid of animals with an open circulatory system. Most arthropods, ascidians, and many molluscs have hemolymph. In these animals, a heart pumps hemolymph at low pressures through vessels to tissue spaces (hemocoel) and sinuses. Generally, the hemolymph volume is high and the circulation slow. In the process of movement, essential gases, nutrients, and wastes are transported.

Many times, hemolymph has noncirculatory functions. For example, in insects, hemolymph pressure assists in molting of the old cuticle and in inflation of the wings. In certain jumping spiders, hydrostatic pressure of the hemolymph provides a hydraulic mechanism for limb extension.

The coelomic fluid, hemolymph, or blood of most animals contains circulating cells called blood cells or **hemocytes.** Some cells contain a respiratory pigment, such as hemoglobin, and are called erythrocytes or red blood cells. These cells are usually present in high numbers to facilitate oxygen transport. Cells that do not contain respiratory pigments have other functions, such as blood clotting.

The number and types of blood cells vary dramatically in different invertebrates. For example, annelid blood contains hemocytes that are phagocytic. The coelomic fluid contains a variety of coelomocytes (amoebocytes, eleocytes, lampocytes, linocytes) that function in phagocytosis, glycogen storage, encapsulation, defense responses, and excretion. The hemolymph of molluscs has two general types of hemocytes (amoebocytes and granulocytes) that have most of the aforementioned functions as well as nacrezation (pearl formation) in some bivalves. Insect hemolymph contains large numbers of various hemocyte types that function in phagocytosis, encapsulation, and clotting (figure 26.2).

TRANSPORT SYSTEMS IN VERTEBRATES

All vertebrates have a closed circulatory system in which the walls of the heart and blood vessels are continuously contracted, and blood never leaves the blood vessels (figure 26.1*g*). Blood moves from the heart, through arteries, arterioles, capillaries, venules, veins, and back to the heart. Exchange between the blood and extracellular fluid only occurs at the capillary level.

CHARACTERISTICS OF VERTEBRATE BLOOD AND BLOOD CELLS

Overall, vertebrate blood transports oxygen, carbon dioxide, and nutrients; defends against harmful microorganisms, cells, and viruses; prevents blood loss through coagulation (clotting); and helps regulate body temperature and pH. Because it is a liquid, vertebrate blood is classified as a specialized type of connective tissue. Like other connective tissues, blood contains a fluid matrix called plasma and cellular elements called formed elements.

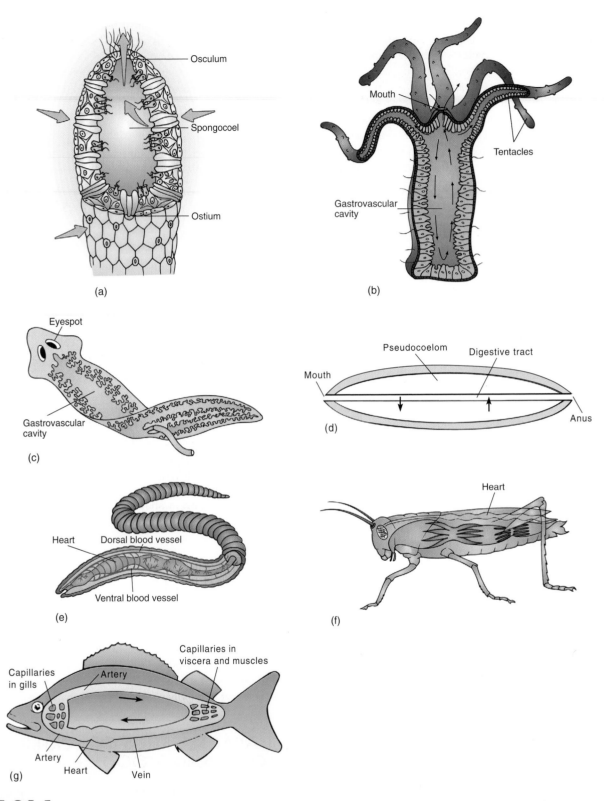

FIGURE 26.1

Some Transport and Circulatory Systems.   (*a*) Sponges use water from the environment as a circulatory fluid by passing it through their bodies (blue arrows). (*b*) Cnidarians, such as this *Hydra,* also use water from the environment and circulate it (black arrows) through the gastrovascular cavity. Cells lining the cavity exchange gases and nutrients from the water and release waste into it. (*c*) The planarian's gastrovascular cavity is branched, allowing for more effective distribution of materials. (*d*) Pseudocoelomates use their body cavity fluid for internal transport from and to the digestive tract as the black arrows indicate. (*e*) The circulatory system of an earthworm contains blood that is kept separate from the coelomic fluid. This is an example of a closed circulatory system. (*f*) The dorsal heart of an arthropod, such as this grasshopper, pumps blood through an open circulatory system. In this example, blood and body cavity (hemocoelic) fluid are one and the same. (*g*) Octopuses, other cephalopod molluscs, annelids, and vertebrates, such as this fish, have closed circulatory systems. In a closed system, the walls of the heart and blood vessels are continuously connected, and blood never leaves the vessels. Black arrows indicate the direction of blood flow.

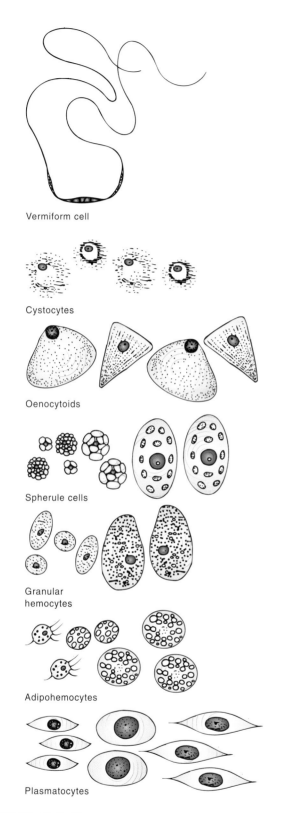

Vermiform cell

Cystocytes

Oenocytoids

Spherule cells

Granular
hemocytes

Adipohemocytes

Plasmatocytes

FIGURE 26.2

Examples of Invertebrate Hemocytes. These hemocytes are representative of those found in an insect. The different cells function in phagocytosis, agglutination, nutrient storage, wound repair, formation of connective tissue cells, and lipid transport.

Plasma

Plasma (Gr., anything formed or molded) is the straw-colored, liquid part of blood. In mammals, plasma is about 90% water and provides the solvent for dissolving and transporting nutrients. A group of proteins (albumin, fibrinogen, and globulins) comprises another 7% of the plasma. The concentration of these plasma proteins influences the distribution of water between the blood and extracellular fluid. Because albumin represents about 60% of the total plasma proteins, it plays important roles with respect to water movement. Fibrinogen is necessary for blood coagulation (clotting), and the globulins include the immunoglobulins and various metal-binding proteins. **Serum** is plasma from which the proteins involved in blood clotting have been removed. The gamma globulin portion functions in the immune response because it consists mostly of antibodies. The remaining 3% of plasma is composed of electrolytes, amino acids, glucose and other nutrients, various enzymes, hormones, metabolic wastes, and traces of many inorganic and organic molecules.

Formed Elements

The **formed-element fraction** (cellular component) of vertebrate blood consists of erythrocytes (red blood cells; RBCs), leukocytes (white blood cells; WBCs), and platelets (thrombocytes) (figure 26.3). White blood cells are present in lower number than are red blood cells, generally being 1 to 2% of the blood by volume. White blood cells are divided into agranulocytes (without granules in the cytoplasm) and granulocytes (have granules in the cytoplasm). The two types of agranulocytes are lymphocytes and monocytes. The three types of granulocytes are eosinophils, basophils, and neutrophils. Fragmented cells are called platelets (thrombocytes). Each of these cell types is now discussed in more detail.

Red Blood Cells Red blood cells (erythrocytes; Gr. *erythros*, red + cells) vary dramatically in size, shape, and number in the different vertebrates (figures 26.4 and 26.5*a*). For example, the RBCs of most vertebrates are nucleated, but mammalian RBCs are enucleated (without a nucleus). Some fishes and amphibians also have enucleated RBCs. Among all vertebrates, the salamander *Amphiuma* has the largest RBC (figure 26.4*a*). Avian RBCs (figure 26.4*c*) are oval-shaped, nucleated, and larger than mammalian RBCs. Among birds, the ostrich has the largest RBC. Most mammalian RBCs are biconcave disks (figure 26.5*a*); however, the camel (figure 26.4*e*) and llama have elliptical RBCs. The shape of a biconcave disk provides a larger surface area for gas diffusion than a flat disk or sphere. Generally, the lower vertebrates tend to have fewer but larger RBCs than the higher invertebrates.

Almost the entire mass of a RBC consists of **hemoglobin** (Gr. *haima*, blood + L. *globulus*, little globe), an iron-containing protein. The major function of an erythrocyte is to pick up oxygen from the environment, bind it to hemoglobin to form **oxyhemoglobin,** and transport it to body tissues. Blood rich in oxyhemoglobin is bright red. As oxygen diffuses into the tissues, blood becomes darker and appears blue when observed through the

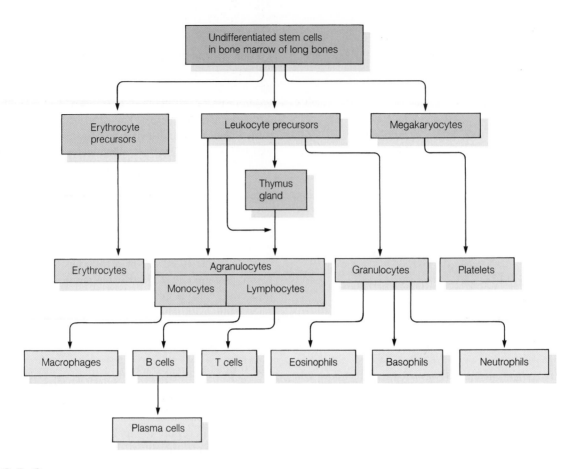

FIGURE 26.3

Cellular Components of Vertebrate Blood. Hematopoiesis is the process of blood cell production. Notice that all blood cells initially begin their lives in the bone marrow of long bones within a vertebrate's body.

blood vessel walls. However, when this less oxygenated blood is exposed to oxygen (such as when a vein is cut and a mammal begins to bleed), it instantaneously turns bright red. Hemoglobin also carries waste carbon dioxide (in the form of **carbaminohemoglobin**) from the tissues to the lungs (or gills) for removal from the body.

White Blood Cells
White blood cells (leukocytes) (Gr. *leukos*, white + cells) are scavengers that destroy microorganisms at infection sites, remove foreign chemicals, and remove debris that results from dead or injured cells. All WBCs are derived from immature cells (called stem cells) in bone marrow by a process called **hematopoiesis** (Gr. *hemato*, blood + *poiein*, to make; *see figure 26.3*).

Among the granulocytes, **eosinophils** are phagocytic and ingest foreign proteins and immune complexes rather than bacteria (figure 26.5b). In mammals, eosinophils also release chemicals that counteract the effects of certain inflammatory chemicals released during allergic reactions. **Basophils** are the least numerous WBC (figure 26.5c). When they react with a foreign substance, their granules release histamine and heparin. Histamine causes blood vessels to dilate and leak fluid at a site of inflammation, and heparin prevents blood clotting. **Neutrophils** are the most numerous of the white blood cells (figure 26.5d). They are chemically attracted to sites of inflammation and are active phagocytes.

The two types of agranulocytes are the **monocytes** and **lymphocytes** (figure 26.5e, f). Two distinct types of lymphocytes are B cells and T cells, both of which are central to the immune response. **B cells** originate in the bone marrow and colonize the lymphoid tissue, where they mature. In contrast, **T cells** are associated with and influenced by the thymus gland before they colonize lymphoid tissue and play their role in the immune response. When B cells are activated, they divide and differentiate to produce **plasma cells.**

Platelets (Thrombocytes)
Platelets (so named because of their platelike flatness), or **thrombocytes** (Gr. *thrombus*, clot + cells), are disk-shaped cell fragments that initiate blood clotting. When a blood vessel is injured, platelets immediately move to the site and clump, attaching themselves to the damaged area, and thereby beginning the process of blood coagulation.

VERTEBRATE BLOOD VESSELS

Arteries are elastic blood vessels that carry blood away from the heart to the organs and tissues of the body. The central canal of an artery (and of all blood vessels) is a lumen. Surrounding the lumen of an artery is a thick wall composed of three layers, or tunicae (L. *tunica*, covering) (figure 26.6a).

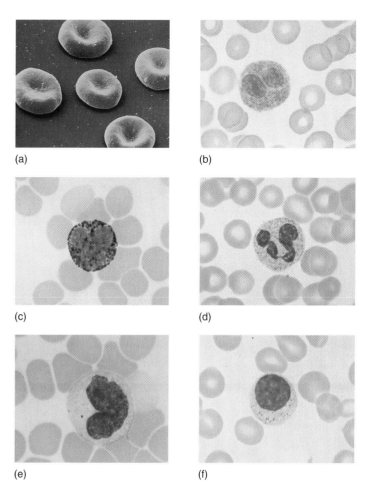

(a)

(b)

(c)

(d)

(e)

(f)

FIGURE 26.5

Blood Cells. (*a*) The biconcave shape of human erythrocytes (SEM ×1,500). (*b*) Red-staining cytoplasmic granules characterize an eosinophil. (*c*) Blue-staining granules characterize a basophil. (*d*) Light pink granules and a multilobed nucleus characterize a neutrophil. Cells in *b–d* are also known as granulocytes. The agranulocytes consist of large monocytes (*e*) and lymphocytes (*f*). (*b–f* are LM ×710.)

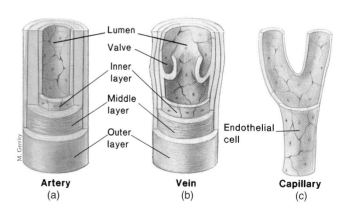

Lumen

Valve

Inner
layer

Middle
layer

Outer
layer

Endothelial
cell

M. Gerrity

Artery
(a)

Vein
(b)

Capillary
(c)

FIGURE 26.6

Structure of Blood Vessels. (*a, b*) The walls of arteries and veins have three layers (tunicae). The outermost layer consists of connective tissue, the middle layer has elastic and smooth muscle tissue, and the inner layer consists of a single layer of smooth endothelial cells (endothelium). Notice that the wall of an artery is much thicker than the wall of a vein. The middle layer is greatly reduced in a vein. (*c*) A capillary consists of a single layer of endothelial cells.

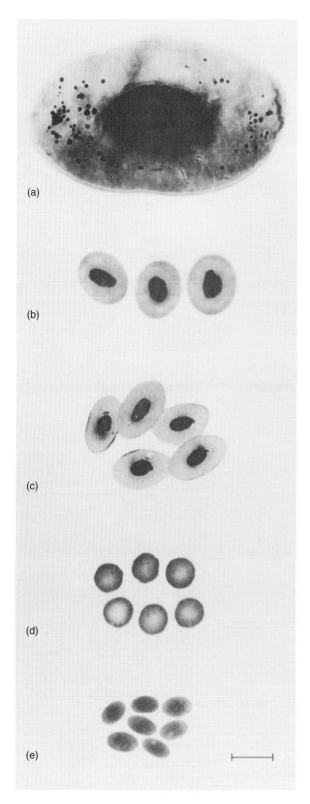

(a)

(b)

(c)

(d)

(e)

FIGURE 26.4

Comparison of Red Blood Cells from a Variety of Vertebrates.
Light micrographs of (*a*) a nucleated cell from a salamander; (*b*) nucleated cells from a snake; (*c*) nucleated cells from an ostrich; (*d*) enucleated cells (biconcave disk) from a red kangaroo; and (*e*) enucleated cells (ellipsoid) from a camel (bar = 10 μm).

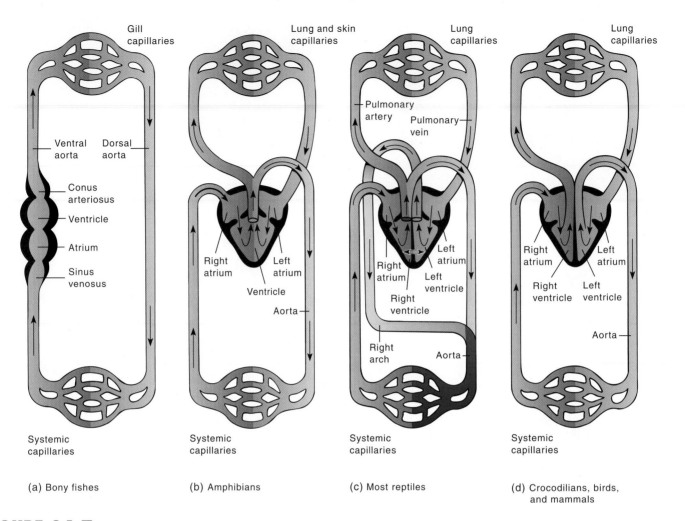

Gill capillaries

Lung and skin capillaries

Lung capillaries

Lung capillaries

Ventral aorta   Dorsal aorta

Conus arteriosus

Ventricle

Atrium

Sinus venosus

Pulmonary artery   Pulmonary vein

Right atrium   Left atrium

Ventricle

Aorta

Right atrium   Left atrium

Right atrium   Left ventricle

Right ventricle

Right arch   Aorta

Right atrium   Left atrium

Right ventricle   Left ventricle

Aorta

Systemic capillaries

Systemic capillaries

Systemic capillaries

Systemic capillaries

(a) Bony fishes

(b) Amphibians

(c) Most reptiles

(d) Crocodilians, birds, and mammals

FIGURE 26.7

Heart and Circulatory Systems of Various Vertebrates. Oxygenated blood is red; less oxygenated blood is blue; a mixture of oxygenated and less oxygenated blood is purple. (*a*) In bony fishes, the heart's two chambers (atrium, ventricle) pump in series. Respiratory and systemic circulations are not separate. (*b*) The amphibian heart has two atria and one ventricle. Blood from the lungs enters the left atrium, and blood from the body enters the right atrium. The blood from both atria empties into one ventricle, which then pumps it into the respiratory and systemic circulations. (*c*) Most reptiles exhibit a greater degree of anatomical division of the ventricle into two halves. (*d*) In crocodilians, birds, and mammals, the ventricle is completely divided, forming a four-chambered heart, with the blood flow through the lungs completely separated from the flow to other tissues. Black arrows indicate the direction of blood flow.

Most **veins** are relatively inelastic, large vessels that carry blood from the body tissues to the heart. The wall of a vein contains the same three layers (tunicae) as arterial walls, but the middle layer is much thinner, and one or more valves are present (figure 26.6*b*). The valves permit blood flow in only one direction, which is important in returning the blood to the heart.

Arteries lead to terminal **arterioles** (those closest to a capillary). The arterioles branch to form **capillaries** (L. *capillus*, hair), which connect to **venules** and then to veins. Capillaries are generally composed of a single layer of endothelial cells and are the most numerous blood vessels in an animal's body (figure 26.6*c*). An abundance of capillaries makes an enormous surface area available for the exchange of gases, fluids, nutrients, and wastes between the blood and nearby cells.

THE HEARTS AND CIRCULATORY SYSTEMS OF BONY FISHES, AMPHIBIANS, AND REPTILES

The heart and blood vessels changed greatly as vertebrates moved from water to land and as endothermy evolved. Examples of these trends are now presented.

The bony fish heart has two chambers—the atrium and ventricle (figure 26.7*a*). Blood leaves the heart via the ventral aorta, which goes to the gills. In the gills, blood becomes oxygenated, loses carbon dioxide, and enters the dorsal aorta. The dorsal aorta distributes blood to all of the body organs, and then blood returns to the heart via the venous system. Because blood only passes through the heart once, this system is called a single circulation circuit. This circuit has the advantage of circulating oxygenated blood from the gills to the systemic capillaries in all organs almost simultaneously. However, the circulation of blood through the gill capillaries offers resistance to flow. Blood pressure and rates of flow to other organs are thus appreciably reduced. This arrangement probably could not support the high metabolic rates present in most birds and mammals.

In amphibians and reptiles, the evolution of a double circulatory circuit, in which blood passes through the heart twice during its circuit through the body, has overcome the slow blood-flow problem. Amphibians and most reptiles (alligators and crocodiles have a four-chambered heart) have hearts that are not fully divided in two. In amphibians, a single ventricle pumps blood both to the lungs and to the rest of the body (figure 26.7*b*). However, because most amphibians absorb more oxygen through their skin than through their lungs or gills, blood returning from the skin also contributes oxygenated blood to the ventricle. The blood pumped out to the rest of the body is thus highly oxygenated.

In the heart of most reptiles, the ventricle is partially divided into a right and left side (figure 26.7*c*). Oxygenated blood from the lungs returns to the left side of the heart via the pulmonary vein and does not mix much with deoxygenated blood in the right side of the heart. When the ventricles contract, blood is pumped out two aortae for distribution throughout the body, as well as to the lungs. The incomplete separation of the ventricles is an important adaptation for reptiles, such as turtles, because it allows blood to be diverted away from the pulmonary circulation during diving and when the turtle is withdrawn into its shell. This conserves energy and diverts blood to vital organs during the time when the lungs cannot be ventilated.

THE HEARTS AND CIRCULATORY SYSTEMS OF BIRDS AND MAMMALS

Even though the physiological separation of blood in left and right ventricles is almost complete in reptiles, the complete anatomical separation of ventricles occurs only in crocodilians, birds, and mammals (figure 26.7d). This facilitates the double circulation required to maintain high blood pressure. High blood pressure is important in the rapid delivery of oxygenated nutrient-rich blood to tissues with high metabolic rates.

Blood circulates throughout the avian and mammalian body in two main circuits: the pulmonary and systemic circuits (figure 26.8). The **pulmonary circuit** supplies the blood only to the lungs. It carries oxygen-poor (deoxygenated) blood from the heart to the lungs, where carbon dioxide is removed and oxygen is added. It then returns the oxygen-rich (oxygenated) blood to the heart for distribution to the rest of the body. The **systemic circuit** supplies all the cells, tissues, and organs of the body with oxygen-rich blood and returns oxygen-poor blood to the heart.

THE HUMAN HEART

The human heart is a hard-working pump that moves blood through the body. It pumps its entire blood volume (about 5 liters) every minute; about 8,000 liters of blood move through 96,000 km

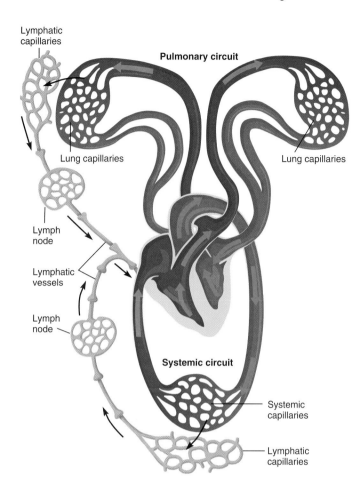

FIGURE 26.8

Circulatory Circuits. The cardiovascular system of a bird or mammal has two major capillary beds and transport circuits: the pulmonary circuit and systemic circuit. The lymphatic system consists of one-way vessels that are also involved in returning tissue fluid, called lymph, to the heart. The black arrows indicate the direction of lymph flow, and the colored (blue and pink) arrows indicate the direction of blood flow.

of blood vessels every day. The heart of an average adult beats about 70 times per minute—more than 100,000 times per day. In a 70-year lifetime, the heart beats more than 2.6 billion times without fatiguing.

Most of the human heart is composed of cardiac muscle tissue called myocardium (Gr. *myo*, muscle). The outer protective covering of the heart, however, is fibrous connective tissue called the epicardium (Gr. *epi*, upon). Connective tissue and endothelium form the inside of the heart, the endocardium (Gr. *endo*, inside). (Endothelium is a single layer of epithelial cells lining the chambers of the heart, as well as the lumen of blood vessels; *see figure 26.6.*)

The left and right halves of the heart are two separate pumps, each containing two chambers (figure 26.9). In each half, blood first flows into a thin-walled atrium (L. *antichamber*, waiting room) (pl., atria), then into a thick-walled ventricle. Valves are between the upper (atria) and lower (ventricles) chambers. The tricuspid valve is between the right atrium and right ventricle, and the bicuspid valve is between the left atrium and left

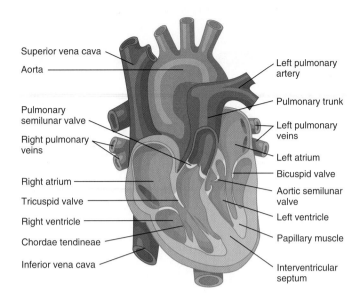

FIGURE 26.9

Structures of the Human Heart. Less oxygenated blood from the tissues of the body returns to the right atrium and flows through the tricuspid valve into the right ventricle. The right ventricle pumps the blood through the pulmonary semilunar valve into the pulmonary circuit, from which it returns to the left atrium and flows through the bicuspid valve into the left ventricle. The left ventricle then pumps blood through the aortic semilunar valve into the aorta. The various heart valves are shown in yellow.

ventricle. (Collectively, these are referred to as the AV valves—atrioventricular valves.) The pulmonary semilunar valve is at the exit of the right ventricle, and the aortic semilunar valve is at the exit of the left ventricle. (Collectively, these are referred to as the semilunar valves.) All of these valves open and close due to blood pressure changes when the heart contracts during each heartbeat. Like the valves in veins, heart valves keep blood moving in one direction, preventing backflow.

The heartbeat is a sequence of muscle contractions and relaxations called the cardiac cycle. A "pacemaker," a small mass of tissue called the sinoatrial node (SA node) at the entrance to the right atrium, initiates each heartbeat (figure 26.10). (Because the pacemaker is in the heart, nervous innervation is not necessary, which is why a heart transplant without connection to nerves is possible.) The SA node initiates the cardiac cycle by producing an action potential that spreads over both atria, causing them to contract simultaneously. The action potential then passes to the atrioventricular node (AV node), near the interatrial septum. From here, the action potential continues through the atrioventricular bundle (bundle of His), at the tip of the interventricular septum. The atrioventricular bundle divides into right and left branches, which are continuous with the Purkinje fibers in the ventricular walls. Stimulation of these fibers causes the ventricles to contract almost simultaneously and eject blood into the pulmonary and systemic circulations.

The action potential moving over the surface of the heart causes current flow, which can be recorded at the surface of the body as an electrocardiogram (ECG or EKG).

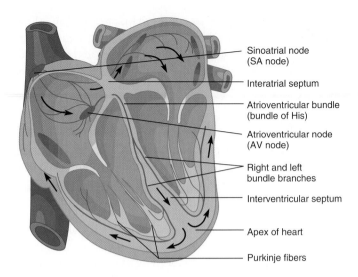

FIGURE 26.10

Electrical Conduction System of the Human Heart. The SA node initiates the depolarization wave, which passes successively through the atrial myocardium to the AV node, the atrioventricular bundle, the right and left bundle branches, and the Purkinje fibers in the ventricular myocardium. Black arrows indicate the direction of the electrical current flow.

During each cycle, the atria and ventricles go through a phase of contraction called **systole** and a phase of relaxation called **diastole.** Specifically, while the atria are relaxing and filling with blood, the ventricles are also relaxed. As more and more blood accumulates in the atria, blood pressure rises, and the atria contract, forcing the AV valves open and causing blood to rush into the ventricles. When the ventricles contract, the AV valves close, and the semilunar valves open, allowing blood to be pumped into the pulmonary arteries and aorta. After blood has been ejected from the ventricles, they relax and start the cycle anew.

BLOOD PRESSURE

Ventricular contraction generates the fluid pressure, called **blood pressure,** that forces blood through the pulmonary and systemic circuits. More specifically, blood pressure is the force the blood exerts against the inner walls of blood vessels. Although such a force occurs throughout the vascular system, the term blood pressure most commonly refers to systemic arterial blood pressure.

Arterial blood pressure rises and falls in a pattern corresponding to the phases of the cardiac cycle. When the ventricles contract (ventricular systole), their walls force the blood in them into the pulmonary arteries and the aorta. As a result, the pressure in these arteries rises sharply. The maximum pressure achieved during ventricular contraction is called the **systolic pressure.** When the ventricles relax (ventricular diastole), the arterial pressure drops, and the lowest pressure that remains in the arteries before the next ventricular contraction is called the **diastolic pressure.**

TABLE 26.1
MAJOR STRUCTURAL AND FUNCTIONAL COMPONENTS OF THE LYMPHATIC SYSTEM IN VERTEBRATES

STRUCTURE	FUNCTION
Lymphatic capillaries	Collect excess extracellular fluid in tissues
Lymphatics	Carry lymph from lymphatic capillaries to veins in the neck, where lymph returns to the bloodstream
Lymph nodes	House the WBCs that destroy foreign substances; play a role in antibody formation
Spleen	Filters foreign substances from blood; manufactures phagocytic lymphocytes; stores red blood cells; releases blood to the body when blood is lost
Thymus gland (in mammals)	Site of antibodies in the newborn; is involved in the initial development of the immune system; site of T cell differentiation
Bursa of Fabricius (in birds)	A lymphoid organ at the lower end of the alimentary canal in birds; the site of B cell maturation

In humans, normal systolic pressure for a young adult is about 120 mm Hg, which is the amount of pressure required to make a column of mercury (Hg) in a sphygmomanometer (sfig″mo-mah-nom′e-ter) rise 120 mm. Diastolic pressure is approximately 80 mm Hg. Conventionally, these readings are expressed as 120/80.

THE LYMPHATIC SYSTEM

The vertebrate **lymphatic system** begins with small vessels called lymphatic capillaries, which are in direct contact with the extracellular fluid surrounding tissues (*see figure 26.8*). The system has four major functions: (1) to collect and drain most of the fluid that seeps from the bloodstream and accumulates in the extracellular fluid; (2) to return small amounts of proteins that have left the cells; (3) to transport lipids that have been absorbed from the small intestine; and (4) to transport foreign particles and cellular debris to disposal centers called lymph nodes. The small lymphatic capillaries merge to form larger lymphatic vessels called lymphatics. Lymphatics are thin-walled vessels with valves that ensure the one-way flow of lymph. **Lymph** (L. *lympha*, clear water) is the extracellular fluid that accumulates in the lymph vessels. These vessels pass through the lymph nodes on their way back to the heart. Lymph nodes concentrate in several areas of the body and play an important role in the body's defense against disease.

In addition to the previously mentioned parts, the lymphatic system of birds and mammals consists of lymphoid organs—the spleen and either the bursa of Fabricius in birds or the thymus gland, tonsils, and adenoids in mammals. Table 26.1 summarizes the major components of the lymphatic system. The lymphatic system is also vital to an animal's defense against injury and attack.

GAS EXCHANGE

To take advantage of the rich source of energy that earth's organic matter represents, animals must solve two practical problems. First, they must break down and digest the organic matter so that it can enter the cells that are to metabolize it (chapter 27 describes this digestive process). Second, they must provide cells with both an adequate supply of oxygen required for aerobic respiration and a way of eliminating the carbon dioxide that aerobic respiration produces. This process of gas exchange with the environment, also called external respiration, is the subject of the rest of this chapter.

RESPIRATORY SURFACES

Protists and animals have five main types of respiratory systems (surfaces): (1) simple diffusion across plasma membranes, (2) tracheae, (3) cutaneous (integument or body surface) exchange, (4) gills, and (5) lungs. Each of these surfaces is now discussed.

INVERTEBRATE RESPIRATORY SYSTEMS

In single-celled protists, such as protozoa, **diffusion** across the plasma membrane moves gases into and out of the organism (figure 26.11*a*). Some multicellular invertebrates either have very flat bodies (e.g., flatworms) in which all body cells are relatively close to the body surface or are thin-walled and hollow (e.g., *Hydra*) (figure 26.11*b*, *c*). Again, gases diffuse into and out of the animal.

Invertebrates such as earthworms that live in moist environments use **integumentary exchange.** Earthworms have capillary networks just under their integument, and they exchange gases with the air spaces among soil particles (*see figure 13.14*).

Most aquatic invertebrates carry out gas exchange with **gills.** The simplest gills are small, scattered projections of the skin, such as the gills of sea stars. Other aquatic invertebrates have their gas exchange structures in more restricted areas. For example, marine and annelid worms have prominent lateral projections called parapodia that are richly supplied with blood vessels and function as gills.

Crustaceans and molluscs have gills that are compact and protected with hard covering devices (*see figure 12.11*). Such gills divide into highly branched structures to maximize the area for gas exchange.

Some terrestrial invertebrates (e.g., insects, centipedes, and some mites, ticks, and spiders) have **tracheal systems** consisting of highly branched chitin-lined tubes called tracheae (figure 26.12*a*). Tracheae open to the outside of the body through spiracles, which

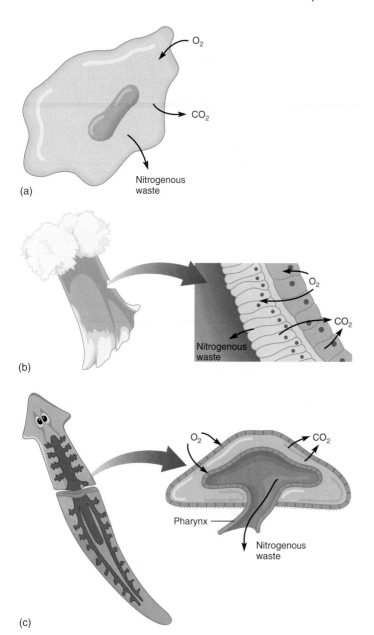

FIGURE 26.11

Invertebrate Respiration: Diffusion through Body Surfaces. The cells of small organisms, such as (a) protozoa, (b) cnidarians, and (c) flatworms, maintain close enough contact with the environment that they have no need for a respiratory system. Diffusion moves gases, as well as waste products, into and out of these organisms.

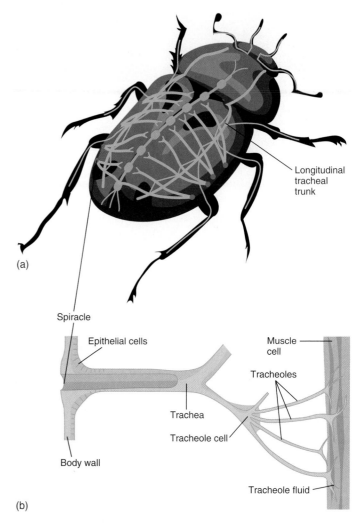

FIGURE 26.12

Invertebrate Respiration: A Tracheal System. (a) Tracheal system of an insect, showing the major tracheal trunks. (b) Tracheoles end at cells, and the terminal portions of tracheoles are fluid filled. The fluid acts as a solvent for gases.

usually have some kind of closure device to prevent excessive water loss. Spiracles lead to branching tracheal trunks that eventually give rise to smaller branches called tracheoles, whose blind ends lie close to all cells of the body. Since no cells are more than 2 or 3 μm from a tracheole, gases move between the tracheole and the tissues of the body by diffusion (figure 26.12b). Most insects have ventilating mechanisms that move air into and out of the trachea. For example, contracting flight muscles of insects alternately compress and expand the large tracheal trunks and thereby ventilate the tracheae.

Arachnids possess tracheae, book lungs, or both. **Book lungs** are paired invaginations of the ventral body wall that are folded into a series of leaflike lamellae (figure 26.13). Air enters the book lung through a slitlike opening called a spiracle and circulates between lamellae. Respiratory gases diffuse between the hemolymph moving along the lamellae and the air in the air chamber. Some ventilation also results from the contraction of a muscle attached to the dorsal side of the air chamber. This contraction dilates the chamber and opens the spiracle, but most gas movement is still by diffusion.

The only other major group of terrestrial invertebrates whose members have distinct air-breathing structures is the molluscan subclass Pulmonata—the land snails and slugs. The gas-exchange structure in these animals is a **pulmonate lung** that opens to the outside via a pore called a **pneumostome** (Gr. *pneumo*, breath + *stoma*, mouth) (figure 26.14). This lung is derived from a feature common to molluscs in general—the mantle cavity—which in other molluscs houses the gills and other

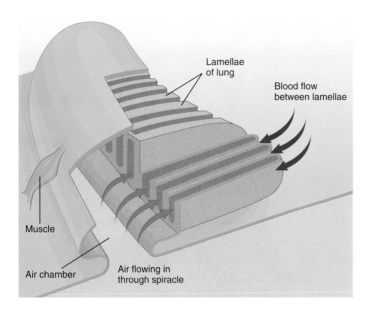

FIGURE 26.13

Invertebrate Respiration: A Book Lung. Structure of an arachnid (spider) book lung. Air enters through a spiracle into the air chamber by diffusion and by ventilation due to muscle contraction. Air diffuses from the air chamber into the lamellar spaces; hemolymph circulates through the blood lamellar spaces that alternate with air lamellar spaces. Small, peglike surface projections hold the lamellae apart. Due to this structural arrangement, air (blue arrows) and blood (purple arrows) move on opposite sides of a lamella in a countercurrent flow, allowing the exchange of respiratory gases by diffusion.

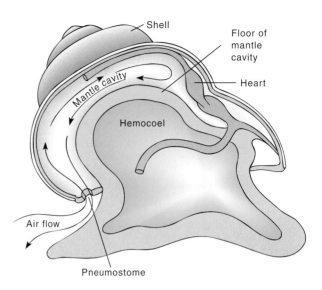

FIGURE 26.14

Invertebrate Respiration: The Pulmonate Lung. The mantle cavity of the pulmonate snail, *Lymnaea*, is highly vascularized and functions as a lung. Downward movement of the floor of the cavity increases the cavity's volume, so that air is drawn into the mantle cavity for respiration. Decreasing the volume of the mantle cavity expels the air. Air flows into and out of the lung through a single pore called the pneumostome. Black arrows indicate the direction of air flow.

organs. Some of the more primitive pulmonate snails are aquatic (freshwater) and close the pneumostome during submergence. When the snail surfaces to breathe air, the pneumostome opens. Most of the higher pulmonates are terrestrial and rely on their lungs for gas exchange. The lung may be ventilated by arching and then flattening the body, but most gas exchange occurs by diffusion through the pneumostome, which is open most of the time.

VERTEBRATE RESPIRATORY SYSTEMS

Aquatic vertebrates (fish, amphibians, and some reptiles) rely on one, or a combination of, the following surfaces for gas exchange: the cutaneous body surface, external filamentous gills, and internal lamellar gills. **Bimodal breathing** is the ability of an organism to exchange respiratory gases simultaneously with both air and water. A bimodal organism (e.g., some salamanders, crabs, barnacles, bivalve molluscs, and fishes [lungfishes]) uses gills for water breathing and lungs for air breathing. However, some gas exchange is always cutaneous, and some bimodal breathers are actually trimodal (skin, gills, and lungs). *B*imodal breathing was an important respiratory adaptation that made possible the evolutionary transition between aquatic and terrestrial habitats. Fundamental changes in the structure and function of the respiratory organs accompanied the transition from water to air breath-

ing. In air-breathing terrestrial vertebrates (reptiles, birds, and mammals) lungs replaced gills. These vertebrate surfaces and transitions are now discussed.

CUTANEOUS EXCHANGE

Some vertebrates that have lungs or gills, such as some aquatic turtles, salamanders with lungs, snakes, fishes, and mammals, use **cutaneous respiration** or integumentary exchange to supplement gas exchange. However, cutaneous exchange is most highly developed in frogs, toads, lungless salamanders, and newts.

Amphibian skin has the simplest structure of all the major vertebrate respiratory organs (*see figure 23.6*). In frogs, a uniform capillary network lies in a plane directly beneath the epidermis. This vascular arrangement facilitates gas exchange between the capillary bed and the environment by both diffusion and convection. A slimy mucous layer that keeps amphibian skin moist and protects against injury aids in this gas exchange. Some amphibians obtain about 25% or more of their oxygen by this exchange, and the lungless salamanders carry out all of their gas exchange through the skin and buccal-pharyngeal region.

GILLS

Gills are respiratory organs that have either a thin, moist, vascularized layer of epidermis to permit gas exchange across thin gill membranes, or a very thin layer of epidermis over highly vascularized dermis. Larval forms of a few fishes and amphibians have external gills projecting from their bodies (figure 26.15). Adult fishes have internal gills.

FIGURE 26.15

Vertebrate Respiration: External Gills.   This axolotl (*Ambystoma tigrinum*) has elaborate external gills with a large surface for gas exchange with the water.

Gas exchange across internal gill surfaces is extremely efficient (figure 26.16). It occurs as blood and water move in opposite directions on either side of the lamellar epithelium. Water passing between gill lamellae first passes the portion of a gill lamella containing blood that is about to leave the lamella. This blood has a relatively high oxygen concentration, having picked up oxygen earlier in the lamella. Because water at this point has lost none of its oxygen, a diffusion gradient still favors movement of more oxygen into the blood. Water then passes by the portion of the vessels bringing blood from deep within the body. This blood is lower in oxygen. Even though the water has already lost some oxygen to the blood earlier in its movement through the gills, there is still a higher concentration of oxygen in the water than the blood. Thus, a diffusion gradient still favors movement of oxygen into the blood. Carbon dioxide also diffuses into the water because its concentration (pressure) is higher in the blood than in the water. This countercurrent exchange mechanism provides efficient gas exchange by maintaining a concentration gradient between the blood and water over the length of the capillary bed.

LUNGS

A **lung** is an internal sac-shaped respiratory organ. The typical lung of a terrestrial vertebrate comprises one or more internal blind pouches into which air is either drawn or forced. The respiratory epithelium of lungs is thin, well vascularized, and divided into a large number of small units, which greatly increase the surface area for gaseous exchange between the lung air and the blood. This blind-pouch construction, however, limits the efficiency with which oxygen and carbon dioxide are exchanged with the atmosphere because only a portion of the lung air is ever replaced with any one breath. Birds are an exception in that they have very efficient lungs with a one-way pass-through system (*see figure 21.10*). For example,

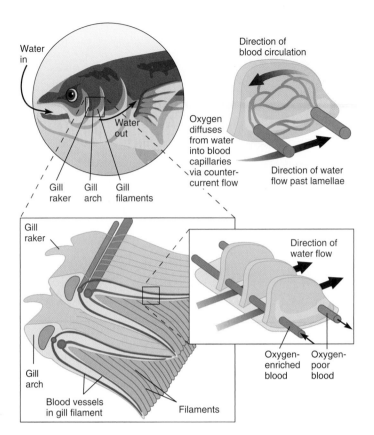

FIGURE 26.16

Vertebrate Respiration: Internal Gills.   Removing the protective operculum exposes the feathery internal gills of this bony fish. Each side of the head has four gill arches, and each arch consists of many filaments. A filament houses capillaries within lamellae. Note that the direction of water flow opposes that of blood flow. This countercurrent flow allows the fish to extract the maximal amount of oxygen from the water.

a mammal removes approximately 25% of the oxygen from air with each breath, whereas a bird removes approximately 90%.

The evolution of the vertebrate lung is related to the evolution of the swim bladder. The swim bladder is an air sac located dorsal to the digestive tract in the body of many modern fishes. Evidence indicates that both lungs and swim bladders evolved from a lunglike structure present in primitive fishes that were ancestors of both present-day fishes and tetrapods (amphibians, reptiles, birds, and mammals). These ancestral fishes probably had a ventral sac attached to their pharynx (*see figure 18.17*). This sac may have served as a supplementary gas-exchange organ when the fishes could not obtain enough oxygen through their gills (e.g., in stagnant or oxygen-depleted water). By swimming to the surface and gulping air into this sac, ancestral fishes could exchange gas through its wall.

Further evolution of this blind sac proceeded in two different directions (see figure 18.17). One adaptation is in the majority of modern bony fishes, where the swim bladder lies dorsal to the digestive tract. The other adaptation is in the form of the lungs, which are ventral to the digestive tract. A few present-day fishes and the tetrapods have ventral lungs. The evolution of the structurally

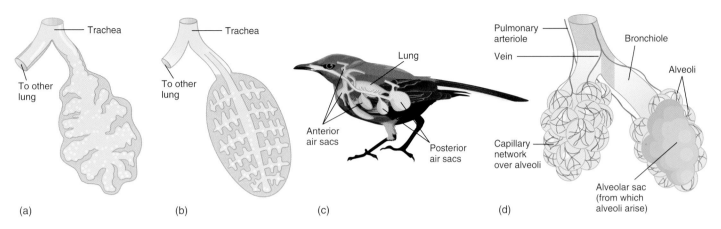

FIGURE 26.17

Vertebrate Respiration: Lungs.   Evolution of the vertebrate lung, showing the increased surface area from (*a*) amphibians and (*b*) reptiles to (*c*) birds and (*d*) mammals. This evolution has paralleled the evolution of larger body size and higher metabolic rates.

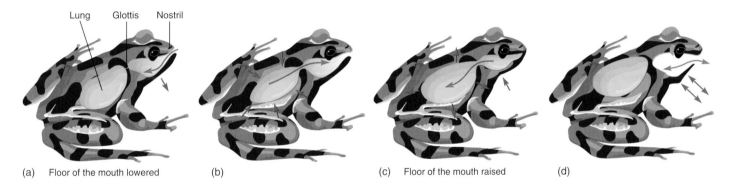

FIGURE 26.18

Ventilation in Amphibians.   The positive pressure pumping mechanism in a frog (*Rana*). The breathing cycle has several stages. (*a*) Air is taken into the mouth and pharynx by lowering the floor of the mouth. Notice that the glottis is closed. (*b*) The glottis is then opened, and air is permitted to escape from the lungs, passing over the air just taken in. (*c*) With the nostrils and mouth firmly shut, the floor of the mouth is raised. This positive pressure forces air into the lungs. (*d*) With the glottis closed, fresh oxygenated air can again be brought into the mouth and pharynx. Some gas exchange occurs in the mouth cavity (buccopharyngeal respiration), and frogs may repeat this "mouth breathing" movement several times before ventilating the lungs again. Red arrows indicate body wall movement, and blue arrows indicate air flow.

complex lung paralleled the evolution of the larger body sizes and higher metabolic rates of endothermic vertebrates (birds and mammals), which necessitated an increase in lung surface area for gas exchange, compared to the smaller body size and lower metabolic rates of ectothermic vertebrates (figure 26.17).

LUNG VENTILATION

Ventilation is based on several physiological principles that apply to all air-breathing animals with lungs:

1. Air moves by bulk flow into and out of the lungs in the process called ventilation.
2. Oxygen and carbon dioxide diffuse across the respiratory surface of the lung tissue from pulmonary capillaries.
3. At systemic capillaries, oxygen and carbon dioxide diffuse between the blood and interstitial fluid in response to concentration gradients.

4. Oxygen and carbon dioxide diffuse between the interstitial fluid and body cells.

Vertebrates exhibit two different mechanisms for lung ventilation based on these physiological principles. Amphibians and some reptiles use a positive pressure pumping mechanism. They push air into their lungs. Most reptiles and all birds and mammals, however, use a negative pressure system; that is, they inhale (breathe in) by suction.

Figure 26.18 shows the positive pressure pumping mechanism of an amphibian. The muscles of the mouth and pharynx create a positive pressure to force air into the lungs.

Most reptiles (e.g., snakes, lizards, crocodilians) expand the body cavity with a posterior movement of the ribs to ventilate the lungs. This expansion decreases pressure in the lungs and draws air into the lungs. Elastic recoil of the lungs and the movement of the ribs and body wall, which compress the lungs, expel air. The ribs of turtles are a part of the shell (*see figure 20.5*); thus, movements of

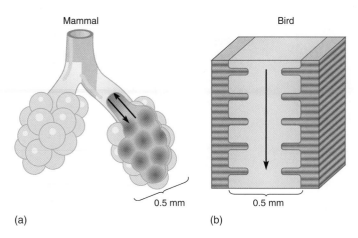

(a)   (b)

FIGURE 26.19

Gas Exchange Surfaces in Mammals and Birds. (*a*) The gas exchange surfaces in a mammal's lung are in saclike alveoli. Ventilation is by an ebb-and-flow mechanism (arrows), and the air inside the alveoli can never be completely replaced. (*b*) The smallest-diameter passages in a bird lung are tubes that are open at both ends. Ventilation is by one-way flow (arrow), and complete replacement of air in the tubes is continuous.

the body wall to which they attach are impossible. Turtles exhale by contracting muscles that force the viscera upward, compressing the lungs. They inhale by contracting muscles that increase the volume of the visceral cavity, creating negative pressure to draw air into the lungs.

Because of the high metabolic rates associated with flight, birds have a greater rate of oxygen consumption than any other vertebrate. Birds also use a negative pressure system to move air into and out of their lungs in an ebb and flow breathing pattern similar to that of mammals. However, birds also have a special lung ventilation mechanism that permits one-way flow over gas exchange surfaces. This mechanism makes bird lungs more efficient than mammalian lungs (figure 26.19). This is also why bird lungs are smaller than the lungs of mammals of comparable body size. Bird lungs have tunnel-like passages called parabronchi, which lead to air capillaries in which gas exchange occurs. The arrangement and functioning of a system of air sacs make one-way flow possible. These air sacs ramify throughout the body cavity, are collapsible, and open and close as a result of muscle contractions around them. Inhaled air bypasses the lungs and enters the abdominal (posterior) air sacs. It then passes through the lungs into the thoracic (anterior) air sacs. Finally, air is exhaled from the thoracic air sacs. This whole process requires two complete breathing cycles (figure 26.20).

HUMAN RESPIRATORY SYSTEM

The structure and function of external respiration in humans are typical of mammals. Thus the human respiratory system is used here to describe those principles that apply to all air-breathing mammals.

Air-Conducting Portion

Figure 26.21 shows the various organs of the human respiratory system. Air normally enters and leaves this system through either

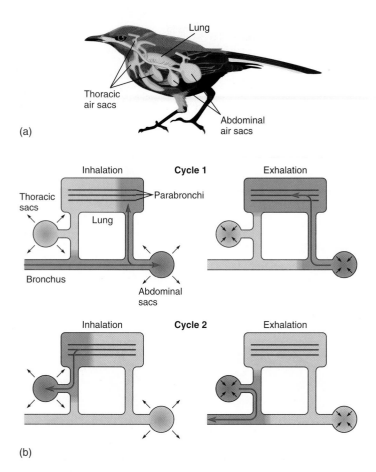

(a)

(b)

FIGURE 26.20

Gas Exchange Mechanism in Birds. (*a*) Birds have a number of large air sacs. Some of them (abdominal) are posterior to the small pair of lungs, and others (thoracic) are anterior to the lungs. The main bronchus (air passageway) that runs through each lung has connections to air sacs, as well as to the lung. In (*b*), abdominal and thoracic air sacs are sketched as single functional units to clarify their relationship to the lung and bronchus. (*b*) Air flow through the bird respiratory system. The darker blue portion in each diagram represents the volume of a single inhalation and distinguishes it from the remainder of the air in the system. Two full breathing cycles are needed to move the volume of gas taken in during a single inhalation through the entire system and out of the body. This system is associated with one-way flow through the gas exchange surfaces in the lungs. Black arrows indicate expansion and contraction of air sacs. Red arrows indicate movement of air.

nasal or oral cavities. From these cavities, air moves into the pharynx, which is a common area for both the respiratory and digestive tracts. The pharynx connects with the larynx (voice box) and with the esophagus that leads to the stomach. The epiglottis is a flap of cartilage that allows air to enter the trachea during breathing. It covers the trachea during swallowing to prevent food or water from entering.

During inhalation, air from the larynx moves into the trachea (windpipe), which branches into a right and left bronchus (pl., bronchi). After each bronchus enters the lungs, it branches into smaller tubes called bronchioles, then even smaller tubes called terminal bronchioles, and finally, the respiratory bronchioles, which are part of the gas-exchange portion of the respiratory system.

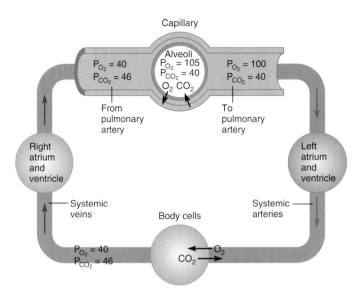

FIGURE 26.22

Gas Exchange between the Lungs and Tissues. Gases diffuse according to partial pressure (P) differences, as the numbers and arrows indicate.

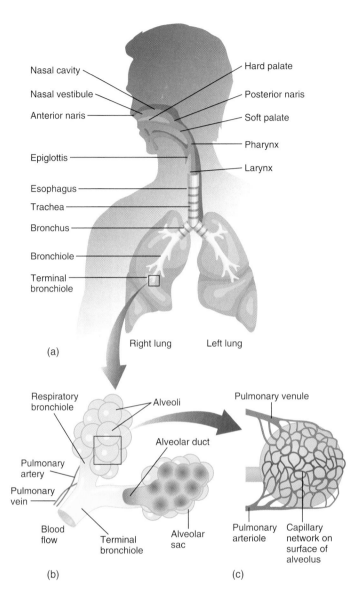

FIGURE 26.21

Organs of the Human Respiratory System. (*a*) Basic anatomy of the respiratory system. (*b, c*) The respiratory tubes end in minute alveoli, each of which is surrounded by an extensive capillary network.

Gas-Exchange Portion

Small tubes called alveolar ducts connect the respiratory bronchioles to grapelike outpouchings called **alveoli** (sing., alveolus) (L. *alveus*, hollow) (figure 26.21*b*). The alveoli cluster to form an alveolar sac. Surrounding the alveoli are many capillaries (figure 26.21*c*). Alveoli are the functional units of the lungs (gas-exchange portion). Passive diffusion, driven by a partial pressure gradient, moves oxygen from the alveoli into the blood and moves carbon dioxide from the blood into the alveoli (figure 26.22). Collectively, the alveoli provide a large surface area for gas exchange. If the alveolar epithelium of a human were removed from the lungs and put into a single layer of cells side by side, the cells would cover the area of a tennis court.

Ventilation

Breathing (also called pulmonary ventilation) has two phases: (1) inhalation, the intake of air; and (2) exhalation, the outflow of air. These air movements result from the rhythmic increases and decreases in thoracic cavity volume. Changes in thoracic volume lead to reversals in the pressure gradients between the lungs and the atmosphere; gases in the respiratory system follow these gradients. The mechanism of inhalation operates in the following way (figure 26.23):

1. Several sets of muscles, the main ones being the diaphragm and intercostal muscles, contract. The intercostal muscles stretch from rib to rib, and when they contract, they pull the ribs closer together, enlarging the thoracic cavity.
2. The thoracic cavity further enlarges when the diaphragm contracts and flattens.
3. The increased size of the thoracic cavity causes pressure in the cavity to drop below the atmospheric pressure. Air rushes into the lungs, and the lungs inflate.

During ordinary exhalation, air is expelled from the lungs in the following way:

1. The intercostal muscles and the diaphragm relax, allowing the thoracic cavity to return to its original, smaller size and increasing the pressure in the thoracic cavity.
2. Abdominal muscles contract, pushing the abdominal organs against the diaphragm, further increasing the pressure within the thoracic cavity.
3. The action in step 2 causes the elastic lungs to contract and compress the air in the alveoli. With this compression, alveolar pressure becomes greater than atmospheric pressure, causing air to be expelled (exhaled) from the lungs.

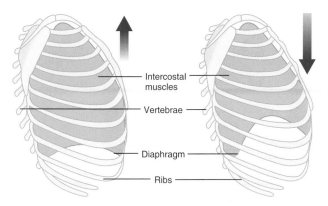

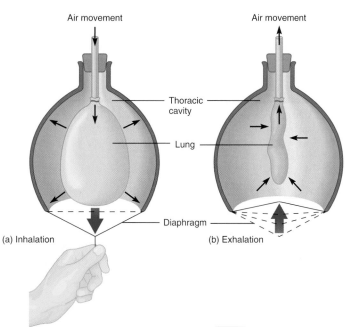

FIGURE 26.23

Ventilation of Human Lungs as an Example of Breathing in Mammals. (*a*) During inhalation, muscle contractions lift the ribs up and out (upper diagram arrows) and lower the diaphragm. These movements increase the size of the thoracic cavity and decrease the pressure around the lungs. This negative pressure causes more air to enter the lungs. (*b*) Exhalation follows the relaxation of the rib cage and diaphragm muscles, as the increased pressure forces the air out of the lungs. Arrows indicate the direction pressure changes take in the thoracic (lower diagrams) cavity during inhalation and exhalation.

GAS TRANSPORT

As noted in the previous discussion, oxygen must be transported from the sites of environmental gas exchange to the cells of an animal's body. Various systems (e.g., tracheae, cutaneous exchange, gills, lungs) help accomplish this transport.

As animals became larger and acquired higher metabolic rates, simple diffusion became increasingly inadequate as a means of delivering oxygen to the tissues. Consequently, in most animals with high metabolic rates and tissues more than a few millimeters from respiratory surfaces, a specialized circulatory system circulates body fluids to aid in the internal distribution of oxygen

(*see figure 26.1*). In general, more active animals have an increased demand for oxygen. However, simply creating a convection of a water-based body fluid does not in itself guarantee internal transport of sufficient oxygen to meet this increased demand. The reason is the low solubility of oxygen in water-based body fluids. *Thus, fluid-borne respiratory pigments specialized for reversibly binding large quantities of oxygen evolved in most phyla.* Respiratory pigments help the various transport systems satisfy this increased oxygen demand. In addition to oxygen transport, respiratory pigments may also function in short-term oxygen storage.

Respiratory pigments are organic compounds that have either metallic copper or iron that binds oxygen. These pigments may be in solution within the blood or body fluids, or they may be in specific blood cells. In general, the pigments respond to a high oxygen concentration by combining with oxygen; they respond to low oxygen concentrations by releasing oxygen. The four most common respiratory pigments are hemoglobin, hemocyanin, hemerythrin, and chlorocruorin.

Hemoglobin is a reddish pigment that contains iron as the oxygen-binding metal. It is the most common respiratory pigment in animals, being found in a variety of invertebrates (e.g., protozoa, platyhelminths, nemerteans, nematodes, annelids, crustaceans, some insects, and molluscs), and with the exception of a few fishes, in all vertebrates. *This wide distribution suggests that hemoglobin evolved very early in the history of animal life.* Hemoglobin may be carried within red blood cells (erythrocytes; *see figure 26.5a*) or simply dissolved in the blood or coelomic fluid.

Hemocyanin is the most commonly occurring respiratory pigment in molluscs and certain crustaceans. Hemocyanin contains metallic copper, has a bluish color when oxygenated, and always occurs dissolved in hemolymph. Unlike most hemoglobin, hemocyanin tends to release oxygen easily and to provide a ready source of oxygen to the tissues as long as concentrations of oxygen in the environment are relatively high.

Hemerythrin contains iron and is pink when oxygenated. It is in nucleated cells, rather than free in body fluids or hemolymph. Sipunculans, priapulids, a few brachiopods, and some polychaetes have hemerythrin.

Chlorocruorin also contains iron but is green when associated with low oxygen concentrations and bright red when associated with high oxygen concentrations. Chlorocruorin occurs in several families of polychaete worms.

As just discussed, respiratory pigments raise the oxygen-carrying capacity of body fluids far above what simple transport in a dissolved state would achieve. Similarly, carbon dioxide concentrations in animal body fluids (and in seawater as well) are much higher than would be expected strictly on the basis of its solubility. The reason for this increased transport is that, in addition to being transported bound to hemoglobin and in the dissolved state, most carbon dioxide is transported in the form of carbonic acid (H_2CO_3) and the bicarbonate ion (HCO_3^-) in a series of reversible reactions:

$$CO_2 + H_2O \rightleftharpoons H_2CO_3 \rightleftharpoons H^+ + HCO_3^-$$

Thus, "tying up" carbon dioxide in other forms lowers its concentration in solution, thereby raising the overall carrying capacity of a body fluid such as blood.

From an evolutionary perspective, the occurrence of respiratory pigments among various taxa has no phylogenetic explanation. Their sporadic distribution suggests that some of the pigments may have evolved more than once through parallel evolution. Interestingly, respiratory pigments are rare in the successful insects. The general absence of respiratory pigments among most insects reflects the fact that most insects do not use blood as a medium for gas transport, but employ extensive tracheal systems to carry gases directly to the tissues (see figure 26.12). In those insects without well-developed tracheae, oxygen is simply carried in solution in the hemolymph.

SUMMARY

1. Any system of moving fluids that reduces the functional diffusion distance that nutrients, wastes, and gases must traverse is an internal transport system or circulatory system.

2. The two basic types of circulatory systems are open and closed. Open systems generally circulate hemolymph, and closed systems circulate blood.

3. Blood is a type of connective tissue made up of blood cells (red blood cells and white blood cells), plasma, and platelets.

4. The heart pumps blood through a series of vessels in the following order: arteries, arterioles, capillaries, venules, veins, and back to the heart.

5. The action of the heart consists of cyclic contraction (systole) and relaxation (diastole). Systolic contraction generates blood pressure that forces blood through the closed system of vessels.

6. The lymphatic system consists of one-way vessels that help return fluids and proteins to the circulatory system.

7. Animals that respire aerobically need a constant supply of oxygen. The process of acquiring oxygen and eliminating carbon dioxide is called external respiration.

8. The exchange of oxygen and carbon dioxide occurs across respiratory surfaces. Such surfaces include gills, cutaneous surfaces, and lungs.

9. The air-conducting portion of the respiratory system of air-breathing vertebrates moves air into (inhalation) and out of (exhalation) this system. This process of air movement is called ventilation.

10. Oxygen and carbon dioxide diffuse from areas of higher concentration to areas of lower concentration.

11. Once in the blood, oxygen diffuses into red blood cells and binds to hemoglobin for transport to the tissues. Carbon dioxide is transported bound to hemoglobin, as well as in the form of the bicarbonate ion and carbonic acid.

12. Respiratory pigments are organic compounds that have either metallic copper or iron that binds oxygen. Examples include hemoglobin, hemocyanin, hemerythrin, and chlorocruorin.

SELECTED KEY TERMS

alveoli (p. 439)
bimodal breathing (p. 435)
lung (p. 436)
lymph (p. 433)
plasma (p. 427)
serum (p. 427)

CRITICAL THINKING QUESTIONS

1. Many invertebrates utilize the body cavity as a circulatory system. However, in humans, the body cavity plays no role whatsoever in circulation. Why?

2. Describe the homeostatic functions of the vertebrate circulatory system. What functions are maintained at relative stability?

3. The area of an animal's respiratory surface is usually directly related to the animal's body weight. What does this tell you about the mechanism of gas exchange?

4. How can seals and whales stay under water for long periods?

ONLINE LEARNING CENTER

Visit our Online Learning Center (OLC) at www.mhhe.com/zoology (click on this book's title) to find the following chapter-related materials:

- CHAPTER QUIZZING
- ANIMATIONS
   Bronchoscopy
   Breathing
   Gas Exchange
   Respiration
   Smoking Risks
   Hemoglobin
   Asthma
   ABO Blood Types
   Cardiac Cycle Blood Flow (Normal Speed)
   Cardiac Cycle Blood Flow (Slow Motion)
   Cardiac Cycle Electrical
   Cardiac Cycle Muscular
   Cardiac Cycle Sounds
   Cardiac Cycle Sounds (Detail)
   Valvular Insufficiency
   Valvular Stenosis
   Myocardial Infarction
   Portal System
   Lymphatic System
   Phagocytic Cells
   Fever
   Complement Proteins
   Antiviral Defense
   Clonal Selection
   T Cell Function
   Vaccination

- RELATED WEB LINKS
    Blood and Vertebrate Circulation
        Vertebrates: Macroscopic Anatomy of Circulatory Elements
        Vertebrates: Microscopic Anatomy of Circulatory Elements
        Human Circulatory System Topics
    Respiratory Systems
        Vertebrates: Macroscopic Anatomy of the Respiratory System
        Vertebrates: Microscopic Anatomy of the Respiratory System
        Human Respiratory Topics

- BOXED READING ON
    How Deep-Diving Marine Mammals Adapt to Oxygen-Poor
    Environments

- SUGGESTED READINGS
- LAB CORRELATIONS
    Check out the OLC to find specific information on these related lab
    exercises in the *General Zoology Laboratory Manual*, 5th edition, by
    Stephen A. Miller:

NUTRITION AND DIGESTION

Outline

Concepts

1. Animals are heterotrophic organisms that use food to supply both raw materials and energy. Nutrition includes all of those processes by which an animal takes in, digests, absorbs, stores, and uses food to meet its metabolic needs.

2. Digestion is the chemical and/or mechanical breakdown of food into particles that individual cells of an organism can absorb. Digestion can occur either inside a cell (intracellular), outside a cell (extracellular), or in both places.

3. Most animals must work for their nutrients. The number of specializations that have evolved for food procurement (feeding) and extracellular digestion are almost as numerous as the number of animal species. Some examples include continuous versus discontinuous feeding, suspension feeding, deposit feeding, herbivory, predation, surface nutrient absorption, and fluid feeding.

4. Intracellular digestion occurs in some invertebrates (e.g., in sponges); others (e.g., some cnidarians and molluscs) utilize both intracellular and extracellular digestion; and most higher invertebrates (e.g., insects) have evolved variations in extracellular digestion that allow them to exploit different food sources.

5. In primitive, multicellular animals, such as cnidarians, the gut is a blind (closed) sac called a gastrovascular cavity. Its one opening is both entrance and exit; thus, it is an incomplete digestive tract. The development of an anus and complete digestive tract in the aschelminths was an evolutionary breakthrough. The many variations of the basic complete digestive tract correlate with different food-gathering mechanisms and diets.

6. Vertebrate digestive systems have evolved into assembly lines where food is first broken down mechanically and then chemically by digestive enzymes. The simple sugars, fats, triglycerides, amino acids, vitamins, and minerals that result are then taken into the circulatory systems for distribution throughout the animal's body and are used in maintenance, growth, and energy production.

Nutrition includes all of those processes by which an animal takes in, digests, absorbs, stores, and uses food (nutrients) to meet its metabolic needs. **Digestion** (L. *digestio*, from + *dis*, apart + *gerere*, to carry) is the chemical and/or mechanical breakdown of food into particles that individual cells of an animal can absorb. This chapter discusses animal nutrition, the different strategies animals use for consuming and using food, and various animal digestive systems.

*T*his chapter contains evolutionary concepts, which are set off in this font.

EVOLUTION OF NUTRITION

Nutrients in the food an animal consumes provide the necessary chemicals for growth, maintenance, and energy production. Overall, the nutritional requirements of an animal are inversely related to its ability to synthesize molecules essential for life. The fewer such biosynthetic abilities an animal has, the more kinds of nutrients it must obtain from its environment. Green plants and photosynthetic protists have the fewest such nutritional requirements because they can synthesize all their own complex molecules from simpler inorganic substances; they are **autotrophs** (Gr. *auto*, self + *trophe*, nourishing). Animals, fungi, and bacteria that cannot synthesize many of their own organic molecules and must obtain them by consuming other organisms or their products are **heterotrophs** (Gr. *heteros*, another or different + *trophe*, nourishing). Animals such as rabbits that subsist entirely on plant material are **herbivores** (L. *herba*, plant + *vorare*, to eat). **Carnivores** (L. *caro*, flesh), such as hawks, are animals that eat only meat. **Omnivores** (L. *omnius*, all), such as humans, bears, raccoons, and pigs, eat both plant and animal matter. **Insectivores,** such as bats, eat primarily arthropods.

Losses of biosynthetic abilities have marked much of animal evolution. Once an animal routinely obtains essential, complex organic molecules in its diet, it can afford to lose the ability to synthesize those molecules. Moreover, the loss of this ability confers a selective advantage on the animal because the animal stops expending energy and resources to synthesize molecules that are already in its diet. Thus, as the diet of animals became more varied, they tended to lose their abilities to synthesize such widely available molecules as some of the amino acids.

THE METABOLIC FATES OF NUTRIENTS IN HETEROTROPHS

The nutrients that a heterotroph ingests can be divided into macronutrients and micronutrients. **Macronutrients** are needed in large quantities and include the carbohydrates, lipids, and proteins. **Micronutrients** are needed in small quantities and include organic vitamins and inorganic minerals. Together, these nutrients make up the animal's dietary requirements. Besides these nutrients, animals require water.

CALORIES AND ENERGY

The energy value of food is measured in terms of calories or Calories. A **calorie** (L. *calor*, heat) is the amount of energy required to raise the temperature of 1 g of water 1° C. A calorie, with a small c, is also called a gram calorie. A **kilocalorie,** also known as a **Calorie** or kilogram calorie (kcal), is equal to 1,000 calories. In popular usage, you talk about calories but actually mean Calories, because the larger unit is more useful for measuring the energy value of food. If an advertisement says that a so-called light beer contains 95 calories per 12 oz, it really means 95,000 calories, 95 Calories, or 95 kcal.

MACRONUTRIENTS

With a few notable exceptions, heterotrophs require organic molecules, such as carbohydrates, lipids, and proteins, in their diets. Enzymes break down these molecules into components that can be used for energy production or as sources for the "building blocks" of life.

Carbohydrates: Carbon and Energy from Sugars and Starches

The major dietary source of energy for heterotrophs is complex carbohydrates (figure 27.1a). Most carbohydrates originally come from plant sources. Various polysaccharides, disaccharides, or any of a variety of simple sugars (monosaccharides) can meet this dietary need. Carbohydrates also are a major carbon source for incorporation into important organic compounds. Many plants also supply cellulose, a polysaccharide that humans and other animals (with the exception of herbivores) cannot digest. Cellulose is sometimes called dietary fiber. It assists in the passage of food through the alimentary canal of mammals. Cellulose may also reduce the risk of cancer of the colon, because the mutagenic compounds that form during the storage of feces are reduced if fecal elimination is more frequent.

Lipids: Highly Compact Energy-Storage Nutrients

Neutral lipids (fats) or triacylglycerols are contained in fats and oils, meat and dairy products, nuts, and some fruits and vegetables high in fats, such as avocados (figure 27.1b). Lipids are the most concentrated source of food energy. They produce about 9 Calories (kcal) of usable energy per gram, more than twice the energy available from an equal mass of carbohydrate or protein (table 27.1).

Many heterotrophs have an absolute dietary requirement for lipids, sometimes for specific types. For example, many animals require unsaturated fatty acids (e.g., linoleic acid, linolenic acid, and arachidonic acid). These fatty acids act as precursor molecules for the synthesis of sterols, the most common of which is cholesterol. The sterols are also required for the synthesis of steroid hormones and cholesterol, which is incorporated into cell membranes. Other lipids insulate the bodies of some vertebrates and help maintain a constant temperature.

Proteins: Basic to the Structure and Function of Cells

Animal sources of protein include other animals and milk. Plant sources include beans, peas, and nuts. Proteins are needed for their amino acids, which heterotrophs use to build their own body proteins (figure 27.1c).

MICRONUTRIENTS

Micronutrients are usually small ions, organic vitamins, inorganic minerals, and molecules that are used repeatedly in enzymatic reactions or as parts of certain proteins (e.g., copper in hemocyanin

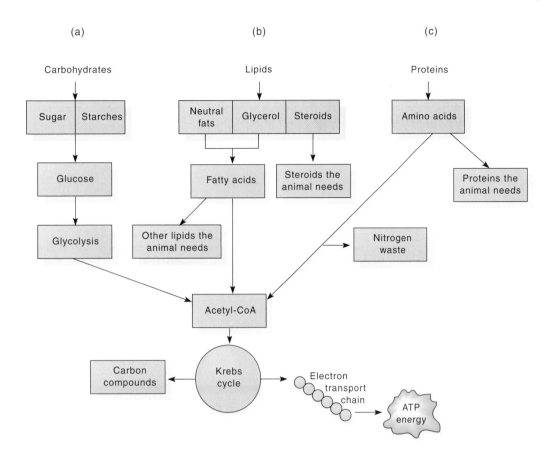

FIGURE 27.1

Macronutrients in the Diet. (*a*) Carbohydrate foods break down to their constituent sugars and starches, and ultimately into glucose. Individual cells use this sugar in glycolysis and aerobic respiration to create new carbon compounds or ATP energy. (*b*) Lipids (fats and oils) in the diet break down to neutral fats, glycerol, and steroids. These molecules can be modified and incorporated into the lipids or steroids the animal needs for storing fat or generating hormones, or they can be converted to acetyl-CoA and enter the Krebs cycle and electron transport chain for ATP production. (*c*) Proteins break down to amino acids, which are incorporated into new proteins or modified to enter the Krebs cycle and electron transport chain to produce ATP energy.

TABLE 27.1
THE AVERAGE CALORIC VALUES OF MACRONUTRIENTS

MACRONUTRIENT	CALORIES PER GRAM
Carbohydrates	4.1
Lipids	9.3
Proteins	4.4

and iron in hemoglobin). Even though they are needed in small amounts, animals cannot synthesize them rapidly (if at all); thus, they must be obtained from the diet.

Minerals

Some minerals are needed in relatively large amounts and are called essential minerals, or macrominerals. For example, sodium and potassium are vital to the functioning of every nerve and

muscle in an animal's body. Animals lose large quantities of these minerals, especially sodium, in the urine every day. Animals that sweat to help regulate body temperature lose sodium in their sweat. A daily supply of calcium is needed for muscular activity and, with phosphorus, for bone formation. Table 27.2 lists the functions of the major essential minerals.

Other minerals are known as trace minerals, trace elements, or microminerals. Animals need these in only very small amounts for various enzymatic functions. Table 27.3 lists the function of some trace minerals.

Vitamins

Normal metabolic activity depends on very small amounts of more than a dozen organic substances called vitamins. **Vitamin** (L. *vita*, life) is the general term for a number of chemically unrelated, organic substances that occur in many foods in small amounts and are necessary for normal metabolic functioning. Vitamins may be water soluble or fat soluble. Most water-soluble vitamins, such as the B vitamins and vitamin C, are coenzymes needed in metabolism (table 27.4). The fat-soluble vitamins have various functions (table 27.5).

TABLE 27.2
PHYSIOLOGICAL ROLES OF THE ESSENTIAL MINERALS (MACROMINERALS) ANIMALS REQUIRE IN LARGE AMOUNTS

MINERAL	MAJOR PHYSIOLOGICAL ROLES
Calcium (Ca)	Component of bone and teeth; essential for normal blood clotting; needed for normal muscle, neuron, and cellular function
Chlorine (Cl)	Principal negative ion in extracellular fluid; important in acid-base and fluid balance; needed to produce stomach HCl
Magnesium (Mg)	Component of many coenzymes; needed for normal neuron and muscle function, as well as carbohydrate and protein metabolism
Phosphorus (P)	Major constituent of bones, blood plasma; needed for energy metabolism; part of DNA, RNA, ATP, energy metabolism
Potassium (K)	Major positive ion in cells; influences muscle contraction and neuron excitability
Sodium (Na)	Principal positive ion in extracellular fluid; important in fluid balance; essential for conduction of action potentials, active transport
Sulfur (S)	Protein structure; detoxification reactions and other metabolic activity

TABLE 27.3
SOME PHYSIOLOGICAL ROLES OF TRACE MINERALS (MICROMINERALS) IN ANIMALS

MINERAL	PHYSIOLOGICAL ROLES
Cobalt (Co)	Component of vitamin B_{12}; essential for red blood cell production
Copper (Cu)	Component of many enzymes; essential for melanin and hemoglobin synthesis; part of cytochromes
Fluorine (F)	Component of bone and teeth; prevents tooth decay
Iodine (I)	Component of thyroid hormones
Iron (Fe)	Component of hemoglobin, myoglobin, enzymes, and cytochromes
Manganese (Mn)	Activates many enzymes; an enzyme essential for urea formation and parts of the Krebs cycle
Molybdenum (Mo)	Constituent of some enzymes
Selenium (Se)	Needed in fat metabolism
Zinc (Zn)	Component of at least 70 enzymes; needed for wound healing and fertilization

The dietary need for vitamin C and the fat-soluble vitamins (A, D, E and K) tends to be limited to the vertebrates. Even in closely related groups, vitamin requirements vary. For example, among vertebrates, humans and guinea pigs require vitamin C, but rabbits do not. Some birds require vitamin A; others do not.

DIGESTION

In some of the simplest forms of life (the protists and sponges), some cells take in whole food particles directly from the environment by diffusion, active transport, and/or endocytosis and break them down with enzymes to obtain nutrients. This strategy is called **intracellular** ("within the cell") **digestion** (figure 27.2a). Intracellular digestion circumvents the need for the mechanical breakdown of food or for a gut or other cavity in which to chemically digest food. At the same time, however, intracellular digestion limits an animal's size and complexity—only very small pieces of food can be used. Intracellular digestion provides all or some of the nutrients in protozoa, sponges, cnidarians, platyhelminths, rotifers, bivalve molluscs, and primitive chordates.

Larger animals have evolved structures and mechanisms for **extracellular digestion**: *the enzymatic breakdown of larger pieces of food into constituent molecules, usually in a special organ or cavity (figure 27.2b).* Nutrients from the food then pass into body cells lining the organ or cavity and can take part in energy metabolism or biosynthesis.

ANIMAL STRATEGIES FOR GETTING AND USING FOOD

As noted earlier, only a few protists and animals can absorb nutrients directly from their external environment via intracellular digestion. Most animals must work for their nutrients. The number of specializations that have evolved for food procurement (feeding) and extracellular digestion are almost as numerous as the number of animal species. What follows is a brief discussion of the major feeding strategies animals use.

CONTINUOUS VERSUS DISCONTINUOUS FEEDERS

One variable related to the structure of digestive systems is whether an animal is a continuous or discontinuous feeder. Many **continuous feeders** are slow-moving or completely sessile animals (they remain permanently in one place). For example, aquatic **suspension feeders,** such as tube worms and barnacles, remain in one place and continuously "strain" small food particles from the water.

Discontinuous feeders tend to be active, sometimes highly mobile, animals. Typically, discontinuous feeders have more digestive specializations than continuous feeders because discontinuous

TABLE 27.4
WATER-SOLUBLE VITAMINS

VITAMIN	CHARACTERISTICS	FUNCTIONS	SOURCES
Thiamin (vitamin B_1)	Destroyed by heat and oxygen, especially in alkaline environment	Part of coenzyme needed for oxidation of carbohydrates, and coenzyme needed in synthesis of ribose	Lean meats, liver, eggs, whole-grain cereals, leafy green vegetables, legumes
Riboflavin (vitamin B_2)	Stable to heat, acids, and oxidation; destroyed by alkalis and light	Part of enzymes and coenzymes needed for oxidation of glucose and fatty acids and for cellular growth	Meats, dairy products, leafy green vegetables, whole-grain cereals
Niacin (nicotinic acid)	Stable to heat, acids, and alkalis; converted to niacinamide by cells; synthesized from tryptophan	Part of coenzymes needed for oxidation of glucose and synthesis of proteins, fats, and nucleic acids	Liver, lean meats, poultry, peanuts, legumes
Vitamin B_6	Group of three compounds; stable to heat and acids; destroyed by oxidation, alkalis, and ultraviolet light	Coenzyme needed for synthesis of proteins and various amino acids, for conversion of tryptophan to niacin, for production of antibodies, and for synthesis of nucleic acids	Liver, meats, fish, poultry, bananas, avocados, beans, peanuts, whole-grain cereals, egg yolk
Pantothenic acid	Destroyed by heat, acids, and alkalis	Part of coenzyme needed for oxidation of carbohydrates and fats	Meats, fish, whole-grain cereals, legumes, milk, fruits, vegetables
Cyanocobalamin (vitamin B_{12})	Complex, cobalt-containing compound; stable to heat; inactivated by light, strong acids, and strong alkalis; absorption regulated by intrinsic factor from gastric glands; stored in liver	Part of coenzyme needed for synthesis of nucleic acids and for metabolism of carbohydrates; plays role in synthesis of myelin	Liver, meats, poultry, fish, milk, cheese, eggs
Folate (folic acid)	Occurs in several forms; destroyed by oxidation in acid environment or by heat in alkaline environment; stored in liver, where it is converted into folinic acid	Coenzyme needed for metabolism of certain amino acids and for synthesis of DNA; promotes production of normal red blood cells	Liver, leafy green vegetables, whole-grain cereals, legumes
Biotin	Stable to heat, acids, and light; destroyed by oxidation and alkalis	Coenzyme needed for metabolism of amino acids and fatty acids and for synthesis of nucleic acids	Liver, egg yolk, nuts, legumes, mushrooms
Ascorbic acid	Closely related to monosaccharides; stable in acids, but destroyed by oxidation, heat, light, and alkalis	Needed for production of collagen, conversion of folate to folinic acid, and metabolism of certain amino acids; promotes absorption of iron and synthesis of hormones from cholesterol	Citrus fruits, citrus juices, tomatoes, cabbage, potatoes, leafy green vegetables, fresh fruits

feeders take in large meals that must be either ground up or stored, or both. Many carnivores, for example, pursue and capture relatively large prey. When successful, they must eat large meals so that they need not spend their time in the continuous pursuit of prey. Thus, carnivores have digestive systems that permit the storage and gradual digestion of large, relatively infrequent meals.

Herbivores spend more time eating than carnivores do, but they are also discontinuous feeders. They need to move from area to area when food is exhausted and, at least in natural environments, must limit their grazing time to avoid excessive exposure to predators. Thus, their digestive systems permit relatively rapid food gathering and gradual digestion.

TABLE 27.5
FAT-SOLUBLE VITAMINS

VITAMIN	CHARACTERISTICS	FUNCTIONS	SOURCES
Vitamin A	Occurs in several forms; synthesized from carotenes; stored in liver; stable in heat, acids, and alkalis; unstable in light	Necessary for synthesis of visual pigments, mucoproteins, and mucopolysaccharides; for normal development of bones and teeth; and for maintenance of epithelial cells	Liver, fish, whole milk, butter, eggs, leafy green vegetables, and yellow and orange vegetables and fruits
Vitamin D	A group of sterols; resistant to heat, oxidation, acids, and alkalis; stored in liver, skin, brain, spleen, and bones	Promotes absorption of calcium and phosphorus; promotes development of teeth and bones	Produced in skin exposed to ultraviolet light; in milk, egg yolk, fish-liver oils, fortified foods
Vitamin E	A group of compounds; resistant to heat and visible light; unstable in presence of oxygen and ultraviolet light; stored in muscles and adipose tissue	An antioxidant; prevents oxidation of vitamin A and polyunsaturated fatty acids; may help maintain stability of cell membranes	Oils from cereal seeds, salad oils, margarine, shortenings, fruits, nuts, and vegetables
Vitamin K	Occurs in several forms; resistant to heat, but destroyed by acids, alkalis, and light; stored in liver	Needed for synthesis of prothrombin; needed for blood clotting	Leafy green vegetables, egg yolk, pork liver, soy oil, tomatoes, cauliflower

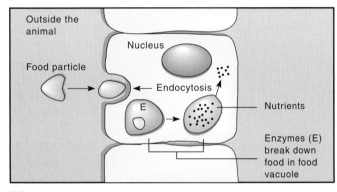

(a)

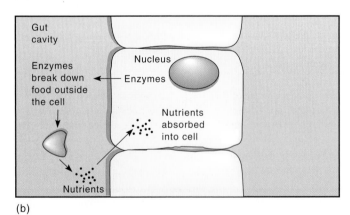

(b)

FIGURE 27.2

Intracellular and Extracellular Digestion.   (*a*) A simple invertebrate, such as a sponge, has no gut and thus carries out intracellular digestion. Tiny food particles are taken into the body wall cells by endocytosis. Digestive enzymes in the vacuole then break the small particles into constituent molecules. (*b*) A dog, for example, has a gut and so can take in and digest (extracellularly) relatively large food particles. Cells lining the gut cavity secrete enzymes into the cavity. There, the enzymes break down food materials into constituent nutrients, and the nearby cells absorb these nutrients.

SUSPENSION FEEDERS

Suspension feeding is the removal of suspended food particles from the surrounding water by some sort of capture, trapping, or filtration structure. This feeding strategy involves three steps: (1) transport of water past the feeding structure, (2) removal of nutrients from the water, and (3) transport of the nutrients to the mouth of the digestive system. Sponges, ascidians, branchiopods, ectoprocts, entoprocts, phoronids, most bivalves, and many crustaceans, polychaetes, gastropods, and some nonvertebrate chordates are suspension feeders.

DEPOSIT FEEDERS

Deposit feeding involves primarily omnivorous animals. These animals obtain their nutrients from the sediments of soft-bottom habitats (muds and sands) or terrestrial soils. Direct deposit feeders simply swallow large quantities of sediment (mud, soil, sand, organic matter). The usable nutrients are digested, and the remains pass out the anus. Direct deposit feeding occurs in many polychaete annelids, some snails, some sea urchins, and in most earthworms. Other direct deposit feeders utilize tentaclelike structures to consume sediment. Examples include sea cucumbers, most sipunculans, certain clams, and several types of polychaetes.

HERBIVORY

Herbivory (L. *herba*, herb + *vorare*, to eat) is the consumption of macroscopic plants. This common feeding strategy requires the ability to "bite and chew" large pieces of plant matter (macroherbivory). *Although biting and chewing mechanisms evolved within the architectural framework of a number of invertebrate lineages, they are often characterized by the development of* *hard surfaces (e.g., teeth) that powerful muscles manipulate.* Invertebrates that evolved macroherbivory include molluscs, polychaete worms, arthropods, and sea urchins.

Many molluscs have a radula. A radula is a muscularized, belt-like rasp armed with chitinous teeth. Molluscs use the radula to scrape algae off rocks or to tear the leaves off terrestrial plants. Polychaetes have sets of large chitinous teeth on an eversible proboscis or pharynx that is used to scrape off algae. This toothed pharynx is also suitable for carnivory when plant material is scarce. Macroherbivory is found in almost every group of arthropods. For example, insects and crustaceans have large, powerful mandibles capable of biting off plant material and subsequently grinding and chewing it before passing the plant material to the mouth.

PREDATION

Predation (L. *praedator,* a plunderer, pillager) is one of the most sophisticated feeding strategies, since it requires the capture of live prey. Only a few generalizations about the many kinds of predation are presented here; discussions of various taxa are presented in their appropriate chapters.

Predators can be classified by how they capture their prey: motile stalkers, lurking predators, sessile opportunists, or grazers. Motile stalkers actively pursue their prey. Examples include ciliate protozoa, nemerteans, polychaete worms, gastropods, octopuses and squids, crabs, sea stars, and many vertebrates. Lurking predators sit and wait for their prey to come within seizing distance. Examples include certain species of praying mantises, shrimp, crabs, spiders, polychaetes, and many vertebrates. Sessile opportunists usually are not very mobile. They can only capture prey when the prey organism comes into contact with them. Examples include certain protozoa, barnacles, and cnidarians. Grazing carnivores move about the substrate picking up small organisms. Their diet usually consists largely of sessile and slow-moving animals, such as sponges, ectoprocts, tunicates, snails, worms, and small crustaceans.

SURFACE NUTRIENT ABSORPTION

Some highly specialized animals have dispensed entirely with all mechanisms for prey capture, ingestion of food particles, and digestive processes. Instead they directly absorb nutrients from the external medium across their body surfaces. This medium may be nutrient-rich seawater, fluid in other animals' digestive tracts, or the body fluids of other animals. For example, some free-living protozoa, such as *Chilomonas,* absorb all of their nutrients across their body surface. The endoparasitic protozoa, cestode worms, endoparasitic gastropods, and crustaceans (all of which lack mouths and digestive systems) also absorb all of their nutrients across their body surface.

A few nonparasitic multicellular animals also lack a mouth and digestive system and absorb nutrients across their body surface. Examples include the gutless bivalves and pogonophoran worms. Interestingly, many pogonophoran worms absorb some nutrients from seawater across their body surface and also supplement their nutrition with organic carbon that symbiotic bacteria fix within the pogonophoran's tissues.

FLUID FEEDERS

The biological fluids of animals and plants are a rich source of nutrients. Feeding on this fluid is called **fluid feeding.** Fluid feeding is especially characteristic of some parasites, such as the intestinal nematodes that bite and rasp off host tissue or suck blood. External parasites (ectoparasites), such as leeches, ticks, mites, lampreys, and certain crustaceans, use a wide variety of mouthparts to feed on body fluids. For example, the sea lamprey has a funnel structure surrounding its mouth (*see figure 27.6a*). The funnel is lined with over 200 rasping teeth and a rasplike tongue. The lamprey uses the funnel like a suction cup to grip its fish host, and then with its tongue, rasps a hole in the fish's body wall. The lamprey then sucks blood and body fluids from the wound.

Insects have the most highly developed sucking structures for fluid feeding. For example, butterflies, moths, and aphids have tubelike mouthparts that enable them to suck up plant fluids. Blood-sucking mosquitoes have complex mouthparts with piercing stylets.

Most pollen- and nectar-feeding birds have long bills and tongues. In fact, the bill is often specialized (in shape, length, and curvature) for particular types of flowers. The tongues of some birds have a brushlike tip or are hollow, or both, to collect the nectar from flowers. Other nectar-feeding birds have short bills; they make a hole in the base of a flower and use their tongue to obtain nectar through the hole.

The only mammals that feed exclusively on blood are the vampire bats, such as *Desmodus,* of tropical South and Central America. These bats attack birds, cattle, and horses, using knife-sharp front teeth to pierce the surface blood vessels, and then lap at the oozing wound. Nectar-feeding bats have a long tongue to extract the nectar from flowering plants, and compared to the blood-feeding bats, have reduced dentition. In like manner, the nectar-feeding honey possum has a long, brush-tipped tongue and reduced dentition.

DIVERSITY IN DIGESTIVE STRUCTURES: INVERTEBRATES

In primitive, multicellular animals, such as cnidarians, the gut is a blind (closed) sac called a **gastrovascular cavity.** It has only one opening that is both entrance and exit (figure 27.3a); thus, it is an incomplete digestive tract. Some specialized cells in the cavity secrete digestive enzymes that begin the process of extracellular digestion. Other phagocytic cells that line the cavity engulf food material and continue intracellular digestion inside food vacuoles. Some flatworms have similar digestive patterns (figure 27.3b).

The development of the anus and complete digestive tract in the aschelminths was an evolutionary breakthrough (figure 27.3c). A complete digestive tract permits the one-way flow of ingested food without mixing it with previously ingested food or waste. Complete digestive tracts also have the advantage of progressive digestive processing in specialized regions along the system. Food can be digested efficiently in a series of distinctly different steps. The many variations of the basic plan of a

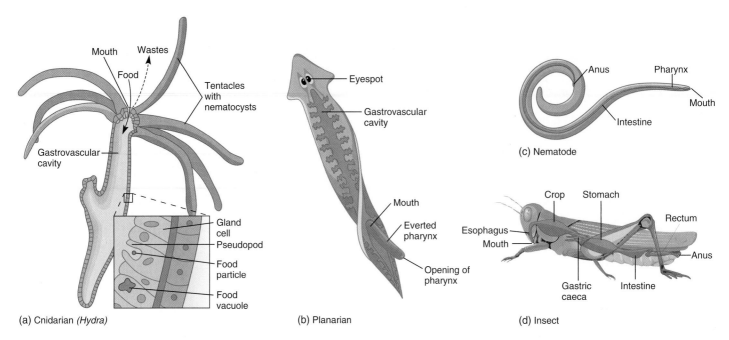

FIGURE 27.3

Various Types of Digestive Structures in Invertebrates. (*a*) The gastrovascular cavity of a cnidarian (*Hydra*) is an incomplete digestive tract because its one opening, a mouth, must serve as the entry and exit point for food and waste. Extracellular digestion occurs in the gastrovascular cavity, and intracellular digestion occurs inside food vacuoles formed when phagocytic cells engulf food particles. (*b*) Even though the gastrovascular cavity in a platyhelminth (planarian) branches extensively, it is also an incomplete digestive tract with only one opening. When a planarian feeds, it sticks its muscular pharynx out of its mouth and sucks in food. (*c*) A nematode (*Ascaris*) has a complete digestive tract with a mouth, pharynx, and anus. (*d*) The complete digestive tract of an insect (grasshopper) has an expanded region called a crop that functions as a food storage organ.

complete digestive tract correlate with different food-gathering mechanisms and diets (figure 27.3*d*). Most of these have been presented in the discussion of the many different protists and invertebrates and are not repeated in this chapter. Instead, three examples further illustrate digestive systems in protozoa and invertebrates: (1) the incomplete digestive system of a ciliated protozoan is an example of an intracellular digestive system; (2) the bivalve mollusc is an example of an invertebrate that has both intracellular and extracellular digestion; and (3) an insect is an example of an invertebrate that has extracellular digestion and a complete digestive tract.

PROTOZOA

As presented in chapter 8, protozoa may be autotrophic, saprozoic, or heterotrophic (ingest food particles). Ciliated protozoa are good examples of protists that utilize heterotrophic nutrition. Ciliary action directs food from the environment into the buccal cavity and cytostome (figure 27.4). The cytostome opens into the cytopharynx, which enlarges as food enters and pinches off a food-containing vacuole. The detached food vacuole then moves through the cytoplasm. During this movement, excess water is removed from the vacuole, the contents are acidified and then made alkaline, and a lysosome adds digestive enzymes. The food particles are then digested within the vacuole and the nutrients absorbed into the cytoplasm. The residual vacuole then excretes its waste products via the cytopyge.

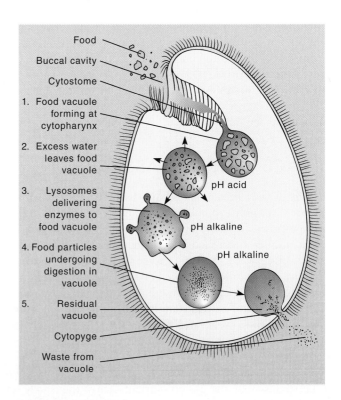

FIGURE 27.4

Intracellular Digestion in a Ciliated Protozoan. Cilia direct food toward the cytostome ("mouth"). The food enters the cytopharynx, where a food vacuole forms and detaches from the cytopharynx. The detached vacuole undergoes acidic and alkaline digestion, and the waste vacuole moves to the cytopyge ("anus") for excretion.

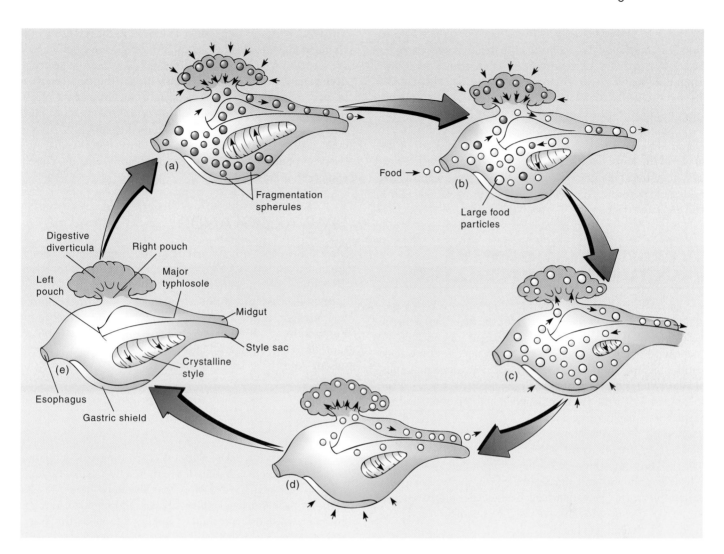

FIGURE 27.5

Extracellular and Intracellular Digestion in a Bivalve Mollusc.   (*a*) Extracellular digestion begins before food ingestion by the dissolving of the crystalline style and the formation of fragmentation spherules in the stomach. (*b*) As food enters the stomach, the rotating style and the enzymes released by the gastric shield mechanically and enzymatically break it down. (*c*) The small food particles then move into the digestive diverticulae for intracellular digestion. (*d*) A progressive passage of food particles from the stomach to the digestive diverticulae follows cessation of feeding. (*e*) During this resting phase, the stomach empties and the style reforms, while intracellular digestion in the diverticulae is completed, and fragmentation spherules begin to form again. The movement of fragmentation spherules starts the next feeding cycle.

BIVALVE MOLLUSCS

Many bivalve molluscs suspension feed and ingest small food particles. The digestive tract has a short esophagus opening into a stomach, midgut, hindgut, and rectum. The stomach contains a crystalline style, gastric shield, and diverticulated region. These diverticulae are blind-ending sacs that increase the surface area for absorption and intracellular digestion. The midgut, hindgut, and rectum function in extracellular digestion and absorption (figure 27.5).

Digestion is a coordination of three cycles: (1) feeding, (2) extracellular digestion, and (3) intracellular digestion. The resting phase is preparative for extracellular digestion. The mechanical and enzymatic breakdown of food during feeding provides the small particles for intracellular digestion. Intracellular digestion releases the nutrients into the blood and produces the fragmenta-

tion spherules that both excrete wastes and lower the pH for optimal extracellular digestion. These three cycles are linked to tidal immersion and emersion of the mollusc.

INSECTS

The grasshopper is a representative insect with a complete digestive tract and extracellular digestion (*see figure 27.3*d). During feeding, the mandibles and maxillae first break up (masticate) the food, which is then taken into the mouth and passed to the crop via the esophagus. During mastication, the salivary glands add saliva to the food to lubricate it for passage through the digestive tract. Saliva also contains the enzyme amylase, which begins the enzymatic digestion of carbohydrates. This digestion continues during food storage in the crop. The midgut secretes other enzymes (carbohydrases, lipases, proteases) that enter the crop. Food passes slowly from the crop to the stomach, where it is mechanically reduced and the nutrient particles sorted. Large particles are returned to the crop for further processing; the small particles enter the gastric cecae, where extracellular digestion is completed. Most nutrient absorption then occurs in the intestine. Undigested food is moved along the intestine and passes into the rectum, where water and ions are absorbed. The solid fecal pellets that form then pass out of the animal via the anus. During this entire feeding process the nervous system, the endocrine system, and the presence of food exert considerable control over enzyme production at various points in the digestive tract.

DIVERSITY IN DIGESTIVE STRUCTURES: VERTEBRATES

The complete vertebrate digestive tract (gut tube) is highly specialized in both structure and function for the digestion of a wide variety of foods. The basic structures of the gut tube include the buccal cavity, pharynx, esophagus, stomach, small intestine, large intestine, rectum, and anus/cloaca. In addition, three important glandular systems are associated with the digestive tract: (1) the salivary glands; (2) the liver, gallbladder, and bile duct; and (3) the pancreas and pancreatic duct.

Because most vertebrates spend the majority of their time acquiring food, feeding is the universal pastime. The oral cavity (mouth), teeth, intestines, and other major digestive structures usually reflect the way an animal gathers food, the type of food it eats, and the way it digests that food. These major digestive structures are now discussed to illustrate the diversity of form and function among different vertebrates.

TONGUES

A tongue or tonguelike structure develops in the floor of the oral cavity in many vertebrates. For example, a lamprey has a protrusible tongue with horny teeth that rasp its prey's flesh (figure 27.6a). Fishes may have a primary tongue that bears teeth that help hold prey; however, this type of tongue is not muscular (figure 27.6b). Tetrapods have evolved mobile tongues for gathering food. Frogs and salamanders and some lizards can rapidly project part of their tongue from the mouth to capture an insect (figure 27.6c). A woodpecker has a long, spiny tongue for gathering insects and grubs (figure 27.6d). Ant- and termite-eating mammals also gather food with long, sticky tongues. Spiny papillae on the tongues of cats and other carnivores help these animals rasp flesh from a bone.

TEETH

With the exception of birds, turtles, and baleen whales, most vertebrates have teeth. Birds lack teeth, probably to reduce body weight for flight. Teeth are specialized, depending on whether an animal feeds on plants or animals, and on how it obtains its food.

The teeth of snakes slope backward to aid in the retention of prey while swallowing (figure 27.7a), and the canine teeth of wolves are specialized for ripping food (figure 27.7b). Herbivores, such as deer, have predominantly grinding teeth, the front teeth of a beaver are used for chiseling trees and branches, and the elephant has two of its upper, front teeth specialized as weapons and for moving objects (figure 27.7c–e). Because humans, pigs, bears, raccoons, and a few other mammals are omnivores, they have teeth that can perform a number of tasks—tearing, ripping, chiseling, and grinding (figure 27.7f).

SALIVARY GLANDS

Most fishes lack salivary glands in the head region. Lampreys are an exception because they have a pair of glands that secrete an anticoagulant needed to keep their prey's blood flowing as they feed. Modified salivary glands of some snakes produce venom that is injected through fangs to immobilize prey. Because the secretion of oral digestive enzymes is not an important function in amphibians or reptiles, salivary glands are absent. Most birds lack salivary glands, while all mammals have them.

ESOPHAGI

The esophagus (pl., esophagi) is short in fishes and amphibians, but much longer in amniotes due to their longer necks. Grain- and seed-eating birds have a crop that develops from the caudal portion of the esophagus (figure 27.8a). Storing food in the crop ensures an almost continuous supply of food to the stomach and intestine for digestion. This structure allows these birds to reduce the frequency of feeding and still maintain a high metabolic rate.

STOMACHS

The stomach is an ancestral vertebrate structure that evolved as vertebrates began to feed on larger organisms that were caught at less frequent intervals and required storage. Some zoologists believe that the gastric glands and their production of hydrochloric acid (HCl) evolved in the context of killing bacteria and helping preserve food. The enzyme pepsinogen may have evolved later because the stomach is not essential for digestion.

GIZZARDS

Some fishes, some reptiles such as crocodilians, and all birds have a gizzard for grinding up food (figure 27.8a). The bird's gizzard develops from the posterior part of the stomach called the ventriculus. Pebbles (grit) that have been swallowed are often retained in the gizzard of grain-eating birds and facilitate the grinding process.

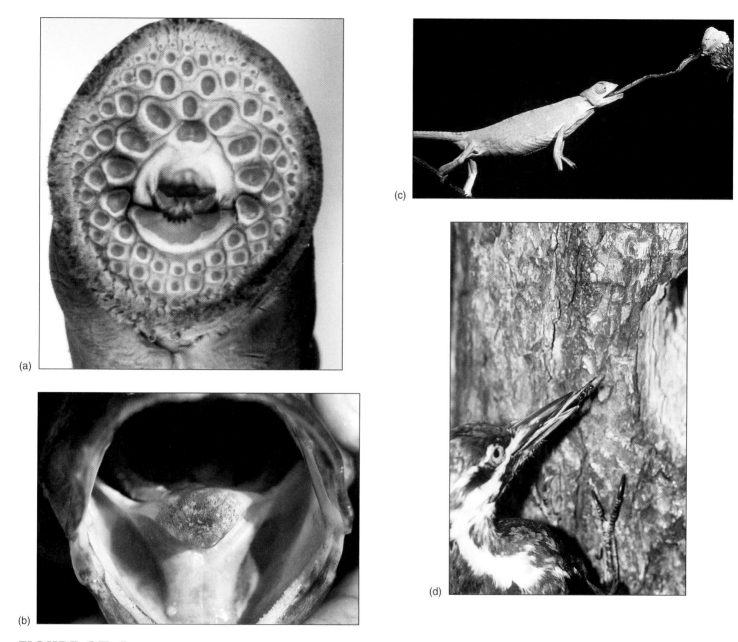

(a)

(b)

(c)

(d)

FIGURE 27.6

Tongues.   (*a*) Rasping tongue and mouth of a lamprey. (*b*) Fish tongue. (*c*) Tongue of a chameleon catching an insect. (*d*) The tongue of a woodpecker extracts insects from the bark of a tree.

RUMENS

Ruminant mammals—animals that "chew their cud," such as cows, sheep, and deer—show some of the most unusual modifications of the stomach. *This method of digestion has evolved in animals that need to eat large amounts of food relatively quickly, but can chew the food at a more comfortable or safer location.* More important, though, the ruminant stomach provides an opportunity for large numbers of microorganisms to digest the cellulose walls of grass and other vegetation. Cellulose contains a large amount of energy; however, animals generally lack the ability to produce the enzyme cellulase for digesting

cellulose and obtaining its energy. Because gut microorganisms can produce cellulase, they have made the herbivorous lifestyle more effective.

In ruminants, the upper portion of the stomach expands to form a large pouch, the rumen, and a smaller reticulum. The lower portion of the stomach consists of a small antechamber, the omasum, with a "true" stomach, or abomasum, behind it (figure 27.9). Food first enters the rumen, where it encounters the microorganisms. Aided by copious fluid secretions, body heat, and churning of the rumen, the microorganisms partially digest the food and reduce it to a pulpy mass. Later, the pulpy mass moves into the reticulum, from which mouthfuls are regurgitated as "cud" (L. *ruminare*, to

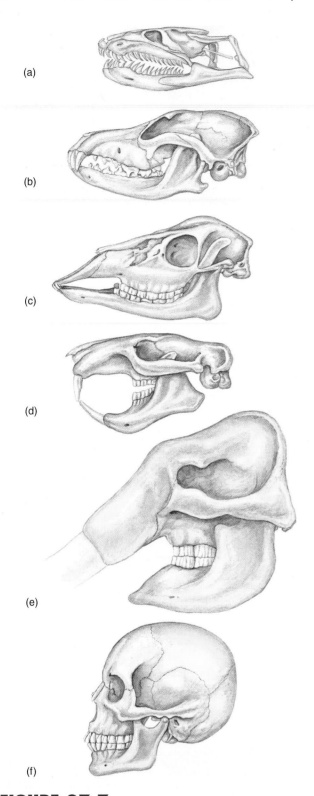

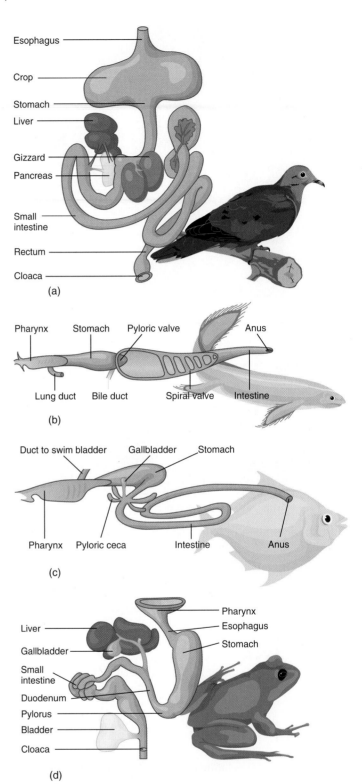

FIGURE 27.7

Arrangement of Teeth in a Variety of Vertebrates.   (*a*) Snake.
(*b*) Wolf. (*c*) Deer. (*d*) Beaver. (*e*) Elephant. (*f*) Human.

FIGURE 27.8

Arrangement of Stomachs and Intestines in a Variety of Vertebrates.
(*a*) Pigeon. (*b*) Lungfish. (*c*) Teleost fish. (*d*) Frog.

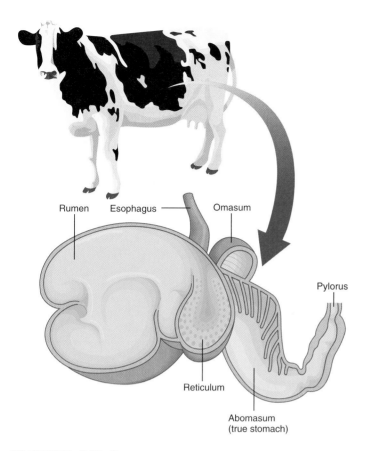

FIGURE 27.9
Ruminant Mammal. Four-chambered stomach of a cow, where symbiotic microorganisms digest cellulose.

chew the cud). At this time, food is thoroughly chewed for the first time. When reswallowed, the food enters the rumen, where it becomes more liquid in consistency. When it is very liquid, the digested food material flows out of the reticulum and into the omasum and then the glandular region, the abomasum. Here the digestive enzymes are first encountered, and digestion continues.

CECA

Microorganisms attack the food of ruminants before gastric digestion, but in the typical nonruminant herbivore, microbial action on cellulose occurs after digestion. Rabbits, horses, and rats digest cellulose by maintaining a population of microorganisms in their unusually large cecum, the blind pouch that extends from the colon (figure 27.10). Adding to this efficiency, a few nonruminant herbivores, such as mice and rabbits, eat some of their own feces to process the remaining materials in them, such as vitamins.

LIVERS AND GALLBLADDERS

In those vertebrates with a gallbladder, it is closely associated with the liver. The liver manufactures bile, which the gallbladder then stores. **Bile** is a fluid containing bile salts and bile pigments.

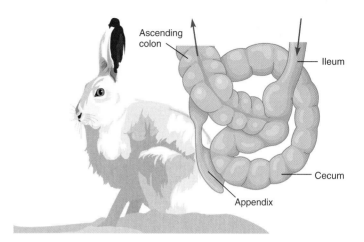

FIGURE 27.10
Extensive Cecum of a Nonruminant Herbivore, Such as a Rabbit. The cecum contains microorganisms that produce digestive enzymes (e.g., cellulase that helps break down cellulose).

Bile salts play an important role in the digestion of fats, although they are not digestive enzymes. They emulsify dietary fat, breaking it into small globules (emulsification) on the surface of which the fat-digesting enzyme lipase can function. Bile pigments result from phagocytosis of red blood cells in the spleen, liver, and red bone marrow. Phagocytosis cleaves the hemoglobin molecule, releasing iron, and the remainder of the molecule is converted into pigments that enter the circulation. These pigments are subsequently extracted from the circulation in the liver and excreted in the bile as bilirubin ("red bile") and biliverdin ("green bile").

Because of the importance of bile in fat digestion, the gallbladder is relatively large in carnivores and vertebrates in which fat is an important part of the diet. It is much reduced or absent in bloodsuckers, such as the lamprey, and in animals that feed primarily on plant food (e.g., some teleosts, many birds, and rats).

PANCREATA

Every vertebrate has a pancreas (pl., pancreata); however, in lampreys and lungfishes it is embedded in the wall of the intestine and is not a visible organ. Both endocrine and exocrine tissues are present, but the cell composition varies. Pancreatic fluid containing many enzymes empties into the small intestine via the pancreatic duct.

INTESTINES

The configuration and divisions of the small and large intestines vary greatly among vertebrates. Intestines are closely related to the animal's type of food, body size, and levels of activity. For example, cyclostomes, chondrichthian fishes, and primitive bony fishes have short, nearly straight intestines that

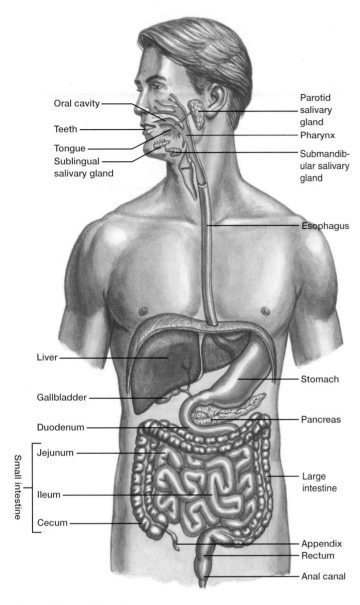

FIGURE 27.11

Major Organs and Parts of the Human Digestive System. Food passes from the mouth through the pharynx and esophagus to the stomach. From the stomach, it passes to the small intestine, where nutrients are broken down and absorbed into the circulatory and lymphatic systems. Nutrients then move to the large intestine, where water is reabsorbed, and feces form. Feces exit the body via the anal canal.

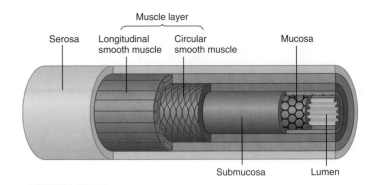

FIGURE 27.12

Mammalian Gastrointestinal Tract. Common structural layers of the gastrointestinal tract. The central lumen extends from the mouth to the anus.

THE MAMMALIAN DIGESTIVE SYSTEM

Humans, pigs, bears, raccoons, and a few other mammals are omnivores. The digestive system of an omnivore has the mechanical and chemical ability to process many kinds of foods. The sections that follow examine the control of gastrointestinal motility, the major parts of the alimentary canal, and the accessory organs of digestion (figure 27.11).

The process of digesting and absorbing nutrients in a mammal includes:

1. Ingestion—eating
2. Peristalsis—the involuntary, sequential muscular contractions that move ingested nutrients along the digestive tract
3. Segmentation—mixing the contents in the digestive tract
4. Secretion—the release of hormones, enzymes, and specific ions and chemicals that take part in digestion
5. Digestion—the conversion of large nutrient particles or molecules into small particles or molecules
6. Absorption—the passage of usable nutrient molecules from the small intestine into the bloodstream and lymphatic system for the final passage to body cells
7. Defecation—the elimination from the body of undigested and unabsorbed material as waste

GASTROINTESTINAL MOTILITY AND ITS CONTROL

As with any organ, the function of the gastrointestinal tract is determined by the type of tissues it contains. Most of the mammalian gastrointestinal tract has the same anatomical structure along its entire length (figure 27.12). From the outside inward is a thin layer of connective tissue called the serosa. (The serosa forms a moist epithelial sheet called the peritoneum. This peritoneum lines the entire abdominal cavity and covers all internal organs. The space it encompasses is the coelom.) Next are the longitudinal smooth-muscle layer and circular smooth-muscle layer.

extend from the stomach to the anus (*see figure 27.8b*). In more advanced bony fishes, the intestine increases in length and begins to coil (*see figure 27.8c*). The intestines are moderately long in most amphibians and reptiles (*see figure 27.8d*) In birds and mammals, the intestines are longer and have more surface area than those of other tetrapods (*see figure 27.8a*). Birds typically have two ceca, and mammals have a single cecum at the beginning of the large intestine. The large intestine is much longer in mammals than in birds, and it empties into the cloaca in most vertebrates.

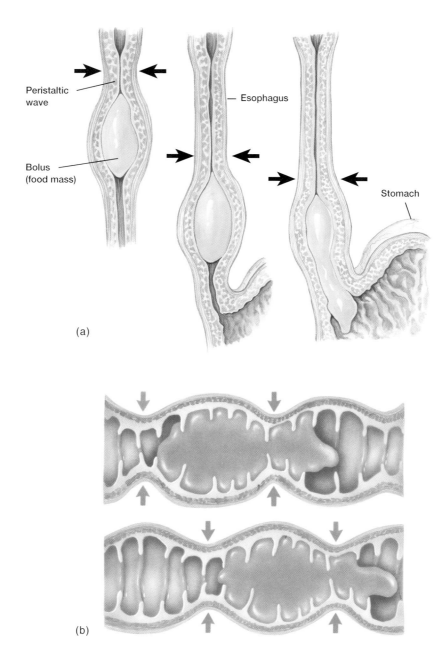

FIGURE 27.13

Peristalsis and Segmentation. (*a*) Peristaltic waves move food through the esophagus to the stomach. (*b*) In segmentation, simultaneous muscular contractions of many sections of the intestine (blue arrows) help mix nutrients with digestive secretions.

Underneath this muscle layer is the submucosa. The submucosa contains connective tissue, blood, and lymphatic vessels. The mucosa faces the central opening, which is called a lumen.

The coordinated contractions of the muscle layers of the gastrointestinal tract mix the food material with various secretions and move the food from the oral cavity to the rectum. The two types of movement involved are peristalsis and segmentation.

During **peristalsis** (Gr. *peri*, around + *stalsis*, contraction), food advances through the gastrointestinal tract when the rings of circular smooth muscle contract behind it and relax in front of it (figure 27.13*a*). Peristalsis is analogous to squeezing icing

from a pastry tube. The small and large intestines also have rings of smooth muscles that repeatedly contract and relax, creating an oscillating back-and-forth movement in the same place, called **segmentation** (figure 27.13*b*). This movement mixes the food with digestive secretions and increases the efficiency of absorption.

Sphincters also influence the flow of material through the gastrointestinal tract and prevent backflow. Sphincters are rings of smooth or skeletal muscle at the beginning or ends of specific regions of the gut tract. For example, the cardiac sphincter is between the esophagus and stomach, and the pyloric sphincter is between the stomach and small intestine.

Control of gastrointestinal activity is based on the volume and composition of food in the lumen of the gut. For example, ingested food distends the gut and stimulates mechanical receptors in the gut wall. In addition, digestion of carbohydrates, lipids, and proteins stimulates various chemical receptors in the gut wall. Signals from these mechanical and chemical stimuli travel through nerve plexuses in the gut wall to control the muscular contraction that leads to peristalsis and segmentation, as well as the secretion of various substances (e.g., mucus, enzymes) into the gut lumen. In addition to this local control, long-distance nerve pathways connect the receptors and effectors with the central nervous system. Either or both of these pathways function to maintain homeostasis in the gut. The endocrine cells of the gastrointestinal tract also produce hormones that help regulate secretion, digestion, and absorption.

ORAL CAVITY

A pair of lips protects the **oral cavity** (mouth). The lips are highly vascularized, skeletal muscle tissue with an abundance of sensory nerve endings. Lips help retain food as it is being chewed and play a role in phonation (the modification of sound).

The oral cavity contains the tongue and teeth (figure 27.14). Mammals can mechanically process a wide range of foods because their teeth are covered with enamel (the hardest material in the body) and because their jaws and teeth exert a strong force. The oral cavity is continuously bathed by **saliva,** a watery fluid that at least three pairs of salivary glands secrete. Saliva moistens food, binds it with mucins (glycoproteins), and forms the ingested food into a moist mass called a bolus. Saliva also contains bicarbonate ions (HCO_3^-), which buffer chemicals in the mouth, and thiocyanate ions (SCN^-) and the enzyme lysozyme, which kill microorganisms. It also contributes an enzyme (amylase) necessary for the initiation of carbohydrate digestion.

PHARYNX AND ESOPHAGUS

Chapter 26 discusses how both air and swallowed foods and liquids pass from the mouth into the **pharynx**—the common passageway for both the digestive and respiratory tracts. The epiglottis temporarily seals off the opening (glottis) to the trachea so that swallowed food does not enter the trachea. Initiation of the swallowing reflex can be voluntary, but most of the time it is involuntary. When swallowing begins, sequential, involuntary contractions of smooth muscles in the walls of the **esophagus** propel the bolus or liquid to the stomach. Neither the pharynx nor the esophagus contribute to digestion.

STOMACH

The mammalian **stomach** is a muscular, distensible sac with three main functions. It (1) stores and mixes the food bolus received from the esophagus, (2) secretes substances (enzymes, mucus, and hydrochloric acid [HCl]) that start the digestion of proteins, and

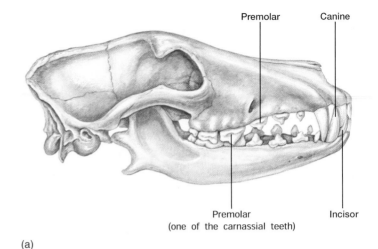
(a)

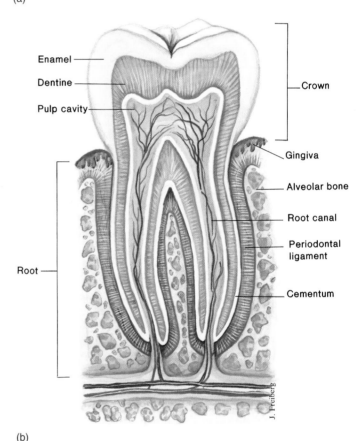
(b)

FIGURE 27.14

Teeth. (*a*) The teeth of a carnivore, such as this wolf, are specialized for slicing, puncturing, tearing, and grinding animal flesh. (*b*) Anatomy of a typical mammalian tooth.

(3) helps control the rate at which food moves into the small intestine via the pyloric sphincter (figure 27.15*a*).

The stomach is made up of an inner mucous membrane containing thousands of gastric glands (figure 27.15*b*). Three types of cells are in these glands. **Parietal cells** secrete a solution containing HCl, and **chief cells** secrete pepsinogen, the precursor of the enzyme pepsin. Both of the cells are in the pits of the gastric glands.

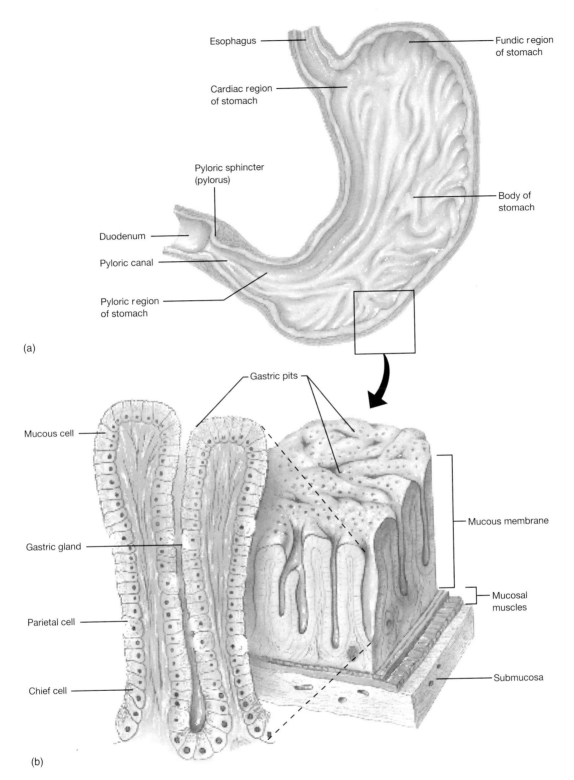

Esophagus

Fundic region
of stomach

Cardiac region
of stomach

Pyloric sphincter
(pylorus)

Body of
stomach

Duodenum

Pyloric canal

Pyloric region
of stomach

(a)

Gastric pits

Mucous cell

Gastric gland

Mucous membrane

Parietal cell

Mucosal
muscles

Chief cell

Submucosa

(b)

FIGURE 27.15

Stomach. (*a*) Food enters the stomach from the esophagus. (*b*) Gastric glands cover the mucosa of the stomach and include mucous cells, parietal cells, and chief cells. Each type produces a different secretion.

The surface of the mucous membrane at the openings of the glands contains numerous **mucous cells** that secrete mucus that coats the surface of the stomach and protects it from the HCl and digestive enzymes. The surfaces of the upper gastrointestinal tract—the esophagus and mouth—have a much thinner mucous-cell layer than the stomach, which is why vomiting can cause a burning sensation in the esophagus or mouth. Endocrine cells in one part of the stomach mucosa release the hormone gastrin, which travels to target cells in the gastric glands, further stimulating them.

When the bolus of food enters the stomach, it distends the walls of the stomach. This distention, as well as the act of eating, causes the gastric pits to secrete HCl (as H^+ and Cl^-) and pepsinogen. The H^+ ions cause pepsinogen to be converted into the active enzyme pepsin. As pepsin, mucus, and HCl mix with and begin to break down proteins, smooth mucosal muscles contract and vigorously churn and mix the food bolus. About three to four hours after a meal, the stomach contents have been sufficiently mixed and are a semiliquid mass called **chyme** (Gr. *chymos*, juice). The pyloric sphincter regulates the release of the chyme into the small intestine.

When the stomach is empty, peristaltic waves cease; however, after about 10 hours of fasting, new waves may occur in the upper region of the stomach. These waves can cause "hunger pangs" as sensory nerve fibers carry impulses to the brain.

SMALL INTESTINE: MAIN SITE OF DIGESTION

Most of the food a mammal ingests is digested and absorbed in the **small intestine.** The human small intestine is about 4 cm in diameter and 7 to 8 m in length (*see figure 27.11*). It is intermediate in length between the small intestines of typical carnivores and herbivores of similar size, and it reflects the human's omnivorous eating habits. The length of the small intestine directly relates to the total surface area available for absorbing nutrients, as determined by the many circular folds and minute projections of the inner gut surface (figure 27.16a). On the circular folds, thousands of fingerlike projections called **villi** (L. *villus*, tuft of hair) (sing. villus) project from each square centimeter of mucosa (figure 27.16b, c). Simple columnar epithelial cells, each bearing numerous microvilli, cover both the circular folds and villi (figure 27.16d). These minute projections are so dense that the inner wall of the human small intestine has a total surface area of approximately 300 m^2—the size of a tennis court.

The first part of the small intestine, called the duodenum, functions primarily in digestion. The next part is the jejunum, and the last part is the ileum. Both function in nutrient absorption.

The duodenum contains many digestive enzymes that intestinal glands in the duodenal mucosa secrete. The pancreas secretes other enzymes. In the duodenum, digestion of carbohydrates and proteins is completed, and most lipids are digested. The jejunum and ileum absorb the end products of digestion (amino acids, simple sugars, fatty acids, glycerol, nucleotides, water). Much of this absorption involves active transport and the sodium-dependent ATPase pump. Sugars and amino acids are absorbed the capillaries of the villi, whereas free fatty acids enter the

epithelial cells of the villi and recombine with glycerol to form triglycerides. The triglycerides are coated with proteins to form small droplets called **chylomicrons,** which enter the lacteals of the villi (figure 27.16c). From the lacteals, the chylomicrons move into the lymphatics and eventually into the bloodstream for transport throughout the body.

Besides absorbing organic molecules, the small intestine absorbs water and dissolved mineral ions. The small intestine absorbs about 9 liters of water per day, and the large intestine absorbs the rest.

LARGE INTESTINE

Unlike the small intestine, the **large intestine** has no circular folds, villi, or microvilli; thus, the surface area is much smaller. The small intestine joins the large intestine near a blind-ended sac, the **cecum** (L. *caecum*, blind gut) (*see figure 27.11*). *The human cecum and its extension, the **appendix,** are storage sites and possibly represent evolutionary remains of a larger, functional cecum, such as is found in herbivores* (see figure 27.10). The appendix contains an abundance of lymphoid tissue and may function as part of the immune system.

The major functions of the large intestine include the reabsorption of water and minerals, and the formation and storage of feces. As peristaltic waves move food residue along, minerals diffuse or are actively transported from the residue across the epithelial surface of the large intestine into the bloodstream. Water follows osmotically and returns to the lymphatic system and bloodstream. When water reabsorption is insufficient, diarrhea (Gr. *rhein*, to flow) results. If too much water is reabsorbed, fecal matter becomes too thick, resulting in constipation.

Many bacteria and fungi exist symbiotically in the large intestine. They feed on the food residue and further break down its organic molecules to waste products. In turn, they secrete amino acids and vitamin K, which the host's gut absorbs. What remains—feces—is a mixture of bacteria, fungi, undigested plant fiber, sloughed-off intestinal cells, and other waste products.

ROLE OF THE PANCREAS IN DIGESTION

The **pancreas** (Gr. *pan*, all + *kreas*, flesh) is an organ that lies just ventral to the stomach and has both endocrine and exocrine functions. Exocrine cells in the pancreas secrete digestive enzymes into the pancreatic duct, which merges with the hepatic duct from the liver to form a common bile duct that enters the duodenum. Pancreatic enzymes complete the digestion of carbohydrates and proteins and initiate the digestion of lipids. Trypsin, carboxypeptidase, and chymotrypsin digest proteins into small peptides and individual amino acids. Pancreatic lipases split triglycerides into smaller, absorbable glycerol and free fatty acids. Pancreatic amylase converts polysaccharides into disaccharides and monosaccharides. Table 27.6 summarizes the major glands, secretions, and enzymes of the mammalian digestive system.

The pancreas also secretes bicarbonate (HCO_3^-) ions that help neutralize the acidic food residue coming from the stomach.

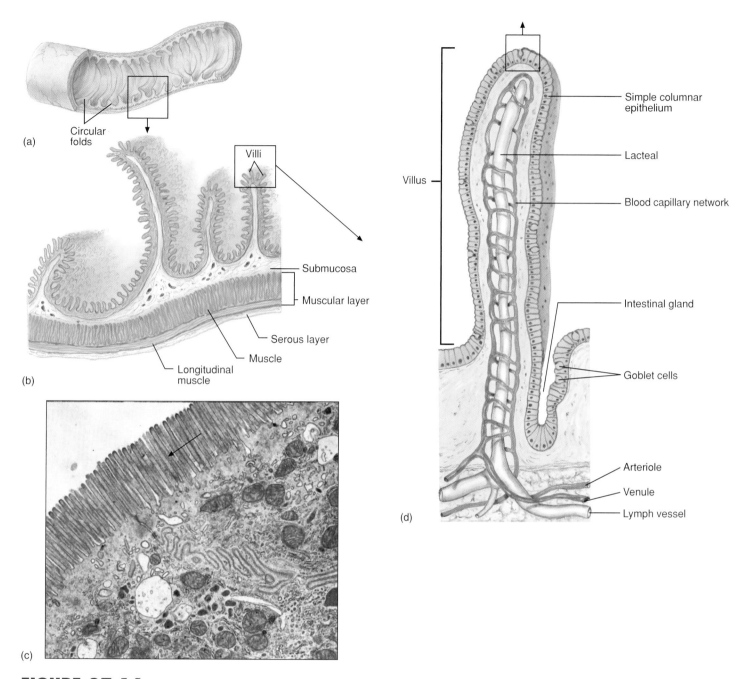

(a) Circular folds

(b) Submucosa
Muscular layer
Serous layer
Muscle
Longitudinal muscle
Villi

(c)

(d) Villus
Simple columnar epithelium
Lacteal
Blood capillary network
Intestinal gland
Goblet cells
Arteriole
Venule
Lymph vessel

FIGURE 27.16

Small Intestine.   The small intestine absorbs food over a large surface area. (*a*) The lining of the intestine has many circular folds. (*b, d*) Fingerlike villi line the intestine. A single villus contains a central capillary network and a lymphatic lacteal, both of which transport nutrients absorbed from the lumen of the intestine. (*c*) The plasma membrane of the simple columnar epithelial cells covering the villi fold into microvilli (arrow), which further increase the surface area facing the lumen.

Bicarbonate raises the pH from 2 to 7 for optimal digestion. Without such neutralization, pancreatic enzymes could not function.

ROLE OF THE LIVER AND GALLBLADDER IN DIGESTION

The **liver,** the largest organ in the mammalian body, is just under the diaphragm (*see figure 27.11*). In the liver, millions of specialized cells called hepatocytes take up nutrients absorbed from

the intestines and release them into the bloodstream. Hepatocytes also manufacture the blood proteins prothrombin and albumin.

In addition, some major metabolic functions of the liver include:

1. Removal of amino acids from organic compounds.
2. Urea formation from proteins and conversion of excess amino acids into urea to decrease body levels of ammonia.
3. Manufacture of most of the plasma proteins, formation of fetal erythrocytes, destruction of worn-out erythrocytes, and

TABLE 27.6
MAJOR DIGESTIVE GLANDS, SECRETIONS, AND ENZYMES IN MAMMALS

PLACE OF DIGESTION	SOURCE	SECRETION	ENZYME	DIGESTIVE FUNCTION
Mouth	Salivary glands	Saliva	Salivary amylase	Begins the digestion of carbohydrates; inactivated by stomach HCl
	Mucous glands	Mucus	—	Lubricates food bolus
Esophagus	Mucous glands	Mucus	—	Lubricates food bolus
Stomach	Gastric glands	Gastric juice	Lipase	Digests lipids into fatty acids and glycerol
			Pepsin	Digests proteins into polypeptides
	Gastric mucosa	HCl	—	Converts pepsinogen into active pepsin; kills microorganisms
	Mucous glands	Mucus	—	Lubricates
Small intestine	Liver	Bile	—	Emulsifies lipids; activates lipase
	Pancreas	Pancreatic juice	Amylase	Digests starch into maltose
			Chymotrypsin	Digests proteins into peptides and amino acids
			Lipase	Digests lipids into fatty acids and glycerol (requires bile salts)
			Nuclease	Digests nucleic acids into mononucleotides
			Trypsin	Digests proteins into peptides and amino acids
	Intestinal glands	Intestinal juice	Enterokinase	Digests inactive trypsinogen into active trypsin
			Lactase	Digests lactose into glucose and galactose
			Maltase	Digests maltose into glucose
			Peptidase	Digests polypeptides into amino acids
			Sucrase	Digests sucrose into glucose and fructose
	Mucous glands	Mucus	—	Lubricates
Large intestine	Mucous glands	Mucus	—	Lubricates

synthesis of the blood-clotting agents prothrombin and fibrinogen from amino acids.

4. Synthesis of nonessential amino acids.
5. Conversion of galactose and fructose to glucose.
6. Oxidation of fatty acids.
7. Formation of lipoproteins, cholesterol, and phospholipids (essential cell membrane components).
8. Conversion of carbohydrates and proteins into fat.
9. Modification of waste products, toxic drugs, and poisons (detoxification).
10. Synthesis of vitamin A from carotene, and with the kidneys, participation in the activation of vitamin D.
11. Maintenance of a stable body temperature by raising the temperature of the blood passing through it. Its many metabolic activities make the liver the major heat producer in a mammal's body.
12. Manufacture of bile salts, which are used in the small intestine for the emulsification and absorption of simple fats, cholesterol, phospholipids, and lipoproteins.
13. Main storage center. The liver stores glucose in the form of glycogen, and with the help of insulin and enzymes, converts glycogen back into glucose as the body needs it. The liver also stores fat-soluble vitamins (A, D, E, and K), and minerals, such as iron, from the diet. The liver can also store fats and amino acids, and convert them into usable glucose as required.

The **gallbladder** (L. *galbinus*, greenish yellow) is a small organ near the liver (*see figure 27.11*). The gallbladder stores the greenish fluid called bile that the liver cells continuously produce. Bile is very alkaline and contains pigments, cholesterol, lecithin, mucin, bilirubin, and bile salts that act as detergents to emulsify fats (form them into droplets suspended in water) and aid in fat digestion and absorption. (Recall that fats are insoluble in water.) Bile salts also combine with the end products of fat digestion to form micelles. **Micelles** are lipid aggregates (fatty acids and glycerol) with a surface coat of bile salts. Because they are so small, they can cross the microvilli of the intestinal epithelium.

SUMMARY

1. Nutrition describes all of those processes by which an animal takes in, digests, absorbs, stores, and uses food (nutrients) to meet its metabolic needs. Digestion is the mechanical and chemical breakdown of food into smaller particles that the individual cells of an animal can absorb.

2. Losses of biosynthetic abilities have marked much of animal evolution. This tendency has led to the evolution of the following nutritional types: insectivores, herbivores, carnivores, and omnivores.

3. The nutrients that a heterotroph ingests can be divided into macronutrients and micronutrients. Macronutrients are needed in large quantities and include the carbohydrates, lipids, and proteins. Micronutrients are needed in small quantities and include the vitamins and minerals.

4. Only a few protists and animals can absorb nutrients directly from their external environment. Most animals must work for their nutrients. Various specializations have evolved for food procurement (feeding) in animals. Some examples include continuous versus discontinuous feeding, suspension feeding, deposit feeding, herbivory, predation, surface nutrient absorption, and fluid feeding.

5. The evolution and structure of the digestive system in various invertebrates and vertebrates reflect their eating habits, their rate of metabolism, and their body size.

6. The digestive system of vertebrates is one-way, leading from the mouth (oral cavity), to the pharynx, esophagus, stomach, small intestine, large intestine, rectum, and anus.

7. The coordinated contractions of the muscle layer of the gastrointestinal tract mix food material with various secretions and move the food from the oral cavity to the rectum. The two types of movement involved are segmentation and peristalsis.

8. Most digestion occurs in the duodenal portion of the small intestine. The products of digestion are absorbed in the walls of the jejunum and ileum. In the process of digestion, bile that the liver secretes makes fats soluble. The liver has many diverse functions. It controls the fate of newly synthesized food molecules, stores excess glucose as glycogen, synthesizes many blood proteins, and converts nitrogenous and other wastes into a form that the kidneys can excrete.

9. The large intestine has little digestive or nutrient absorptive activity. It functions principally to absorb water, to compact the material left over from digestion, and to be a storehouse for microorganisms.

SELECTED KEY TERMS

autotrophs (p. 444)
carnivores (p. 444)
digestion (p. 443)
herbivores (p. 444)
heterotrophs (p. 444)
macronutrients (p. 444)

micronutrients (p. 444)
nutrition (p. 443)
omnivores (p. 444)
suspension feeding (p. 448)
vitamin (p. 445)

CRITICAL THINKING QUESTIONS

1. What advantages are there to digestion? Would it not be simpler for a vertebrate to simply absorb carbohydrates, lipids, and proteins from its food and to use these molecules without breaking them down?

2. What might have been some evolutionary pressures acting on animals that led to the internalization of digestive systems?

3. Many digestive enzymes are produced in the pancreas and released into the duodenum. Why, then, has the mammalian stomach evolved the ability to produce pepsinogen?

4. Human vegetarians, unlike true herbivores, have no highly specialized fermentation chambers. Why?

5. Trace the fate of a hamburger from the mouth to the anus, identifying sites and mechanisms of digestion and absorption.

ONLINE LEARNING CENTER

Visit our Online Learning Center (OLC) at www.mhhe.com/zoology (click on this book's title) to find the following chapter-related materials:

- CHAPTER QUIZZING
- ANIMATIONS
    Digestion
    Plaque Formation
    Dental Caries
    Endoscopy
    Digestion (Mouth to Stomach)
    Digestion (Stomach)
    Stomach Digestion
    Ulcers
    Formation of Gallstones
    Small Intestine Digestion
    Digestion (Stomach to Small Intestine)

- RELATED WEB LINKS
    Digestion and Nutrition
        Vertebrates: Macroscopic Anatomy of the Digestive System
        Vertebrates: Microscopic Anatomy of the Digestive System
        Nutrition
        Human Digestive Topics

- BOXED READINGS ON
    Suspension Feeding in Invertebrates and Nonvertebrate Chordates
    Filter Feeding in Vertebrates

- SUGGESTED READINGS
- LAB CORRELATIONS

Check out the OLC to find specific information on these related lab exercises in the *General Zoology Laboratory Manual,* 5th edition, by Stephen A. Miller:

Exercise 22   *Vertebrate Digestion*

CHAPTER 28

TEMPERATURE AND BODY FLUID REGULATION

Outline

Concepts

1. Thermoregulation is a complex and important physiological process that maintains, to varying degrees, an animal's body temperature, despite variations in environmental temperature. Based on this regulation, animals can be categorized as endotherms or ectotherms, and homeotherms or heterotherms.

2. For osmoregulation, some invertebrates have contractile vacuoles, flame-cell systems, antennal (green) glands, maxillary glands, coxal glands, nephridia, or Malpighian tubules.

3. A vertebrate's urinary system functions in osmoregulation and excretion, both of which are necessary for internal homeostasis. Osmoregulation governs water and salt balance, and excretion eliminates metabolic wastes. In fishes, reptiles, birds, and mammals, the kidneys are the primary osmoregulatory structures.

The earth's environments vary dramatically in temperature and amount of water present. In the polar regions, high mountain ranges, and deep oceans, the temperature remains near or below 0° C (32° F) throughout the year. Temperatures exceeding 40° C (103° F) are common in equatorial deserts. Between these two extremes, in the earth's temperate regions, temperatures fluctuate widely. The temperate regions have varying amounts of water, as well as varied habitats—freshwater, saltwater, wetlands, mountains, and grasslands.

Animals have successfully colonized these varied places on earth by possessing homeostatic mechanisms for maintaining a relatively constant internal environment, despite fluctuations in the external environment. This chapter covers three separate but related homeostatic systems that enable animals to survive the variations in temperature, water availability, and salinity (salt concentration) on the earth. The thermoregulatory system maintains an animal's body temperature and/or its responses to shifts in environmental temperature. The osmoregulatory system maintains the level and concentration of water and salts in the body. And the urinary system eliminates metabolic wastes from the body and functions in osmoregulation.

HOMEOSTASIS AND TEMPERATURE REGULATION

The temperature of a living cell affects the rate of its metabolic processes. An animal can grow faster and respond to the environment more rapidly if its cells are kept warm. *In fact, zoologists believe that the ability of some higher animals to maintain a constant*

This chapter contains evolutionary concepts, which are set off in this font.

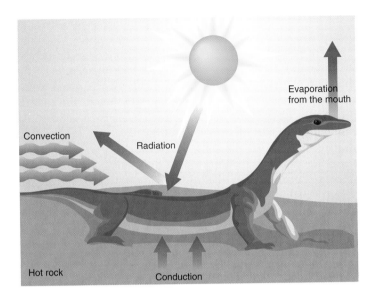

FIGURE 28.1

Heat Gain and Loss for a Terrestrial Reptile in a Typical Terrestrial Environment.   Heat is either gained or lost by objects in direct contact with the animal (conduction), by air currents (convection), by exhaled air (evaporation), or by electromagnetic waves (radiation).

(homeostatic), relatively high body temperature is a major reason for their evolutionary success. This ability to control the temperature of the body is called **thermoregulation** ("heat control") and involves the nervous, endocrine, respiratory, and circulatory systems in higher animals.

THE IMPACT OF TEMPERATURE ON ANIMAL LIFE

Every animal's physiological functions are inexorably linked to temperature, because metabolism is sensitive to changes in internal temperature. Thus, temperature has been a strong source of selective pressure on all animals. The rate of cellular respiration increases with temperature up to a point. When the temperature rises above the temperature optima at which enzymes most efficiently catalyze their chemical reactions, the rates decline as the enzymes begin to denature. The chemical interactions holding the enzymes in their three-dimensional shape are also disrupted. *Thus, the results of enzyme evolution have frequently been enzymes with temperature optima that reflect an animal's habitat.* For example, a digestive enzyme in a trout might function optimally at 10° C, whereas another enzyme in the human body that catalyzes the same reaction functions best at 37° C. Higher temperatures cause the proteins in nucleic acids to denature, and lower temperatures may cause membranes to change from a fluid to a solid state, which can interfere with many cellular processes, such as active-transport pumps.

Animals can guard against these damaging effects of temperature fluctuations by balancing heat gains and heat losses with their environment.

HEAT GAINS AND LOSSES

Animals produce heat as a by-product of metabolism and either gain heat from, or lose it to, the environment. The total body temperature is a result of an interaction of these factors and can be expressed as:

Body temperature = heat produced metabolically
+ heat gained from the environment
− heat lost to the environment

Animals use four physical processes to exchange heat with the environment: conduction, convection, evaporation, and radiation (figure 28.1). **Conduction** is the direct transfer of thermal motion (heat) between molecules of the environment and those on the body surface of an animal. This transfer is always from an area of higher temperature to one of lower temperature because heat moves down thermal gradients. For example, when you sit on the cold ground, you lose heat, and when you sit on warm sand, you gain heat.

Convection is the movement of air (or a liquid) over the surface of a body; it contributes to heat loss if the air is cooler than the body or heat gain if the air is warmer than the body. On a cool day, your body loses heat by convection because your skin temperature is higher than the surrounding air temperature.

Evaporation is loss of heat from a surface as water molecules escape in the form of a gas. It is useful only to terrestrial animals. For example, humans and some other mammals (chimpanzees and horses) have sweat glands that actively move watery solutions through pores to the skin surface. When skin temperature is high, water at the surface absorbs enough thermal energy to break the hydrogen bonds holding the individual water molecules together, and they depart from the surface, carrying heat with them. As long as the environmental humidity is low enough to permit complete evaporation, sweating can rid the mammalian body of excess heat; however, the water must evaporate. Sweat dripping from a mammal has no cooling effect at all.

Radiation is the emission of electromagnetic waves that objects, such as another animal's body or the sun, produce. Radiation can transfer heat between objects that are not in direct contact with each other, as happens when an animal suns itself (figure 28.2).

SOME SOLUTIONS TO TEMPERATURE FLUCTUATIONS

Animals cope with temperature fluctuations in one of three basic ways. (1) They can occupy a place in the environment where the temperature remains constant and compatible with their physiological processes; (2) their physiological processes may have adapted to the range of temperatures in which the animals are capable of living; or (3) they can generate and trap heat internally to maintain a constant body temperature, despite fluctuations in the temperature of the external environment.

Animals can be categorized as ectotherms or endotherms, based on whether their source of body heat is from internal

FIGURE 28.2

Radiation Warms an Animal.   After a cold night in its den on the Kalahari Desert, a meerkat (*Suricata suricatta*) stands at attention, allowing the large surface area of its body to absorb radiation from the sun.

processes or derived from the environment. **Ectotherms** (Gr. *ectos*, outside) derive most of their body heat from the environment rather than from their own metabolism (figure 28.3). They have low rates of metabolism and are poorly insulated. In general, reptiles, amphibians, fishes, and invertebrates are ectotherms, although a few reptiles, insects, and fishes can raise their internal temperature. Ectotherms tend to move about the environment and find places that minimize heat or cold stress to their bodies.

Birds and mammals are called **endotherms** (Gr. *endos*, within) because they obtain their body heat from cellular processes. A constant source of internal heat allows them to maintain a nearly constant core temperature, despite the fluctuating environmental temperature. ("Core" refers to the body's internal temperature as opposed to the temperature near its surface.)

Most endotherms have bodies insulated by fur or feathers and a relatively large amount of fat. This insulation enables them to retain heat more efficiently and to maintain a high core temperature. Endothermy allows animals to stabilize their core temperature so that biochemical processes and nervous system functions can proceed at steady, high levels. *Endothermy allows some animals to colonize habitats denied to ectotherms.*

Another way of categorizing animals is based on whether they maintain a constant or variable body temperature. Although most endotherms are **homeotherms** (maintain a relatively constant body temperature), and most ectotherms are **heterotherms** (have a variable body temperature), there are many exceptions.

Some endotherms vary their body temperatures seasonally (e.g., hibernation); others vary it on a daily basis.

For example, some birds (e.g., hummingbirds) and mammals (e.g., shrews) can only maintain a high body temperature for a short period because they usually weigh less than 10 g and have a body mass so small that they cannot generate enough heat to compensate for the heat lost across their relatively large surface area. Hummingbirds must devote much of the day to locating and sipping nectar (a very high-calorie food source) as a constant energy source for metabolism. When not feeding, hummingbirds rapidly run out of energy unless their metabolic rates decrease considerably. At night, hummingbirds enter a sleep-like state, called **daily torpor,** and their body temperature approaches that of the cooler surroundings. Some bats also undergo daily torpor to conserve energy.

Some ectotherms can maintain fairly constant body temperatures. Among these are a number of reptiles that can maintain fairly constant body temperatures by changing position and location during the day to equalize heat gain and loss.

In general, ectotherms are more common in the tropics because they do not have to expend as much energy to maintain body temperature there, and they can devote more energy to food gathering and reproduction. Indeed, in the tropics, amphibians are far more abundant than mammals. Conversely, in moderate to cool environments, endotherms have a selective advantage and are more abundant. Their high metabolic rates and insulation allow them to occupy even the polar regions (e.g., polar bears). In fact, the efficient circulatory systems of birds and mammals can be thought of as adaptations to endothermy and a high metabolic rate.

TEMPERATURE REGULATION IN INVERTEBRATES

As previously noted, environmental temperature is critical in limiting the distribution of all animals and in controlling metabolic reactions. Many invertebrates have relatively low metabolic rates and have no thermoregulatory mechanisms; thus, they passively conform to the temperature of their external environment. These invertebrates are termed **thermoconformers.**

Evidence indicates that some higher invertebrates can directly sense differences in environmental temperatures; however, specific receptors are either absent or unidentified. What zoologists do know is that many arthropods, such as insects, crustaceans, and the horseshoe crab (*Limulus*), can sense thermal variation. For example, ticks of warm-blooded vertebrates can sense the "warmth of a nearby meal" and drop on the vertebrate host.

Many arthropods have unique mechanisms for surviving temperature extremes. For example, temperate-zone insects avoid freezing by reducing the water content in their tissues as winter approaches. Other insects can produce glycerol or other glycoproteins that act as an antifreeze. Some moths and bumblebees warm up prior to flight by shivering contractions of their thoracic flight muscles. *Most large, flying insects have evolved*

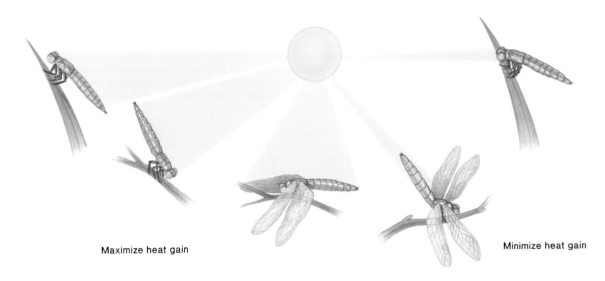

Maximize heat gain

Minimize heat gain

FIGURE 28.3

Heat Gain in an Insect.   Postures a dragonfly adopts to either maximize or minimize heat gain.

a mechanism to prevent overheating during flight; blood circulating through the flight muscles carries heat from the thorax to the abdomen, which gets rid of the heat—much as coolant circulating through an automobile engine passes through the radiator. Certain cicadas (Diceroprocta apache) that live in the Sonoran Desert have independently evolved the complete repertoire of evaporative cooling mechanisms that vertebrates use. When threatened with overheating, these cicadas extract water from their blood and transport it through large ducts to the surface of their body, where it passes through sweat pores and evaporates. In other words, these insects can sweat.

Body posture and orientation of the wings to the sun can markedly affect the body temperature of basking insects. For example, perching dragonflies and butterflies can regulate their radiation heat gain by postural adjustments (figure 28.3).

To prevent overheating, many ground-dwelling arthropods (*Tenebrio* beetles, locusts, scorpions) raise their bodies as high off the ground as possible to minimize heat gain from the ground. Some caterpillars and locusts orient with reference to both the sun and wind to vary both radiation heat gain and convective heat loss. Some desert-dwelling beetles can exude waxes from thousands of tiny pores on their cuticle. These "wax blooms" prevent dehydration and also are an extra barrier against the desert sun.

Color has a significant effect on thermoregulation since 50% of the radiant energy from the sun is in the visible spectrum. A black surface reflects less radiant energy than a white surface. Thus, many black beetles may be more active earlier in the day because they absorb more radiation and heat faster. Conversely, white beetles are more active in the hotter parts of the day because they absorb less heat.

The previous examples of invertebrate temperature regulation give clues to how thermoregulation may have evolved in vertebrates. The endothermic temperature regulation of active

insects apparently evolved because locomotion produced sufficient metabolic heat that thermoregulatory strategies could evolve. An increased locomotor metabolism could well have preceded the evolution of thermoregulation in vertebrates.

TEMPERATURE REGULATION IN FISHES

The temperature of the surrounding water determines the body temperature of most fishes. Fishes that live in extremely cold water have "antifreeze" materials in their blood. Polyalcohols (e.g., sorbitol, glycerol) or water-soluble peptides and glycopeptides lower the freezing point of blood plasma and other body fluids. These fishes also have proteins or protein-sugar compounds that stunt the growth of ice crystals that begin to form. These adaptations enable these fishes to stay flexible and swim freely in a supercooled state (i.e., at a temperature below the normal freezing temperature of a solution).

Some active fishes maintain a core temperature significantly above the temperature of the water. Bluefin tuna and the great white shark have major blood vessels just under the skin. Branches deliver blood to the deeper, powerful, red swimming muscles, where smaller vessels are arranged in a countercurrent heat exchanger called the **rete mirabile** ("miraculous net") (figure 28.4). The heat that these red muscles generate is not lost because it is transferred in the rete mirabile from venous blood passing outward to cold arterial blood passing inward from the body surface. This arrangement of blood vessels enhances vigorous activity by keeping the swimming muscles several degrees warmer than the tissue near the surface of the fish. This system has been adaptive for these fishes. *Their muscular contractions can have four times as much power as those of similar muscles in fishes with cooler bodies. Thus, they can swim faster and range more widely through various depths in search of prey than can other predatory fishes more limited to given water depths and temperatures.*

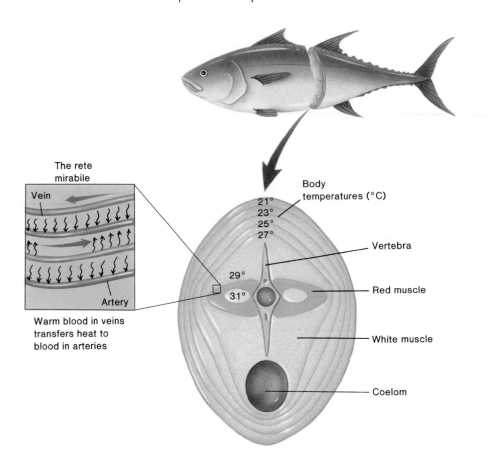

The rete mirabile

Vein

Warm blood in veins transfers heat to blood in arteries

Artery

Body temperatures (°C)

21°
23°
25°
27°
29°
31°

Vertebra

Red muscle

White muscle

Coelom

FIGURE 28.4

Thermoregulation in Large, Active Fishes.   In the bluefin tuna, the rete mirabile of arteries and veins acts as a countercurrent exchange system that helps reduce the loss of body heat. The cross section through the body shows that the temperature is highest around the red swimming muscles.

TEMPERATURE REGULATION IN AMPHIBIANS AND REPTILES

Animals, such as amphibians and reptiles, that have air rather than water as a surrounding medium face marked daily and seasonal temperature changes. Most of these animals are ectotherms. They derive heat from their environment, and their body temperatures vary with external temperatures.

Most amphibians have difficulty in controlling body heat because they produce little of it metabolically and rapidly lose most of it from their body surfaces. However, as previously noted, behavioral adaptations enable them to maintain their body temperature within a homeostatic range most of the time. Amphibians have an additional thermoregulatory problem because they must exchange oxygen and carbon dioxide across their skin surface, and this moisture layer acts as a natural evaporative cooling system. This problem of heat loss through evaporation limits the habitats and activities of amphibians to warm, moist areas. Some amphibians, such as bullfrogs, can vary the amount of mucus they secrete from their body surface—a physiological response that helps regulate evaporative cooling.

Reptiles have dry rather than moist skin, which reduces the loss of body heat through evaporative cooling of the skin. They also have an expandable rib cage, which allows for more powerful and efficient ventilation. Reptiles are almost completely ectothermic. They have a low metabolic rate and warm themselves by behavioral adaptations. In addition, some of the more sophisticated regulatory mechanisms found in mammals are first found in reptiles. For example, diving reptiles (e.g., sea turtles, sea snakes) conserve body heat by routing blood through circulatory shunts into the center of the body. These animals can also increase heat production in response to the hormones thyroxine and epinephrine. In addition, tortoises and land turtles can cool themselves through salivating and frothing at the mouth, urinating on the back legs, moistening the eyes, and panting.

TEMPERATURE REGULATION IN BIRDS AND MAMMALS

Birds and mammals are the most active and behaviorally complex vertebrates. They can live in habitats all over the earth because they are homeothermic endotherms; they can maintain body temperatures between 35 and 42° C with metabolic heat.

Various cooling mechanisms prevent excessive warming in birds. Because they have no sweat glands, birds pant to lose heat

(a)

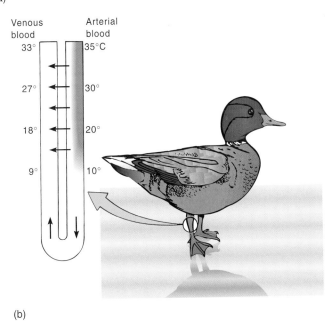

(b)

FIGURE 28.5

Insulation and Countercurrent Heat Exchange. (a) A thick layer of down feathers keeps these Chinstrap penguins (*Pygocelis antarctia*) warm. Their covering of short, stiff feathers interlocks to trap air, forming the ornithological equivalent of a diver's suit. (b) The countercurrent heat exchanger in a bird foot. Some aquatic birds, such as this duck, possess countercurrent systems of arteries and veins (rete mirabile) in their legs that reduce heat loss. The arteries carry warm blood down the legs to warm the cooler blood in the veins, so the heat is carried back to the body rather than lost through the feet that are in contact with a cold surface.

through evaporative cooling. Some species have a highly vascularized pouch (gular pouch) in their throat that they can flutter (a process called **gular flutter**) to increase evaporation from the respiratory system.

Some birds possess mechanisms for preventing heat loss. Feathers are excellent insulators for the body, especially downy-type feathers that trap a layer of air next to the body to reduce

FIGURE 28.6

Temperature Regulation. This antelope jackrabbit (*Lepus alleni*) must get rid of excess body heat. Its huge, thin, highly vascularized ears have a large surface area for heat exchange.

heat loss from the skin (figure 28.5a). (This mechanism explains why goose down is such an excellent insulator and is used in outdoor vests and coats for protection from extreme cold.) Aquatic species, who lose heat from their legs and feet, have peripheral countercurrent heat exchange vessels called a rete mirabile in their legs to reduce heat loss (figure 28.5b). Mammals that live in cold regions, such as the arctic fox and barren-ground caribou, also have these exchange vessels in their extremities (e.g., legs, tails, ears, nose). Animals in hot climates, such as jackrabbits, have mechanisms (e.g., large ears) to rid the body of excess heat (figure 28.6).

Thick pelts and a thick layer of insulating fat called **blubber** just under the skin help marine animals, such as seals and whales, to maintain a body temperature of around 36 to 38° C. In the tail and flippers, which have no blubber, a countercurrent system of arteries and veins helps minimize heat loss.

Birds and mammals also use behavioral mechanisms to cope with external temperature changes. Like ectotherms, they sun themselves or seek shade as the temperature fluctuates. Many animals huddle to keep warm; others share burrows for protection from temperature extremes. Migration to warm climates and hibernation enable many different birds and mammals to survive the harsh winter months. Others, such as the desert camel, have a multitude of evolutionary adaptations for surviving in some of the hottest and driest climates on earth.

HEAT PRODUCTION IN BIRDS AND MAMMALS

In endotherms, heat generation can warm the body as it dissipates throughout tissues and organs. Birds and mammals can generate heat (**thermogenesis**) by muscle contraction, ATPase pump enzymes, oxidation of fatty acids in brown fat, and other metabolic processes.

Every time a muscle cell contracts, the actin and myosin filaments sliding over each other and the hydrolysis of ATP molecules generate heat. Both voluntary muscular work (e.g., running, flying, jumping) and involuntary muscular work (e.g., shivering) generate heat. Heat generation by shivering is called **shivering thermogenesis.**

Birds and mammals have a unique capacity to generate heat by using specific enzymes of ancient evolutionary origin—the ATPase pump enzymes in the plasma membranes of most cells. When the body cools, the thyroid gland releases the hormone thyroxine. Thyroxine increases the permeability of many cells to sodium (Na^+) ions, which leak into the cells. The ATPase pump quickly pumps these ions out. In the process, ATP is hydrolyzed, releasing heat energy. The hormonal triggering of heat production is called **nonshivering thermogenesis.**

Brown fat is a specialized type of fat found in newborn mammals, in mammals that live in cold climates, and in mammals that hibernate (figure 28.7). The brown color of this fat comes from the large number of mitochondria with their iron-containing cytochromes. Deposits of brown fat are beneath the ribs and in the shoulders. A large amount of heat is produced when brown fat cells oxidize fatty acids because little ATP is made. Blood flowing past brown fat is heated and contributes to warming the body.

The basal metabolic rate of birds and mammals is high and also produces heat as an inadvertent but useful by-product.

In amphibians, reptiles, birds, and mammals, specialized cells in the hypothalamus of the brain control thermoregulation. The two hypothalamic thermoregulatory areas are the heating center and the cooling center. The heating center controls vasoconstriction of superficial blood vessels, erection of hair and fur, and shivering or nonshivering thermogenesis. The cooling center controls vasodilation of blood vessels, sweating, and panting. Overall, feedback mechanisms (with the hypothalamus acting as a thermostat) trigger either the heating or cooling of the body and thereby control body temperature (figure 28.8). Specialized neuronal receptors in the skin and other parts of the body sense temperature changes. Warm neuronal receptors excite the cooling center and inhibit the heating center. Cold neuronal receptors have the opposite effects.

During the winter, various endotherms (e.g., bats, woodchucks, chipmunks, ground squirrels) go into **hibernation** (L. *hiberna,* winter). During hibernation, the metabolic rate slows, as do the heart and breathing rates. Mammals prepare for hibernation by building up fat reserves and growing long winter pelts. All hibernating animals have brown fat. Decreasing day length stimulates both increased fat deposition and fur growth. Another physiological state characterized by slow metabolism and inactivity is **aestivation** (L. *aestivus,* summer), which allows certain mammals to survive long periods of elevated temperature and diminished water supplies.

Some animals, such as badgers, bears, opossums, raccoons, and skunks, enter a state of prolonged sleep in the winter. Since their body temperature remains near normal, this is not true hibernation.

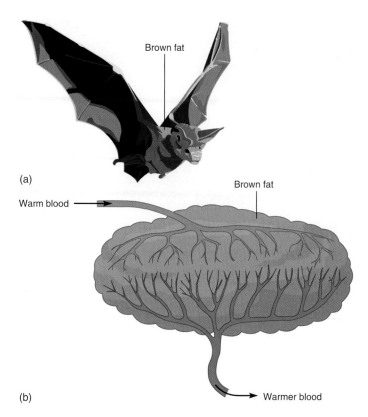

(a)

Brown fat

Warm blood

Brown fat

(b)

Warmer blood

FIGURE 28.7

Brown Fat.   (*a*) Many mammals, such as this bat, have adipose tissue called brown fat between the shoulder blades. (*b*) The area of brown fat is much warmer than the rest of the body. Blood flowing through the brown fat is warmed.

CONTROL OF WATER AND SOLUTES (OSMOREGULATION AND EXCRETION)

Excretion (L. *excretio,* to eliminate) can be defined broadly as the elimination of metabolic waste products from an animal's body. These products include carbon dioxide and water (which cellular respiration primarily produces), excess nitrogen (which is produced as ammonia from the deamination of amino acids), and solutes (various ions). Chapter 26 covers the excretion of respiratory carbon dioxide.

The excretion of nitrogenous wastes is usually associated with the regulation of water and solute (ionic) balance by a physiological process called **osmoregulation.** Osmoregulation is necessary for animals in all habitats. If the osmotic concentration of the body fluids of an animal equals that of the medium (the animal's environment), the animal is an **osmoconformer.** When the osmotic concentration of the environment changes, so does that of the animal's body fluids. Obviously, this type of osmoregulation is not efficient and has limited the distribution of those animals using it. In contrast, an animal that maintains its body fluids at a different osmotic concentration from that of its surrounding environment is an **osmoregulator.**

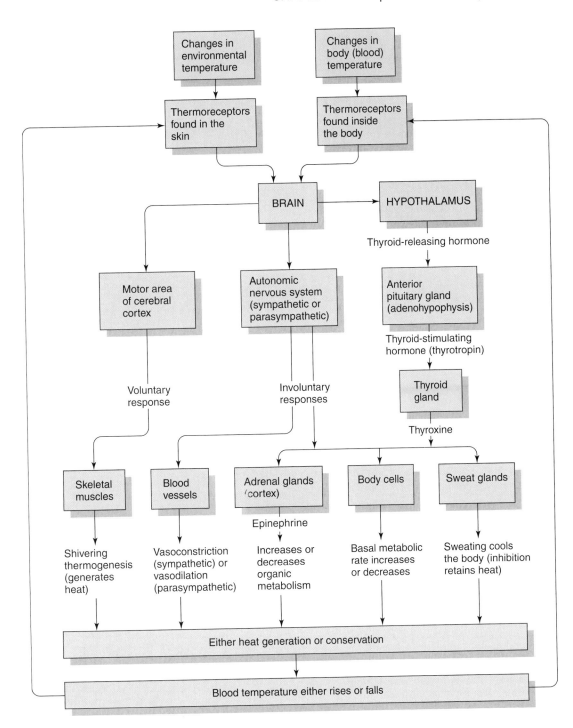

FIGURE 28.8

Thermoregulation. Overview of the feedback pathways that control the core body temperature of a mammal. Arrows show the major control pathways.

Animals living in seawater have body fluids with an osmotic concentration that is about a third less (hypoosmotic) than the surrounding seawater, and water tends to leave their bodies continually. *To compensate for this problem, mechanisms evolved in these animals to conserve water and prevent dehydration. Freshwater animals have body fluids that are hyperosmotic with respect to their environment, and water tends to continually enter their bodies. Mechanisms evolved in these animals that excrete water and prevent fluid accumulation.* Land animals have a higher concentration of water in their fluids than in the surrounding air. They tend to lose water to the air through evaporation and may use considerable amounts of water to dispose of wastes.

The form and function of organs or systems associated with excretion and osmoregulation are related both to environmental conditions (saltwater, freshwater, terrestrial) and to body size (especially the surface-to-volume ratio).

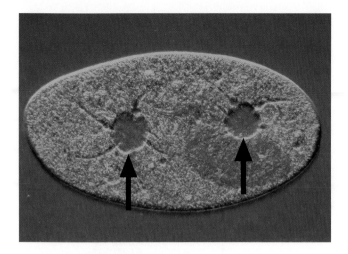

FIGURE 28.9

Contractile Vacuoles. A photomicrograph (×100) showing the location of two contractile vacuoles (black arrows) in a stained *Paramecium*. Notice the small tubules surrounding each vacuole. These tubes collect water and deliver it to the contractile vacuole, which expels the fluid through a pore.

INVERTEBRATE EXCRETORY SYSTEMS

Aquatic invertebrates occur in a wide range of media, from freshwater to markedly hypersaline water (e.g., salt lakes). Generally, marine invertebrates have about the same osmotic concentration as seawater (i.e., they are osmoconformers). This avoids any need to osmoregulate. Most water and ions are gained across the integument, via gills, by drinking, and in food. Ions and wastes are mostly lost by diffusion via the integument, gills, or urine.

Freshwater invertebrates are strong osmoregulators because it is impossible to be isosmotic with dilute media. Any water gain is usually eliminated as urine.

A number of invertebrate taxa have more or less successfully invaded terrestrial habitats. The most successful terrestrial invertebrates are the arthropods, particularly the insects, spiders, scorpions, ticks, mites, centipedes, and millipedes. Overall, the water and ion balance of terrestrial invertebrates is quite different from that of aquatic animals because terrestrial invertebrates face limited water supplies and water loss by evaporation from their integument. Some of the invertebrate excretory mechanisms and systems are now discussed.

CONTRACTILE VACUOLES

Some protists and marine invertebrates (e.g., protozoa, cnidarians, echinoderms, sponges) do not have specialized excretory structures because wastes simply diffuse into the surrounding isoosmotic water. In some freshwater species, cells on the body surface actively pump ions into the animal. Many freshwater species (protozoa, sponges), however, have contractile vacuoles that pump out excess water. **Contractile vacuoles** are energy-requiring devices that expel excess water from individual cells exposed to hypoosmotic environments (figure 28.9).

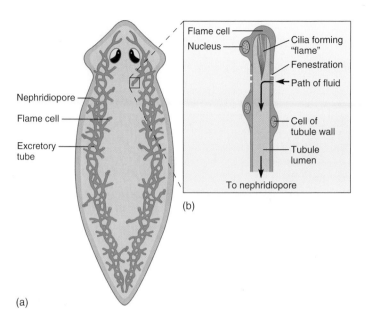

(a)

(b)

FIGURE 28.10

Protonephridial (Excretory) System in a Turbellarian. (*a*) The system lies in the mesenchyme and consists of a network of fine tubules that run the length of the animal on each side and open to the surface by minute excretory pores called nephridiopores. (*b*) Numerous fine side branches from the tubules originate in the mesenchyme in enlargements called flame cells.

PROTONEPHRIDIA

Although a few groups of metazoan invertebrates possess no known excretory structures, most have **nephridia** (Gr. *nephros*, kidney) (sing., nephridium) that serve for excretion, osmoregulation, or both. **Probably the earliest type of nephridium to appear in the evolution of animals was the protonephridium (Gr. protos, first + nephridium).**

Among the simplest of the protonephridia are flame-cell systems, such as those in rotifers, some annelids, larval molluscs, and some flatworms (figure 28.10). The protonephridial excretory system is composed of a network of excretory canals that open to the outside of the body through excretory pores. Bulblike **flame cells** are located along the excretory canals. Fluid filters into the flame cells from the surrounding interstitial fluid, and beating cilia propel the fluid through the excretory canals and out of the body through the excretory pores. Flame-cell systems function primarily in eliminating excess water. Nitrogenous waste simply diffuses across the body surface into the surrounding water.

METANEPHRIDIA

A more advanced type of excretory structure among invertebrates is the **metanephridium** (Gr. *meta*, beyond + nephridium) (pl., metanephridia). Protonephridia and metanephridia have critical structural differences. Both open to the outside, but metanephridia (1) also open internally to the body fluids and (2) are multicellular.

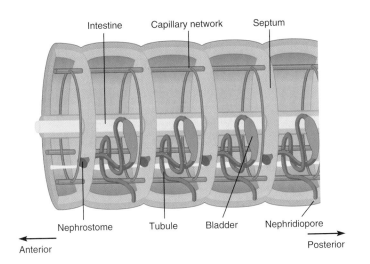

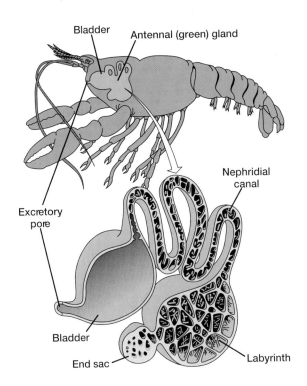

FIGURE 28.11

Earthworm Metanephridium. The metanephridium opens by a ciliated nephrostome into the cavity of one segment, and the next segment contains the nephridiopore. The main tubular portion of the metanephridium is coiled and is surrounded by a capillary network. Waste can be stored in a bladder before being expelled to the outside. Most segments contain two metanephridia.

Most annelids (such as the common earthworm) and a variety of other invertebrates have a metanephridial excretory system. Recall that the earthworm's body is divided into segments and that each segment has a pair of metanephridia. Each metanephridium begins with a ciliated funnel, the nephrostome, that opens from the body cavity of a segment into a coiled tubule (figure 28.11). As beating cilia move the fluid through the tubule, a network of capillaries surrounding the tubule reabsorbs and carries away ions. Each tubule leads to an enlarged bladder that empties to the outside of the body through an opening called the nephridiopore.

The excretory system of molluscs includes protonephridia in larval stages and metanephridia in adults.

ANTENNAL (GREEN) AND MAXILLARY GLANDS

In those crustaceans that have gills, nitrogenous wastes are removed by simple diffusion across the gills. Most crustaceans release ammonia, although they also produce some urea and uric acid as waste products. Thus, the excretory organs of freshwater species may be more involved with the reabsorption of ions and elimination of water than with the discharge of nitrogenous wastes. The excretory organs in some crustaceans (crayfish, crabs) are **antennal glands** or **green glands** because of their location near the antenna and their green color (figure 28.12). Fluid filters into the antennal gland from the hemocoel. Hemolymph pressure from the heart is the main driving force for filtration. Marine crustaceans have a short nephridial canal and produce urine that is isoosmotic to their hemolymph. The nephridial canal is longer in freshwater crustaceans, which allows more surface area for ion transport.

In other crustaceans (some malacostracans [crabs, shrimp, pillbugs]), the excretory organs are near the maxillary segments and

FIGURE 28.12

Antennal (Green) Gland of the Crayfish. The antennal gland, which lies in front of and to both sides of the esophagus, is divided into an end sac, where fluid collects by filtration, and a labyrinth. The labyrinth walls are greatly folded and glandular and appear to be an important site for reabsorption. The labyrinth leads via a nephridial canal into a bladder. From the bladder, a short duct leads to an excretory pore.

are termed **maxillary glands.** In maxillary glands, fluid collects within the tubules from the surrounding blood of the hemocoel, and this primary urine is modified substantially by selective reabsorption and secretion as it moves through the excretory system and rectum.

MALPIGHIAN TUBULES

Insects have an excretory system made up of the gut and **Malpighian tubules** attached to the gut (figure 28.13). Excretion involves the active transport of potassium ions into the tubules from the surrounding hemolymph and the osmotic movement of water that follows. Nitrogenous waste (uric acid) also enters the tubules. As fluid moves through the Malpighian tubules, some of the water and certain ions are recovered. All of the uric acid passes into the gut and out of the body.

COXAL GLANDS

Coxal (L. *coxa*, hip) **glands** are common among arachnids (spiders, scorpions, ticks, mites). These spherical sacs resemble annelid nephridia (figure 28.14). Wastes are collected from the surrounding hemolymph of the hemocoel and discharged through pores on from one to several pairs of appendages near the proximal joint (coxa) of the leg. Recent evidence suggests that the coxal glands may also function in the release of pheromones.

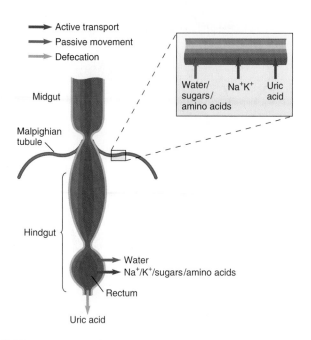

FIGURE 28.13

Malpighian Tubules.   Malpighian tubules remove nitrogenous wastes (uric acid) from the hemocoel. Various ions are actively transported across the outer membrane of the tubule. Water follows these ions into the tubule and carries amino acids, sugars, and some nitrogenous wastes along passively. Some water, ions, and organic compounds are reabsorbed in the basal portion of the Malpighian tubules and the hindgut; the rest are reabsorbed in the rectum. Uric acid moves into the hindgut and is excreted.

Other arachnid species have Malpighian tubules instead of, or in addition to, the coxal glands. In some of these species, however, Malpighian tubules seem to function in silk production rather than in excretion.

VERTEBRATE EXCRETORY SYSTEMS

Vertebrates face the same problems as invertebrates in controlling water and ion balance. Generally, water losses are balanced precisely by water gains (table 28.1). Vertebrates gain water by absorption from liquids and solid foods in the small and large intestines, and by metabolic reactions that yield water as an end product. They lose water by evaporation from respiratory surfaces, evaporation from the integument, sweating or panting, elimination in feces, and excretion by the urinary system.

Solute losses also must be balanced by solute gains. Vertebrates take in solutes by the absorption of minerals from the small and large intestines, through the integument or gills, from secretions of various glands or gills, and by metabolism (e.g., the waste products of degradative reactions). They lose solutes in sweat, feces, urine, and gill secretions, and as metabolic wastes. The major metabolic wastes that must be eliminated are ammonia, urea, uric acid.

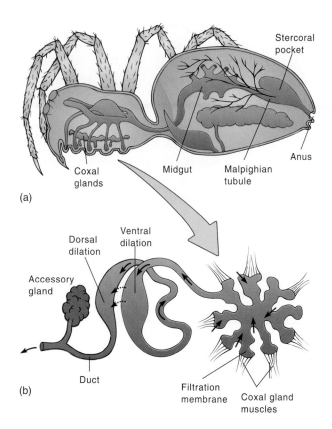

FIGURE 28.14

Coxal Glands in Arachnids.   (*a*) The gut and excretory systems of a spider. (*b*) Coxal gland muscles attach to the thin saccular filtration membrane. These muscles promote filtration and fluid flow (black arrows) by contracting and relaxing along the tubular duct. Water and solutes are reabsorbed along the tubular duct.

Vertebrates live in saltwater, freshwater, and on land; each of these environments presents different water and solute problems that vertebrates have solved in different ways. The next section discusses how vertebrates avoid losing or gaining too much water and, in turn, how they maintain a homeostatic solute concentration in their body fluids. The disposal (excretion) of certain metabolic waste products is also coupled with osmotic balance and is discussed with the urinary system.

HOW VERTEBRATES ACHIEVE OSMOREGULATION

A variety of mechanisms have evolved in vertebrates to cope with their osmoregulatory problems, and most of them are adaptations of the urinary system. As presented in chapter 26, vertebrates have a closed circulatory system containing blood that is under pressure. This pressure forces blood through a membrane filter in a kidney, where the following three key functions take place:

1. Filtration, in which blood passes through a filter that retains blood cells, proteins, and other large solutes but lets small molecules, ions, and urea pass through

TABLE 28.1
AVERAGE WATER GAIN AND LOSS IN A HUMAN AND KANGAROO RAT

VERTEBRATE	WATER GAIN (ML)		WATER LOSS (ML)	
Human (daily)	Ingested in solid food	1,200	Feces	100
	Ingested as liquids	1,000	Urine	1,500
	Metabolically produced	350	Skin and lungs	950
	Total	2,550		2,550
Kangaroo rat (over 4 weeks)	Ingested in solid food	6	Feces	3
	Ingested in liquids	0	Urine	13
	Metabolically produced	54	Skin and lungs	44
	Total	60		60

2. Reabsorption, in which selective ions and molecules are taken back into the bloodstream from the filtrate
3. Secretion, whereby select ions and end products of metabolism (e.g., K^+, H^+, NH_3) that are in the blood are added to the filtrate for removal from the body

VERTEBRATE KIDNEY VARIATIONS

Vertebrates have two kidneys that are in the back of the abdominal cavity, on either side of the aorta. Each kidney has a coat of connective tissue called the renal capsule (L. *renes*, kidney). The inner portion of the kidney is called the medulla; the region between the capsule and the medulla is the cortex.

The structure and function of vertebrate kidneys differ, depending on the vertebrate groups and the developmental stage. Overall, there are three kinds of vertebrate kidneys: the pronephros, mesonephros, and metanephros. The **pronephros** appears only briefly in many vertebrate embryos, and not at all in mammalian embryos (figure 28.15*a*). In some vertebrates, the pronephros is the first osmoregulatory and excretory organ of the embryo (tadpoles and other amphibian larvae); in others (hagfishes), it remains as the functioning kidney. During the embryonic development of amniotes, or during metamorphosis in amphibians, the mesonephros replaces the pronephros (figure 28.15*b*). The **mesonephros** is the functioning embryonic kidney of many vertebrates and also adult fishes and amphibians. The mesonephros gives way during embryonic development to the **metanephros** in adult reptiles, birds, and mammals (figure 28.15*c*).

The physiological differences between these kidney types are primarily related to the number of blood-filtering units they contain. The pronephric kidney forms in the anterior portion of the body cavity and contains fewer blood-filtering units than either the mesonephric or metanephric kidneys. The larger number of filtering units in the latter has allowed vertebrates to face the rigorous osmoregulatory and excretory demands of freshwater and terrestrial environments.

What follows is a presentation of how a few vertebrates maintain their water and solute concentrations in different habitats—in the seas, in freshwater, and on land (table 28.2).

Sharks

Sharks and their relatives (skates and rays) have mesonephric kidneys and have solved their osmotic problem in ways different from the bony fishes (figure 28.15*b*). Instead of actively pumping ions out of their bodies through the kidneys, they have a **rectal gland** that secretes a highly concentrated salt (NaCl) solution. To reduce water loss, they use two organic molecules—urea and trimethylamine oxide (TMO)—in their body fluids to raise the osmotic pressure to a level equal to or higher than that of the seawater.

Urea denatures proteins and inhibits enzymes, whereas TMO stabilizes proteins and activates enzymes. Together in the proper ratio, they counteract each other, raise the osmotic pressure, and do not interfere with enzymes or proteins. This reciprocity is termed the **counteracting osmolyte strategy.**

A number of other fishes and invertebrates have evolved the same mechanism and employ pairs of counteracting osmolytes to raise the osmotic pressure of their body fluids.

Teleost Fishes

Most teleost fishes have mesonephric kidneys. Because the body fluids of freshwater fishes are hyperosmotic relative to freshwater (*see table 28.2*), water tends to enter the fishes, causing excessive hydration or bloating (figure 28.16*a*). At the same time, body ions tend to move outward into the water. To solve this problem, freshwater fishes usually do not drink much water. Their bodies are coated with mucus, which helps stem inward water movement. They absorb salts and ions by active transport across their gills. They also excrete a large volume of water as dilute urine.

Although most groups of animals probably evolved in the sea, many marine bony fishes probably had freshwater

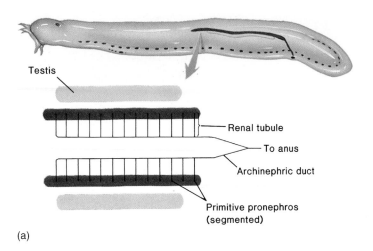

(a)

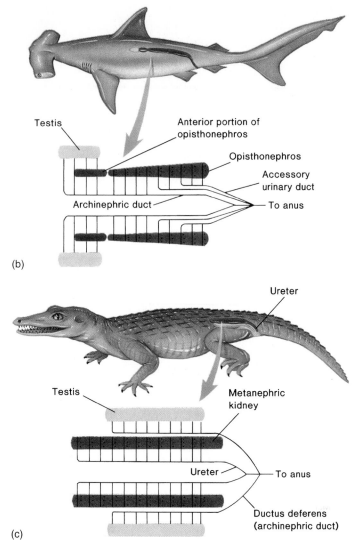

(b)

(c)

FIGURE 28.15

Types of Kidneys in Vertebrates and Their Association with the Male Reproductive System. The brown portions of the drawings represent the mesoderm that forms both the kidneys and gonads. Notice that it extends much of the length of the body during development. (*a*) The primitive pronephric kidney is found in adult hagfishes and embryonic fishes and amphibians. It is anterior in the body and contains segmental renal tubules that lead from the body of the pronephros to the archinephric duct. Notice that the testes are separated from the kidneys. (*b*) The mesonephros is the functional kidney in the amniote embryo, adult fishes, and amphibians. It is structurally similar to the nonsegmented opisthonephric (advanced mesonephric) kidney of most nonamniote vertebrates, such as sharks. The anterior portion of the opisthonephros functions in blood cell formation and secretion of sex hormones. Notice that the testes occupy the position of the anterior opisthonephros, and the archinephric duct carries both sperm and urine. (*c*) The metanephric kidney of adult amniotes (reptiles, birds, and mammals) is the most advanced kidney. Notice the separate ureters (new ducts) for carrying urine. The archinephric duct becomes the ductus deferens for carrying sperm. The kidney is more compact and located more caudally in the body.

TABLE 28.2
HOW VARIOUS VERTEBRATES MAINTAIN WATER AND SALT BALANCE

ORGANISM	ENVIRONMENTAL CONCENTRATION RELATIVE TO BODY FLUIDS	URINE CONCENTRATION RELATIVE TO BLOOD	MAJOR NITROGENOUS WASTE	KEY ADAPTATION
Freshwater fishes	Hypoosmotic	Hypoosmotic	Ammonia	Absorb ions through gills
Saltwater fishes	Hyperosmotic	Isoosmotic	Ammonia	Secrete ions through gills
Sharks	Isoosmotic	Isoosmotic	Ammonia	Secrete ions through rectal gland
Amphibians	Hypoosmotic	Very hypoosmotic	Ammonia and urea	Absorb ions through skin
Marine reptiles	Hyperosmotic	Isoosmotic	Ammonia and urea	Secrete ions through salt gland
Marine mammals	Hyperosmotic	Very hyperosmotic	Urea	Drink some water
Desert mammals	No comparison	Very hyperosmotic	Urea	Produce metabolic water
Marine birds	No comparison	Weakly hyperosmotic	Uric acid	Drink seawater and use salt glands
Terrestrial birds	No comparison	Weakly hypersmotic	Uric acid	Drink freshwater

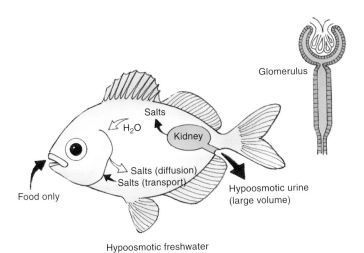

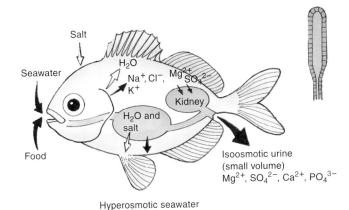

(a) Freshwater teleosts (hypertonic blood)

(b) Marine teleosts (hypotonic blood)

FIGURE 28.16

Osmoregulation. Osmoregulation by (*a*) freshwater and (*b*) marine fishes. Large black arrows indicate passive uptake or loss of water or ions. Small black and white arrows indicate active transport processes at gill membranes and kidney tubules. Insets of kidney nephrons depict adaptations within the kidney. Water, ions, and small organic molecules are filtered from the blood at the glomerulus of the nephron. Essential components of the filtrate can be reabsorbed within the tubule system of the nephron. Marine fishes conserve water by reducing the size of the glomerulus of the nephron, and thus reducing the quantity of water and ions filtered from the blood. Ions can be secreted from the blood into the kidney tubules. Marine fishes can produce urine that is isoosmotic with the blood. Freshwater fishes have enlarged glomeruli and short tubule systems. They filter large quantities of water from the blood, and tubules reabsorb some ions from the filtrate. Freshwater fishes produce a hypoosmotic urine.

ancestors, as presented in chapter 18. Marine fishes face a different problem of water balance—their body fluids are hypoosmotic with respect to seawater (*see table 28.2*), and water tends to leave their bodies, resulting in dehydration (figure 28.16*b*). To compensate, marine fishes drink large quantities of water, and they secrete Na^+, Cl^-, and K^+ ions through secretory cells in their gills. Channels in plasma membranes of their kidneys actively

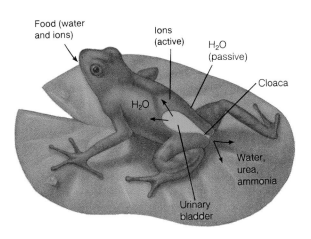

FIGURE 28.17

Water and Ion Uptake in an Amphibian. Water can enter this frog via food, through its highly permeable skin, or from the urinary bladder. The skin also actively transports ions such as Na^+ and Cl^- from the environment. The kidney forms a dilute urine by reabsorbing Na^+ and Cl^- ions. Urine then flows into the urinary bladder, where most of the remaining ions are reabsorbed.

transport the multivalent ions that are abundant in seawater (e.g., Ca^{2+}, Mg^{2+}, SO_4^{2-}, and PO_4^{3-}) out of the extracellular fluid and into the nephron tubes. The ions are then excreted in a concentrated urine.

Some fishes encounter both fresh- and saltwater during their lives. Newborn Atlantic salmon swim downstream from the freshwater stream of their birth and enter the sea. Instead of continuing to pump ions in, as they have done in freshwater, the salmon must now rid their bodies of salt. Years later, these same salmon migrate from the sea to their freshwater home to spawn. As they do, the pumping mechanisms reverse themselves.

Amphibians

The amphibian kidney is identical to that of freshwater fishes (figure 28.16*a*), which is not surprising, because amphibians spend a large portion of their time in freshwater, and when on land, they tend to seek out moist places. Amphibians take up water and ions in their food and drink, through the skin that is in contact with moist substrates, and through the urinary bladder (figure 28.17). This uptake counteracts what is lost through evaporation and prevents osmotic imbalance (*see table 28.2*).

The urinary bladder of a frog, toad, or salamander is an important water and ion reservoir. For example, when the environment becomes dry, the bladder enlarges for storing more urine. If an amphibian becomes dehydrated, a brain hormone causes water to leave the bladder and enter the body fluid.

Reptiles, Birds, and Mammals

Reptiles, birds, and mammals all possess metanephric kidneys (*see figure 28.15c*). Their kidneys are by far the most complex animal kidneys, well suited for these animals' high rates of metabolism.

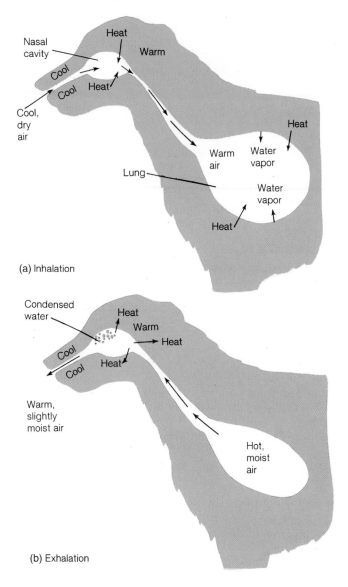

(a) Inhalation

(b) Exhalation

FIGURE 28.18

Water Retention by Countercurrent Heating and Cooling in a Mammal. (*a*) When this animal inhales, the cool, dry air passing through its nose is heated and humidified. At the same time, its nasal tissues are cooled. (*b*) When the animal exhales, it gives up heat to the previously cooled nasal tissue. The air carries less water vapor, and condensation occurs in the animal's nose.

In most reptiles, birds, and mammals, the kidneys can remove far more water than can those in amphibians, and the kidneys are the primary regulatory organs for controlling the osmotic balance of the body fluids. Some desert and marine reptiles and birds build up high salt (NaCl) concentrations in their bodies because they consume salty foods or seawater, and they lose water through evaporation and in their urine and feces. To rid themselves of excess salt, these animals also have salt glands near the eye or in the tongue that remove excess salt from the blood and secrete it as tearlike droplets.

A major site of water loss in mammals is the lungs. To reduce this evaporative loss, many mammals have nasal cavities that

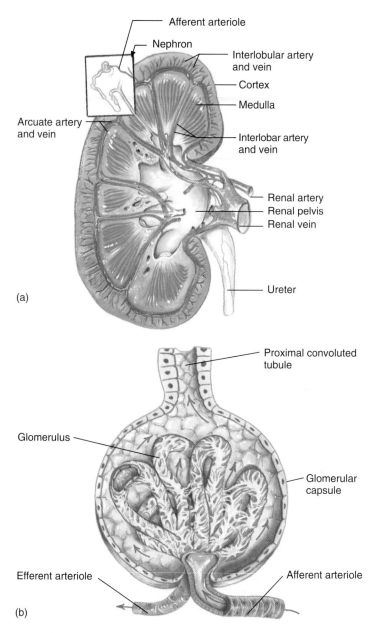

(a)

(b)

FIGURE 28.19

Filtration Device of the Metanephric Kidney. (*a*) Interior of a kidney, showing the positioning of the nephron and the blood supply to and from the kidney. (*b*) Glomerular capsule. Red arrows show that high blood pressure forces water and ions through small perforations in the walls of the glomerular capillaries to form the glomerular filtrate.

act as countercurrent exchange systems (figure 28.18). When the animal inhales, air passes through the nasal cavities and is warmed by the surrounding tissues. In the process, the temperature of this tissue drops. When the air gets deep into the lungs, it is further warmed and humidified. During exhalation, as the warm, moist air passes up the respiratory tree, it gives up its heat to the nasal cavity. As the air cools, much of the water condenses on the nasal surfaces and does not leave the body. This mechanism explains why a dog's nose is usually cold and moist.

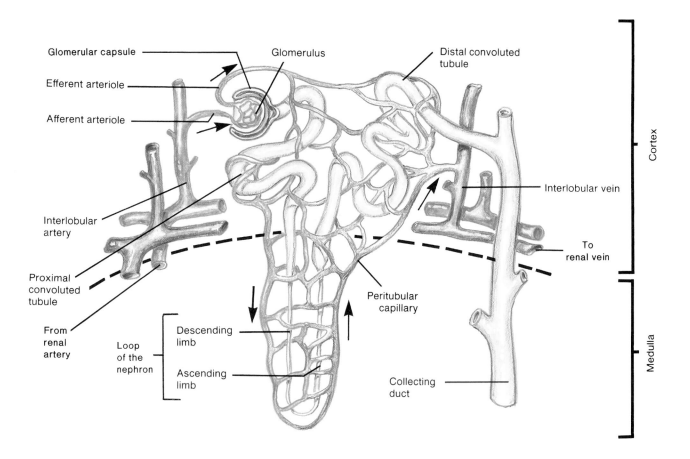

FIGURE 28.20

Metanephric Nephron.   The proximal convoluted tubule reabsorbs glucose and some ions. The distal convoluted tubule reabsorbs other ions and water. Final water reabsorption takes place in the collecting duct. Black arrows indicate the direction of movement of materials in the nephron.

HOW THE METANEPHRIC KIDNEY FUNCTIONS

The filtration device of the metanephric kidney consists of over one million individual filtration, secretion, and absorption structures called **nephrons** (Gr. *nephros*, kidney + *on*, neuter) (figure 28.19*a*). At the beginning of the nephron is the filtration apparatus called the glomerular capsule (formerly Bowman's capsule), which looks rather like a tennis ball that has been punched in on one side (figure 28.19*b*). The capsules are in the cortical (outermost) region of the kidney. In each capsule, an afferent ("going to") arteriole enters and branches into a fine network of capillaries called the **glomerulus.** The walls of these glomerular capillaries contain small perforations called filtration slits that act as filters. Blood pressure forces fluid through these filters. The fluid is now known as glomerular filtrate and contains small molecules, such as glucose, ions (Ca^{2+}, PO_4^{3-}), and the primary nitrogenous waste product of metabolism—urea or uric acid. Because the filtration slits are so small, large proteins and blood cells remain in the blood and leave the glomerulus via the efferent ("outgoing") arteriole. The efferent arteriole then divides into a set of capillaries called the peritubular capillaries that wind profusely around the tubular portions of the nephron (figure 28.20).

Eventually, they merge to form veins that carry blood out of the kidney.

Beyond the glomerular capsule are the proximal convoluted tubule, the loop of the nephron (formerly the loop of Henle), and the distal convoluted tubule. At various places along these structures, the glomerular filtrate is selectively reabsorbed, returning certain ions (e.g., Na^+, K^+, Cl^-) to the bloodstream. Both active (ATP-requiring) and passive procedures are involved in the recovery of these substances. Potentially harmful compounds, such as hydrogen (H^+) and ammonium (NH_4^+) ions, drugs, and various other foreign materials are secreted into the nephron lumen. In the last portion of the nephron, called the collecting duct, final water reabsorption takes place so that the urine contains an ion concentration well above that of the blood. Thus, the filtration, secretion, and reabsorption activities of the nephron do not simply remove wastes. They also maintain water and ion balance, and therein lies the importance of the homeostatic function of the kidney.

Mammalian, and to a lesser extent avian and reptilian, kidneys can remove far more water from the glomerular filtrate than can the kidneys of amphibians. For example, human urine is four times as concentrated as blood plasma, a camel's urine is eight times as concentrated, a gerbil's is 14 times as concentrated, and some desert rats and mice have urine more than 20 times as concentrated as their plasma. This concentrated waste enables them to live in dry or desert environments, where little water is available for them to drink. Most of their water is metabolically produced from the oxidation of carbohydrates, fats, and proteins in the seeds that they eat (*see table 28.1*). **M**ammals and, to a lesser extent, birds achieve this remarkable degree of water conservation by a unique, yet simple, evolutionary adaptation: the bending of the nephron tube into a loop. By bending, the nephron can greatly increase the salt concentration in the tissue through which the loop passes and use this gradient to draw large amounts of water out of the tube.

Countercurrent Exchange

The loop of the nephron increases the efficiency of reabsorption by a countercurrent flow similar to that in the gills of fishes or in the legs of birds, but with water and ions being reabsorbed instead of oxygen or heat. Generally, the longer the loop of the nephron, the more water and ions that can be reabsorbed. It follows that desert rodents (e.g., the kangaroo rat) that form highly concentrated urine have very long nephron loops (figure 28.21). Similarly, amphibians that are closely associated with aquatic habitats have nephrons that lack a loop.

Figure 28.22 shows the countercurrent flow mechanism for concentrating urine. The process of reabsorption in the proximal convoluted tubule removes some salt (NaCl) and water from the

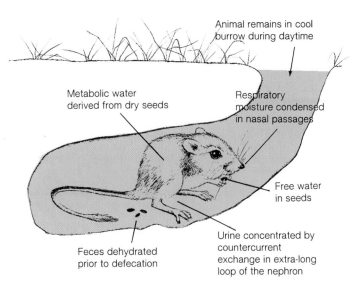

FIGURE 28.21

Kangaroo Rat (*Dipodomys ordii*), a Master of Water Conservation. Its efficient kidneys can produce an ion concentration in its urine that is 20 times that of its blood plasma. As a result, these kidneys, as well as other adaptations, prevent unnecessary water loss to the environment.

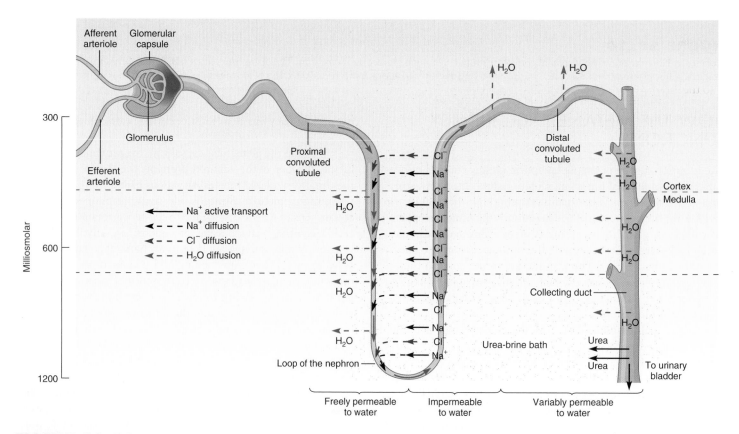

FIGURE 28.22

Countercurrent Exchange. Movement of materials in the nephron and collecting duct. Solid arrows indicate active transport; dashed arrows indicate passive transport. The shading at intervals along the tubules illustrates the relative concentration of the filtrate in milliosmoles.

glomerular filtrate and reduces its volume by approximately 25%. However, the concentrations of salt and urea are still isosmotic with the extracellular fluid.

As the filtrate moves to the descending limb of the loop of the nephron, it becomes further reduced in volume and more concentrated. Water moves out of the tubule by osmosis due to the high salt concentration (the "brine-bath") in the extracellular fluid.

Notice in figure 28.22 that the highest urea-brine bath concentration is around the lower portion of the loop of the nephron. As the filtrate passes into the ascending limb, sodium (Na^+) ions are actively transported out of the filtrate into the extracellular fluid, with chloride (Cl^-) ions following passively. Water cannot flow out of the ascending limb because the cells of the ascending limb are impermeable to water. Thus, the salt concentration of the extracellular fluid becomes very high. The salt flows passively into the descending loop, only to move out again in the ascending loop, creating a recycling of salt through the loop and the extracellular fluid. Because the flows in the descending and ascending limbs are in opposite directions, a countercurrent gradient in salt is set up. The osmotic pressure of the extracellular brine bath is made even higher because of the abundance of urea that moves out of the collecting ducts.

Finally, the distal convoluted tubule empties into the collecting duct, which is permeable to urea, and the concentrated urea in the filtrate diffuses out into the surrounding extracellular fluid. The high urea concentration in the extracellular fluid, coupled with the high concentration of salt, forms the urea-brine bath that causes water to move out of the filtrate by osmosis as it moves down the descending limb. Finally, the many peritubular capillaries surrounding each nephron collect the water and return it to the systemic circulation.

The renal pelvis of the mammalian kidney is continuous with a tube called the **ureter** that carries urine to a storage organ called the **urinary bladder** (figure 28.23). Urine from two ureters (one from each kidney) accumulates in the urinary bladder. The

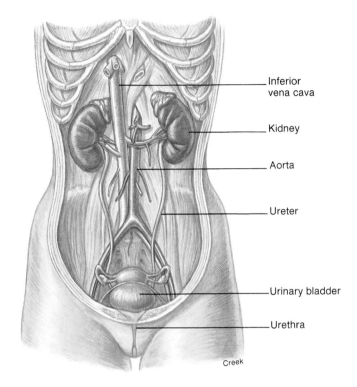

FIGURE 28.23

Component Parts of the Human Urinary System.   The positions of the kidneys, ureters, urinary bladder, and urethra.

urine leaves the body through a single tube, the **urethra,** which opens at the body surface at the end of the penis (in human males) or just in front of the vaginal entrance (in human females). As the urinary bladder fills with urine, tension increases in its smooth muscle walls. In response to this tension, a reflex response relaxes sphincter muscles at the entrance to the urethra. This response is called urination. The two kidneys, two ureters, urinary bladder, and urethra constitute the urinary system of mammals.

SUMMARY

1. Thermoregulation is a complex and important physiological process for maintaining heat homeostasis despite environmental changes.

2. Ectotherms generally obtain heat from the environment, whereas endotherms generate their own body heat from metabolic processes.

3. Homeotherms generally have a relatively constant core body temperature, while heterotherms have a variable body temperature.

4. The high, constant body temperature of birds and mammals also depends on insulation, panting, sweating, specific behaviors, vasoconstriction or vasodilation of peripheral blood vessels, and in some species, a rete mirabile system.

5. Thermogenesis involves mainly shivering, enzymatic activity, brown fat, and high cellular metabolism.

6. The hypothalamus is the temperature-regulating center that functions as a thermostat with a fixed set point. This set point can either rise or fall during hibernation or torpor.

7. Some invertebrates have contractile vacuoles, flame-cell systems, antennal (green) glands, maxillary glands, coxal glands, nephridia, or Malpighian tubules for osmoregulation.

8. The osmoregulatory system of vertebrates governs the concentration of water and ions; the excretory system eliminates metabolic wastes, water, and ions from the body.

9. Freshwater animals tend to lose ions and take in water. To avoid hydration, freshwater fishes rarely drink much water, have impermeable body surfaces covered with mucus, excrete a dilute urine, and take up ions through their gills.

10. Marine animals tend to take in ions from the seawater and to lose water. To avoid dehydration, they frequently drink water, have

relatively permeable body surfaces, excrete a small volume of concentrated urine, and secrete ions from their gills.

11. Amphibians can absorb water across the skin and urinary bladder wall. Desert and marine reptiles and birds have salt glands to remove and secrete excess salt (NaCl).

12. In reptiles, birds, and mammals, the kidneys are important osmoregulatory structures. The functional unit of the kidney is the nephron, composed of the glomerular capsule, proximal convoluted tubule, loop of the nephron, distal convoluted tubule, and collecting duct. The loop of the nephron and the collecting duct are in the kidney's medulla; the other nephron parts lie in the kidney's cortex. Urine passes from the pelvis of the kidney to the urinary bladder.

13. To make urine, kidneys produce a filtrate of the blood and reabsorb most of the water, glucose, and needed ions, while allowing wastes to pass from the body. Three physiological mechanisms are involved: filtration of the blood through the glomerulus, reabsorption of the useful substances, and secretion of toxic substances. In those animals with a loop of the nephron, salt (NaCl) and urea are concentrated in the extracellular fluid around the loop, allowing water to move by osmosis out of the loop and into the peritubular capillaries.

SELECTED KEY TERMS

brown fat (p. 470)
ectotherm (p. 466)
endotherm (p. 466)
heterotherm (p. 466)
hibernation (p. 470)
homeotherm (p. 466)
nephron (p. 479)
osmoconformer (p. 470)
osmoregulation (p. 470)
thermoregulation (p. 465)

CRITICAL THINKING QUESTIONS

1. Reptiles are said to be behavioral homeotherms. Explain what this means.
2. Why do very small birds and mammals go into a state of torpor at night?
3. How does the countercurrent mechanism help regulate heat loss?
4. In endotherms, what controls the balance between the amount of heat lost and the amount gained?
5. If marooned on a desert isle, do not drink seawater; it is better to be thirsty. Why is this true?

ONLINE LEARNING CENTER

Visit our Online Learning Center (OLC) at www.mhhe.com/zoology (click on this book's title) to find the following chapter-related materials:

- CHAPTER QUIZZING
- ANIMATIONS
    Kidney Function (two animations; same name)
- RELATED WEB LINKS
    Homeostasis
        Water and Osmotic Regulation in Aquatic Organisms
        Water and Osmotic Regulation in Terrestrial Organisms
        Vertebrates: Macroscopic Anatomy of Excretory Organs
        Vertebrates: Microscopic Anatomy of Excretory Organs
        Human Excretory System Topics
        Temperature Regulation
- BOXED READINGS ON
        How the One-Humped Dromedary (Camel) of the Arabian and African Deserts Thrives in Some of the Hottest and Driest Climates on Earth
        Waste Products of Nitrogen Metabolism
        Extrarenal Secretion by Avian and Reptilian Salt Glands
- SUGGESTED READINGS
- LAB CORRELATIONS

    Check out the OLC to find specific information on these related lab exercises in the *General Zoology Laboratory Manual*, 5th edition, by Stephen A. Miller:

Exercise 23   *Vertebrate Excretion*

REPRODUCTION AND DEVELOPMENT

Outline

Concepts

1. All animals have the capacity to reproduce. The simplest form of reproduction is asexual. Asexual reproduction produces new individuals from one parent, but it does not produce new genetic combinations among offspring, as does sexual reproduction.
2. Almost all animals reproduce sexually, at least sometimes. Sexual reproduction involves mechanisms that bring sperm and egg together for fertilization and ensure that the fertilized egg has a suitable place to develop until the new animal is ready to function on its own.
3. Sexual reproduction evolved in aquatic environments, and its modification for organisms living on dry land entailed evolutionary innovations to prevent the gametes and embryos from drying out.
4. Mammalian fertilization and embryonic development are both internal.
5. Hormones coordinate the reproductive functions in both males and females. In female mammals, hormones also maintain pregnancy, and after childbirth, stimulate the production and letdown of milk from the mammary glands.

Reproduction is a basic attribute of all forms of life. Chapter 3 describes the general features of animal development and the control processes that allow a genotype to be translated into its phenotype. Although in modern zoology, development is "the center stage" in reproduction, the whole process includes the behavior, anatomy, and physiology of adults—whether protists, invertebrates, or vertebrates. This chapter begins with a comparative focus on the different reproductive strategies observed in protists, invertebrates, and the five major groups of vertebrates. The chapter concludes with a discussion of human reproduction, not only because of the subject's basic interest to everyone, but because scientists know more about the biochemistry, hormones, anatomy, and physiology of human reproduction than they do about those of any other species.

ASEXUAL REPRODUCTION IN INVERTEBRATES

In the biological sense, reproduction means producing offspring that may (or may not) be exact copies of the parents. Reproduction is part of a life cycle, a recurring frame of events in which animals grow, develop, and reproduce according to a program of instruction encoded in the DNA they inherit from their parents. One of the two major types of reproduction in the biological world is asexual reproduction.

This chapter contains evolutionary concepts, which are set off in this font.

The first organisms to evolve probably reproduced by pinching in two, much like the simplest organisms that exist today do. This is a form of **asexual reproduction,** which is reproduction without the union of gametes or sex cells. *In the first two billion years or more of evolution, forms of asexual reproduction were probably the only means by which the primitive organisms could increase their numbers. While asexual reproduction effectively increases the numbers of a species, those species reproducing asexually tend to evolve very slowly, because all offspring of any one individual are alike, providing less genetic diversity for evolutionary selection.*

Asexual reproduction is common among the protozoa, as well as among lower invertebrates, such as sponges, jellyfishes, flatworms, and many segmented worms. Asexual reproduction is rare among the higher invertebrates. The ability to reproduce asexually often correlates with a marked capacity for regeneration.

In the lower invertebrates, the most common forms of asexual reproduction are fission, budding (both internal and external), and fragmentation. Parthenogenesis, which is comparatively uncommon, also occurs in a few invertebrates.

FISSION

Protists and some multicellular animals (cnidarians, annelids) may reproduce by fission. **Fission** (L. *fissio,* the act of splitting) is the division of one cell, body, or body part into two (figure 29.1*a*). In this process, the cell pinches in two by an inward furrowing of the plasma membrane. Binary fission occurs when the division is equal; each offspring contains approximately equal amounts of protoplasm and associated structures. Binary fission is common in protozoa; for some, it is their only means of reproduction.

In fission, the plane of division may be asymmetrical, transverse, or longitudinal, depending on the species. For example, the multicellular, free-living flatworms, such as the common planarian, reproduce by longitudinal fission (figure 29.1*b*). Some flatworms and annelids reproduce by forming numerous constrictions along the length of the body; a chain of daughter individuals results (figure 29.1*c*). This type of asexual reproduction is called multiple fission.

BUDDING

Another method of asexual reproduction found in lower invertebrates is **budding** (L. *bud,* a small protuberance). For example, in the cnidarian *Hydra* and many species of sponges, certain cells divide rapidly and develop on the body surface and form an external bud (figure 29.1*d*). The bud cells proliferate and form a cylindrical structure, which develops into a new animal, usually breaking away from the parent. If the buds remain attached to the parent, they form a colony. A **colony** is a group of closely associated individuals of one species. Internal budding (as in the freshwater sponges) produces gemmules, which are collections of many cells surrounded by a body wall. When the body of the parent dies and degenerates, each gemmule gives rise to a new individual.

FRAGMENTATION

Fragmentation is a type of asexual reproduction whereby a body part is lost and then regenerates into a new organism. Fragmentation occurs in some cnidarians, platyhelminths, rhynchocoels, and echinoderms. For example, in sea anemones, as the organism moves, small pieces break off from the adult and develop into new individuals (figure 29.1*e*).

PARTHENOGENESIS

Certain flatworms, rotifers, roundworms, insects, lobsters, some lizards, and some fishes can reproduce without sperm and normal fertilization. These animals carry out what is called **parthenogenesis** (Gr. *parthenos,* virgin + *genesis,* production). (However, most parthenogenetic animals also can reproduce sexually at some point in their life history.) Parthenogenesis is a spontaneous activation of a mature egg, followed by normal egg divisions and subsequent embryonic development. In fact, mature eggs of species that do not undergo parthenogenesis can sometimes be activated to develop without fertilization by pricking them with a needle, by exposing them to high concentrations of calcium, or by altering their temperature.

Because parthenogenetic eggs are not fertilized, they do not receive male chromosomes. The offspring would thus be expected to have only a haploid set of chromosomes. In some animals, however, meiotic division is suppressed, so the diploid number is conserved. In other animals, meiosis occurs, but an unusual mitosis returns the haploid embryonic cells to the diploid condition.

Overall, animals that reproduce parthenogenetically have substantially less genetic variability than do animals with chromosome sets from two parents. This condition may be an advantage for animals that are well adapted to a relatively stable environment. However, in meeting the challenges of a changing environment, parthenogenetic animals may have less flexibility, which may explain why this form of reproduction is relatively uncommon.

Parthenogenesis also plays an important role in social organization in colonies of certain bees, wasps, and ants. In these insects, large numbers of males (drones) are produced parthenogenetically, whereas sterile female workers and reproductive females (queens) are produced sexually.

ADVANTAGES AND DISADVANTAGES OF ASEXUAL REPRODUCTION

The predominance of asexual reproduction in protists and some invertebrates can be partially explained by the environment in which they live. The marine environment is usually very stable. Stable environments may favor this form of reproduction because a combination of genes that matches the relatively unchanging environment is an advantage over a greater number of gene combinations, many of which do not match the environment. In other habitats, asexual reproduction is seasonal. The season during which asexual reproduction occurs coincides with the period when the environment is predictably hospitable. Under such

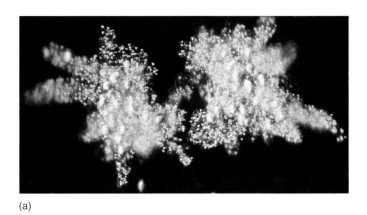

(a)

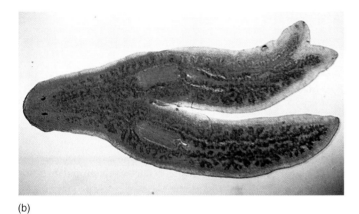

(b)

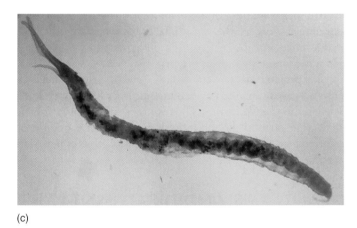

(c)

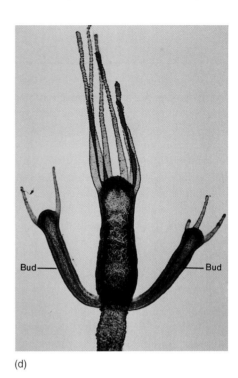

Bud — | — Bud

(d)

(e)

FIGURE 29.1

Asexual Reproduction.   (*a*) An amoeba (a protist) undergoes fission to form two individual organisms. (*b*) Planarian worms undergoing longitudinal fission. (*c*) The annelid, *Autolysis*, undergoing various constrictions. (*d*) A *Hydra* with developing buds. (*e*) Small sea anemones produced by fragmentation.

conditions, it is advantageous for the animal to produce asexually a large number of progeny with identical characteristics. A large number of animals, well adapted to a given environment, can be produced even if only one parent is present.

Without the tremendous genetic variability bestowed by meiosis and sexual processes, however, a population of genetically identical animals stands a greatly increased chance of being devastated by a single disease or environmental insult, such as a long drought. A given line of asexually reproducing animals can cope with a changing environment only through the relatively rare spontaneous mutations (alterations in genetic material) that prove to be beneficial. Paradoxically, however, most mutations are detrimental or lethal, and herein lies one of the greatest disadvantages of asexual reproduction. All such mutations are passed on to every offspring along with the normal, unmutated genes. Consequently, the typical asexual animal may have only one "good" copy of each hereditary unit (gene); the one on the homologous chromosome may be a mutated form that is nonfunctional or potentially lethal.

SEXUAL REPRODUCTION IN INVERTEBRATES

In **sexual** (L. *sexualis*, pertaining to sex) **reproduction**, the offspring have unique combinations of genes inherited from the two parents. Offspring of a sexual union are somewhat different from their parents and siblings—they have genetic diversity. Each new individual represents a combination of traits derived from two parents because syngamy, or fertilization, unites one gamete from each parent.

Sexual reproductive strategies and structures in the invertebrates are overwhelming. What follows is an overview of some principles of reproductive structure and function. The coverage of each invertebrate phylum in chapters 9 through 17 provides more specific details.

EXTERNAL FERTILIZATION

Many invertebrates (e.g., sponges and corals) simply release their gametes into the water in which they live **(broadcast spawning),** allowing external fertilization to occur. In these invertebrates, the gonads are usually simple, often transient structures for releasing the gametes from the body through various arrangements of coelomic ducts, metanephridia, sperm ducts, or oviducts.

INTERNAL FERTILIZATION

Other invertebrates (from flatworms to insects) utilize internal fertilization to transfer sperm from male to female and have structures that facilitate such transfer (figure 29.2).

In the male, sperm are produced in the testes and transported via a sperm duct to a storage area called the seminal vesicle. Prior to mating, some invertebrates (e.g., arrow worms, leeches, some insects) incorporate many sperm into packets termed spermatophores. Spermatophores provide a protective casing for sperm and facilitate the transfer of large numbers of sperm with minimal loss. Some spermatophores are even motile and act as independent sperm carriers. Sperm or the spermatophores are then passed into an ejaculatory duct to a copulatory organ (e.g., penis, cirrus, gonopore). The copulatory organ is used as an intromittent structure to introduce sperm into the female's system. Various accessory glands (e.g., seminal vesicle) may be present in males that produce seminal fluid or spermatophores.

In the female, ova (eggs) are produced in the ovaries and transported to the oviduct. Sperm move up the oviduct, where they encounter the ova and fertilize them. Accessory glands (e.g., those that produce egg capsules or shells) may also be present in females.

As noted earlier, sexual reproduction usually involves the fusion of gametes from a male and female parent. However, some sexually reproducing animals occasionally depart from this basic reproductive mode and exhibit variant forms of sexual reproduction.

Hermaphroditism (Gr. *hermaphroditos*, an organism with the attributes of both sexes) occurs when an animal has both functional male and female reproductive systems. This dual sexuality is sometimes called the monoecious (Gr. *monos*, single + *oikos*, house) condition. Although some hermaphrodites fertilize themselves,

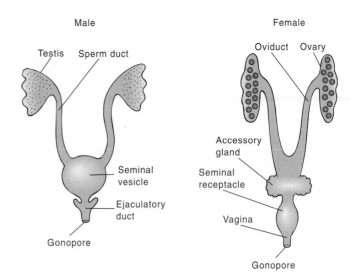

FIGURE 29.2

Stylized Male and Female Reproductive Systems in Invertebrates. Sexual reproduction is possible via these systems.

most also mate with another member of the same species (e.g., earthworms and sea slugs). When this occurs, each animal serves as both male and female—donating and receiving sperm. Hermaphroditism is especially beneficial to sessile (attached) animals (e.g., barnacles) that may only occasionally encounter the opposite sex.

Another variation of hermaphroditism—**sequential hermaphroditism**—occurs when an animal is one sex during one phase of its life cycle and the opposite sex during another phase. Hermaphrodites are either **protogynous** (Gr. *protos*, first + *gyne*, women) or **protandrous** (Gr. *protos*, first + *andros*, man; **protandry**). In protandry, an animal is a male during its early life history and a female later in the life history. The reverse is true for protogynous animals. A change in the sex ratio of a population is one factor that can induce sequential hermaphroditism, which is common in oysters.

ADVANTAGES AND DISADVANTAGES OF SEXUAL REPRODUCTION

New combinations of traits can arise more rapidly in sexually reproducing animals because of genetic recombinations. The resulting genetic diversity or variability increases the chances of the species surviving sudden environmental changes. Furthermore, variation is the foundation for evolution. In contrast to the way asexually reproducing populations tend to retain mutations, sexually reproducing populations tend to eliminate deleterious and lethal mutations.

Sexual reproduction also has some disadvantages. For example, an animal that cannot reproduce asexually can never bequeath its own exact set of genetic material to its progeny. Sexual reproduction bestows on the progeny a reassortment of maternal and paternal chromosomes. Thus, the same mixing processes that create the adaptive gene combinations in the adult work to dismantle it partially in the offspring. In addition, many of the gametes that are released are not fertilized, leading to a significant waste of metabolic effort.

SEXUAL REPRODUCTION IN VERTEBRATES

Since the evolution of the first animals, the basic use of male and female gametes has been preserved. Vertebrate evolution has also given rise to the close link between reproductive biology and sexual behavior. The strong drive to mate or reproduce dominates the lives of many vertebrates, as illustrated by the salmon's fateful spawning run or the rutting of bull elk. Females of most mammal species come into heat or **estrus** (Gr. *oistros*, a most vehement desire; the period of sexual receptivity) about the same time each year. Genetic, hormonal, and nervous system controls usually time estrus so that the young are born when environmental conditions make survival most likely.

SOME BASIC VERTEBRATE REPRODUCTIVE STRATEGIES

Fishes are well known for their high potential fecundity, with most species releasing thousands to millions of eggs and sperm annually. Fish species have reproductive methods, structures, and an attendant physiology that have allowed them to adapt to a great variety of aquatic conditions.

The reproductive strategies in amphibians are much more diverse than those observed in other groups of vertebrates. In each of the three living orders of Amphibia (caecilians, salamanders, anurans) are trends toward terrestriality. The variety of these adaptations is especially noteworthy in anurans. These reproductive adaptations have been viewed as pioneering evolutionary experiments in the conquest of terrestrial environments by vertebrates. Noteworthy is the evolution of direct development of terrestrial eggs, ovoviviparity, and viviparity that have been important in the successful invasion of mountainous environments by amphibians.

The reproductive adaptations of reptiles, birds, and early mammals foreshadow changes evident in the reproductive systems of later mammals, including humans. The reptilian system includes shelled, desiccation-resistant eggs. These eggs had the three basic embryonic membranes that still characterize the mammalian embryo, as well as a flat embryo that developed and underwent gastrulation atop a huge yolk mass. The same process of gastrulation is still seen in mammalian embryos, even though the massive yolk mass has been lost.

The mechanisms for maintaining the developing embryo within the female for long periods of time evolved in the early mammals. During **gestation** (L. *gestatio*, from + *gestare*, to bear), the embryo was nourished with nutrients and oxygen, yet it was protected from attack by the female's immune system. After birth, the first mammals nourished their young with milk from the mammary glands, just as primates do today.

Female apes and monkeys are asynchronous breeders. Mating and births can take place over much of the year. Females mate only when in estrus, increasing the probability of fertilization. Human females show a less distinctive estrus phase and can reproduce throughout the year. They can also engage in sexual activity without reproductive purpose; no longer is sexual behavior precariously tied to ovulation. The source of this important reproductive adaptation may be physiological or a result of concomitant evolution of the brain—a process that gave humans some conscious control over their emotions and behaviors that hormones, instincts, and the environment control in other animals. *This separation of sex from a purely reproductive function has evolved into the long-lasting pair bonds between human males and females (e.g., marriage) that further support the offspring. This type of behavior has also resulted in the transmission of culture—a key to the evolution and success of the human species.*

With this background, the reproductive anatomy and physiology of selected vertebrate classes is now presented.

EXAMPLES OF REPRODUCTION AMONG VARIOUS VERTEBRATE CLASSES

Almost all vertebrates reproduce sexually; only a few lizards and fishes normally reproduce parthenogenetically. *Sexual reproduction evolved among aquatic animals and then spread to the land as animals became terrestrial.*

FISHES

All fishes reproduce in aquatic environments. In bony fishes, fertilization is usually external, and eggs contain only enough yolk to sustain the developing fish for a short time. After this yolk is consumed, the growing fish must seek food. Although many thousands of eggs are produced and fertilized, few survive and grow to maturity. Some succumb to fungal and bacterial infections, others to siltation, and still others to predation. Thus, for reproduction to be successful, the fertilized egg must develop rapidly, and the young must achieve maturity within a short time.

AMPHIBIANS

The vertebrate invasion of land meant facing for the first time the danger of drying out or desiccating; the tiny gametes were especially vulnerable. The gametes could not simply be released near one another on the land because they would quickly desiccate.

The amphibians were the first vertebrates to invade the land. They have not, however, become adapted to a completely terrestrial environment; their life cycle is still inextricably linked to water. Among most amphibians, fertilization is still external, just as it is among the fishes. Among the frogs and toads, the male grasps the female and discharges fluid containing sperm onto the eggs as she releases them into the water (figure 29.3a).

The developmental period is much longer in amphibians than in fishes, although the eggs do not contain appreciably more yolk. *An evolutionary adaptation present in amphibians is the*

(a)

(b)

(c)

(d)

FIGURE 29.3

Vertebrate Reproductive Strategies.   (*a*) A male wood frog (*Rana sylvatica*) clasping the female in amplexus, a form of external fertilization. As the female releases eggs into the water, the male releases sperm over them. (*b*) Reptiles, such as these turtles, were the first terrestrial vertebrates to develop internal fertilization. (*c*) Birds are oviparous. Their shelled eggs have large yolk reserves, and the young develop and hatch outside the mother's body. Birds may show advanced parental care. (*d*) A placental mammal. This female dog is nursing her puppies.

presence of two periods of development: larval and adult stages. The aquatic larval stage develops rapidly, and the animal spends much time eating and growing. After reaching a sufficient size, the larval form undergoes a developmental transition called metamorphosis into the adult (often terrestrial) form.

REPTILES

The reptiles were the first group of vertebrates to completely abandon the aquatic habitat because of adaptations that permitted sexual reproduction on land. *A crucial adaptation first found in reptiles is internal fertilization (figure 29.3b). Internal fertilization protects the gametes from drying out, freeing the animals from returning to the water to breed.*

Many reptiles are **oviparous** (L. *ovum*, egg + *parere*, to bring forth), and the eggs are deposited outside the body of the female. Others are **ovoviviparous** (L. *ovum*, egg + *vivere*, to live, + *parere*, to bring forth). They form eggs that hatch in the body of the female, and the young are born alive.

The shelled egg and extraembryonic membranes, also first seen in reptiles, constitute two other important evolutionary adaptations to life on land. These adaptations allowed reptiles to lay eggs in dry places without danger of desiccation. As the embryo develops, the extraembryonic chorion and amnion help protect it, the latter by creating a fluid-filled sac for the embryo. The allantois permits gas exchange and stores excretory products. Complete development can occur within the eggshell. When the

animal hatches, it has developed to the point that it can survive on its own or with some parental care.

BIRDS

Birds have retained the important adaptations for life on land that evolved in the early reptiles. With the exception of most waterfowl, birds lack a penis. Males simply deposit semen against the cloaca for internal fertilization. Sperm then migrate up the cloaca and fertilize the eggs before hard shells form. This method of mating occurs more quickly than the internal fertilization that reptiles practice. All birds are oviparous, and the eggshells are much thicker than those of reptiles. Thicker shells permit birds to sit on their eggs and warm them. This brooding, or incubation, hastens embryo development. When many young birds hatch, they are incapable of surviving on their own. Extensive parental care and feeding of young are more common among birds than among fishes, amphibians, or reptiles (figure 29.3*c*).

MAMMALS

The most primitive mammals, the monotremes (e.g., the duck-billed platypus and spiny anteater), lay eggs, as did the reptiles from which they evolved. All other mammals are viviparous.

Mammalian viviparity was another major evolutionary adaptation, and it has taken two forms. The marsupials developed the ability to nourish their young in a pouch after a short gestation inside the female. The other, much larger group—the placentals—retain the young inside the female, where the mother nourishes them by means of a placenta. Even after birth, mammals continue to nourish their young. *Mammary glands are a unique mammalian adaptation that permit the female to nourish the young with milk that she produces (figure 29.3d).* Some mammals nurture their young until adulthood, when they are able to mate and fend for themselves. **A**s noted at the beginning of this section, mammalian reproductive behavior also contributes to the transmission and evolution of culture that is the key to the evolution of the human species.

THE HUMAN MALE REPRODUCTIVE SYSTEM

The reproductive role of the human male is to produce sperm and deliver them to the vagina of the female. This function requires the following structures:

1. Two testes that produce sperm and the male sex hormone, testosterone.
2. Accessory glands and tubes that furnish a fluid for carrying the sperm to the penis. This fluid, together with the sperm, is called semen.
3. Accessory ducts that store and carry secretions from the testes and accessory glands to the penis.
4. A penis that deposits semen in the vagina during sexual intercourse.

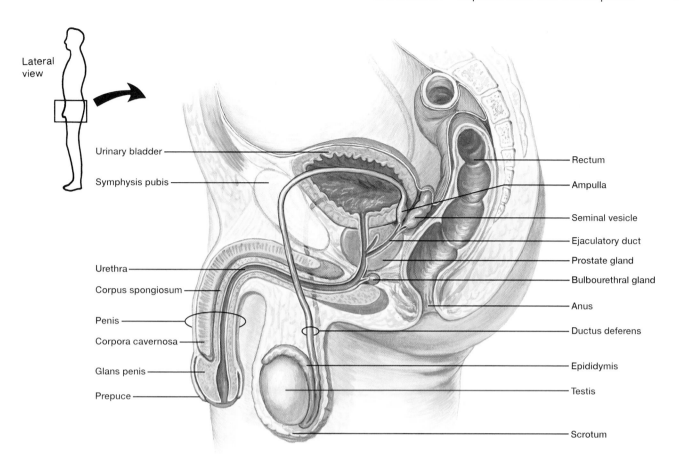

Lateral
view

Urinary bladder

Symphysis pubis

Urethra

Corpus spongiosum

Penis

Corpora cavernosa

Glans penis

Prepuce

Rectum

Ampulla

Seminal vesicle

Ejaculatory duct

Prostate gland

Bulbourethral gland

Anus

Ductus deferens

Epididymis

Testis

Scrotum

FIGURE 29.4

Lateral View of the Human Male Reproductive System.   There are two each of the following structures: testis, epididymis, ductus deferens, seminal vesicle, ejaculatory duct, and bulbourethral gland.

PRODUCTION AND TRANSPORT OF SPERM

The paired **testes** (sing., testis) (L. *testis*, witness; the paired testes were believed to bear witness to a man's virility) are the male reproductive organs (gonads) that produce sperm (figure 29.4). Shortly after birth, the testes descend from the abdominal cavity into the **scrotum** (L. *scrautum*, a leather pouch for arrows), which hangs between the thighs. Because the testes hang outside the body, the temperature inside the scrotum is about 34° C compared to a 38° C core temperature. The lower temperature is necessary for active sperm production and survival. Muscles elevate or lower the testes, depending on the outside air temperature.

Each testis contains over eight hundred tightly coiled **seminiferous tubules** (figure 29.5*a, b*), which produce thousands of sperm each second in healthy young men. The walls of the seminiferous tubules are lined with two types of cells: spermatogenic cells, which give rise to sperm, and sustentacular cells, which nourish the sperm as they form and which also secrete a fluid (as well as the hormone inhibin) into the tubules to provide a liquid medium for the sperm. Between the seminiferous tubules are clusters of endocrine cells, called interstitial cells (Leydig cells), that secrete the male sex hormone testosterone.

A system of tubes carries the sperm that the testes produce to the penis. The seminiferous tubules merge into a network of tiny tubules called the rete testis (L. *rete*, net), which merges into a coiled tube called the epididymis. The epididymis has three main functions: (1) it stores sperm until they are mature and ready to be ejaculated, (2) it contains smooth muscle that helps propel the sperm toward the penis by peristaltic contractions, and (3) it serves as a duct system for sperm to pass from the testis to the ductus deferens. The ductus deferens (formerly called the vas deferens or sperm duct) is the dilated continuation of the epididymis. Continuing upward after leaving the scrotum, the ductus deferens passes through the lower part of the abdominal wall via the inguinal canal. If the abdominal wall weakens at the point where the ductus deferens passes through, an inguinal hernia may result. (In an inguinal hernia, the intestine may protrude downward into the scrotum.) The ductus deferens then passes around the urinary bladder and enlarges to form the ampulla (*see figure 29.4*). The ampulla stores some sperm until they are ejaculated. Distal to the ampulla, the ductus deferens becomes the ejaculatory duct. The urethra is the final section of the reproductive duct system.

After the ductus deferens passes around the urinary bladder, several accessory glands add their secretions to the sperm as they are propelled through the ducts. These accessory glands are the seminal

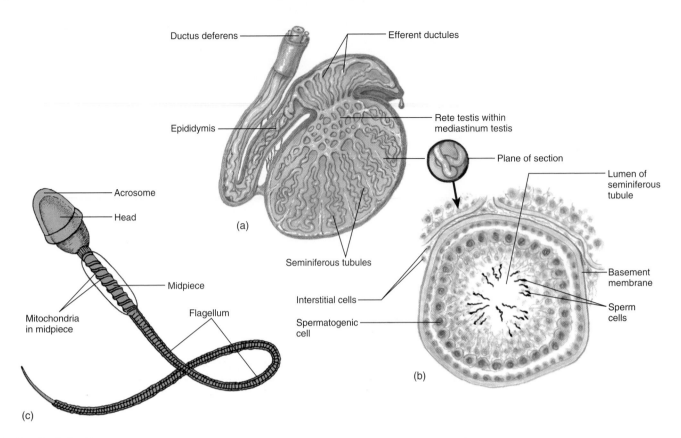

FIGURE 29.5

Human Male Testis. (*a*) Sagittal section through a testis. (*b*) Cross section of a seminiferous tubule, showing the location of spermatogenesis. (*c*) A mature sperm.

vesicles, prostate gland, and bulbourethral glands (*see figure 29.4*). The paired **seminal vesicles** secrete water, fructose, prostaglandins, and vitamin C. This secretion provides an energy source for the motile sperm and helps to neutralize the natural protective acidity of the vagina. (The pH of the vagina is about 3 to 4, but sperm motility and fertility are enhanced when it increases to about 6.) The **prostate gland** secretes water, enzymes, cholesterol, buffering salts, and phospholipids. The **bulbourethral glands** secrete a clear, alkaline fluid that lubricates the urethra and facilitates the ejaculation of semen and lubricates the penis prior to sexual intercourse. The fluid that results from the combination of sperm and glandular secretions is **semen** (L. *seminis*, seed). The average human ejaculation produces 3 to 4 ml of semen and contains 300 to 400 million sperm.

The penis has two functions. It carries urine through the urethra to the outside during urination, and it transports semen through the urethra during ejaculation. In addition to the urethra, the penis contains three cylindrical strands of erectile tissue: two corpora cavernosa and the corpus spongiosum (*see figure 29.4*). The corpus spongiosum extends beyond the corpora cavernosa and becomes the expanded tip of the penis called the glans penis. The loosely fitting skin of the penis folds forward over the glans to form the prepuce or foreskin. **Circumcision** is the removal of the prepuce for religious or health reasons. Today, many circumcisions are performed in the belief that they lessen the likelihood of cancer of the penis.

A mature human sperm consists of a head, midpiece, and tail (figure 29.5c). The head contains the haploid nucleus, which is mostly DNA. The acrosome, a cap over most of the head, contains an enzyme called acrosin that assists the sperm in penetrating the outer layer surrounding a secondary oocyte. The sperm tail contains an array of microtubules that bend and produce whiplike movements. The spiral mitochondria in the midpiece supply the ATP necessary for these movements.

HORMONAL CONTROL OF MALE REPRODUCTIVE FUNCTION

Before a male can mature and function sexually, special regulatory hormones must come into play (table 29.1). Male sex hormones are collectively called **androgens** (Gr. *andros*, man + *gennan*, to produce). The hormones that travel from the brain and pituitary gland to the testes (and ovaries in the female) are called **gonadotropins**. As previously noted, the interstitial cells produce the male sex hormone **testosterone.** Figure 29.6 shows the negative feedback mechanisms that regulate the production and secretion of testosterone, as well as its actions. When the level of testosterone in the blood decreases, the hypothalamus is stimulated to secrete GnRH (gonadotropin-releasing hormone). GnRH stimulates the secretion of FSH (follicle-stimulating hormone) and LH (luteinizing hormone), also called ICSH (interstitial cell–stimulating hormone), into the bloodstream. (FSH and LH were first named for their functions in females, but their molecular structure is exactly the same in males.) FSH causes the spermatogenic cells in the

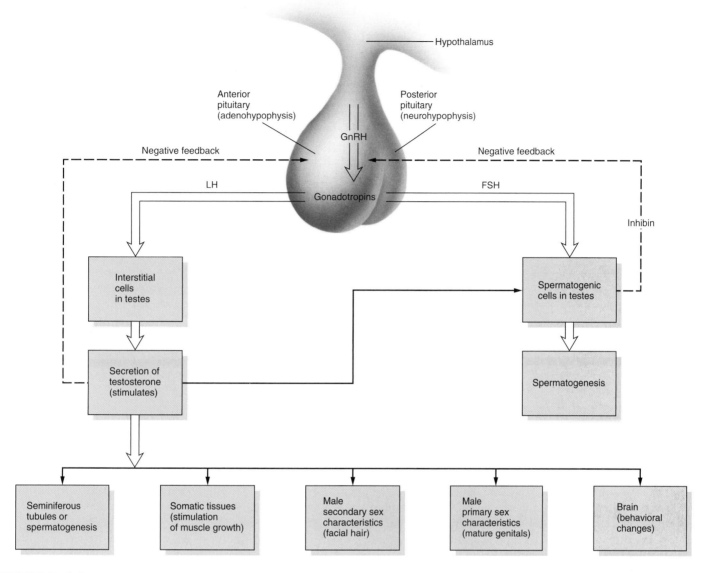

FIGURE 29.6

Hormonal Control of Reproductive Function in Adult Human Males.   Negative feedback mechanisms by which the hypothalamus controls sperm maturation and the development of male secondary sexual characteristics. (GnRH = gonadotropin-releasing hormone; LH = luteinizing hormone; FSH = follicle-stimulating hormone.)

TABLE 29.1
MAJOR HUMAN MALE REPRODUCTIVE HORMONES IN AN ADULT

HORMONE	FUNCTIONS	SOURCE
FSH (follicle-stimulating hormone)	Aids sperm maturation; increases testosterone production	Pituitary gland
GnRH (gonadotropin-releasing hormone)	Controls pituitary secretion	Hypothalamus
Inhibin	Inhibits FSH secretion	Sustentacular cells in testes
LH (luteinizing hormone) or ICSH (interstitial cell–stimulating hormone)	Stimulates testosterone secretion	Pituitary gland
Testosterone	Increases sperm production; stimulates development of male primary and secondary sexual characteristics; inhibits LH secretion	Interstitial cells in testes

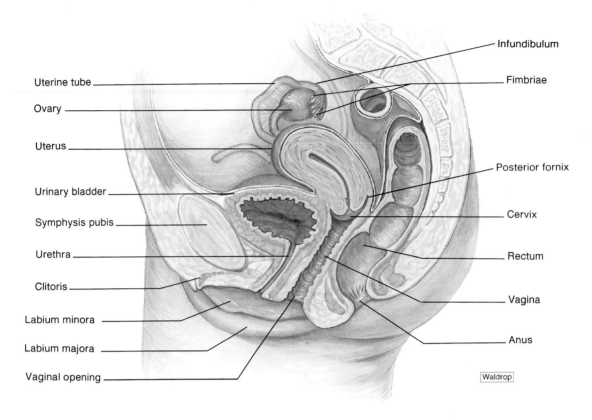

Uterine tube

Ovary

Uterus

Urinary bladder

Symphysis pubis

Urethra

Clitoris

Labium minora

Labium majora

Vaginal opening

Infundibulum

Fimbriae

Posterior fornix

Cervix

Rectum

Vagina

Anus

Waldrop

FIGURE 29.7
Lateral View of the Human Female Reproductive System. Two uterine tubes lead into the uterus and two ovaries.

seminiferous tubules to initiate spermatogenesis, and LH stimulates the interstitial cells to secrete testosterone. The cycle is completed when testosterone inhibits the secretion of LH, and another hormone, inhibin, is secreted. Inhibin inhibits the secretion of FSH from the anterior pituitary. This cycle maintains a constant rate (homeostasis) of spermatogenesis.

THE HUMAN FEMALE REPRODUCTIVE SYSTEM

The reproductive role of human females is more complex than that of males. Not only do females produce gametes (eggs or ova), but after fertilization, they also nourish, carry, and protect the developing embryo. After the offspring is born, the mother may nurse it for a time. Another difference between the sexes is the monthly rhythmicity of the female reproductive system.

The female reproductive system consists of a number of structures with specialized functions (figure 29.7):

1. Two ovaries produce eggs and the female sex hormones estrogen and progesterone.
2. Two uterine tubes, one from each ovary, carry eggs from the ovary to the uterus. Fertilization usually occurs in the upper third of a uterine tube.
3. If fertilization occurs, the uterus receives the blastocyst and houses the developing embryo.

4. The vagina receives semen from the penis during sexual intercourse. It is the exit point for menstrual flow and is the canal through which the baby passes from the uterus during childbirth.
5. The external genital organs have protective functions and play a role in sexual arousal.
6. The mammary glands, contained in the paired breasts, produce milk for the newborn baby.

PRODUCTION AND TRANSPORT OF THE EGG

The female gonads are the paired **ovaries** (L. *ovum*, egg), which produce eggs and female hormones. The ovaries are located in the pelvic part of the abdomen, one on each side of the uterus. A cross section of an ovary reveals rounded vesicles called follicles, which are the actual centers of egg production (oogenesis) (figure 29.8). Each follicle contains an immature egg called a primary oocyte, and follicles are always present in several stages of development. After the release of a secondary oocyte (commonly called an egg) in the process called **ovulation,** the lining of the follicle grows inward, forming the corpus luteum ("yellow body"), which serves as a temporary endocrine tissue and continues to secrete the female sex hormones estrogen and progesterone.

The paired tubes that receive the secondary oocyte from the ovary and convey it to the uterus are called either the **uterine**

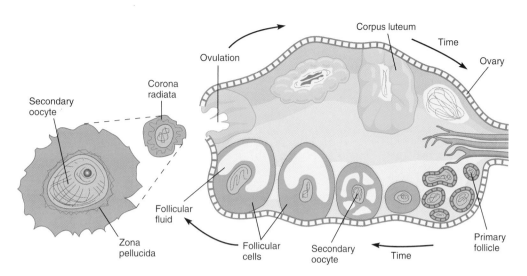

FIGURE 29.8

Cross Section Through a Human Ovary.   The events in the ovarian cycle proceed from the growth and maturation of the primary follicle, through ovulation (rupture of a mature follicle with the concurrent release of a secondary oocyte), through the formation and maintenance (during pregnancy) or degeneration (no pregnancy) of an endocrine structure called the corpus luteum. The positions of the oocyte and corpus luteum are varied for illustrative purposes only. An oocyte matures at the same site, from the beginning of the cycle to ovulation.

tubes or **fallopian tubes** (*see figure 29.7*). Feathery fimbriae fringe the part of the uterine tube that encircles the ovary. Each month, as a secondary oocyte is released, the motion of the fimbriae sweep it across a tiny space between the uterine tube and the ovary into the tube.

Unlike sperm, the secondary oocyte cannot move on its own. Instead, the peristaltic contractions of the tube and the waving motions of the cilia in the mucous membrane of the tube carry the secondary oocyte along (figure 29.9). Fertilization usually occurs in the uppermost third of the uterine tube. A fertilized oocyte (zygote) continues its journey toward the uterus, where it will implant. The journey takes four to seven days. If fertilization does not occur, the secondary oocyte degenerates in the uterine tube.

The uterine tubes terminate in the **uterus,** a hollow, muscular organ in front of the rectum and behind the urinary bladder (figure 29.10). The uterus terminates in a narrow portion called the cervix, which joins the uterus to the vagina. The uterus has three layers of tissues. The outer layer (perimetrium) extends beyond the uterus to form the two broad ligaments that stretch from the uterus to the lateral walls of the pelvis. The middle muscular layer (myometrium [Gr. *myo,* muscle + *metra,* womb]) makes up most of the uterine wall. The endometrium is the specialized mucous membrane that contains an abundance of blood vessels and simple glands.

The cervix leads to the **vagina,** a muscular tube 8 to 10 cm long. The wall of the vagina is composed mainly of smooth muscle and elastic tissue.

The external genital organs, or genitalia, include the mons pubis, labia majora, labia minora, vestibular glands, clitoris, and vaginal opening (*see figure 29.7*). As a group, these organs are called the **vulva.** In most young women, the vaginal opening is partially covered by a thin membrane, the hymen, which may be ruptured during normal strenuous activities or may be stretched or broken during sexual activity.

FIGURE 29.9

Cilia Lining the Uterine Tubes.   The tiny, beating cilia on the surfaces of the uterine tube cells propel the secondary oocyte downward and perhaps the sperm upward (EM ×1000).

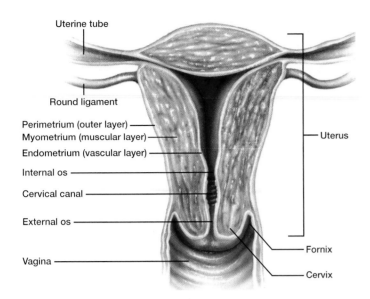

FIGURE 29.10

Human Female Uterus. This frontal section of a uterus shows the three major tissue layers. The outer layer is the perimetrium. The middle myometrium makes up the bulk of the uterine wall. It is composed of smooth muscle fibers. The innermost layer is composed of a specialized mucous membrane called the endometrium, which is deep and velvety in texture. Breakdown of the endometrium comprises part of the menstrual flow.

The **mammary glands** (L. *mammae*, breasts) are modified sweat glands that produce and secrete milk. They contain varying amounts of adipose tissue. The amount of adipose tissue determines the size of the breasts, but the amount of mammary tissue does not vary widely from one woman to another.

HORMONAL CONTROL OF FEMALE REPRODUCTIVE FUNCTION

The male is continuously fertile from puberty to old age, and throughout that period, sex hormones are continuously secreted. The female, however, is fertile only during a few days each month, and the pattern of hormone secretion is intricately related to the cyclical release of a secondary oocyte from the ovary.

The cyclical production of hormones controls the development of a secondary oocyte in a follicle (figure 29.11; table 29.2). Gonadotropin-releasing hormone (GnRH) from the hypothalamus acts on the anterior pituitary gland, which releases follicle-stimulating hormone (FSH) and luteinizing hormone (LH) to bring about the oocyte's maturation and release from the ovary. These hormones regulate the **menstrual cycle,** which is the cyclic preparation of the uterus to receive a fertilized egg, and the **ovarian cycle,** during which the oocyte matures and ovulation occurs. This monthly preparation of the uterine lining for the fertilized egg normally begins at puberty. When a female reaches 45 to 55 years of age, the ovaries lose their sensitivity to FSH and LH, they stop making normal amounts of progesterone and estrogen, and the monthly menstrual cycle

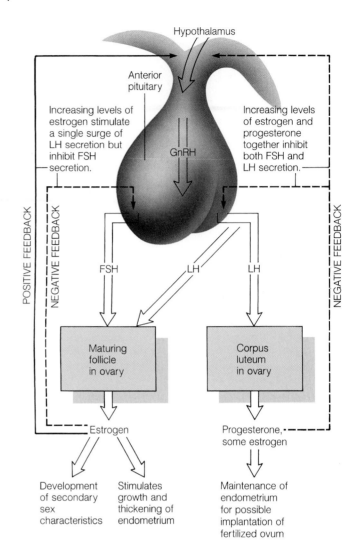

FIGURE 29.11

Hormonal Control of Reproductive Functions in an Adult Human Female. Feedback loops involve the hypothalamus, anterior pituitary, and ovaries. Gonadotropin-releasing hormone (GnRH) stimulates the release of both follicle-stimulating hormone (FSH) and luteinizing hormone (LH). Two negative feedback systems and a positive feedback system (one of the few in the human body) control the ovarian cycle.

ceases in what is called the **menopause** (Gr. *men*, month + *pausis*, cessation).

One way to understand the hormonal pattern in the normal monthly cycle is to follow the development of the oocyte and the physical events in the menstrual cycle (figure 29.12; table 29.3). On average, it takes 28 days to complete one menstrual cycle, although the range may be from 22 to 45 days. During this time, the following events take place:

1. The controlling center for ovulation and menstruation is the hypothalamus. It releases, on a regular cycle, GnRH, which stimulates the anterior pituitary to secrete FSH and LH (*see figure 29.11*).
2. FSH promotes the development of the oocyte in one of the immature ovarian follicles.

TABLE 29.2
MAJOR HUMAN FEMALE REPRODUCTIVE HORMONES

HORMONE	FUNCTIONS	SOURCE
Estrogen	Stimulates thickening of uterine wall, maturation of oocyte and development of female sexual characteristics; inhibits FSH secretion; increases LH secretion	Ovarian follicle, corpus luteum
FSH (follicle-stimulating hormone)	Causes immature oocyte and follicle to develop; increases estrogen secretion; stimulates new gamete formation and development of uterine wall after menstruation	Pituitary gland
GnRH (gonadotropin-releasing hormone)	Controls pituitary secretion	Hypothalamus
hCG (human chorionic gonadotropin)	Prevents corpus luteum from disintegrating; stimulates corpus luteum to secrete estrogen and progesterone	Embryonic membranes and placenta
Inhibin	Inhibits secretion of FSH from the anterior pituitary gland	Ovaries
LH (luteinizing hormone)	Stimulates further development of oocyte and follicle; stimulates ovulation; increases progesterone secretion; aids in development of corpus luteum	Pituitary gland
Oxytocin	Stimulates uterine contractions during labor and milk release during nursing	Pituitary gland
Prolactin	Promotes milk secretion by mammary glands after childbirth	Pituitary gland
Progesterone	Stimulates thickening of uterine wall	Corpus luteum
Relaxin	Increases flexibility of pubic symphysis during pregnancy and helps dilate uterine cervix during labor and delivery	Placenta and ovaries

TABLE 29.3
SUMMARY OF THE MENSTRUAL CYCLE EVENTS

PHASE	EVENTS	DURATION IN DAYS*
Follicular	Follicle matures in the ovary; menstruation (endometrium breaks down); endometrium rebuilds	1–5
Ovulation	Ovary releases secondary oocyte	6–14
Luteal	Corpus luteum forms; endometrium thickens and becomes glandular	15–28

*Using a 28-day menstrual cycle as an example.

3. The follicles produce estrogen, causing a buildup and proliferation of the endometrium, as well as the inhibition of FSH production.
4. The elevated estrogen level about midway in the cycle triggers the anterior pituitary to secrete LH, which causes the mature follicle to enlarge rapidly and release the secondary oocyte (ovulation). LH also causes the collapsed follicle to become another endocrine tissue, the corpus luteum.

5. The corpus luteum secretes estrogen and progesterone, which act to complete the development of the endometrium and maintain it for 10 to 14 days.
6. If the oocyte is not fertilized, the corpus luteum disintegrates into a corpus albicans, and estrogen and progesterone secretion cease.
7. Without estrogen and progesterone, the endometrium breaks down, and **menstruation** occurs. The menstrual flow is composed mainly of sloughed-off endometrial cells, mucus, and blood.
8. As progesterone and estrogen levels decrease further, the pituitary renews active secretion of FSH, which stimulates the development of another follicle, and the monthly cycle begins again.

HORMONAL REGULATION IN THE PREGNANT FEMALE

Pregnancy sets into motion a new series of physiological events. The ovaries are directly affected because, as the embryo develops, the cells of the embryo and placenta release the hormone human chorionic gonadotropin (hCG), which keeps the corpus luteum from disintegrating. The progesterone that it secretes is necessary to maintain the uterine lining. After a time, the placenta takes over progesterone production, and the corpus luteum degenerates. By the end of two weeks following implantation, the concentration of hCG is so high in the female's blood, and in her urine as

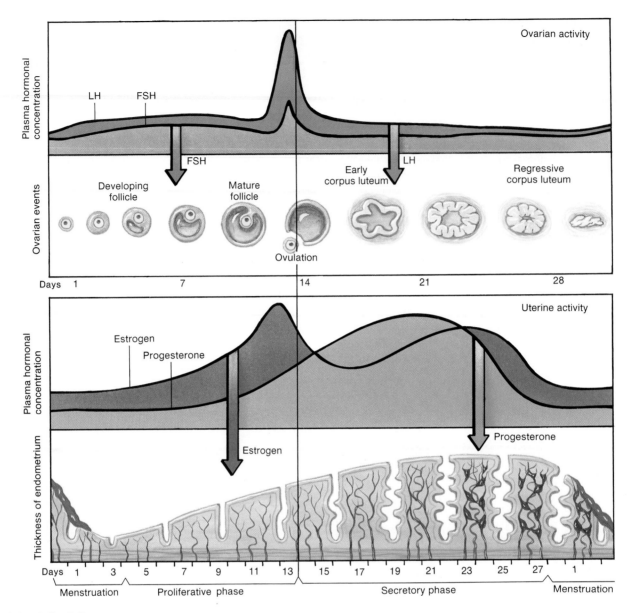

FIGURE 29.12

Major Events in the Female Ovarian and Menstrual Cycles.   The two charts correlate the gonadotropins, ovarian hormones, follicle development, ovulation, and changes in uterine anatomy during the cycles.

well, that an hCG immunological test can check for pregnancy. As the embryo develops, other hormones are secreted. For example, prolactin and oxytocin induce the mammary glands to secrete and eject milk after childbirth. Oxytocin and prostaglandins also stimulate the uterine contractions that expel the baby from the uterus during childbirth.

PRENATAL DEVELOPMENT AND BIRTH

This section covers the main event in reproduction—the nine-month pregnancy period, during which time the human female's body carries, nourishes, and protects the embryo as it grows to a full-term baby.

EVENTS OF PRENATAL DEVELOPMENT: FROM ZYGOTE TO NEWBORN

The development of a human being can be divided into prenatal ("before birth") and postnatal ("after birth") periods. During the prenatal period, the developing individual begins life as a zygote, then becomes a ball of cells called a morula, and eventually becomes a blastocyst that implants in the endometrium. From two weeks after fertilization until the end of the eighth week of its existence, the individual is called an embryo. From nine weeks until birth, it is a fetus. During or after birth, it is called a newborn, or baby.

Pregnancy is arbitrarily divided into three-month periods called trimesters. The first trimester begins at fertilization, and during this time most of the organs are formed. The next two trimesters are mainly periods of growth for the fetus.

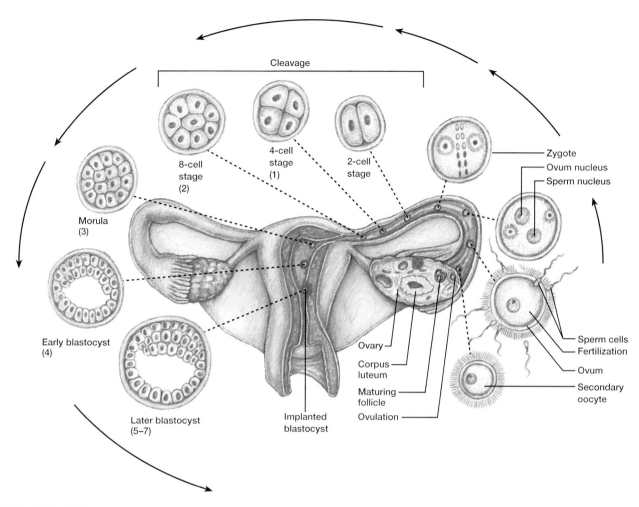

FIGURE 29.13

Early Stages in Human Development. The numbers in parentheses indicate the days after fertilization. The secondary oocyte is fertilized in the upper third of the uterine tube, undergoes cleavage while traveling down the tube, and finally implants in the endometrium of the uterus.

The First Trimester

After fertilization, usually in the upper third of the uterine tube, the zygote goes through several cleavages as it moves down the tube (figure 29.13). It eventually becomes a solid ball of cells called a **morula,** and by the fourth day, it develops into a 50- to 120-cell blastula stage called a **blastocyst.**

The next stage of development occurs when the blastocyst adheres to the uterine wall and implants. During implantation, the outer cells of the blastocyst, called the trophoblast, invade the endometrium. Implantation is usually completed 11 to 12 days after fertilization; from then on, the female is considered to be pregnant.

One of the unique features of mammalian development is that most of the cells of the early embryo make no contribution to the embryo's body, giving rise instead to supportive and protective membranes. Only the inner cell mass gives rise to the embryonic body. Eventually, these cells arrange in a flat sheet that undergoes a gastrulation similar to that of reptiles and birds.

Once gastrulation is completed, the rest of the first trimester is devoted to organogenesis and growth (figure 29.14). Regulatory events and inductive-tissue interactions shape most of the organ systems. By the middle of the first trimester, all of the major body systems have begun to develop.

The Second Trimester

In the second trimester (fourth month), fetal growth is spectacular. By now, the pregnant mother is aware of fetal movements. The heartbeat can be heard with a stethoscope. During the sixth month, the upper and lower eyelids separate, and the eyelashes form. During the seventh month, the eyes open. During this period, the bones begin to ossify.

The Third Trimester

The third trimester extends from the seventh month until birth. During this time, the fetus has developed sufficiently (with respect to the circulatory and respiratory systems) to potentially survive if born prematurely. During the last month, fetal weight doubles.

THE PLACENTA: EXCHANGE SITE AND HORMONE PRODUCER

The lengthy pregnancy characteristic of mammals is possible, in part, because of the embryonic membranes that originated in the reptiles: the amnion, yolk sac, chorion, and allantois. The latter two

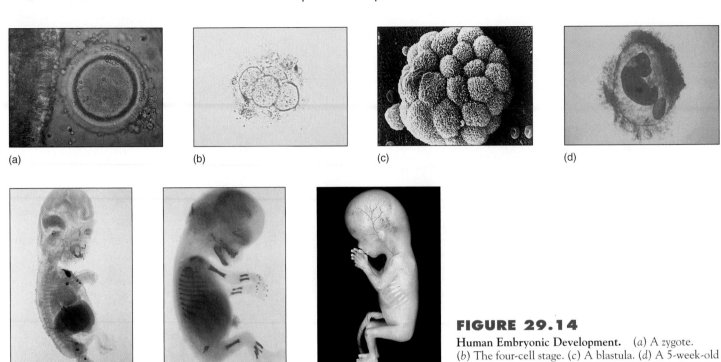

(a)   (b)   (c)   (d)

(e)   (f)   (g)

FIGURE 29.14

Human Embryonic Development. (*a*) A zygote. (*b*) The four-cell stage. (*c*) A blastula. (*d*) A 5-week-old embryo. (*e*) The embryo at 8 weeks. (*f*) The embryo at 10 weeks. (*g*) The fetus at about 3 months.

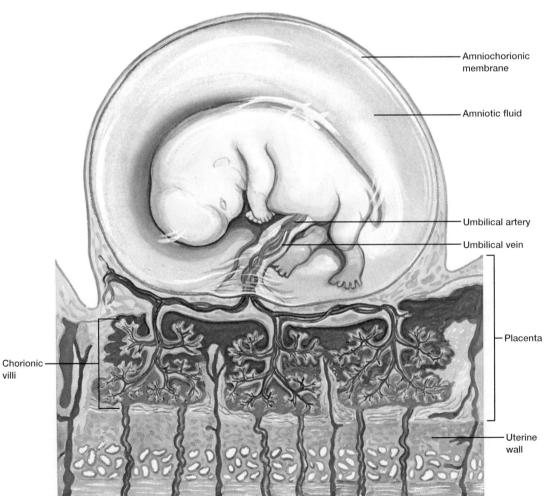

Amniochorionic membrane

Amniotic fluid

Umbilical artery

Umbilical vein

Placenta

Chorionic villi

Uterine wall

FIGURE 29.15

Fetus and Placenta at Seven Weeks. The circulations of mother and fetus are in close contact at the site of the chorionic villi, but they do not actually mix. Branches of the mother's arteries in the wall of her uterus open into pools near the chorionic villi. Oxygen and nutrients from the mother's blood diffuse into the fetal capillaries of the placenta. The fetal capillaries lead into the umbilical vein, which is enclosed within the umbilical cord. From here, the fresh blood circulates through the fetus's body. Blood that the fetus has depleted of nutrients and oxygen returns to the placenta in the umbilical arteries, which branch into capillaries, from which waste products diffuse to the maternal side.

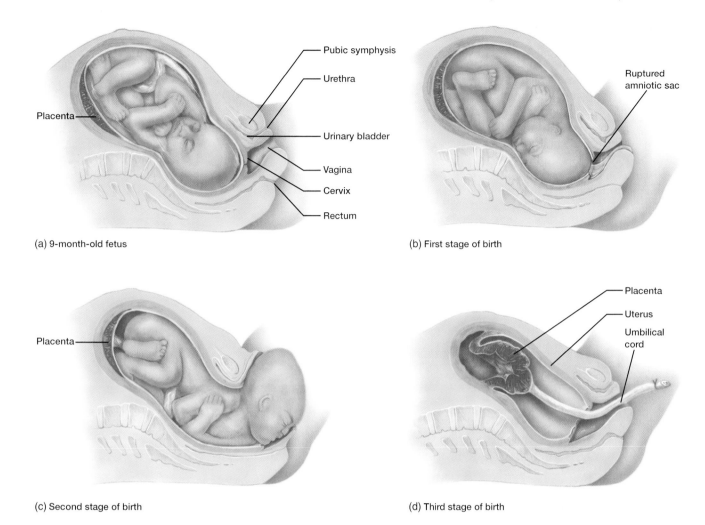

Pubic symphysis

Urethra

Placenta

Urinary bladder

Vagina

Cervix

Rectum

(a) 9-month-old fetus

Ruptured
amniotic sac

(b) First stage of birth

Placenta

(c) Second stage of birth

Placenta

Uterus

Umbilical
cord

(d) Third stage of birth

FIGURE 29.16

Stages of Labor and Parturition.   (*a*) The position of the fetus prior to labor. (*b*) The ruptured amniotic sac and early dilation of the cervix. (*c*) The expulsion stage of parturition. (*d*) Passage of the afterbirth.

gave rise to the embryonic parts of the placenta. The **placenta** is the organ that sustains the embryo and fetus throughout the pregnancy and through which gases, nutrients, and wastes are exchanged between the maternal and fetal systems (figure 29.15). The tiny, fingerlike projections that were sent out from the blastocyst during implantation develop into numerous chorionic villi, which contain embryonic blood vessels. These blood vessels do not merge with those of the mother; the two bloodstreams remain separate throughout the pregnancy. The **umbilical cord** connects the placenta to the abdomen of the fetus. Two fetal umbilical arteries and one fetal umbilical vein spiral about each other in the umbilical cord.

BIRTH: AN END AND A BEGINNING

About 266 days after fertilization, or 280 days from the beginning of the last menstrual period, the human infant is born. The birth process is called **parturition** (L. *parturire*, to be in labor). During parturition, the mother's uterine muscles begin to contract, and the cervix begins to dilate, or open. The hormone relaxin, pro-

duced by the ovaries and placenta, causes the mother's pelvic bones to slightly separate so that the baby can pass through the birth canal.

Changing hormone levels initiate parturition. When it is time for the baby to be born, its pituitary gland secretes adrenocorticotropic hormone (ACTH), which stimulates the adrenal glands to secrete steroids. These steroids stimulate the placenta to produce prostaglandins that, along with the hormone oxytocin from the mother's pituitary, cause the uterus to begin powerful muscular contractions. The contractions build in length and increase in frequency over a period that usually lasts from 2 to 18 hours. During that time, the cervix becomes fully dilated, and the amniotic sac ruptures. Usually within an hour of these events, the baby is expelled from the uterus (figure 29.16*a*–*c*). After the baby emerges, uterine contractions continue to expel the placenta and membranes commonly called the **afterbirth** (figure 29.16*d*). The umbilical cord is severed, and the newborn embarks on its nurtured existence in the outside world. (In mammals other than humans, the female bites through the cord to sever it.)

MILK PRODUCTION AND LACTATION

Lactation (L. *lactare*, to suckle) includes both milk secretion (production) by the mammary glands and milk release from the breasts. *(Mammary glands, a unique characteristic of mammals, evolved from sweat glands in the skin.)* During pregnancy, the breasts enlarge in response to increasing levels of the hormone prolactin. Before birth, placental secretions of estrogen and progesterone inhibit milk secretion from the breasts. After the placenta has been expelled from the uterus, the concentrations of estrogen and progesterone drop, and the breasts begin to produce copious amounts of milk.

The mother's breasts do not actually release milk until one to three days after the baby is born. During these first days, the suckling baby receives **colostrum,** a high-protein fluid present in the breast at birth. Colostrum contains an abundance of maternal antibodies and thus helps strengthen the baby's immune system. It also functions as a laxative, removing fetal wastes, called **meconium,** retained in the intestines. After about three days, the prolactin secreted from the pituitary stimulates milk production. The newborn's suckling stimulates the pituitary to release oxytocin as well as prolactin. Oxytocin triggers milk release from the mammary glands.

SUMMARY

1. Asexual reproductive processes do not involve the production and subsequent fusion of haploid cells, but rely solely on vegetative growth through mitosis.

2. Sexual reproduction involves the formation of haploid cells through meiosis and the subsequent fusion of pairs of those cells to produce a diploid zygote.

3. The forms of asexual reproduction in invertebrates are binary fission, budding (both external and internal), and fragmentation. Parthenogenesis also occurs in a few invertebrates.

4. Sexual reproductive strategies and structures in the invertebrates are numerous and varied.

5. Sexual reproduction first evolved in aquatic animals. The invasion of land meant facing the danger of the gametes and embryos desiccating. The five major groups of vertebrates have reproductive adaptations for the environment in which they reproduce.

6. The reproductive role of the human male is to produce sperm and deliver them to the vagina of the female. This function requires different structures. The testes produce sperm and the male sex hormone, testosterone. Accessory glands furnish a fluid, called semen, for carrying the sperm to the penis. Accessory ducts store and carry secretions from the testes and accessory glands to the penis. The penis deposits semen in the vagina during sexual intercourse.

7. Before a human male can mature and function sexually, special regulatory hormones (FSH, GnRH, inhibin, LH, and testosterone) must function.

8. The reproductive roles of the human female are more complex than those of the male. Not only do females produce eggs, but after fertilization, they also nourish, carry, and protect the developing embryo. They may also nourish the infant for a time after it is born. The female reproductive system consists of two ovaries, two uterine tubes, a uterus, vagina, and external genitalia. The mammary glands contained in the paired breasts produce milk for the newborn baby.

9. The human female is fertile for only a few days each month, and the pattern of hormone secretion is intricately related to the cyclical release of a secondary oocyte from the ovary. Various hormones regulate the menstrual and ovarian cycles.

10. Pregnancy sets a new series of physiological events into motion that are directed to housing, protecting, and nourishing the embryo.

11. The development of a human may be divided into prenatal and postnatal periods. Pregnancy is arbitrarily divided into trimesters.

12. The placenta is the organ that sustains the embryo and fetus throughout the pregnancy. The birth process is called parturition and occurs about 266 days after fertilization.

13. Lactation includes both milk secretion (production) by the mammary glands and milk release from the breasts.

SELECTED KEY TERMS

asexual reproduction (p. 484)
broadcast spawning (p. 486)
budding (p. 484)
estrus (p. 487)
fission (p. 484)
fragmentation (p. 484)

gestation (p. 487)
parthenogenesis (p. 484)
sequential hermaphroditism (p. 486)
sexual reproduction (p. 486)

CRITICAL THINKING QUESTIONS

1. Is the fertility of a woman affected by the length of a given menstrual cycle or whether the cycles are regular or irregular? Explain.

2. Looking at a variety of animals, what are the advantages to restricting reproduction to a limited time period? Why do so many animals have a sharply defined reproductive season during the year?

3. In most sexual species, males produce far more gametes than do females. Why does this occur when, in most cases, only one male gamete can fertilize one female gamete?

4. Why are the accessory glands of the male so important in reproduction?

ONLINE LEARNING CENTER

Visit our Online Learning Center (OLC) at www.mhhe.com/zoology (click on this book's title) to find the following chapter-related materials:

- CHAPTER QUIZZING
- ANIMATIONS
    - Spermatogenesis
    - Penile Erection
    - Circumcision
    - Vasectomy
    - Menstruation
    - Female Reproductive Cycle
    - Oogenesis
    - Tubal Ligation
- RELATED WEB LINKS
    - Reproduction
        - Asexual Reproduction
        - Sexual Reproduction
    - Animal Reproduction
    - Development of Male Gametes, and Structures Involved
    - Invertebrate Reproduction
    - Vertebrate Reproduction
    - Development of Female Gametes, and Structures Involved
    - Fertilization and Development of the Embryo
    - Hormonal Control of Events During Gestation
    - Principles of Development/Embryology in Invertebrates
    - Principles of Development/Embryology in Vertebrates
- BOXED READING ON
    - The Participation of the Fetus During Childbirth
- SUGGESTED READINGS
- LAB CORRELATIONS

Check out the OLC to find specific information on these related lab exercises in the *General Zoology Laboratory Manual*, 5th edition, by Stephen A. Miller:

Exercise 24   *Vertebrate Reproduction*

GLOSSARY

In this pronunciation system, long vowels that stand alone or occur at the end of a syllable are unmarked. When a long vowel occurs in the middle of a syllable, it is marked with a macron (ˉ).

Short vowels that occur in the middle of a syllable are unmarked. When a short vowel occurs at the end of a syllable, or when it stands alone, it is marked with a breve (˘). The following are examples: cape (cāp), cell (sel), vacation (va-ka'shun), and vegetal (vej"ĕ-tal').

The vowel heard in *foot* and *book* is indicated as "fŏŏt" and "bŏŏk." The vowel heard in *loose* and *too* is unmarked and is indicated as "loos" and "too."

abdomen (ab'do-men)   1. The portion of a tetrapod's body between the thorax and pelvic girdle. 2. The region of an arthropod's body behind the thorax. It contains the visceral organs.

aboral (ab-or'al)   The end of a radially symmetrical animal opposite the mouth.

acanthella (a-kan'thel-a)   Developing acanthocephalan larva between an acanthor and a cystacanth, in which the definitive organ systems are developed. Develops in the intermediate host.

Acanthocephala (a-kan"tho-sef'a-lah)   The phylum of aschelminths commonly called the spiny-headed worms.

acanthor (a-kan'thor)   Acanthocephalan larva (first larval stage) that hatches from the egg. The larva has a rostellum with hooks that penetrate host tissues.

accommodation (ah-kom"o-da'shun)   The adjustment of the eye for focusing at various distances.

acetabulum (as"e-tab'u-lum)   Sucker. The ventral sucker of a fluke. A sucker on the scolex of a tapeworm.

Acetospora (ah-sēt-o-spor'ah)   The protozoan phylum whose members have multicellular spores. All parasitic in invertebrates. Examples: acetosporans (*Paramyxa, Halosporidium*).

acetylcholine (as-e-tel-ko'lēn)   A neurotransmitter that certain neurons liberate. It is excitatory at neuromuscular junctions and inhibitory at other synapses.

acid   A substance that ionizes in water to release hydrogen ions (H^+).

acid deposition   The combination of sulfur dioxide and nitrogen oxides with water in the atmosphere. This combination produces acidic precipitation called acid deposition. The burning of fossil fuels is a major contributor to acid deposition.

acoelomate (a-se'lah-māt)   Without a body cavity.

acrosome (ak'ro-sōm")   The enzyme-filled cap on the head of a sperm. Used in egg penetration.

actin (ak'tin)   A protein in a muscle fiber that, together with myosin, is responsible for contraction and relaxation.

action potential   The sequence of electrical changes when a nerve cell membrane is exposed to a stimulus that exceeds its threshold.

activational effects of hormones   Occur when an external stimulus triggers a hormonally mediated response by the organism.

active transport   A process that requires an expenditure of ATP energy to move molecules across a cell membrane. Usually moved against the concentration gradient with the aid of specific carrier (transport) proteins.

adaptation   Structures or processes that increase an organism's potential to successfully reproduce in a specified environment.

adaptive radiation   Evolutionary change that results in the formation of a number of new characteristics from an ancestral form, usually in response to the opening of new habitats.

Adenophorea (a-den"o-for'e-ah)   The class of nematodes formerly called Aphasmidia. Examples: *Trichinella, Trichuris*.

adenosine diphosphate (ADP) (ah-dēn'o-sēn di-fos'fāt)   A nucleotide composed of the pentose sugar D-ribose, adenine, and two phosphates. Formed by hydrolysis of ATP.

adenosine monophosphate (AMP) (ah-dēn'o-sēn mon-o-fos'fāt)   Molecule created when the terminal phosphate is lost from a molecule of adenosine diphosphate.

adenosine triphosphate (ATP) (ah-dēn'o-sēn tri-fos'fāt)   Molecule that stores and releases energy for use in cellular processes. A nucleotide composed of adenine, ribose, and three phosphate groups.

adhesive gland   Attachment glands in turbellarians that produce a chemical that attaches part of the turbellarian to a substrate.

adipose tissue (ad'i-pōs)   Fat-storing tissue.

adrenal gland (ah-dre'nal)   The endocrine gland on top of the kidney.

aerobic (a"er-o'bik)   Having molecular oxygen present. An oxygen-dependent form of respiration.

aerobic respiration   *See* **cellular respiration.**

aestivation (es'te-va-shun)   The condition of dormancy or torpidity during the hot, dry summer months.

afterbirth (af'ter-berth)   The placental and fetal membranes expelled from the uterus after the birth of a mammal.

age structure   The proportion of a population that is in prereproductive, reproductive, and postreproductive classes.

Agnatha (ag-nath'ah)   A paraphyletic group of vertebrates whose members lack jaws and paired appendages and possess a cartilaginous skeleton and a persistent notochord. Lampreys and hagfishes.

airfoil   A surface, such as a wing, that provides lift by using currents of air it moves through.

allantois (ah-lan'tois)   One of the extraembryonic membranes in the embryo of an amniote. Forms as a ventral outgrowth of the gut, enlarges during development, and functions in waste (uric acid) storage and gas exchange.

alleles (ah-lēlz')   Alternate forms of a gene that occur at the same locus of a chromosome. For example, in humans, three alleles (I^A, I^B, and i) that occur at the same locus encode the A, B, and O blood types.

allopatric speciation (al"o-pat'rik spe"se-a'shun)   Speciation that occurs in populations separated by geographical barriers.

all-or-none law   The phenomenon in which a muscle fiber contracts completely when it is exposed to a stimulus of threshold strength. Also, the principle that states that a neuron will "fire" at full power or not at all.

altricial (al-trish'al)   An animal that is helpless at hatching or birth.

altruism (al'troo-iz"em)   The principle or practice of unselfish concern for, or devotion to, the welfare of others.

alula (al′u-lah)   A group of feathers on the wing of a bird that is supported by the bones of the medial digit. The alula reduces turbulent air flow over the upper surface of the wing.

alveolus (al-ve′o-lus)   An air sac of a lung. A saclike structure.

ambulacral groove (am″byul-ac′ral)   The groove along the length of the oral surface of a sea star arm. Ambulacral grooves contain tube feet.

ametabolous metamorphosis (a″me-tab′a-lus met″ah-mor′fe-sis)   Development in which the number of molts is variable. Immature stages resemble adults, and molting continues into adulthood.

amictic eggs (e-mik′tic)   Pertaining to female rotifers that produce only diploid eggs that cannot be fertilized. The eggs develop directly into amictic females.

amino acid (ah-me′no)   A relatively small organic compound that contains an amino group ($-NH_2$) and a carboxyl group ($-COOH$). The structural unit of a protein molecule.

ammonotelic excretion (ah-mo″no-tel′ik)   Having ammonia as the chief excretory product of nitrogen metabolism. Occurs in freshwater fishes.

amnion (am′ne-on)   One of the extraembryonic membranes of the embryos of reptiles, birds, and mammals. The amnion encloses the embryo in a fluid-filled sac.

amniote lineage (am′ne-ōt lin′e-ij)   The evolutionary lineage of vertebrates leading to modern reptiles, birds, and mammals.

amniotic egg (am″ne-ot′ik)   The egg of reptiles, birds, and mammals. It has extraembryonic membranes that help prevent desiccation, store wastes, and promote gas exchange. These adaptations allowed vertebrates to invade terrestrial habitats.

amoeboid movement (ah-me′boid)   A form of movement similar to that found in amoebae. Fluid endoplasm (plasmasol) flows forward inside the cell and changes state to viscous ectoplasm (plasmagel) on reaching the tip of a pseudopodium. At the opposite end of the cell, ectoplasm changes into endoplasm.

Amphibia (am-fib′e-ah)   The class of vertebrates whose members are characterized by skin with mucoid secretions, which serves as a respiratory organ. Developmental stages are aquatic and are usually followed by metamorphosis to an amphibious adult. Frogs, toads, and salamanders.

amphid (am′fed)   One of a pair of chemosensory organs on the anterior end of certain nematodes.

amplexus (am-plek′sus)   The positioning of a male amphibian dorsal to a female amphibian, his forelimbs around her waist. During amplexus, the male releases sperm as the female releases eggs.

ampullary organ (am′pu-la″re)   A receptor that can detect electrical currents. These electroreceptors are found in most fishes, some amphibians, and the platypus. *See also* **electroreceptor.**

anabolism (an-nab′o-lizm)   Any constructive metabolic process by which organisms convert substances into other components of the organism's chemical architecture. System of biosynthetic reactions in a cell by which large molecules are made from smaller ones.

anaerobic (an-a″er-o′bik)   The phase of cellular respiration that occurs in the absence of oxygen. Lacking oxygen.

analogous (a-nal′a-ges)   Describes structures that have similar functions in two organisms but that have not evolved from a common ancestral form. Analogous structures often arise through convergent evolution.

anaphase (an′ah-fāz)   The stage in mitosis and meiosis, following metaphase, in which the centromeres divide and the chromatids, lined up on the mitotic spindle, begin to move apart toward the poles of the spindle to form the daughter chromosomes.

anatomy (ah-nat′o-me)   The study of the structure of an organism and its parts.

androgen (an′dro-jen)   Any substance that contributes to masculinization, such as the hormone testosterone.

aneuploidy (an′u-ploid″e)   An addition or deletion of one or more chromosomes. Aneuploidy may be represented by $2N + 1$, $2N - 1$, etc.

animal behavior   Activities animals perform during their lifetime.

Animalia (an″i-mal′e-ah)   The kingdom of organisms whose members are multicellular, eukaryotic, and heterotrophic. The animals.

animal pole   The region of a fertilized egg where meiosis is completed. It contains less yolk and is more metabolically active than the opposite vegetal pole.

Annelida (ah-nel′i-dah)   The phylum of triploblastic, coelomate animals whose members are metameric (segmented) and wormlike. Annelids have a complete digestive tract and a ventral nerve cord.

annuli (an′u-li)   Secondary divisions of each body segment of a leech (phylum Annelida, class Hirudinea).

antennal gland (an-ten′al)   The excretory organ in some crustaceans (crayfish).

Called antennal gland because of its location near the base of each second antenna and its green color. Also called **green gland.**

anterior   The head end. Usually, the end of a bilateral animal that meets its environment.

Anthozoa (an″tho-zo′ah)   The class of cnidarians whose members are solitary or colonial polyps. Medusae are absent. Gametes originate in the gastrodermis. Mesenteries divide the gastrovascular cavity. Sea anemones and corals.

anthropomorphism (an″thro-po-mor′fizm)   The attribution of human characteristics to nonhuman beings and objects.

anticodon (an′ti-ko″don)   A sequence of three bases on transfer RNA that pairs with codons of messenger RNA to position amino acids during protein synthesis.

antiparallel (an″ti-par′ah-lel″)   Refers to opposing strands of DNA that are oriented in opposite directions.

Apicomplexa (a″pi-kom-plex′ah)   The protozoan phylum characterized by members having an apical complex for penetrating host cells. Cilia and flagella are lacking, except in certain reproductive stages. Examples include the gregarines (*Monocystis*), coccidians (*Eimeria, Isospora, Sarcocystis, Toxoplasma*), *Pneumocystis*, and *Plasmodium*.

Aplacophora (a″pla-kof′o-rah)   The class of molluscs whose members lack a shell, mantle, and foot. Wormlike, burrowing animals with poorly developed head. Some zoologists divide this group into two classes: Caudofoveata and Solenogasters.

aposematic coloration (ah″pos-mat′ik)   Sharply contrasting colors of an animal that warn other animals of unpleasant or dangerous effects.

appendicular skeleton (ap″en-dik′u-lar)   The bones of the upper and lower extremities. Includes the shoulder and pelvic girdles.

appendix (a-pen′diks)   Refers to the appendix vermiformis of the colon.

Arachnida (ah-rak′nǐ-dah)   The class of chelicerate arthropods whose members are mostly terrestrial, possess book lungs or tracheae, and usually have four pairs of walking legs as adults. Spiders, scorpions, ticks, mites, and harvestmen.

arachnoid (ah-rak′noid)   The weblike middle covering (meninx) of the central nervous system.

Archaea (ahr-kay′e-ah)   A domain of organisms that includes prokaryotic microbes that live in harsh anaerobic environments. Probably the most primitive life-forms known.

archenteron (ar-ken′te-ron″)   The embryonic digestive tract that forms during gastrulation.

Aristotle's lantern   The series of ossicles making up the jawlike structure of echinoid echinoderms.

arteriole (ar-te′re-ōl)   A minute arterial branch, especially just proximal to a capillary.

artery   A vessel that transports blood away from the heart.

Arthropoda (ar″thrah-po′dah)   The phylum of animals whose members possess metamerism with tagmatization, a jointed exoskeleton, and a ventral nervous system. Includes insects, crustaceans, spiders, and related animals.

aschelminth (ask′hel-minth)   Any animal in the phyla Rotifera, Kinorhyncha, Nematoda, Nematomorpha, Acanthocephala, Loricitera, or Priapulida.

Ascidiacea (as-id″eas′e-ah)   A class of urochordates whose members are sessile as adults, and solitary or colonial.

ascon (as′kon)   The simplest of the three sponge body forms. Ascon sponges are vaselike, with choanocytes directly lining the spongocoel.

asexual reproduction   Having no sex; not sexual; not pertaining to sex. Reproduction of an organism without fusion of gametes by fission, budding, or some other method not involving the fertilization of gametes.

aster (as′ter)   The star-shaped structure in a cell during prophase of mitosis. Composed of a system of microtubules arranged in astral rays around the centrosome. May emanate from a centrosome or from a pole of a mitotic spindle.

Asteroidea (as″te-roi′de-ah)   The class of echinoderms whose members have rays that are not sharply set off from the central disk, ambulacral grooves with tube feet, and suction disks on tube feet. Sea stars.

asymmetry (a-sim′i-tre)   Without a balanced arrangement of similar parts on either side of a point or axis.

asynchronous flight   *See* **indirect (asynchronous) flight.**

atom   Smallest particle of an element that has the same properties. The basic unit of an element that can enter into chemical combinations. The smallest unit of matter that is unique to a particular element.

atomic mass   A mass unit determined by arbitrarily assigning the carbon-12 isotope a mass of 12 atomic mass units.

atomic number   A value equal to the number of protons in the nucleus of an atom. It differs for each element.

ATP   *See* **adenosine triphosphate.**

auricle (aw′re-kl)   The portion of the external ear not connected within the head. Also used to designate an atrium of a heart. In the class Turbellaria, the sensory lobes that project from the side of the head.

autosome (au′te-sōm″)   Chromosome other than sex chromosomes.

autotomy (au-tot′o-me)   The self-amputation of an appendage. For example, the casting off of a section of a lizard's tail caught in the grasp of a predator. The autotomized appendage is usually regenerated.

autotroph (aw′to-trōf)   An organism that uses carbon dioxide as its sole or principal source of carbon and makes its organic nutrients from inorganic raw materials.

autotrophic (au″to-trōf′ic)   Having the ability to synthesize food from inorganic compounds.

Aves (a′vez)   A class of vertebrates whose members are characterized by scales modified into feathers for flight, endothermy, and amniotic eggs. The birds.

axial filament   *See* **axoneme.**

axial skeleton (ak′se-al)   Portion of the skeleton that supports and protects the organs of the head, neck, and trunk.

axon (ak′son)   A fiber that conducts a nerve impulse away from a neuron cell body. Action potentials move rapidly, without alteration, along an axon. Their arrival at an axon ending may trigger the release of neurotransmitter molecules that influence an adjacent cell.

axoneme (ak′so-nēm)   The axial thread of the chromosome in which is located the axial combination of genes. The central core of a cilium or flagellum, consisting of a central pair of filaments surrounded by nine other pairs. Also called axial filament.

axopodium (ak′se-pōd-e-um)   Fine, needlelike pseudopodium that contains a central bundle of microtubules. Also called axopod. Found in certain sarcodine protozoa. Pl., axopodia.

B

balanced polymorphism (pol″e-morf′izm)   Occurs when different phenotypic expressions are maintained at a relatively stable frequency in a population.

baleen plate (ba-lēn′)   A keratined growth in toothless whales.

bare sensory nerve endings   Nerve endings that are sensitive to pain.

baroreceptor (bar″o-re-sep′tor)   A specialized nerve ending that is stimulated by changes in pressure.

basal body (ba′sel)   A centriole that has given rise to the microtubular system of a cilium or flagellum and is located just beneath the plasma membrane. Serves as a nucleation site for the growth of the axoneme.

base   A substance that ionizes in water to release hydroxyl ions (OH⁻) or other ions that combine with hydrogen ions.

basophil (ba′so-fil)   White blood cell characterized by the presence of cytoplasmic granules that become stained blue-purple by a basophilic dye.

B cell   A type of lymphocyte derived from bone marrow stem cells that matures into an immunologically competent cell under the influence of the bursa of Fabricius in the chicken, and the bone marrow in non-avian species. Following interaction with antigen, it becomes a plasma cell, which synthesizes and secretes antibody molecules involved in humoral immunity. Also called B lymphocyte.

Bdelloidea (del-oid′e-ah)   A class of rotifers without males. Anterior end is retractile and bears two disks. Mastax is adapted for grinding. Paired ovaries. Cylindrical body. Example: *Rotaria.*

behavioral ecology   The scientific study of all aspects of animal behavior as related to the environment.

bilateral symmetry (bi-lat′er-al sim′i-tre)   A form of symmetry in which only the midsagittal plane divides an organism into mirror images. Bilateral symmetry is characteristic of actively moving organisms that have definite anterior (head) and posterior (tail) ends.

bile (bīl)   A fluid secreted by the liver, stored in the gallbladder, and released into the small intestine via the bile duct. Emulsifies fats.

bimodal breathing (bi-mo′del)   The ability of an organism to exchange respiratory gases simultaneously with both air and water, usually using gills for water breathing and lungs for air breathing.

binary fission (bi′ne-re fish′en)   Asexual reproduction in protists in which cytoplasmic division follows mitosis to produce two new protists.

binomial nomenclature (bi-no′me-al no′men-kla′cher)   A system for naming in which each kind of organism (a species) has a two-part name: the genus and the species epithet.

biochemistry   The chemistry of living organisms and of vital processes. Also known as physiological or biological chemistry. The study of the molecular basis of life.

biodiversity The variety of organisms in an ecosystem.

biogeochemical cycles The cycling of elements between reservoirs of inorganic compounds and living matter in an ecosystem.

biogeography The study of the geographic distribution of life on earth. Biogeographers attempt to explain the factors that influence where species of plants and animals live on the earth.

biological magnification The concentration of substances in animal tissues as the substances pass through ecosystem food webs.

biosonar *See* **sonar.**

biotic potential (bi-ot'ik) The capacity of a population to increase maximally. Also called **intrinsic rate of growth.**

biramous appendages (bi-ra'mus ah-pen'dij-ez) Appendages having two distal processes that a single proximal process connects to the body.

Bivalvia (bi'val ve"ah) The class of molluscs whose members are enclosed in a shell consisting of two dorsally hinged valves, lack a radula, and possess a wedge-shaped foot. Clams, mussels, oysters.

bladder worm The unilocular hydatid cyst of a tapeworm. *See also* **cysticercus.**

blastocoel (blas'to-sēl) The fluid-filled cavity of the blastula.

blastocyst (blas'to-sist) An early stage of embryonic development consisting of a hollow ball of cells. The product of cleavage.

blastoderm (blas'to-derm) A small disk of cells at the animal end of a reptile or bird embryo that results from early cleavages.

blastomeres (blas'to-merz) Any of the cells produced by cleavage of a zygote.

blastopore (blas'to-por) The point of invagination at which cells on the surface of the blastula move to the interior of the embryo during gastrulation.

blastula (blas'tu-lah) An early stage in the development of an embryo. It consists of a sphere of cells enclosing a fluid-filled cavity (blastocoel).

blood A type of connective tissue with a fluid matrix called plasma in which blood cells are suspended. The fluid that circulates through the heart, arteries, capillaries, and veins.

blood pressure The force (energy) with which blood pushes against the walls of blood vessels and circulates throughout the body when the heart contracts.

blubber (blub'er) The fat found between the skin and muscle of whales and other cetaceans, from which oil is made. Functions in insulation.

bone cells The hard, rigid form of connective tissue constituting most of the skeleton of vertebrates. Composed chiefly of calcium salts. Also called bone (osseous) tissue.

book gill Modifications of a horseshoe crab's exoskeleton into a series of leaflike plates that are a surface for gas exchange between the arthropod and the water (phylum Arthropoda, class Merostomata).

book lung Modification of the arthropod exoskeleton into a series of internal plates that provide surfaces for gas exchange between the blood and air. Found in spiders.

bottleneck effect Changes in gene frequency that result when numbers in a population are drastically reduced, and genetic variability is reduced as a result of the population being built up again from relatively few surviving individuals.

Brachiopoda (bra-ke-op'o-dah) A phylum of marine animals whose members possess a bivalved calcareous and/or chitinous shell that a mantle secretes and that encloses nearly all of the body. Unlike the molluscs, the valves are dorsal and ventral. Possess a lophophore. Lampshells.

broadcast spawning The release of gametes into the water for external fertilization.

brood patch The patch of feathers birds use to incubate eggs. Also known as incubation patch.

brown fat Fat deposits with extensive vascularization, mitochondria, and enzyme systems for oxidation. Found in small, specific deposits in a few mammals and used for nonshivering thermogenesis.

buccal pump (buk'el) The mechanism by which lung ventilation occurs in amphibians. Muscles of the mouth and pharynx create positive pressure to force air into the lungs.

buccopharyngeal respiration (buk"o-fah-rin'je-al res"pah-ra'shun) The diffusion of gases across moist linings of the mouth and pharynx of amphibians.

budding (bud'ing) The process of forming new individuals asexually in many different invertebrates. Offspring arise as outgrowths from the parent and are smaller than the parent. When the offspring do not separate from the parent, a colony forms.

buffer (buf'er) A substance that can react with a strong acid or base to form a weaker acid or base and thus resist a change in pH. A buffer accepts H^+ ions from or donates H^+ ions to solutions.

bulb of Krause A sensory receptor in the skin believed to be the sensor for touch-pressure. Also called bulbous corpuscle.

bulbourethral gland (bul"bo-u-re'thral) Gland that secretes a viscous fluid into the male urethra during sexual excitement.

bursa (bur'sah) A membranous sac that invaginates from the oral surface of ophiuroid echinoderms. Functions in diffusion of gases and waste material.

bursa of Fabricius (bur'sah of fah-bris'e-us) The lymphoid organ of birds that, like the thymus, develops as an outpouching of the gut near the cloaca rather than the foregut. It is responsible for the maturation of lymphocytes.

C

Calcarea (kal-kar'e-ah) The class of sponges whose members are small and possess monaxon, triaxon, or tetraaxon calcium carbonate spicules.

calcitonin (kal"si-to'nin) A thyroid hormone that lowers calcium and phosphate levels in the blood. Also called thyrocalcitonin.

calorie A unit used in the measurement of heat energy and the energy value of foods. The amount of heat energy required to raise the temperature of 1 g of water 1°C.

Calorie Amount of heat energy required to raise the temperature of 1,000 g of water 1°C.

calyx (ka'liks) 1. A boat-shaped or cuplike central body of an entoproct. The body and tentacles of an entoproct. 2. A cuplike set of ossicles that support the crown of a sea lily or feather star (class Crinoidea, phylum Echinodermata).

camouflage Color patterns in an animal that help to hide it from other animals.

capillary (kap'i-lar-e) A small blood vessel that connects an arteriole and a venule. The functional unit of the circulatory system. The site of exchange between the blood and tissue cells.

carapace (kar'ah-pās) The dorsal portion of the shell of a turtle. Formed from a fusion of vertebrae, ribs, and dermal bone.

carbaminohemoglobin (kar"bah-me'no-he'mo-glo"bin) Compound formed by the union of carbon dioxide and hemoglobin.

carbohydrate (kahr-bo-hi'drāt) An organic compound that contains carbon, hydrogen, and oxygen. Carbohydrates consist of simple single-monomer sugars (monosaccharides), disaccharides, or other multi-unit sugars (polysaccharides).

cardiac muscle Specialized type of muscle tissue found only in the heart.

cardiovascular system *See* **circulatory system.**

carnivore (kar′ne-vor)   One of the flesh-eating animals of the order Carnivora. Also, any organism that eats flesh.

carrying capacity   The maximum population size that an environment can support.

cartilage (kar′ti-lij)   Type of connective tissue in which cells are within lacunae and separated by a semisolid matrix. Provides a site for muscle attachment, aids in movement of joints, and provides support.

caste (kast)   One of the distinct kinds of individuals in a colony of social insects (e.g., queens, drones, and workers in a honeybee colony).

catabolism (kah-tă′bo-lizm)   Metabolic process by which large molecules are broken down into smaller ones. Catabolic metabolism. Intermediates in these reactions are called catabolites.

catalysis (kah-tal′i-sis)   An increase in the velocity of a chemical reaction or process produced by the presence of a substance that is not consumed in the net chemical reaction or process.

catalyst (kat′ah-list)   A substance that increases the rate of a chemical reaction but is not permanently altered by the reaction. Enzymes are protein catalysts.

Caudofoveata (kaw′do-fo″ve-ah″ta)   The class of molluscs characterized by a wormlike, shell-less body and scalelike, calcareous spicules. Lack eyes, tentacles, statocysts, crystalline style, foot, and nephridia. Deep-water marine burrowers. *Chaetoderma.*

cecum (se′kum)   1. Each arm of the blind-ending, Y-shaped digestive tract of trematodes (phylum Platyhelminthes). 2. A region of the vertebrate digestive tract where fermentation can occur. It is at the proximal end of the large intestine.

cell body   Portion of a nerve cell that includes a cytoplasmic mass and a nucleolus, and from which the nerve fibers extend.

cell cycle   The regular sequence of events (including interphase and the mitotic phase) during which a eukaryotic cell grows, prepares for division, duplicates its contents, and divides to form two daughter cells.

cellular organization   Pertaining to or made up of cellular organelles; an organized cellular structure.

cellular respiration   Process by which energy is released from organic compounds within cells. The aerobic process that involves glycolysis, the Krebs cycle, the electron transport chain, and chemiosmosis.

central nervous system   The brain and spinal cord.

centriole (sen′tre-ol)   A cellular organelle (composed of microtubules) that organizes the mitotic spindle during mitosis. A pair of centrioles is usually in the center of a microtubule-organizing center in animal cells.

centromere (sen′tro-mer)   Constricted region of a mitotic chromosome that holds sister chromatids together. Also, the site on the DNA where the kinetochore forms and then captures microtubules from the mitotic spindle.

Cephalaspidomorphi (sef″a-las″pe-do-morf′e)   The class of vertebrates characterized by the absence of paired appendages and the presence of sucking mouthparts with teeth and a rasping tongue. Lampreys.

cephalic (se-fal′ik)   Having to do with, or toward, the head of an animal.

cephalization (sef″al-iz-a′shun)   The development of a head with an accumulation of nervous tissue into a brain.

Cephalocarida (sef″ah-lo-kar′ĭ-dah)   A class of marine crustaceans characterized by uniform, leaflike, triramous appendages.

Cephalochordata (sef″a-lo-kor-dat′ah)   The subphylum of chordates whose members possess a laterally compressed, transparent body. They are fishlike and possess all four chordate characteristics throughout life. Amphioxus.

Cephalopoda (sef″ah-lop′o-dah)   The class of molluscs whose members have a foot modified into a circle of tentacles and a siphon. A shell is reduced or absent, and the head is in line with the elongate visceral mass. Octopuses, squid, and cuttlefish.

cephalothorax (sef″al-o-thor′aks)   The fused head and thoracic regions of crustaceans and some arachnids.

cercaria (ser-kar′e-a)   Juvenile digenetic trematode produced by asexual reproduction within a sporocyst or redia. Cercariae are free-swimming and have a digestive tract, suckers, and a tail. They develop into metacercariae.

cerebellum (ser″e-bel′um)   Portion of the brain that coordinates skeletal muscle movement. Part of the metencephalon, it consists of two hemispheres and a central vermis.

cerebrum (sah-re′brum)   The main portion of the brain, occupying the upper part of the cranial cavity. Its two hemispheres are united by the corpus callosum. It forms the largest part of the central nervous system in mammals.

Cestoidea (ses-toid′e-ah)   The class of platyhelminthes with members that are all parasitic with no digestive tract. Have great reproductive potentials. Tapeworms.

chaetae (ke′tah)   *See* **setae.**

Chaetognatha (ke″tog-nath′ah)   A phylum of planktonic, marine deuterostomes whose body consists of a head, trunk, and tail. Their streamlined shape and darting locomotion are reflected in their common name—arrowworms.

character   Any animal attribute that has a genetic basis and that can be measured. Used by systematists in investigating relatedness among animal groups.

chelicerae (ke-lis′er-ae)   One of the two pairs of anterior appendages of arachnids. May be pincerlike or modified for piercing and sucking or other functions.

Chelicerata (ke-lis″er-ah′tah)   The subphylum of arthropods whose members have a body that is divided into prosoma and opisthosoma. The first pair of appendages are feeding appendages called chelicerae. Spiders, scorpions, mites, and ticks.

chemical synapse (sin′apse)   A synapse at which neurotransmitters that one neuron releases diffuse across an extracellular gap to influence a second neuron's activity.

chemiosmosis (kem″e-os-mo′sis)   The process whereby electron transport generates a proton gradient and an electrochemical gradient that drive ATP synthesis by oxidative phosphorylation.

chemistry   The science dealing with the elements, atomic relations of matter, and various elemental compounds. The study of the properties of substances and how substances react with one another.

chemoreceptor (ke″mo-re-sep′tor)   A receptor that is stimulated by the presence of certain chemical substances.

chiasma (ki-as′mah)   A decussation or X-shaped crossing. The places where pairs of homologous chromatids remain in contact during late prophase to anaphase of the first meiotic division. The chiasma indicates where nonsister chromatids have exchanged homologous segments by crossing-over.

chief cell   Cell of a gastric gland that secretes various digestive enzymes, including pepsinogen.

Chilopoda (ki″lah-pod′ah)   The class of uniramous arthropods whose members have one pair of legs per segment and whose body is oval in cross section. Centipedes.

chitin (ki′tin)   The polysaccharide in the exoskeleton of arthropods.

chloragogen tissue (klo″rah-gog′en)   Cells covering the dorsal blood vessel and digestive tract of annelids. Functions in glycogen and fat synthesis and urea formation.

chlorocruorin (klo″ro-kroo′e-ren)   The greenish iron-containing respiratory pigment dissolved in the blood plasma of certain marine polychaetes.

choanocytes (ko-an′o-sītz) Cells of sponges that create water currents and filter food.

Chondrichthyes (kon-drik′thi-es) The class of vertebrates whose members are fishlike, possess paired appendages and a cartilaginous skeleton, and lack a swim bladder. Skates, rays, and sharks.

chordamesoderm (kor″dah-mez′o-derm) Tissue in the amphibian gastrula that forms between ectoderm and endoderm in the dorsal lip region of the blastopore. Develops into the mesoderm and notochord.

Chordata (kor-dat′ah) A phylum of animals whose members are characterized by a notochord, pharyngeal slits, a dorsal tubular nerve cord, and a postanal tail.

chorion (kor′e-on) The outermost extraembryonic membrane of the embryo of an amniote. Becomes highly vascular and aids in gas exchange.

chromaffin tissue (kro′maf-in) Specialized endocrine cells near the kidneys of various vertebrates. Produces various steroid hormones.

chromatid (kro′mah-tid) One copy of a eukaryotic chromosome formed by DNA replication.

chromatin (kro′mah-tin) Nuclear material that gives rise to chromosomes during mitosis. Complex of DNA, histones, and nonhistone proteins.

chromatophores (kro-mah′tah-forz) Cells containing pigment that, through contraction and expansion, produce temporary color changes.

chromosome (kro′mo-sōm) Rodlike structure that appears in the nucleus of a cell during mitosis. Contains the genes responsible for heredity. Structure composed of a long DNA molecule and associated proteins that carries part (or all) of the hereditary information of an organism.

chrysalis (kris′ah-lis) The pupal case of a butterfly that forms from the exoskeleton of the last larval instar.

chylomicron (ki″lo-mi′kron) A particle of the class of lipoproteins responsible for the transport of cholesterol and triglycerides from the small intestine to tissue after meals.

chyme (kīm) Semifluid mass of food materials that passes from the stomach to the small intestine. Consists of partially digested food and gastric juice.

cilia (sil′e-ah) Microscopic, hairlike processes on the exposed surfaces of certain eukaryotic cells. Cilia contain a core bundle of microtubules and can perform repeated beating movements. They are also responsible for the swimming of many single-celled organisms.

ciliary creeping The principal means of nemertine locomotion.

Ciliophora (sil″e-of′or-ah) The protozoan phylum whose members have simple or compound cilia at some stage in their life history. Heterotrophs with a well-developed cytostome and feeding organelles. At least one macronucleus and micronucleus present. Examples: *Paramecium, Stentor, Vorticella, Balantidium*.

circadian rhythms (sur-ka′de-an) Daily cycles of activity. Circadian rhythms are usually based on photoperiods.

circulatory system (ser′ku-lah-to″re) Pertaining to the circulation.

circumcision (ser″kum-sizh′un) The removal of all or part of the prepuce or foreskin.

cirri (ser′i) Any of various slender or filamentous, usually flexible appendages, such as one of the compound organelles composed of groups of fused cilia in certain peritrichious ciliate protozoa and used for locomotion. An eversible penis in flatworms. A fingerlike projection of a polychete parapodium.

Cirripedia (sir″ĭ-pēd′e-ah) The class of crustaceans whose members are sessile and highly modified as adults. Enclosed by calcium carbonate valves. Barnacles.

citric acid cycle (sit′rik) A series of chemical reactions in the mitochondrion by which various molecules are oxidized and energy is released. Krebs cycle. TCA or tricarboxylic acid cycle.

clade (klād) A subset of organisms in a phylogenetic group that share a certain synapomorphy (derived character).

cladistics (klad-is′tiks) *See* **phylogenetic systematics.**

cladogram (klad′o-gram) Diagram depicting the evolutionary history of taxa. Derived from phylogenetic systematics (cladistics).

class A level of classification between phylum and order.

classical conditioning A type of learning in which positive or negative reinforcement influences later responses of an animal to a stimulus.

claw The sharp, usually curved, nail on the foot of an animal or insect. The pincerlike extremity of specific limbs of certain arthropods (e.g., lobster claws).

cleavage (kle′vij) The early mitotic and cytoplasmic divisions of an embryo.

climax community A final, relatively stable stage in an ecological succession.

Clitellata (kli-te′lah-tah) The class of Annelida characterized by no parapodia, few or no setae, a clitellum that functions in cocoon formation, and monoecious reproduction. Earthworms and leeches.

clitellum (kli-te′lum) The region of an annelid responsible for secreting mucus around two worms in copula and for secreting a cocoon to protect developmental stages.

cloaca (klo-a′kah) A common opening for excretory, digestive, and reproductive systems.

closed circulatory system A circulatory system in an animal (vertebrates) in which blood is confined to vessels throughout its circuit.

clouds of electrons The distribution of electrons in space around the atomic nucleus.

clutch The number of eggs laid and chicks produced by a female bird.

cnida (ni′dah) An organelle characteristic of the cnidaria that is used in defense, food gathering, and attachment.

Cnidaria (ni-dar′e-ah) The phylum of animals whose members are characterized by radial or biradial symmetry, diploblastic organization, a gastrovascular cavity, and nematocysts. Jellyfish, sea anemones, and their relatives.

cnidocytes (ni′do-sītz) The cells that produce and discharge the cnidae in members of the phylum Cnidaria.

cocoon The protective covering of a resting or developmental stage. Sometimes refers to both the covering and the contents.

codominance (ko-dom′ah-nens) An interaction of two alleles such that both alleles are expressed in a phenotype.

codon (ko′don) A sequence of three bases on messenger RNA that specifies the position of an amino acid in a protein.

coelom (se′lom) A fluid-filled body cavity lined by mesoderm.

coelomic fluid (se-lom′ic) The fluid within the body cavity of triploblastic animals.

coenzyme (ko-en′zīm) An organic nonprotein molecule, frequently a phosphorylated derivative of a water-soluble vitamin, that binds with the protein molecule (apoenzyme) to form the active enzyme (holoenzyme). Examples include biotin, NAD^+, and coenzyme A.

coevolution (ko″ev″ah-lu′shun) The evolution of ecologically related species such that each species exerts a strong selective influence on the other.

cofactor (ko′fak-tor) A metal ion or inorganic ion with which an enzyme must unite to function. *See also* **coenzyme.**

colloblasts (kol′ah-blasts) Adhesive cells on ctenophoran tentacles for capturing prey.

colonial hypothesis A hypothesis formulated to explain the origin of multicellularity from protist ancestors. Animals may have been derived when protists associated and cells became specialized and interdependent.

colony An aggregation of organisms. Usually a group of closely associated individuals of one species.

colostrum (ko-los′trum) The first secretion of the mammary glands following the birth of a mammal.

comb rows Rows of cilia that are the locomotor organs of ctenophorans.

commensalism (kah-men′sal-izm) Living within or on an individual of another species without harm. The commensal benefits, but the host is unharmed.

communication An act on the part of one organism (or cell) that alters the probability of patterns of behavior in another organism (or cell) in an adaptive fashion. The transfer of information from one animal to another.

community The different kinds of organisms living in an area.

community diversity The number of different kinds of organisms living in an area.

comparative anatomy The study of animal structure to deduce evolutionary pathways in particular animal groups.

comparative embryology (em″bre-ol′o-je) The study of animal development in an attempt to deduce evolutionary pathways in particular animal groups.

comparative psychology Study of the genetic, neural, and hormonal bases of animal behavior.

competitive exclusion principle The idea that two species with identical niches cannot coexist.

complex camera eye The type of image-forming eye found in squids and octopuses.

compound A substance composed of atoms of two or more elements joined by chemical bonds and chemically united in fixed proportions.

compound eye An eye consisting of many individual lens systems (ommatidia). Present in many members of the phylum Arthropoda.

Concentricycloidea (kon-sen″tri-si-kloi′de-ah) The group of echinoderms whose members are characterized by two concentric water-vascular rings encircling a disklike body; no digestive system; and internal brood pouches. Formerly considered a class, they are now believed to be highly modified asteroideans. Sea daisies.

conduction (ken-duk′shun) The conveyance of energy, such as heat, sound, or electricity. The direct transfer of thermal motion (heat) between molecules of the environment and those on the body surface of an animal.

cone cell A color-sensitive photoreceptor cell concentrated in the retina.

conjugation (kon″ju-ga′shun) A form of sexual union that ciliates use to mutually exchange haploid micronuclei.

connective tissue A basic type of tissue that includes bone, cartilage, and various fibrous tissues. Connective tissue supports and binds.

continental drift The breakup and movement of landmasses of the earth. The earth had a single landmass about 250 million years ago. This mass broke apart into continents, which have moved slowly to their present positions.

continuous feeder Usually slow-moving, sessile animals that feed all of the time.

contour feathers Feathers that cover the body, wings, and tail of a bird. Contour feathers provide flight surfaces and are responsible for plumage colors.

contractile vacuole (kon-trak′til vak′u-ōl″) An organelle that collects and discharges water in protists and a few lower metazoa in a cyclical manner to accomplish osmoregulation and some excretion.

convection (ken-vek′shun) The act of conveying or transmission. The movement of air (or a liquid) over the surface of a body that contributes to heat loss (if the air is cooler than the body) or heat gain (if air is warmer than the body).

convergent evolution Evolutionary changes that result in members of one species resembling members of a second unrelated (or distantly related) species.

Copepoda (ko″pe-pod′ah) A class of crustaceans characterized by maxillipeds modified for feeding and antennae modified for swimming. The copepods.

coracidium (kor″ah-sid′e-um) Larva with a ciliated epithelium hatching from the egg of certain cestodes. A ciliated, free-swimming oncosphere.

coralline algae (kor′ah-lin al′je) Any red alga that is impregnated with calcium carbonate. Coralline algae often contribute to coral reefs.

coral reef Association of stony coral organisms and algae that forms one of the most highly productive ecosystems in the world.

corona (ko-ro′nah) A crown. An encircling structure. The ciliated organ at the anterior end of rotifers used for swimming or feeding.

cortisol (kor′ti-sol) A glucocorticoid that the adrenal cortex secretes that helps regulate overall metabolism and the concentration of blood sugar.

counteracting osmolyte strategy (os-mo-lyt) An osmolyte (ion) that counteracts another ion.

countercurrent exchange mechanism The passive exchange of something between fluids moving in opposite directions past each other.

countershading Contrasting coloration that helps conceal the animal (e.g., the darkly pigmented top and lightly pigmented bottom of frog embryos).

covalent bond (ko-va′lent) Chemical bond created when two atoms share a pair of electrons.

coxal glands (koks′el) An organ of excretion in some arthropods (spiders) that empties through a pore near the proximal joint (coxa) of the leg.

Craniata (kra″ne-ah′tah) A group of chordates characterized by a skull that surrounds the brain, olfactory organs, ears, and eyes. They possess a unique neural crest tissue. Hagfishes and vertebrates.

Crinoidea (krin-oi′de-ah) The class of echinoderms whose members are attached by a stalk of ossicles or are free-living. Possess a reduced central disk. Sea lilies and feather stars.

crossing-over The exchange of material between homologous chromosomes, during the first meiotic division, resulting in a new combination of genes.

Crustacea (krus-tās′e-ah) The subphylum of mandibulate arthropods whose members are characterized by having two pairs of antennae, one pair of mandibles, two pairs of maxillae, and biramous appendages. Crabs, crayfish, lobsters.

cryptic coloration (kript′ik) An animal taking on color patterns of its environment.

crystalline style A proteinaceous, rodlike structure in the digestive tract of a bivalve (Mollusca) that rotates against a gastric shield and releases digestive enzymes.

Ctenophora (te-nof′er-ah) The phylum of animals whose members are characterized by biradial symmetry, diploblastic organization, colloblasts, and meridionally arranged comb rows. Comb jellies.

Cubozoa (ku″bo-zo′ah) The class of cnidarians whose members have prominent cuboidal medusae with tentacles that hang from the corner of the medusa. Small polyp, gametes gastrodermal in origin. *Chironex.*

cutaneous respiration (kyoo-ta′ne-us) Gas exchange across thin, moist surfaces of the skin. Also called cutaneous exchange or integumentary exchange.

cuticle (kyoo′ti-kel) A noncellular, protective, organic layer secreted by the external epithelium (hypodermis) of many invertebrates. Refers to the epidermis or skin in higher animals and usually consists of cornified dead cells.

Cycliophora (si″kle-o-for′ah)   A phylum of animals whose members are marine, acoelomate, and bilaterally symmetrical. They live in association with the mouthparts of lobsters. The body consists of a buccal funnel, a trunk, and a stalk that ends in an adhesive disk. The buccal funnel contains the mouth and is surrounded by a ring of compound cilia. It is the most recently described animal phylum.

cystacanth (sis′ta-kanth)   Juvenile acanthocephalan that is infective to its definitive host.

cysticercosis (sis″ti-ser-ko′-sis)   Disease in humans that results from cysticercus development in body tissues following infection by *Taenia solium*.

cysticercus (sis″ti-ser′kus)   Metacestode developing from the oncosphere in most Cyclophyllidea. Usually has a tail and a well-formed scolex and is characterized by a fluid-filled oval body with an invaginated scolex. Cysticercoid.

cytokinesis (si″ta-kin-e′sis)   The division of the cytoplasm of a cell into two parts, as distinct from the division of the nucleus (which is mitosis).

cytomembrane system (si″to-mem′brăn)   Organelles, functioning as a system, to modify, package, and distribute newly formed proteins and lipids. Endoplasmic reticulum, Golgi apparatus, lysosomes, and a variety of vesicles and vacuoles are its components.

cytopharynx (si″to-far′inks)   A region of the plasma membrane and cytoplasm of some ciliated and flagellated protists specialized for endocytosis. A permanent oral canal.

cytoplasm (si′to-plazm)   The contents of a cell surrounding the nucleus. Consists of a semifluid medium and organelles.

cytopyge (si′te-pij)   Found in some protozoa; functions as a site for the expulsion of wastes.

cytoskeleton (si″to-skel′e-ton)   In the cytoplasm of eukaryotic cells, an internal protein framework of microtubules, microfilaments, and intermediate filaments by which organelles and other structures are anchored, organized, and moved about.

cytosol (si′to-sol)   Contents of the main compartment of the cytoplasm, excluding membrane-bound organelles.

D

daily torpor (tor′por)   Daily sluggishness that some animals experience. A period of inactivity that is normally induced by cold. Observed in numerous ectotherms and also endotherms. Often used to describe the specific physiological state of endotherms with a circadian cycle of lowered body temperature and depressed metabolic rate.

daughter sporocysts   In digenetic trematodes, the embryonic cells that develop from sporocysts and give rise to rediae.

deamination reaction (de-am″i-na′shun)   A reaction in which an amino group, $-NH_2$, is enzymatically removed from a compound.

decomposer   Mostly heterotrophic bacteria and fungi that obtain organic nutrients by breaking down the remains or products of other organic compounds. Their activities help cycle the simple compounds back to the autotrophs.

definitive host   The host in the life cycle of a parasite that harbors the adult stage or sexual stage of the parasite. If no sexual reproduction takes place, then the definitive host is the host in which the symbiont becomes mature and reproduces.

degeneracy (de-jen′er-ah-se)   The genetic code is said to be degenerate because more than one three-base sequence in DNA can code for one kind of amino acid.

delayed fertilization   When fertilization of an egg does not occur immediately following coitus but is delayed for weeks or months.

deme (deem)   A small, local subpopulation. Isolated subpopulations sometimes display genetic changes that may contribute to evolutionary change in the subpopulation.

Demospongiae (de-mo-spun′je-a)   The class of poriferans whose members have monaxon or tetraaxon siliceous spicules or spongin. Leucon body forms are present and vary in size from a few centimeters to 1 m in height.

denaturation (de-na″chur-a′shun)   Disruption of bonds holding a protein in its three-dimensional form such that its polypeptide chain(s) unfolds partially or completely. Changes in pH, salt concentration, or environmental temperature can cause denaturation.

dendrite (den′drīt)   Nerve fiber that transmits impulses toward a neuron cell body. Dendrites compose most of the receptive surface of a neuron.

density-dependent factors   Environmental parameters that are more severe when population density is high (or sometimes very low) than they are at other densities. Disease, predation, and parasitism are density-dependent factors.

density-independent factors   Environmental parameters that influence the number of animals in a population without regard to the number of animals per unit space (density). Weather conditions and human influences often have density-independent effects on animal populations.

dental formula   A notation that indicates the number of incisors, canines, premolars, and molars in the upper and lower jaws of a mammal.

denticle (den′ti-kl)   A small, toothlike process.

deoxyribonucleic acid (DNA) (de-oks′e-ri″bo-nu-kla′ik)   A polymer of deoxyribonucleotides in the form of a double helix. DNA is the genetic molecule of life in that it codes for the sequence of amino acids in proteins. Contains nucleotide monomers with deoxyribose sugar and nitrogenous bases adenine (A), cytosine (C), guanine (G), and thymine (T). *See also* **gene.**

depolarization (de-po″lar-i-za′shun)   The loss of an electrical charge or polarity on the surface of a membrane.

deposit feeding   The type of feeding whereby an animal obtains its nutrients from the sediments of soft-bottom habitats (mud or sands) or terrestrial soils. Examples include polychaete annelids, some snails, some sea urchins, and most earthworms.

derived character   Character that has arisen since common ancestry with an outgroup. *See also* **synapomorphies.**

dermal branchiae (der′mal branch′e-a)   Thin folds of the body wall of a sea star that extend between ossicles and function in gas exchange and other exchange processes.

dermis (der′mis)   The layer of the skin deep to the epidermis, consisting of a dense bed of vascular connective tissue.

deuterostomes (du′te-ro-stōms″)   Animals in which the anus forms from, or in the region of, the blastopore. Often characterized by enterocoelous coelom formation, radial cleavage, and the presence of a dipleurula-like larval stage.

diabetes (di″ah-be′tez)   Condition characterized by a high blood glucose level and the appearance of glucose in the urine due to a deficiency of insulin or the inability of body cells to respond to insulin. Diabetes mellitus.

diaphragm (di′ah-fram)   The domed respiratory muscle between thoracic and abdominal compartments of mammals.

diastole (di-as′to-le)   Phase of the cardiac cycle during which a heart chamber wall relaxes. *See also* **diastolic pressure.**

diastolic pressure (di″ah-stol′ik)   The blood pressure measurement during the interval between heartbeats. It is the second number in a blood-pressure reading.

differentiation (dif″ah-ren″she-a′shun)   The development of embryonic structures from a nondescript form in the early embryo to their form in the adult.

diffusion (dĭ-fu′zhun)   The random movement of molecules from one location to another because of random thermal molecular motion. Net diffusion always occurs from a region of higher concentration to a region of lower concentration.

digestion (di-jest′yun)   The process by which mechanical and chemical means break down larger molecules of food into smaller molecules that the digestive system can take up. Hydrolysis.

dihybrid cross (di-hi′brid)   A mating between individuals heterozygous for two traits. Dihybrid crosses usually result in a phenotypic ratio of 9 dominant, dominant : 3 dominant, recessive : 3 recessive, dominant : 1 recessive, recessive in the offspring.

dioecious (di-e′shus)   Having separate (male and female) sexes. Male and female organs are in separate individuals.

diploblastic (dip″lo-blas′tik)   Animals whose body parts are organized into layers that are derived embryologically from two tissue layers: ectoderm and endoderm. Animals in the phyla Cnidaria and Ctenophora are diploblastic.

diploid (di′ploid)   Having two sets of chromosomes. The *2N* chromosome number.

Diplopoda (dip″lah-pod′ah)   The class of arthropods whose members are characterized by two pairs of legs per apparent segment and a body that is round in cross section. Millipedes.

direct (synchronous) flight   Insect flight that is accomplished by flight muscles acting on wing bases and in which a single nerve impulse results in a single wing cycle. *See also* **indirect (asynchronous) flight.**

directional selection   Natural selection that occurs when individuals at one phenotypic extreme have an advantage over individuals with more common phenotypes.

disaccharide (di-sak′ah-rīd)   A sugar produced by the union of two monosaccharide molecules as a result of a dehydration synthesis.

discontinuous feeder   An animal that does not feed all the time. Instead, it generally eats large meals sporadically and does not spend time in the continuous pursuit of prey. Most carnivores are discontinuous feeders.

disruptive selection   Natural selection that occurs when individuals of the most common phenotypes are at a disadvantage. Produces contrasting subpopulations.

distal   Away from the point of attachment of a structure on the body (e.g., the toes are distal to the knee).

diving reflex   The reflex certain animals have to stay underwater for prolonged periods of time.

domain (do-mān′)   The broadest taxonomic grouping. Recent evidence from molecular biology indicates that there are three domains: Archaea, Eubacteria, and Eukarya.

dominant   Describes a gene that masks one or more of its alleles. For a dominant trait to be expressed, at least one member of the gene pair must be the dominant allele. *See also* **recessive.**

dominant species   A species that exerts an overriding influence in determining the characteristics of a community.

dorsal (dor′sal)   The back of an animal. Usually the upper surface. Synonymous with posterior for animals that walk upright.

down feathers   Feathers that provide insulation for adult and immature birds.

dura mater (du′rah ma′ter)   The outermost and toughest meninx covering the brain and spinal cord.

dyad (di′ad)   A double chromosome resulting from the halving of a tetrad in the first meiotic division.

E

ecdysis (ek-dis′is)   1. The shedding of the arthropod exoskeleton to accommodate increased body size or a change in morphology (as may occur in molting from immature to adult). 2. The shedding of the cuticle in aschelminths in order to grow. 3. The shedding of the epidermal layers of the skin of a reptile. Also called **molting.**

Echinodermata (i-ki″na-dur′ma-tah)   The phylum of coelomate animals whose members are pentaradially symmetrical as adults and possess a water-vascular system and an endoskeleton covered by epithelium. Sea stars, sea urchins, sea cucumbers, sea lilies.

Echinoidea (ek″i-noi′de-ah)   The class of echinoderms whose members are globular or disk shaped, possess movable spines, and have a skeleton of closely fitting plates. Sea urchins and sand dollars.

Echiura (ek-e-yur′e-ah)   A phylum of protostomate, marine animals whose members burrow in mud or sand or live in rock crevices. They possess a spatula-shaped proboscis and are 15 to 50 cm in length. Spoon worms.

echolocation (ek″o-lo-ka′shun)   A method of locating objects by determining the time required for an echo to return and

the direction from which the echo returns. As in bat echolocation.

ecological niche (ek′o-loj′i-kal nich)   The role of an organism in a community.

ecology (e-kol′-ah-je)   The study of the relationships between organisms and their environment.

ecosystem (ek′o-sis″tem)   All of the populations of organisms living in a certain area plus their physical environment.

ectoderm (ek′ta-durm)   The outer embryological tissue layer. Gives rise to skin epidermis and glands, also hair and nervous tissues in some animals.

ectoplasm (ek′to-plaz-em)   The outer, viscous cytoplasm of a protist. Contrasts with **endoplasm.**

Ectoprocta (ek-to-prok′tah)   A phylum of animals whose members are colonial and freshwater or marine. Anus ends outside a ring of tentacles. Lophophore used in feeding. Moss animals or bryozoans.

ectotherm (ek′to-therm)   Having a variable body temperature derived from heat acquired from the environment. Contrasts with **endotherm.**

egestion vacuole (e-jes′chen vak′u-ol)   Within the cytoplasm of a protist, a membrane-bound vacuole that expels wastes.

electrical synapse (sin′aps)   A synapse at which local currents resulting from electrical activity flow between two neurons through gap junctions joining them.

electrolyte (e-lek′tro-līt)   A substance that dissociates into ions when fused or in solution and thus becomes capable of conducting electricity. An ionic solute.

electron (e-lek′tron)   A small, negatively charged particle that orbits the nucleus of an atom. It has a very low mass. Atoms can gain, lose, or share electrons with other atoms.

electron transport chain   In the cristae of a mitochondrial membrane, electron carriers and enzymes positioned in an organized array that enhance oxidation-reduction reactions. Such systems function in the release of energy that is used in ATP formation and other reactions.

electroreception (i-lek″tro-re-sep′shun)   The ability to detect weak electrical fields in the environment.

electroreceptor (i-lek″tro-re-sep′tor)   A receptor that senses changes in an electrical current, usually in the surrounding water. Also called an **ampullary organ.**

element   A basic chemical substance. A substance that cannot be separated into simpler substances by ordinary chemical means. Scientists recognize 92 naturally occurring elements.

elephantiasis (el″e-fan-ti′ah-sis)   A chronic filarial disease most commonly occurring in the tropics due to infection of the lymphatic vessels with the nematode *Wuchereria* spp.

embryology (em″bre-ol′a-je)   The study of animal development from the fertilized egg to the formation of all major organ systems.

embryonic diapause (em″bre-on′ik di′ah-pauz)   The arresting of early development to allow young to hatch, or be born, when environmental conditions favor survival.

endangered species   A species that is in imminent danger of extinction throughout its range.

end bulb   A tiny swelling on the terminal end of telodendria at the distal end of an axon. Also called the synaptic bouton.

endergonic (end″er-gon′ik)   Characterized by the absorption of energy. Said of chemical reactions that require energy to proceed.

endocrine gland (en′do-krin)   Ductless, hormone-producing gland (e.g., pituitary, hypothalamus, thyroid) that is part of the endocrine system.

endocrinology (en″do-kri-nol′o-je)   The study of the endocrine system and its role in the physiology of an animal.

endocytosis (en″do-si-to′sis)   Physiological process by which substances move across a plasma membrane into the cell via membranous vesicles or vacuoles.

endoderm (en′do-durm)   The innermost embryological tissue layer. Gives rise to the inner lining of the gut tract, digestive gland, and the inner lining of the respiratory system.

endoplasm (en′do-plaz-em)   The inner, fluid cytoplasm of a protist. Contrasts with **ectoplasm.**

endoplasmic reticulum (ER) (en-do-plaz′mic re-tik′u-lum)   Cytoplasmic organelle composed of a system of interconnected membranous tubules and vesicles. Rough ER has ribosomes attached to the side of the membrane facing the cytoplasm, and smooth ER does not. Rough ER functions in protein synthesis, while smooth ER functions in lipid synthesis.

endopodite (end-op′o-dīt)   The medial ramus of the biramous appendages of crustaceans and trilobites (phylum Arthropoda).

endoskeleton (end″o-skel′ĕ-ton)   A skeleton that lies beneath the surface of the body (e.g., the bony skeleton of vertebrates and the calcium carbonate skeleton of echinoderms).

endostyle (en′do-stīl″)   A ciliated tract within the pharynx of some chordates that is used in forming mucus for filter feeding.

endosymbiont (en′do-sim″be-ont)   A free-living bacterium or other organism that becomes engulfed by a larger cell.

endosymbiont theory (en′do-sim″be-ont ther-e)   The idea that eukaryotic cells evolved from larger prokaryotic cells that engulfed once–free-living bacteria or other organisms.

endotherm (en′do-therm)   Having a body temperature determined by heat derived from the animal's own metabolism. Contrasts with **ectotherm.**

end-product (feedback) inhibition   The inhibition of the first enzyme in a pathway by the end product of that pathway.

energy   An ability to cause matter to move, and thus to do work.

energy budget   An accounting of the way in which organisms of an ecosystem process and lose energy from the sun.

energy-level shell   The distribution of electrons around the nucleus of an atom.

Enteropneusta (ent″er-op-nus′tah)   A class of hemichordates whose members live in burrows in shallow marine water. Their bodies are divided into three regions: proboscis, collar, and trunk. Acorn worms.

Entognatha (en″to-na′tha)   The class of Hexapoda whose members possess mouth appendages hidden within the head, mandibles with a single articulation, and legs with one undivided tarsus. Collembolans, proturans, and diplurans.

Entoprocta (en′to-prok-tah)   A phylum of aschelminths commonly called entoprocts.

entropy (en′tro-pe)   A measure of the degree of disorganization of a system. How much energy in a system has become so dispersed (usually as heat) that it is no longer available to do work. The higher the entropy, the more the disorder.

environmental resistance   The constraints that climate, food, space, and other environmental factors place on a population.

enzyme (en′zīm)   A protein that is synthesized by a cell and acts as a catalyst in a specific cellular reaction. The substances that each type of enzyme acts upon are called its substrates.

enzyme-substrate complex (ES) (en′zīm sub′strāt)   The binding of a substrate molecule to the active site of an enzyme.

eosinophil (e″o-sīn′o-fil)   White blood cell characterized by the presence of cytoplasmic granules that become stained red by an acid dye (eosin).

ephyra (e-fi′rah)   Miniature medusa produced by asexual budding of a scyphistoma (class Scyphozoa, phylum Cnidaria). Ephyrae mature into sexually mature medusae.

epiblast (ep′ĭ-blast)   An outer layer of cells in the embryo of an amniote that forms from the proliferation and movement of cells of the blastoderm.

epidermis (ep″ĭ-dur-mis)   A sheet of cells covering the surface of an animal's body. In invertebrates, a single layer of ectodermal epithelium.

epigenesis (ep″i-jen′ĭ-sis)   The mistaken belief that the egg contains all the materials from which the embryo is constructed.

epithelial tissue (ep″ĭ-the′le-al)   The cellular covering of internal and external surfaces of the body. Consists of cells joined by small amounts of cementing substances. Epithelium is classified into types based on the number of layers deep and the shape of the superficial cells.

epitoky (ep′ĭ-to″ke)   The formation of a reproductive individual (epitoke) that differs from the nonreproductive (atoke) form of that species.

esophagus (e-sof′ah-gus)   The passage extending from the pharynx to the stomach.

estrus (es′trus)   The recurrent, restricted period of sexual receptivity in female mammals (other than primates) marked by intense sexual urges. Also known as "being in heat."

estrus cycle (es′trus)   A recurrent series of changes in the reproductive physiology of female mammals other than primates. Females are receptive, physiologically and behaviorally, to the male only at certain times in this cycle.

ethology (e-thol′o-je)   The study of whole patterns of animal behavior in natural environments, stressing the analysis of adaptation and the evolution of the patterns.

Eubacteria (u″bak-teer′e-ah)   The domain that includes the true bacteria.

euchromatic regions (u″kro-mat′ik)   Less densely staining regions of chromosomes that contain active genes.

Eukarya (u″kar′e-ah)   The domain that includes all eukaryotic organisms: protists, fungi, plants, and animals.

eukaryote (u-kar′e-ōt)   Having a true nucleus. A cell that has membranous organelles, most notably the nucleus. All organisms except bacteria are composed of eukaryotic cells. Compare to **prokaryote.**

eutely (u′te-le)   Condition where the body is composed of a constant number of somatic cells or nuclei in all adult members of a species (e.g., rotifers, some nematodes, and acanthocephalans).

evaporation (e-vap″o-ra′shun)   The act or process of evaporating. Heat that is lost from a surface as water molecules escapes in the form of a gas. The conversion of a liquid or solid into a vapor.

evolution (ev″o-lu′shun)   Change over time. Organic or biological evolution is a series of changes in the genetic composition of a population over time. *See also* **natural selection** and **punctuated equilibrium model.**

evolutionary conservation   The slowness of change in a characteristic of an animal over time. Evolutionary conservation usually indicates that the characteristic is vital for normal functions and that change is not tolerated.

evolutionary systematics   The study of the classification of, and evolutionary relationships among, animals. Evolutionary systematists attempt to reconstruct evolutionary pathways based on resemblances between animals that result from common ancestry.

excretion (eks-kre′shun) The act, process, or function of excreting. The elimination of metabolic waste products from an animal's body.

exergonic (ek″ser-gon′ik)   Characterized or accompanied by the release of energy. Said of chemical reactions that release energy so that the products have a lower free energy than the reactants.

exocrine gland (ek′so-krin)   A gland (e.g., mammary, salivary, sweat) that secretes its product to an epithelial surface or body cavity directly or through ducts.

exocytosis (eks″o-si-to′sis)   The process by which substances move out of a cell. The substances are transported in the cytoplasmic vesicles, the surrounding membrane of which merges with the plasma membrane in such a way that the substances are dumped outside.

exopodite (eks-op′o-dīt)   The lateral ramus of the biramous appendages of crustaceans and trilobites (phylum Arthropoda).

exoskeleton (eks″o-skel′ĕ-ton)   A skeleton that forms on the outside of the body (e.g., the exoskeleton of an arthropod).

exponential growth (ek″spo-nen′shal)   Population growth in which the number of individuals increases by a constant multiple in each generation.

extracellular digestion (eks″tra-sel′u-ler)   Digestion that occurs outside the cell, usually in a special organ or cavity.

facilitated diffusion (fah-sil′i-tāt″id di-fu′zhun)   Diffusion in which a substance is moved across a membrane from a region of higher concentration to a region of lower concentration (down its concentration gradient) by carrier protein molecules.

fallopian tube (fal-lo′pe-an)   *See* **uterine tube.**

family   The level of classification between order and genus.

fermentation (fer″men-ta′shun)   Degradative pathway that begins with glycolysis and ends with the electrons being transferred back to one of the breakdown products or intermediates. Does not require molecular oxygen.

fertilization membrane   A membrane that rises off the surface of an egg after sperm penetration. Prevents multiple fertilization.

fibrillar flight muscle (fi′bra-lar)   Insect flight muscle responsible for indirect flight. A single nerve impulse results in many cycles of flight muscle contraction and relaxation.

fibrocartilage (fi″bro-kar′ti-lij)   The type of cartilage made up of parallel, thick, compact bundles, separated by narrow clefts containing typical cartilage cells (chondrocytes).

fibrous connective tissue (fi′brus)   The tissue made up of densely packed fibers (e.g., tendons and ligaments).

filoplume feather (fil′o-ploom)   A small, thin feather that probably has sensory functions in birds (pinfeather).

filopodium (fi″lo-po′de-um)   Pseudopodium that is slender, clear, and sometimes branched. Pl., filopodia.

filtration (fil-tra′shun)   Movement of material across a membrane as a result of hydrostatic pressure.

first law of thermodynamics (thur″mo-di-nam′-iks)   The total amount of energy in the universe remains constant. More energy cannot be created, and existing energy cannot be destroyed. Energy can only undergo conversion from one form to another.

fission (fish′un)   Asexual reproduction in which the cell divides into two (binary fission) or more (multiple fission) daughter parts, each of which becomes an individual organism.

flagella (flah-jel′ah)   Relatively long, motile processes that extend from the surface of a cell. Eukaryotic flagella are longer versions of cilia. Flagellar undulations drive a cell through a fluid medium. Like cilia, flagella have a 9 + 2 arrangement of microtubules covered by the cell's plasma membrane.

flame cell   Specialized, hollow excretory or osmoregulatory structure consisting of one to several cells containing a tuft of cilia (the "flame") and located at the end of a minute tubule. Flame bulb.

flavin adenine dinucleotide (FAD) (fla′vin ad′e-nēn di″nuc′le-o-tīd)   A coenzyme that is a condensation product of riboflavin phosphate and adenylic acid. It forms the prosthetic group of certain enzymes.

fluid feeding   The process by which an animal feeds on fluid. Examples include some parasites, leeches, ticks, mites, lampreys, and certain crustaceans.

fluke (flook)   Any trematode worm. A member of the class Trematoda or class Monogenea.

food chain   A linear sequence of organisms through which energy is transferred in an ecosystem from producers through several levels of consumers.

food vacuole (vak′u-ōl)   A cell organelle that functions in intracellular digestion.

food web   A sequence of organisms through which energy is transferred in an ecosystem. Rather than being linear, a food web has highly branched energy pathways.

forebrain (for′brān)   Consists of the diencephalon and telencephalon. Also known as the prosencephalon.

formed-element fraction   The cellular component of vertebrate blood.

fossil   Any remains, impressions, or traces of organisms of a former geological age.

founder effect   Changes in gene frequency that occur when a few individuals from a parental population colonize new habitats. The change is a result of founding individuals not having a representative sample of the parental population's genes.

fragmentation (frag″men-ta′shun)   Division into smaller units. A type of asexual reproduction whereby a body part is lost and then regenerates into a new organism.

Fungi (fun′ji)   The kingdom of life whose members are eukaryotic, multicellular, and saprophytic (mushrooms, molds).

furcula (fur′kyah-lah)   The fused clavicles of a bird. The furcula is one point of attachment for flight muscles. The wishbone.

Galápagos Islands (gah-lah′pĕ-gōs″)   An archipelago on the equator in the Pacific Ocean about 900 km west of Ecuador. Charles Darwin's observations of the plant and animal life of these islands were important in the formulation of the theory of evolution by natural selection.

gallbladder (gawl′blad-der)   The pear-shaped reservoir for bile under the right lobe of the liver.

gamete (gam'ēt)   Mature haploid cell (sperm or egg) that functions in sexual reproduction. The union of two gametes of opposite sex (fertilization) produces a zygote.

gametogenesis (gam″e-to-jen′ĕ-sis)   Gamete formation by way of meiosis.

gametogony (ga′mēt-o-gony)   Multiple fission that forms gametes that fuse to form a zygote. Also called gamogony. Occurs in the class Sporozoea.

ganglion (gang′gle-on)   A group of nerve cell bodies outside the central nervous system.

gastric shield   A chitinized plate in the stomach of a bivalve (phylum Mollusca) on which the crystalline style rotates.

gastrodermis (gas-tro-derm′is)   The endodermally derived lining of the gastrovascular cavity of Cnidaria.

Gastropoda (gas-trop′o-dah)   The class of molluscs characterized by torsion. A shell, when present, is usually coiled. Snails.

Gastrotricha (gas-tro-tri′kah)   A small phylum of marine and freshwater species of gastrotrichs that inhabit the spaces between bottom sediments.

gastrovascular cavity (gas″tro-vas′ku-lar kav′ĭ-te)   The large central cavity of cnidarians and flatworms that receives and digests food. Has a single opening serving as both mouth and anus.

gastrozooid (gas″tro-zo′oid)   A feeding polyp in a colonial hydrozoan (phylum Cnidaria).

gastrulation (gast″ru-la′shun)   The embryological process that results in the formation of the gastrula. Results in the formation of the embryonic gut, ectoderm, and endoderm.

gemmule (jem′yool)   Resistant, overwintering capsule formed by freshwater, and some marine, sponges that contains masses of mesenchyme cells. Amoeboid mesenchyme cells are released and organize into a sponge.

gene   A heritable unit in a chromosome. A series of nucleotide bases on the DNA molecule that codes for a single polypeptide.

gene flow   Changes in gene frequency in a population that result from emigration or immigration.

gene pool   The sum of all genes in a population.

generator potential   A graded potential that travels only a short distance along the plasma membrane of a sensory cell.

genetic drift (je-net′ik)   Occurs when chance events influence the frequency of genes (evolutionary change). Also called **neutral selection.**

genetic recombination (je-net′ik re-kom-bĕ-na′shun)   Crossing-over. A major source of genetic variation in a population or a given species.

genetics (je-net′iks)   The study of the mechanisms of transmission of genes from parents to offspring.

genotype (je′no-tīp)   The specific gene combinations that characterize a cell or an individual.

genotypic ratio (je″no-tip′-ik)   The relative numbers of progeny in each genotypic category that a genetic cross produces.

genus (je′nus)   The level of classification between species and family.

georeceptor (je″o-re-cep′tor)   A specialized nerve ending that responds to the force of gravity.

gerontology (jer″on-tol′o-je)   Scientific study of the problems of aging in all their aspects, including clinical, biological, and sociological.

gestation (jes-ta′shun)   Period of development of the young in viviparous animals, from the time of fertilization of the ovum until birth.

giardiasis (je″ar-di′ah-sis)   A common infection of the lumen of the small intestine with the flagellated protozoan *Giardia lamblia.* Spread via contaminated food and water and by direct person-to-person contact.

gill   An aquatic respiratory organ for obtaining oxygen and getting rid of carbon dioxide. Found in fishes, molluscs, and many arthropods.

gill arches   Bony or cartilaginous gill support of some vertebrates. Also called **visceral arches.**

gill filament   A thin-walled, fleshy extension of a gill arch that contains vessels carrying blood to and from gas exchange surfaces.

gill rakers   Processes extending from pharyngeal arches of some fishes that trap food and protect gill filaments from mechanical damage.

glochidium (glo-kid′e-um)   A larval stage of freshwater bivalves in the family Unionidae. It lives as a parasite on the gills or fins of fishes.

glomerulus (glo-mer′u-lus)   A capillary tuft (coiled mass of capillaries) within the capsule (Bowman's) of a nephron. Forms the glomerular filtrate.

glycocalyx (gli′ko-kal′iks)   The glycoprotein and glycolipid covering (cell coat) that surrounds many eukaryotic cells.

glycolysis (gli-kol′ĭ-sis)   The conversion of glucose to pyruvic acid (pyruvate) with the release of some energy in the form of ATP. Occurs in the cytosol. Literally, "sugar splitting."

Gnathostomata (na″tho-sto′ma-tah)   A group of vertebrates whose members possess hinged jaws and paired appendages. A vertebral column may have replaced the notochord.

Golgi apparatus (gol′je ap″ah-ră′tus)   The membrane-bound cytoplasmic organelle where the proteins and lipids made in the endoplasmic reticulum are modified and stored.

gonad (go′nad)   A gamete-producing gland. Ovary or testis.

gonadotropin (go-nad″o-trōp′in)   A hormone that stimulates activity in the gonads.

gonangium (go′nanj″e-um)   *See* **gonozooid.**

gonozooid (gon′o-zo″id)   A polyp of a hydrozoan cnidarian that produces medusae.

Gordian worm   *See* **horsehair worms.**

gravid (grav′id)   Containing developing young. Pregnant.

gray crescent   A dark, arching band that forms on the surface of the amphibian zygote opposite the point of sperm penetration. Forms in the region where gastrulation will occur.

green gland   *See* **antennal gland.**

greenhouse effect   The warming of a global climate due to the accumulation of carbon dioxide in the atmosphere from burning fossil fuels.

gular flutter (gu′lar)   A type of breathing in some birds. Rapid movement of the throat region promotes evaporative water loss.

gustation (gus-ta′shun)   The act of tasting or the sense of taste.

H

habitat   The native environment of an organism.

habitat selection   The choice of an animal's place to live. Habitat selection involves the interaction of physiological and psychological factors.

habituation (hab-bich″u-a′shun)   The gradual decrease in a response to a stimulus or to the environment.

hair   A long, slender filament. Applied especially to such filamentous appendages of the skin.

haploid (hap′loid)   Having one member of each pair of homologous chromosomes. Haploid cells are the product of meiosis and are often gametes.

Hardy-Weinberg theorem (har′de wīn′berg)   A theorem stating that the frequency of genes in a population does not change from one generation to another if specified conditions are met.

head-foot   The body region of a mollusc that contains the head and is responsible for locomotion as well as retracting the visceral mass into the shell.

heartworm disease   A parasitic infection in dogs caused by the nematode *Dirofilaria immitis*.

hectocotylus (hek"to-kot'ĭ-lus)   A modified arm of some male cephalopods that is used in sperm transfer.

hemal system (he'mal)   Strands of tissue found in echinoderms. The hemal system is of uncertain function. It may aid in the transport of large molecules or coelomocytes, which engulf and transport waste particles within the body.

hematopoiesis (hem"ah-to-poi-e'sis)   The formation and development of blood cells.

hemerythrin (hem"e-rith'rin)   The red iron-containing respiratory pigment in the blood plasma of some polychaetes, sipunculids, priapulids, and brachiopods.

Hemichordata (hem"e-kor-da'tah)   The phylum of marine, wormlike animals whose members have an epidermal nervous system and pharyngeal slits. Acorn worms and pterobranchs.

hemimetabolous metamorphosis (hem"e-met-ab'ol-us met-a-morf'ō-sis)   A type of insect metamorphosis involving a species-specific number of molts between egg and adult stages, during which immatures gradually take on the adult form.

hemizygous (hem"e-zi'gus)   An individual having one member of a pair of genes. Males of most animal species are hemizygous for traits carried on the X chromosome.

hemocoel (he'mo-sēl)   Large tissue space within an arthropod that contains blood. Derived from the blastocoel of the embryo.

hemocyanin (he"mo-si'ah-nin)   A non-heme, blue respiratory pigment in the plasma of many molluscs and arthropods. Composed of monomers, each of which contains two atoms of copper and can bind one molecule of oxygen.

hemocyte (he'mo-sīt)   Any blood corpuscle or formed element of the blood in animals with an open circulatory system.

hemoglobin (he'mo-glo"bin)   An iron-containing respiratory pigment of red blood cells responsible for the transport of oxygen and carbon dioxide. Occurs in vertebrate red blood cells and in the plasma of many invertebrates.

hemolymph (he'mo-limf)   The fluid in the coelom or hemocoel of some invertebrates that represents the blood and lymph of higher animals. Found in animals with an open circulatory system.

herbivore (her'bĭ-vor)   A plant-eating animal. Any organism that subsists on plants.

herbivory (her-bĭ'vor-e)   The process of existing by eating macroscopic plants. Examples include molluscs, polychaete worms, arthropods, and sea urchins.

hermaphroditism (her-maf'ro-di"tizm)   A state characterized by the presence of both male and female reproductive organs in the same animal. Monecious.

heterochromatic regions (het"er-o-krōm-ă'tik)   Having inactive genes. Inactive regions of chromosomes are heterochromatic.

heterodont (het'er-o-dont)   Having a series of teeth specialized for different functions.

heterotherm (het'e-ro-therm)   An animal whose body temperature fluctuates markedly. "Cold-blooded." An animal that derives essentially all of its body heat from the environment.

heterotrophic (het"er-o-tro'fik)   The type of nutrition in which organisms derive energy from the oxidation of organic compounds either by consumption or absorption of other organisms.

heterotrophs (het'er-o-trofs)   Organisms that obtain both inorganic and organic raw materials from the environment to live. Animals, fungi, many protists, and most bacteria are heterotrophs.

heterozygous (het"er-o-zi'ges)   Having different expressions of a gene on homologous chromosomes.

Hexactinellida (hex-act"in-el'id-ah)   The class of sponges whose members are characterized by triaxon siliceous spicules, which sometimes form an intricate lattice. Cup or vase shaped. Sycon body form. Glass sponges.

Hexapoda (hex"sah-pod'ah)   The subphylum of mandibulate arthropods whose members are characterized by a body consisting of a head, thorax, and abdomen; five pairs of head appendages; and three pairs of uniramous appendages on the thorax.

hibernation (hi"ber-na'shun)   Condition of mammals that involves passing the winter in a torpid state in which the body temperature drops to nearly freezing and the metabolism drops close to zero. May last weeks or months.

hindbrain (hīnd'brān)   Includes the medulla oblongata, cerebellum, and pons. Also known as the rhombencephalon.

Hirudinea (hi"ru-din'e-ah)   The class of annelids whose members are characterized by bodies with 34 segments, each of which is subdivided into annuli. Anterior and posterior suckers are present. Leeches.

holoblastic (hol"o-blas'tik)   Division of a zygote that results in separate blastomeres.

holometabolous metamorphosis (hol"o-met-ab'ol-us met-ah-morf'a-sis)   A type of insect metamorphosis in which immatures, called larvae, are different in form and habitats from the adult. The last larval molt results in the formation of a pupa. Radical cellular changes in the pupal stage end in adult emergence.

homeostasis (ho"me-o-sta'sis)   A state of equilibrium in which the internal environment of the body of an animal remains relatively constant with respect to the external environment.

homeotherm (ho'me-o-therm)   Having nearly uniform body temperature, regulated independently of the environmental temperature. "Warm-blooded."

homodont (ho'mo-dont)   Having a series of similar, unspecialized teeth.

homologous (ho-mol'o-ges)   Describes structures, processes, or molecules that have a common evolutionary origin. The wing of a bat and the arm of a human are homologous, since each can be traced back to a common ancestral appendage.

homologous chromosomes (ho-mol'o-ges kro'mo-sōmz)   Chromosomes that carry genes for the same traits. One of two copies of a particular chromosome in a diploid cell, each copy being derived from a different parent.

homozygous (ho"mo-zi'ges)   Having the same expression of a gene on homologous chromosomes.

hormone (hor'mōn)   A chemical secreted by an endocrine gland that is transmitted by the bloodstream or body fluids to a target cell or tissue.

horns   The paired growths on the heads of certain ungulate animals. The median growth of hair on the snout of the rhinoceros.

horsehair worms   Pseudocoelomate animals that belong to the phylum Nematomorpha. Also known as Gordian worms or hairworms (Gordius is the name of an ancient king who tied an intricate knot).

host   An animal or protist that harbors or nourishes another organism (parasite).

humoral immunity (hu'mor-al ĭ-mu'nĭ-te)   The type of immunity that results from the presence of antibodies that are soluble in blood and lymph.

hyaline cartilage (hī'ah-lin kar'tĭ-lij)   The type of cartilage with a glassy, translucent appearance.

hydranth (hi′dranth) *See* **gastrozooid.**

hydraulic skeleton (hi-drol-ĭk) The use of body fluids in open circulatory systems to give support and facilitate movement. Muscles contracting in one part of the body force body fluids into some distant tissue space, thus causing a part of the body to extend or become turgid.

hydrocarbon (hi′dro-kar″bon) An organic molecule that contains only carbon and hydrogen and has its carbons bonded in a linear fashion.

hydrogen bond (hi′dro-jen) A weak to moderate attractive force between a hydrogen atom bonded to an electronegative atom and one pair of electrons of another electronegative atom.

hydrological cycle (hi″dro-loj′ĭ-kal) The cycling of water between reservoirs in oceans, lakes, and groundwater, and the atmosphere.

hydrostatic skeleton (hi″dro-stat′ik) The use of body cavity fluids, confined by the body wall, to give support (e.g., the hydrostatic skeleton of nematodes and annelids). Also called hydroskeleton.

hydrothermal vents (hi″dro-thur′mal) Deep, oceanic regions where the tectonic plates in the earth's crust are moving apart. They are characterized by occasional lava flows and hot water springs. These vents support a rich community by chemolithotrophy.

Hydrozoa (hi″dro-zo′ah) The class of cnidarians whose members have epidermally derived gametes, mesoglea without wandering amoeboid cells, and gastrodermis without nematocysts. Medusae, when present, with a velum. *Hydra, Obelia, Physalia.*

hygroreceptor (hi″gro-re-sep′tor) A receptor in insects that detects the water content of air.

Hyperotreti (hi″per-ot′re-te) The subphylum of craniates whose skull consists of cartilaginous bars, jawless, no paired appendages, four head tentacles, and olfactory sacs that open to the mouth cavity. Hagfishes.

hypertonic (hi″per-ton′ik) A solution having a greater number of solute particles than another solution to which it is compared.

hypoblast (hi′po-blast) An inner layer of cells that results from the proliferation and movement of cells in the blastoderm of an avian or reptilian embryo.

hypodermis (hi″po-der′mis) The layer of integument below the cuticle. The outer cellular layer of the body of invertebrates that secretes the cuticular exoskeleton.

hypothalamus (hi″po-thal′ah-mus) A structure within the diencephalon and below the thalamus that functions as an autonomic center and regulates the pituitary gland.

hypotonic (hi″po-ton′ik) A solution having a lesser number of solute particles than another solution to which it is compared.

I

imprinting (im′print-ing) A young animal developing an attachment toward an animal or object.

incomplete dominance An interaction between alleles in which both alleles are expressed more or less equally, and the phenotype of the heterozygote is different from either homozygote.

indirect (asynchronous) flight Insect flight accomplished by flight muscles acting on the body wall. Changes in shape of the thorax cause wing movements. A single nerve impulse results in many cycles of the wings. *See also* **direct (synchronous) flight.**

induced fit The precise fit between an enzyme and its substrate.

inferior Below a point of reference (e.g., the mouth is inferior to the nose in humans).

inorganic molecules (compounds) Compounds that are not hydrocarbons or their derivatives. Compounds other than organic compounds.

Insecta (in-sekt′ah) Class of Hexapoda with exposed mouth appendages, mandibles with two points of articulation, and well-developed Malpighian tubules. Insects.

insectivore (in-sek′tĕ-vor) An animal or plant that consumes insects. Any mammal of the order Insectivora, comprising the moles, shrews, and Old World hedgehogs.

insight learning The use of cognitive or mental processes to associate experiences and solve problems.

instrumental conditioning (in″strĕ-men′tal kon-dish′en-ing) Trial-and-error learning. The reinforcement of certain behaviors in animals leads to an animal repeating the behavior.

integument (in-teg′u-ment) A covering (e.g., the skin).

integumentary exchange (in-teg″u-men′tah-re) Gas exchange through the integument. Also called cutaneous exchange.

intercalated disk (in-ter″kah-lāt′ed) Membranous boundary between adjacent cardiac muscle cells.

intermediate filament The chemically heterogeneous group of protein fibers, the specific proteins of which vary with cell type. One of the three most prominent types of cytoskeletal filaments. Made of fibrous proteins.

intermediate host The organism in the life cycle of a parasite that harbors an immature stage of the parasite and where asexual reproduction usually occurs.

intermediate lobe The area in the pituitary gland between the anterior and posterior lobes. Also called the pars intermedia. Produces melanophore-stimulating hormone.

interneuron (in″ter-nu′ron) A neuron between a sensory neuron and a motor neuron. Interneurons function as integrating centers.

interphase (in′ter-fāz) Period between two cell divisions when a eukaryotic cell is carrying on its normal functions. Long period of the cell cycle between one mitosis and the next. Includes G_1 phase, S phase, and G_2 phase. DNA replication occurs during interphase.

intracellular digestion Digestion that occurs inside a cell, as in many protozoans, sponges, cnidarians, flatworms, rotifers, bivalve molluscs, and primitive chordates.

intraspecific competition Competition among members of the same species for environmental resources.

intrinsic rate of growth *See* **biotic potential.**

introvert (in′tro-vert) The anterior narrow portion that can be withdrawn (introverted) into the trunk of a sipunculid worm, a loriciferan, or a bryozoan.

involuntary (visceral or autonomic) nervous system Stimulates smooth and cardiac muscle and glands of the body.

involution (in″vo-lu′shun) The rolling of superficial cells over the dorsal lip of the blastopore during gastrulation of amphibian embryos.

ion (i′on) An atom or group of atoms with an electrical charge. The charged particle formed when a neutral atom or group of atoms gains (negative) or loses (positive) one or more electrons.

ionic bond (i-on′ik) An association between ions of opposite charges. The electrostatic force that holds ions together in an ionic compound.

isomer (i′so-mer) Organic compounds with the same molecular formula but different structures, functions, and properties.

isotonic (i″so-ton′ik) In a comparison of two solutions, both have equal concentrations of solutes.

isotope (i′sĕ-tōp)   Any of two or more forms of a chemical element having the same number of protons in the nucleus or the same atomic number, but having different numbers of neutrons in the nucleus, or different atomic masses. Isotopes of a single element possess almost identical chemical properties.

J

Jacobson's (vomeronasal) organ   Olfactory receptor present in most reptiles. Blind-ending sacs that open through the secondary palate into the mouth cavity. Used to sample airborne chemicals.

Johnston's organ   Mechanoreceptor (auditory receptor) found at the base of the antennae of male mosquitoes, midges, and most other insects.

K

keratin (ker′ă-tin)   A tough, water-resistant protein found in the epidermal layers of the skin of reptiles, birds, and mammals. Found in hair, feathers, hooves, nails, claws, bills, etc.

kilocalorie (kil′o-kal″o-re)   A unit of heat equal to 1,000 calories. Used to measure the energy content of food. *See also* **Calorie.**

kinetic energy (kĭ-net′ik en′er-je)   The energy associated with a body by virtue of its motion. The energy of a mass of matter that is moving.

kinetochore (kĭ-nĕ′to-kor)   A specialized group of proteins and DNA at the centromere of a chromosome. Is an attachment site for the microtubules of the mitotic spindle and plays an active part in the movement of chromosomes to the pole.

kingdom (king′dom)   The level of classification above phylum. The traditional classification system includes five kingdoms: Monera, Protista, Fungi, Plantae, and Animalia. Recent evidence from molecular biological studies indicates that these five kingdoms may not be monophyletic lineages.

Kinorhyncha (kin″o-rink′ah)   The phylum of aschelminths that contains members called kinorhynchs. Small, elongate worms found exclusively in marine environments, where they live in mud and sand.

kin selection   The idea that natural selection acting on related animals can affect the fitness of an individual. When genes are common to related animals, an individual's fitness is based on the genes the individual passes on and on those common genes that relatives pass on. Kin selection

is thought to explain how altruism could evolve in a population. *See also* **altruism.**

Krebs cycle   *See* **citric acid cycle.**

krill (kril)   Any of the small, pelagic, shrimp-like crustaceans. Krill are an important source of energy in Antarctic food webs.

L

labial palp (la′be-al palp)   1. Chemosensory appendage found on the labium of insects (Arthropoda). 2. Flaplike lobe surrounding the mouth of bivalve molluscs that directs food toward the mouth.

labium (la′be-um)   The posterior mouthpart of insects. It is often referred to as the "lower lip," is chemosensory, and was derived evolutionarily from paired head appendages (Hexapoda, Arthropoda).

lactation (lak-ta′shun)   The production of milk by the mammary glands.

large intestine   That part of the digestive system between the ileocecal valve of the small intestine and the anus. Removes salt and water from undigested food and releases feces through the anus.

larva (lar′vah)   The immature feeding stage of an insect that undergoes holometabolous metamorphosis. The immature stage of any animal species in which adults and immatures are different in body form and habitat.

Larvacea (lar-vas′e-ah)   The class of urochordates whose members are planktonic and whose adults retain a tail and notochord with a gelatinous covering of the body.

larval instars (lar′val in′starz)   Any of the different immature feeding stages of an insect metamorphosis.

latent learning (lāt′ent)   Exploratory learning. Latent learning occurs when an animal makes associations without immediate reinforcement or reward.

lateral (lat′er-al)   Away from the plane that divides a bilateral animal into mirror images.

lateral-line system (lat′er-al)   1. A line of sensory receptors along the side of some fishes and amphibians used to detect water movement (phylum Chordata). 2. The external manifestation of a lateral excretory canal of nematodes (phylum Nematoda).

learning   Changes in the behavior of an individual due to experience.

leucon (lu′kon)   The sponge body form that has an extensively branched canal system. The canals lead to chambers lined by choanocytes.

light microscope (LM)   The type of microscope in which the specimen is viewed under ordinary illumination.

limiting factor   A nutrient or other component of an organism's environment that is in relatively short supply and, therefore, restricts the organism's ability to reproduce successfully.

lipid (lip′id)   A fat, oil, or fatlike compound that usually has fatty acids in its molecular structure. An organic compound consisting mainly of carbon and hydrogen atoms linked by nonpolar covalent bonds. Examples include fats, waxes, phospholipids, and steroids that are insoluble in water.

Lissamphibia (lis′am-fib′e-ah)   The group of amphibians comprised of living species. Members of the orders Anura, Caudata, and Gymnophiona.

liver (liv′er)   A large, dark-red gland that carries out many vital functions, such as the formation of urea, manufacture of plasma proteins, synthesis of amino acids, synthesis and storage of glycogen, and many others.

lobopodium (lo″bo-po′de-um)   A blunt, lobelike pseudopodium that is commonly tubular and is composed of both ectoplasm and endoplasm. Pl., lobopodia.

local chemical messenger   A chemical that acts on nearby cells.

locus (lo′kus)   The position of a gene in a chromosome.

logistic population growth   The population growth pattern that occurs when environmental resistance limits exponential growth.

looping movement   The type of locomotion exhibited by leeches and some insect larvae whereby they alternate temporary points of attachment to move forward.

loose connective tissue   The type of tissue in which the matrix contains strong, flexible fibers of the protein collagen interwoven with fine, elastic, and reticular fibers.

lophophore (lof′a-for)   Tentacle-bearing ridge or arm within which is an extension of the coelomic cavity in lophophorate animals (e.g., brachiopods, ectoprocts, phoronids).

lorica (lo-ri′kah)   The protective external case in rotifers and some protozoa. It is formed by a thickened cuticle.

Loricifera (lor″a-sif′er-ah)   A phylum of aschelminths. A recent animal phylum to be described. Members are commonly called loriciferans.

lotic ecosystems (lo′tik ē′ko-sis″temz)   Flowing water ecosystems. They include brooks, streams, and rivers.

lung   An organ of the respiratory system in which gas exchange occurs between body fluids (e.g., blood) and air.

lymph (limf)   Fluid that the lymphatic vessels transport.

lymphatic system (lim-fat′ik)   The one-way system of lymphatic vessels, lymph nodes, and other lymphoid organs and tissues. Drains excess tissue fluid from the extracellular space and provides a site for immune surveillance.

lymphocyte (lim′fo-sīt)   A type of white blood cell that provides protection to an animal.

lysosome (li′so-sōm)   Cytoplasmic, membrane-bounded organelle that contains digestive and hydrolytic enzymes, which are typically most active at the acid pH found in the lumen of lysosomes.

M

macroevolution (mak″ro-ev″o-lu′shun)   Large-scale evolutionary changes that result in extinction and the formation of new species.

macronucleus (mak-ro-nu′kle-us)   A large nucleus found within the Ciliata (Protista) that regulates cellular metabolism. Directly responsible for the phenotype of the cell.

macronutrient (mak″ro-noo′tre-ent)   An essential nutrient for which an animal has a large minimal daily requirement (greater than 100 mg). For example, calcium, phosphorus, magnesium, potassium, sodium, and chloride, along with carbohydrates, lipids, and proteins.

Malacostraca (mal-ah-kos′trah-kah)   The class of crustaceans whose members are characterized by having appendages modified for crawling along the substrate, as in lobsters, crayfish, and crabs. Alternatively, the abdomen and body appendages may be used in swimming, as in shrimp.

Malpighian tubules (mal-pig′e-an tu′būlz)   The blind-ending excretory and osmoregulatory tubules that join the midgut of insects and some other arthropods. Secrete waste products and form urine.

Mammalia (mah-ma′le-ah)   The class of vertebrates whose members are at least partially covered by hair, have specialized teeth, and are endothermic. Young are nursed from mammary glands. The mammals.

mammary gland (mam′ar-e)   The breast. In female mammals, the mammary glands produce and secrete milk to nourish developing young.

mandible (man′dib-el)   1. The lower jaw of vertebrates. 2. The paired, grinding and tearing arthropod mouthparts, derived from anterior head appendages.

mantle (man′tel)   The outer fleshy tissue of molluscs that secretes the shell. The mantle of cephalopods may be modified for locomotion.

mantle cavity (man′tel kav′ĭ-te)   The space between the mantle and the visceral mass of molluscs.

manubrium (mah-nu′bre-um)   A structure that hangs from the oral surface of a cnidarian medusa and surrounds the mouth.

mass   A measure of the quantity of matter in an object.

mastax (mas′tax)   The pharyngeal apparatus of rotifers used for grinding ingested food.

Mastigophora (mas″ti-gof′o-rah)   The protozoan subphylum whose members possess one or more flagella for locomotion. Autotrophic, heterotrophic, or saprozoic.

matter   Anything that has mass and occupies space.

maturation (mach″oo-ra′shun)   To complete the natural development of an animal system. The performance of behavior improves as parts of the nervous system and other structures complete development.

maxilla (maks′il-ah)   One member of a pair of mouthparts just posterior to the mandibles of many arthropods.

maxillary gland (mak′si-ler″e)   In malacostracan crustaceans, the excretory organ located near the maxillary segments. Helps regulate ion concentrations.

Maxillopoda (maks″il-ah-pod′ah)   The class of Crustacea characterized by five head, six thoracic, and four abdominal somites plus a telson. Abdominal segments lack typical appendages. Abdomen often reduced. Barnacles and copepods.

mechanoreceptor (mek″ah-no-re-sep′tor)   A sensory receptor that is sensitive to mechanical stimulation, such as changes in pressure or tension.

meconium (mĭ-ko′ne-um)   A dark green mucilaginous material in the intestine of the full-term fetus, being a mixture of the secretions of the intestinal glands and some amniotic fluid.

medial (me′de-al)   On or near the plane that divides a bilateral animal into mirror images. Also median.

median (parietal) eye (me′de-an)   A photoreceptor located middorsally on the head of some vertebrates (reptiles). It is associated with the vertebrate epithalamus.

medulla oblongata (me-dul′ah ob″lon-gah′tah)   Inferior-most portion of the brain stem between the pons and the spinal cord.

medusa (me-du′sah)   Usually, the sexual stage in the life cycle of cnidarians. The jellyfish body form.

meiosis (mi-o′sis)   Process of cell division by which egg and sperm cells form, involving a diminution in the amount of genetic material. Comprises two successive nuclear divisions with only one round of DNA replication, which produces four haploid daughter cells from an initial diploid cell.

melatonin (mel″ah-to′nin)   The hormone that the pineal gland secretes. Regulates photoperiodicity.

meninges (me-nin′jēz)   A group of three membranes that covers the brain and spinal cord.

menopause (men′o-pawz)   Termination of the menstrual cycle. Period of life in a female during which hormonal changes cause ovulation and menstruation to cease.

menstrual cycle (men′stroo-al)   The period of regularly recurring physiologic changes in the endometrium that culminates in its shedding (menstruation).

menstruation (men″stroo-a′shun)   Loss of blood and tissue from the uterus at the end of a female primate's reproductive cycle.

meroblastic (mer″ah-blas′tik)   The division of a zygote in which large quantities of yolk prevent complete cleavages.

merogony (mĕ-rog′ah-ne)   Schizogony resulting in the production of merozoites.

Merostomata (mer″o-sto′mah-tah)   The class of arthropods whose members are aquatic and possess book gills on the opisthosoma. Eurypterids (extinct) and horseshoe crabs.

mesenchyme (mez′en-kīm)   Undifferentiated mesoderm. It eventually develops into muscle, blood vessels, skeletal elements, and (other) connective tissues.

mesenchyme cells (mez′en-kīm)   Amoeboid cells within the mesohyl of a sponge. Mesenchyme cells are specialized for reproduction, secreting skeletal elements, transporting food, storing food, and forming contractile rings around openings in the sponge wall.

mesoderm (mez′ah-durm)   The embryonic tissue that gives rise to tissues between the ectoderm and endoderm (e.g., muscle, skeletal tissues, and excretory structures).

mesoglea (mez-o-gle′ah)   A gel-like matrix between the epidermis and gastrodermis of cnidarians.

mesohyl (mez-o-hīl′)   A jellylike layer between the outer (pinacocyte) and inner (choanocyte) layers of a sponge. Contains wandering amoeboid cells.

mesonephros (mez″o-nef′rōs)   The middle of three pairs of embryonic renal organs in vertebrates. The functional kidney of fishes and amphibians. Its collecting duct is a wolffian duct.

mesothorax (mez″o-thor′aks) The middle of the three thoracic segments of an insect. Usually contains the second pair of legs and the first pair of wings.

Mesozoa (mes″o-zo′ah) A phylum of animals whose members are parasites of marine invertebrates. Two-layered body organization. Dioecious, complex life histories. Orthonectids and dicyemids.

messenger RNA (mRNA) A single-stranded polyribonucleotide. Formed in the nucleus from a DNA template and carries the transcribed genetic code to the ribosome, where the genetic code is translated into protein.

metabolism (me-tab′o-lizm) All of the chemical changes and processes within cells. All controlled, enzyme-mediated chemical reactions by which cells acquire and use energy.

metacercaria (mĕ″tă-ser-ka′re-ah) Stage between the cercaria and adult in the life cycle of most digenetic trematodes. Usually encysted and quiescent.

metamerism (mĕt-tăm′a-riz″em) A segmental organization of body parts. Metamerism occurs in the Annelida, Arthropoda, and other smaller phyla.

metamorphosis (met″ah-mor′fo-sis) Change of shape or structure, particularly a transition from one developmental stage to another, as from larva to adult form.

metanephridium (met″ah-nĕ-frid′e-um) An excretory organ in many invertebrates. It consists of a tubule that has one end opening at the body wall and the opposite end in the form of a funnel-like structure that opens to the body cavity.

metanephros (me″tah-nef′rōs) The renal organs of some vertebrates arising behind the mesonephros. The functional kidney of reptiles, birds, and mammals. It is drained by a ureter.

metaphase (met′ah-fāz) Stage in mitosis when chromosomes align in the middle of the cell and firmly attach to the mitotic spindle but have not yet segregated toward opposite poles.

metathorax (met″ah-thor′aks) The posterior of the three segments of an insect thorax. It usually contains the third pair of walking legs and the second pair of wings (Arthropoda).

micelle (mi-sēl′) Lipid aggregates with a surface coat of bile salts. A stage in the digestion of lipids in the small intestine.

microevolution (mi″kro-ev′o-lu′shun) A change in the frequency of alleles in a population over time.

microfilament (mi″kro-fil′ah-ment) Component of the cytoskeleton. Involved in cell shape, motion, and growth. Helical protein filament formed by the polymerization of globular actin molecules.

microfilaria (mi″kro-fi-lar′e-ah) The prelarval stage of filarial worms. Found in the blood of humans and the tissues of the vector.

micronucleus (mi″kro-nu′kle-us) A small body of DNA that contains the hereditary information of ciliates (Protista). Exchanged between protists during conjugation. It undergoes meiosis before functioning in sexual reproduction.

micronutrient (mi″kro-nu′tre-ent) A dietary element essential in only small quantities. For example, iron, chlorine, copper, and vitamins.

microscopy (mi-kros′ko-pe) Examination with a microscope.

Microspora (mi″cro-spor′ah) The protozoan phylum whose members have unicellular spores. Intracellular parasites in nearly all major animal groups. Example: microsporeans (*Nosema*).

microtubule (mi″kro-tu′būl) A hollow cylinder of tubulin subunits. Involved in cell shape, motion, and growth. Functional unit of cilia and flagella. It is one of three major classes of filaments of the cytoskeleton.

microtubule-organizing center (MTOC) (mi″kro-tu′būl) Region in a cell, such as a centrosome or a basal body, from which microtubules grow.

mictic eggs (mik′tik) Pertaining to the haploid eggs of rotifers. If not fertilized, the egg develops parthenogenetically into a male. If fertilized, mictic eggs secrete a heavy shell and become dormant, hatching in the spring into amictic females.

midbrain The portion of the brain between the pons and forebrain (diencephalon).

migration Periodic round trips of animals between breeding and nonbreeding areas or to and from feeding areas.

mimicry (mim′ik-re) When one species resembles one or more other species. Often, protection is afforded the mimic species.

miracidium (mi-rah-sid′e-um) The ciliated, free-swimming, first-stage larva of a digenean trematode that undergoes further development in the body of a snail.

mitochondrion (mi″to-kon′dre-on) Membrane-bound organelle that specializes in aerobic respiration (oxidative phosphorylation) and produces most of the ATP in eukaryotic cells.

mitosis (mi-to′sis) Nuclear division in which the parental number of chromosomes is maintained from one cell generation to the next. Basis of reproduction of single-cell eukaryotes. Basis of physical growth (through cell divisions) in multicellular eukaryotes.

mitotic spindle (mi-tŏ′tic spin′dl) Collectively, the asters, spindle, centrioles, and microtubules of a dividing cell.

modern synthesis The combination of the principles of population genetics and Darwinian evolutionary theory.

molecular biology The study of the biochemical structure and function of organisms.

molecular genetics The study of the biochemical structure and function of DNA.

molecule A particle composed of two or more atoms bonded together. An aggregate of at least two atoms in a definite arrangement held together by special forces.

Mollusca (mol-lus′kah) The phylum of coelomate animals whose members possess a head-foot, visceral mass, mantle, and mantle cavity. Most molluscs also possess a radula and a shell. The molluscs. Bivalves, snails, octopuses, and related animals.

molting The periodic renewal of feathers of birds by shedding and replacement. In arthropods and other invertebrates, the shedding of the exoskeleton or other body covering is called molting or ecdysis. *See also* **ecdysis.**

Monera (mon′er-ah) The kingdom of life whose members have cells that lack a membrane-bound nucleus, as well as other internal, membrane-bound organelles (they are prokaryotic). Bacteria.

monocyte (mon′o-sīt) A type of white blood cell that functions as a phagocyte and develops into macrophages in tissues. A monocyte has a kidney-shaped nucleus and gray-blue cytoplasm.

monoecious (mah-ne′shus) An organism in which both male and female sex organs occur in the same individual. Hermaphroditic.

monogamous (mah-nog′ah-mus) Having one mate at a time.

Monogenea (mon″oh-gen′e-ah) The class of Platyhelminthes with members that are called monogenetic flukes. Most are ectoparasites on vertebrates (usually on fishes, occasionally on turtles, frogs, copepods, squids). One life-cycle form in only one host. Bear an opisthaptor. Examples: *Disocotyle, Gyrodactylus, Polystoma.*

Monogononta (mon″o-go-non′tah) A class of rotifers with members that possess one ovary. Mastax not designed for grinding. Produce mictic and amictic eggs. Example: *Notommata.*

monohybrid cross (mon-o-hi′brid)   A mating between two individuals heterozygous for one particular trait. Monohybrid crosses usually result in a phenotypic ratio of 3 dominant : 1 recessive in the offspring.

monophyletic group (mon″o-fi-let′ik)   A group of organisms descended from a single ancestor.

Monoplacophora (mon″o-pla-kof′o-rah)   The class of molluscs whose members have a single, arched shell; a broad, flat foot; and certain serially repeated structures. Example: *Neopilina*.

monosaccharide (mon″o-sak′ah-rīd)   A simple sugar, such as glucose or fructose, that represents the structural unit of a carbohydrate. Monosaccharides are building blocks of more complex sugars (disaccharides) and polysaccharides.

morphogenesis (mor″fo-gen′ĕ-sis)   The evolution and development of form, as the development of the shape of a particular organ or part of the body.

morula (mor′u-lah)   A stage in the embryonic development of some animals that consists of a solid ball of cells (blastomeres).

mosaic evolution (mo-za′ik ev″ah-loo′shun)   A change in a portion of an organism (e.g., a bird wing) while the basic form of the organism is retained.

motor (effector or efferent) neuron or nerve   A neuron or nerve that transmits impulses from the central nervous system to an effector, such as a muscle or gland.

motor unit   A motor neuron and the muscle fibers associated with it.

mucous cell   A glandular cell that secretes mucus.

Müller's larva   A free-swimming, ciliated larva that resembles a modified ctenophore. Characteristic of many marine polyclad turbellarians.

multiple alleles (mul′tĭ-pl ah-lēlz′)   The presence of more than two alleles in a population.

multiple fission (mul′tĭ-pl fish′un)   Asexual reproduction by the splitting of a cell or organism into many cells or organisms. *See also* **schizogony.**

muscle fiber   The contractile unit of a muscle. A muscle cell.

muscle tissue   The type of tissue that allows movement. The three kinds are skeletal, smooth, and cardiac. Tissue made of bundles of long cells called muscle fibers.

musk gland   *See* **scent gland.**

mutation pressure (myoo-ta′shun presh′er)   A measure of the tendency for gene frequencies to change through mutation.

mutualism (myoo′choo-ah-liz-em)   A relationship between two species that benefits both members. The association is necessary to both species.

myelin (mi′ĕ-lin)   Lipoprotein material that forms a sheathlike covering around some nerve fibers.

myofibril (mi″o-fi′bril)   Contractile fibers within muscle cells.

myoglobin (mi″o-glo′bin)   The oxygen-transporting pigment of muscle tissue.

myomere (mi′o-mēr)   The muscle plate or portion of a somite that develops into voluntary muscle.

myosin (mi′ah-sin)   A protein that, together with actin, is responsible for muscular contraction and relaxation.

Myriapoda (mir″e-ă-pod′ah)   The subphylum of arthropods characterized by a body consisting of two tagmata and uniramous appendages.

Myxozoa (mix″o-zo′ah)   The protozoan phylum whose members have spores of multicellular origin. The myxozoans.

N

naiad (na′ad)   The aquatic immature stage of any hemimetabolous insect.

nail   The horny, cutaneous plate of the dorsal surface of the distal end of a finger or toe.

natural selection   A theory, conceived by Charles Darwin and Alfred Wallace, of how some evolutionary changes occur. The idea that some individuals in a population possess variations that make them less able to survive and/or reproduce. These individuals die or fail to reproduce. Their genes are less likely to be passed into the next generation. Therefore, the genetic composition of the population in the next generation changes (i.e., evolution occurs).

nematocyst (nĕ-mat′ah-sist)   A cnidarian cnida usually armed with spines or barbs and containing a venom that is injected into a prey's flesh.

Nematoda (nem-ă-to′dah)   The phylum of aschelminths that contains members commonly called either roundworms or nematodes. Triploblastic, bilateral, vermiform, unsegmented, and pseudocoelomate.

Nematomorpha (nem″ă-to-mor′fah)   The phylum of aschelminths commonly called horsehair worms.

Nemertea (nem-er′te-ah)   The phylum that has members commonly called the proboscis worms. Elongate, flattened worms in marine mud and sand. Triploblastic. Complete digestive tract with anus. Closed circulatory system.

neo-Darwinism (ne″o-dar′wĭ-niz″um)   *See* **modern synthesis.**

nephridiopore (nĕ-frid′e-o-por)   The opening to the outside of a nephridium. An excretory opening in invertebrates.

nephridium (nĕ-frid′e-um)   The excretory organ of the embryo. The embryonic tube from which the kidney develops. Tubular osmoregulatory and excretory organ of many invertebrates. Functions in excretion, osmoregulation, or both.

nephron (nef′ron)   The functional unit of a vertebrate kidney, consisting of a renal corpuscle and a renal tubule.

nerve   A bundle of neurons or nerve cells outside the central nervous system.

nerve net   A diffuse, two-dimensional plexus of bi- or multipolar neurons found in cnidarians. The simplest pattern of invertebrate nervous systems.

nervous tissue   The type of tissue composed of individual cells called neurons and supporting neuroglial cells.

neuroendocrine system (nur′o-en′do-krin)   The combination of the nervous and endocrine systems.

neurofibril node (nur″o-fib′ril nōd)   Regular gaps in a myelin sheath around a nerve fiber. Formerly called **node of Ranvier.**

neurohormone (nur″o-hor′mōn)   A chemical transmitter that nervous tissue produces. Uses the bloodstream or other body fluids for distribution to its target site.

neurolemmocyte (nur″o-lem′o-sīt)   The cell that surrounds a fiber of a peripheral nerve and forms the neurolemmal sheath and myelin. Formerly called **Schwann cell.**

neuromast (nur′o-mast)   The hair-cell mechanoreceptor within pits of the lateralline system. Used to detect water currents and the movements of other animals.

neuromuscular junction (nur″o-mus′ku-lar jungk′shun)   The junction between nerve and muscle. Myoneural junction or neuromuscular cleft.

neuron (nu′ron)   A nerve cell that consists of a cell body and its processes. The basic functional unit of communication in nervous systems.

neuropeptide (nur″o-pep′tīd)   A hormone (neurohormone) that secretory nervous tissue produces.

neurotoxin (nur″o-tok′sin)   A toxin that is poisonous to or destroys nerve tissue.

neurotransmitter (nur″o-trans-mit′er)   Chemical substance that the terminal end of an axon secretes that either stimulates or inhibits a muscle fiber contraction or an impulse in another neuron.

neurulation (noor″yah-la′shun)   External changes along the upper surface of a chordate embryo that form the neural tube.

neutral selection   *See* **genetic drift.**

neutron (nu′tron)   A neutral particle with a mass approximately the same as a proton that exists in atomic nuclei.

neutrophil (nu′tro-fil)   A type of phagocytic white blood cell with a multilobed nucleus and inconspicuous cytoplasmic granules.

nicotinamide adenine dinucleotide (NAD⁺) (nik″ah-tēn′ah-mīd ad′e-nēn di-nu′kle-o-tīd)   A local electron carrier that transfers hydrogen atoms and electrons within metabolic pathways. A free-moving carrier. Not membrane bound in a transport system.

nictitating membrane (nik′tĭ-tāt-ing mem′brān)   The thin, transparent lower eyelid of amphibians and reptiles.

nociceptor (no″se-sep′tor)   A sensory receptor responding to potentially harmful stimuli. Produces a sensation of pain.

node of Ranvier (nōd ov Ran′ve-a)   A constriction of myelinated nerve fibers at regular intervals at which the myelin sheath is absent and the axon is enclosed only by sheath cell processes. Also known as a **neurofibril node.**

nomenclature (no′men-kla-cher)   The study of the naming of organisms in the fashion that reflects their evolutionary relationships.

nonamniote lineage (non-am′ne-ōt lin′e-ij)   The lineage of vertebrate tetrapods leading to modern amphibians.

nondisjunction (non″dis-junk′shun)   The failure of homologous chromosomes to separate during meiosis. Nondisjunction results in gametes with a deficiency of one chromosome or an extra chromosome.

nonpolar covalent bond (non-po′lar co-va′lent)   The bond formed when electrons spend as much time orbiting one nucleus as the other. Thus, the distribution of charges is symmetrical.

nonshivering thermogenesis (non-shiv′er-ing ther-mo-gen′esis)   The hormonal triggering of heat production. A thermogenic process in which enzyme systems for fat metabolism are activated, breaking down and oxidizing conventional fats to produce heat.

norepinephrine (nor″ep-i-nef′rin)   A catecholamine neurotransmitter released from the axon ends of some nerve fibers. Noradrenaline. Also, an adrenal medulla hormone.

notochord (no′tah-kord)   A rodlike, supportive structure that runs along the dorsal midline of all larval chordates and many adult chordates.

nuclear envelope   Double membrane forming the surface boundary of a eukaryotic nucleus. Consists of outer and inner membranes perforated by nuclear pores.

nucleic acid (nu-kle′ik)   A substance composed of bonded nucleotides. DNA and RNA are the two nucleic acids.

nucleolus (nu-kle′o-lus)   A small structure within the nucleus of a cell that transcribes ribosomal RNA and assembles ribosomal subunits.

nucleosome (nu-kle′ah-sōm)   An association of DNA and histone proteins that makes up chromatin. Many nucleosomes are linked in a chromatin strand.

nucleotide (nu′kle-o-tīd)   A component of a nucleic acid molecule consisting of a sugar, a nitrogenous base, and a phosphate group. Nucleotides are the building blocks of nucleic acids.

nucleus (nu′kle-us)   1. Cell nucleus. A spheroid body within a cell, contained in a double membrane, the nuclear envelope, and containing chromosomes and one or more nucleoli. The genetic control center of a eukaryotic cell. 2. The cell bodies of nerves within the central nervous system. 3. An atom's central core, containing protons and neutrons.

Nuda (nu′dah)   The class of ctenophorans whose members lack tentacles and have a flattened body with a highly branched gastrovascular cavity.

numerical taxonomy (nu-mer′ĭ-kal takson′ah-me)   A system of classification that does not attempt to distinguish true and false similarities.

nutrition (nu-trĭ′shun)   The study of the sources, actions, and interactions of nutrients. The study of foods and their use in diet and therapy.

nymph (nimf)   The immature stage of a paurometabolous insect. Resembles the adult but is sexually immature and lacks wings (Arthropoda).

O

ocellus (o-sel′as)   A simple eye or eyespot in many invertebrates. A small cluster of photoreceptors.

odontophore (o-dont′o-for″)   The cartilaginous structure that supports the radula of molluscs.

olfaction (ol-fak′shun)   The act of smelling. The sense of smell.

Oligochaeta (ol″ĭ-go-kēt′ah)   The subclass of annelids whose members are characterized by having few setae and no parapodia. Monecious with direct development. The earthworm (*Lumbricus*) and *Tubifex*.

ommatidia (om″ah-tid′e-ah)   The sensory unit of the arthropod compound eye.

omnivore (om′nĭ-vor)   Subsisting upon both plants and animals. An animal that obtains its nutrients by consuming plants and other animals.

onchosphere (ong′ko-sfēr)   The larva of the tapeworm contained within the external embryonic envelope and armed with six hooks and cilia. Typically referred to as a coracidium when released into the water.

oncomiracidium (on″ko-mir-ă-sid′e-um)   Ciliated larva of a monogenetic trematode.

Onychophora (on-e-kof′o-rah)   A phylum of terrestrial animals with 14 to 43 pairs of unjointed legs, oral papillae, and two large antennae. Onycophorans live in humid, tropical areas of the world. Their ancestor may have been an evolutionary transition between annelids and arthropods. Velvet worms or walking worms.

oogenesis (o″o-jen′e-sis)   The process by which an egg cell forms from an oocyte.

Opalinata (op″ah-li-not′ah)   The protozoan subphylum whose members are cylindrical and covered with cilia. Examples: *Opalina*, *Zelleriella*.

open circulatory system   A circulatory system found in insects and some other invertebrates in which blood is not confined to vessels in part of its circuit. Blood bathes tissues in blood sinuses.

operculum (o-per′ku-lum)   A cover. 1. The cover of a gill chamber of a bony fish (Chordata). 2. The cover of the genital pore of a horseshoe crab (Meristomata, Arthropoda). 3. The cover of the aperture of a snail shell (Gastropoda, Mollusca).

Ophiuroidea (o-fe-u-roi′de-ah)   The class of echinoderms whose members have arms sharply set off from the central disk and tube feet without suction disks. Brittle stars.

opisthaptor (ah″pis-thap′ter)   Posterior attachment organ of a monogenetic trematode.

opisthosoma (ah″pis-tho-so′mah)   The portion of the body of a chelicerate arthropod that contains digestive, reproductive, excretory, and respiratory organs.

oral   Having to do with the mouth. The end of an animal containing the mouth.

oral cavity   The cavity within the mouth.

oral sucker   The sucker on the anterior end of a tapeworm, fluke, or leech.

order   The level of classification between class and family.

organ   A structure consisting of a group of specialized tissues that performs a specialized function.

organelle (or″gah-nel′)   A part of a cell that performs a specific function. A membrane-bound sac or compartment.

organic evolution   The change in an organism over time. A change in the sum of all genes in a population.

organic molecule   A molecule that contains one or more carbon atoms.

organizational effects of hormones   Changes resulting from the presence of hormones at critical time periods such that specific developmental pathways for specific brain regions and developing gonadal tissues are influenced to become either femalelike or malelike.

organ level   Animals whose body functions are carried out by organs that are not organized into organ systems.

organ of Ruffini   Sensory receptor in the skin believed to be a sensor for touch-pressure, position sense of a body part, and movement. Also known as **corpuscle of Ruffini.**

organ system   A set of interconnected or interdependent parts that function together in a common purpose or produce results that cannot be achieved by one of them acting alone.

osmoconformer (oz″mo-con-form′er)   An organism whose body fluids have the same or very similar osmotic pressure as that of its aquatic environment. A marine organism that does not utilize energy in osmoregulation.

osmoregulation (oz″mo-reg″u-la′shun)   The maintenance of osmolarity by an organism or body cell with respect to the surrounding medium.

osmoregulator (oz″mo-reg′u-la-ter)   An organism that regulates its internal osmolarity with respect to the environment.

osmosis (oz-mo′sis)   Net movement of water across a selectively permeable membrane driven by a difference in concentration of solute on either side. The membrane must be permeable to water but not to the solute molecules.

Osteichthyes (os-te-ik′the-es)   The class of fishes whose members are characterized by a bony skeleton, a swim bladder, and an operculum. Bony fishes.

outgroup   In cladistic studies, an outgroup is a group outside of a study group that shares an ancestral characteristic with the study group.

ovarian cycle (o-va′re-an)   The cycle in the ovary during which the oocyte matures and ovulation occurs.

ovary (o′var-e)   The primary reproductive organ of a female. Where eggs (ova) are produced.

oviparous (o-vip′er-us)   Organisms that lay eggs that develop outside the body of the female.

ovipositor (ov-ĭ-poz′it-or)   A modification of the abdominal appendages of some female insects that is used for depositing eggs in or on some substrate (Arthropoda, Hexapoda).

ovoviviparous (o′vo-vi-vip′er-us)   Organisms with eggs that develop within the female reproductive tract and that are nourished by food stored in the egg.

ovulation (ov″vu-la′shun)   The release of an egg (female gamete) from a mature ovarian follicle or ovary.

oxidation (ok″sĭ-da′shun)   The loss of electrons from a compound.

oxidation-reduction reaction   An electron transfer from one atom or molecule to another. Often, hydrogen is transferred along with the electron(s).

oxygen debt   The amount of oxygen that must be supplied following physical exercise to convert accumulated lactic acid (lactate) to glucose.

oxyhemoglobin (ok″sĭ-he′mo-glo″bin)   Compound formed when oxygen combines with hemoglobin.

P

Pacinian corpuscle (Pah-sin′e-an kor′pus-l)   A sensory receptor in skin, muscles, body joints, body organs, and tendons that is involved with the vibratory sense and firm pressure on the skin. Also called a lamellated corpuscle.

paedomorphosis (pe′dah-mor′fo-sis)   The development of sexual maturity in the larval body form.

pain receptor   A modified nerve ending that, when stimulated, gives rise to the sense of pain.

paleontology (pa″le-on-tol′o-je)   The study of early life-forms on earth.

pancreas (pan′kre-as)   Glandular organ in the abdominal cavity and behind the stomach that secretes hormones and digestive enzymes.

pancreatic islet (pan″kre-at′ic i′let)   An island of special tissue in the pancreas. Secretes insulin or glucagon.

papulae (pap′yoo-lah)   See **dermal branchiae.**

parabronchi (par″ah-brong′ke)   The tiny air tubes within the lung of a bird that lead to air capillaries where gases are exchanged between blood and air.

parapatric speciation (par″ah-pat′rik spe″she-a′shun)   Speciation that occurs in small, local populations, called **demes.**

paraphyletic group (par″ah-fi-let′ik)   A group that includes some, but not all, members of

a lineage. Paraphyletic groups result from insufficient knowledge of the group.

parapodia (par″ah-pod′e-ah)   Paired lateral extensions on each segment of polychaetes (Annelida). May be used in swimming, crawling, and burrowing.

parasitism (par′ah-si″tizm)   A relationship between two species in which one member (the parasite) lives at the expense of the second (the host).

parasympathetic nervous system (par″ah-sim″pah-thet′ik ner′vus sis′tem)   Portion of the autonomic nervous system that arises from the brain and sacral region of the spinal cord.

parathyroid gland (par″ah-thi′roid)   One of the small glands within a lobe of the thyroid gland. Produces the hormone parathormone.

parenchyma (pă-ren′kă-mă)   A spongy mass of mesenchyme cells filling spaces around viscera, muscles, or epithelia in acoelomate animals. Depending on the species, parenchyma may provide skeletal support, nutrient storage, motility, reserves of regenerative cells, transport of materials, structural interactions with other tissues, modifiable tissue for morphogenesis, oxygen storage, and perhaps other functions that have yet to be determined.

parietal cell (pah-ri′ĕ-tal)   Cell of a gastric gland that secretes hydrochloric acid and intrinsic factor.

parietal eye (pah-ri′ĕ-tal)   See **median (parietal) eye.**

pars intermedia   See **intermediate lobe.**

parthenogenesis (par″thĕ-no-jen′ĕ-sis)   A modified form of sexual reproduction by the development of a gamete without fertilization, as occurs in some bees, wasps, certain lizards, and a few other animals.

parturition (par″tu-rish′un)   The process of childbirth.

Pauropoda (por″o-pod′ah)   A class of arthropods whose bodies are small, soft, and divided into 11 segments, and who have 9 pairs of legs.

pebrine (pă-brēn′)   An infectious disease of silkworms caused by the protozoan *Nosema bombicis.*

pedal locomotion (ped′el)   The type of locomotion exhibited by flatworms, some cnidarians, and the gastropod molluscs. This locomotion involves waves of activity in the muscular system, which are applied to the substratum.

pedicellariae (ped″e-sel-ar′i-a)   Pincerlike structures found on the body wall of many echinoderms. They are used in cleaning and defense.

pedipalps (ped′ĕ-palps)   The second pair of appendages of chelicerate arthropods. These appendages are sensory in function.

pellicle (pel′ik-el)   A thin, frequently non-cellular covering of an animal (e.g., the protective and supportive pellicle of protists occurs just below the plasma membrane). May be composed of a cell membrane, cytoskeleton, and other organelles.

pentaradial symmetry (pen″tah-ra′de-al sim′ĭ-tre)   A form of radial symmetry in the echinoderms in which body parts are arranged in fives around an oral-aboral axis.

Pentastomida (pen-tah-stom′id-ah)   A phylum of worms that are all endoparasites in the lungs or nasal passageways of carnivorous vertebrates. Tongue worms.

peptide bond   The covalent bond that joins individual amino acids. Formed by dehydration synthesis.

peripheral nervous system (pĕ-rif′er-al)   The nerves and ganglia of the nervous system that lie outside of the brain and spinal cord.

peristalsis (per″ĭ-stal′sis)   Rhythmic waves of muscular contraction in the walls of various tubular organs that move material through the organs.

peristomium (per″i-stōm′e-um)   The segment of an annelid body that surrounds the mouth.

perspiration (per″spĭ-ra′shun)   Sweating. The function of sweat secretion.

phagocytosis (fag″o-si-to′sis)   Process (endocytosis) by which a cell engulfs bacteria, foreign proteins, macromolecules, and other cells and digests their substances. Cellular eating.

phagolysosome (fag″o-li′so-sōm)   The organelle that forms when a lysosome combines with a vesicle.

pharyngeal lamellae   Thin plates of tissue on gill filaments that contain the capillary beds across which gases are exchanged.

pharyngeal slit (far-in′je-al)   One of several openings in the pharyngeal region of chordates. Pharyngeal slits allow water to pass from the pharynx to the outside of the body. In the process, water passes over gills, or suspended food is removed in a filter-feeding mechanism.

pharynx (far′inks)   The passageway posterior to the mouth that is common to respiratory and digestive systems.

phasmid (faz′mid)   Sensory pit on each side near the end of the tail of nematodes of the class Secernentea (formerly Phasmidea).

phenotype (fe′no-tīp)   The expression that results from an interaction of one or more gene pairs and the environment.

phenotypic ratio (fe″no-tip′ik)   The relative numbers of progeny in each phenotypic category that a genetic cross produces.

pheromone (fer′o-mōn)   A chemical that is synthesized and secreted to the outside of the body by one organism and that is perceived (as by smell) by a second organism of the same species, releasing a specific behavior in the recipient.

phonoreceptor (fo″no-re-sep′tor)   A specialized nerve ending that responds to sound.

Phoronida (fo-ron′ĭ-dah)   A phylum of marine animals whose members live in permanent, chitinous tubes in muddy, sandy, or solid substrates. They feed via an anterior lophophore with two parallel rings of long tentacles.

photoreceptor (fo″to-re-sep′tor)   A nerve ending that is sensitive to light energy.

pH scale   The numerical scale that measures acidity and alkalinity, ranging from 0 (most acidic) to 14 (most basic). pH stands for potential hydrogen and refers to the concentration of hydrogen ions (H^+).

phyletic gradualism (fi-let′ik graj′oo-el-izm)   The idea that evolutionary change occurs at a slow, constant pace over millions of years.

phylogenetic systematics (fi-lo-je-net′ik sistem-at′iks)   The study of the phylogenetic relationships among organisms in which true and false similarities are differentiated. Cladistics.

phylogeny (fi′-loj′en-e)   The evolutionary relationships among species.

phylum (fi′lum)   The level of classification between kingdom and class. Members are considered a monophyletic assemblage derived from a single ancestor.

physiology   The branch of science that deals with the function of living organisms.

Phytomastigophorea (fi″to-mas-ti-go-for′ah)   The protozoan class whose members usually have chloroplasts. Mainly autotrophic, some heterotrophic. Examples: *Euglena*, *Volvox*, *Chlamydomonas*.

pia mater (pi′ah mā′ter)   The innermost meninx that is in direct contact with the brain and spinal cord.

pilidium larva (pi-lid-e-um lar′va)   Free-swimming, hat-shaped larva of nemertean worms characterized by an apical tuft of cilia.

pinacocyte (pin″ah-ko′sīt)   Thin, flat cell covering the outer surface, and some of the inner surface, of poriferans.

pineal gland (pin′e-al)   A small gland in the midbrain of vertebrates that converts a

signal from the nervous system into an endocrine signal. Also called the pineal body.

pinfeather   *See* **filoplume feather.**

pinocytosis (pin″o-si-to′sis)   Cell drinking. The engulfment into the cell of liquid and of dissolved solutes by way of small membranous vesicles. A type of endocytosis.

pioneer community   The first community to become established in an area.

pit organ   Receptor of infrared radiation (heat) on the heads of some snakes (pit vipers).

pituitary gland (pĭ-tu′ĭ-tar″e)   Endocrine gland that attaches to the base of the brain and consists of anterior and posterior lobes. Hypophysis.

placenta (plah-sen′tah)   Structure by which an unborn child or animal attaches to its mother's uterine wall and through which it is nourished.

placid (plas′id)   Plates on the Kinorhyncha.

Placozoa (plak″o-zo′ah)   A phylum of small, flattened, marine animals that feed by forming a temporary digestive cavity. Example: *Tricoplax adherans.*

planktonic (plangk-ton′ik)   Describes small organisms that passively float or drift in a body of water.

Plantae (plant′a)   One of the five kingdoms of life. Characterized by being eukaryotic and multicellular, and having rigid cell walls and chloroplasts.

planula (plan′u-lah)   A ciliated, free-swimming larva of most cnidarians. The planula develops following sexual reproduction and metamorphoses into a polyp.

plasma (plaz′mah)   The fluid or liquid portion of circulating blood within which formed elements and various solutes are suspended.

plasma cell (plaz′mah)   A mature, differentiated B lymphocyte chiefly occupied with antibody synthesis and secretion. A plasma cell lives for only five to seven days.

plasma membrane (plaz′mah mem′brān)   Outermost membrane of a cell. Its surface has molecular regions that detect changes in external conditions and act as a selective barrier to ions and molecules passing between the cell and its environment. Consists of a phospholipid bilayer in which are embedded molecules of protein and cholesterol. The external covering of a protozoan.

plastron (plas′tron)   The ventral portion of the shell of a turtle. Formed from bones of the pectoral girdle and dermal bone.

platelet (plāt′let)   Cytoplasmic fragment formed in the bone marrow that functions in blood coagulation. Also called **thrombocyte.**

plate tectonics (tek-ton′iks) The study of the movement of the earth's crustal plates. These movements are called continental drift.

Platyhelminthes (plat″e-hel-min′thēz) The phylum of flatworms. Bilateral acoelomates.

plerocercoid larva (ple″ro-ser′koid) Metacestode that develops from a procercoid larva. It usually shows little differentiation.

pneumatic sacs (nu-mat′ik) Gas-filled sacs that arise from the esophagus, or another part of the digestive tract, of fishes. Pneumatic sacs are used in buoyancy regulation (swim bladders) or gas exchange (lungs).

pneumostome (nu′mo-stōm) The outside opening of the gas exchange structure (lung) in land snails and slugs (Pulmonata).

Pogonophora (po″go-nof′e-rah) A phylum of protostomate, marine animals that are distributed throughout the world's oceans. Live in secreted, chitinous tubes in cold water at depths exceeding 100 m. Lack a mouth and digestive tract. Nutrients absorbed across the body wall and from endosymbiotic bacteria that they harbor. Beard worms.

point mutations A change in the structure of a gene that usually arises from the addition, deletion, or substitution of one or more nucleotides.

polar covalent bond (po′-lar co-va′lent) The bond formed by asymmetrically moving electrons.

polyandrous (pol″e-an′drus) Having more than one male mate. Polyandry is advantageous when food is plentiful, but because of predation or other factors, the chances of successfully rearing young are low.

Polychaeta (pol″e-kēt′ah) The class of annelids whose members are mostly marine and are characterized by a head with eyes and tentacles and a body with parapodia. Parapodia bear numerous setae. Examples: *Nereis, Arenicola.*

polygynous (pah′-lij′ah-nus) Having more than one female mate. Polygyny tends to occur in species whose young are relatively independent at birth or hatching.

polyp (pol′ip) The attached, usually asexual, stage of a cnidarian.

polyphyletic (See **polyphyletic group**)

polyphyletic group (pol-e-fi-let′ik) An assemblage of organisms that includes multiple evolutionary lineages. Polyphyletic assemblages usually reflect insufficient knowledge regarding the phylogeny of a group of organisms.

Polyplacophora (pol″e-pla-kof′o-rah) The class of molluscs whose members are elongate and dorsoventrally flattened, and have a shell consisting of eight dorsal plates.

polyploidy (pol′e-ploi-de) Having more than two sets of chromosomes ($3N$, $4N$, and so forth).

polysaccharide (pol″e-sak′ah-rīd) A carbohydrate composed of many monosaccharide molecules joined by dehydration synthesis reactions.

pons (ponz) A portion of the brain stem above the medulla oblongata and below the midbrain.

population A group of individuals of the same species that occupy a given area at the same time and share a unique set of genes.

population genetics The study of genetic events in gene pools.

Porifera (po-rif′er-ah) The animal phylum whose members are sessile and either asymmetrical or radially symmetrical. Body is organized around a system of water canals and chambers. Cells are not organized into tissues or organs. Sponges.

porocytes (por′o-sītz) Tubular cells in sponge body wall that create a water channel to an interior chamber.

postanal tail (post-an′al) A tail that extends posterior to the anus. One of the four unique characteristics of chordates.

postmating isolation (post-māt′ing i-sah-la′shun) Isolation that occurs when fertilization and development are prevented, even though mating has occurred.

potential energy The energy matter has by virtue of its position. Stored energy.

preadaptation (pre-ă-dap-ta′shun) Occurs when a structure or a process present in members of a species proves useful in promoting reproductive success when an individual encounters new environmental situations.

precocial (pre-ko′shel) Having developed to a high degree of independence at the time of hatching or birth.

predation (pre-da′shun) The derivation of an organism of elements essential for its existence from organisms of other species that it consumes and destroys. The ingestion of prey by a predator for energy and nutrients.

preformation (pre-for-ma′shun) The erroneous idea that gametes contain miniaturized versions of all of the elements present in an adult.

premating isolation (pre-māt′ing i-sah-la′shun) Isolation that occurs when behaviors or other factors prevent animals from mating.

preoptic nuclei Neurons in the hypothalamus that produce various neuropeptides (hormones).

Priapulida (pri″ă-pyu′-li-da) A phylum of aschelminths commonly called priapulids.

primary consumer A plant-eating animal (an herbivore) that obtains organic molecules by eating producers (autotrophs) or their products.

primary germ layers Blocks or layers of embryonic cells that give rise to tissues and organs of animals. *See also* **ectoderm, mesoderm,** and **endoderm.**

primary producer An autotrophic organism that is able to build its own complex organic molecules from simple inorganic substances (CO_2, H_2O) in the environment.

Primates The order of mammals whose members include humans, monkeys, apes, lemurs, and tarsiers.

primitive streak A medial thickening along the dorsal margin of an amniote embryo that forms during the migration of endodermal and mesodermal cells into the interior of the embryo.

principle of independent assortment One of Mendel's observations on the behavior of hereditary units (genes) during gamete formation. A modern interpretation of this principle is that genes carried in one chromosome are distributed to gametes without regard to the distribution of genes in nonhomologous chromosomes.

principle of segregation One of Mendel's observations on the behavior of hereditary units (genes) during gamete formation. A modern interpretation of this principle is that genes exist in pairs, and during gamete formation, members of a pair of genes are distributed into separate gametes.

procercoid larva (pro-ser′koid lar′vă) Cestode developing from a coracidium in some orders. It usually has a posterior cercomer. Developmental stage between oncosphere and plerocercoid.

proglottid (pro-glot′id) One set of reproductive organs in a tapeworm strobila. Usually corresponds to a segment. One of the linearly arranged segmentlike sections that make up the strobila of a tapeworm.

prokaryote (pro-kar′e-ot) A cell that lacks nuclei and other organelles. Eubacteria and Archaea. (Same as prokaryotic)

prokaryotic See **prokaryote.**

pronephros (pro-nef′res) Most anterior of three pairs of embryonic renal organs of vertebrates. Functional only in larval amphibians and fishes, and in adult hagfishes. Vestigial in mammalian embryos.

prophase (pro'fāz)   The stage of mitosis during which the chromosomes become visible under a light microscope. The first stage of mitosis during which the chromosomes are condensed but not yet attached to a mitotic spindle.

proprioceptor (pro″pre-o-sep'tor)   A sensory nerve terminal that gives information concerning movements and positions of the body. They occur chiefly in the muscles, tendons, and the labyrinth of the mammalian ear and in the joints of arthropods.

prosoma (pro-so'mah)   A sensory, feeding, and locomotor tagma of chelicerate arthropods.

prostate gland (pros'tāt)   Gland around the male urethra below the urinary bladder that adds its secretions to seminal fluid during ejaculation.

prostomium (pro-stōm'e-um)   A lobe lying in front of the mouth, as found in the Annelida.

protandrous (pro-tan'drus)   *See* **protandry.**

protandry (pro-tan'dre)   The condition in a monoecious (hermaphroditic) organism in which male gonads mature before female gametes. Prevents self-fertilization.

protein (pro'tēn)   Nitrogen-containing organic compound composed of amino acid molecules joined by peptide bonds. The unique sequence of amino acids in the protein is the basis for the protein's three-dimensional structure and chemical behavior.

prothorax (pro-thor'aks)   The first of the three thoracic segments of an insect. Usually contains the first pair of walking appendages.

Protista (pro-tist'ah)   The kingdom whose members are eukaryotic and unicellular or colonial.

protogynous (pro-toj'ĭ-nus)   Hermaphroditism in which the female gonads mature before the male gonads.

proton (pro'ton)   A positively charged particle in the atomic nucleus. The mass of a proton is about 1,840 times that of an electron.

protonephridium (pro-to-nĕ-frid'e-um)   Primitive osmoregulatory or excretory organ composed of a tubule terminating internally with a flame cell or solenocyte. The unit of a flame-cell system. Protonephridia are specialized for ultrafiltration.

protoplasmic (pro″to-plas'mic)   Referring to living matter. The organized living substance; cytoplasm and nucleoplasm of the cell.

protoplasmic organization   Organisms whose functions are carried out within the boundaries of a single cell.

protopodite (pro″to-po'dīt)   The basal segment of a biramous appendage of a crustacean.

protostome (pro'to-stōm″)   Animal in which the embryonic blastopore becomes the mouth. Often possesses a trochophore larva, schizocoelous coelom formation, and spiral embryonic cleavage.

protostyle (pro'to-stīl″)   A rotating mucoid mass into which food is incorporated in the gut of a gastropod (phylum Mollusca).

protozoa (pro″to-zo'ah)   A subkingdom (formerly a phylum) comprising the simplest organisms called protista. Divided into seven phyla.

protozoologist (pro-to-zo-ol'-o-jist)   A person who studies protozoa.

proximal (proks'em-al)   Toward the point of attachment of a structure on an animal (e.g., the hip is proximal to the knee).

pseudocoel   *See* **pseudocoelom.**

pseudocoelom (soo″do-se'lom)   A body cavity between the mesoderm and endoderm. A persistent blastocoel that is not lined with peritoneum. Also called **pseudocoel.**

pseudocoelomate (soo′do-sēl'o-māt)   Animals having a pseudocoelom, as the aschelminths.

pseudopodia (soo-dah-po'de-ah)   Temporary cytoplasmic extensions of amoebas or amoeboid cells that are used in feeding and locomotion. Sing., pseudopodium.

Pterobranchia (ter″o-brang'ke-ah)   The class of hemichordates whose members lack gill slits and have two or more arms. Colonial, living in externally secreted encasements.

pulmonary circuit (pul'mo-ner″e)   The system of blood vessels from the right ventricle of the heart to the lungs, transporting deoxygenated blood and returning oxygenated blood from the lungs to the left atrium of the heart.

pulmonate lung (pul'mon-āt)   The gas exchange structure in the Pulmonata—the land snails and slugs.

punctuated equilibrium model (pungk'choo-āt″ed e″kwĕ-lib're-ahm)   The idea that evolutionary change can occur rapidly over thousands of years and that these periods of rapid change are interrupted by periods of constancy (stasis).

Punnett square   A tool geneticists use to predict the results of a genetic cross. Different kinds of gametes that each parent produces are placed on each axis of the square. Combining gametes in the interior of the square gives the results of random fertilization.

pupa (pu'pah)   A nonfeeding immature stage in the life cycle of holometabolous insects. It is a time of radical cellular changes that result in a change from the larval to the adult body form.

puparium (pu-par'e-um)   A pupal case formed from the last larval exoskeleton. *See also* **pupa.**

purine (pyoor'ēn)   A nitrogen-containing organic compound that contributes to the structure of a DNA or RNA nucleotide. Uric acid is also derived from purines.

Pycnogonida (pik″no-gon'ĭ-dah)   The class of chelicerate arthropods whose members have a reduced abdomen and four to six pairs of walking legs. Without special respiratory or excretory structures. Sea spiders.

pygostyle (pig'o-stīl)   The fused posterior caudal vertebrae of a bird. Helps support tail feathers that are important in steering.

pyrenoid (pi-re'noid)   Part of the chloroplast that synthesizes and stores polysaccharides.

pyrimidine (pi-rim'i-dēn)   A nitrogen-containing organic compound that is a component of the nucleotides making up DNA and RNA.

R

radial symmetry   A form of symmetry in which any plane passing through the oral-aboral axis divides an organism into mirror images.

radiation (ra″de-a'shun)   A form of energy that includes visible light, ultraviolet light, and X rays. Means by which body heat is lost in the form of infrared rays.

radioisotope (ra″de-o-i'so-tōp)   An isotope that is radioactive (i.e., has an unstable nucleus), which gives it the property of decay by one or more of several processes. Also called a radioactive isotope.

radula (raj'oo-lah)   The rasping, tonguelike structure of most molluscs that is used for scraping food. Composed of minute chitinous teeth that move over a cartilaginous odontophore.

ram ventilation   The movement of water across gills as a fish swims through the water with its mouth open.

range of optimum   The range of values for a condition in the environment that is best able to support survival and reproduction of an organism.

receptor-mediated endocytosis (re-cep'tor-me'de-āt-ed en″do-si-to'sis)   The type of endocytosis that involves a specific receptor on the plasma membrane that recognizes an extracellular molecule and binds with it.

recessive A gene whose expression is masked when it is present in combination with a dominant allele. For a recessive gene to be expressed, both members of the gene pair must be the recessive form of the gene.

rectal gland (rek′tel) The excretory organ of elasmobranchs and the coelacanth near the rectum. It excretes a hyperosmotic salt (NaCl) solution.

red blood cell (erythrocyte) The type of blood cell that contains hemoglobin and no nucleus. During their formation in mammals, erythrocytes lose their nuclei. Those of other vertebrates retain the nuclei.

redia (re′de-ah) A larval, digenetic trematode produced by asexual reproduction within a miracidium, sporocyst, or mother redia.

redox reaction *See* **oxidation-reduction reaction.**

reduction The gain of electrons by a compound.

refractory period (re-frak′ter-e) Time period following stimulation during which a neuron or muscle fiber cannot respond to a stimulus.

releaser gland A gland in turbellarians that secretes a chemical that dissolves the organism's attachment to a substrate.

Remipedia (re-mi-pe′de-ah) A class of crustaceans whose members possess about 30 body segments and uniform, biramous appendages. This class contains a single species of cave-dwelling crustaceans from the Bahamas.

renette (re′net) An excretory structure in some worms.

repolarized (re-po′ler-īzd) The reestablishment of polarity, especially the return of plasma membrane potential to resting potential after depolarization.

reproductive isolation Occurs when individuals are prevented from mating, even though they may occupy overlapping ranges. *See also* **premating** and **postmating isolation.**

Reptilia (rep-til′e-ah) The class of vertebrates whose members have dry skin with epidermal scales and amniotic eggs that develop in terrestrial environments. Snakes, lizards, and alligators.

resilin (rez′ĭ-lin) The elastic protein in some arthropod joints that stores energy and functions in jumping.

respiratory pigment Organic compounds that have either metallic copper or iron with which oxygen combines.

respiratory tree A pair of tubules attached to the rectum of a sea cucumber that branch through the body cavity and function in gas exchange.

resting membrane potential The potential difference that results from the separation of charges along the plasma membrane of a neuron or other excitable cell. The steady voltage difference across the plasma membrane.

rete mirabile (re′te ma-rab′a-le) A network of small blood vessels arranged so that the incoming blood runs countercurrent to the outgoing blood and, thus, makes possible efficient exchange of heat or gases between the two bloodstreams. An extensive countercurrent arrangement of arterial and venous capillaries.

reticulopodium (re-tik″u-lo-po′de-um) A pseudopodium that forms a threadlike branched mesh and contains axial microtubules. Pl., reticulopodia.

rhabdite (rab′dīt) A rodlike structure in the cells of the epidermis or underlying parenchyma in certain turbellarians that is discharged in mucous secretions, possibly in response to attempted predation or desiccation.

rhodopsin (ro-dop′sin) Light-sensitive substance in the rods of the retina. Visual purple.

rhopalium (ro-pal′e-um) A sensory structure at the margin of the scyphozoan medusa. It consists of a statocyst and a photoreceptor (phylum Cnidaria).

rhynchocoel (ring′ko-sēl) In nemerteans, the fluid-filled coelomic cavity that contains the inverted proboscis.

ribonucleic acid (RNA) (ri″bo-nu-kle′ik) A single-stranded polymer of ribonucleotides. RNA forms from DNA in the nucleus and carries a code for proteins to ribosomes, where proteins are synthesized. Contains the nitrogenous bases adenine (A), cytosine (C), guanine (G), and uracil (U).

ribosomal RNA (rRNA) (ri″bo-sōm′al) A form of ribonucleic acid that makes up a portion of ribosomes.

ribosome (ri′-bo-sōm) Cytoplasmic organelle that consists of protein and RNA, and functions in protein synthesis.

ribozyme (ri′bo-zīm) An RNA enzyme.

rod cell A type of light receptor that is sensitive to dim light.

Rotifera (ro-tif′er-ah) The phylum of aschelminths that has members with a ciliated corona surrounding a mouth. Muscular pharynx (mastax) present with jawlike features. Nonchitinous cuticle. Parthenogenesis common. Both freshwater and marine species.

 s

saliva (sah-li′vah) The enzyme-containing secretion of the salivary glands.

saltatory conduction (sal′tah-tor-e kon-duk′shun) A type of nerve impulse conduction in which the impulse seems to jump from one neurofibril node to the next.

salt gland An orbital gland of many reptiles and birds that secretes a hyperosmotic NaCl or KCl solution. An important osmoregulatory organ, especially for marine species.

Sarcodina (sar″ko-din′ah) The protozoan subphylum whose members have pseudopodia for movement and food gathering. Naked or with shell or test. Mostly free living.

sarcolemma (sar″ko-lem′ah) The plasma membrane of a muscle fiber.

Sarcomastigophora (sar″ko-mas-ti-gof′o-rah) The protozoan phylum whose members possess flagella, pseudopodia, or both for locomotion and feeding. Single type of nucleus.

sarcomere (sar′ko-mer) The contractile unit of a myofibril. The repeating units, delimited by the Z bands along the length of the myofibril.

scale A thin, compacted, flaky fragment. One of the thin, flat, horny plates forming the covering of certain animals (snakes, fishes, lizards).

scalid (sca′lid) A set of complex spines found on the kinorhynchs, loriciferans, priapulans, and larval nematomorphs, with sensory, locomotor, food capture, or penetrant function.

scanning electron microscope (SEM) The type of microscope in which an electron beam, instead of light, forms a three-dimensional image for viewing, allowing much greater magnification and resolution.

scanning tunneling microscope (STM) The type of microscope that uses a needle probe and electrons to determine the surface features of specimens.

Scaphopoda (ska-fop′o-dah) A class of molluscs whose members have a tubular shell that is open at both ends. Possess tentacles but no head. Example: *Dentalium.*

scent gland A gland around the feet, face, or anus of many mammals. Secretes pheromones, which may be involved with defense, species and sex recognition, and territorial behavior. Musk gland.

schizogony (skiz-og′on-e) A form of fission (asexual reproduction) involving multiple nuclear divisions and the formation of many individuals from the parental organism. Occurs in the class Sporozoea. *See also* **multiple fission** and **merogony.**

Schwann cell *See* **neurolemmocyte.**

scolex (sko′leks)   The attachment or holdfast organ of a tapeworm, generally considered the anterior end. It is used to adhere to the host.

scrotum (skro′tum)   A pouch of skin that encloses the testes.

scyphistoma (si-fis′to-mah)   The polyp stage of a scyphozoan (phylum Cnidaria). Develops from a planula and produces ephyrae by budding.

Scyphozoa (si″fo-zo′ah)   A class of cnidarians whose members have prominent medusae. Gametes are gastrodermal in origin and are released to the gastrovascular cavity. Nematocysts are present in the gastrodermis. Polyps are small. Example: *Aurelia.*

sebaceous (oil) gland (se-ba′shus)   Gland of the skin that secretes sebum. Oil gland.

sebum (se′bum)   Oily secretion from the sebaceous gland.

Secernentea (ses-er-nen′te-ah)   The class of nematodes formerly called Phasmidea. Examples: *Ascaris, Enterobius, Necator, Wuchereria.*

secondary consumer   An animal that preys on and eats a primary consumer.

secondary palate (pal′et)   A plate of bone that separates the nasal and oral cavities of mammals and some reptiles.

second law of thermodynamics (ther″mo-di-nam′iks)   Physical and chemical processes proceed in such a way that the entropy of the universe (the system and its surroundings) increases to the maximum possible. The spontaneous direction of energy flow is from organized (high-quality) to less organized (low-quality) forms.

segmentation (seg″men-ta′shun)   1. In many animal species, a series of body units that may be externally similar to, or quite different from, one another. 2. The oscillating back-and-forth movement in the small intestine that mixes food with digestive secretions and increases the efficiency of absorption.

Seisonidea (sy″son-id′e-ah)   A class of rotifers with members that are commensals of crustaceans. Large and elongate body with rounded corona. Example: *Seison.*

selection pressure   The tendency for natural selection to occur. Natural selection occurs whenever some genotypes are more fit than other genotypes.

selective permeability (si-lek′tiv per″me-ah-bil′i-te)   The ability of the plasma membrane to let some substances in and keep others out.

semen (se′men)   The thick, whitish secretion of the reproductive organs in the male; composed of sperm and secretions from the prostate, seminal vesicles, and various other glands and ducts. Fluid containing sperm.

seminal receptacle (sem′i-nal ri-sep′tah-kl)   A structure in the female reproductive system that stores sperm received during copulation (e.g., in many insects and annelids).

seminal vesicle (sem′i-nal ves′i-kl)   1. One of the paired accessory glands of the reproductive tract of male mammals. It secretes the fluid medium for sperm ejaculation (phylum Chordata). 2. A structure associated with the male reproductive tract that stores sperm prior to its release (e.g., in earthworms—phylum Annelida).

seminiferous tubule (se″mi-nif′er-us tu′būl)   The male duct that produces sperm. Highly convoluted tube in testis.

sensilla (sen-sil′ah)   Modifications of the arthropod exoskeleton that, along with nerve cells, form sensory receptors.

sensory (receptor or afferent) neuron or **nerve**   A neuron or nerve that conducts an impulse from a receptor organ to the central nervous system.

sequential hermaphroditism (si-kwen′shal her-maf′ro-di-tizm)   The type of hermaphroditism that occurs when an animal is one sex during one phase of its life cycle and an opposite sex during another phase.

seral stage (ser′al)   A successional stage in an ecosystem.

sere (ser)   An entire successional sequence in an ecosystem (e.g., the sequence of stages in the succession of a lake to a climax forest).

serially homologous (ser′e-al-e ho-mol′o-ges)   Metameric structures that have evolved from a common form. The biramous appendages of crustaceans are serially homologous.

serum (se′rum)   The fluid portion of coagulated blood. The protein component of plasma.

setae (se′tah)   1. Hairlike modifications of an arthropod's exoskeleton that may be set into a membranous socket. Displacement of a seta initiates a nerve impulse in an associated cell. 2. Hairlike bristles found on parapodia of polychaete annelids and on the body wall of oligochaete annelids. Also **chaetae.**

sex chromosome   A chromosome that carries genes determining the genetic sex of an individual.

sexual reproduction   The generation of a new cell or organism by the fusion of two haploid cells so that genes are inherited from each parent.

shell   The calcium carbonate outer layer of cnidarians, molluscs, and other animals. Produced by mucous glands.

shivering thermogenesis (ther″mo-jen′e-sis)   The generation of heat by shivering, especially within the animal body.

Simphyla (sim-fi′lah)   A class of arthropods whose members are characterized by having long antennae, 10 to 12 pairs of legs, and centipede-like bodies. Occupy soil and leaf mold.

simple diffusion (di-fu′zhun)   The process of molecules spreading out randomly from where they are more concentrated to where they are less concentrated until they are evenly distributed.

siphon (si′fon)   A tubular structure through which fluid flows. Siphons of some molluscs allow water to enter and leave the mantle cavity.

Sipuncula (si-pun′kyu-lah)   A phylum of protostomate worms whose members burrow in soft marine substrates throughout the world's oceans. Range in length from 2 mm to 75 cm. Peanut worms.

sister chromatid (kro′mah-tid)   One of the two identical parts of a duplicated chromosome in a eukaryotic cell. Sister chromatids are exact copies of a long, coiled DNA molecule with associated proteins. Sister chromatids join at the centromere of a duplicated chromosome.

skeletal muscle   Type of muscle tissue in muscles attached to skeletal parts.

skin   The outer integument or covering of an animal body, consisting of the dermis and the epidermis and resting on the subcutaneous tissues.

small intestine   The part of the digestive system consisting of the duodenum, jejunum, and ileum.

smooth muscle   Type of muscle tissue in the walls of hollow organs. Visceral muscle.

society   A stable group of individuals of the same species that maintains a cooperative social relationship.

sociobiology (so″se-o-bi-ol′o-je)   The study of the evolution of social behavior.

sodium-potassium ATPase pump   The active transport mechanism that concentrates sodium ions on the outside of a plasma membrane and potassium ions on the inside of the membrane.

somatic cell (so-mat′ik)   Ordinary body cell. Pertaining to or characteristic of a body cell. Any cell other than a germ cell or germ-cell precursor.

somites (so′mīts)   Segmental thickenings along the side of a vertebrate embryo that result from the development of mesoderm.

sonar (so′nar) or **biosonar**   A system that uses sound at sonic or ultrasonic frequencies to detect and locate objects.

speciation (spe″she-a′shun)   The process by which two or more species form from a single ancestral stock.

species   A group of populations in which genes are actually, or potentially, exchanged through multiple generations. Numerous problems with this definition make it difficult to apply in all circumstances.

species diversity   *See* **community diversity.**

spermatogenesis (sper″mah-to-jen′ĕ-sis)   The production of sperm cells.

spermatophores (sper-mat′ah-fors)   Encapsulated sperm that a male can deposit on a substrate for a female to pick up or that a male can transfer directly to a female.

sphincter (sfingk′ter)   A ringlike band of muscle fibers that constricts a passage or closes a natural orifice. Acts as a valve.

spicules (spik′ūlz)   Skeletal elements that some mesenchyme cells of a sponge body wall secrete. May be made of calcium carbonate or silica.

spiracle (spi′rah-kel)   An opening for ventilation. The opening(s) of the tracheal system of an arthropod or an opening posterior to the eye of a shark, skate, or ray.

spongin (spun′jin)   A fibrous protein that makes up the supportive framework of some sponges.

sporocyst (spor′o-sist)   1. Stage of development of a sporozoan protozoan, usually with an enclosing membrane, the oocyst. 2. An asexual stage of development in some digenean trematodes that arises from a miracidium and gives rise to rediae.

sporogony (spor-og′ah-ne)   Multiple fission that produces sporozoites after zygote formation. Occurs in the class Sporozoea.

stabilizing selection   Natural selection that results in the decline of both extremes in a phenotypic range. Results in a narrowing of the phenotypic range.

statocyst (stat′o-sist)   An organ of equilibrium and balance in many invertebrates. Statocysts usually consist of a fluid-filled cavity containing sensory hairs and a mineral mass called a statolith. The statolith stimulates the sensory hairs, which helps orient the animal to the pull of gravity.

statolith   *See* **statocyst.**

steroid (ster′oid)   An organic substance (lipid) whose molecules include four complex rings of carbon and hydrogen atoms. Examples are estrogen, cholesterol, and testosterone.

stigma (stig′mah)   1. The mass of bright red photoreceptor granules found in certain flagellated protozoa (*Euglena*) that shields the photoreceptor. 2. The spiracle of certain terrestrial arthropods.

stimulus (stim′u-lus)   Any form of energy an animal can detect with its receptors.

stimulus filter (stim′u-lus)   The ability of the nervous system to block incoming stimuli that are unimportant for the animal.

stomach   The expansion of the alimentary canal between the esophagus and duodenum.

strobila (stro-bi′lah)   The chain of proglottids constituting the bulk of the body of adult tapeworms.

substrate (sub′strāt)   The substance or molecule upon which an enzyme acts. A reactant or precursor molecule for a metabolic reaction.

substrate-level phosphorylation (sub′strāt-lev′el fos″for-ĭ-la′shun)   ATP generation by coupling strongly exergonic reactions with ATP synthesis from ADP and phosphate.

succession   The sequence of community types during the maturation of an ecosystem.

sudoriferous gland (su″do-rif′er-us)   A sweat gland.

superior   Above a point of reference (e.g., the neck is superior to the chest of humans).

suspension feeder   The type of feeding whereby an animal obtains its nutrients by removing suspended food particles from the surrounding water by some sort of capturing, trapping, or filtering structure. Examples include tube worms and barnacles. Also, **suspension feeding.**

swim bladder   A gas-filled sac, usually along the dorsal body wall of bony fishes. It is an outgrowth of the digestive tract and regulates the buoyancy of a fish.

sycon (si′kon)   A sponge body form characterized by choanocytes lining radial canals.

symbiosis (sim′bi-os′es)   The living together of two different species in an intimate relationship. In this relationship, the symbiont always benefits; the host may benefit, may be unaffected, or may be harmed as in mutualism, commensalism, and parasitism, respectively.

symmetry (sim′ĭ-tre)   A balanced arrangement of similar parts on either side of a common point or axis.

sympathetic nervous system (sim″pah-thet′ik)   Portion of the autonomic nervous system that arises from the thoracic and lumbar regions of the spinal cord. Also called thoracolumbar division.

sympatric speciation (sim-pat′rik spe″she-a′shun)   Speciation that occurs in populations that have overlapping ranges.

symplesiomorphies (sim-ples″e-o-mor′fēz)   Taxonomic characters that are common to all members of a group of organisms. These characters indicate common ancestry but cannot be used to describe relationships within the group.

synapomorphies (sin-ap′o-mor″fēz)   Characters that have arisen within a group since it diverged from a common ancestor. Synapomorphies indicate degrees of relatedness within a group. Also called **shared, derived characters.**

synapse (sin′aps)   The junction between the axon end of one neuron and the dendrite or cell body of another neuron or effector cell.

synapsis (sĭ-nap′sis)   The time in reduction division when the pairs of homologous chromosomes lie alongside each other in the first meiotic division.

synaptic cleft (sĭ-nap′tik)   The narrow space between the terminal ending of a neuron and the receptor site of the postsynaptic cell.

synchronous flight   *See* **direct (synchronous) flight.**

syncytial hypothesis (sin-sit′e-al hi-poth′ĕ-sis)   The idea that multicellular organisms could have arisen by the formation of plasma membranes within a large, multinucleate protist.

synsacrum (sin-sak′rum)   The fused posterior thoracic vertebrae, all lumbar and sacral vertebrae, and anterior caudal vertebrae of a bird. Helps maintain proper flight posture.

syrinx (sir′ingks)   The vocal apparatus of a bird. Located near the point where the trachea divides into two bronchi.

system level   Animals whose body functions are carried out by organ systems.

systematics (sis-tem-at′iks)   The study of the classification and phylogeny of organisms. *See also* **taxonomy.**

systemic circuit (sis-tem′ik)   The portion of the circulatory system concerned with blood flow from the left ventricle of the heart to the entire body and back to the heart via the right atrium.

systole (sis′to-le)   Phase of the cardiac cycle during which a heart chamber wall contracts. *See also* **systolic pressure.**

systolic pressure (sis-tol′ik)   The portion of blood pressure measurement that represents the highest pressure reached during ventricular ejection. It is the first number shown in a blood-pressure reading.

T

tactile (touch) receptor (tak′tīl rĭ-sep′ter)   A sensory receptor in the skin that detects light pressure. Formerly called Meissner's corpuscle.

tagmatization (tag″mah-ti-za′shun) The specialization of body regions of a metameric animal for specific functions. The head of an arthropod is specialized for feeding and sensory functions, the thorax is specialized for locomotion, and the abdomen is specialized for visceral functions.

Tardigrada (tar-di-gra′dah) A phylum of animals whose members live in marine and freshwater sediments and in water films on terrestrial lichens and mosses. Possess four pairs of unsegmented legs and a proteinaceous cuticle. Water bears.

target cell The cell a specific hormone influences.

taste buds The receptor organs in the tongue that are stimulated and give rise to the sense of taste.

taxis (tak′sis) The movement of an organism in a particular direction in response to an environmental stimulus.

taxon (tak′son) A group of organisms that are genetically (evolutionarily) related.

taxonomy (tak″son′ah-me) The description of species and the classification of organisms into groups that reflect evolutionary relationships. *See also* **phylogenetic systematics, evolutionary systematics,** and **numerical taxonomy.** *Also* **systematics.**

T cell A type of lymphocyte, derived from bone marrow stem cells, that matures into an immunologically competent cell under the influence of the thymus. T cells are involved in a variety of cell-mediated immune reactions. Also known as a T lymphocyte.

tegument (teg′u-ment) The external epithelial covering in cestodes, trematodes, and acanthocephalans. Once called a cuticle.

telophase (tel′o-fāz) Stage in mitosis during which daughter cells become separate structures. The two sets of separated chromosomes decondense, and nuclear envelopes enclose them.

tendon (ten′den) A cord or bandlike mass of white, fibrous connective tissue that connects a muscle to a bone or to another muscle.

Tentaculata (ten-tak′u-lah-tah) The class of ctenophorans with tentacles that may or may not be associated with sheaths into which tentacles can be retracted. Example: *Pleurobranchia.*

test A shell or hardened outer covering, typically covered externally by cytoplasm or living tissue.

testis (tes′tis) Primary reproductive organ of a male. A sperm-cell producing organ.

testosterone (tes-tos′tĕ-rōn) Male sex hormone that the interstitial cells of the testes secrete.

tetrad (tet′rad) A pair of homologous chromosomes during synapsis (prophase I of meiosis). A tetrad consists of four chromatids.

tetrapod (tĕ′trah-pod) A taxonomic designation that refers to extant amphibians, reptiles, birds, and mammals and their closest common ancestor.

thalamus (thal′ah-mus) An oval mass of gray matter within the diencephalon that is a sensory relay area.

Thaliacea (tal″e-as′e-ah) A class of urochordates whose members are planktonic. Adults are tailless and barrel shaped. Oral and atrial openings are at opposite ends of the tunicate. Water currents are produced by muscular contractions of the body wall and result in a weak form of jet propulsion.

Thecostraca (the′ko-stra′se-ah) The subclass of maxillopod crustaceans that includes the barnacles.

theory of evolution by natural selection A theory conceived by Charles Darwin and Alfred Russel Wallace on how some evolutionary changes occur. *See also* **natural selection.**

thermoconformer (ther″mo-con-form′er) To conform to the temperature of the external environment.

thermodynamics (ther″mo-di-nam′iks) The branch of science that deals with heat, energy, and the interconversion of these. The study of energy transformations.

thermogenesis (ther″mo-jen′ah-sis) The metabolic generation of heat by muscle contraction or brown fat metabolism.

thermoreceptor (ther″mo-re-sep′tor) A sensory receptor that is sensitive to changes in temperature. A heat receptor.

thermoregulation (ther″mo-reg′u-la′shun) Heat regulation. Involves the nervous, endocrine, respiratory, and circulatory systems in higher animals.

threatened species A species that is likely to become endangered in the near future. *See also* **endangered species.**

thrombocyte *See* **platelet.**

thymus gland (thi′mus) A ductless mass of flattened lymphoid tissue behind the top of the sternum. It forms antibodies in the newborn and is involved in the development of the immune system.

thyroid gland (thi′roid) An endocrine gland in the neck of vertebrates and involved with the metabolic functions of the body. Produces thyroxins.

tissue (tish′u) A group of similar cells that performs a specialized function.

tissue level An aggregation of similarly specialized cells united in the performance of a specific function.

tolerance range The range of variation in an environmental parameter that is compatible with the life of an organism.

tonicity (to-nis′ĭ-te) The state of tissue tone or tension. In body fluid physiology, the effective osmotic pressure equivalent.

tornaria (tor-nar′e-ah) The ciliated larval stage of an acorn worm (class Enteropneusta, phylum Hemichordata).

torpor (tor′per) A time of decreased metabolism and lowered body temperature in daily activity cycles.

torsion (tor′shun) A developmental twisting of the visceral mass of a gastropod mollusc that results in an anterior opening of the mantle cavity and a twisting of nerve cords and the digestive tract.

tracheae (tra′che-e) The small tubes that carry air from spiracles through the body cavity of an arthropod. Arthropod tracheae are modifications of the exoskeleton.

tracheal system *See* **tracheae.**

tract A bundle of nerve fibers within the central nervous system.

transcription (tran-skrip′shun) The formation of a messenger RNA molecule that carries the genetic code from the nucleus to the cytoplasm of a cell.

transducer (trans-du′ser) A receptor that converts one form of energy into another.

transfer RNA (tRNA) (trans′fer) A single-stranded polyribonucleotide that carries amino acids to a ribosome and positions those amino acids by matching the tRNA anticodon with the messenger RNA codon.

transgenic (tranz-gen′ik) An animal that develops from a cell that received a foreign gene.

translation (trans-la′shun) The production of a protein based on the code in messenger RNA.

transmission electron microscope (TEM) The type of microscope that produces highly magnified images of ultra-thin tissue sections or other specimens by using electron beams.

Trematoda (trem″ah-to′dah) The class of Platyhelminthes with members that are all parasitic. Several holdfast devices are present. Complicated life cycles involve both sexual and asexual reproduction.

trichinosis (trik″ĭ-no′sis) A disease resulting from infection by *Trichinella spiralis* (Nematoda) larvae from eating undercooked meat. Characterized by muscular pain, fever, edema, and other symptoms.

trichocyst (trik′o-sist) A protective structure in the ectoplasm of some ciliates. A bottle-shaped extrusible organelle of the ciliate pellicle.

Trilobitamorpha (tri″lo-bit″a-mor′fah) The subphylum of arthropods whose members had bodies divided into three longitudinal lobes. Head, thorax, and abdomen were present. One pair of antennae and biramous appendages. Entirely extinct.

triploblastic (trip″lo-blas′tik) Animals whose body parts are organized into layers that are derived embryologically from three tissue layers: ectoderm, mesoderm, and endoderm. Platyhelminthes and all coelomate animals are triploblastic.

trochophore larva (trok′o-for lar′va) A larval stage characteristic of many molluscs, annelids, and some other protostomate animals.

trophic level (trōf′ik) The feeding level of an organism in an ecosystem. Green plants and other autotrophs function at producer trophic levels. Animals function at the consumer trophic levels.

tube feet Muscular projections from the water-vascular systems of echinoderms that are used in locomotion, gas exchange, feeding, and attachment.

tubular nerve cord A hollow nerve cord that runs middorsally along the back of chordates. One of four unique chordate characteristics. Also called the neural tube and, in vertebrates, the spinal cord.

Turbellaria (tur″bel-lar′e-ah) The class of Platyhelminthes with members that are mostly free living and aquatic. External surface usually ciliated. Predaceous. Possess rhabdites. Protrusable proboscis. Mostly hermaphroditic. Examples: *Convoluta, Notoplana, Dugesia.*

tympanic (tympanal) organs (tim-pan′ik) Auditory receptors present on the abdomen or legs of some insects.

U

ultimobranchial gland (ul″ti-mo-bronk′e-el) In jawed fishes, primitive tetrapods, and birds, the small gland(s) that forms ventral to the esophagus. Produces the hormone calcitonin that helps regulate calcium concentrations.

umbilical cord (um-bil′ĭ-kal) Cordlike structure that connects the fetus to the placenta.

umbo (um′bo) The rounded prominence at the anterior margin of the hinge of a bivalve (Mollusca) shell. It is the oldest part of the shell.

unicellular (cytoplasmic) organization The life-form in which all functions are carried out within the confines of a single plasma membrane. Members of the kingdom Protista display unicellular organization. Also called **cytoplasmic organization.**

uniformitarianism (u″nah-for″mĭ-tar′e-an-ism) The idea that forces of wind, rain, rivers, volcanoes, and geological uplift shape the earth today, just as they have in the past.

ureotelic excretion (u″re-o-tel′ik eks-kre′shun) Having urea as the chief excretory product of nitrogen metabolism. Occurs in mammals.

ureter (u-re′ter) The tube that conveys urine from the kidney to the bladder.

urethra (u-re′thrah) The tube that conveys urine from the bladder to the exterior of the body.

uricotelic excretion (u′ri-ko-tel″ik eks-kre′shun) Having uric acid as the chief excretory product of nitrogen metabolism. Occurs in reptiles and birds.

urinary bladder (u′rĭ-ner″e blad′er) The storage organ for urine.

Urochordata (u″ro-kor-da′tah) The subphylum of chordates whose members have all four chordate characteristics as larvae. Adults are sessile or planktonic and enclosed in a tunic that usually contains cellulose. Sea squirts or tunicates.

urophysis (u″ro-fi′sis) A discrete structure in the spinal cord of the fish tail that produces neuropeptides (hormones) that help control water and ion balance, blood pressure, and smooth muscle contractions.

uterine tube (u′ter-in) The tube that leads from the ovary to the uterus and transports the ovum or egg(s). Also called **fallopian tube.**

uterus (u′ter-us) The hollow, muscular organ in female mammals in which the fertilized ovum normally becomes embedded and in which the developing embryo/fetus is nourished.

V

vacuole (vak′u-ōl) Any small membrane-bound space or cavity formed in the protoplasm of a cell. Functions in either food storage or water expulsion.

vagina (vah-ji′nah) Tubular organ that leads from the uterus to the vestibule of the female reproductive tract. Receptacle for the penis during copulation.

valves 1. Devices that permit a one-way flow of fluids through a vessel or chamber. 2. The halves of a bivalve (Mollusca) shell.

vault A newly discovered organelle that is shaped like an octagonal barrel that may function as a "cellular truck."

vegetal pole (vĕ′jĕ-tl) The lower pole of an egg. Usually more dense than the animal pole because it contains more yolk.

vein A vessel that carries blood toward the heart from the various organs.

veliger larva (vel′ĭ-jer lar′va) The second free-swimming larval stage of many molluscs. Develops from the trochophore and forms rudiments of the shell, visceral mass, and head-foot before settling to the substrate and undergoing metamorphosis.

ventral The belly of an animal. Usually the lower surface. Synonymous with anterior for animals that walk upright.

venule (ven′yūl) A small blood vessel that collects blood from a capillary bed and joins a vein.

Vertebrata (ver″te-bra′tah) The subphylum of chordates whose members are characterized by cartilaginous or bony vertebrae surrounding a nerve cord. The skeleton is modified anteriorly into a skull for protection of the brain.

vibrissa (vi-bris′ah) A long, coarse hair, such as those occurring about the nose (muzzle) of a dog or cat. A sensor for mechanical stimuli.

villus (vil′us) Tiny, fingerlike projection that extends outward from the inner lining of the small intestine and increases the surface area for absorption.

visceral arches *See* **gill arches.**

visceral mass (vis′er-al) The region of a mollusc's body that contains visceral organs.

vitamin An organic substance other than a carbohydrate, lipid, or protein that is needed for normal metabolism but that the body cannot synthesize in adequate amounts.

viviparous (vi-vip′er-us) Organisms with eggs that develop within the female reproductive tract and are nourished by the female.

voluntary (somatic) nervous system That part of the nervous system that relays commands to skeletal muscles.

vomeronasal organ *See* **Jacobson's (vomeronasal) organ.**

vulva (vul′vah) The external genital organs in the female.

W

water expulsion vacuoles (*See* **contractile vacuoles**)

water-vascular system (wah′ter vas′ku-ler) A series of water-filled canals and muscular tube feet in echinoderms. Provides the basis for locomotion, food gathering, and attachment.

white blood cell (leukocyte)   A type of blood cell involved with body defenses.

winter sleep   A period of inactivity in which a mammal's body temperature remains near normal and the mammal is easily aroused.

work   The exertion or effort needed to accomplish something.

Y

yolk plug   The large, yolk-filled cells that protrude from beneath the ectoderm at the blastopore of the amphibian gastrula.

yolk sac   The stored food reserve (yolk) and its surrounding membranes. Found in embryonic reptiles, birds, and mammals.

Z

zonite (zo′nĭt)   The individual body unit of a member of the phylum Kinorhyncha.

zooid (zo′oid)   An individual member of a colony of animals, such as colonial cnidarians and ectoprocts, produced by incomplete budding or fission.

zoology (zo-ol′o-je)   The study of animals.

Zoomastigophorea (zo″o-mas-ti-go-for′ah)   The protozoan class whose members lack chloroplasts. Heterotrophic or saprozoic. Examples: *Trypanosoma, Trichonympha, Trichomonas, Giardia.*

zooxanthellae (zo″o-zan-thel′e)   A group of dinoflagellates that live in mutualistic relationships with some cnidarians. They promote high rates of calcium carbonate deposition in coral reefs.

zygote (zi′gōt)   Diploid cell produced by the fusion of an egg and sperm. Fertilized egg cell.

CREDITS

12.24: © James Cutler/Visuals Unlimited; **Box Figure 12.2:** Courtesy of Kevin S. Cummings, Illinois Natural History Survey, Champaign, IL; **Box Figure 12.3:** © M.C. Barnhart, 1995.

CHAPTER 13

Figure 13.1a: © Stan Combs; **13.1b:** © Peter Craig, National Park of American Samoa; **13.2:** © Marty Snyderman/Visuals Unlimited; **13.5:** © R. DeGoursey/Visuals Unlimited; **13.6, 13.9-13.12, 13.7 line art:** From A LIFE OF INVERTEBRATES, © 1979 W.D. Russell-Hunter.

CHAPTER 14

Figure 14.1: © Roger Klocek/Visuals Unlimited; **14.4 line art:** From A LIFE OF INVERTEBRATES © 1979 W.D. Russell-Hunter; **14.6:** © John Cancalosi/Tom Stack & Assoc.; **14.8a:** © Francois Gohier/Photo Researchers, Inc.; **14.11a:** © SS 11/PhotoDisc; **14.12 line art:** (a) After Sherman and Sherman; (b) After the Kastons; **14.13:** © Ken Highfill/Photo Researchers, Inc.; **14.14a:** © S. Masiowski/Visuals Unlimited; **14.14b:** © Larry Miller/Photo Researchers, Inc.; **14.15:** © David Scharf/Peter Arnold, Inc ; **14.16a:** © R. Calentine/ Visuals Unlimited; **14.16b:** © Scott Camazine/ Photo Researchers, Inc.; **14.17:** © Heather Angel; **14.19:** © Andrew J. Martinez/ Photo Researchers, Inc.; **14.22a:** © John Cunningham/ Visuals Unlimited; **14.22b:** © William Jorgensen/Visuals Unlimited; **14.23a:** © David M. Dennis/Tom Stack & Assoc.; **14.23b:** © Alan Desbonnet/Visuals Unlimited; **14.25b:** © Tom Stack/Tom Stack & Assoc.; **Box Figure 14.1:** © Jason Gunter.

CHAPTER 15

Figure 15.1: © Patti Murray/Animals Animals/ Earth Scenes; **15.2a:** © Bill Beatty/Visuals Unlimited; **15.2b:** © Glenn M. Oliver/Visuals Unlimited; **15.10a:** © S.L. Flegler/Visuals Unlimited; **15.14a-c:** © Treat Davidson/Photo Researchers, Inc.; **Box Figure 15.1:** © A. Rider/Photo Researchers, Inc.

CHAPTER 16

Figure 16.1: © Carl Roessler/Tom Stack & Assoc.; **16.3a:** © Michael Dispezio; **16.8:** Courtesy of the Museum of New Zealand Te Papa Tongarewa, Alan M. Baker, photographer; **16.9a:** © Robert Dunne/Photo Researchers, Inc.; **16.9b:** © 2004 Harold W. Pratt/BPS; **16.11a:** © C. McDaniel/ Visuals Unlimited; **16.11b:** © Bruce Iverson/Visuals Unlimited; **16.13:** © Daniel W. Gotshall/Visuals Unlimited; **Box Figure 16.2a,b:** © Jonathan Green/Senderos Naturales.

CHAPTER 17

Figure 17.1: © Michael DiSpezio; **17.7a:** © William Jorgenson/Visuals Unlimited; **17.7b:** © Daniel W. Gotshall/Visuals Unlimited; **Box Figure 17.1:** © Dr. Desmond Collins, Royal Ontario Museum.

CHAPTER 18

Figure 18.1: © Vol. 53/Corbis; **18.2b:** © The Natural History Museum, London; **18.6:** © Russ Kinne/Photo Researchers, Inc.; **18.9a:** © Vol. 53/Corbis; **18.9b,c:** © Daniel W. Gotshall/Visuals Unlimited; **18.10a:** © Science VU/NOAA/Visuals Unlimited; **18.10b line art:** From R. Steale, SHARKS OF THE WORLD © 1985 Cassel PLC, London; **18.11:** © John Cunningham/Visuals Unlimited; **18.13a:** © Patrice Ceisel/Visuals Unlimited; **18.12:** © Peter Scoones/SPL/Photo Researchers, Inc.; **18.13b:** © Patrice Ceisel/Visuals Unlimited; **18.14a:** © R. DeGoursey/Visuals Unlimited; **18.14b:** © Vol. 6/PhotoDisc; **18.14c:** © Vol. 64/Corbis; **18.16b:** © Fred Hossler/Visuals Unlimited; **18.18b:** © Tom McHugh/Photo Researchers, Inc.; **18.19:** © Daniel W. Gotshall/ Visuals Unlimited; **18.21:** © Vol. 6/Corbis; **Box Figure 18.2:** © NEBRASKALAND Magazine/ Nebraska Game and Parks Commission photo; **18.22c:** © Ken Lucas/Visuals Unlimited.

CHAPTER 19

Figure 19.1: © Joe McDonald/Visuals Unlimited; **19.4:** © John Cunningham/Visuals Unlimited; **19.5a,b:** © Dwight Kuhn; **19.5c:** © Joel Arrington/ Visuals Unlimited; **19.6:** © Tom McHugh/Photo Researchers, Inc.; **19.7:** © John Serrao/Visuals Unlimited; **19.13 line art:** From N. Heisler, JOURNAL OF EXPERIMENTAL BIOLOGY © 1983 The Company of Biologists Limited, Cambridge, UK. Used by permission; **19.15a:** © Nada Pecnik/Visuals Unlimited; **19.15b:** © Edward S. Ross; **19.16:** © Nick Bergkessel/Photo Researchers, Inc.; **19.17b:** © Dan Kline/Visuals Unlimited; **19.18:** © M.J. Tyler, The University of Adelaid, South Australia; **19.19a-d:** © Jane Burton/Bruce Coleman, Inc ; **Box Figure 19.1:** © D. Dennis/Tom Stack & Assoc.

CHAPTER 20

Figure 20.1: © E.R. Degginger/Photo Researchers, Inc.; **20.6:** © Valorie Hoegson/Visuals Unlimited; **20.7:** © Nathan Cohen/Visuals Unlimited; **20.8:** © Joe McDonald/Visuals Unlimited; **20.9:** © Thomas Gula/Visuals Unlimited; **20.10:** © John Cunningham/Visuals Unlimited; **20.12:** © Stephen Dalton/Animals Animals/Earth Scenes; **20.13a:** © Joe McDonald/Visuals Unlimited; **20.14 line art:** From N. Heisler, JOURNAL OF EXPERIMENTAL BIOLOGY © 1983 The Company of Biologists Limited, Cambridge, UK. Used by permission; **20.15:** © Zigmund Leszcynski/Animals Animals/ Earth Scenes; **20.16:** © Robert Hernes/Photo Researchers, Inc.; **Box Figure 20.1:** © Tom McHugh/Photo Researchers, Inc.

CHAPTER 21

Figure 21.1a: © Tim Hauf/Visuals Unlimited; **21.2a:** © John Cunningham/Visuals Unlimited; **21.2b:** Neg. #K-set3 #29. Courtesy of Library Services, American Museum of Natural History; **21.6e:** © Vol. 44/PhotoDisc; **21.7:** © Robert A Lubeck/Animals Animals/Earth Scenes; **21.8a:** © Vol. 46/Corbis; **21.8b:** © Kirtley-Perkins/Visuals Unlimited; **21.8c:** © Brian Parker/Tom Stack & Assoc.; **21.10c:** Courtesy of Dr. Hans-Rainer Duncker; **21.11:** © Vol. 6/Corbis; **21.14:** © John Cunningham/Visuals Unlimited; **21.15a:** © Karl Maslowski/Visuals Unlimited; **21.15b:** © John Cunningham/Visuals Unlimited; **Box Figure 21.1:** © C.C. Lockwood/Animals Animals/Earth Scenes.

CHAPTER 22

Figure 22.1: © Vol. 6/Corbis; **22.4a:** © Tom McHugh/Photo Researchers, Inc.; **22.4b:** © J. Alcock/ Visuals Unlimited; **22.4c:** © Vol. 8/Corbis; **22.5a:** © Walt Anderson/Visuals Unlimited; **22.5b:** © William J. Weber/Visuals Unlimited; **22.6:** © Vol. 6/Corbis; **22.11:** © Vol. 44/PhotoDisc; **22.15a:** © Dennis Schmidt/Valan Photos; **22.16:** © Walt Anderson/Visuals Unlimited; **Box Figure 22.2:** © Thomas Kitchin/Tom Stack & Assoc.; **22.17a,b:** © John Cunningham/Visuals Unlimited.

CHAPTER 23

Figure 23.10: © Gary Robinson/Visuals Unlimited; **23.11a:** © Richard Walters/Visuals Unlimited; **23.12b:** Copyright Richard G. Kessel and Randy H. Kardon, Tissues and Organs: A Text Atlas of Scanning Electron Microscopy, 1979, W.H. Freeman and Company. All rights reserved. **23.2, 23.17, 23.18 line art:** From A LIFE OF INVERTEBRATES © 1979 W.D. Russell-Hunter.

CHAPTER 24

Figure 24.10 line art: Redrawn from Cecie Starr and Ralph Taggart, BIOLOGY: THE UNITY AND DIVERSITY OF LIFE, 4th ed. © 1987 Wadsworth Publishing Company, Belmont, CA; **24.14 line art:** Redrawn with permission from Richard C. and Gary J. Brusca, INVERTEBRATES, © 1990 Sinauer Associates, Inc. Sunderland, MA; **24.16c:** © Thomas Eisner/Cornell University; **24.24:** © Milton Tierney/Visuals Unlimited.

CHAPTER 25

Figure 25.5 line art: Redrawn with permission from Richard C. Brusca and Gary J. Brusca, INVERTEBRATES © 1990 Sinauer Associates, Inc., Sunderland, MA; **25.9:** Courtesy of Dr. Aaron Lerner, Yale University; **Table 25.1:** Barrett/ Abramoff/Kumaran/Millington, BIOLOGY, © 1986, p. 383. Adapted by permission of Prentice Hall, Inc., Englewood Cliffs, New Jersey.

CHAPTER 26

Figure 26.4: Courtesy of Dr. Philip C. Withers; **26.5a:** © Stanley Flegler/Visuals Unlimited; **26.5b-f:** © John Cunningham/Visuals Unlimited; **26.15:** © Victor Hutchison/Visuals Unlimited.

CHAPTER 27

Figure 27.6a: © R. Calentine/Visuals Unlimited; **27.6b:** © Daniel Gotshall/Visuals Unlimited; **27.6c:** © Dwight Kuhn; **27.6d:** © Science VU/ Visuals Unlimited; **27.16c:** © Visuals Unlimited.

CHAPTER 28

Figure 28.2: © Joe McDonald/Visuals Unlimited; **28.5a:** © G. Prance/Visuals Unlimited; **28.6:** © Dan Kline/Visuals Unlimited; **28.9:** © Cabisco/ Visuals Unlimited; **28.12 line art:** Redrawn with permission from W.K. Purves and G.H. Orians, LIFE: THE SCIENCE OF BIOLOGY, 2nd Edition. © 1987 Sinauer Associates, Inc., Sunderland, MA; **28.21 line art:** From ANIMAL PHYSIOLOGY by Roger Eckert, David Randall, and George Augustine. © 1998 by W. H. Freeman and Company. Used with permission.

CHAPTER 29

Figure 29.1a: © Cabisco/Visuals Unlimited; **29.1b:** © John Cunningham/Visuals Unlimited; **29.1c:** © Mike Mulligan/Visuals Unlimited; **29.1d:** © John Cunningham/Visuals Unlimited; **29.1e:** © Daniel Gotshall/Visuals Unlimited; **29.2 line art:** Redrawn with permission from Richard C. Brusca and Gary J. Brusca, INVERTEBRATES © 1990 Sinauer Associates, Inc., Sunderland, MA;

29.3a: © John Serrao/Visuals Unlimited; **29.3b:** © A. Gurmankin/Visuals Unlimited; **29.3c:** © Karl Maslowski/Visuals Unlimited; **29.3d:** © SIU/ Visuals Unlimited; **29.9:** © David Phillips/Visuals Unlimited; **29.14a:** © Petit Format/Photo Researchers, Inc.; **29.14b:** © A. Tsiaras/Photo Researchers, Inc.; **29.14c:** © Petit Format/Photo Researchers, Inc.; **29.14d-f:** © Science/Visuals Unlimited; **29.14g:** © Cabisco/Visuals Unlimited.

INDEX

Lymph, 433
Lymphatic capillaries, 433, 433t
Lymphatics, 433, 433t
Lymphatic system
  in amphibians, 303
  in vertebrates, 433, 433t
Lymph nodes, 433, 433t
Lymphocytes, 427, 428
Lysine, 42f
Lysosomes, 11f, 16, 17f, 18t,
    19–20, 20f

M

Macaque, 389f
MacArthur, Robert, 84, 84f
Macracanthorhynchus hirudinaceus,
    175
Macroevolution, 58–59
Macrogamete, 116
Macronucleus, in ciliates, 123
Macronutrients, 444, 445f, 445t
Macropus. See Kangaroo
Macrostomida, digestive system
    of, 151f
Madagascar day gecko, 323f
Madreporite, 252–253, 254f
Magnesium, 446t
Magnetic compass, 339
Maintenance functions. *See specific*
    functions
Malacostraca, 226–229
  classification of, 217t, 226
Malaria, 120–121
  caused by *Plasmodium*, 120, 121
  sickle-cell anemia and, 73
Malawi, Lake, cichlids in, 4, 4f
Malleus, 398, 398f
Malpighian tubules, 221, 243,
    473, 474f
Maltase, 462t
Malthus, Thomas, 55, 57
Mammalia, 342–358
  behavior of, 354–355
  circulatory system in, 350–352,
    351f, 352f, 427, 431–432
  classification of, 267t, 313–314,
    344t, 345
  development in, 357–358, 488
  digestive system in, 350,
    456–462
  diversity of, 343–345
  endocrine system of, 417–422
  evolution of, 342–343, 345, 345f
  excretory system in, 354,
    479–480
  external structure of, 347–350
  eyes of, 402–403
  feeding in, 350
  gas exchange in, 352, 438–439,
    438f
  hearing in, 354, 398
  hormones of, 357
  kidneys in, 477–478
  lymphatic system in, 433, 433t
  muscles in, 347, 350
  nervous system in, 354, 387f

olfaction in, 354, 400
osmoregulation in, 354,
    477–478, 478f
parental care in, 488
reproduction in, 355–357, 487,
    488, 488f
sensory system of, 354
skeleton of, 350
skin of, 347, 364–365, 364f
skull of, 348–350
study of. *See* Mammalogy
teeth of, 348–350, 458, 458f
territoriality in, 355
thermoregulation in, 352–354,
    353f, 364–365, 466,
    468–470, 469f, 471f
tongue in, 402, 452
vision in, 354, 402–403
Mammalogy, 3t
Mammary glands, 347–348, 348f,
    488, 494, 500
Mandibles
  of crayfish, 226
  of insects, 240, 240f
Manganese, 446t
Manta hamiltoni, 282f
Manta ray, 282f, 283
Mantle, 182, 182f
  of bivalves, 186, 188
  of cephalopods, 191, 192
Mantle cavity, 182, 182f, 184, 187,
    434–435, 435f
Mantodea, classification of, 237t
Mantophasmatodea, 238
Manubrium, 137
Margulis, Lynn, 28
Marsupials
  classification of, 345
  development in, 357, 488
Marsupium, 345
Mastax, 167, 167f
Mastigias quinquecirrha. See Stinging
    nettle
Mastigophora, 114, 115t
Matrix
  of cartilage, 28
  of mitochondrion, 20, 20f
Maxillae
  of crayfish, 226
  of insects, 240, 240f
Maxillary gland, 228, 473
Maxillopoda, 231
  classification of, 217t, 231
Mechanoreceptors, 396f, 397,
    400, 401f
Meconium, 500
Median eye, 322, 402, 403f
Medulla, 475
Medulla oblongata, 388
Medusa, 135, 136f
Meerkat, 466f
Megantherium, 54f
Mehlis' gland, 154f
Meiomenia, 196f
Meiosis, 35–37, 37f
  errors in, 41, 43f
  first division, 35–36, 36f

independent assortment during,
    45, 47f
  second division, 36–37, 36f
Meissner's corpuscles, 400, 401f
Melanism, industrial, 72, 72f
Melanocyte-stimulating hormone
    (MSH), 414t
Melatonin, 413, 419
Memory, in insects, 242
Mendel, Gregor, 32
Meninges, 387
Meningoencephalitis, primary
    amebic, 112f
Menopause, 494
Menstrual cycle, 357, 494–495,
    495t, 496f
Menstruation, 495
"Mermaid's pennies," 119
Merogony, 120
Merostomata, 219–220
  body structure of, 219, 220f
  classification of, 215, 217t, 219
Merozoites, 120
Mesenchyme, 105
Mesenchyme cells, 130, 165
Mesenteries, 141, 165
Mesoderm, 106
Mesoglea, 105, 134
Mesohyl, 130
Mesonephros kidney, 475, 476f
Mesothorax, in insects, 239
Mesozoa, 107, 107t
Mesozoans, classification of, 107, 107t
Mesozoic era, 61f
Messenger RNA (mRNA), 17, 39
  DNA transcribed into, 40, 41f
  in translation, 40–41, 42f
Metabolism
  in heterotrophs, 444–446
  rate of, in birds, 335
Metacercaria, 155, 156f, 157
Metamerism, 201
  in annelids, 201–202, 202f
  in arthropods, 215–216
  evolution of, 201, 215
Metamorphosis, 218
  ametabolous, 245
  in amphibians, 308, 308f, 488
  in arthropods, 218
  in frogs, 308, 308f, 414, 416f
  hemimetabolous, 245, 245f
  holometabolous, 245, 245f
  in insects, 245
  in salamanders, 298, 300f
  in urochordates, 271–272, 271f
Metanephric kidneys, 322, 475,
    476f, 478f, 479–481
Metanephridium, 205, 206f, 209,
    472–473, 473f
Metaphase, in mitosis, 34f, 35
Metaphase I, in meiosis, 35, 36f
Metaphase II, in meiosis, 36, 36f
Metatheria, classification of
    344t, 345
Metathorax, in insects, 239
Methionine, 42f
Mexican beaded lizard, 317

Micelles, 462
Microevolution, 58, 68
Microfilaments, 9, 18t, 20, 21f
Microfilariae, 173, 174f
Microgamete, 116
Micronucleus, in ciliates, 123
Micronutrients, 444–446
Microspora, 114, 115t, 121
Microtubule(s), 9, 18t, 20, 21f
Microtubule doublet, 21, 21f
Microtubule-organizing centers, 11f,
    18t, 22, 34f
Mictic eggs, 168–169, 168f
Midbrain, 388, 388f
Middle ear, 62, 63f, 398, 398f
Migration, 339
  of birds, 326, 327f, 339
  diadromous, 291
  of fishes, 289, 291, 400
  and gene flow, 71
MIH. *See* Molt-inhibiting hormone
Milk production, 500
Millipedes, 235–238, 235f
Mimicry, 85
Mineralization, 60f
Mineralized tissues, in invertebrates,
    366
Mineralocorticoids, 415t, 420
Minerals, 445, 446t
Miracidium, 155, 156f, 157f, 158
Mirounga angustirostris. See
    Northern elephant seal,
    diversity of
Mites, 224
Mitochondrial DNA, 98
Mitochondrion (mitochondria),
    11f, 18t, 20, 20f
Mitosis, 33–35, 33f, 34f
Mitotic spindle, 35
Mnemiopsis, 143f
Mobile-receptor mechanism,
    409, 409f
Modern synthesis, 64
Molars, 349
Molecular biology, 62–63
  definition of, 3t
  in taxonomy, 98–99
Molecular evolution, 76
Molecular genetics, 32, 37
Mollusca, 179–197
  body structure of, 180–183, 182f
  circulatory system in, 425
  classification of, 107t, 181t,
    196f, 197
  digestive system in, 451, 451f
  evolution of, 179–180, 181f,
    196f, 197
  feeding in, 449
  gas exchange in, 434–435
  hormones of, 410
  nervous system of, 410
Molting. *See also* Ecdysis
  in aschelminths, 166
  in birds, 331
  in crustaceans, 228
  in mammals, 347
  in reptiles, 363

THE HISTORY OF THE EARTH: GEOLOGICAL ERAS, PERIODS, AND MAJOR BIOLOGICAL EVENTS

ERA	PERIOD	AGE AT BEGINNING OF THE PERIOD (MILLIONS OF YEARS)	MAJOR BIOLOGICAL EVENTS
CENOZOIC	Quaternary	0.01	Subtropical forests gave way to cooler forests and grassland areas.
CENOZOIC	Tertiary	65	Modern orders of mammals evolved. Humans evolved in the last 5 million years.
MESOZOIC	Cretaceous	135	Continental seas and swamps spread. Extinction of ancient birds and reptiles.
MESOZOIC	Jurassic	195	Climate warm and stable. High reptilian diversity. Birds first appeared.
MESOZOIC	Triassic	240	Climate warm. Extensive deserts. Dinosaurs replace mammal-like reptiles. First true mammals.
PALEOZOIC	Permian	285	Climate cold early, but then warmed. Mammal-like reptiles common. Widespread extinction of amphibians.
PALEOZOIC	Carboniferous	375	Warm and humid with extensive coal-producing swamps. Arthropods and amphibians common. First reptiles appeared.
PALEOZOIC	Devonian	420	Land high and climate cool. Freshwater basins developed. Fish diversified. Early amphibians appeared.
PALEOZOIC	Silurian	450	Extensive shallow seas. Warm climate. First terrestrial arthropods. First jawed fish.
PALEOZOIC	Ordovician	520	Shallow extensive seas. Climate warmed. Many marine invertebrates. Jawless fish widespread.
PALEOZOIC	Cambrian	570	Extensive shallow seas and warm climate. Trilobites and brachiopods common. Earliest vertebrates found late in the Cambrian.
PROTEROZOIC		2,000	Multicellular organisms appeared and flourished. Many invertebrates. Eukaryotic organisms appeared (1,500 million years ago). Oxygen accumulated in the atmosphere.
ARCHEAN		4,600	Prokaryotic life appeared (3,500 million years ago). Origin of the earth (4,600 million years ago).